Springer Collected Works in Mathematics

More information about this series at http://www.springer.com/series/11104

Luis Antonio Santaló in his office

(Courtesy of Santalo's family)

Luis Antonio Santaló

Selected Works

Edited by
Antonio M. Naveira and Agusti Reventós

In Collaboration with
Graciela S. Birman and Ximo Gual-Arnau

Preface by
Simon K. Donaldson

Reprint of the 2009 Edition

Author
Luis Antonio Santaló
(1911–2001)

Editors
Antonio M. Naveira
Facultad de Matemáticas
Universitat de València
Burjassot, Valencia
Spain

Agusti Reventós
Facultat de Ciències
Universitat Autonoma de Barcelona
Bellaterra
Spain

Collaborators
Graciela S. Birman
Universidad Nacional del Centro de la Provincia de Buenos Aires
Tandil
Argentina

Ximo Gual-Arnau
Departament de Matemàtiques
Universitat Jaume I
Castelló de la Plana
Spain

Simon K. Donaldson
Department of Mathematics
Imperial College London
London
UK

ISSN 2194-9875
Springer Collected Works in Mathematics
ISBN 978-3-662-53237-9 (Softcover)

Library of Congress Control Number: 2012954381

Printed on acid-free paper

This Springer imprint is published by Springer Nature
The registered company is Springer-Verlag GmbH Germany
The registered company address is: Heidelberger Platz 3, 14197 Berlin, Germany

Luis Antonio Santaló

Selected Works

Editors
Antonio M. Naveira • Agustí Reventós

in collaboration with
Graciela S. Birman • Ximo Gual

Preface by
Simon K. Donaldson

Author

Luis Antonio Santaló
(Spain 1911 – Argentina 2001)

Editors

Antonio M. Naveira
Departamento de Geometria y Topologia
Facultad de Matemáticas
University of Valencia
Avda. Andrés Estellés 1
46100 Burjassot, Valencia
Spain
naveira@uv.es

Agustí Reventós
Departament de Matemàtiques
Universitat Autònoma de Barcelona
08193 Bellaterra, Catalunya
Spain
agusti@mat.uab.cat

In collaboration with

Graciela S. Birman
CONICET y Departamento de matemática
Facultad de Ciencias Exactas
Universidad Nacional del Centro de la Provincia de Buenos Aires (UNCPBA)
Campus Universitario
Paraje Arrojo Seco
7000 Tandil
Argentina
gbirman@exa.unicen.edu.ar

Ximo Gual-Arnau
Departament de Matemàtiques
Campus Riu Sec
Universitat Jaume I
8029AP Castelló
Spain
gual@mat.uji.es

Preface by

Simon K. Donaldson
Huxley Building
Department of Mathematics
Imperial College
Queen's Gate
London SW7 2AZ
UK
s.donaldson@imperial.ac.uk

ISBN: 978-3-540-89580-0 e-ISBN: 978-3-540-89581-7

Library of Congress Control Number: 2009920723

Cover design: WMX Design GmbH, Heidelberg

Printed on acid-free paper

9 8 7 6 5 4 3 2 1

springer.com

Contents

Preface

The word "geometry" can cover many different things. The paths that can be traced from the most ancient concepts to sophisticated modern abstractions form one of the charms of the subject. We can all agree that the study of lines in three dimensional Euclidean space is a part of geometry. It is a wonderful idea that the set of all lines can itself be considered as a 'space', which has in turn its own geometry. Of course, this leap into abstraction may seem commonplace now – the set of lines is an example of a four-dimensional manifold and of a homogeneous space, for the group of Euclidean isometries. Modern differential geometry provides the language and tools for doing calculus on such spaces and in particular for *integration*. Then we can talk about the volume of a set of lines with a particular property, the mean value of a function on the space of lines and so on. Likewise if, instead of lines, we take some other distinguished objects, such as spheres, or if we begin with a non-Euclidean space. This is the theory of "Integral Geometry", to which Luis Santaló contributed so much.

The span of Santaló's life and career was large: in time, covering nearly all the twentieth century, geographically, and in his phenomenal research output. There is much to gain from a study of his papers. We can see the development and interaction of different themes, points of view and schools; the evolution of the mathematical ideas, language and notation. We see the shift to the modern abstract points of view, involving general theories of Lie groups, homogeneous spaces and differential forms. But we can also find marvellous, and probably little-known, gems of geometry, such as Santaló's proof of the isoperimetric inequality in surfaces of negative curvature.

Nowadays, if one has access to MathSciNet, it takes only a few moments at a computer to produce a complete list of any mathematicians published work, but a volume such as this *Selecta* goes far beyond that. There is the careful choice of the most significant papers, some of which appeared in journals which are not easy to find, grouped under different themes and with comments from leading experts. There is also a biographical outline, fascinating as a story of an eventful life, as a slice of mathematical history and for the insights it provides into the times. We should heartily thank the Editors for their tremendous labour of love in the production of this volume, which conveys, especially to those of us who did not have the good fortune to know him, such a picture of Santaló, both the mathematician and the man.

Simon K. Donaldson

1 A Short Biography of L. A. Santaló

Some Biographical Data

L. A. Santaló was born in Girona on October 9, 1911, and died in Buenos Aires on November 22, 2001. He began studying at the *Scholastic Group*, where his father was a teacher. He then moved to the *Institute*, of which he preserved a long-lasting memory. At the age of 16 he went to study in Madrid, where he was later to obtain his degree at the Faculty of Mathematics in 1934. Here he entered into scientific relations with professors Julio Rey Pastor and Esteve Terradas, who both exercised a decisive influence on him. Indeed, it was on their advice that he went to Hamburg to work with Wilhelm Blaschke. While he was there he also met Chern, Wu, Varga, Petkantschin and others. At that time, Blaschke was engaged in the study of geometric probabilities, thereby originating what he himself was to term *Integralgeometrie*.

In 1936, sponsored by Pedro Pineda, Santaló obtained his Ph. D. with a thesis on Integral Geometry, [36.1].

After the Spanish Civil War, Santaló moved to France, and with the assistance of Rey Pastor obtained a position at the University of Litoral, Rosario, Argentina. It was at that time that the *Instituto de Matemática* under the direction of Beppo Levy was created at the University of Litoral, and Santaló was named as Assistant Director, a position he held until 1949, during which time he helped to create the mathematical journal *Mathematicae Notae*.

In 1948 was awarded a grant from the Guggenheim Foundation to study in Princeton and Chicago. In 1949 he obtained a post at the University of La Plata and supervised his first doctoral thesis. In 1957 he became a full professor of the Faculty of Exact and Natural Sciences at the University of Buenos Aires, where he remained until his retirement. Beginning in 1955, he visited Spain several times, maintaining in particular a close relationship with Vidal-Abascal, who was responsible for introducing the study of the Integral Geometry into Spain.

More Important Distinctions for His Educational and Research Work

In Spain

- *Corresponding Academic* of the Royal Academy of Exact, Physical and Natural Sciences of Madrid, 1955.

- *Corresponding Academic* of the Royal Academy of Sciences and Arts of Barcelona, 1970.
- Doctor *Honoris Causa* from the Polytechnical University of Catalonia, 1977.
- *Corresponding Member* of the Institute of Catalan Studies, 1977.
- *Prince of Asturias* Prize for Scientific and Technological Research, 1983.
- *Narcís Monturiol* Medal for Science and Technology from the Government of Catalonia, 1984.
- Doctor *Honoris Causa* from the Autonomous University of Barcelona, 1986.
- Doctor *Honoris Causa* from the University of Sevilla, 1990.
- Awarded the Medal of the University of Valencia, 1993.
- *Corresponding Academic* of the Academy of Sciences of Canarias, 1993.
- Cross of Sant Jordi, from the Government of Catalonia, 1994.
- *Cross of Alfonso X*, granted by King Juan Carlos and given by the ambassador of Spain in Argentina.
- *Honorary Member* of the Royal Spanish Mathematical Society, 1999.
- The University of Girona creates the Santaló's Chair, 2000.
- *Honorary Member* of the Catalan Mathematical Society, 2000.

In Argentina

- National Prize of Culture, 1954.
- Argentina Scientific Society Award, 1959.
- *Academic* of the National Academy of Exact, Physical and Natural Sciences of Buenos Aires, 1960. Secretary (1972–1976). Vice-president (1976–1980). President (1980–1984), Honorary Academic (1997).
- *Corresponding Academic* of the National Academy of Sciences of Cordoba, 1961.
- Doctor *Honoris Causa* from the National University of Nordeste, Argentina, 1977.
- Severe Vaccaro Foundation Prize, 1977.
- *Honorary Professor* from the National University of La Plata, 1979.
- Doctor *Honoris Causa* from the National University of Misiones, 1982.
- Doctor *Honoris Causa* from the National University of Tucumán, 1983.
- Inter-American *Prize of Sciences, B. A. Houssay,* 1986.
- *Academic* of the National Academy of Educatión, Buenos Aires, 1989.
- Doctor *Honoris Causa* from the University of San Juan, 1991.
- *National Consecration Prize* of the Culture Secretariat, Buenos Aires, 1992.
- Doctor *Honoris Causa* from the University CAECE of Buenos Aires, 1992.
- Doctor *Honoris Causa* from the University of Buenos Aires, 1992.
- Doctor *Honoris Causa* from the University of Morón, 1992.
- *Honorary Member* of the Argentina Scientific Society, 1994.
- *Gold Rose Prize* of the Foundation Tapia by the *trajectory in benefit of the education of excellence*, Buenos Aires, 1994.
- *Jose Manuel Estrada Prize* for *Teacher of Argentine Sciences*, Arquidiocesana Commission for Culture, Buenos Aires, 1995.
- *Konex of Honor Prize* (posthumous), 2003.

Other Distinctions

- *Corresponding Academic* of the National Academy of Exact, Physical and Natural Sciences of Lima (Perú), 1945.
- Member of the International Statistical Institute, Holland, 1979.
- *Honorary Member* of the Mathematical Society of Paraguay, 1982.
- *Honorary Member* of the Academy of Sciences of Latin America, Venezuela, 1983.
- *Honorary Member* of the Royal Statistical Society of London, 1984.
- *Honorary Member* of the Latin American Society of the History of Sciences and the Technology, 1984.
- *Corresponding Academic* of the Academy of Sciences of Chile, 1986.
- *Honorary Life Member* of the International Society of Stereology, Germany, 1994.
- *Badge Tribute* for persistence and dedication to the Mathematical Education, Blumenau, Brazil, 1994.
- *Academic* of the New York Academy of Sciences, USA, 1997.

2 Scientific Work of L. A. Santaló

Analysis of Santaló's scientific work reveals certain fundamental characteristics; his considerable powers of abstraction, his brilliant geometric intuition, his surprising clarity of exposition and his outstanding gifts as a disseminator of science. It is therefore hardly surprising that his work should have had such profound repercussions in the specialized scientific community and in society in general. The international mathematical community regards Santaló as one of the great geometers of the 20th century, comparable with Blaschke, Chern and Hadwiger, among others. His contribution to the development of Integral Geometry throughout the 20th century is exceptional, both from the mathematical perspective and from the point of view of the applications of his theories to the applied sciences. One may state with complete certainty that men such as Santaló will always be necessary for the development of Mathematics, its applications and its dissemination.

Santaló was a person who always showed a lively interest in the problems surrounding him. We the Editors of Professor Santaló's Selected Papers had the honour and the pleasure of knowing him personally and of enjoying his friendship. To each one of us, albeit on different occasions, he conveyed his personal and scientific experiences, both while he was in Spain and during his prolonged sojourn in Argentina, the country where he carried out practically the whole of his activity as a professor and researcher.

This activity covered various scientific fields; among them, research into Differential Geometry, Integral Geometry and Theoretical Physics. He was also greatly drawn to the Teaching of Mathematics at all levels (in particular, at Secondary School and University level), and his cherished wish was to bring Mathematics closer to society in order to make them understandable to the scientific community in general. Close analysis of his research papers shows that from the very beginning that he was concerned with making clear the applications of Mathematics in general, and Integral Geometry in particular, to other applied sciences, especially to the Theory of Geometric Probability, to Statistics and to Physics.

His long career as a researcher (more than fifty years), his global vision of the problems that aroused his interest, his professional gifts, his vocation, his perseverance and his enormous capacity for work enabled him to produce a body of scientific work of great value. According to the references that appear in Zentralblatt MATH and Mathematical Reviews, he wrote scientific papers that alone amount to approximately 2000 pages. In addition, he authored twenty-five books, all with an exceptional educational value, both from the teaching and learning points of view. As a complement to his professional activity, and beginnng in the 1970s, he wrote many articles devoted to mathematical dissemination and education. Every time he participated in meetings of this nature, he was greatly contented and radiated optimism to all those who attended his seminars and talks.

The content of his publications covers a wide range. We believe that it is very difficult, not to say impossible, to arrive at an objective classification of his work according to fields and specialities, since very often many of his papers could very well be assigned to several of them at once. Nevertheless, with the aim of showing the different lines of research to which he devoted his work, we have divided these Selected Papers into five main Chapters (see Section 6 in Contents). We also believe it is useful to accompany the contents of each of these sections by a short commentary, each one drawn up by a specialist. However, as a possible Historical overview, we also believe it apposite to situate briefly that part of his work dealing with invariant measures within the scientific context in which they were produced, since the concept of kinematic density is present in much of that work. In addition to the articles and papers that figure in this Selection, Santaló also wrote many others, some of which are of great interest.

Analysis of the history of the development of Integral Geometry enables us to state that the origin of this speciality is to be found in the so-called "*Buffon's Needle problem*", posed by this naturalist in his famous *Essai d'Arithmetique Morale*, written in the late 18th century. In this regard, Crofton's formulas, contained in his well-known article *On the theory of local probability*, should also be taken into consideration.

This type of problem, which at first appeared to be simply a game or a mathematical pastime, led to the development of a mathematical theory in its own right, due mainly to the work of Poincaré, Blaschke and his school. Among the most prestigious pupils belonging to Blaschke's school, we may mention, among others, Chern, Santaló, Wu, Petkanschin and Varga, but most of all the first two. All of them turned up for some time at Hamburg. Thus it was that in the first third of the 20th century, when Integral Geometry emerged as a specialized field of mathematics with its own content, Blaschke and his school, with Chern and Santaló in particular, employed Cartan's moving reference method in their work. This would turn out to be the most powerful tool in the development of the whole of Santaló's mathematical work, and in much of Chern's work too. Written in this language, these works on Integral Geometry have provided the basis for this discipline right down to our own time, the study of which remains obligatory for all those reseachers who are interested in this speciality.

When in a problem of probabilities the set of possible cases has the power of the continuous, this is known as "*Geometric Probability*". This definition coincides with the classical one, the number of cases being replaced by the "*measure*", which is required to be defined for each type of set, and which, furthermore, can vary if the measure is changed. By analysing the quotient (favourable cases / possible cases) in the problem of Buffon's Needle we find that infinite possibilities exist. These possibilities can be parametrized and identified again as points in the plane, so that one has as many positions as points, and thus it seems quite obvious that the area can be used to "*measure*" or "*count*" the number of points. As Santaló states in [76.1], "*in order to apply the idea of probability to randomly given elements that are geometric objects (such as points, lines, geodesics, congruent sets, movements of affinities) it is first necessary to define a measure for such sets of elements*".

Thus in the problems posed by geometric probabilities, it is towards the geometric interest in itself that the mathematician feels drawn, and he addresses these problems without concerning himself with whether or not there may be an underlying concept of probability. The discussion about what measure it is necessary to chose is related with the group that determines the geometry of the problem in the sense of F. Klein's Erlangen programme. This is the mathematical reason why in Santaló's work the properties of Lie groups and homogeneous spaces figure so explicitly. E. Cartan, by looking for differential invariants for the Lie transformation groups rather than delving into problems of geometric probabilities, was the first to justify the concept of density and to generalize it to sets of straight lines and planes in space. He proved that the property of such a density must be that its value does not change when all the lines and planes considered are subject to a movement in space. Obviously, density depends on the group used at each moment. Thus, Cartan paved the way from the theory of continuous groups towards the theory of geometric probabilities. It should be noted that invariance with respect to the group of movements is equivalent to stating, for example, that all straight lines have the same behaviour.

According to Santaló, the basis of Integral Geometry consists fundamentally of four words: *probability, measure, group and geometry*. Indeed, some of his most important results come from measuring directly in the group. For instance, he identifies all the positions of a particular figure in the plane with the movements that carry an initial given figure to each of the possible positions. The formulas found thereby are known as kinematic formulas, because they express this idea of movement, even though the group may not specifically be the group of movements. Kinematic density in Euclidean space was introduced by Poincaré. In modern terminology, this is the Haar measure for the group of movements. One of the basic problems in Integral Geometry is to obtain explicit formulas for the integrals of geometric quantities with respect to kinematic density in terms of well-known integral invariants. As Santaló pointed out in the introduction to [67.5], Integral Geometry consists of three steps that constitute Integral Geometry in Blaschke's original sense:

1. The definition of measure for sets of geometric objects with certain properties of invariance;
2. The evaluation of this measure for some particular sets;
3. The application of the result obtained to arrive at statements of geometric interest.

In 1976, Santaló published an outstanding book entitled "*Integral Geometry and Geometric Probability*", [76.1], which we refer to as *The Encyclopedia*, since it contained almost all the basic existing bibliographic information on this speciality available at the time of publication. This book, which is regarded as a masterpiece of Classical Integral Geometry, was translated into Russian and Chinese. In order to understand much of Santaló's work, a basic knowledge of the theory of differenciable manifolds and of Lie groups is required. This is because that all Integral Geometry in Blaschke's, Chern's and Santaló's sense is grounded on the analysis of invariant properties under the effect of groups that act as homogeneous spaces, since as is well-known these spaces always admit a differentiable manifold structure.

In accordance with the ideas of Poincaré, Cartan and Blaschke, Santaló defines and analyses densities and measures in homogeneous spaces. To this end, he uses very simple mathematical foundations; let G be a Lie group and H a closed subgroup of G; then G/H admits a differentiable manifold structure. Chern gives a necessary and sufficient condition for the existence of an invariant measure on G/H in terms of the constants of the structure of G. Santaló extends this result, giving conditions for the existence of a differentiable non-null form on G/H and whenever it is G-invariant. He defines such a form as a "*density*" on G/H, and by integration it leads to an "*invariant measure*" on the homogeneous space: under the hypothesis that G acts transitively on a manifold, and the invariant density (if it exists) is unique except for a constant factor. Basically, Integral Geometry in Santaló's and Chern's sense rests on the following basic result: "*A necessary and sufficient condition for a form ω, verifying $\omega(X, \ldots) = 0$ for any X belonging to the Lie algebra of H, to be a density for G/H is that its exterior differential vanishes*".

Santaló employed this result repeatedly to construct densities and invariant measures on different homogeneous spaces, which enabled him to measure geometric objects with a high degree of generalization. This theorem was used and generalized in various contexts by various authors.

Following Hadwiger, Bonnesen and Fenchel, Santaló introduced the quermass-integrals[1]. Among other results, Santaló obtained the measure for sets of r-planes intersecting a convex set. This measure enabled him to obtain interesting formulas on the geometric probability of subspaces intersecting a convex set in terms of the integrals of the mean curvatures, thereby generalizing the results of Blaschke and his school to Euclidean spaces of arbitrary dimension.

Thus, the combination of problems arising from processes involving geometric elements together with probabilistic questions paved the way to Stochastic Geometry, thus called by Kendall and Harding in 1974, which is currently applied to different fields in Pure Mathematics (such as the theory of convex bodies) and in Applied Mathematics. Santaló also expressed his interest in this speciality in a series of intersting articles, such as that he wrote on random mosaics.

Perhaps Santaló's most important contribution to the applied sciences was his having laid the mathematical foundations for the creation of a new applied science: *Stereology*. This he did with a series of papers published over many years. However, it appears that the origin of Stereology is to be found in [43.2]: *On the probable distribution of corpuscles in a body, deduced from the distribution of its sections and analogous problems*.

In the 1970s, Elias put forward the following definition: "*Stereology deals with a set of methods for the exploration of 3-dimensional space, when only sections of two dimensions are known, by solid sets or their projections*". According to Santaló, the basic objective of Stereology is "*to determine the measure of the distribution of convex*

[1] Averaged measures, which, by the Steiner Formula, can be indentified with the integrals of mean curvatures of a convex set in Euclidean space.

particles distributed randomly in 3-dimensional Euclidean space from the measure of the distribution of its sections with random figures of known form (for example, a convex body, a cylinder, a plane, a band or a line). Stereology is an interdisciplinary science that draws together subjects as seemingly disparate as Biology, Engineering, Minerology, Metallurgy, Biomedicine, Geometry and Statistics".

By applying the techniques of this speciality, volumes, areas, lengths, number of particles, forms of bodies, etc., can all be estimated. This enables Stereology to be successfully applied in other apparently disparate sciences. Santaló was also interested in Integral Geometry in spaces of constant curvature. He wrote many interesting papers on this subject, perhaps the most fundamental being [52.4]. He extends many of the properties of Euclidean Integral Geometry to these spaces; in particular, one might mention Steiner's formula, its generalizations and properties, as well as the Cauchy and Kubota formulas. Furthermore, Santaló defines new concepts; for example, for the case of negative curvature, convexity by subsets that behave like Euclideans sets; specifically, he introduces convex sets by horospheres.

Integral Geometry in Riemannian spaces of non-constant curvature cannot be based on the same principle as that use for constructing Integral Geometry in spaces of constant curvature, because in general such spaces do not admit a transitive group of transformations that preserve the metric, neither do they apply geodesics in geodesics. Therefore, density for a set of geodesics cannot be defined by the property of being invariant under a certain group of transformations. Nevertheless, as Santaló points out, one may adopt another approach and define a measure for sets of geodesics and sets of points, which, although it is not invariant under a group, has properties of invariance that make it interesting from the geometric point of view. Santaló defines an invariant measure for geodesics and analyses its properties and consequences.

The term Integral Geometry was also used in the mathematical bibliography in a sense apparently different from that of the Blaschke, Santaló and Chern school. In the early 20th century, J. Radon proved that a differentiable function on 3-dimensional Euclidean space can be explicitly determined by means of its integrals on the planes.

The mathematical theory of Radon's Transform was built around this idea, and is related to the Fourier transform of a function. The classical intepretation of X-rays can also be regarded as an attempt to reconstruct properties of a 3-dimensional body using the projection of such rays onto a plane. The modern geometric interpretation of the action of X-rays, as well as many other elements currently employed in Medicine, can be performed using Radon's transform.

While the two concepts of Integral Geometry (the schools of Blaschke, Santaló and Chern on the one hand, and those of Gelfand and Helgason on the other) appear not to be related, Guillemin points to the fact that they are. Smith, Solmon and Wagner (Bull. Amer. Math. Soc. 83 (1977), 1227–1270) published an exceptional article entitled: *Mathematical and practical aspects of the problem of reconstructing objects from radiographies*. The mathematics employed in this article is fundamentally the theory of Integral Geometry in Gelfand and Helgason's sense. Thus, it is possible to explain all the medical activities of Tomography from a mathematical point of view. This branch of Analysis and of Geometry has undergone considerable development in recent years. At present, it appears to be one of the most interesting lines of research in the field of Integral Geometry. During the last years of his life, Santaló was wont to attend all the congresses held on this subject and invited us to study and explore this new technique more deeply.

Perhaps as a result of the time he spent in Princeton and Chicago, Santaló pubished a series of interesting articles on the Theory of Relativity and Theoretical Physics which we have decided to omit from this Selection, since they depart from the main lines of his mathematical research.

THE EDITORS

3 Some Photographs of L. A. Santaló

Dean Fac. San Juan, Santaló, J. Rey Pastor, E. Corominas.
San Juan (Argentina) 1941.
(Courtesy of Instituto de España (Madrid))

Front row: A. Sagastume, A. Durañona, G. D. Birkhoff, J. Rey Pastor, L. A. Santaló.
Buenos Aires, July 25, 1942.
(Courtesy of Instituto de España (Madrid))

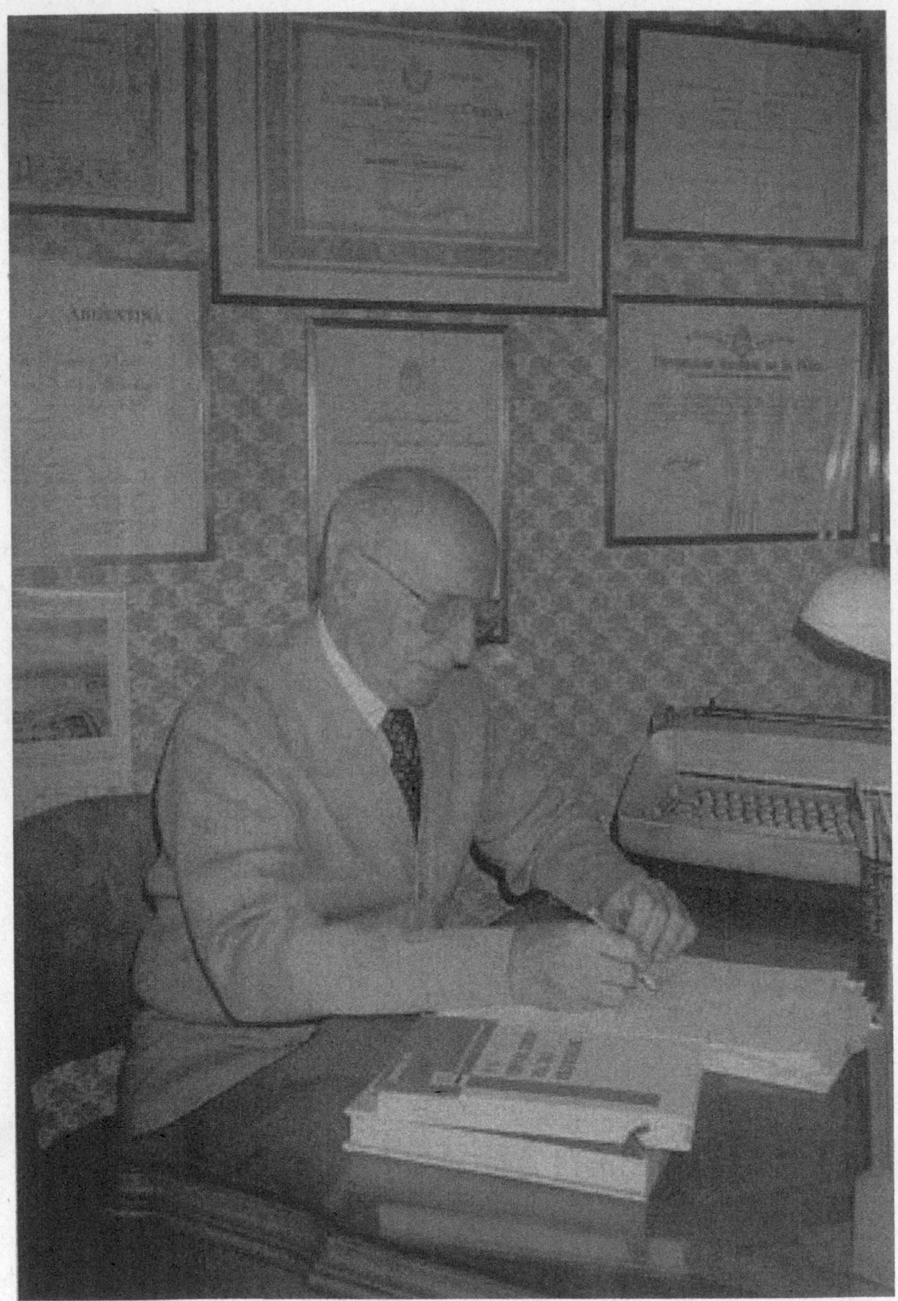

Santaló at home (1988).
(Courtesy of Santalo's family)

With Vidal-Abascal in Santiago de Compostela. 2nd International Colloquium on Differential Geometry. Santaló presents the work [67.1]. (Courtesy of Vidal-Abascal's family)

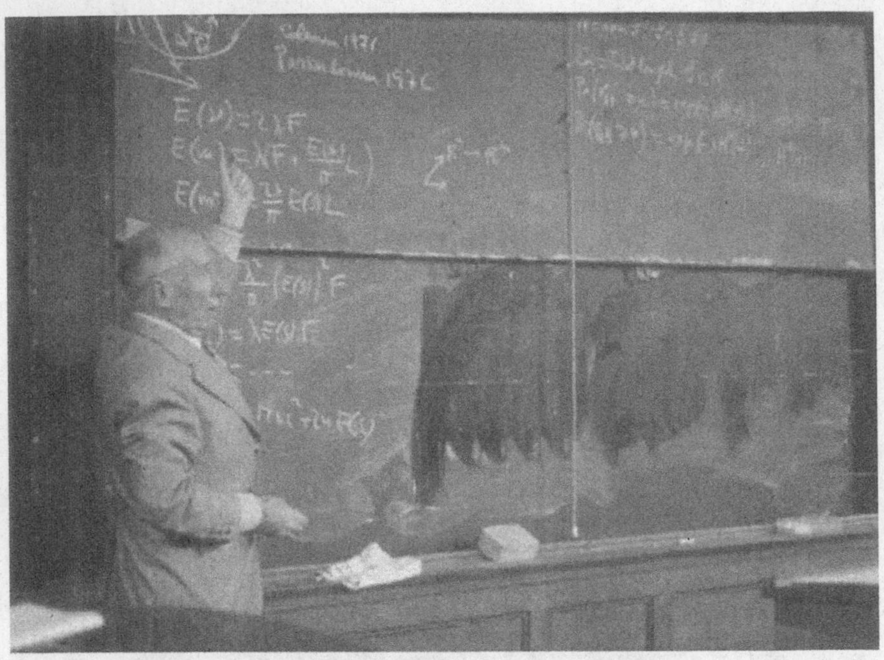

Explaining [78.3] at the Buffon Bicentenary Symposium. Paris, June 1977.
(Courtesy of L. M. Cruz-Orive)

Buffon Bicentenary Symposium. Jardin des Plantes, París, June 1977.
(Courtesy of L. M. Cruz-Orive)

With Cruz-Orive. Buffon Bicentenary Symposium.
Jardin des Plantes, París, June 1977.
(Courtesy of L. M. Cruz-Orive)

With S. M. King Juan Carlos I, receiving the Prince of Asturias Prize in Technical and Scientific Research, 1983.
(Courtesy of Santalo's family and Casa Real of Spain)

This price was established in 1981 and is one of the most prestigious given in Spain to members of the internacional scientific community. Santaló is the only one that had received this Prize for his contribution to pure mathematics.

Giving a talk.
(Courtesy of Santalo's family)

With Naveira in Santaló's office. Fac. Ciencias Matemáticas, Buenos Aires, September 1997.
(Courtesy of A. M. Naveira)

Santaló and his wife Hilda Rossi in the celebration
in the anniversary of its daughter Claudia, 1997.
(Courtesy of Santalo's family)

4 L. A. Santaló's Published Work

1934

[34.1.r][2] Some combinatorial problems, (Spanish); *Matemática Elemental* **3** (1934), 21–22.

[34.2.r] Area generated by a segment which moves keeping itself normal to a line and describing a developable surface, (Spanish); *Rev. Mat. Hisp.-Amer.* **9** (1934), 101–107.

[34.3.r] Developable surfaces through a straight line, (Spanish); *Las Ciencias*, **I** (1934), 1–7.

1935

[35.1.r] Some properties of spherical curves and a characteristic of the sphere, (Spanish); *Rev. Mat. Hisp.-Amer.* **10** (1935), 9–12.

[35.2.r] An integral formula for convex figures in the plane and the space, (Spanish); *Rev. Mat. Hisp.-Amer.* **10** (1935), 209–216.

1936

[36.1.r]* Integral Geometry 7: New applications of the concept of the kinematic measure in the plane and the space, (Spanish); *Rev. Acad. Ci. Exact. Fis. Nat.* (Ph. D. thesis), Madrid **33**, (1936), 451–504.

[36.2.r] Some problems referring to geometrical probabilities, (Spanish); *Rev. Mat. Hisp.- Amer.* **11** (1936), 87–97.

[36.3.r] Integral Geometry 4: On the kinematic measure in the plane, (Spanish); *Abh. Math. Sem. Univ. Hamburg* **11** (1936), 222–236.

[36.4.r] Integral Geometry 5: On the kinematic measure in space, (Spanish); *Actualités Sci. Industr. Hermann* **357**, París, 1936.

[36.5.r] Curves on a surface which hold the condition $\delta \int(k, \tau)ds = 0$, (Spanish); *Rev. Mat. Hisp.-Amer.* **11** (1937), 129–138.

1937

[37.1.r] Integral geometrie 15: Fundamental formula of the kinematic measure for cylinders and moving parallel planes, *Abh. Math. Sem. Univ. Hamburg* **12** (1937), 38–41.

---

[2] The notation $[\alpha.\beta.\gamma]$ denotes: α the year of the publication, β the order in that year and γ means book (b), research (r), general interest (g) or education (e). Also the references indicated with (*) are those chosen for the "Selecta".

4 L. A. Santaló's Published Work

1939

[39.1.r] Integral Geometry of unlimited figures, (Spanish); *Publ. Inst. Mat. Univ. Nac. Litoral*, Rosario **1** (1939), 1–58.

1940

[40.1.r]* On some problems of geometric probabilities, (French); *Tôhoku Math. J.* **47** (1940), 159–171.

[40.2.r]* A theorem on sets of parallelepipeds with parallel edges, (Spanish); *Publ. Inst. Mat. Univ. Nac. Litoral*, Rosario **2** (1940), 49–60.

[40.3.r] Integral Geometry 31: On mean values and geometric probabilities, (Spanish); *Abh. Math. Sem. Univ. Hamburg* **13** (1940), 284–294.

[40.4.r] Integral Geometry 32: Some integral formulae in the plane and in the space; *Abh. Math. Sem. Univ. Hamburg* **13** (1940), 344–356.

[40.5.r] A demonstration of the isoperimetric property of the circle, (Spanish); *Publ. Inst. Mat. Univ. Nac. Litoral*, Rosario **2** (1940), 37–46.

[40.6.g] On some geometrical problems concerning aviation, (Spanish); *Boletín Matemático*, Buenos Aires **13** (1940), 66-71.

[40.7.g] On continuous probabilities, (Spanish); *Ciencia*, Méjico **1**, 1940.

1941

[41.1.r]* A generalization of a theorem of Kubota on ovals, (German); *Tôhoku Math. J.* **48** (1941), 64–67.

[41.2.r]* Proof of a theorem of Bottema on ovals, (German); *Tôhoku Math. J.* **48** (1941), 221–224.

[41.3.r]* A theorem and an inequality referring to rectificable curves, *Amer. J. Math.* **63** (1941), 635–644.

[41.4.r]* Curves of extremal total torsion and *D*-curves, (Spanish); *Publ. Inst. Mat. Univ. Nac. Litoral*, Rosario **3** (1941), 131–156.

[41.5.r] Some infinitesimal properties of plane curves, (Spanish); *Math. Notae* **1** (1941), 129–144.

[41.6.r] Generalization of a problem of geometrical probability, (Spanish); *Rev. Un. Mat. Argentina* **7** (1941), 129–132.

[41.7.r] Nicolas Tartaglia and the resolution of the equation of third order, (Spanish); *Math. Notae* **1** (1941), 26–33.

[41.8.r] A system of mean values in the theory of geometric probabilities, (Spanish); *Revista Ci. Lima* **43** (1941), 147-154.

[41.9.g] The mathematics and language, (Spanish); *Asoc. Cult. Conferencias*, Rosario, 1941.

[41.10.g] Probability and its several applications, (Spanish); *Asoc. Cult. Conferencias*, Rosario, 1941.

[41.11.r] The mean value of the number of parts into which a convex domain is divided by *n* arbitrary straight lines, (Spanish); *Rev. Un. Mat. Argentina* **7** (1941), 33–37.

1942

[42.1.r]* Integral formulas in Crofton's style on the sphere and some inequalities referring to spherical curves, *Duke Math. J.* **9** (1942), 707–722.

[42.2.r]* On the isoperimetric inequality for surfaces of constant negative curvature, (Spanish); *Univ. Nac. Tucumán Rev.* **3** (1942), 243–259.

[42.3.r]* Supplement to the note: A theorem on sets of parallelepipeds with parallel edges, (Spanish); *Publ. Inst. Mat. Univ. Nac. Litoral*, Rosario **3** (1942), 202–210.

[42.4.r] On the concept of curvature of a surface, (Spanish); *Math. Notae* **2** (1942), 165–184.

[42.5.r] An integral formula concerning convex figures, (Spanish); *Rev. Un. Mat. Argentina* **8** (1942), 165–169.

[42.6.r] On certain varieties of the type of a developable in Euclidean space of four dimensions, (Spanish); *Publ. Inst. Mat. Univ. Nac. Litoral*, Rosario **4** (1942), 3–42.

[42.7.r] Some mean values and inequalities relating to curves on the sphere, (Spanish); *Rev. Un. Mat. Argentina* **8** (1942), 113—125.

[42.8.r] Isaac Newton and the binomial theorem, (Spanish); *Math. Notae* **2** (1942), 61–72.

[42.9.r] Some properties of the twisted curves in the affine differential geometry, (French); *Portugaliae Mat.* **3** (1942), 63–68.

[42.10.r] Surfaces of constant negative curvature, (Spanish); *Univ. Nac. Tucumán Rev.* **3** (1942), 243–259.

[42.11.r] (With Cosnita, Thebault and Court). Problems and solutions. Advanced problems: Problems for solution: 4036–4039. *Amer. Math. Monthly* **49** (1942), 340–341.

[42.12.g] Possibilities of interplanetary flight, (Spanish); *Rev. Ingeniería y Arquitectura*, Rosario, 1942.

[42.13.e] What must be done for the progress of mathematics in Argentina?, (Spanish); *Publ. Facultad Ci. Mat., Físico-Químicas y Nat. Apl. a la Industria*. Univ. Nac. Litoral **34** (1942), 41-45.

[42.14.g] Probability and its several applications. Conference published by *Asoc. Cult. Conferencias*, Rosario, 1942.

1943

[43.1.r]* Integral Geometry on surfaces of constant negative curvature, *Duke Math. J.* **10** (1943), 687–709.

[43.2.r]* On the probable distribution of corpuscles in a body, deduced from the distribution of its sections, and analogous problems, (Spanish); *Rev. Un. Mat. Argentina* **9** (1943), 145–164.

[43.3.r] (With Fritz). Problems and solutions: Advanced problems: Solutions: 4036. *Amer. Math. Monthly* **50** (1943), 397–399.

[43.4.r] A characteristic property of the circle, (Spanish); *Math. Notae* **3** (1943), 142–147.

[43.5.r] Some inequalities between the elements of a triangle, (Spanish); *Math. Notae* **3** (1943), 65–73.

[43.6.r] On the osculating conic section at an ordinary point of a plane curve, (Spanish); *Rev. Un. Mat. Argentina* **9** (1943), 53–60.

[43.7.r] On some systems of linear equations and their determinants, (Spanish); *Math. Notae* **3** (1943), 129–184.

[43.8.g] Brief history and current state of some chimera and fantasies of human being, (Spanish); *Rev. Centr. Estud., Fac. Ciencias Matemáticas*, Rosario, 1943.

[43.9.r] Solution to the question n° 8 (on a question of geometry), (Spanish); *Math. Notae* **3** (1943), 105–111.

1944

[44.1.r]* Note on convex spherical curves, *Bull. Amer. Math. Soc.* **50** (1944), 528–534.

[44.2.r] (Whith Frink, Jr., Thebault and Dulmage). Problems and solutions. Elementary problems: Problems for solutions E646–E650. *Amer. Math. Montly* **51** (1944), 586–587.

[44.3.r] (With Erdos, Brauer and Cohen). Problems and solutions: Advanced problems: Solutions: 4070. *Amer. Math. Monthly* **51** (1944), 234–236.

[44.4.r] Origin and development of Integral Geometry, (Spanish); *Rev. Univ. Católica Perú* **12** (1944), 205–230.

[44.5.r] Area bounded by the curve generated by the end of a segment whose other end traces a fixed curve, and application to the derivation of some theorems on ovals, (Spanish); *Math. Notae* **4** (1944), 213–226.

[44.6.r] Bounds for the length of a curve or for the number of points necessary for an approximate covering of a domain, (Spanish); *An. Acad. Brasil. Ci.* **16** (1944), 111–121.

[44.7.r] Properties of convex figures on a sphere, (Spanish); *Math. Notae* **4** (1944), 11–40.

[44.8.r] Solution to the question n° 11 (on a question of probability), (Spanish); *Math. Notae* **4** (1944), 105–111.

1945

[45.1.r]* Mean value of the number of regions into which a body in space is divided by n arbitrary planes, (Spanish); *Rev. Un. Mat. Argentina* **10** (1945), 101–108.

[45.2.r]* Note on convex curves on the hyperbolic plane, *Bull. Amer. Math. Soc.* **51** (1945), 405–412.

[45.3.r] (With Rosenbaum). Problems and solutions: Elementary problems: Solutions: E665. *Amer. Math. Montly* **52** (1945), 521–522.

[45.4.r] (With Fine and Eves). Problems and solutions: Elementary Problems: Solutions: E649. *Amer. Math. Montly* **52** (1945), 344–345.

[45.5.r] (With Eves, Thebault, Kaplansky and Browne). Problems and solutions: Elementary problems: Problems for solution: E661-E665, E637. *Amer. Math. Montly* **52** (1945), 159.

[45.6.r] Some properties of skew curves in projective differential geometry, (Spanish); *Actas Acad. Ci. Lima* **8** (1945), 203–216.

[45.7.r] Addendum to the note "On a Diophantine problem", (Spanish); *Math. Notae* **5** (1945), 162–171.

[45.8.r] Surfaces whose *D*-curves are geodesics or isogonal trajectories of the lines of curvature, (Spanish); *Publ. Inst. Mat. Univ. Nac.* Litoral **5** (1945), 255–267.

[45.9.r] On the circle of maximum radius contained in a domain, (Spanish); *Rev. Un. Mat. Argentina* **10** (1945), 155–162.

[45.10.r] A theorem on conformal mapping, (Spanish); *Math. Notae* **5** (1945), 29–40.

[45.11.g] On the problem of the radius of action of the airplanes, (Spanish); *Rev. Centr. Estud., Facultad Ciencias Matemáticas*, Rosario, 1945.

[45.12.g] Geometric probabilities and Integral Geometry, (Spanish); *Bol. Fac. Ingeniería*, Montevideo **3**, (1945), 91–113.

[45.13.g] Origin and evolution of some mathematical theories, (Spanish); *Rev. Ingeniería*, Montevideo, 1945.

[45.14.g] Contribution of the aeronautic to the progress of the sciences, (Spanish); *Asoc. Cult. Conferencias*, Rosario, 1945.

1946

[46.1.r]* Convex regions on the *n*-dimensional spherical surface, *Ann. of Math.* **47** (1946), 448–459.

[46.2.r]* On convex bodies of constant width in E^n, (Spanish); *Portugaliae Math.* **5** (1946), 195–201.

[46.3.r]* A geometrical characterization for the affine differential invariants of a space curve, *Bull. Amer. Math. Soc.* **52** (1946), 625–632.

[46.4.r] Some integral formulas referring to convex bodies, (Spanish); *Rev. Un. Mat. Argentina* **12** (1946), 78-87.

[46.5.r] On the length of a space curve as mean value of the lengths of its orthogonal projections, (Spanish); *Math. Notae* **6** (1946), 158–166.

[46.6.b] History of aeronautics, (Spanish); *Espasa-Calpe Argentina*, 1946.

[46.7.r] (With Stewart). Problems and solutions: Advanced problems: Solutions 4151. *Amer. Math. Montly* **53** (1946), 342–344.

[46.8.r] On plane hyperconvex figures, (Spanish); *Summa Brasil. Math.* **1** (1946), 221-239.

[46.9.r] (With Levi and De María). Enumerative studies on the varieties of contact of the surfaces in a space of *n*-dimensions, (Spanish); *Fac. Ci. Mat. Univ. Nac. Litoral. Publ. Inst. Mat*, Rosario. **8** (1946), 3–72.

[46.10.r] On a linear complex related to a closed space curve, (Spanish); *Math. Notae* **6** (1946), 45–56.

[46.11.g] Contribution of aviation to the progress of science, (Spanish); *Asoc. Cult. Conferencias* Rosario **1** (1946), 3–15.

1947

[47.1.r]* Affine invariants of certain pairs of curves and surfaces, *Duke Math. Journal* **14** (1947), 559–574.

[47.2.r]* On the first two moments of the measure of a random set, *Ann. Math. Statistics* **18** (1947), 37–49.

[47.3.r]* *D*-curves on cones, (Spanish); *Math. Notae* **7** (1947), 179–190.

[47.4.r] (With Bellman, Thebault, Dyson and Eves). Advanced problems and solutions: Problems for solution: 4259–4263, 4248. *Amer. Math. Montly* **54** (1947), 418–419.

[47.5.r] A characteristic property of the quadrics of revolution and of cylinders whose cross section is a logarithmic spiral, (Spanish); *Math. Notae* **7** (1947), 81–90.

[47.6.r] On the measure of the set of congruent figures contained in the interior of a rectangle or of a triangle, (Spanish); *Actas Acad. Ci. Lima* **10** (1947), 103–116.

1948

[48.1.r] An affine invariant for closed convex plane curves, (Spanish); *Math. Notae* **8** (1948), 103–111.

[48.2.r] On the distribution of planes in space, (Spanish); *Rev. Un. Mat. Argentina* **13** (1948), 120–124.

1949

[49.1.r]* Integral Geometry in three-dimensional spaces of constant curvature, (Spanish); *Math. Notae* **9** (1949), 1–28.

[49.2.r]* Integral Geometry on surfaces, *Duke Math. J.* **16** (1949), 361–375.

[49.3.r]* An affine invariant for convex bodies in *n*-dimensional space, (Spanish); *Portugaliae Math.* **8** (1949), 155–161.

[49.4.r] Some inequalities between the elements of a tetrahedron in non-Euclidean geometry, (Spanish); *Math. Notae* **9** (1949), 113–117.

[49.5.r] Advanced problems and solutions: Solutions: 4262. *Amer. Math. Montly* **56** (1949), 270–271.

1950

[50.1.r]* On parallel hypersurfaces in the elliptic and hyperbolic *n*-dimensional space, *Proc. Amer. Math. Soc.* **1** (1950), 325–330.

[50.2.r]* Integral Geometry in general spaces, *Proceedings of the International Congress of Mathematiciens, Cambridge Mass. Amer. Math. Soc.* Providence, R. I. **1** (1950), 482–489.

[50.3.r]* Integral Geometry in projective and affine spaces, *Ann. of Math.* **51** (1950), 739–755.

[50.4.r] On some integral formulas and mean values concerning movable convex figures in the plane, (Spanish); *Publ. Fac. Ci. Exact. Fis. Nat. Univ. Buenos Aires* **1** (1950), 23–45.

[50.5.r] Some integral formulas and a definition of q-dimensional area for a set of points, (Spanish); *Univ. Nac. Tucumán Rev.* **7** (1950), 271–282.

[50.6.r] Integral Geometry in 3-dimensional spaces of constant curvature, (Spanish); *Math. Notae* **9** (1950), 1–28.

[50.7.r] Remarks on surfaces and inscript polyhedrons, (Spanish); *Las Ciencias*, Madrid **15**, 1950.

[50.8.r] Applications and current problems of some mathematical theories, (Spanish); *An. Soc. Científica Argentina* **150**, 1950, 136–154.

1951

[51.1.b] (With Rey Pastor). Integral Geometry, (Spanish); *Espasa-Calpe*, Buenos Aires, 1951.

[51.2.r] Two characteristic properties of circles on a spherical surface, (Spanish); *Math. Notae* **11** (1951), 73–78.

[51.3.r] Probability in geometrical constructions, (Spanish); *An. Soc. Ci. Argentina* **152** (1951), 203–229.

[51.4.r] On permanent vector-varieties in n-dimensions, *Portugaliae Math.* **10** (1951), 125–127.

[51.5.r] Generalization of an inequality of H. Hornich to spaces of constant curvature, (Spanish); *Rev. Un. Mat. Argentina* **15** (1951), 62–66.

[51.6.r] On pairs of convex figures, *Gaz. Mat.*, Lisboa **12** (1951), 7–10.

1952

[52.1.r]* Measure of sets of geodesics in a Riemannian space and applications to integral formulas in elliptic and hyperbolic spaces, *Summa Brasil. Math.* **3** (1952), 1–11.

[52.2.r]* Integral Geometry in Hermitian spaces, *Amer. J. Math.* **74** (1952), 423–434.

[52.3.r] Some mean values on the hemisphere, (Spanish); *Math. Notae* **12–13** (1952), 32–37.

[52.4.r] Integral Geometry in spaces of constant curvature, (Spanish); *Rep. Argentina. Publ. Comisión Nac. Energía Atómica, Serie Matemática* **1** (1952), 1–68.

[52.5.r] Problems of Integral Geometry, (Spanish); *Symposium on "Some mathematical problems which are being studied in Latin America"*, Punta del Este, Uruguay, December 1951. Cent. Coop. Ci. UNESCO for Latin America, Montevideo, (1952), 23–40.

[52.6.g] New problems posed to mathematics by other sciences, (Spanish); Bol. Centro Coop. Cient. UNESCO, Montevideo, 1952.

1953

[53.1.b] Introduction to Integral Geometry. *Act. Sci. Ind., 1198. Publ. Inst. Mat. Univ. Nancago, II, Hermann*, París, 1953.

[53.2.r] Algebraic curves and analytic curves, (Spanish); *Univ. Nac. Eva Perón Publ. Fac. Ci. Físico-Mat.* **4** (1953), 493–506.

[53.3.r] Correction to the article: " On pairs of convex figures". *Gaz. Mat.* Lisboa **14** (1953), 6.

[53.4.g] The problem of the unification of fields: the last theory of Einstein, (Spanish); *Mundo Atómico*, Buenos Aires, **4** 1953.

[53.5.g] The last theory of the unified field by Einstein, (Spanish); *Ciencia e Investigación* Buenos Aires, **9** 1953.

1954

[54.1.r]* Questions of differential and Integral Geometry in spaces of constant curvature, (Spanish); *Univ. e Politec. Torino. Rend. Sem. Mat.* **14** (1954), 277–295. (1955).

[54.2.r] Questions of the affine differential geometry of surfaces, (Spanish); *Second Symposium on "Some mathematical problems which are studying in Latin America"*, Cent. Coop. Ci. UNESCO for Latin America, Villavicencio, Argentina (1954), 21–33. (1955).

[54.3.r] On a theorem of Holditch and analogous in non-Euclidean geometry, (Spanish); *Math. Notae* **14** (1954), 32–49.

[54.4.r] On some tensors analogous to that of curvature in spaces with a non-symmetrical affine connection, (Spanish); *Univ. Nac. Tucumán Rev.* **10** (1954), 19–26.

[54.5.r] Some generalizations of the four-vertex theorem, (Spanish); *Math. Notae* **11** (1954), 69–78.

[54.6.r] On the kinematic formula in spaces of constant curvature, *Proceedings of the International Congress of Mathematicians*, Amsterdam **2** (1954), 251–252.

[54.7.r] Generalization of a geometric inequality of Feller, (Spanish); *Rev. Un. Mat. Argentina* **16** (1954), 78–81.

[54.8.r] Probability in non-Euclidean geometries, (Spanish); *Estocástica* **2**, 1954.

[54.9.g] Modern aspects in the field of geometry, (Spanish); *Ciencia y Tecnología*, OEA, **4**, Unión Panamericana, Washington 6. D. C., 1954.

1955

[55.1.r]* On the distribution of sizes of particles contained in a body given the distribution in its sections or projections, (Spanish); *Trabajos Estadist.* **6** (1955), 181–196.

[55.2.r] On geometry of numbers, *J. Math. Soc. Japan.* **7** (1955), 208–213.

[55.3.b] (With Rey Pastor and Balanzat). Analytical geometry, *Editorial Kapelusz*, Buenos Aires, 1955.

[55.4.b] Probability and its applications, (Spanish); *Editorial Iberoamericana*, Buenos Aires, 1955.

[55.5.r] On the chords of a convex curve, (Spanish); *Rev. Un. Mat. Argentina* **17** (1955), 217–222.

[55.6.r] On the uniqueness of the elementary vector operators, (Spanish); *Math. Notae* **14** (1955), 120–132. (1956).

[55.7.g] The work by Einstein in the mathematical field, *Ciencia e Investigación*, Buenos Aires, 1955.

1956

[56.1.r]* On the mean curvatures of a flattened convex body, *Rev. Fac. Sci., Univ. Istanbul* **21** (1956), 189–194.

[56.2.b] Introduction to Integral Geometry, (Russian); *Izdat. Insotr. Lit.*, Moscow, 1956.

[56.3.r] Curves on a surface which are extremals of a function of the curvature and the torsion, (Spanish); *Abh. Math. Sem. Univ. Hamburg*, **20** (1956), 216–222.

[56.4.r] On the measure of the linear spaces which cut a convex body and related problems, (conference on questions of reality in geometry), Liège, (French); *Masson & Cie*, París (1955), 177–190.

1957

[57.1.r] Affine differential geometry and convex bodies, (Spanish); *Math. Notae* **16** (1957), 20–42.

[57.2.r] Some properties of the local conformal representation of one surface on another, (Spanish); *Rev. Un. Mat. Argentina* **18** (1957), 45–52.

[57.3.r] A new affine invariant of plane and solid convex bodies, (Spanish); *Math. Notae* **16** (1957), 78–91. (1958).

[57.4.r] Some inequalities referring to convex figures of the plane and the space, (Spanish); *Actas Reunión Un. Mat. Argentina*, Bahía Blanca, 1957.

1959

[59.1.r]* On complete systems of inequalities between three elements of a plane convex figure, (Spanish); *Math. Notae* **17** (1959), 82–104. (1961).

[59.2.r] On Einstein's unified field equations, (Spanish); *Univ. Nac. Tucumán Rev.* **12** (1959), 31–55.

[59.3.g] Professor Dr. Alberto Sagastume Berra (obituary), (Spanish); *Math. Notae* **17** (1959), 57–58.

1960

[60.1.r]* Steiner's formula for parallel surfaces in affine geometry, (Spanish); *Univ. Nac. Tucumán Rev.* **13** (1960), 194–208.

[60.2.r]* Two applications of the Integral Geometry in affine and projective spaces, *Publ. Math. Debrecen* **7** (1960), 226–237.

[60.3.r] On unified field theories, (Spanish); *Rev. Un. Mat. Argentina* **19** (1960), 196–206.

[60.4.g] Analitic geometry and synthetical geometry, (Spanish); *An. Acad. Ci. Exact. Fís. y Nat.* **15**, (1960), 9–31.

[60.5.g] Perspective for the development of mathematics in Latin America, (Spanish); *Rev. Un. Mat. Argentina* **20** (1960), 23–32.

1961

[61.1.b] Non-Euclidean geometries, (Spanish); *EUDEBA*, Buenos Aires, 1961. 64 pp. Reedited in 1961, 1963, 1966, 1969.

[61.2.b] Vectors and tensors, with applications, (Spanish); *EUDEBA*, Buenos Aires, 1961. 383 pp.

[61.3.g] Mathematics in Argentina, (Spanish); *Rev. Universitaria de Buenos Aires*, **5**, 1961.

[61.4.g] Analytic geometry and synthetical geometry, (Spanish); *Ciencia e Investigación*, Buenos Aires **17** (1961), 145–154.

1962

[62.1.r]* On the Gauss-Bonnet formula for polyhedra in spaces of constant curvature, (Spanish); *Rev. Un. Mat. Argentina* **18** (1962), 79–91.

[62.2.r]* On the measure of sets of parallel linear subspaces in affine space, *Canad. J. Math.* **14** (1962), 313–319.

[62.3.r]* On the fundamental kinematic formula of Integral Geometry in spaces of constant curvature, (Spanish); *Math. Notae* **18** (1962), 79–94.

[62.4.r] On some characteristic properties of the sphere (homage to Terracini and Cernuschi), (Spanish); *Univ. Nac. Tucumán Rev.* **14** (1962), 287–297.

[62.5.r] (With Babini and González Domínguez). Julio Rey Pastor, *Rev. Un. Math. Argentina* **21** (1962), 3–22.

[62.6.g] The scientific work of Beppo Levi, (Spanish); *Math. Notae* **18** (1962), 23–28.

[62.7.e] (With Valeiras). The formation of teachers in mathematics, (Spanish); *Educación Matemática en las Américas I*, Columbia University, Teachers College, 1962.

[62.8.e] New trends in the teaching of geometry, (Spanish); *Instituto Superior del Profesorado, Buenos Aires, 1962, 25 pp.*, Columbia University, Teachers College, 1962.

1963

[63.1.r]* A relation between the mean curvatures of parallel convex bodies in spaces of constant curvature, (Spanish); *Rev. Un. Mat. Argentina* **21** (1963), 131–137.

[63.2.g] (With R. Carranza). Finite geometries and its applications, (Spanish); *Ciencia e investigación*, Buenos Aires **19** (1963), 49–62.

[63.3.e] The teaching of sciences in the high school: The mathematic, (Spanish); *Ciencia e investigación* **19** (1963), 245–252.

1965

[65.1.b] Introduction to the differential geometry of differentiable manifolds, (Spanish); *Cursos y Seminarios de Matemática, Universidad de Buenos Aires*, Buenos Aires **19**, 1965, 153 pp.

[65.2.r] Integral Geometry of the projective groups on the plane depending on more than three parameters, *An. Sti.. Univ. "Al. I. Cuza" Iasi* **11** (1965), 307–335.

[65.3.g] Vector spaces and analytical geometry, (Spanish); *Washington, Monografías de la OEA*, 1965, 56 pp.

[65.4.e] Modern mathematics in the primary and high school, (Spanish); *La Educación*, OEA Washington **37** (1965), 25–44.

[65.5.g] Algebra and geometry: their relationships, (Spanish); *An. Acad. Nac. Ci. Exact. Fís. y Naturales* Buenos Aires **20** (1965), 47–63.

[65.6.g] Premium "Sociedad científica argentina" and seminary of mathematic "Claro C. Dassen", (Spanish); *Bol. Inf. De la Soc. Cient. Argentina*, Buenos Aires, **30** (1965), 1–16.

1966

[66.1.r]* Groups of the plane with respect to which sets of points and lines admit an invariant measure, (Spanish); *Rev. Un. Mat. Argentina* **23** (1966), 119–148. (1967).

[66.2.r]* Average values for polygons formed by random lines in the hyperbolic plane, (Spanish); *Univ. Nac. Tucumán Rev.* **16** (1966), 29–43.

[66.3.b] Projective geometry, (Spanish); *EUDEBA*, Buenos Aires, 1966. 371 pp.

[66.4.r] On Einstein's unified field theory, *Perspectives in geometry and relativity. Essays in Honor of V. Hlavaty*, Indiana Univ. Press, Bloomington (1966), 343–352.

[66.5.b] Mathematics in the secondary school, (Spanish); *EUDEBA*, Buenos Aires, 1966. 64 pp.

[66.6.e] Problems founded for the reform of the teaching of mathematic in Latin America referring to teachers and programs, (Spanish); *Ed. Mat. en las Américas II*, (1966), 23–29.

[66.7.e] (With Voelker). Preparation of mathematical teachers for teaching in high school, (Spanish); *Ed. Mat. en las Américas II*, (1966), 189–196.

1967

[67.1.r]* Total absolute curvatures of manifolds contained in Euclidean space, (Spanish); *Acta Ci. Compostelana* **4** (1967), 149–158.

[67.2.r]* Horocycles and convex sets in the hyperbolic plane, *Arch. Math. (Basel)* **18** (1967), 529–533.

[67.3.r] Spaces with two affine connections, *Bull. Calcutta Math. Soc.* **59** (1967), 3–8.

[67.4.r] On the converse of a theorem of Jacobi relative to space curves, (Spanish); *Univ. Nac. Tucumán Rev. Ser. A* **17** (1967), 83–89.

[67.5.r] Integral Geometry, *Studies in global geometry and analysis*, Edited by S. S. Chern, *The Math. Assoc. Amer., Prentice Hall*, (1967), 147–193.

1968

[68.1.r]* Horospheres and convex bodies in hyperbolic space. *Proc. Amer. Math. Soc.* **19** (1968), 390–395.

[68.2.r] Alessandro Terracini (1889–1968), *Rev. Un. Mat. Argentina* **23** (1968), 149–151.

[68.3.r] (With Dieulefait and Ferrari). Prof. Juan Olguin, α: 1.I.1902, ω: 6.II.1968, (Spanish); *Math. Notae* **21** (1968), 1–8.

1969

[69.1.r]* Convexity in the hyperbolic plane, (Spanish); *Univ. Nac. Tucumán Rev.* **19** (1969), 173–183.

[69.2.r]* On some geometric inequalities in the style of Fary, *Amer. J. Math.* **91**, (1969), 25–31.

[69.3.r] Some problems of differential geometry, (Spanish); *Notas Ci. Ser. M Mat.* **7** (1969), 27–34.

1970

[70.1.r]* Mean values and curvatures, *Izv. Akad. Nauk. Armejan*, SSR, Ser. Math. **5** (1970), 286–295. Reprinted in the book *Stochastic Geometry*, (tribute memory Rollo Davidson), Ed. Harding-Kendall, Wiley, London (1974), 165-174.

[70.2.r] Probabilities on convex bodies and cylinders, (Spanish); *Rev. Un. Mat. Argentina* **25** (1970), 95–104. (1971)

[70.3.g] Probability and statistical inference, (Spanish); *Washington, Monografías de la OEA*, 1970. 132 pp.

[70.4.e] Mathematics in the Faculty of Exact and Natural Sciences of the University of Buenos Aires in the period 1865–1930, (Spanish); *Boletín Academia de Ciencias*, Córdoba, Buenos Aires (1970), 255–273.

1972

[72.1.r]* (With Yáñez). Averages for polygons formed by random lines in Euclidean and hyperbolic planes, (Spanish); *J. Appl. Probab.* **9** (1972), 140–157.

[72.2.r] Unified field theories of Einstein's type deduced from a variational principle: conservation laws. Commemoration volumes for Prof. Dr. Akitsugu Kawaguchi's seventieth birthday, Vol. II.*Tensor* **25** (1972), 383–389.

[72.3.r] On certain asymmetric unified field theories, (Spanish); *Rev. Acad. Ci. Exact. Fis. Nat.* Madrid **66** (1972), 395–425.

[72.4.e] Mathematics and Education, (Spanish); *Office of Science of UNESCO for Latin America*, Montevideo, 1972.

[72.5.g] Editorials on Scientific Policy, (Spanish); *Ciencia e Investigación*, 1968–72 (period in which he was co-director).

1973

[73.1.e] Mathematics and its teaching at the elementary, medium and superior level, (Spanish); *Actas del IV Congreso Bolivariano de Matemáticas*, Panamá, 1973. (Published also in *Conceptos de Matemática* **27**, 1973).

1974

[74.1.r]* Total curvatures of compact manifolds immersed in Euclidean space, *Symposia Math., Inst. Naz. di Alta Matematica, Academic Press*, Roma **14** (1974), 363–390.

[74.2.r] Curves and quaternions, (Spanish); *Rev. Un. Mat. Argentina* **27** (1974), 41–52. (1975).

[74.3.e] The applications of mathematics in the Secondary School: Role of statistics and probability, (Spanish); In the book "The applications of teaching and learning of mathematics in High School (Spanish)", *Office of Science of UNESCO for Latin America*, Montevideo, 1974.

[74.4.g] Panorama of mathematics in Latin America in 1974, (Spanish); *Office of Science of UNESCO for Latin America*, Montevideo **8**, 1974.

1975

[75.1.r]* The kinematic formula in Integral Geometry for cylinders, *Ann. Mat. Pura Appl.* **103** (1975), 71–79.

[75.2.r] Geodesics in Gödel-Synge spaces (homage to Lora Tamayo), (Spanish); *Rev. R. Acad. Ci. Ex. Fís. Nat.* Madrid (1975), 51–69.

[75.3.b] [With other authors] Evolution of the Sciences in the República Argentina, 1923–1972. Vol. I: Mathematics, (Spanish); *Soc. Ci. Argentina*, 1972. 243 pp.

[75.4.b] Mathematical education today, (Catalan); *Teide*, Barcelona, 1975. 108 pp.

[75.5.e] The theory of sets and the teaching of mathematics, (Spanish); *Conceptos de Matemática* **34** (1975), 4–10.

1976

[76.1.b] Integral Geometry and Geometric Probability, (with a foreword by Marc Kac), *Enciclopedia of Mathematics and its Applications, Massachussets, Addison-Wesley, Reading*, 1976. 404 pp.

[76.2.b] Spinor geometry, (Spanish); Courses in Mathematics, No. 2, Consejo nacional de investigaciones científicas y técnicas, *Instituto argentino de matemática*, Buenos Aires, 1976. 130 pp.

[76.3.r] Random segments in E^n, (Spanish); *Univ. Nac. Tucumán Rev.* **26** (1976), 229–238. (1981).

[76.4.e] The third international symposium about mathematical education, (Spanish); *Conceptos de Matemática* **40** (1976), 19–24.

[76.5.b] (With Deulofeu and Galloni). History of the "Academia Nacional de Ciencias Exactas, Físicas y Naturales", (Spanish); Buenos Aires 1976. 126 pp. 19–24.

1977

[77.1.r] Sets of segments and graphs in the plane, (Spanish); *Dep. Mat., Fac. Ci. Ex. Fís. Nat., Univ.* Buenos Aires **39** (1977), 1–37.

[77.2.e] The teaching of mathematics: from Plato to modern mathematics, (Spanish); *Rev. Inst. Investigaciones Educativas (IIE)*, Buenos Aires **13** (1977), 3–26.

[77.3.e] The current debate on modern mathematics, (Spanish); *Rev. Inst. Investigaciones Educativas (IIE)*, Buenos Aires **14** (1977), 3–22.

[77.4.g] Geometry and Physics, (Spanish); *Conceptos de Matemática* **43** (1977), 24–34.

[77.5.e] (With other authors). The modulus in the teaching and learning of mathematics in the high school, (Spanish); *Office of Science of UNESCO for Latin America*, Montevideo, 1977.

1978

[78.1.r]* Sets of segments on surfaces, (Spanish); *Math. Notae* **26** (1978), 63–72.

[78.2.r]* (With Fava). Plate and line segment processes, *J. Appl. Prob.* **15** (1978), 494–501.

[78.3.r] Random processes of linear segments and graphs, *Lect. Not. on Biomathematics*, Springer **23** (1978), 279–294.

[78.4.r] Integral Geometry: history and perspectives, (Spanish); *Proc. IV Int. Colloq. on Diff. Geometry*, Santiago de Compostela, (1978), 1–48. (1979).

[78.5.g] Mathematical thinking: mathematics in tecnique and in art, (Spanish); *Revista Hitos*, Buenos Aires, 1978.

[78.6.g] Information and culture, (Spanish); *Rev. Nac. Cultura*, Buenos Aires **1** (1978), 75–81.

1979

[79.1.r]* (With Fava). Random processes on manifolds in $\mathbf{R}^n$, *Z. Wahrsch. Verw. Gebiete* **50** (1979), 85–96.

[79.2.g] (With Rios and Balanzat). Julio Rey Pastor, mathematician, (Spanish); *Instituto de España, Colección Cultura y Cencia*, Madrid, 1979.

[79.3.g] Mathematics and the humanistical sciences, (Spanish); In the book *Quantitative methods in social sciences(Spanish) (in memoriam of Dr. Barral Souto)*, Macchi, Buenos Aires, (1979), 137–152.

1980

[80.1.r]* Notes on the Integral Geometry in the hyperbolic plane, Special issue in honor of Antonio Monteiro. *Portugaliae Math.* **39** (1980), 239–249 (1985).

[80.2.r]* Cauchy and Kubota formulas for convex bodies in elliptic n-space, *Rend. Sem. Mat.Univ. Torino* **38** (1980), 51–58.

[80.3.r]* Random lines and tesselations in a plane, *Stochastica* **4**, (1980), 3–13.

[80.4.r] Geometric probabilities, Integral Geometry and Stochastic Geometry, (Spanish); *An. Acad. Nac. Ci. Ex., Fís. Nat.*, Buenos Aires **32** (1980), 65–93.

[80.5.g] Mathematics and society, (Spanish); *Docencia*, Buenos Aires, 1980. 43 pp.

[80.6.e] Applications of mathematics in the elementary and high school (1st. part), (Spanish); *L'Escaire*, Dep. Mat. Arquitectura, UPC, Barcelona **5** (1980), 44–58.

[80.7.e] Teaching statistics in Argentina, *Teaching Statistics* **2** (1980).

[80.8.e] Situation of the teaching of geometry in front of the new trends in mathematical education, (Spanish); *Revista de Bachillerato*, Ministerio Educación y Ciéncia, Madrid, Addendum to number **13** (1980), 23–38.[3] Talk given in "Quinta Conferencia Iberoamericana sobre educación matemática", Campinas, Brasil 1979.

[80.9.r] Commentary to Vidal's works on Classical Differential Geometry and Integral Geometry of Enrique Vidal-Abascal, (Spanish); *Selecta of Scientific Jubilee of Prof. Enrique Vidal-Abascal*, University of Santiago de Compostela, Spain. Pag. VII-XI of the Preface, 1980.

1981

[81.1.g] Influence of Einstein in the mathematical field, (Spanish); In the book *Homage to Einstein on the centenary of his birthday*, (Spanish). Edited by Univ. Nac. Tucumán, (1981), 27–46.

[81.2.e] Applications of mathematics in the elementary and high school (2nd part), (Spanish); *L'Escaire*, Dep. Mat. Arquitectura, UPC, Barcelona **6** (1981), 29–44.

[81.3.b] The teaching of mathematics in the high school (Spanish); *Docencia*, Buenos Aires 1981, 144 pp.

1982

[82.1.r] Geodesics in Gödel-Synge spaces. *Tensor* (N.S.) **37** (1982), 173–178.

[82.2.e] Computation and probability in education, (Spanish); *Rev. Inst. Investigaciones Educativas* **37** (1982), 3–14.

[82.3.g] Argentinian science: Its history and its philosophy, (Spanish); *Actas primeras jornadas de historia del pensamiento científico argentino*, Buenos Aires (1982), 13–19.

[82.4.g] Carlos Encina (1838–1882), (Spanish); Web page of *Acad. Nac. Ci. Exact. Fís. y Naturales*. www.ancefn.org.ar/institucional/presidentes/encina.htm.

1983

[83.1.r]* An inequality between the parts into which a convex body is divided by a plane section, *Rend. Circ. Mat. Palermo* **32** (1983), 124–130.

[83.2.b] Integral Geometry and Geometric Probability (transl. of [76.1] from the English by Maksimov), (Russian); *Nauka*, Moscow.

[83.3.r] Some current problems of stocastic geometry, (Spanish); *Not. First Intern. Symposium*, Barcelona, 1983.

---

[3] Also "Causes and effects of the present tendencies in the education of geometry" (Spanish) *Lecturas Matemáticas* Sociedad Colombiana de matemáticas **2** (1981) 299–318.

[83.4.e] The cultivation of scientific attitudes for an integral education of the human being, (Spanish); *El sistema educativo hoy*, *CINAE*, Buenos Aires **1** (1983), 15–26.

[83.5.g] Science and technique in contemporary society, (Spanish); *CONICET*, Buenos Aires, 1983.

[83.6.g] The work of Terradas in Argentina, *Rev. R. Acad. Ci. Exact., Fís. Nat.* Madrid, 1983.

1984

[84.1.r]* Mixed random mosaics. *Math. Nachr.* **117** (1984), 129-133.

[84.2.g] (With Balanzat). J. Babini, (Spanish); *Rev. Un. Mat. Argentina* **31** (1984), 160–158. (1963).

[84.3.r] Integral Geometry in the affine plane, (Spanish); *Publicación CRM* Barcelona **2**, 1984.

[84.4.g] Integral Geometry, stereology and computerized tomography, *Ciencia Hoje, Statistics*, Rio de Janeiro **2** (1984), 26–32.

[84.5.g] Science, technique and society at the end of the second millennium, (Spanish); *Rev. Escuela Superior de Guerra*, 1984.

[84.6.g] The role of the university in scientific and technological development, (Spanish); *Educación Superior*, *UNESCO*, Caracas **16**, 1984.

1985

[85.1.r] On some invariants under similitudes for convex bodies. *Discrete geometry and convexity (N. Y. 1982), Ann. New York Acad. Sci.* (1985), 128–131.

[85.2.r] A simple problem of decision and a waiting time problem, (Spanish); *Trab. Estad. Invest. Oper.*, Madrid **36** (1985), 269–279.

[85.3.g] Information and its influence on science and philosophy, (Spanish); *Estudios dedicados a Luis Farré*, FEPAI (1985), 75–79.

[85.4.e] The teaching of the geometry in the secondary cycle, (Spanish); *La enseñanza de la matemática a debate*, Ministerio de Educación y Ciencia, Madrid (1985), 11–23.

[85.5.g] Science and technique at the end of XX century, (Spanish); *Ingeniería Militar* **1** (1985), 5–8.

[85.6.g] The work of Wilhelm Blaschke on Integral Geometry, *Blashcke Gesammelte Werke*, Thales Verlag, Essen **2** (1985), 211–218.

1986

[86.1.r]* On the measure of line segments entirely contained in a convex body, *Aspects of Math. and its Appl., Elsevier Science Publ., North-Holland Math. Library* **34** (1986), 677–687.

[86.2.r] Fausto Ismael Toranzos (1908–1996), *Rev. Un. Mat. Argentina* **32** (1986), 220–221. (1987).

[86.3.e] Mathematics and education, (Spanish); *Elementos de Matemática*, Universidad CAECE, **I** Buenos Aires, 1986.

[86.4.g] The engineer José Babini, mathematician and historian of science, (Spanish); *Actas de las segundas jornadas de historia del pensamiento científico argentino*, Buenos Aires (1986), 135–138.

[86.5.e] The teaching of the sciences in high school (a contribution to the pedagogical congress), (Spanish); *An. Soc. Científica* **216**, 1986.

[86.6.g] What are Mathematics (Catalan); Publications of the Autonomous University of Barcelona. Lecture given in Girona, on the occasion of his investiture as doctor Honoris Causa, June 1986.

1987

[87.1.r] On the superposition of random mosaics, (Spanish); *Acta Stereol.* **6** (1987), 141–145.

[87.2.g] Deterministic thinking, probabilistic thinking, informatic thinking, (Spanish); *An. I. Congr. Arg. Informática Educativa*, Buenos Aires, (1987), 5–10.

1988

[88.1.r]* Affine Integral Geometry and convex bodies, *J. of Microscopy* **152** (1988), 229–233.

[88.2.r] Fundamental formulas of stereology using sections by nonlinear manifolds, (Spanish); *Rev. Un. Mat. Argentina* **34**, (1988), 56–68. (1990).

[88.3.r] Random mosaics, (Spanish). *Rev. R. Acad. Ci. Ex. Fis. Nat.* Madrid **82** (1988), 483–522.

[88.4.g] Influence of science in literature, (Spanish); *Literatura y Ciencia*, Fundación Casa de la Cultura de Córdoba (1988), 29–33.

[88.5.e] Statistic and probability in the high school, (Spanish); *Elementos de matemática, II*, **17** (1988), 16–26.

[88.6.e] Probability in the high school: Random use of tables, (Spanish); *Épsilon (Rev. Soc. Mat. Thales)* **10** (1988), 9–22.

[88.7.g] Julio Rey Pastor, on the centennial of his birthday, (Spanish); *Ciencia e investigación*, Buenos Aires **42** (1988), 326–330.

[88.8.e] Proportionality and probability, (Spanish); *Rev. Soc. Can. Profesores de Matemáticas* **18** (1988), 7–17.

[88.9.g] Rey Pastor in Latin America, (homage to Rey-Pastor on the centenary of his birthday), (Spanish); *R. Acad. Ci. Ex., Fís. Nat.*, Madrid (1988), 27–34.

[88.10.g] Scientific research, profession of our century, (Spanish); *Acad. Ci. Ex., Fís. Nat., Buenos Aires*, 1988.

1989

[89.1.g] The works of Rey Pastor in geometry and topology, (Spanish); *Rev. Un. Mat. Argentina* **35** (1989), 3–12. (1991).

[89.2.r] Integral Geometry, *Global differential geometry, Math. Ass. Amer. Stud. Math.* **27** (1989), 303–350.

[89.3.e] Probability in the high school, simulation of games, (Spanish); *Rev. Educación Matemática* **4** (1989), 4–17.

[89.4.g] Science on the threshold of the third millennium, (Spanish); *Ingeniería militar*, Buenos Aires **9** (1989), 4–11.

1990

[90.1.g] Experimental method and mathematical method, (Spanish); *Elementos de Matemàtica* **5**, 1990.

[90.2.e] Mathematics for non-mathematicians, (Spanish); *Mem. Congreso Ibero-Americano, UNESCO*, Sevilla, 1990.

[90.3.g] Scientific Essay, *Fundación Casa de Cultura de Córdoba*, (1990), 16–21.

1991

[91.1.r] Differential Geometry, Integral Geometry and Stochastic Geometry, (Spanish); *Noticiero de la Un. Mat. Argentina*, 1991.

[91.2.e] Mathematical olympiads, (Spanish); *Revista de educación matemática* **6** (1991), 21-36.

[91.3.g] Fractals and chaotic systems, (Spanish); *Rev. Inst. Tecnológico de Buenos Aires* **14**, 1991.

[91.4.e] (With other authors). Miss: What is statistics?, (Spanish); *República Argentina, Editorial COPEA*, Buenos Aires, 1991.

[91.5.e] Mathematic in the high school, (Spanish); *Pensar y repensar la educación*, Academia Nacional de Educación, Buenos Aires, (1991) 529–556. Lecture given when he was elected member of the Academia Nacional de Educación, April 3, 1989.

1992

[92.1.g] Ernest Corominas (1913–1992), (Spanish); *Rev. Un. Mat. Argentina* **38** (1992), 157–158. (1993).

[92.2.g] Mathematics and general culture, (Spanish); *SUMA*, Huelva **10** (1992), 4–8.

[92.3.g] Fractal sets, (Spanish); *Elementos de Matemática* **1** (1992), 5–26.

[92.4.g] Science in Spanish America throughout the five hundred years since its discovery, (Spanish); *An. Acad. Ci. Ex. Fis. Nat.*, Buenos Aires **44**, 1992.

1993

[93.1.b] Mathematics I for basic general teaching (initiation to creativity) (Spanish); *Kapelusz*, Buenos Aires, 1993.

[93.2.e] Geometry in the forming of teachers, (Spanish); *Red Olímpica*, Buenos Aires, 1993. 115 pp.

1994

[94.1.b] Mathematics: A philosophy and a technique, (Catalan); *Vic: Eumo editorial*, 1993; (Spanish), *Ariel*, Barcelona, 1994.

[94.2.b] Mathematics II for basic general teaching (initiation to creativity), (Spanish); *Kapelusz*, Buenos Aires, 1994.

[94.3.e] The teaching of mathematics in intermediate education (directed by Victor Garcia Hoz), (Spanish); *Rialp*, Madrid, **14**.

[94.4.e] (With other authors). Focus on a humanistic didactics of mathematics, (Spanish); *Troquel*, Buenos Aires, 1994, 245 pp. Contains: "Contributions to continue teaching mathematics", "Probabilities, chance and statistics", "The Euler graphs, the Escher mosaics and the Moebius band".

[94.5.e] (With Palacios and Giordano). About education and Statistics, (Spanish); *Serie Eureka, Kapelusz*, Buenos Aires 1994. 111 pp.

1995

[95.1.b] Mathematics IIII for basic general teaching (initiation to creativity), (Spanish); *Kapelusz*, Buenos Aires, 1995.

[95.2.r] Foundations of stereology in three-dimensional hyperbolic spaces, (Spanish); *Rev. Acad. Canaria Cienc.* **7** (1995), 117-134.

[95.3.g] Manuel Balanzat (1912–1994), (Spanish); *Rev. Un. Mat. Argentina* **39** (1995), 235–239.

1997

[97.1.g] (With López de Aráoz). Dr. Raúl Ernesto Luccioni (1927–1997), (Spanish); *Rev. Un. Mat. Argentina* **40** (1997), 159–160.

[97.2.g] José Babini, mathematician, (Spanish); *Saber y tiempo*, Buenos Aires **3**, 1997.

[97.3.e] Mathematic for teachers, (Spanish); *Épsilon, Rev. Soc. Mat. Thales*, SAEM, Sevilla **38**, 1997.

2002

[02.1.g] History of the Unión Matemática Argentina, 1936–1996, (Spanish); *Rev. Un. Mat. Argentina* **43** (2002), 1–38.

2004

[04.1.b] Integral Geometry and Geometric Probability. 2nd ed., (with a foreword by Marc Kac), *Cambridge University Press*, 2004.

He also contributed to many works by other authors. For example, he wrote the prologue to the following: "Philosophy of the natural laws (Spanish)" by D. Papp, *Troquel* 2^{a} ed., Buenos Aires, 1980; "Doing geometry (Spanish)" by P. Marbah and L. Saidón, *Centro de Investigación Babbage*, Buenos Aires, 1997; "Traveling in space-time" (Spanish) by I. Hernaiz and C. Ottolengui, *Troquel*, Buenos Aires, 1990. He also collaborated on "Probabilities and statistic: its teaching (Spanish) by J. Foncuberta, *Conicet*, Buenos Aires, 1996; "Matemática moderna, matemática viva" by A. Revuz *OCDL*, Buenos Aires, 1965; "Problems in the teaching of mathematics" by J. Bosch et al. *Conceptos de Matemática*, Buenos Aires 1980.

5 List of Santaló's Papers Included in the Selection

Part I. Differential Geometry

[45.2] Note on convex curves on the hyperbolic plane, *Bull. Amer. Math. Soc.* **51** (1945), 405–412.

[50.1] On parallel hypersurfaces in the elliptic and hyperbolic n-dimensional space, *Proc. Amer. Math. Soc.* **1** (1950), 325–330.

[62.1] On the Gauss-Bonnet formula for polyhedra in spaces of constant curvature, (Spanish); *Rev. Un. Mat. Argentina* **18** (1962), 79–91.

[63.1] A relation between the mean curvatures of parallel convex bodies in spaces of constant curvature, (Spanish); *Rev. Un. Mat. Argentina* **21** (1963), 131–137.

[67.1] Total absolute curvatures of manifolds contained in Euclidean space, (Spanish); *Acta Ci. Compostelana* **4** (1967), 149–158.

[67.2] Horocycles and convex sets in the hyperbolic plane, *Arch. Math. (Basel)* **18** (1967), 529–533.

[68.1] Horospheres and convex bodies in hyperbolic space. *Proc. Amer. Math. Soc.* **19** (1968), 390–395.

[69.1] Convexity in the hyperbolic plane, (Spanish); *Univ. Nac. Tucumán Rev.* **19** (1969), 173–183.

[69.2] On some geometric inequalities in the style of Fary, *Amer. J. Math.* **91**, (1969), 25–31.

[70.1] Mean values and curvatures, *Izv. Akad. Nauk. Armejan*, SSR, S. Math. **5** (1970), 286–295.

[74.1] Total curvatures of compact manifolds immersed in Euclidean space, *Symposia Math., Inst. Naz. di Alta Matematica, Academic Press*, Roma **14** (1974), 363–390.

Part II. Integral Geometry

[36.1] Integral Geometry 7: New applications of the concept of the kinematic measure in the plane and space, (Spanish); *Rev. Acad. Ci. Ex. Fis. Nat.* Madrid **33** (1936), 3–50.

[41.1] Proof of a theorem of Bottema on ovals, (German); *Tohöku Math. J.* **48** (1941), 221–224.

[41.2] A generalization of a theorem of T. Kubota on ovals, (German); *Tohöku Math. J.* **48** (1941), 64–67.

[41.3] A theorem and an inequality referring to rectifiable curves, *Amer. J. Math.* **63** (1941), 635–644.

[41.4] Curves of extremal total torsion and *D*-curves, (Spanish); *Publ. Inst. Mat. Univ. Nac. Litoral*, Rosario **3** (1941), 131–156.

[42.1] Integral formulas in Crofton's style on the sphere and some inequalities referring to spherical curves, *Duke Math. J.* **9** (1942), 707–722.

[42.2] On the isoperimetric inequality for surfaces of constant negative curvature, (Spanish); *Univ. Nac. Tucumán Rev.* **3** (1942), 243–259.

[43.1] Integral Geometry on surfaces of constant negative curvature, *Duke Math. J.* **10** (1943), 687–704.

[44.1] Note on convex spherical curves, *Bull. Amer. Math. Soc.* **50** (1944), 528–534.

[45.1] Mean value of the number of regions into which a body in space is divided by *n* arbitrary planes, (Spanish); *Rev. Un. Mat. Argentina* **10** (1945), 101–108.

[47.3] *D*-curves on cones, (Spanish); *Math. Notae* **7** (1947), 179–190.

[49.1] Integral Geometry in three-dimensional spaces of constant curvature, (Spanish); *Math. Notae* **9** (1949), 1–28.

[49.2] Integral Geometry on surfaces, *Duke Math. J.* **16** (1949), 361–375.

[50.2] Integral Geometry in general spaces, *Proceedings of the International Congress of Mathematiciens, Cambridge Mass. Amer. Math. Soc. Providence, R. I.* **1** (1950), 482–489.

[52.1] Measure of sets of geodesics in a Riemannian space and applications to integral formulas in elliptic and hyperbolic spaces, *Summa Brasil. Math.* **3** (1952), 1–11.

[52.2] Integral Geometry in Hermitian spaces, *Amer. J. Math.* **74** (1952), 423–434.

[54.1] Questions of Differential and Integral Geometry in spaces of constant curvature, (Spanish); *Univ. e Politec. Torino. Rend. Sem. Mat.* **14** (1954), 277–295. (1955).

[56.1] On the mean curvatures of a flattened convex body, *Rev. Fac. Sci., Univ. Istanbul* **21** (1956), 189–194.

[60.3] Two applications of the Integral Geometry in affine and projective spaces, *Publ. Math. Debrecen* **7** (1960), 226–237.

[62.3] On the fundamental kinematic formula of Integral Geometry in spaces of constant curvature, (Spanish); *Math. Notae* **18** (1962), 79–94.

[66.1] Groups of the plane with respect to which sets of points and lines admit an invariant measure, (Spanish); *Rev. Un. Mat. Argentina* **23** (1966), 119–148. (1967).

[66.2] Average values for polygons formed by random lines in the hyperbolic plane, (Spanish); *Univ. Nac. Tucumán Rev.* **16** (1966), 29–43.

[75.1] The kinematic formula in Integral Geometry for cylinders, *Ann. Mat. Pura Appl.* **103** (1975), 71–79.

[78.1] Sets of segments on surfaces, (Spanish); *Math. Notae* **26** (1978), 63–72.

[80.1] Notes on the Integral Geometry in the hyperbolic plane, *Portugaliae Math.* **39** (1980), 239–249.

[80.2] Cauchy and Kubota formulas for convex bodies in elliptic *n*-space, *Rend. Sem. Mat. Univ. Torino* **38** (1980), 51–58.

Part III. Convex Geometry

[40.2] A theorem on sets of parallelepipeds with parallel edges, (Spanish); *Publ. Inst. Mat. Univ. Nac. Litoral*, Rosario **2** (1940), 49–60.

[42.3] Supplement to the note: A theorem on sets of parallelepipeds with parallel edges, (Spanish); *Publ. Inst. Mat. Univ. Nac. Litoral*, Rosario **3** (1942), 202–210.

[46.1] Convex regions on the n-dimensional spherical surface, *Ann. of Math.* **47** (1946), 448–459.

[46.2] On the convex bodies of constant width in E^n, (Spanish); *Portugaliae Math.* **5** (1946), 195–201.

[49.3] An affine invariant for convex bodies in n-dimensional space, (Spanish); *Portugaliae Math.* **8** (1949), 155–161.

[59.1] On complete systems of inequalities between elements of a plane convex figure, (Spanish); *Math. Notae* **17** (1959), 82–104. (1961).

[83.1] An inequality between the parts into which a convex body is divided by a plane section, *Rend. Circ. Mat. Palermo* **32** (1983), 124–130.

[86.1] On the measure of line segments entirely contained in a convex body, *Aspects of Math. And its Appl., Elsevier Science Publ., North-Holland Math. Library* **34** (1986), 677–687.

[88.1] Affine Integral Geometry and convex bodies, *J. of Microscopy* **152** (1988), 229–233.

Part IV. Affine Geometry

[46.3] A geometrical characterization for the affine differential invariants of a space curve, *Bull. Amer. Math. Soc.* **52** (1946), 625–632.

[47.1] Affine invariants of certain pairs of curves and surfaces, *Duke Math. Journal* **14** (1947), 559–574.

[50.3] Integral Geometry in projective and affine spaces, *Ann. of Math.* **51** (1950), 739–755.

[60.1] Steiner's formula for parallel surfaces in affine geometry, (Spanish); *Univ. Nac. Tucumán Rev.* **13** (1960), 194–208.

[62.2] On the measure of sets of parallel linear subspaces in affine space, *Canad. J. Math.* **14** (1962), 313–319.

Part V. Statistic and Stereology

[40.1] Sur quelques problèmes de probabilités géométriques, (French); *Tohöku Math. J.* **47** (1940), 159–171.

[43.2] On the probable distribution of corpuscles in a body, deduced from the distribution of its sections, and analogous problems, (Spanish); *Rev. Un. Mat. Argentina* **9** (1943), 145–164.

[47.2] On the first two moments of the measure of a random set, *Ann. Math. Statistics* **18** (1947), 37–49.

[55.1] On the distribution of sizes of particles contained in a body given the distribution in its sections or projections, (Spanish); *Trabajos Estadist.* **6** (1955), 181–196.

[72.1] (With Yáñez). Averages for polygons formed by random lines in Euclidean and hyperbolic planes, (Spanish); *J. Appl. Probab.* **9** (1972), 140–157.

[78.2] (With Fava). Plate and line segment processes, *J. Appl. Prob.* **15** (1978), 494–501.

[79.1] (With Fava). Random processes on manifolds in $\mathbf{R}^n$, *Z. Wahrsch. Verw. Gebiete* **50** (1979), 85–96.

[80.3] Random lines and tesselations in a plane, *Stochastica* **4**, (1980), 3–13.

[84.1] Mixed random mosaics. *Math. Nachr.* **117** (1984), 129–133.

6 Selected Papers of L. A. Santaló

Part I. Differential Geometry, with comments by E. Teufel

The comprehensive scientific opus of L. A. SANTALÓ contains numerous contributions to Differential Geometry. Most of them belong to the classical Differential Geometry of curves and surfaces, in the spirit of the "Erlanger Program" of F. KLEIN (1872) ordered into Euclidean, non-Euclidean, Affine, Conformal Differential Geometry. Many of them join Differential Geometry and Convexity or Integral Geometry respectively.

The publications selected for this Part I can be divided according to four key words: *1) Horocycles and horospheres in hyperbolic geometry*, *2) Parallel surfaces, Gauß-Bonnet formula*, *3) Convexity in Hyperbolic spaces* and *4) Total (absolute) curvatures in Euclidean spaces*.

1) Horocycles and horospheres in hyperbolic geometry. In Integral Geometry the invariant density with respect to the group of motions for r-planes in Euclidean spaces is well known, M. W. CROFTON (1868), W. BLASCHKE (1936), S. S. CHERN (1952). In non-Euclidean spaces the invariant density for totally geodesic subspaces was investigated by T.-J. WU (1938) and L. A. SANTALÓ (1949) ([49.1] Part II).

By [67.2] and [68.1] L. A. SANTALÓ opened up Integral Geometry in hyperbolic spaces using horocycles and horospheres. He derived explicitly the invariant densities for horocycles and horospheres, and he used them to obtain the Cauchy-Crofton formula for the length of a curve, Crofton's formula for chords, and the measure of horospheres intersecting a h-convex body (i.e. convex with respect to horocycles).

He pointed out that, in order to extend topics from Euclidean to hyperbolic geometry, in some cases the natural analogue of the Euclidean lines and planes are horocycles and horospheres respectively. The same can be found, for instance in the work of I. M. GELFAND and M. I. GRAEV (1962) on the application of Integral Geometry to group representations.

2) Parallel surfaces, Gauß-Bonnet formula. The classical Steiner formula for the length of a plane curve parallel to a given curve was extended by G. HERGLOTZ (1943) and C. B. ALLENDOERFER (1948) for the area $M_0(\lambda)$ of parallel hypersurfaces $S^{n-1}(\lambda)$ at distance λ of a smooth hypersurface S^{n-1} in an n-dimensional space of constant curvature. In 1950 L. A. SANTALÓ [50.1] further extended this for the total ith mean curvatures $M_i(\lambda)$ of $S^{n-1}(\lambda)$ $(i = 0, \ldots, n-1)$: $M_i(\lambda)$ is equal to a linear function of the total jth mean curvatures M_j, $j = 0, \ldots, n-1$, of S^{n-1} with coefficients depending on λ.

Especially for polar hypersurfaces ($\lambda = \pi/2$) in the elliptic case, this gave relations between the total mean curvatures of the hypersurface and its polar hypersurface, extending results of W. BLASCHKE (1936) in dimension 3.

Moreover, he investigated hypersurfaces of constant width. Here the general formula showed relations between the total mean curvatures and the width, extending results of W. BLASCHKE (1915), and related to investigations of L. A. SANTALÓ (1946) on convex bodies of constant width in Euclidean spaces ([46.2] Part III).

In 1963 he addressed this topic especially for convex bodies Q with smooth boundaries in spaces of constant curvature K [63.1]. His main result was on the first derivative $M_i'(\lambda)$ of the total ith mean curvature $M_i(\lambda)$ of the exterior parallel hypersurface $\partial Q(\lambda)$: $M_i'(\lambda) = -iKM_{i-1}(\lambda) + (n-i-1)M_{i+1}(\lambda)$. From this, inter alia, he reached an inequality between the area and the total mean curvature of convex bodies in 3-dimensional spaces, which was conjectured by W. BLASCHKE (1938) as a generalization of a classical Minkowski inequality in Euclidean space.

The classical Gauß-Bonnet formula was extended to manifolds by C. B. ALLENDOERFER and A. WEIL (1943), S. S. CHERN (1944/45), G. HERGLOTZ (1943). In [62.1] L. A. SANTALÓ looked closer into this for (convex) polyhedrons Q in n-dimensional spaces of constant curvature. Using the smooth version of the Gauß-Bonnet formula applied to the exterior parallel hypersurface $\partial Q(\lambda)$, he obtained through the limit $\lambda \to 0$ the Gauß-Bonnet formula for Q. The formula is written in terms of the h-dimensional edges of Q and the $(n-h-1)$-dimensional polar angles there.

In particular, in case of even-dimensional ambient spaces he obtained the volume of Q explicitly in terms of the polar angles of Q, related to the work of L. SCHLÄFLI (1852), H. POINCARÉ (1905), H. HOPF (1925), E. PESCHL (1956) on the volume of simplices. He explicitly worked out the computations in dimensions 2, 3, and 4. For instance in dimension 4, he reached a volume formula, which for simplices reduces to the Poincaré-Hopf formula.

3) Convexity in hyperbolic spaces. In [45.2] and [69.1] L. A. SANTALÓ investigated convexity in hyperbolic plane.

In the first publication he used the usual convexity with respect to geodesics. He introduced a new breadth function for smooth convex curves, defined for pairs of support geodesics having a common perpendicular geodesic which is also a normal of the curve, as the length of this common perpendicular segment. The new definition was necessary because unlike the Euclidean case, there is no parallelity for geodesics in the hyperbolic plane. In particular he considered convex curves of constant breadth. As for constant width in the Euclidean case, constant breath is characterized by the fact that each normal of the curve is a double normal. He obtained the relation between the curvatures in opposite points, and a generalization of the classical Barbier formula which writes the circumference in terms of the area and the breadth. This supplemented his work on convex spherical curves (1944) ([44.1] Part II), generalizing classical Euclidean results.

In the second publication he used h-convexity, i.e. convexity with respect to horocycles (cf. also [67.2]). Among many basic results on h-convexity, he introduced a new h-breadth function, defined as the distance between "parallel" support horocycles. In particular, he analyzed h-convex curves of constant h-breadth. And amongst others, he justified that under all h-convex curves of a given constant h-

breadth, the Releaux triangle has the minimal area. In the Euclidean plane this minimum property of the Releaux triangle was proved by H. L. LEBESGUE (1914) and W. BLASCHKE (1915).

4) Total (absolute) curvatures in Euclidean spaces. The Gauß-Bonnet formula acts as a classical bridge between Geometry and Topology. For submanifolds in Euclidean spaces orthogonal projections onto lines ("height functions") are tools to apply Morse theory and to obtain deep ongoing relations between the curvature and the topology of the submanifold, S. S. CHERN and R. K. LASHOF (1957/58), N. H. KUIPER (1969, 1970), et al.

In this spirit, perhaps inspired by the classical projection and intersection formulas for convex bodies in Euclidean space, L. A. SANTALÓ, [67.1], [70.1], [74.1], introduced by a global differential topological definition a series of new rth total (absolute) curvatures of order q for smooth submanifolds X in Euclidean spaces. In the case $q = 1$, they are mean values of the volumes of the critical value set of orthogonal projections of X onto r-planes. For this case $q = 1$ he derived the reproductive property for the new curvatures, and moreover he obtained local representations in terms of curvatures of X. He analyzed many special cases, for instance the new curvatures contain the Lipschitz-Killing case, and moreover, e.g. for surfaces in 4-dimensional Euclidean space, they are related to the "normal curvatures" and "tangent curvatures" of the surfaces, T. OTSUKI (1961), B. Y. CHEN (1970), S. S. CHERN (1945).

In [69.2] he again combined Differential Geometry and Integral Geometry in order to obtain geometric inequalities in the style of Fáry for hypersurfaces in Euclidean space, relating total mean curvatures and the radius of a ball enclosing the hypersurface, generalizing results of I. FÁRY (1950) and G. D. CHAKERIAN (1962/64). See the comment on this paper by G. Solanes, page 82.

The scientific work of L. A. SANTALÓ covered topics of immediate interest in their time and developed them further. In many aspects his works served as an inspiration and a starting point for mathematical research to the present day.

EBERHARD TEUFEL (Stuttgart).

NOTE ON CONVEX CURVES ON THE HYPERBOLIC PLANE

L. A. SANTALÓ

1. Introduction. In a previous note [5][1] we have obtained some properties referring to convex curves on the sphere. Following an analogous way our purpose is now to obtain the same properties for convex curves on a surface of constant negative curvature $K=-1$, or, what is equivalent, for convex curves on the hyperbolic plane.

In §§6 and 7 we consider the curves of constant breadth, for which we obtain the formula (7.3) which relates the length L and area F with the breadth α.

For the curves which are not of constant breadth the formula (4.5), which contains (7.3) as a particular case, holds. But (4.5) is true only if we suppose that the curve has in all its points geodesic curvature κ_g greater than one.

2. Definitions. A closed curve C on a surface of constant negative curvature $K=-1$ is said to be convex when it cannot be cut by any geodesic in more than two points, except that a complete arc of geodesic may belong to the curve. Any closed convex curve C has a finite length L and bounds a finite area F. In the following, unless otherwise specified, we shall suppose that C is composed of a finite number of arcs each with continuous geodesic curvature κ_g.

Let ω_i be the exterior angles which these arcs form at the vertices of C. Then we have the Gauss-Bonnet formula [3, p. 191],

$$\int_C \kappa_g ds + \sum \omega_i = 2\pi + F. \tag{2.1}$$

If a point O on C is taken as origin, any point A of C can be determined by the length of the arc $OA=s$ or by the angle τ defined by

$$\tau = \int_0^s \kappa_g ds + \sum_s \omega_i \tag{2.2}$$

where $\sum_s \omega_i$ is extended over all the vertices of C contained in the arc OA.

Any geodesic with only one common point or with a complete arc in common with C is called a "geodesic of support" of C. In each

Received by the editors November 18, 1944.

[1] Numbers in brackets refer to the references cited at the end of the paper.

point of C for which there exists a tangent geodesic the geodesic of support coincides with this. Any geodesic which passes through a vertex without crossing C is also a geodesic of support. To any geodesic of support of C corresponds a value of the angle τ (2.2).

Let g be a geodesic of support of C and let A be its point of support. Let g' be the orthogonal geodesic to g at the point A and g_1 another geodesic of support of C which is also orthogonal to g'. The curve C will be contained between g and g_1. If A' is the point in which g_1 cuts g', we shall call *breadth* α of C corresponding to the point A the length of the arc AA' of geodesic g'. The breadth α is a function of the angle τ or the arc s corresponding to the point A.

3. Closed convex curves with $\kappa_g>1$ at any point. Let us suppose that C has $\kappa_g>1$ at any point. We shall call *pseudospherical osculating circle* of C at the point A the limit of the geodesic circle determined by the points A, A_1, A_2 of C when $A_1, A_2 \to A$.

If we suppose $\kappa_g>1$, the radius R of the pseudospherical osculating circle (which we shall call "radius of pseudospherical curvature") has a finite value and is related to the geodesic curvature κ_g by

$$\kappa_g = \coth R. \tag{3.1}$$

This equality is obtained by applying the Gauss-Bonnet formula to a geodesic circle and using the following formulas for its length and area:

$$L = 2\pi \sinh R, \qquad F = 2\pi(\cosh R - 1). \tag{3.2}$$

The center O of pseudospherical curvature is the limiting position of the point in which the orthogonal geodesic to C at $A(s)$ cuts the orthogonal geodesic at $A(s+\Delta s)$ when $\Delta s \to 0$.

The condition $\kappa_g>1$, which is equivalent to $R<\infty$, will then be necessary and sufficient for two sufficiently close orthogonal geodesics to C to intersect each other.

Before proceeding it is necessary that we prove the following lemma.

LEMMA. *Let g_1 and g_2 be two geodesics which are orthogonal to the geodesic g. Let MN be an arc of the curve C with $\kappa_g>1$ at each point and which is tangent to g_1 and g_2 at the ends M, N respectively. We affirm that: on the arc MN there is only one point with the property that the orthogonal geodesic to C which passes through it is also orthogonal to g.*

We suppose the arc MN is composed of a finite number of arcs with continuous geodesic curvature, and a geodesic will be considered orthogonal to MN at a corner if it is orthogonal to a geodesic of support through the corner.

PROOF. The angle which the geodesics of support of the arc MN form with the orthogonal geodesic to g through their contact points increases from 0 to π. Hence there is a point at which this angle equals $\pi/2$. It remains to be proved that there is no other point with this property.

Let us consider the curvilinear coordinate system formed by the orthogonal geodesics to g as curves $v=\text{const.}$ and their orthogonal trajectories as curves $u=\text{const.}$, $u=0$ being the geodesic g. Then the element of length is given by [3, p. 282]

$$ds^2 = du^2 + \cosh^2 u\,dv^2. \tag{3.3}$$

If $u=u(s)$, $v=v(s)$ are the equations of the curve C, calling ϕ the angle which C forms at each point with the corresponding geodesic $v=\text{const.}$, we have $\tan\phi=\cosh uv'/u'$, hence

$$d\tan\phi/ds = [(u'v'' - u''v')\cosh u + u'^2v'\sinh u]u'^{-2}. \tag{3.4}$$

The geodesic curvature of the curve $u=u(s)$, $v=v(s)$ is given by [3, p. 187]

$$\kappa_g = (u'v'' - u''v')\cosh u + 2u'^2v'\sinh u + v'^3\cosh^2 u\sinh u. \tag{3.5}$$

From this and from (3.3) we deduce

$$(u'v'' - u''v')\cosh u + u'^2v'\sinh u = \kappa_g - v'\sinh u. \tag{3.6}$$

From $u'^2+v'^2\cosh^2 u=1$ and $\cosh^2 u-\sinh^2 u=1$ we have $v'^2\sinh^2 u=1-u'^2-v'^2$, that is, $v'^2\sinh^2 u\leqq 1$. Hence, from (3.4) and (3.6) under the assumption that $\kappa_g>1$, we get (if $u'\neq 0$)

$$d\tan\phi/ds = (\kappa_g - v'\sinh u)u'^{-2} > 0.$$

The angle ϕ is then always increasing from M to N, that is, from 0 to π. Consequently at only one point will $\phi=\pi/2$, which proves our lemma.

4. Principal formula. Let g be a geodesic which cuts the closed convex curve C and let O be a fixed point on the surface with $K=-1$ that contains C. Let w be the distance from O to g and θ the angle which the orthogonal geodesic from O to g makes with a fixed direction at O. Then it is known [4, p. 687] that the measure of a set of geodesics is the integral of the expression $dg=\cosh w\,d\theta\,dw$ extended to the set.

Consider the set of "oriented" geodesics which cut C; it is known [4, p. 691] that

$$\int_{C,g\neq 0} dg = \int \cosh w d\theta dw = 2L, \tag{4.1}$$

where L is the length of C.

According to the former lemma if we suppose that $\kappa_g > 1$ in each point of C, on each side of g there will be only one point A in which the orthogonal geodesic to C will also be orthogonal to g. Consequently the geodesic g can be determined by the point A (that is, by the corresponding value of s or τ) and the distance a from A to g (Fig. 1). We wish to express the density $dg = \cosh w d\theta dw$ in terms of τ and a. It is known that the differential expression $dg = \cosh w d\theta dw$

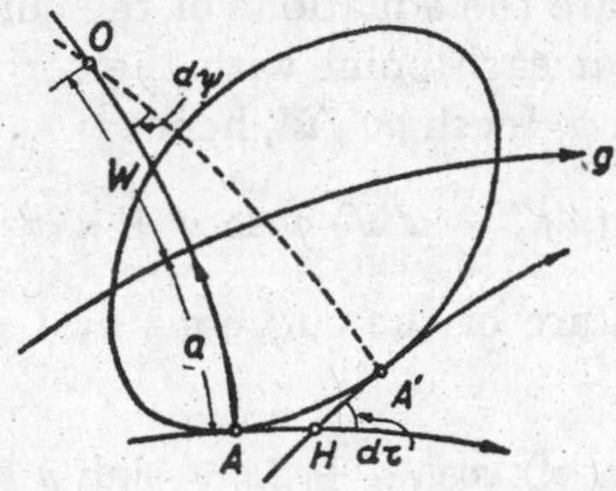

FIG. 1

does not depend on the point O or the direction origin of the angles θ. Consequently we can consider for a moment that the point O is the pseudospheric center of curvature of C at A, that is, the point in which the orthogonal geodesic to C at $A(\tau)$ is intersected by the orthogonal geodesic at $A' = A(\tau + d\tau)$. Hence $OA = R$, $w = R - a$. Let H be the point in which the geodesic of support of C at A intersects the geodesic of support at A'. From the Gauss-Bonnet theorem we deduce that the area of the geodesic quadrilateral $OAHA'$ has the value

$$(\pi/2 + d\tau + \pi/2 + \pi - d\psi) - 2\pi = d\tau - d\psi, \tag{4.2}$$

$d\tau$ being the angle which the geodesic of support at A forms with the geodesic of support at A' and $d\psi$ the angle AOA'. But save for infinitesimals of second order the same area equals the area of the sector of geodesic circle AOA' which has the value $(\cosh R - 1)d\psi$. Consequently

$$d\tau = \cosh R d\psi. \tag{4.3}$$

Since R is independent of w, from $w = R - a$, $d\theta = d\psi$, and (4.3) we deduce

$$|\partial(a, \tau)/\partial(\theta, w)| = \cosh R.$$

Hence $dg = \cosh w d\theta dw = (\cosh (R - a)/\cosh R) da d\tau$ or, since $d\tau = \kappa_g ds$, we have

$$dg = \cosh a da d\tau - \sinh a da ds. \tag{4.4}$$

Let us substitute this expression (4.4) in (4.1). For each value of s (or τ) the arc a can vary from 0 to the breadth α of C corresponding to the point s (or τ). Therefore

$$\int_{C, g\neq 0} dg = \int_C d\tau \int_0^\alpha \cosh a da - \int_C ds \int_0^\alpha \sinh a da$$

or, in accordance with (4.1),

$$L = \int_C \sinh \alpha d\tau - \int_C \cosh \alpha ds. \tag{4.5}$$

This is our principal formula from which we wish to obtain some consequences. The formula (4.5) is analogous to that obtained for convex spherical curves in a previous paper [5] and holds for any convex curve with $\kappa_g > 1$ at each point on a surface of constant negative curvature $K = -1$.

5. **Consequences.** (a) Let Δ be the minimum breadth of C, that is to say, the minimum value of α, and δ the maximum value of α, that is, the diameter of C. From (4.5), (2.1), and (2.2) we deduce

$$\sinh \Delta/(1 + \cosh \delta) \leqq L/(2\pi + F) \leqq \sinh \delta/(1 + \cosh \Delta). \tag{5.1}$$

Therefore: *on a surface of constant negative curvature $K = -1$, for any convex curve C with $\kappa_g > 1$ the inequalities* (5.1) *are verified.*

(b) If C has a continuous geodesic curvature, from (4.5) we deduce

$$\int_C \sinh \alpha \kappa_g ds - \int_C (1 + \cosh \alpha) ds = 0$$

or

$$\int_C \cosh \frac{\alpha}{2}\left(\kappa_g \sinh \frac{\alpha}{2} - \cosh \frac{\alpha}{2}\right) ds = 0. \tag{5.2}$$

From this and according to (3.1), we have: *In any closed convex curve C on a surface of constant negative curvature $K = -1$, with continuous goedesic curvature $\kappa_g > 1$, there are at least two points for which* *the radius R of pseudospherical curvature equals $\alpha/2$, where α is the breadth of K corresponding to the point considered.*

6. Convex curves of constant breadth. In §2 we have defined the breadth α of a closed convex curve C on the surface with $K=-1$, corresponding to a point A of C. When α is constant, C is called a *curve of constant breadth.*

The geodesic circles of finite radius R are the first examples of curves of constant breadth $\alpha=2R$. Another class of curves of constant breadth is the generalization on the surfaces of constant negative curvature of the Reuleaux polygons [2, p. 130].

Let us consider a geodesic circle of radius R; we divide it into $2n+1$ equal parts and through the division points we draw the tangent geodesics. If two consecutive tangents intersect each other we shall have a geodesic regular polygon of an odd number of sides. Taking each vertex as center let us draw the arc of the geodesic circle which joins the two opposite vertices. These arcs form a Reuleaux polygon of $2n+1$ sides, and it is easily seen that this polygon has constant breadth.

The necessary relation can easily be found between the radius R of the geodesic circle and the number $2n+1$ of sides so that two consecutives tangent geodesics intersect. Let A_i and A_{i+1} be two consecutive division points and OH the geodesic which halves the angle A_iOA_{i+1}. The condition that the geodesic OH be cut for the tangent geodesic at A_i is that the angle $A_iOH=\pi/2n+1$ be smaller than the angle of parallelism corresponding to the tangent geodesic at A_i and to the center O, hence [1, p. 621]

$$\tan\,(\pi/(2n+1)) < 2e^R/(e^{2R}-1).$$

That is, to have a Reuleaux polygon of $2n+1$ sides we must start from a geodesic circle with a radius R which satisfies the inequality

$$R < \log \frac{1}{\tan\,(\pi/2(2n+1))}.$$

7. Properties of the convex curves of constant breadth. The convex curves of constant breadth on the surfaces of constant negative curvature (or on the hyperbolic plane) have analogous properties to the curves of constant breadth on the plane. For example it is easily seen that the constant breadth α equals the diameter δ. Therefore if A is a point of C the orthogonal geodesic g to C at A will cut C at the point A' and at this point g will also be orthogonal to C. The points A and A' can be called opposite points.

We shall prove that if C is of constant breadth any orthogonal geodesic to C, say AA', will intersect a neighbouring orthogonal geodesic at a point contained in the arc AA'.

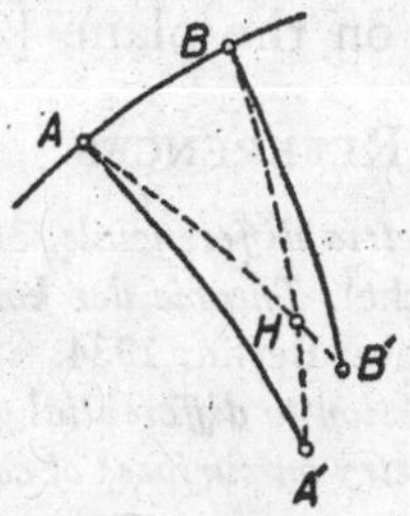

FIG. 2

Let $B \equiv A(\tau + d\tau)$ and BB' be the orthogonal geodesic to C at B (Fig. 2). If BB' does not intersect the arc AA' the point B' must have the position indicated in Fig. 2. But in this position we have

$$(7.1) \qquad AA' < AH + HA', \qquad BB' < BH + HB'$$

and by addition

$$(7.2) \qquad AA' + BB' < AB' + BA'.$$

Since C is of constant breadth we have $AA' = BB' > AB', BA'$ and therefore $AA' + BB' > AB' + BA'$ which gives a contradiction with (7.2).

From this we deduce that the pseudospherical radius of curvature R is always finite and not greater than α. Moreover since the orthogonal geodesics at A and B are also orthogonal geodesics at A' and B' the pseudospherical center of curvature will be the same at A and A'. Hence: *the sum of the pseudospherical radii of curvature corresponding to opposite points equals the breadth* α.

Since $R < \infty$, we shall have $\kappa_g > 1$; therefore we can apply the formula (4.5) which gives, according to (2.1),

$$L = (2\pi + F) \sinh \alpha - L \cosh \alpha,$$

that is,

$$(7.3) \qquad L = (2\pi + F) \tanh (\alpha/2).$$

We conclude the following theorem: *For any convex curve of constant breadth on the surface of constant negative curvature* $K = -1$, *the equation* (7.3) *holds.*

If $K = -1/a^2$ instead of -1, (7.3) gives

$$L/a = (2\pi + F/a^2) \tanh (\alpha/2a).$$

Muliplying both sides by a and letting $a \to \infty$ we find

$$L = \pi\alpha,$$

which is a well known relation between the length and breadth of the curves of constant breadth on the plane [2, p. 131].

References

1. L. Bianchi, *Lezioni di geometria differenziale*, 3d ed., vol. 1, Bologna, 1927.

2. T. Bonnesen, and W. Fenchel, *Theorie der konvexen Körper*, Ergebnisse der Mathematik und ihrer Grenzgebiete, Berlin, 1934.

3. L. P. Eisenhart, *An introduction to differential geometry*, Princeton, 1940.

4. L. A. Santaló, *Integral geometry on surfaces of constant negative curvature*, Duke Math. J. vol. 10 (1943).

5. ———, *Note on convex spherical curves*, Bull. Amer. Math. Soc. vol. 50 (1944).

Universidad Nacional del Litoral, Rosario

ON PARALLEL HYPERSURFACES IN THE ELLIPTIC AND HYPERBOLIC n-DIMENSIONAL SPACE

L. A. SANTALÓ

1. Introduction. Let S^{n-1} be a hypersurface of class C^3 in the elliptic or hyperbolic n-dimensional space, which is closed and bounding and whose principal curvatures with respect to an inside normal are all positive. Let $S^{n-1}(\lambda)$ be the hypersurface parallel to S^{n-1} at distance λ.

If $\rho_1, \rho_2, \cdots, \rho_{n-1}$ are the principal radii of curvature of S^{n-1} at a point P and dP denotes the element of area at P, the mean curvatures of S^{n-1} are defined by

$$(1.1) \qquad M_i = \int_{S^{n-1}} \left(\sum \frac{1}{\rho_{\nu_1}\rho_{\nu_2} \cdots \rho_{\nu_i}} \right) dP, \qquad i = 0, 1, \cdots, n-1,$$

where the sum is extended to the $C_{n-1,i}$ combinations of ith order of the indices $1, 2, \cdots, n-1$. In particular, M_0 coincides with the area A of S^{n-1}.

Herglotz [6][1] and, from a more general point of view, Allendoerfer [1] have obtained the area $A(\lambda)$ and volume $V(\lambda)$ of the parallel hypersurface $S^{n-1}(\lambda)$, which can be expressed as linear functions of the mean curvatures M_i of S^{n-1} with coefficients depending upon λ. For this purpose it is enough to find the expression of $A(\lambda)$, that is, $M_0(\lambda)$, because $V(\lambda)$ is then given by

$$(1.2) \qquad V(\lambda) = V + \int_0^\lambda A(\lambda) d\lambda.$$

The purpose of the present note is to extend these results to the evaluation of all mean curvatures $M_i(\lambda)$ of $S^{n-1}(\lambda)$. The resulting formulae are also linear with respect to M_i; they are (2.9) for the elliptic case, and (3.2) for the hyperbolic case. For $i=0$, they give the value of $A(\lambda)$ obtained by Herglotz and Allendoerfer.

As a consequence, in the elliptic case we obtain the relation (4.2) between the mean curvature of an S^{n-1} and those of its polar hypersurface. Finally we obtain the equations (5.3) which hold for the mean curvatures of convex surfaces of "constant width" in the elliptic or hyperbolic n-dimensional spaces.

In all these questions, in order to obtain simplifications in the re-

Received by the editors January 8, 1949.

[1] Numbers in brackets refer to the bibliography at the end of the paper.

sulting formulas the generalized Gauss-Bonnet formula as obtained by Allendoerfer-Weil [2] plays a fundamental role. In our particular case of the elliptic and hyperbolic space, this formula can be written (see [1]):

For $n-1$ even

$$C_{n-1}M_{n-1} + C_{n-3}M_{n-3} + \cdots + C_1M_1 + K^{n/2}V = -\omega^n\chi'/2 \tag{1.3}$$

and for $n-1$ odd

$$C_{n-1}M_{n-1} + C_{n-3}M_{n-3} + \cdots + C_0M_0 = \omega^n\chi'/2 \tag{1.4}$$

where ω^j is the surface area of a j-dimensional unit sphere ($\omega^0=2$) and

$$C_i = \frac{\omega^n}{\omega^i\omega^{n-1-i}} K^{(n-1-i)/2},$$

being $K=1$ in the elliptic and $K=-1$ in the hyperbolic case. χ' is the inner characteristic of the volume bounded by S^{n-1}; if S^{n-1} is a topologic sphere it is $\chi'=-1$ for $n-1$ even and $\chi'=1$ for $n-1$ odd.

2. **The elliptic case.** Let C_i ($i=1, 2, \cdots, n-1$) be the lines of curvature of S^{n-1} which pass through the point P and let ds_i be the element of arc of C_i at P. The element of area of S^{n-1} at P will be

$$dP = ds_1 ds_2 \cdots ds_{n-1}. \tag{2.1}$$

If ρ_i is the principal radius of curvature at P corresponding to C_i and R_i represents the distance from P to the contact point of the normal to S^{n-1} at P with the envelope of the normals to S^{n-1} along C_i, the relation (see, for instance, [5, p. 214])

$$\rho_i = \tan R_i \tag{2.2}$$

is well known.

Furthermore if $d\alpha_i$ is the angle between two infinitely near normals to S^{n-1} along C_i at their intersection point,

$$ds_i = \sin R_i d\alpha_i. \tag{2.3}$$

From (2.1) and (2.3) we deduce

$$dP = \prod_{i=1}^{n-1} \sin R_i d\alpha_i. \tag{2.4}$$

Applying (2.4) to the hypersurface $S^{n-1}(\lambda)$, we have

$$dP(\lambda) = \prod_{i=1}^{n-1} \sin (R_i + \lambda) d\alpha_i \tag{2.5}$$

or, according to (2.4),

$$(2.6)\qquad \begin{aligned} dP(\lambda) &= \prod_{i=1}^{n-1} (\sin R_i \cos \lambda + \cos R_i \sin \lambda) d\alpha_i \\ &= \prod_{i=1}^{n-1} (\cos \lambda + \sin \lambda / \tan R_i) dP. \end{aligned}$$

From the definition (1.1) and from (2.2) we deduce

$$(2.7)\quad M_i(\lambda) = \int_{S^{n-1}(\lambda)} \left(\sum \frac{1}{\tan (R_{\nu_1} + \lambda) \cdots \tan (R_{\nu_i} + \lambda)} \right) dP(\lambda),$$

or, according to (2.5) and (2.4)

$$(2.8)\qquad \begin{aligned} M_i(\lambda) &= \int \sum \left(\prod_{j=1}^{i} \cos (R_{\nu_j} + \lambda) \right. \\ &\qquad \left. \cdot \prod_{j=i+1}^{n-1} \sin (R_{\nu_j} + \lambda) \right) d\alpha_1 \cdots d\alpha_{n-1} \\ &= \int_{S_{n-1}} \sum \left(\prod_{j=1}^{i} \left(\frac{\cos \lambda}{\tan R_{\nu_j}} - \sin \lambda \right) \right. \\ &\qquad \left. \cdot \prod_{j=i+1}^{n-1} \left(\cos \lambda + \frac{\sin \lambda}{\tan R_{\nu_j}} \right) \right) dP. \end{aligned}$$

The sums are always extended over all combinations of ith order of the indices $1, 2, \cdots, n-1$.

If we take into account (2.2) and the definition (1.1) of M_i, from the last equality results[2]

$$(2.9)\qquad M_i(\lambda) = \sum_{k=0}^{n-1} M_k \phi_{ik}(\lambda)$$

where

$$(2.10)\quad \phi_{ik}(\lambda) = \sum_{h=p}^{q} (-1)^{i-h} C_{n-1-k,i-h} C_{k,h} \sin^{i+k-2h} \lambda \cos^{n-1-i-k+2h} \lambda,$$

where the sum is extended over all values of h for which the combina-

[2] The combinatory coefficients which appear in (2.10) are easily obtained if we observe that the number of terms in the sum (2.8) with k factors $1/\tan R_{\nu_j}$ and coefficient $\sin^{i+k-2h}\lambda \cos^{n-1-i-k+2h}\lambda$ is $C_{i,h}C_{n-1-i,k-h}C_{n-1,i}$ and the number of terms in the sum (1.1) which gives M_k is $C_{n-1,k}$. Therefore the product $M_k \sin^{i+k-2h}\lambda \cos^{n-1-i-k+2h}\lambda$ appears a number of times equal to the quotient of the two foregoing combinatory numbers, which is equal to $C_{n-1-k,i-h}C_{k,h}$.

tory symbols have a sense, that is

$$(2.11) \qquad p = \max(0, i + k - n + 1), \qquad q = \min(i, k).$$

Formulas (2.9) and (2.10) solve our problem for the elliptic case.

3. **The hyperbolic case.** For the case of a hypersurface S^{n-1} in the hyperbolic n-dimensional space, formulas (2.2) and (2.3) must be replaced respectively by

$$(3.1) \qquad \rho_i = \tanh R_i, \qquad ds_i = \sinh R_i d\alpha_i.$$

Exactly the same calculation as before gives now

$$(3.2) \qquad M_i(\lambda) = \sum_{k=0}^{n-1} M_k \phi_{ik}(\lambda)$$

with

$$(3.3) \quad \phi_{ik}(\lambda) = \sum_{h=p}^{q} (-1)^{i-h} C_{n-1-k,i-h} C_{k,h} \sinh^{i+k-2h} \lambda \cosh^{n-1-i-k+2h} \lambda,$$

where p, q are given by (2.11).

4. **Polar surfaces.** In the elliptic case it is interesting to consider the polar surface $S^{n-1}(\pi/2)$ to the given S^{n-1}.

Applying (2.9), (2.10) for $\lambda = \pi/2$ we obtain

$$(4.1) \qquad M_i(\pi/2) = (-1)^i M_{n-1-i}.$$

If M_i^P denotes the ith mean curvature of the polar surface, we have $M_i^P = (-1)^i M_i(\pi/2)$ and consequently

$$(4.2) \qquad M_i^P = M_{n-1-i}.$$

For $i = 0$

$$A^P = M_{n-1},$$

which is a result due to Allendoerfer [1, formula (30)]. For $n = 3$ we get $A^P = M_2$, $M_1^P = M_1$ or, applying the Gauss-Bonnet formula (1.3)

$$M_1^P = M_1, \qquad A^P + A = -4\pi\chi'.$$

The last formula is due to Blaschke [4].

5. **Hypersurfaces of constant width.** Let us assume S^{n-1} to be a topological sphere such that the inward drawn normal at every point P cuts S^{n-1} beside P at only one opposite point P^*. Let Δ be the distance PP^* measured along the normal. If Δ is constant for every point

P, S^{n-1} is said to be a hypersurface of "constant width."

In such a case the normal at P^* coincides with P^*P. Indeed, if Q is a point of S^{n-1} such that the distance PQ is a maximum (P fixed, Q variable on S^{n-1}), QP must be normal to S^{n-1} at Q and therefore, by assumption, distance $QP=\Delta$; on the other hand, if P^*P were not normal to S^{n-1} at P^*, the distance PP^* would not be a maximum, thus distance $PP^*<$distance $PQ=\Delta$, contrary to the assumption.

Furthermore, according to the definition of the radii R_i and the assumption that they are not negative (see §1 and (2.2), (3.1)), the point of contact of the normal PP^* with the envelope of the normals along each line of curvature through P does lie inside the segment PP^*; therefore for the hypersurfaces of constant width, between the corresponding radii R_i, R_i^* at opposite points, the relation

$$R_i + R_i^* = \Delta, \qquad i = 1, 2, \cdots, n-1, \tag{5.1}$$

holds.

We have also $dP=(-1)^{n-1}dP^*$, and consequently (2.7) gives

$$M_i(-\Delta) = (-1)^{n-1-i}M_i \tag{5.2}$$

which holds the same in both elliptic and hyperbolic cases.

Therefore, taking into account the relations (2.9) and (3.2) we get: *Between the mean curvatures M_i of a hypersurface of constant width Δ in the elliptic or hyperbolic n-dimensional space, the relations*

$$M_i = (-1)^{n-1-i}\sum_{k=0}^{n-1} M_k\phi_{ik}(-\Delta), \qquad i = 0, 1, 2, \cdots, n-1, \tag{5.3}$$

hold, where ϕ_{ik} are given by (2.10) *in the elliptic case and by* (3.3) *in the hyperbolic case.*

Furthermore, if V is the volume enclosed by S^{n-1}, we have $V(-\Delta) =(-1)^n V$ and therefore (1.2) and (2.9), (3.2) give the following relation

$$V = (-1)^n V + (-1)^n \sum_{k=0}^{n-1} M_k \int_0^{-\Delta} \phi_{0k}(\lambda)d\lambda, \tag{5.4}$$

which must be added to the preceding ones (5.3).

The obtained relations (5.3) and (5.4) are, in general, not independent, as the following examples will show.

EXAMPLE 1. If $n=2$, (5.3) and (5.4) are equivalent to the unique relation

$$M_0 \sin\Delta - M_1(1-\cos\Delta) = 0 \text{ (elliptic case)},$$

$$M_0 \sinh \Delta - M_1(1 - \cosh \Delta) = 0 \text{ (hyperbolic case).}$$

If L is the length and A the area enclosed by S^1, $M_0 = L$ and the Gauss-Bonnet formula gives $M_1 = 2\pi \pm A$; therefore the foregoing relations may be written respectively

$$L = (2\pi - A) \tan (\Delta/2), \qquad L = (2\pi + A) \tanh (\Delta/2). \tag{5.5}$$

Example 2. For $n=3$, if we set $M_0 = A$ and take into account (1.3) which gives $M_2 = 4\pi \pm A$, the relations (5.3) become equivalent to

$$\begin{aligned} M_1 \cos \Delta &= 2(2\pi - A) \sin \Delta \text{ (elliptic case),} \\ M_1 \cosh \Delta &= 2(2\pi + A) \sinh \Delta \text{ (hyperbolic case).} \end{aligned} \tag{5.6}$$

(5.4) gives

$$2V = 2\pi\Delta - (M_1/2) \sin^2 \Delta - (2\pi - A) \sin \Delta \cos \Delta \text{ (elliptic case),}$$

$$2V = -2\pi\Delta - (M_1/2) \sinh^2 \Delta + (2\pi + A) \sinh \Delta \cosh \Delta \text{ (hyperbolic case).}$$

If we take into account (5.6), the last relations can be written respectively

$$4V = 4\pi\Delta - M_1, \qquad 4V = M_1 - 4\pi\Delta. \tag{5.7}$$

(5.5) and (5.7) are due to Blaschke [**3**]. For the analogous questions in the n-dimensional euclidean space, see [**7**].

Bibliography

1. C. B. Allendoerfer, *Steiner formulae on a general* S^{n+1}, Bull. Amer. Math. Soc. vol. 54 (1948) pp. 128–135.

2. C. B. Allendoerfer and A. Weil, *The Gauss-Bonnet theorem for Riemannian polyhedra*, Trans. Amer. Math. Soc. vol. 53 (1943) pp. 101–129.

3. W. Blaschke, *Einige Bemerkungen ueber Kurven und Flächen von konstanter Breite*, Leipziger Berichte vol. 67 (1915) pp. 290–297.

4. ———, *Integralgeometrie* 22: *Zur elliptischen Geometrie*, Math. Zeit. vol. 41 (1936) pp. 785–786.

5. E. P. Eisenhart, *Riemannian geometry*, Princeton, 1926.

6. G. Herglotz, *Ueber die Steinersche Formel für Parallelflächen*, Abh. Math. Sem. Hansischen Univ. vol. 15 (1943) pp. 165–177.

7. L. A. Santaló, *Sobre los cuerpos convexos de anchura constante en* E_n, Portugaliae Mathematica vol. 5 (1946) pp. 195–201.

The Institute for Advanced Study and
Facultad de Ciencias Matematicas, Rosario, Argentina

On the Gauss-Bonnet formula for Polyhedra in spaces of constant curvature

L. A. Santaló

Abstract

For spaces S_n of constant curvature K, the generalized formula of Gauss-Bonnet for a closed, orientable surface S of class ≥ 3 which is the boundary of a body Q, takes the forms (4) and (5) according to the parity of n. In this paper we consider the case in which Q is a polyhedron. Then the formula takes the form (20), (21), where the terms $\alpha_{n-h-1}L_h$ are the sums (18) between the h-dimensional measures L_h^i of the h-dimensional measures L_h^i of the h-dimensional edges of Q and the $(n-h-1)$-dimensional polyhedral angles "polar" of those formed by the faces of Q which arc incident in L_h^i.

If Q is a simplex of S_n these formulae must contain Poincaré's relations between the angles of spherical simplexes [11], [8], [10], [15]. This is computed for $n = 2, 3, 4$ in N.° 4 and 5. However the computation in the general case seems to be far from obvious.

1 The Gauss-Bonnet formula in spaces of constant curvature

The classical Gauss-Bonnet formula of surface theory was generalized to multidimensional varieties by Allendoefer-Weil in 1943 [1] Shortly afterwards, a further simple proof was given by S. S. Chern [3], [4]. For the particular case of hypersurfaces in a space of constant curvature, the generalized Gauss-Bonnet formula is contained in results obtained independently and almost at the same time by Herglotz [7]. In this case, the elements in the Gauss-Bonnet formula have a very precise geometric meaning. Let us now recall this formula, a proof of which can be seen in [13]. Let $S_n(K)$ be the space of constant curvature K of n-dimensions; that is, n-dimensional non-Euclidean space. Let S be a closed hypersurface of the class $n \geq 3$ that is the boundary of a body Q. At each point of S, if $R_1, R_2, \ldots, R_{n-1}$ are the principle radii of curvature, the mean curvatures are then defined[1]

$$m_i = \left\{ \frac{1}{R_{\alpha_1}} \frac{1}{R_{\alpha_2}} \cdots \frac{1}{R_{\alpha_i}} \right\}, \quad m_0 = 1,\ i = 1, 2, \ldots, n-1, \tag{1}$$

[1] In paper [12], as usual we define the mean curvatures by

$$m_i' = \quad n-1 \quad {}^{-1} \quad m_i$$
$$\qquad\quad i$$

This modifies the following formulae by a combinatorial factor. In the present case, the definition adopted here is preferable.

where the brackets indicate the elementary symmetric function of order i formed by the principle curvatures $1/R_i$. Hence we may deduce the *mean curvature integrals*

$$M_i = \int_S m_i dF \quad (i = 0, 1, 2, \ldots, n-1) \tag{2}$$

where dF indicates the area element ($(n-1)$-dimensional) of S. In particular, this is

$$M_0 = F = \text{area of } S. \tag{3}$$

With these notations, the generalized Gauss-Bonnet formula for spaces of constant curvature K can be written:

$$c_{n-1}M_{n-1} + c_{n-3}M_{n-3} + \cdots + c_1 M_1 + K^{n/2}V = \frac{1}{2}O_n \chi(Q), \tag{4}$$

and for n odd

$$c_{n-1}M_{n-1} + c_{n-3}M_{n-3} + \cdots + c_2 M_2 + c_0 F = \frac{1}{2}O_n \chi(Q), \tag{5}$$

where the constants c_i are

$$c_i = \frac{O_n}{O_i O_{n-1-i}} K^{(n-1-i)/2}, \tag{6}$$

where O_h is the area of the Euclidean unit sphere and dimension h; that is

$$O_h = \frac{2\pi^{(h+1)/2}}{\Gamma((h+1)/2)}. \tag{7}$$

$\chi(Q)$ indicates the Euler-Poincaré characteristic of the body Q bounded by S. If Q splits into simplexes and a_i is the number of these simplexes of dimension i, we have

$$\chi(Q) = a_0 - a_1 + a_2 - \cdots + (-1)^n a_n. \tag{8}$$

For the boundary S of Q, this is

$$\chi(Q) = 0 \text{ for } n \text{ even}, \; \chi(S) = 2\chi(Q) \text{ for } n \text{ odd}. \tag{9}$$

For the case of a topological sphere, it is

$$\chi(Q) = 1. \tag{10}$$

2 Polar Bodies

Let us consider in particular the n-dimensional unit sphere of Euclidean space E_{n+1}; this is space S_n of constant curvature $K = 1$. Given in this space a hypersurface S that bounds a body Q, the parallel exterior hypersurface at distance $\pi/2$ is called the "dual" or "polar" of S, which we denote by S^*. The body Q^* bounded by S^* on the side that does not contain Q will be the dual or polar of Q. Obviously, $S^{**} = S$ and $Q^{**} = Q$.

The polar hypersurface S^* can also be obtained as the extremal geometric location of the radii of the unit sphere E_{n+1} normal to the diametric hyperplanes tangent to S.

If we call the volumes of Q and Q^*, V and V^*, and the respective areas of S and S^*, F and F^*, the following formulae hold (see [13]): For n even

$$(11)\quad \begin{cases} c_{n-1}F + c_{n-3}M_2 + \cdots + c_1 M_{n-2} + V^* = \frac{1}{2}O_n\chi(Q), \\ c_{n-1}F^* + c_{n-3}M_{n-3} + \cdots + c_1 M_1 + V = \frac{1}{2}O_n\chi(Q) \end{cases}$$

and for n odd

$$(12)\quad \begin{cases} c_{n-1}F + c_{n-3}M_2 + \cdots + c_1 M_{n-2} + c_0 F^* = \frac{1}{2}O_n\chi(Q), \\ c_{n-2}M_{n-2} + c_{n-4}M_{n-4} + \cdots + c_1 M_1 + V + V* = (1 - \frac{1}{2}\chi(Q))O_n. \end{cases}$$

In the latter cases of n odd, according to (9), $\chi(Q) = \frac{1}{2}\chi(S)$ can be substituted. The constants c_i are the same (6).

Furthermore, in all cases, the following formula holds

$$(13)\quad M_i^* = M_{n-1-i}.$$

We therefore have, for example, the following particular cases:

For $n = 2$, if we denote by F the area of a domain Q and by L the length of its boundary (which will be the above-mentioned V and F, respectively), we arrive at the following elementary formulae

$$(14)\quad L + F^* = 2\pi\chi(Q), \quad L^* + F = 2\pi\chi(Q).$$

For $n = 3$, we have

$$(15)\quad \begin{cases} F + F^* = 2\pi\chi(S), \\ \frac{1}{2}M_1 + V + V^* = (2 - \chi(Q))\pi^2. \end{cases}$$

For $n = 4$, we have

$$(16)\quad \begin{cases} 2F + M_2 + 6V^* = 4\pi^2\chi(S), \\ 2F^* + M_1 + 6V = 4\pi^2\chi(Q). \end{cases}$$

3 The Case of Polyhedra

Let us consider the case in which Q is a polyhedron. We wish to find the value which in this case must be attributed to the integrals of the mean curvature M_i, since they cannot be defined directly by the formulae (1) and (2) due to the fact that they have edges on which some R_i vanish.

Let us consider the body Q_ϵ parallel exterior to Q at distance ϵ (set of points $S_n(K)$ whose distance from Q is $\leq \epsilon$; its principle curvatures are $1/(R_i+\epsilon)$, and therefore its mean curvatures are

$$m_i(Q_\epsilon) = \left\{\frac{1}{R_{\alpha_1}+\epsilon}, \frac{1}{R_{\alpha_2}+\epsilon}, \dots, \frac{1}{R_{\alpha_i}+\epsilon}\right\}.$$

The parts of Q_ϵ that correspond to the faces of Q are also flat, and therefore its curvature is null. The part of Q_ϵ obtained from an edge L_h^s of dimension h $(0 \leq h \leq n-2)$ is the sector of a cylinder whose cross section is a simplex of dimension $n-h-1$ of the sphere of radius ϵ and dimension $n-h-1$ contained within the plane normal to L_h^s. For this cylindrical sector, the principle curvatures are $1/R_1 = 1/R_2 = \cdots = 1/R_{n-h-1} = 1/\epsilon$, $1/R_{n-h} = \cdots = 1/R_{n-1} = 0$ and the area element is $\epsilon^{n-h-1} dO_{n-h-1} dL_{h^\epsilon}$, where dO_{n-h-1} is the area element of the $(n-h-1)$-dimensional unit sphere and dL_h^s is the h-dimensional volume element of L_h^s.

If we calculate the integral of mean curvature M_i corresponding to the cylindrical sector of axis L_h^s, and make $\epsilon \to 0$, then this result is zero if $i < n-h-1$ or if $i > n-h-1$, and only the value $\alpha_{n-h-1}^s L_h^s$ remains for the case $i = n-h-1$, where α_{n-h-1}^s is the integral of dO_{n-h-1}, and denoting by the same letters L_h^s the h-dimensional volume of the edge L_h^s. For $h=0$, L_0^s is a vertex of Q for whose measure it is necessary to take the unit (that is, we must write $L_{0^\epsilon} = 1$), since the area element of Q_ϵ corresponding to the vertices does not contain the factor dL_0^s.

Adding together the values obtained for all the edges L_h^s of the same dimension h, we have that for a polyhedron Q

$$M_{n-h-1} = \sum_s \alpha_{n-h-1}^s L_h^s,$$

which it is appropriate to write as

$$M_h = \sum_s \alpha_h^s L_{n-h-1}^s, \tag{17}$$

where:

L_{n-h-1}^s = the $(n-h-1)$-dimensional volume of the edge L_{n-h-1}^s of dimension $n-h-1 (h = 1, 2, \dots, n-1)$.

α_h^s = the measure of the h-dimensional polyhedral angle "polar" of the polyhedral angle formed by the $h+1$ faces of Q that meet atL_{n-h-1}^s.

The sum with respect to s refers to all the $(n-h-1)$-dimensional edges of Q.

Let us recall the meaning of "polar" polyhedral angle. To each edge L^s_{n-h-1} there corresponds a polyhedral angle of dimension h interior to Q; let α^s_h*. The measure of this angle is that of a simplex of dimension h on the unit sphere and dimension h; this simplex has a polar in the direction of N° 2, and the volume of this polar simplex is precisely the previous value α^ϵ_h. In the interests of brevity, this may be written as

$$(18)\qquad \sum_s \alpha^s_h L^s_{n-h-1} = \alpha_h L_{n-h-1},$$

whereby (17) may be written as

$$(19)\qquad M_h = \alpha_h L_{n-h-1}.$$

In particular, for $h = n-1$ we have $L^s_0 = 1$, and therefore

$$M_{n-1} = \sum \alpha^s_h = \alpha_h$$

is the sum of the polar angles of those corresponding to the vertices of Q.

Substitution of the values (19) in (4) and (5) gives the generalized Gauss-Bonnet formula for polyhedra in n-dimensional space of constant curvature K the following form: For n even:

$$(20)\; c_{n-1}\alpha_{n-1} + c_{n-3}\alpha_{n-3}L_2 + \cdots + c_1\alpha_1 L_{n-2} + K^{n/2}V = \frac{1}{2}O_n\chi(Q),$$

and for n odd

$$(21)\quad c_{n-1}\alpha_{n-1} + c_{n-3}\alpha_{n-3}L_2 + \cdots + c_2\alpha_2 L_{n-3} + c_0 F = \frac{1}{2}O_n\chi(Q).$$

Let us now see some consequencies of these formulae:

a) For n even, formula (20) enables us to calculate the volume V of a polyhedron from its angles and from the measures of its edges of even dimension. However, since these edges are also polyhedra in a space of constant curvature of a lesser dimension, their measures can again be expressed in terms of their angles. By proceeding successively, we have the following classical result.

For n even, the volume V of a polyhedron in a space of constant curvature $K \neq 0$ can be expressed in terms of the angles of the same polyhedron of different dimensions. For n odd, these angles are constrained by a linear relation, although with these angles V cannot be expressed.

This theorem, which may already have been known to L. Schläfli in 1852 for the case of spherical polyhedra[2], was reencountered by Poincaré in 1905 [11],

[2]Schläfli's original manuscripts remained unknown in the Schweizerischen Landesbibliothek in Berne until 1901, when they were published in Vol.38.I of the *Denkschriften der Schweizerischen Naturforschenden Gesellschaft.* See [14].

and was later extended to spaces of constant negative curvature by Hopf [9]. In general, these works are confined to the case of the n-dimensional simplex, considering that all polyhedra may split into simplexes. Recent works on this subject are those by Peschl [10] and Höhn [8], in which further bibliography can be found.

b) For n odd, formula (21) is none other than (20) applied to the faces of the polyhedron Q, which are polyhedra of dimension $n-1$, and therefore even, and subsequently adding to all the faces. In the following section this can be seen clearly in the particular case $n=3$.

In general, given a polyhedron Q of dimension n, contained in $S_n(K)$, formula (20) can be applied to all the edges of even dimension. If, for example, it is applied to the edges of dimension $h = 2m \leq n-1$, one obtains a relation between the measures of the edges of dimension $2, 4, \dots, h$ and the polyhedral angles formed by the faces by which they are bounded, these edges being considered as polyhedra (con $\chi = 1$) of dimensions $2, 4, \dots, h$. For the case of a simplex, these formulae should coincide with those given by Poincaré in the aforementioned work [11] and studied in depth by Peschl [10]. However, the effective computation of this equivalence is far from easy in the general case. For $n = 3, 4$ this can be seen in the following Nos. 4 and 5.

c) The following problem appears more interesting and we will return to it on a future occasion. The relations already referred to in b), translated to the general case of any body Q of $S_n(K)$, bounded by a surface S of class ≥ 3, must be relations between the mean curvatures M_i analogous to those of the Gauss-Bonnet-Herglotz theorem (4), (5), but in which only the M_i for $i \leq h \leq n-1 (h = 2, 4, \dots)$ appear. These relations in turn must be a particular case, corresponding to the case of spaces of constant curvature K, of integral formulae of the Gauss-Bonnet type generalized by Allendoerfer-Weil-Chern, and valid for any closed, orientable hypersurface of a Riemann space. One may therefore assume the existence of these new integral formulae, which it would undoubtedly be interesting to study. They could well coincide or be related with those given by Chern ([4], formula (20)), although this is not entirely certain.

4 Cases $n = 2, 3$

In order to understand them better, let us now see the form taken by the formulae (20) and (21) for the most simple cases of dimension 2 and 3.

For $n = 2$, we have $c_1 = 1$, and formula (20) yields

$$\alpha_1 + KV = 2\pi\chi(Q).$$

In this case, V denotes the area of the polygon Q; it is preferable to denote it by F, thereby giving

$$\alpha_1 + KF = 2\pi\chi(Q). \tag{22}$$

If A_s are the interior angles of the polygon Q and m is the number of vertices, we obtain

$$\alpha_1 = \sum_{s=1}^{m} (\pi - A_s) = m\pi - \sum_{1}^{m} A_s, \tag{23}$$

which yields the elementary formula

$$KF = \sum_{1}^{m} A_s - (m - 2\chi(Q))\pi, \tag{24}$$

which gives the area of a polygon over a surface of constant curvature $K \neq 0$. For a simply connected polygon this is $\chi(Q) = 1$.

For $n = 3$, the volume of a polyhedron can no longer be calculated in an elementary way. Even for the case of the tetrahedron, complicated trascendent functions appear; see, for example, Coxeter [5] and Böhm [2].

Let us now see what relation formula (21) yields.

This takes the form $c_2\alpha_2 + c_0F = \frac{1}{2}O_n\chi(Q)$, or according to (6)

$$\alpha_2 + KF = 4\pi\chi(Q), \tag{25}$$

where α_2 is the sum of the polar polyhedral angles of those corresponding to the vertices of Q. If we wish to introduce these same polyhedral angles (recalling that for $n = 2$ between the area α^s of a spherical polygon and the length λ_s* of the boundary of its polar, relation (14) holds, which is now written $\alpha^s + \lambda_s* = 2\pi$), one may write $\alpha_2 = 2\pi m - \sum_1^m \lambda_s*$ where m is the number of vertices of Q. The sum of the λ_s* is precisely the sum of the angles of the faces of Q, which according to (24) is for each face

$$KF_i + (m_i - 2)\pi,$$

where m_i is the number of vertices of the face of area F_i (which we assume to be simply connected and therefore $\chi = 1$). Adding for all the faces, and denoting the number of faces by c and the number of edges by a, we obtain

$$\sum_s \lambda_s* = KF + 2\pi a - 2\pi c.$$

However, in every tri-dimensional polyhedron, between the number of edges, faces and vertices of the surface by which it is bounded, the relations (8) and (9) hold, which in current notation give $m - a + c = 2\chi(Q)$. Therefore

$$\alpha_2 = 4\pi\chi(Q) - KF,$$

and by substituting in (25) an identity is obtained. In other words, relation (21), knowing (20), and which can be applied to each of the faces of Q, yields nothing new.

5 The case $n = 4$

In this case, formula (20) enables us to calculate the volume of a polyhedron of $S_4(K)$ in terms of its angles (polyhedral angles of dimensions 1 and 3). Let us now see what result this gives.

By writing (20) and substituting the coefficients c_i for its values, we obtain

$$2\alpha_3 + \alpha_1 L_2 K + 3K^2 V = 4\pi^2\chi(Q). \tag{26}$$

The expression of V, solely in terms of the angles of Q, acquires a rather complicated expression in the case of the general polyhedron. Let us confine ourselves to the simplest case in which Q is a simplex, in order to see how (26) contains the afore-mentioned Poincaré-Hopf formula.

If Q is a simplex, then$\chi(Q) = 1$.

Let $P_i (i = 1, 2, 3, 4, 5)$ be the vertices of Q. Each vertex has 4 tri-dimensional faces. The normals to these faces towards the exterior of Q determine a spherical tetrahedron T_i of dimension 3 on the unit sphere of centre P_i; the sum of the volumes of these tetrahedrons is precisely α_3. Instead of these tetrahedrons, it is convenient to introduce their polars, which are those measuring the solid interior angles of Q on the vertices P_i. If we denote the above-mentioned spherical tetrahedrons and their volumes by the same letter T_i, and both the polar tetrahedrons and their volumes by A_i, then according to (15) we have

$$T_i = \pi^2 - \frac{1}{2}M_1^i - A_i \quad (i = 1, 2, \ldots, 5), \tag{27}$$

where M_1^i is the mean curvature of the spherical tetrahedron A_i, and therefore, according to (17),

$$M_1^i = \sum_{s=1}^{6} \alpha_{1i}^s L_{1i}^s, \tag{28}$$

where L_{1i}^s are the lengths of the 6 edges of A_i, and α_{1i}^s are the polar angles of the corresponding dihedra. Representing these dihedra of A_i by β_{1i}^s, we obtain

$$M_1^i = \sum_{s=1}^{6} (\pi - \beta_{1i}^s) L_{1i}^s. \tag{29}$$

Let us now see how β_{1i}^s and L_{1i}^s can be substituted by elements of the simplex Q. The lengths L_{1i}^s of the edges of A_i measure the angles of the 2-faces (faces of dimension 2) of Q incident upon P_i. Therefore, the sum total of L_{1i}^s for $s = 1, 2, \ldots, 6$ and $i = 1, 2, \ldots, 5$ will be the sum of all these angles; that is, denoting the area of the 2-faces of Q $(h = 1, 2, \ldots, 10)$ by f_h and the total area of all the faces by $f = f_1 + f_2 + \cdots + f_{10}$, and according to (24), we have

$$\sum_{i=1}^{5}\sum_{s=1}^{6} L_{1i}^s = Kf + 10\pi. \tag{30}$$

Furthermore, the dihedra β^s_{1i} of A_i measure the angles of Q formed by the two 3-faces (tri-dimensional faces) that meet at one 2-face (projection of L^s_{1i} from P_i). Therefore, denoting these dihedral angles by β^h_1, we obtain

$$\sum_{i,s} \beta^s_{1i} L^s_{1i} = \sum_{h=1}^{10} \beta^h_1 (K f_h + \pi) = K \sum_{1}^{10} \beta^h_1 f_h + \pi \beta_1, \tag{31}$$

where

$\beta_1 =$ the sum of the angles formed by the two 3-faces incident upon each of the ten 2-faces of Q.

In short, adding (27) for all the 5 vertices of the simplex Q, we have

$$2\alpha_3 = K(\sum_{i=1}^{10} \beta^h_1 f_h - \pi f) + \pi\beta_1 - 2A, \tag{32}$$

where A represents the sum of the solid interior angles of the 5 vertices of Q.

For the second, by adding (26) and according to (18), we arrive at

$$\alpha_1 L_2 = \sum_{h=1}^{10} (\pi - \beta^h_1) f_h = \pi f - \sum_{h=1}^{10} \beta^h_1 f_h, \tag{33}$$

and therefore by substituting (32) and (33) in (26), we obtain

$$3K^2 V + \pi\beta_1 - 2A = 4\pi^2. \tag{34}$$

This is the Poincaré-Hopf formula for the volume V of a simplex in the space of constant curvature K of 4 dimensions in terms of its 3-angles and its 1-angles. It was also given through direct calculation by M. Dehn [6].

(34) is sometimes written in a slightly different form according to Poincaré. This consists in taking the volume of the sphere O_h as a unit of measure for angles of dimension h.

By doing the same for the volume V, in (34) one may introduce

$$\frac{V}{O_4} = V', \quad \frac{\beta_1}{O_1} = \beta_1', \quad \frac{A}{O_3} = A',$$

and thus

$$2K^2 V' = A' - \frac{1}{2}\beta_1' + 1, \tag{35}$$

which is the formula found, for example, in Pesch ([10], p.331, (7.5a)).

6 Added in the proofs

(October 20th, 1961). If we wish to obtain the expression for the volume of any polyhedron $S_4(K)$ in terms of its angles, we may proceed in the following way, which is completely analogous to the previous one for the case of the simplex.

Let

$$\begin{aligned} s_2 &= \text{the number of 2-faces of the polyhedron } Q. \\ n_i &= \text{the number of sides of the 2-face } i. \end{aligned}$$

One arrives at (27) in the same way as for the simplex. In the present case, since the 2-faces of Q may not be triangles, the general formula (24) is applied for $K = 1$, which gives

$$\sum_{i,s} L_{1i}^s = Kf + \sum_1^{s_2} (n_i - 2)\pi,$$

which substitutes (27).

Analogously, (31) will now be

$$\begin{aligned} \sum_{i,s} \beta_{1i}^s L_{1i}^s &= \sum_{h=1}^{s_2} \beta_1^h (K f_h + (n_h - 2)\pi) \\ &= K \sum_1^{s_2} \beta_1^h f_h + \pi \sum_1^{s_2} (n_h - 2)\beta_1^h, \end{aligned}$$

and thus, by adding the expressions (27) and denoting the number of vertices of Q by s_0, we have

$$\begin{aligned} \alpha_3 = \sum_1^{s_0} (\pi^2 - A_i) &+ \frac{\pi}{2} \sum_1^{s_2} (n_h - 2)(\beta_1^h - \pi) \\ &+ \frac{1}{2} K \sum_1^{s_2} (\beta_1^h - \pi) f_h. \end{aligned}$$

Formula (33) is now written

$$\alpha_1 L_2 = \sum_1^{s_2} (\pi - \beta_1^h) f_h.$$

Thereby, formula (26) gives

$$\begin{aligned} 3K^2 V &+ \pi \sum_1^{s_2} (n_h - 2)(\beta_1^h - \pi) + 2 \sum_1^{s_0} (\pi^2 - A_i) \\ &= 4\pi^2 \chi(Q). \end{aligned}$$

Isolating V, we arrive at the required formula. For the case of the simplex, this is

$$n_h = 3, \quad s_2 = 10, \quad s_0 = 5, \quad \chi(Q) = 1,$$

and as expected (34) is the result.

H. Knothe[3] has recently published a direct way of obtaining (36) for the case of convex polyhedra (for which $\chi(Q) = 1$).

References

[1] C. B. Allendoerfer and A. Weil, *The Gauss-Bonnet theorem for Riemannian polyhedron*, Trans. Am. Math. Soc. **53** (1943), 101–129.

[2] J. Bohm, *Untersuchung des Simplexinhaltes in Raumen konstanter Krümmung beliebiger Dimension*, J. Reine und Ang. Math. (Crelle) **202** (1959), 16–51.

[3] S. S. Chern, *A simple intrinsic proof of the Gauss-Bonnet formula for closed Riemannian manifolds*, Ann. of Math. **45** (1944), 747–752.

[4] ———, *On the curvatura integra in a riemannian manifold*, Ann. of Math. **46** (1945), 674–684.

[5] H. S. M. Coxeter, *Non-euclidean Geometry*, University of Toronto Press, 1957, 3a. ed. Toronto.

[6] M. Dehn, *Die Eulersche Formel im Zusammenhang mit dem Inhalt in der Nicht-Euklidischen Geometrie*, Math. Annalen **61** (1905), 561–586.

[7] G. Herglotz, *Ueber die Steinersche Formel für Parallelflachen*, Hamburg Abh. **XV** (1943), 165–177.

[8] W. Hohn, *Winkel und Winkelsumme in n-dimensionalen euklidischen Simplex*, Eidgenossische Technische Hochschule, 1953, Thesis.

[9] H. Hopf, *Die curvatura integra Clifford-Kleinscher Raumformen*, Nachr. Ges. Wiss. Gottingen, Math. Phys. (1925), 131–141.

[10] E. Peschl, *Winkelrelationen am Simplex und die Eulersche Charakteristik*, Bayer. Akad. Wiss. Math. Nat. Kl. Sitz. (1955), 319–345.

[11] H. Poincaré, *Sur la généralisation d'un théorème élémentaire de géométrie*, C. R. Acad. Sc. Paris **I** (1905), 113–117.

[12] L. A. Santaló, *On parallel hypersurfaces in the elliptic and hyperbolic n-dimensional space*, Proc. Am. Math. Soc. **1** (1950), 325–330.

[13] ———, *Questions of differential and integral geometry in spaces of constant curvatura, (Spanish)*, Rendiconti del Sem. Mat. di Torino **14** (1954-55), 277–295.

[14] L. Schlafli, *Gesammelte Mathematische Abhanlungen*, vol. 1, Basel, 1950, specially pag. 240.

[3] *On Polyhedra in Spaces of Constant Curvature*, Michigan Mathematical Journal, 7, 1960, 251-255.

[15] D. M. Y. Sommerville, *The relations connecting the angle-sums and volume of a polytope in space of n-dimensions*, Proc. Roy. Soc. London **Serie A, 115** (1927), 103–119.

A relation between mean curvatures of parallel convex bodies in spaces of constant curvature

L. A. Santaló

Abstract

Let Q be a closed convex hypersurface of class C^3 of the n-dimensional space of constant curvature $K = k^2$. Let $Q(\lambda)$ be the hypersurface parallel to Q at distance λ. Let $M_i(\lambda)$ denote the i-th mean curvature of $Q(\lambda)$ (defined by (1.3)). We prove the identities (1.9) where $M_i'(\lambda) = dM_i(\lambda)/d\lambda$. From these identities we deduce some consequences for convex curves and surfaces ($n = 2, 3$); among them we get a proof of the inequality (3.14) announced by Blaschke [2] and first proved by Knothe [5] in a very different way.

1 Definitions and fundamental identity

Let us consider the space of constant curvature K. In order to simplify the notation, we write

(1.1) $$K = k^2,$$

whereby k is imaginary if $K < 0$, but the formulae will always have a real meaning by virtue of the equalities

(1.2) $$\sin ix = i \sinh x, \quad \cos ix = \cosh x.$$

Let Q be a closed convex hypersurface of class C^3 of the space. If ρ_i, $(i = 1, 2, \ldots, n-1)$ are the principle radii of curvature of Q at the point P, and dP denotes the area element at P, the mean curvatures of Q are defined by

(1.3) $$M_i = \frac{1}{\binom{n-1}{i}} \int_Q \Big(\sum \frac{1}{\rho_{h1}\rho_{h2}\cdots\rho_{hi}}\Big) dP, \quad i = 1, 2, \ldots, n-1,$$

where the sum is extended to the $\binom{n-1}{i}$ combinations of order i of the indices $1, 2, \ldots, n-1$. In particular, $M_0 =$ the area of Q.

If c_i is the line of curvature corresponding to ρ_i, and R_i denotes the geodesic distance from P to the point of contact of the normal to Q at P with the envelope of the normals to Q along c_i, then we have

(1.4) $$\rho_i = \frac{\tan kR_i}{k},$$

as can be seen in [[4], p. 214].

Furthermore, if $d\alpha_i$ is the angle between two infinitely close normals along c_i at P, and ds_i is the corresponding arc element of c_i, we have

$$ds_i = \frac{\sin kR_i}{k} d\alpha_i. \tag{1.5}$$

Hence,

$$dP = \frac{1}{k^{n-1}} \prod_{i=1}^{n-1} \sin kR_i \, d\alpha_i. \tag{1.6}$$

For the parallel hypersurface $Q(\lambda)$ exterior to Q at distance λ, we have

$$R_i(\lambda) = R_i + \lambda, \quad ds_i(\lambda) = k^{-1} \sin k(R_i + \lambda)\, d\alpha_i. \tag{1.7}$$

Therefore, on substituting (1.6) and (1.7) in (1.3), for the i-th mean curvature of $Q(\lambda)$ we obtain the following expression,

$$M_i(\lambda) = \frac{k^{i-n+1}}{\binom{n-1}{i}} \int \sum \left(\prod_{j=1}^{i} \cos k(R_{hj} + \lambda) \prod_{j=i+1}^{n-1} \sin k(R_{hj} + \lambda) \right) d\alpha_1 d\alpha_2 \ldots d\alpha_{n-1}, \tag{1.8}$$

where the integration is extended to the $(n-1)$-dimensional unit sphere; that is, to all the values of $\alpha_1, \alpha_2, \ldots, \alpha_{n-1}$ corresponding to points P different from Q.

In a previous work [6], we applied formula (1.8) in order to find the value of $M_i(\lambda)$ in terms of the mean curvatures M_i of Q and λ. Our aim here is to note that, without needing to calculate $M_i(\lambda)$, some interesting relations can be obtained among their derivatives. Indeed, by deriving (1.8) with respect to λ, we obtain

$$M_i'(\lambda) = \frac{k^{i-n+1}}{\binom{n-1}{i}} \left[-(n-i) \binom{n-1}{i-1} k^{n-i+1} M_{i-1}(\lambda) + (i+1) \binom{n-1}{i+1} k^{n-i-1} M_{i+1}(\lambda) \right],$$

or, by simplifying,

$$M_i'(\lambda) = -ik^2 M_{i-1}(\lambda) + (n-i-1) M_{i+1}(\lambda). \tag{1.9}$$

Observe that this formula also holds for the extreme cases $i = 0$, $i = n-1$. In other words, formula (1.9) holds for $i = 0, 1, \ldots, n-1$.

Moreover, this formula can be completed as follows: if $V(\lambda)$ is the volume of the body limited by $Q(\lambda)$, the derivative $V'(\lambda)$ is the area of $Q(\lambda)$; that is, equal to $M_0(\lambda)$. In other words, we also have

$$V'(\lambda) = M_0(\lambda). \tag{1.10}$$

For $K > 0$, the relation between the mean curvatures of two "dual" hypersurfaces Q and Q^* is also important. The exterior parallel hypersurface at distance $\pi/2k$ is known as the dual hypersurface Q^* of Q; that is $Q^* = Q(\pi/2k)$. Denoting by M_i^* the i-th mean curvature of Q^*, that is, $M_i(\pi/2k)$, from (1.8) we deduce the following important formula (valid for $i = 0, 1, \ldots, n-1$)

$$M_i^* = (-1)^i k^{2i-n+1} M_{n-i-1} \tag{1.11}$$

which was obtained in [6]. Observe the relation

$$M_i^{**} = (-1)^{n-1} M_i. \tag{1.12}$$

2 Application to the case $n = 2$

For $n = 2$, instead of V we have the area F limited by the convex curve Q, and instead of M_0 the length L of the said convex curve. For $i = 0, 1$, formulae (1.9) give

$$L'(\lambda) = M_1(\lambda), \quad M_1'(\lambda) = -k^2 L(\lambda). \tag{2.1}$$

Hence,

$$L''(\lambda) + k^2 L(\lambda) = 0, \quad M_1''(\lambda) + k^2 M_1(\lambda) = 0. \tag{2.2}$$

The first equation gives

$$L(\lambda) = A \cos k\lambda + B \sin k\lambda. \tag{2.3}$$

The constants A, B are determined in terms of $L(0) = L$ and $L'(0) = M_1$, yielding

$$L(\lambda) = L \cos k\lambda + k^{-1} M_1 \sin k\lambda. \tag{2.4}$$

Taking into account (1.10), which here is written $F'(\lambda) = L(\lambda)$, integrating (2.4) and determining the constant of integration by the value $F(0) = F$, we obtain

$$F(\lambda) = F + k^{-1} L \sin k\lambda + k^{-2} M_1 (1 - \cos k\lambda). \tag{2.5}$$

We thus have the well-known formulae (2.4) and (2.5), which give the length and the area of $Q(\lambda)$. See, for example, Bonnesen [[3], p. 81], Vidal Abascal [7] and Allendoerfer [1].

Analogously to (2.4), from the second formula (2.2) we obtain

$$M_1(\lambda) = M_1 \cos k\lambda - kL \sin k\lambda. \tag{2.6}$$

From (2.4) and (2.6), it follows that $L^2(\lambda) + (k^{-1}M_1(\lambda))^2$ is independent of λ. Hence, for the case of positive curvature the isoperimetric property of the circle [[3], p. 81] can be deduced. Indeed, let $Z(\lambda)$ be the circle circumscribed to $Q(\lambda)$; by taking λ such that $Z(\lambda)$ is the maximum circle of radius k^{-1} and length $2\pi k^{-1}$, the length of $Q(\lambda)$ is equal to or greater than this length (since the arcs between the points of contact are geodesics for $Z(\lambda)$); that is, $L(\lambda) \geq 2\pi k^{-1}$. Therefore,

$$L^2 + (M_1/k)^2 \geq 4\pi^2 k^{-2}, \tag{2.7}$$

since the first term on the right-hand side is independent of λ.

In order to introduce the area F instead of M_1, it is necessary to note that from (2.1) and from $F'(\lambda) + M_1'(\lambda)k^{-2} = 0$, we deduce $F(\lambda) + M_1(\lambda)k^{-2} = C =$ constant (independent of λ). The value of this constant is not given by the foregoing, but by the Gauss-Bonnet formula for the theory of surfaces, and thus $C = 2\pi/k^2$. Thus we have $M_1/k = 2\pi/k - kF$, and the inequality (2.7) is

$$L^2 - 4\pi F + k^2 F^2 \geq 0, \tag{2.8}$$

which is the classical isoperimetric inequality for Spherical curves, or in general for curves of surfaces of constant curvature k^2. The equals sign holds only when $Q(\lambda)$ coincides with $Z(\lambda)$ for the value of λ in which $Z(\lambda)$ is the maximum circle; that is, only if Q is a circumference.

Also observe that for the case $k^2 > 0$, we have the relations of duality (1.11), which for the case $n = 2$, $i = 0, 1$ are written as follows:

$$L^* = k^{-1}M_1, \quad M_1^* = -kL, \tag{2.9}$$

whereby formulae (2.4), (2.5) and (2.6) can be written

$$L(\lambda) = L \cos k\lambda + L^* \sin k\lambda, \tag{2.10}$$

$$F(\lambda) = F + k^{-1}L \sin k\lambda + k^{-1}L^*(1 - \cos k\lambda), \tag{2.11}$$

$$k^{-1}M_1(\lambda) = L^* \cos k\lambda - L \sin k\lambda. \tag{2.12}$$

3 The case $n = 3$

For $n = 3$, formula (1.10) and formulae (1.9), for $i = 0, 1, 2$ take the form

$$\begin{aligned} V'(\lambda) &= M_0(\lambda) \\ M_0'(\lambda) &= 2M_1(\lambda) \\ M_1'(\lambda) &= -k^2 M_0(\lambda) + M_2(\lambda) \\ M_2'(\lambda) &= -2k^2 M_1(\lambda). \end{aligned} \tag{3.1}$$

Formulae (1.11) referring to the duality, for the case of positive curvature, are (for $i = 0, 1, 2$)

$$M_0^* = k^{-2} M_2, \quad M_1^* = -M_1, \quad M_2^* = k^2 M_0. \tag{3.2}$$

A first consequence of (3.1) is $M_2'(\lambda) + k^2 M_0'(\lambda) = 0$, hence

$$M_2(\lambda) + k^2 M_0(\lambda) = C, \tag{3.3}$$

where C is independent of λ. According to (3.2), this relation can be written as follows:

$$M_0^* + M_0 = C k^{-2}. \tag{3.4}$$

Since M_0^* and M_0 are the areas of Q and Q^*, a theorem by Blaschke (see [2], formula (32)) tells us that $C = 4\pi$. Thus we have

$$M_2(\lambda) + k^2 M_0(\lambda) = 4\pi. \tag{3.5}$$

From (3.1), we may also immediately deduce the relations

$$\begin{aligned} M_0''(\lambda) + 4k^2 M_0(\lambda) &= C_0 \\ M_2''(\lambda) + 4k^2 M_2(\lambda) &= C_2, \end{aligned} \tag{3.6}$$

where C_0 and C_2 are independent of λ.

On substituting the dual values (3.2) in the first equation (3.6), and comparing it with the second equation (3.6), then $C_2 = k^2 C_0$. Furthermore, from (3.6) and (3.5) we deduce that $C_2 + k^2 C_0 = 16k^2\pi$; therefore

$$C_0 = 8\pi, \quad C_2 = 8k^2\pi. \tag{3.7}$$

Thereby, integrating equations (3.6) and taking into account the initial values $M_0(0) = M_0$, $M_0'(0) = 2M_1(0) = 2M_1$, $M_2(0) = M_2$, $M_2'(0) = -2k^2 M_1(0) = -2k^2 M_1$, we obtain

$$2k^{-2}\pi - M_0(\lambda) = (2k^{-2}\pi - M_0)\cos 2k\lambda - k^{-1} M_1 \sin 2k\lambda, \tag{3.8}$$

$$M_2(\lambda) - 2\pi = (M_2 - 2\pi)\cos 2k\lambda - k M_1 \sin 2k\lambda. \tag{3.9}$$

Moreover, from (3.1) we also deduce that

$$M_1'' + 4k^2 M_1 = 0, \tag{3.10}$$

and hence, with the initial conditions $M_1(0) = M_1, M_1'(0) = k^2 M_0 + M_2 = 4\pi - 2k^2 M_0$ (the latter according to (3.5)), we arrive at

$$k^{-1} M_1(\lambda) = k^{-1} M_1 \cos 2k\lambda + (2\pi k^{-2} - M_0)\sin 2k\lambda. \tag{3.11}$$

From (3.8) and (3.11), we are able to deduce that the expression

$$D = (2\pi k^{-2} - M_0(\lambda))^2 + k^{-2}M_1^2(\lambda) \tag{3.12}$$

is independent of λ.

Let us consider the case of the Spherical space (positive constant curvature $k^2 > 0$). Having assumed that $Q = Q(0)$ is convex, the volume of $Q(\lambda)$ increases monotonically with λ. As λ increases, a moment arrives when this volume is equal to half the total volume of the space; let us assume that this occurs for $\lambda = \lambda_0$. At that moment, by the isoperimetric inequality in the Spherical space, the area $M_0(\lambda_0)$ is equal to or greater than the area of the maximum sphere $4\pi k^{-2}$, which limits the volume itself. Therefore, for $\lambda = \lambda_0$, the value of D is $D \geq 4\pi^2 k^{-4}$; since D is independent of λ, this bound holds for all λ, in particular for $\lambda = 0$ we have

$$(2\pi k^{-2} - M_0)^2 + k^{-2}M_1^2 \geq 4\pi^2 k^{-4}, \tag{3.13}$$

which by writing $M_0 = F$ to denote the area with its customary notation, may be written as follows:

$$M_1^2 + k^2F^2 - 4\pi F \geq 0. \tag{3.14}$$

This is the inequality that generalizes Minkowski's classical inequality $M^2 - 4\pi F \geq 0$ for the Euclidean case ($k = 0$) to the space of positive constant curvature.

Inequality (3.14) can be found without proof in Blaschke [[2], formula (1)]; the only known proof was given by H. Knothe [5] and is very different from that given here.

It would be interesting to see if inequality (3.14) for spaces of negative curvature could be obtained from the general identities (1.9), as well as its generalization to the case of $n > 3$ dimensions.

References

[1] C. B. Allendoerfer, *Steiner's formulae on a general* S^{n+1}, Bull. Am. Math. Soc. **54** (1948), 128–135.

[2] Wilhelm Blaschke, *Über eine geometrische Frage von Euklid bis heute*, (Hamburg. Math. Einzelschriften. 23) Leipzig, Berlin: B. G. Teubner, 1938.

[3] T. Bonnesen, *Les problèmes des isopérimètres et des isépiphanes*, Collection de monographies sur la théorie des fonctions, Gauthier-Villars, Paris, 1929.

[4] E. P. Eisenhart, *Riemannian Geometry*, Princeton, 1926.

[5] H. Knothe, *Zur Theorie der konvexen Körper im Raum konstanter positiver Krümmung*, Revista de la Facultad de Ciencias de Lisboa **II, Serie A** (1952), 336–348, París.

[6] L. A. Santaló, *On parallel hypersurfaces in the elliptic and hyperbolic n-dimensional space*, Proc. Am. Math. Soc. **1** (1950), 325–330.

[7] E. Vidal-Abascal, *A generalization of Steiner's formulae*, Bull. A. Math. Soc. **53** (1947), 841–844.

Total absolute curvatures of manifolds contained in a Euclidean space

L. A. Santaló

Abstract

LetX^n be a compact (without boundary) differentiable manifold of dimension n and class $\mathcal{C}^2$ contained in the Euclidean space E^{n+N}. We define the following total absolute curvatures $K_r(X^n)$.

a) The case $1 \leq r \leq n$. Let $T_n(p)$ be the tangent space of X^n at the point p. Let $L_{n+N-r}(O)$ be a $(n+N-r)$-dimensional linear subspace through the fixed point O, and let $dL_{n+N-r}(O)$ be the density for sets of $L_{n+N-r}(O)$ (volume element of the Grassman manifold $G_{n+N-r,r}$). Let Γ_r denote the set of all linear subspaces L_r of E^{n+N} which *are contained* in some T_np, pass through p and are orthogonal to $L_{n+N-r}(O)$. Then we define $K_r(X^n)$ by the formulae (1.1), (1.2).

b) The case $n \leq r \leq n+N-1$. Using the same notation as above, and denoting by Γ_r the set of all linear subspaces L_r *containing* some $T_n(p)$ and that are orthogonal to $L_{n+N-r}(O)$, the total absolute curvature $K_r(X^n)$ is defined by the same formulae (1.1) and (1.2).

We prove the following properties of these curvatures: 1. In the case $1 \leq r \leq n$ we have $K_r(X^n)$ if, and only if, $n \geq rN$; 2. For $r \geq n$, the only absolute curvature which is $\neq 0$ is $K_{n+N-1}(X^n)$, and it coincides with Chern-Lashof's curvature; 3. The case $N = 1$ generalizes to compact manifolds the well-known mean curvatures of convex hyperspaces; 4. The case $n = rN$ (formula (2.9)) is particularly interesting; we consider in detail the case $n = 2, N = 2, r = 1$. In Section 4 we state some inequalities among the absolute curvatures $K_r(X^n)$.

1 Definitions

Let X^n be a compact manifold without boundary of class $\mathcal{C}^2$ contained in the Euclidean space E^{n+N}. At each point p of M^n we have the n-dimensional tangent linear space $T_n(p)$. Throughout this paper, L_r denotes the r-dimensional linear subspace in E^{n+N}, and $L_r(O)$ denotes an r-dimensional linear subspace that passes through the point O.

In order to define the total curvatures of X^n, we proceed as follows:

a) Let r be a natural number such that $1 \leq r \leq n$. By the fixed point O of E^{n+N} we take an oriented linear subspace $L_{n+N-r}(O)$ of dimension $n+N-r$. We consider all the L_r of E^{n+N} that *are contained* in some $T_n(p)$ passing through the point p and that *are orthogonal* to $L_{n+N-r}(O)$; we denote by Γ_r the set of these L_r. Since each L_r intersect L_{n+N-r} at a point, the intersection $\Gamma_r \cap L_{n+N-r}$ will be a certain manifold whose dimension will be calculated in Section 2. Let $\mu(\Gamma \cap L_{n+N-r}(O))$ be its measure (if the intersection consists of a finite number of points, the measure is this number; if it is a manifold of dimension h, the measure is understood as being its h-dimensional volume as a submanifold of the Euclidean space $L_{n+N-r}(O)$). Let $dL_{n+N-r}(O)$ be the density in space of all the oriented $L_{n+N-r}(O)$ of E^{n+N} (the Grassmanian volume element $G_{n+N-r,r}$) (see section 3).

We define the rth *total absolute curvature of* $X^n \subset E^{n+N}$ by the integral (r = 1, 2, ... , n)

$$K_r(X^n) = \frac{O_1 O_2 \cdots O_{r-1}}{O_{n+N-2} \cdots O_{n+N-r-1}} \cdot \int_{G_{n+N-r,r}} \mu(\Gamma_r \cap L_{n+N-r}(O))\, dL_{n+N-r}(O), \tag{1.1}$$

where O_i denotes the area of the i-dimensional unit sphere; that is,

$$O_i = \frac{2\pi^{\frac{i+1}{2}}}{\Gamma(\frac{i+1}{2})} \tag{1.2}$$

and the factor preceding the integral sign has been taken to normalize convincingly.

b) Now let $n \leq r \leq n+N-1$. Using the same notation as above, we now consider the set Γ_r of all the L_r that *contain* some $T_n(p)$ and that *are orthogonal* to a fixed $L_{n+N-r}(O)$. The same formula (1.1) then defines the total absolute curvatures of $X^n \subset E^{n+N}$ for $r = n, n+1, \ldots, n+N-1$.

Note that, according to the definition, since the measures cannot be negative, we have always that $K_r \geq 0$.

2 First properties

We now find the relations that should exist between n, N and r in order for the curvature $K_r(X^n)$ to be different from zero.

a) Let us consider the first case $1 \leq r \leq n$.

The set of all the L_r of E^{n+N} constitutes the Grassmanian whose dimension is $(r+1)(n+N-r)$. The set of L_r contained in a fixed $T_n(p)$ and passing through p constitute the Grassmanian $G_{r,n-r}$ of dimension $r(n-r)$, and therefore the set of all the L_r contained in some $T_n(p)$ and passing through the corresponding point have dimension $r(n-r)+n$. Furthermore, the set of the L_r orthogonal to a fixed $L_{n+N-r}(O)$ is of dimension $n+N-r$. Thus, the dimension of the intersection of the last two sets (previously denoted by Γ_r) is as follows:

$$r(n-r)+n+n+N-r-(r+1)(n+N-r) = n - rN. \tag{2.1}$$

Consequently: *in the case* $1 \leq r \leq n$, *in order for* $K_r(X^n) \neq 0$, *the following condition must be fulfilled:*

$$n \geq rN. \tag{2.2}$$

b) Let the case now be $n \leq r \leq n+N-1$.

The set of the L_r of E^{n+N} that *contains* a fixed $T_n(p)$ constitutes the Grassmanian $G_{r-n,n+N-r}$ (it suffices to intersect by an L_N orthogonal to $T_n(p)$ and to consider the intersection), and therefore the set of all the L_r containing some $T_n(p)$ have dimension $(r-n)(n+N-r)$. The remaining dimensions are the same as before, so that the dimension of the set Γ_r, the intersection of the L_r containing some $T_n(p)$ with the set of the L_r orthogonal to a fixed $L_{n+N-r}(O)$, is equal to

$$(r-n)(n+N-r)+n+n+N-r-(r+1)(n+N-r) \\ = rn-n^2-nN+n. \tag{2.3}$$

In order for this dimension not to be negative, we have

$$r \geq n+N-1. \tag{2.4}$$

Hence:
For $r \geq n$, the only non-vanishing curvature $K_r(X^n)$ is $K_{n+N-1}(X^n)$.

Examples: 1. The case $r = n+N-1$. In this case, $L_{n+N-r}(O)$ is an oriented straight line $L_1(O)$, and dL_1 is the area element of the $(n+N-1)$-dimensional unit sphere corresponding to the direction of this line (which we denoted by $dO_{n+N-1} = dL_1(O)$). The total volume of the Grassmanian $G_{1,n+N-1}$ is O_{n+N-1}. Expression (1.1) takes the form

$$K_{n+N-1}(X^n) = \frac{1}{2}\int_{O_{n+N-1}} \nu\, dL_1(O), \tag{2.5}$$

where ν is the number of hyperplanes L_{n+N-1} containing some $T_n(p)$ and that are orthogonal to the line $L_1(O)$.

This total curvature $K_{n+N-1}(X^n)$, except for the factor $\frac{1}{2}$, coincides with that considered by Chern-Lashof in several of their works [6], [7].

2. Let the case now be $N = 1$. According to (2.2), we then have all the total absolute curvatures $K_1, K_2, \ldots, K_n$. The manifold $\Gamma_r \cap L_{n+N-r}$, according to (2.1), has dimension $n-r$, that is, it is a hypersurface of the subspace $L_{n+1-r}(O)$. Expression (1.1) takes the form

$$K_r(X^n) = \frac{O_1O_2\cdots O_{r-1}}{O_{n-1}O_{n-2}\cdots O_{n-r}} \cdot \\ \cdot \int_{G_{n+1-r,r}} \mu(\Gamma_r \cap L_{n+1-r}(O))\, dL_{n+1-r}(O). \tag{2.6}$$

If X^n is the boundary of a convex domain of E^{n+1} (a convex hyperspace of class C^2), we know that this $K_r(X^n)$ coincides with the rth *mean curvature of* X^n, defined by

$$M_r(X^n) = \frac{1}{\binom{n}{r}} \int_{X^n} S_r\, d\sigma, \tag{2.7}$$

where S_r is the rth elementary symmetric function of the principal n curvatures of X^n, and $d\sigma$ is the n-dimensional area element of this hypersurface, the integral being extended to the whole hypersurface (see [16]).

If X^n is not convex, $K_r(X^n)$ is the rth absolute mean curvature; that is, it is equal to

$$K_r(X^n) = \frac{1}{\binom{n}{r}} \int_{X^n} |S_r|\, d\sigma, \tag{2.8}$$

where $|S_r|$ denotes the absolute value of S_r. For this reason the total curvatures $K_r(X^n)$ are known as *absolute.*

3. The case $n = rN$. This case is particularly interesting because, according to (2.2), here the intersection $\Gamma_r \cap L_{n+1-r}(O)$ has null dimension; that is, the integrand of (1.1) is equal to the natural number ν of linear subspaces L_r contained in some tangent space $T_n(p)$ and orthogonal to $L_{n+N-r}(O)$. In other words, K_r can be written

$$K_r(X^n) = \frac{O_1 O_2 \cdots O_{r-1}}{O_{n+N-2} \cdots O_{n+N-r-1}} \cdot \int_{G_{n+N-r,r}} \nu\, dL_{n+N-r}(O). \tag{2.9}$$

For $N = 1, r = n$, we once again have the case (2.5).

For $r < n, n = rN$, if X^n is contained within a L_{n+N-1}, the tangent spaces $T_n(p)$ will also be within this L_{n+N-1}, and therefore $\nu = 0$, except for a set of directions of zero measure (those parallel to L_{n+N-1}). Reciprocally, if X^n is not contained by any subspace of E^{n+N}, the dimensional calculation that leads to (2.2) proves that in general $\nu > 0$ and therefore $K_r(X^n)$. In other words, referring always to compact manifolds X^n of class $\mathcal{C}^2$ in E^{n+N}, we have:

If the relation $n = rN$ is satisfied, with $N > 1$, the condition $K_r(X^n) = 0$ is necessary and sufficient for X^n to be contained in a subspace of dimension $n + N - 1$.

This proves that the curvature $K_r(X^n)$ is different from $K_{n+N-1}(X^n)$, since this latter (Chern-Lashof's curvature) does not depend on the dimension of the ambient space ([12], [7]).

4. The case $n = 2, r = 1, N = 2$.

This case is interesting because it is the simplest in which the condition $n = rN$ is fulfilled, with $N > 1$. This is the case of a surface X^2 in E^4.

According to (2.9), $K_1(X^2)$ is written thus,

$$k_1(X^2) = \frac{1}{O_2} \int_{O_3} \nu_1\, dL_1(O), \tag{2.10}$$

where $dL_1(O)$ replaces $dL_3(O)$ due to the duality $dL_1(O) = dL_3(O)$, which can be seen in the following section. Recall that ν_1 is the number of straight lines parallel to the direction $L_1(O)$ that are contained in some $T_2(p)$.

The curvature $K_3(X^2)$, according to (2.5), can in this case be written thus,

$$K_3(X^2) = \frac{1}{2} \int_{O_3} \nu_3\, dL_3(O), \tag{2.11}$$

where the foregoing duality is employed and where ν_3 is now the number of hyperplanes parallel to $L_3(O)$ that contain some $T_2(p)$.

According to the inequalities (2.2) and (2.4), these two curvatures K_1 and K_3 are the only curvatures of X^2 not always null. The condition for X^2 to be contained by a L_3 is $K_1(X^2) = 0$. On the other hand, $K_3(X^2) \geq O_3 = 2\pi^2$ is always fulfilled, since it is obviously always $\nu_3 \geq 2$.

In order to provide a convenient interpretation of these two curvatures, one may proceed in the following manner: let us consider the mapping $\varphi : T_2(p) \rightarrow L_2(O)$, where $L_2(O)$ is the plane parallel to $T_2(p)$ through the fixed point O. Moreover, by intersecting the set of subspaces $L_i(O)\,(i = 1, 2, 3)$ by a 3-dimensional unit sphere centered at O, and identifying the points diametrically opposed to this unit sphere in order to have an elliptical space S_3, we obtain the map $\phi : L_i(O) \rightarrow S_{i-1}$, denoting the points of S_3 by S_0, the lines by S_1, and the planes by S_2. The composition $\phi \circ \varphi$ maps the set of the tangent planes $T_2(p)$ in a congruence of lines T of S_3, and in this congruence ν_1 denotes the number of lines of the congruence passing through the point S_0, and ν_3 denotes the number of lines of the congruence contained in the plane S_2. The densities $dL_1(O)$ and $dL_3(O)$ that appear in (2.10) and (2.11) are the densities dS_0 and dS_2 of points and planes in the elliptical space S_3 (see [15]). In the cases wherever ν_1 and ν_3 are constants, $K_1(X^2)$ indicates the *order* of the congruence T, and $K_3(X^2)$ the *class* of this congruence (except for a constant factor).

3 Expression of the densities $dL_h(O)$

The expressions of the densities $dL_h(O)$ $(h = 1, 2, \ldots, n + N - 1)$, which appear in (1.1) for h-dimensional oriented linear spaces of E^{n+N} passing through a fixed point O, are known (see, for example, [16] or [5]). Let is now quickly recall them for the sake of convenience.

Let $(O; e_1, e_2, \ldots, e_{n+N})$ be an orthogonal reference of E^{n+N} with vertex O (set of $n + N$ mutually orthogonal unit vectors that pass through the point O). In the space of all the orthogonal references with vertex O, the following differential forms are defined thus,

$$\omega_{im} = -\omega_{mi} = e_m\, de_i. \tag{3.1}$$

Let us consider $L_h(O)$ determined by $e_1, e_2, \ldots, e_h$. Its density is

$$dL_h(O) = \bigwedge_{i,m} \omega_{im} \tag{3.2}$$

for

$$i = 1, 2, \ldots, h; \quad m = h + 1, h + 2, \ldots, n + N,$$

where on the right-hand side of (3.2) Λ means the exterior product.

The L_{n+N-h} normal to $L_h(O)$ is determined by the vectors $e_{h+1}, \ldots, e_{n+N}$, and applying the same definition (3.2) we obtain the "duality"

$$dL_h(O) = dL_{n+N-h}(O). \tag{3.3}$$

The measure of the set of all the oriented $L_h(O)$ (volume of the Grassmanian $G_{h,n+N-h}$) is easily calculated, either directly [16], or by recalling that $G_{h,n+N-h}$ is the quotient space $SO(n+N)/SO(h) \times SO(n+N-h)$, [5], and is

$$(3.4) \qquad \int_{G_{h,n+N-h}} dL_h(O) = \frac{O_{n+N-1} \cdots O_{n+N-h}}{O_1 O_2 \cdots O_{h-1}},$$

where O_i has the value (1.2).

4 Inequalites between total absolute curvatures

a) The case $r = n + N - 1$.

As already stated, in this case, $K_{n+N-1}(X^n)$, given by (2.5), coincides with the Chern-Lashof curvature and thus satisfies the inequality [6]

$$(4.1) \qquad K_{n+N-1}(X^n) \geq O_{n+N-1},$$

which is easily deducible from (2.5), observing that $\nu \geq 2$. Other properties can be found in [6], [7].

In the case $n = 1$, curves in E^{N+1} in particular have been studied. In this case, the total absolute curvature $K_N(X^1)$ is expressed as

$$(4.2) \qquad K_N(X^1) = \frac{O_N}{O_1} \int_{X^1} |\kappa| \, ds,$$

where $|\kappa|$ is the absolute value of the curvature of X^1, and ds is the arc element.

This formula (4.2) arises from (2.5) by the following reasoning: consider the tangential spherical image of the curve X^1; that is, the curve Γ obtained in O_N by tracing radii parallel to the tangents of X^1 through the centre of O_N (note that here we denote by O_N the N-sphere of radius one, not its area, as was done before). The number ν is the number of points at which Γ is intersected by the maximum $(N-1)$-sphere whose pole corresponds to the area element $dL_1(O)$ of (2.5). Therefore, we have that the integral (2.5) is equal to the factor $2O_N/O_1$ along the length of Γ (see [15], p. 27), which is precisely the right-hand side of (4.2).

In this case, the inequality (4.1) is written thus;

$$\int_{X^1} |\kappa| ds \geq 2\pi,$$

which for curves in the space E^3 is due to Fenchel [10].

If we consider curves contained in a sphere of radius R in the space E^{n+N}, the curvature K_N is lower bounded by the length L of X^1 according to the inequality

$$(4.3) \qquad L(X^1) \leq \frac{O_1}{O_N} R K_N(X^1),$$

which for $N = 2$ was given by Fáry [9] and subsequently improved and generalized to any dimension by Chakerian [3], [4].

b) The case $N = 1$.

This is the case of hypersurfaces X^n in E^{n+1}. For the case where X^n is a *convex* hypersurface, the K_i are mutually related by classical quadratic inequalities, which for $n = 2$ are due to Minkowski (see [1] and [2]). Further inequalities (for X^n convex and closed) have been obtained by Fenchel, Alexandrov and Hadwiger, which can be condensed into the following (see [11], p. 282),

$$K_{\alpha-1}^{\beta-\gamma} K_{\beta-1}^{\gamma-\alpha} K_{\gamma-1}^{\alpha-\beta} \geq 1, \quad 0 \leq \alpha < \beta < \gamma \leq n+1, \tag{4.4}$$

where it is convenient to write $K_0 =$ n-dimensional area of X^n and $K_{-1} = (n+1)V$, and where V is the volume of the domain limited by X^n.

Fáry type [9] inequalities, [8], also exist in this case; assuming that the compact hypersurface X^n of E^{n+N} (not necessarily convex) is contained within a Euclidean sphere of radius R, the following inequalities hold:

$$K_r(X^n) \leq R^{n-r} K_n(X^n), \quad r = 1, 2, \ldots, n, \tag{4.5}$$

and furthermore, if F is the n-dimensional area of X^n,

$$F(X^n) \leq R^n K_n(X^n). \tag{4.6}$$

$$F(X^n) \leq \frac{O_{n+1-r} O_{r-1}}{r O_{n+1}} R^r K_r(X^n), \quad r = 1, 2, \ldots, n. \tag{4.7}$$

For $r = 1$, this last inequality can be improved thus, $F(X^n) \leq R K_1(X^n)$, in accordance with Chakerian [4]. These inequalities (4.5), (4.6) and (4.7) were obtained elsewhere [17] by the present author.

c) The case $N > 1, r < n$.

As already observed in Section 2, the most interesting case appears to be $n = rN$, due to the fact that K_r is expressed by the formula (2.9), in which the integrand ν is a natural number. No equalities are known in this case. If the set of the ν points $p \in X^n$ in which $L_r \subset T_n(p)$ is decomposed according to whether or not they are critical points of X^n, and if they are critical, according to their index k, the curvature $K_r(X^n)$ can be decomposed into a sum of "curvatures of index k", in the manner of those introduced by Kuiper for the case $r = n+N-1$ [13], [14], which would also be interesting to study in this case.

References

[1] T. Bonnesen and W. Fenchel, *Theorie der konvexen Körper*, Ergebnisse der Math. Springer, Berlin, 1934.

[2] Herbert Busemann, *Convex surfaces*, Interscience Tracts in Pure and Applied Mathematics, no. 6, Interscience Publishers, Inc., New York, 1958.

[3] G. D. Chakerian, *An inequality for closed space curves*, Pacific J. Math. **12** (1962), 53–57.

[4] ———, *On some geometric inequalities*, Proc. Amer. Math. Soc. **15** (1964), 886–888.

[5] Shiing-shen Chern, *On the kinematic formula in integral geometry*, J. Math. Mech. **16** (1966), 101–118.

[6] Shiing-shen Chern and Richard K. Lashof, *On the total curvature of immersed manifolds*, Amer. J. Math. **79** (1957), 306–318.

[7] ———, *On the total curvature of immersed manifolds. II*, Michigan Math. J. **5** (1958), 5–12.

[8] István Fáry, *Sur la courbure totale d'une courbe gauche faisant un nœud*, Bull. Soc. Math. France **77** (1949), 128–138.

[9] ———, *Sur certaines inégalites géométriques*, Acta Sci. Math. Szeged **12** (1950), no. Leopoldo Fejer et Frederico Riesz LXX annos natis dedicatus, Pars A, 117–124.

[10] W. Fenchel, *Uber krummung und windung geschlossener raumkurven*, Math. Annalen **101** (1929), 238–252.

[11] H. Hadwiger, *Vorlesungen über Inhalt, Oberfläche und Isoperimetrie*, Springer-Verlag, Berlin, 1957.

[12] N. H. Kuiper, *Immersions with minimal total absolute curvature*, Colloque Géom. Diff. Globale (Bruxelles, 1958), Centre Belge Rech. Math., Louvain, 1959, pp. 75–88.

[13] ———, *La courbure d'indice k et les applications convexes*, Séminaire Ehresmann. Topologie et géométrie différentielle **2** (1960), no. 14, 1–15.

[14] ———, *Der Satz von Gauss-Bonnet für Abbildungen in E^N und damit verwandte Probleme*, Jber. Deutsch. Math.-Verein. **69** (1967), 77–88.

[15] L. A. Santaló, *Geometría Integral en espacios de curvatura constante*, Publicaciones de la Com. Nac. de Energía Atómica **1** (1952), 1–68.

[16] ———, *Sur la mesure des espaces linéaires qui coupent un corps convexe et problèmes qui s'y rattachent*, Colloque sur les questions de réalité en géométrie, Liège, 1955, Georges Thone, Liège, 1956, pp. 177–190.

[17] ———, *On some geometric inequalities in the style of Fáry*, Amer. J. Math. **91** (1969), 25–31, (to appear when the present paper was written).

Sonderabdruck aus
ARCHIV DER MATHEMATIK
Vol. XVIII, 1967 BIRKHÄUSER VERLAG, BASEL UND STUTTGART Fasc. 5

Horocycles and Convex Sets in Hyperbolic Plane

By

L. A. SANTALÓ

1. Introduction. The study of the integral geometry in hyperbolic plane was carried out in [2]; see also [3]. However, the basic elements there considered were only points, lines and sets of congruent figures. The horocycles were not considered, in spite of the important role they play in hyperbolic plane geometry. The purpose of the present paper is to fill this gap, by defining a density for sets of horocycles and obtaining the integral formulae (3.2), (4.1), (5.8), (5.11) which generalize to horocycles certain known formulae for convex sets and straight lines. As usual in integral geometry, we call "density" any differential form whose integral gives an invariant measure under the group of hyperbolic motions.

2. Density for horocycles. In terms of the polar coordinates r, φ the line element of the hyperbolic plane has the form

$$ds^2 = dr^2 + \sinh^2 r \, d\varphi^2 \tag{2.1}$$

and the area element is

$$df = \sinh r \, dr \wedge d\varphi . \tag{2.2}$$

Let C be a circle of radius R and center $C_1(r, \varphi)$. Denoting by ϱ the distance from the origin of coordinates O to C, the area element corresponding to C_1 will be $dC_1 = \sinh(\varrho + R)\, d\varrho \wedge d\varphi$ if O is exterior to C and $dC_1 = \sinh(R - \varrho)\, d\varrho \wedge d\varphi$ if O is interior to C.

By fixed R, the product $f(R)\, dC_1$ is invariant, for any $f(R)$, by the group of hyperbolic motions and therefore it can be taken as a density for sets of circles of radius R. As $R \to \infty$ the circle C tends to the horocycle $H(\varrho, \varphi)$ and in order that $f(R)\, dC_1$ approaches to a limit $(\neq 0, \infty)$ we must take $f(R)$ such that $f(R)\, e^R \to a$, a being a constant which for simplicity we assume equal to 2. Then, if we denote by dH_+ the density for horocycles which turn the convexity towards O and by dH_- the density for horocycles which turn the convexity towards the opposite sense, we have

$$dH_+ = e^{\varrho}\, d\varrho \wedge d\varphi , \quad dH_- = e^{-\varrho}\, d\varrho \wedge d\varphi . \tag{2.3}$$

This density, that we will denote indistinctly by dH, is uniquely determined, i.e. it is unique, up to a constant factor, which is invariant under the group of hyperbolic motions, as follows from the way we have obtained it.

3. Horocycles which intersect a curve. Let Q be a rectifiable curve of length L. Then, the so-called Poincaré's formula of the integral geometry applied to Q and to the circle C of radius R writes [2] (having into account that the length of C is $2\pi \sinh R$)

$$\int n\, dC_1 = 4L \sinh R \tag{3.1}$$

where n is the number of intersection points of Q and C, the integral extended over the whole hyperbolic plane, n being zero if Q and C do not intersect. Multiplying (3.1) by $2e^{-R}$ and letting R tend to infinity, we get

$$\int n\, dH = 4L \tag{3.2}$$

where n is now the number of intersection points of Q and the horocycle H, the integral being extended over all horocycles of the plane.

Notice that (3.2) does not change if horocycles are substituted by oriented lines [2].

4. Horocycles which intersect a *h*-convex set. A set of points K in the hyperbolic plane is said to be convex if for each pair of points A, B belonging to K, the entire segment of straight line AB also belongs to K.

A set of points K is said to be h-convex or convex with respect to horocycles, if for each pair of points A, B belonging to K, the entire segments of the two horocycles AB also belong to K.

Two points A, B of the hyperbolic plane determine two horocycles H, H' which contain these points. If K is h-convex the whole lune bounded by H and H' belongs to K and therefore the line segment AB also belongs to K, i.e. any h-convex set is convex. The converse is not true, as is immediately shown by any convex set containing a line segment in its boundary. Since the curvature of the horocycles is equal to 1, it is clear that any convex set bounded by a smooth curve of curvature greater or equal than 1 at every point is h-convex.

Since the set of support horocycles (likewise as the set of support lines) of a h-convex set is a set of measure zero, by applying (3.2) to the boundary of K we will have $n = 2$ up to a set of zero measure, and therefore we get

$$\int_{H \cap K \neq \emptyset} dH = 2L. \tag{4.1}$$

Thus we have: *the measure of the set of horocycles which intersect a h-convex set K is equal to 2L, where L is the length of the boundary of K.*

5. Density for pairs of points and integral formula for chords. Let K be a convex set of the hyperbolic plane and let dG represent the density for lines. If σ denotes the length of the chord that G determines on K, i.e. the length of the intersection $G \cap K$, the following formulae are known [2],

$$\int_{G \cap K \neq \emptyset} \sigma\, dG = \pi F, \quad \int_{G \cap K \neq \emptyset} \sinh \sigma\, dG = \pi F + \tfrac{1}{2} F^2 \tag{5.1}$$

where F is the area of K.

In the euclidean plane the first formula (5.1) holds without change, while the second gives rise to the so-called Crofton's formula for chords, which writes [2]

$$\int_{G \cap K \neq \emptyset} \sigma^3 \, dG = 3 F^2 . \tag{5.2}$$

We wish now to see what happens in the formulae (5.1) when lines are substituted by horocycles. In order to do that we need a formula which gives the product $dP_1 \wedge dP_2$ of the densities of two points P_1, P_2 (area elements at P_1, P_2) in terms of the density dH of a horocycle H determined by those points and the differentials dt_1, dt_2 of the abscissae t_1, t_2 of P_1, P_2 on H.

Let us consider first a circle C (of center $C_1(\varrho, \varphi)$ and radius R) which passes through $P_1(r_1, \varphi + \psi_1)$, $P_2(r_2, \varphi + \psi_2)$ (Fig. 1).

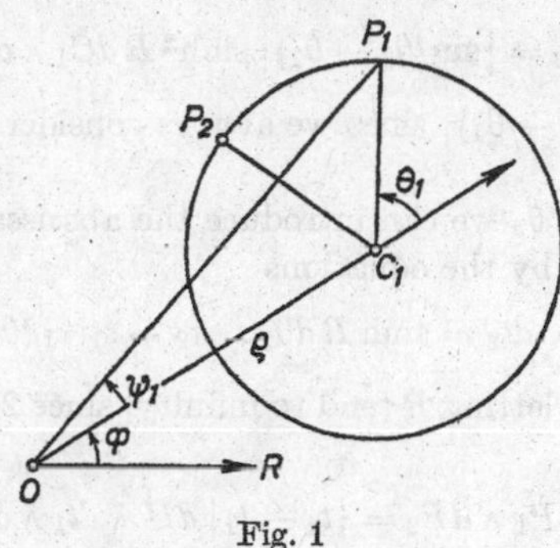

Fig. 1

If θ_1 denotes the angle which forms the radius $C_1 P_1$ with the line $O C_1$ and ψ_1 the angle between $O C_1$ and $O P_1$, by well known formulae of hyperbolic geometry [1, p. 237] we have

$$\cosh r_1 = \cosh R \cosh \varrho + \sinh R \sinh \varrho \cos \theta_1 ,$$
$$\sin \psi_1 \sinh r_1 = \sinh R \sin \theta_1 .$$

Differentiating we get

$$\begin{aligned} \sinh r_1 \, dr_1 &= (\cosh R \sinh \varrho + \sinh R \cosh \varrho \cos \theta_1) \, d\varrho - \sinh R \sinh \varrho \sin \theta_1 \, d\theta_1 , \\ \cos \psi_1 \sinh r_1 \, d\psi_1 &+ \sin \psi_1 \cosh r_1 \, dr_1 = \sinh R \cos \theta_1 \, d\theta_1 . \end{aligned} \tag{5.3}$$

The last equation can be written

$$\begin{aligned} &\cos \psi_1 \sinh r_1 \, d(\psi_1 + \varphi) + \sin \psi_1 \cosh r_1 \, dr_1 = \\ &= \sinh R \cos \theta_1 \, d\theta_1 + \cos \psi_1 \sinh r_1 \, d\varphi . \end{aligned} \tag{5.4}$$

Exterior multiplication of (5.3) and (5.4), putting

$$dP_1 = \sinh r_1 \, dr_1 \wedge d(\psi_1 + \varphi) = \text{area element at } P_1 ,$$

gives

$$\sinh r_1 \cos \psi_1 \, dP_1 =$$
$$= (\cosh R \sinh \varrho + \sinh R \cosh \varrho \cos \theta_1) (\sinh R \cos \theta_1 \, d\varrho \wedge d\theta_1 + \cos \psi_1 \sinh r_1 \, d\varrho \wedge d\varphi)$$
$$- \sinh R \sinh \varrho \sin \theta_1 \cos \psi_1 \sinh r_1 \, d\theta_1 \wedge d\varphi .$$

35*

An analogous formula holds for P_2 and by exterior multiplication we get

$$(5.5)\qquad \sinh r_1 \sinh r_2 \cos\psi_1 \cos\psi_2 \, dP_1 \wedge dP_2 =$$
$$= [\sin\theta_1 \cos\theta_2 \cos\psi_1 \sinh r_1 (\cosh R \sinh\varrho + \sinh R \cosh\varrho \cos\theta_2) -$$
$$- \sin\theta_2 \cos\theta_1 \cos\psi_2 \sinh r_2 (\cosh R \sinh\varrho + \sinh R \cosh\varrho \cos\theta_1)] \times$$
$$\times \sinh^2 R \, dC_1 \wedge d\theta_1 \wedge d\theta_2 ,$$

where $dC_1 = \sinh\varrho \, d\varrho \wedge d\varphi$ is the area element corresponding to C_1.

By a known formula of hyperbolic trigonometry we have, for $i = 1, 2$,

$$\sinh r_i \cos\psi_i = \cosh R \sinh\varrho + \sinh R \cosh\varrho \cos\theta_i$$

and therefore, (5.5) gives

$$(5.6)\qquad dP_1 \wedge dP_2 = |\sin(\theta_2 - \theta_1)| \sinh^2 R \, dC_1 \wedge d\theta_1 \wedge d\theta_2$$

where we have put $|\sin(\theta_2 - \theta_1)|$ since we always consider the densities in absolute value.

Instead of the angles θ_1, θ_2 we can introduce the abscissae t_1, t_2 of P_1, P_2 on the circumference of C related by the equations

$$dt_1 = \sinh R \, d\theta_1 , \quad dt_2 = \sinh R \, d\theta_2 , \quad t_2 - t_1 = (\theta_2 - \theta_1) \sinh R .$$

Substituting in (5.6) and letting R tend to infinity, since $2e^{-R} dC_1 \to dH =$ density for horocycles, we get

$$(5.7)\qquad dP_1 \wedge dP_2 = |t_2 - t_1| \, dH \wedge dt_1 \wedge dt_2 .$$

This formula is the same as the formula for pairs of points in euclidean plane [3] and therefore, integrating over all pairs of points inside a h-convex set K, we obtain

$$(5.8)\qquad \int_{H \cap K \neq \emptyset} \sigma^3 \, dH = 6 F^2$$

where σ is the length of the chord $H \cap K$, assumed K h-convex. This formula (5.8) differs from the formula (5.2) for chords in euclidean plane by a factor 2, due to the fact that two points determine two horocycles.

A kind of dual formula is obtained from (5.6) if we consider P_1, P_2 as centers of two circles of radius R and let $R \to \infty$. Then we get two horocycles H_1, H_2 which intersect at the point C_1 under the angle $\theta_2 - \theta_1$ and the differential formula

$$(5.9)\qquad dH_1 \wedge dH_2 = |\sin(\theta_2 - \theta_1)| \, dC_1 \wedge d\theta_1 \wedge d\theta_2$$

holds.

This formula (5.9) does not change if horocycles are substituted by lines [2] and has the same form as in euclidean plane [3].

In order to generalize the first formula (5.1) to horocycles, let us integrate both sides of (5.9) over all the pairs of horocycles which intersect each other in the interior of a domain K (not necessarily convex) of area F. Since there are two horocycles tangent to a given direction at a point, the right side gives

$$(5.10)\qquad 4 \int_{C_1 \in K} dC_1 \int_0^\pi \int_0^\pi |\sin(\theta_2 - \theta_1)| \, d\theta_1 \wedge d\theta_2 = 8\pi F .$$

In this computation, if the horocycles H_1, H_2 have two intersection points in K, the pair H_1, H_2 has been counted two times. Therefore if we call σ_1 the length of the arc of H_1 which belongs to K, in accordance with (3.2) the integral of dH_2 over all H_2 which cut H_1 in a point of K, is $4\sigma_1$. Thus the integral of the left of (5.9) is $4\int \sigma_1 dH_1$. Equating to (5.10) and writing σ and H in place of σ_1 and H_1, we get

$$\int\limits_{H \cap K \neq \emptyset} \sigma \, dH = 2\pi F \tag{5.11}$$

which is the generalization we wish to obtain. Note that in (5.11) K is not necessarily convex.

Bibliography

[1] H. S. M. Coxeter, Non-euclidean geometry. 3rd ed., Toronto 1957.
[2] L. A. Santaló, Integral geometry on surfaces of constant negative curvature. Duke Math. J. 10, 687–704 (1943).
[3] L. A. Santaló, Introduction to Integral Geometry. Paris 1953.

Eingegangen am 18. 11. 1966

Anschrift des Autors:
L. A. Santaló
Facultad de Ciencias Exactas
Universidad de Buenos Aires
Buenos Aires, Argentina

HOROSPHERES AND CONVEX BODIES IN HYPERBOLIC SPACE

L. A. SANTALÓ

1. Introduction. In extending certain topics from euclidean to hyperbolic geometry one finds that in certain cases the euclidean planes transfer into hyperbolic planes (geodesic surfaces), whilst in other cases the natural analogue for hyperbolic space of the euclidean planes are the horospheres (limit spheres). For instance, in the work of Gelfand and Graev on the application of the integral geometry to group representations [5], in passing from euclidean to hyperbolic space the natural analogue of euclidean planes are the horospheres. In the present note we show that the same happens with certain integral formulae on convex bodies. If K is a convex body in euclidean 3-dimensional space, it is well known that the measure of all planes meeting K is equal to the integral of mean curvature M of the boundary of K (assumed of class C^2) (see, for instance, Kendall-Moran [6, p. 80]). In hyperbolic space the same measure is equal to $M-V$, where V is the volume of K [7]. However if instead of planes we consider the set of horospheres which intersect K, we shall prove that the measure is again M. We also prove, in passing, the formulae (3.4) and (4.4) referring to horospheres which intersect a fixed curve or a fixed surface of hyperbolic space.

2. Horospheres in hyperbolic space. We shall first review a few notions on surfaces in hyperbolic space which will be useful for our purposes.

In a system ρ, θ, ϕ of geodesic polar coordinates the arc element has the form (Cartan [2, p. 240]),

$$ds^2 = d\rho^2 + \sinh^2 \rho(d\theta^2 + \sin^2 \theta d\phi^2) \tag{2.1}$$

and the volume element is

$$dV = \sinh^2 \rho d\rho \wedge d\omega \tag{2.2}$$

where $d\omega = \sin \theta d\theta \wedge d\phi$ represents the element of solid angle corresponding to the direction θ, ϕ.

Between the principal radii of normal curvature r_i $(i=1, 2)$ at a point P of a surface Σ and the distances R_i from P to the contact point of the normal to Σ at P with the envelope of the normals to Σ along the corresponding line of curvature, the relations

Received by the editors October 31, 1966.

$$r_i = \tanh R_i \tag{2.3}$$

hold (Eisenhart [4, p. 214]).

The integral of the mean curvature M of a closed surface Σ (of class C^2) is defined by

$$M = \frac{1}{2}\int_\Sigma \left(\frac{1}{r_1} + \frac{1}{r_2}\right) d\sigma \tag{2.4}$$

where $d\sigma$ denotes the area element of Σ.

If k_a denotes the absolute or intrinsic (Gaussian) curvature and $k_r = 1/r_1 r_2$ the relative curvature of Σ at P, since the curvature of the hyperbolic space is -1, we have (Cartan [2, p. 194])

$$k_a = k_r - 1. \tag{2.5}$$

For the planes (geodesic surfaces) of the hyperbolic space, we have $k_r = 0$ and therefore $k_a = -1$. The horospheres (limit spheres) result from spheres (R_i = constant), when $R_i \to \infty$; therefore, for horospheres we have $r_1 = r_2 = 1$ and $k_r = 1$, $k_a = 0$.

Let us consider a surface Σ and a horosphere H which intersect in the curve Γ. Let P be a point of Γ and N_Σ, N_H the normals to Σ, H at P, and N the principal normal of Γ at P. Let α be the angle between N_Σ and N, α_1 the angle between N_H and N and call $\theta = \alpha + \alpha_1$ = angle between N_Σ and N_H. Meusnier's theorem gives

$$\rho = r \cos \alpha = r' \cos \alpha_1 \tag{2.6}$$

where ρ denotes the radius of curvature of Γ at P and r, r' are respectively the radii of normal curvature of Σ, H at P corresponding to the direction tangent to Γ (Eisenhart [4, p. 152], Cartan [2, p. 224]).

On the other hand, if ρ_g^Σ and ρ_g^H denote the radii of geodesic curvature of Γ considered respectively as a curve of Σ and a curve of H, we have (Eisenhart [4, p. 152], Cartan [2, p. 225])

$$\rho = \rho_g^\Sigma \sin \alpha = \rho_g^H \sin \alpha_1. \tag{2.7}$$

Since H is a horosphere, we have $r' = 1$ and from (2.6) and (2.7) we deduce

$$\rho_g^H = (r \sin \theta)/(1 - r \cos \theta). \tag{2.8}$$

3. Density for horospheres. Let S be a sphere of center C and radius R in hyperbolic space. Assume that the origin of coordinates 0 is exterior to S and let $\rho + R$, θ, ϕ be the geodesic polar coordinates of C, ρ being the distance from 0 to S. In order to have a measure for sets of spheres of radius R which are invariant under the group of hyperbolic motions, we may take any density of the form $f(R)dV$, where $f(R)$ is an arbitrary function of R and $dV=\sinh^2(\rho+R)d\rho\wedge d\omega$ is the volume element corresponding to the center C. When $R\to\infty$, the spheres S pass to horospheres H and in order to obtain a not constant density for sets of horospheres, we take $f(R)$ such that $f(R)e^{2R}\to\text{constant}\neq 0, \infty$. Then, the density for horospheres results, up to a constant factor,

$$dH_+ = e^{2\rho}d\rho\wedge d\omega \tag{3.1}$$

where ρ denotes the distance from the origin 0 to the horosphere H.

The way we have obtained this density shows that, up to a constant factor, it is the unique one which is invariant with respect to the group of hyperbolic motions of the space. The notation H_+ indicates that the horosphere H has its convexity towards the origin of coordinates 0.

Analogously, if we start from spheres which contain 0 in its interior, we will have $dV=\sinh^2(R-\rho)d\rho\wedge d\omega$, so proceeding as before we get

$$dH_- = e^{-2\rho}d\rho\wedge d\omega \tag{3.2}$$

which applies to horospheres with the concavity towards 0. We shall write dH for the density of horospheres, with the convention of taking dH_+ or dH_- according as the convexity of H is turned towards 0 or in the opposite sense.

As an application, consider a fixed curve Γ of length L and a moving sphere of constant radius R. If ν is the number of intersection points between Γ and S, the integral formula

$$\int \nu dV = 2\pi L\sinh^2 R \tag{3.3}$$

is known where dV is the volume element corresponding to the center C of S and the integral is extended over all positions of S for which $\nu\neq 0$. Formula (3.3) is well known in the euclidean case (in which $\sinh R$ must be replaced by R) [8] and it holds with a similar proof in hyperbolic space taking into account that the surface area of the sphere of radius R is $4\pi\sinh^2 R$ in this case. Multiplying (3.3) by $4e^{-2R}$ and making $R\to\infty$, we get

$$\int \nu dH = 2\pi L \tag{3.4}$$

where ν means now the number of intersection points of Γ with the horosphere H and the integral may be considered as extended over all horospheres, ν being zero for the horospheres which do not intersect Γ.

4. A differential formula on densities. We need a formula on densities analogous to a known formula of the integral geometry of euclidean space.

Let Σ_0, Σ_1 be two surfaces (of class C^2) in euclidean space, with Σ_0 fixed and Σ_1 moving with the kinematic density $d\Sigma_1$, which intersect in a curve Γ. We denote by θ the angle between the normals to Σ_0, Σ_1 and by ds the arc element of Γ at a point P. Then Blaschke proved the following formula [**1**],

$$ds \wedge d\Sigma_1 = \sin^2 \theta d\theta \wedge d\sigma_0 \wedge d\phi_0 \wedge d\sigma_1 \wedge d\phi_1 \tag{4.1}$$

where $d\sigma_0$, $d\sigma_1$ denote the area elements of Σ_0, Σ_1 at P and ϕ_0, ϕ_1 denote rotations about the respective normals to Σ_0, Σ_1 at P. This formula (4.1) has been generalized to n-dimensional euclidean space by S. S. Chern [**3**], to elliptic space by Ta-Jen Wu [**9**] and it holds with analogous proof in hyperbolic space [**7**].

Let us consider the case in which Σ_1 is a sphere of radius R and center C. Then $d\Sigma_1 = dV \wedge d\omega \wedge d\tau$, where dV is the element of volume at C, $d\omega$ is the element of solid angle corresponding to the direction CP and $d\tau$ is the rotation element about the line CP. We have also $d\sigma_1 = \sinh^2 R d\omega$, $d\tau = d\phi_1$ and therefore (4.1) gives

$$ds \wedge dV = \sin^2 \theta \sinh^2 R \, d\theta \wedge d\sigma_0 \wedge d\phi_0. \tag{4.2}$$

We recall that, as usual in integral geometry, we always consider the densities in absolute value, so there is no question of sign. Multiplying both sides of (4.2) by $4e^{-2R}$ and making $R \to \infty$, we get (according to §3),

$$ds \wedge dH = \sin^2 \theta d\theta \wedge d\sigma_0 \wedge d\phi_0 \tag{4.3}$$

which is the formula on densities we want to obtain.

We get a first application of (4.3) by integrating over all positions of H in which it intersects a fixed surface Σ_0 of area F_0 ($0 \leqq \theta \leqq 2\pi$, $0 \leqq \phi_0 \leqq \pi$). We get

$$\int_{H \cap \Sigma_0 \neq 0} \lambda dH = \pi^2 F_0 \tag{4.4}$$

where λ is the length of the intersection curve Γ of Σ_0 and H.

5. Horospheres which intersect a convex body. Let us apply the formula (4.3) to the case of a closed surface Σ_0 of class C^2 intersected by a horosphere H. Let ρ_g^H be the radius of geodesic curvature of the intersection curve Γ, considered as a curve of the horosphere H. According to (2.8) and the Euler theorem which gives the normal curvature $1/r$ of Σ_0 at P in terms of the principal normal curvatures $1/r_1$, $1/r_2$ (a theorem which holds in hyperbolic space with the same form as in euclidean space, Eisenhart [4, p. 154]), we have

$$\overset{H}{\kappa_g} = 1/\overset{H}{\rho_g} = (1/\sin\theta)((\cos^2\phi_0)/r_1 + (\sin^2\phi_0)/r_2) - \cot\theta.$$

Multiplying both sides of (4.3) by κ_g^H and integrating over all positions of H with $H\cap\Sigma_0\neq 0$, we get:

(a) At the left side we have

$$\int_{H\cap\Sigma_0\neq 0}\left(\int_\Gamma \overset{H}{\kappa_g}\, ds\right) dH = 2\pi\int_{H\cap\Sigma_0\neq 0} n\, dH$$

where n is the Euler characteristic of the 2-dimensional domain on H which is interior to Σ_0 and whose boundary is Γ. We have applied $\int_\Gamma \kappa_g^H ds = 2\pi n$, which follows from the fact that the absolute curvature of H is 0 and therefore its intrinsic geometry coincides with that of the euclidean plane.

(b) At the right side we have

$$\int_0^{2\pi} |\sin\theta|\, d\theta \int_{\Sigma_0} d\sigma_0 \int_0^\pi \left(\frac{\cos^2\phi_0}{r_1} + \frac{\sin^2\phi_0}{r_2}\right) d\phi_0$$

$$- \pi F_0 \int_0^{2\pi} |\sin\theta|\cos\theta\, d\theta = 4\pi M_0.$$

Consequently, we have

$$\int_{H\cap\Sigma_0\neq 0} n\, dH = 2M_0.$$

If we call h-convex the surfaces which are closed and $n=1$ for any horosphere (i.e. the intersection curve $\Gamma = H\cap\Sigma_0$ is simply connected for any H), we have the

THEOREM. *The measure of the horospheres which intersect an h-convex surface Σ_0 is equal to $2M_0$, where M_0 is the integral of mean curvature of Σ_0.*

Note that every h-convex surface is convex; the converse is clearly not true.

Bibliography

1. W. Blaschke, *Integralgeometrie* 17. *Ueber Kinematick*, Bull. Soc. Math. Grèce 17 (1936), 1–12.

2. E. Cartan, *Leçons sur la géomētrie des espaces de Riemann*, Gauthier-Villars, Paris, 1946.

3. S. S. Chern, *On the kinematic formula in the euclidean space of n dimensions*, Amer. J. Math. 74 (1952), 227–236.

4. L. Eisenhart, *Riemannian geometry*, Princeton Univ. Press, Princeton, N. J., 1949.

5. I. M. Gelfand and M. I. Graev, *An application of the horosphere method to the spectral analysis of functions in real and imaginary Lobatchewsky spaces*, Trudy Moskov. Mat. Obšč. 11 (1962), 243–308.

6. M. G. Kendall and P. A. P. Moran, *Geometrical probability*, Hafner, New York, 1963.

7. L. A. Santaló, *Geometria integral en los espacios tridimensionales de curvatura constante*, Math. Notae 9 (1949), 1–28.

8. ———, *A theorem and an inequality referring to rectifiable curves*, Amer. J. Math. 63 (1941), 635–644.

9. Ta-Jen Wu, *Ueber elliptische Geometrie*, Math. Z. 43 (1938), 495–521.

Facultad de Ciencias Exactas y Naturales, Universidad de Buenos Aires

[69.1] Convexity in the hyperbolic plane

APARTADO DE LA REVISTA
MATEMATICA Y FISICA TEORICA
VOL. XIX 1969 Nos. 1 y 2

CONVEXIDAD EN EL PLANO HIPERBOLICO

Por

L. A. SANTALO

FACULTAD DE CIENCIAS EXACTAS Y NATURALES. UNIVERSIDAD DE BUENOS AIRES, ARGENTINA

(Recibido el 24 de mayo de 1969)

SUMMARY

A set of points Q in the hyperbolic plane is said to be h-convex (convex with respect to horocycles) if for each pair of points A, B belonging to Q, the entire segments of the two horocycles determined by A, B belong to Q. In this paper we give the following properties for h-convex sets: 1. If Q is h- convex, the whole horocyclic lentil bounded by the horocycles O, O' determined by two points A, B of Q belongs to Q and we get the inequalities (2 . 8) between the length L of the boundary ∂Q, the area F, the width Δ and the diameter D of Q . 2. If κ_g denotes the geodesic curvature of the boundary ∂Q of a set Q, then Q is h-convex if and only if $\kappa_g > 1$. 3. For each point A of the boundary ∂Q of a h-convex set Q we consider the two tangent horocycles O^- and O^+ and define the breadth of Q with respect to each of them. These h-breadths α^-, α^+ satisfy the formulae (5.3) and (5.4). 4. The sets of constant breadth are considered specially. For them the formula (6.1) holds The last theorem states that the Reuleaux triangle has minimal length and minimal area among all the sets of the same constant breadth of the hyperbolic plane.

1. INTRODUCCION

En el plano hiperbólico hay dos familias de curvas que en cierto modo tienen las propiedades de las rectas del plano euclidiano, a saber: las rectas propiamente dichas (o geodésicas) y los oriciclos. La principal analogía de los oriciclos con las rectas del plano euclidiano es que ellos son trayectorias ortogonales de un haz de rectas paralelas o bien, en otras palabras, que las rectas del plano hiperbólico ortogonales a un mismo oriciclo son paralelas entre sí. La diferencia esencial es que por dos puntos del plano hiperbólico pasan dos oriciclos, los cuales forman lo que llamaremos un "huso oricíclico".

Interpretando el plano hiperbólico como superficie de curvatura constante $K = -1$, las rectas están caracterizadas por tener la curvatura geodésica nula, κ_g (rectas) $= 0$ y los oriciclos por tener esta curvatura, en valor absoluto, igual a la unidad, o sea $|\kappa_g$ (oriciclos) $| = 1$.

La definición usual de convexidad es la siguiente: un conjunto de puntos Q del plano hiperbólico se dice que es *convexo* si para todo par de puntos A, B pertenecientes a Q, los puntos del segmento de recta de extremos A y B pertenecen todos a Q.

Al sustituir las rectas por oriciclos tendremos la siguiente

Definición. Un conjunto de puntos Q del plano hiperbólico se dice que es *h-convexo* (o convexo respecto de oriciclos), si para todo par de puntos A, B pertenecientes a Q, todos los puntos de los dos segmentos de oriciclo determinados por A y B pertenecen a Q. El contorno de un conjunto h-convexo acotado se llama una *curva h-convexa* cerrada.

Los dos arcos de oriciclo de extremos A, B determinan un *huso oricíclico* el cual, si Q es h-convexo, deberá estar totalmente contenido en Q y como el segmento de recta AB es interior al huso (es eje de simetría), resulta que *todo conjunto h-convexo es convexo.* El recíproco no es cierto: basta considerar un conjunto convexo que contenga en su contorno un segmento de recta (por ejemplo un polígono geodésico) para ver inmediatamente que es convexo pero no h-convexo.

Por consiguiente, todas las propiedades de los conjuntos convexos subsisten para los h-convexos, pero hay algunas propiedades exclusivas de estos últimos, algunas de las cuales queremos estudiar en este trabajo. En un trabajo anterior ya vimos algunas de ellas [7] y recientemente H. Larcher [4] ha dado otras que mencionaremos más adelante.

Tanto los conjuntos convexos como los h-convexos los supondremos siempre acotados y cerrados, vale decir, con el contorno incluído. Además, puesto que la curvatura geodésica de una curva convexa tiene signo constante, *supondremos que este signo es positivo.*

2. ALGUNOS VALORES REFERENTES A "HUSOS ORICICLICOS"

Sean A, B dos puntos del plano hiperbólico y sea D_0 su distancia, o sea, la longitud del segmento AB. Por A y B pasan dos oriciclos, los cuales forman el huso oricíclico de extremos A y B. Queremos calcular el perímetro, el área y el espesor de este huso.

Una manera elemental consiste en utilizar la representación de Poincaré del plano hiperbólico, como los puntos con $y > 0$ y métrica $ds^2 = (dx + dy^2)/y^2$. Como el huso depende únicamente de D_0 podemos suponerlo colocado en la posición de la fig. 1, con A y B de la misma ordenada β. Introduciendo los ángulos $\varphi_A = AOX$, $\varphi_B = BOX$ es sabido (y se calcula fácilmente) que es

$$D_0 = \log [\tan (\varphi_B/2) : \tan (\varphi_A/2)] \tag{2.1}$$

y como $\varphi_B = \pi - \varphi_A$ resulta también

$$D_0 = \log \cot^2 (\varphi_A/2) \tag{2.2}$$

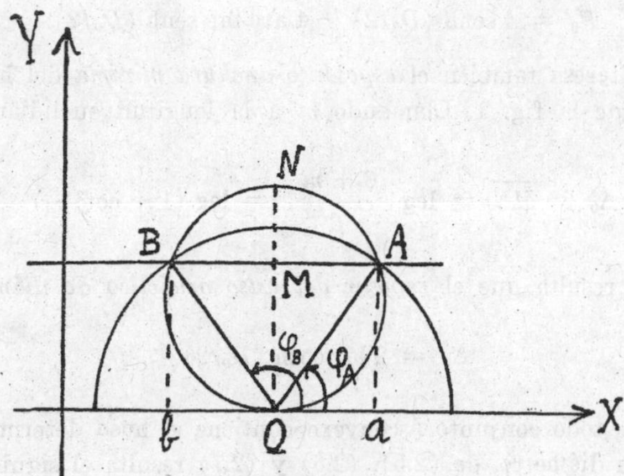

Fig. 1

de donde

$$\tan \frac{\varphi_A}{2} = \exp(-D_0/2). \tag{2.3}$$

Obsérvese que en la representación de Poincaré el segmento AB es el arco de la circunferencia que pasa por AB y tiene el centro en el eje x.

Por otra parte, la longitud del arco de oriciclo AB, siendo $A(a, \beta)$, $B(b, \beta)$ es

$$L_0 = \int_b^a \frac{dx}{y} = \frac{a-b}{\beta} = \frac{2}{\tan \varphi_A} = \frac{1 - \tan^2(\varphi_A/2)}{\tan(\varphi_A/2)}$$

o sea, según (2.3)

$$L_0 = 2 \operatorname{senh} \frac{D_0}{2}. \tag{2.4}$$

Por simetría, la longitud del otro arco de oriciclo AB tendrá el mismo valor. Por tanto: *la longitud del huso oricíclico de diámetro D_0 vale*

$$L_0 = 4 \operatorname{senh}(D_0/2) \tag{2.5}$$

Para calcular el área podemos utilizar las mismas notaciones de la fig. 1. Será

$$\frac{1}{4} F_0 = \int \frac{dx \wedge dy}{y^2} = \int_0^{r\cos\varphi_A} \left(\frac{1}{\beta} - \frac{1}{\sqrt{r^2 - x^2}}\right) dx$$

$$= \cot \varphi_A - \operatorname{arc\,sen} \cos \varphi_A.$$

Teniendo en cuenta (2.3), de donde se deduce $\cot \varphi_A = \operatorname{senh}(D_0/2)$ y $\operatorname{arc\,sen} \cos \varphi_A = \pi/2 - \varphi_A = \operatorname{ar\,tan} \cot \varphi_A$, resulta que *el área del huso oricíclico del diámetro D_0 vale*

$$F_0 = 4 \operatorname{senh} (D_0/2) - 4 \operatorname{arc} \tan \operatorname{senh} (D_0/2). \tag{2.6}$$

A veces interesa también el *espesor* o *anchura mínima* del huso, o sea, el segmento MN de la fig. 1. Llamando m a la longitud euclidiana de MN es

$$\Delta_0 = \overline{MN} = \log \frac{\beta + m}{\beta} = \log (1 + \cot^2 \varphi_A)$$

y según (2.3) resulta que el *espesor del huso oricíclico de diámetro* D_0 *vale*

$$\Delta_0 = 2 \log \cosh (D_0/2). \tag{2.7}$$

Puesto que todo conjunto h-convexo contiene el huso determinado por los extremos de un diámetro, de (2.5), (2.6) y (2.7) resulta el siguiente teorema:

Entre la longitud L del contorno, el área F y el espesor Δ de un conjunto h-convexo y el diámetro D del mismo, valen las siguientes desigualdades

$$L \geqslant 4 \operatorname{senh} (D/2)$$

$$F \geqslant 4 \operatorname{senh} (D/2) - 4 \operatorname{arc} \tan \operatorname{senh} (D/2)$$

$$\Delta \geqslant 2 \log \cosh (D/2).$$

donde las igualdades valen únicamente para los husos oricíclicos.

3. CURVAS h-CONVEXAS CON CURVATURA GEODESICA CONTINUA

Sea Q un conjunto h-convexo y A un punto de su contorno ∂Q. Supongamos que en A la curva convexa ∂Q tenga curvatura $\kappa_g(A)$ continua. En un entorno de A, los dos oriciclos tangentes a ∂Q en A deben quedar exteriores a Q o coincidir, uno de ellos, con ∂Q. En efecto, si un arco de oriciclo quedase interior a Q, considerando un oriciclo paralelo suficientemente próximo, se podría tener un arco de oriciclo con los extremos en Q y no contenido totalmente en Q, contra la hipótesis de ser Q h-convexo. Esto hace que la curvatura $\kappa_g(A)$ deba ser igual o mayor que la curvatura del oriciclo en A, cuyo valor absoluto es 1. Una demostración precisa de este hecho puede verse en H. Karcher [4]. Recíprocamente, si una curva convexa está compuesta de arcos con curvatura continua $\kappa_g \geqslant 1$, ella es h-convexa.

Por tanto:

Una condición necesaria y suficiente para que una curva convexa y cerrada del plano hiperbólico con curvatura continua sea h-convexa, es que sea

$$\kappa_g \geqslant 1 \tag{3.1}$$

en todo punto.

Obsérvese que consideramos la curvatura geodésica de una curva *h-convexa* siempre *positiva*. En caso contrario habría que poner en (3.1) el valor absoluto.

Si R representa el radio del círculo osculador a una curva C en un punto A (= posición límite del círculo geodésico determinado por tres puntos A, A_1, A_2 de C, cuando A_1 y A_2 tienden a A), su relación con κ_g (suponiendo la existencia de κ_g continua) es

$$\kappa_g = \coth R. \tag{3.2}$$

Esta igualdad se obtiene aplicando la fórmula de Gauss-Bonnet a un círculo de radio R y teniendo en cuenta que su longitud y área están dados por las fórmulas (ver Coxeter [3])

$$L = 2\pi \operatorname{senh} R \quad ; \quad F = 2\pi(\cosh R - 1) \tag{3.3}$$

La condición (3.1) equivale por tanto a la siguiente:

Una condición necesaria y suficiente para que una curva convexa con curvatura continua en todo punto, sea h-convexa, es que el radio R de sus círculos osculadores sea finito en todo punto ($R < \infty$).

Hay varias propiedades referentes a curvas convexas o h-convexas y sus círculos inscritos y circunscritos. Algunas de ellas son las mismas que para el plano euclidiano. Otras son diferentes. Por ejemplo, en el plano euclidiano, si C es una curva convexa y cerrada con curvatura continua, el círculo osculador a C de radio mínimo está contenido en el interior de C (Blaschke [1]). Para curvas convexas del plano hiperbólico esta propiedad puede dejar de valer. En cambio sigue valiendo para las curvas h-convexas. Este estudio ha sido hecho por H. Karcher [4].

4. h-*CURVATURA*

Sea una curva C del plano hiperbólico que tenga curvatura continua no nula en un punto A. Se sabe que esta curvatura κ_g puede definirse por

$$d\tau = \kappa_g \, ds \tag{4.1}$$

siendo ds el elemento de arco y $d\tau$ el elemento de ángulo entre dos rectas tangentes a C (ambos elementos tomados en el punto A). Además, si C es una curva cerrada que limita un área F, vale la fórmula de Gauss-Bonnet

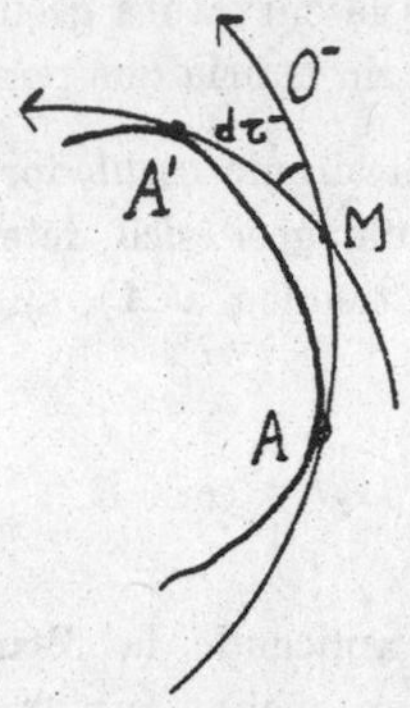

Fig. 2

$$\int_C \kappa_g \, ds = 2\pi + F - \sum_1^n \omega_i \tag{4.2}$$

donde ω_i son los ángulos exteriores de los puntos angulosos, si existen. Si la curva tiene curvatura continua en todo punto, esta suma no aparece en la fórmula (4.2).

Si en vez de rectas (geodésicas) se toman oriciclos tangentes, se tienen dos posibilidades: o bien se toman los oriciclos con la concavidad del mismo lado de C (figura análoga a fig. 2) o bien se toman los oriciclos con la concavidad en sentido opuesto (fig. 3). Aplicando el teorema de Gauss-Bon-

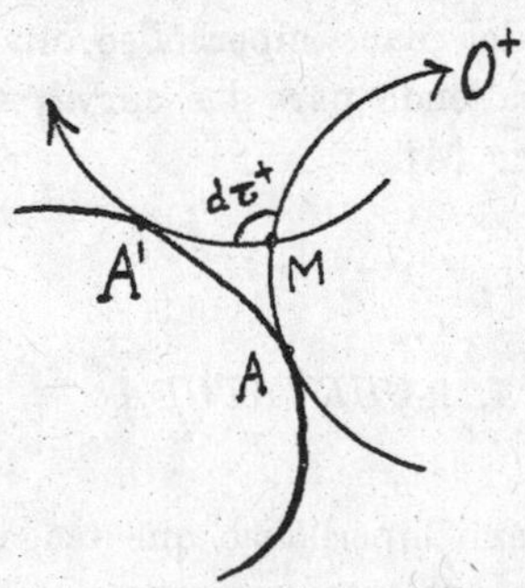

Fig. 3

net al triángulo AMA′ y teniendo en cuenta que en el primer caso la curvatura de los oriciclos tangentes es $+1$ (considerando la curvatura de C como positiva) y en el segundo caso es -1, resulta, para el primer caso la relación

$$d\tau_H^- = d\tau - ds \tag{4.3}$$

y para el segundo caso

$$d\tau_H^+ = d\tau + ds \tag{4.4}$$

siendo $d\tau_H^-$, $d\tau_H^+$ los elementos de ángulo, en cada caso, de los oriciclos tangentes. Para obtener (4.3) y (4.4) se ha tenido en cuenta que el área del triángulo AMA′ es infinitésimo de orden superior al de los elementos que figuran en (4.3) y (4.4).

Las variaciones totales de los ángulos τ_H^+, τ_H^-, para una curva cerrada C que limita un área F y tiene longitud L, resulta valer (aplicando (4.2))

$$\int_C d\tau_H^- = 2\pi + F - L \quad , \quad \int_C d\tau_H^+ = 2\pi + F + L \tag{4.5}$$

5. *h*-ANCHURA DE UNA CURVA CONVEXA

Sea ∂Q una curva convexa cerrada, contorno del conjunto convexo Q. Sea A un punto de ∂Q. De los dos oriciclos tangentes a ∂Q en A, representemos por $O^-(A)$ el que tiene la convexidad del mismo sentido que ∂Q y por $O^+(A)$ el que tiene la convexidad de sentido contrario. Sea g la recta normal a ∂Q en A. Consideremos los oriciclos ortogonales a g que tienen el mismo

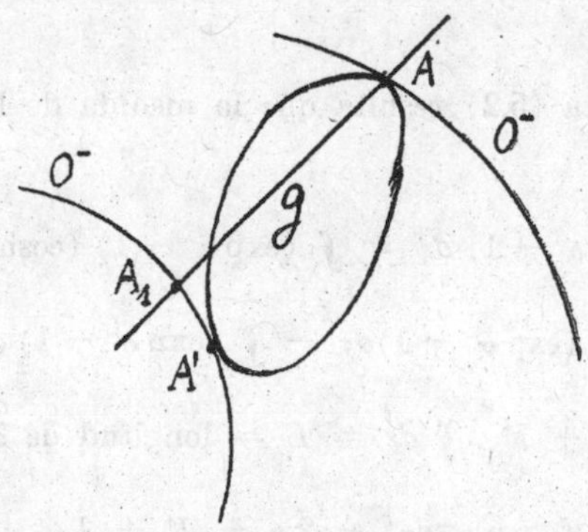

Fig. 4

punto impropio que $O^-(A)$. De estos oriciclos sea $O^-(A')$ el extremo superior de los que tienen punto común con Q $(A' \varepsilon Q)$. Si A_1 es el punto de g correspondiente a $O^-(A')$, se llama *h-anchura* $(-)$ *de* Q *correspondiente al punto* A, a la distancia AA_1. Se representa por $\alpha^-(A)$. Obsérvese que A_1 y A' serán en general puntos distintos (A' es el punto de apoyo de $O^-(A')$), fig. 4.

Para las curvas convexas del plano euclidiano, el valor medio de la anchura, salvo un factor constante, es igual a la longitud de la curva. Que remos obtener algo análogo para las curvas h-convexas del plano hiperbólico.

Para ello debemos recordar que la densidad para medir conjuntos de oriciclos vale

$$dO^{-} = \exp(-r)\, dr \wedge d\psi \tag{5.1}$$

siendo r su distancia a un punto fijo P, situado en la región convexa, no acotada, cuyo contorno es O^{-} y ψ el ángulo de la normal por P al oriciclo con una dirección fija (ver [7]). Además, se sabe que la medida de todos los oriciclos que cortan a una curva convexa cerrada es igual a dos veces su longitud. Como la densidad (5.1) es independiente del punto P, tomemos por un momento el centro de curvatura de C en A. Si R es el radio de curvatura y $d\tau$, ds tienen los significados de antes, aplicando el teorema de Gauss-Bonnet al triángulo elemental P, $A(s)$, $A(s+ds)$ y teniendo en cuenta (3.2) resulta

$$d\psi = \frac{d\tau}{\cosh R} = \frac{ds}{\operatorname{senh} R} \tag{5.2}$$

Cuando los oriciclos ortogonales a g estén entre el centro de curvatura P y A_1, hay que tomar por densidad $\exp(r)\, dr \times d\psi$. Por tanto, fijando ψ o integrando r, se tiene

$$\int_P^A \exp(-r)\, dr + \int_P^{A_1} \exp(r)\, dr = \exp(-R)\,(\exp(\alpha-) - 1).$$

Teniendo en cuenta (5.2) resulta que la medida de los oriciclos que cortan a Q vale

$$\int \exp(-R)\,(\exp\alpha^{-} - 1)\, d\psi = \int (\exp\alpha^{-} - 1)\,(\cosh R - \operatorname{senh} R)\, d\psi$$

$$= \int (\exp\alpha^{-} - 1) d\tau - \int (\exp\alpha^{-} - 1)\, ds\,,$$

y como $\int d\tau = 2\pi + F$, $\int ds = L =$ longitud de ∂Q , resulta

$$\int \exp\alpha^{-}\, d\tau_H^{-} - 2\pi - F + L\,.$$

Como este resultado debe ser igual a $2L$, resulta

$$\int_{\partial Q} \exp(\alpha^{-})\, d\tau_H^{-} = 2\pi + F + L \tag{5.3}$$

Si el centro de curvatura P es exterior al segmento AA_1 el razonamiento es completamente análogo y se llega a la misma fórmula (5.3).

En el punto A se puede tomar también el oriciclo $O^{+}(A)$ que tiene su convexidad en sentido contrario a la de ∂Q. Se puede entonces definir de manera análoga la anchura α^+ y un razonamiento análogo al anterior conduce a la fórmula

$$\int_{\partial Q} \exp(-\alpha^+)\, d\tau^+_H = 2\pi + F - L \tag{5.4}$$

En resumen:

Sea Q un conjunto h-convexo del plano hiperbólico. Llamando α^- y α^+ a las anchuras respecto de los oriciclos tangentes en cada punto, valen las fórmulas (5.3) y (5.4).

Una fórmula análoga para curvas convexas (no necesariamente h-convexas) fue dada en [5].

6. CONJUNTOS CONVEXOS DE ANCHURA CONSTANTE

Si α^- es constante, el mismo razonamiento conocido del caso del plano euclidiano prueba que $A' \equiv A_1$ (notaciones de la fig. 4). Por tanto A y A' son puntos tales que también las rectas tangentes en ellos y los oriciclos O^+ son ortogonales a la recta AA'. Es decir: *si α^- es constante también lo es α^+ y su valor común es la anchura ordinaria del conjunto Q.* En otras palabras: todo conjunto de h-anchura constante es también de anchura constante.

Además, igual que para el plano euclidiano, resulta que si Q es de anchura constante α, los radios de curvatura en puntos opuestos cumplen la condición $R(A) + R(A') = \alpha$. Por tanto R es finito, o sea: *todo conjunto convexo de anchura constante es h-convexo.*

Si Q es de anchura constante α, de (5.3) y (4.3) se deduce

$$\exp \alpha = \frac{2\pi + F + L}{2\pi + F - L} \tag{6.1}$$

Es decir: *para todo conjunto convexo del plano hiperbólico de anchura constante α, vale la igualdad* (6.1). Una fórmula equivalente a la (6.1) ya fue obtenida en [5].

7. POLIGONOS DE REULEAUX

Los conjuntos convexos de anchura constante más simples son los polígonos de Reuleaux. Igual que para el caso del plano, se obtienen dividiendo una circunferencia en $2n+1$ partes iguales por los puntos $A_0, A_1, \ldots, A_{2n}$ y uniendo cada dos vértices consecutivos por un arco de circunferencia cuyo centro es el vértice opuesto. Así, los vértices A_n y A_{n+1} se unen por un arco de circunferencia de centro A_0. Resulta así el polígono de Reuleaux de $2n+1$ lados. La distancia $A_0 A_n = \alpha$ es la anchura constante del polígono.

Llamando $\theta =$ ángulo $A_n A_0 A_{n+1}$ la longitud del polígono es

$$L_R = (2n+1)\,\theta\,\mathrm{senh}\,\alpha \tag{7.1}$$

y la longitud del contorno del círculo de la misma anchura vale

$$L_C = 2\,\pi\,\mathrm{senh}\,(\alpha/2) \tag{7.2}$$

Vamos a demostrar que $L_R < L_C$ para cualquier n. En efecto, si O es el centro de la circunferencia que pasa por los vértices $A_0, A_1, \ldots, A_{2n}$, el triángulo $A_n\,O\,A_0$ es isósceles y por una fórmula elemental de trigonometría hiperbólica se tiene

$$\cosh\frac{\alpha}{2} = \frac{\cos\dfrac{n\pi}{2n+1}}{\mathrm{sen}\dfrac{\theta}{2}}. \tag{7.3}$$

De aquí, o por simple observación de la figura, se deduce que α crece a medida que $n \to \infty$ y $\theta \to 0$. Para θ muy pequeño y n grande es

$$\frac{\cos\dfrac{n\pi}{2n+1}}{\mathrm{sen}\dfrac{\theta}{2}} \sim \frac{\dfrac{\pi}{2} - \dfrac{n\pi}{2n+1}}{\dfrac{\theta}{2}} = \frac{\pi}{(2n+1)\,\theta}: \tag{7.4}$$

Por otra parte, según (7.1) y (7.3)

$$\frac{L_R}{L_C} = \frac{(2n+1)\,\theta}{\pi}\cosh\,(\alpha/2) \tag{7.5}$$

De (7.1), (7.2), (7.4) y (7.5) se deduce que $L_R/L_C < 1$.

Recordando que en geometría elíptica (equivalente en pequeño a la geometría esférica) los polígonos de Reuleaux tienen la longitud mayor que los círculos de igual anchura [6], y que en el plano euclidiano la longitud es la misma, resulta:

Entre la longitud L_R de un polígono de Reuleaux de cualquier número de lados y la longitud L_C del círculo de la misma anchura, vale $L_R = L_C$ en el plano euclidiano, $L_R > L_C$ en el plano elíptico y $L_R < L_C$ en el plano hiperbólico.

Un teorema de Blaschke afirma que todo conjunto convexo de anchura constante se puede aproximar en tanto como se quiera por polígonos de

Reuleaux [2] (aunque Blaschke se refiere al caso del plano euclidiano, la demostración es fácilmente trasladable al plano hiperbólico). Según esto y observando que L_R crece con n y por tanto es mínima para el triángulo, se puede enunciar:

Entre los conjuntos convexos del plano hiperbólico de anchura constante, el triángulo de Reauleux tiene longitud mínima.

Debiéndose cumplir la relación (6.1), se tiene también, como consecuencia:

Entre los conjuntos convexos del plano hiperbólico, el triángulo de Reauleaux es el que tiene área mínima.

Para comprobar algunos resultados anteriores puede ser útil tener a mano los valores de la longitud y el área de los polígonos de Realeaux de $2n + 1$ lados. La longitud está dada por (7.1). En cuanto al área, un cálculo fácil conduce a la expresión

$$F_R = (2n + 1)\,\theta \cosh \alpha + (2n + 1)\,\theta - 2\pi$$

BIBLIOGRAFIA

[1] BLASCHKE, W., *Kreis und Kungel*, 2ª edición, Walter de Gruyter, Berlín, 1956.

[2] BLASCHKE, W., *Convexe Bereiche gegebener konstanter Breite und kleinstein Inhalt*, Math. Ann., Vol. 76, 1915, 504-513.

[3] COXETER, H. S. M., *Non-euclidean Geometry*, University of Toronto Press, 1957.

[4] KARCHER, H., *Umkreise und Inkreise konvexer Kurven in der sphärischen und der hyperbolischen Geometrie*, Math. Annalen, 177, 1968, 122-132.

[5] SANTALÓ, L. A., *Note on convex curves on the hyperbolic plane*, Bull. Am. Math. Soc. 51. 1945, 405-412.

[6] SANTALÓ, L. A., *Propiedades de las figuras convexas sobre la esfera*, Math. Notae, Rosario, año IV, 1944, 11-40.

[7] SANTALÓ, L. A., *Horocycles and convex sets in hyperbolic plane*, Archiv. der Mathematik, XVIII, 1957, 529-533.

ON SOME GEOMETRIC INEQUALITIES IN THE STYLE OF FÁRY.

By L. A. Santaló.

Introduction. Let Σ_h be a compact differentiable manifold of class C^n and dimension h enclosed by a sphere S of radius r in Euclidean n-space E^n. The problem arises as to finding inequalities connecting the area, the total curvature and the total absolute mean curvatures (to be defined below) of Σ_h.

For $n=3$, $h=1,2$ this kind of inequalities were considered by I. Fáry [6] and for $h=1$ and any n by G. D. Chakerian [3], [4]. Our purpose is to consider the case of hypersurfaces $(h=n-1)$. We obtain the inequalities (25), (26) where the sign of equality holds for the sphere of radius r and the inequalities (31) where the equality sign is probably never attained.

1. Densities for linear subspaces in E^n. In this section we review some known results of integral geometry which will be useful in the following sections.

Let $(P; e_1, e_2, \cdots, e_n)$ be an orthonormal frame in E^n so that P is the origin of the frame and $e_1, e_2, \cdots, e_n$ are unit orthogonal vectors. In the space of all frames we define the differential forms

$$\omega_i^m = -\omega_m^i = e_m \cdot de_i. \tag{1}$$

To define the density $dE^h(E^q)$ for h-dimensional linear subspaces about E^q in E^n $(q<h<n)$ we consider the frames such that $P \in E^q$, $e_1, e_2, \cdots, e_q$ span E^q and $e_{q+1}, \cdots, e_h$ are contained in E^h. Then we have (see, for instance, Chern [5])

$$dE^h(E^q) = \bigwedge_{i,m} \omega_i^m \qquad (i=q+1,\cdots,h; m=h+1,\cdots,n) \tag{2}$$

where the right side is the exterior product of the forms ω_i^m. For $q=0$ (2) becomes the density for subspaces E^h through a fixed point P,

$$dE^h(P) = \bigwedge_{i,m} \omega_i^m \qquad (i=1,\cdots,h; m=h+1,\cdots,n). \tag{3}$$

Notice the duality

Received July 18, 1967.

$$dE^h(P) = dE^{n-h}(P). \tag{4}$$

To define the density for linear subspaces E^h in E^n we consider a $E^{n-h}(P)$ orthogonal to E^h through a fixed point P and then, if dP_{n-h} denotes the element of volume of $E^{n-h}(P)$ at the intersection point $E^h \cap E^{n-h}(P)$, we have (see [9])

$$dE^h = dP_{n-h} \wedge dE^{n-h}(P). \tag{5}$$

We consider always the densities in absolute value, so that is no question of sign.

Let $d0_i$ denote the element of area of the i-dimensional unit sphere which contains the end points of the vectors $e_1, e_2, \cdots, e_{i+1}$ at the end point of e_{i+1}. Then the formula

$$dE^h(P) \wedge d0_{h-1} \wedge \cdots \wedge d0_1 = d0_{n-1} \wedge \cdots \wedge d0_{n-h} \tag{6}$$

holds good ([9, formula (2.5)]). It follows that the total measure of all oriented h-dimensional linear subspaces through P in E^n is ([5], [9])

$$\int dE^h(P) = \frac{0_{n-1} \cdots 0_{n-h}}{0_{h-1} \cdots 0_1} \tag{7}$$

where 0_i is the area of the unit i-dimensional sphere, given by

$$0_i = 2\pi^{(i+1)/2}/\Gamma\left(\frac{i+1}{2}\right). \tag{8}$$

More generally, the total measure of all oriented h-dimensional linear subspaces about a fixed E^q in E^n is ([8, formula (2.3)])

$$\int dE^h(E^q) = \frac{0_{n-q-1} \cdots 0_{n-h}}{0_{h-q-1} \cdots 0_1} \tag{9}$$

The density for pairs of linear subspaces $E^h(P) + E^{n-m}(P)$ through P such that $E^h(P) \subset E^{n-m}(P)$ is given by any one of the two equivalent differential forms

$$dE^h(P) \wedge dE^{n-m}(E^h) = dE^{n-m}(P) \wedge dE^h_{n-m}(P) \tag{10}$$

where $dE^h_{n-m}(P)$ indicates the density for $E^h(P)$ in E^{n-m}.

Finally we recall that given an hypersurface Σ_{n-1} of area F in E^n, if N denotes the number of points in which the straight line E^1 cuts Σ_{n-1}, the integral formula

$$\int N\, dE^1 = \frac{0_n}{\pi} F \tag{11}$$

holds, where E^1 is considered oriented and dE^1 is the density (5) for $h=1$. The integral is extended over all lines of E^n, N being 0 if $\Sigma_{n-1}\cap E^1=0$ (see, for instance, Hadwiger [7, p. 229]).

2. Total absolute mean curvatures of a compact hypersurface in E^n. Let Σ_{n-1} be a compact differentiable manifold of class C^n and dimension $n-1$ in E^n. Let us consider the set of all m-dimensional linear subspaces which are tangent to Σ_{n-1} and are orthogonal to a fixed E^{n-m}. The intersection of E^{n-m} with these m-spaces is an $(n-m-1)$-dimensional variety Σ_{n-m-1} (of class C^{n-1}); let $F(\Sigma_{n-m-1})$ be its $(n-m-1)$-dimensional area (for $m=n-1$, $F(\Sigma_0)$ means the number of hyperplanes tangent to Σ_{n-1} which are orthogonal to a fixed line E^1). Then the m-th *total absolute mean curvature* of Σ_{n-1} is defined by the equation $(1\leqq m\leqq n-1)$ (see [9])

$$M_m(\Sigma_{n-1})=\frac{0_{m-1}\cdots 0_1}{0_{n-2}\cdots 0_{n-m-1}}\int F(\Sigma_{n-m-1})\,dE^{n-m}(P). \tag{12}$$

In particular, for $m=n-1$, M_{n-1} becomes the *total absolute curvature* K of Σ_{n-1}. In this case (12) can be written

$$K(\Sigma_{n-1})=\tfrac{1}{2}\int \nu\, d0_{n-1} \tag{13}$$

where ν denotes the number of tangent hyperplanes orthogonal to the direction defined by the point of the unit sphere which element of area is $d0_{n-1}$. If Σ_{n-1} is convex, the total absolute mean curvatures coincide with the integral of the symmetric functions of the principal curvatures (Bonnesen-Fenchel [1, n. 38]). If Σ_{n-1} is not convex, M_m are the integrals of the absolute values of these symmetric functions (see the footnote on page 122 of the paper of Fáry [6]).

We will now give certain relations for the curvatures M_m. By applying the definition (12) to Σ_{n-m-1} as a manifold in E^{n-m}, we have (for $1\leqq h\leqq n-m-1$)

$$M_{n-m-h}(\Sigma_{n-m-1})=\frac{0_{n-m-h-1}\cdots 0_1}{0_{n-m-2}\cdots 0_{h-1}}\int F(\Sigma_{h-1})\,dE^h{}_{n-m}(P) \tag{14}$$

where Σ_{h-1} is the intersection of $E^h(P)\subset E^{n-m}$ with all the $E^{n-m-h}\subset E^{n-m}$ which are orthogonal to $E^h(P)$ and are tangent to Σ_{n-m-1}. The differential form $dE^h{}_{n-m}(P)$ is the density for E^h through P in E^{n-m}. On the other hand, since each E^{n-m-h} tangent to Σ_{n-m-1} together with the corresponding E^m tangent to Σ_{n-1} and orthogonal to E^{n-m} determines a E^{n-h} tangent to Σ_{n-1} and orthogonal to E^h, we have

$$(15) \qquad M_{n-h}(\Sigma_{n-1}) = \frac{0_{n-h-1}\cdots 0_1}{0_{n-2}\cdots 0_{h-1}} \int F(\Sigma_{h-1})\, dE^h(P).$$

From (14) and (10) we deduce

$$(16) \qquad \int M_{n-m-h}(\Sigma_{n-m-1})\, dE^{n-m}(P)$$
$$= \frac{0_{n-m-h-1}\cdots 0_1}{0_{n-m-2}\cdots 0_{h-1}} \int F(\Sigma_{h-1})\, dE^h(P) \wedge dE^{n-m}(E^h)$$
$$= \frac{0_{n-m-h-1}\cdots 0_1}{0_{n-m-2}\cdots 0_{h-1}} \frac{0_{n-2}\cdots 0_{h-1}}{0_{n-h-1}\cdots 0_1} M_{n-h}(\Sigma_{n-1}) \int dE^{n-m}(E^h).$$

The value of the last integral is given by (9) and therefore we have

$$(17) \qquad M_{n-h}(\Sigma_{n-1}) = \frac{0_{m-1}\cdots 0_1}{0_{n-2}\cdots 0_{n-m-1}} \int M_{n-m-h}(\Sigma_{n-m-1})\, dE^m(P).$$

If Σ_{n-1} is convex, the formula (17) is a consequence of the so called formula of Kubota ([1, p. 49] and [9]).

For $h=1$, (17) gives

$$(18) \qquad K(\Sigma_{n-1}) = \frac{0_{m-1}\cdots 0_1}{0_{n-2}\cdots 0_{n-m-1}} \int K(\Sigma_{n-m-1})\, dE^m(P),$$

and for $h=1$, $m=1$ (since $dE^1(P) = d0_{n-1}$),

$$(19) \qquad K(\Sigma_{n-1}) = \frac{1}{0_{n-2}} \int K(\Sigma_{n-2})\, d0_{n-1}$$

which holds for $n > 2$.

3. Inequalities between the total mean curvatures. Let Σ_{n-1} be a compact differentiable manifold of class C^n and dimension $n-1$ in euclidean space E^n, which is contained in a ball S of radius r. Let $(P; e_1, e_2, \cdots, e_n)$ be a orthonormal frame and let Σ_{n-2} be the $(n-2)$-dimensional variety determined in $E^{n-1}(e_2, e_3, \cdots, e_n)$ by the tangent lines to Σ_{n-1} parallel to e_1. Analogously, let Σ_{n-3} be the $(n-3)$-dimensional variety determined in $E^{n-2}(e_3, \cdots e_n)$ by the tangent lines to Σ_{n-2} parallel to e_2. In general, let Σ_i be the i-dimensional variety in $E^{i+1}(e_{n-i}, \cdots, e_n)$ obtained by repeating the process above $(i = n-2, \cdots, 1)$.

Let $N(e_i)$ be the number of points in which the straight line which contains e_i intersects Σ_{n-i}. The intersection of Σ_{n-i} with the plane (e_i, e_{i+1}) is a set of closed plane curves which are divided by the line which contains e_i in a number of arcs $\geqq N(e_i)$ and therefore there are at least $N(e_i)$ tangent to Σ_{n-i} parallel to e_i. Consequently we have $N(e_i) \leqq N(e_{i+1})$. In particular, we have $N(e_1) \leqq N(e_{n-1})$. On the other hand if ν is the number of tangent lines to Σ_1 parallel to e_{n-1}, for an analogous reason, we have $N(e_{n-1}) \leqq \nu$ and therefore

$$N(e_1) \leqq \nu. \tag{20}$$

Since ν is also equal to the number of hyperplanes tangent to Σ_{n-1} orthogonal to e_n, it has the same meaning as in the equation (13).

Assume that P varies over the $(n-1)$-dimensional ball S_{n-1} obtained by orthogonal projection of the given ball S onto the hyperplane $(e_2, \cdots, e_n)$ and denote by dP_{n-1} the volume element in this hyperplane at the point P. Let $d0_{n-i}$ denote the area element of the $(n-i)$-dimensional unit sphere in the linear subspace $(e_i, e_{i+1}, \cdots, e_n)$ at the end point of e_i and take the integral

$$I = \int N(e_1)\, d0_{n-1} \wedge d0_{n-2} \wedge \cdots \wedge d0_1 \wedge dP_{n-1} \tag{21}$$

extended over all positions of the frame $(P; e_1, e_2, \cdots, e_n)$ with $P \in S_{n-1}$.

If E^1 is the straight line which contains e_1, we have $dE^1 = d0_{n-1} \wedge dP_{n-1}$ and therefore, (21) and (11) give

$$I = \frac{0_n}{\pi} 0_{n-2} \cdots 0_1 F(\Sigma_{n-1}). \tag{22}$$

On the other hand, if $d0'_i$ denotes the area element of the i-dimensional unit sphere in the subspace $(e_1, e_2, \cdots, e_{i+1})$ at the end point of e_i, we have

$$d0_{n-1} \wedge \cdots \wedge d0_1 = d0'_1 \wedge \cdots \wedge d0'_{n-1} \tag{23}$$

since both sides of the equality are the so called kinematic density about P. Therefore, having into account (20) and (13) we have

$$I \leqq \int \nu\, d0'_1 \wedge \cdots \wedge d0'_{n-1} \wedge dP_{n-1} = 2K 0_1 \cdots 0_{n-2} V_{n-1}(r), \tag{24}$$

where $V_{n-1}(r) = r^{n-1} 0_{n-2}/(n-1)$ is the volume of the $(n-1)$-dimensional ball of radius r.

From (22) and (24) (having into account that $2\pi 0_{n-2} = (n-1) 0_n$), we have

$$F(\Sigma_{n-1}) \leqq r^{n-1} K(\Sigma_{n-1}). \tag{25}$$

This is the first inequality we wished to prove. From (25), (12) and (18) we deduce

$$M_m(\Sigma_{n-1}) \leqq r^{n-m-1} K(\Sigma_{n-1}). \tag{26}$$

In (25) and (26) the equality sign holds for the sphere of radius r. We have established the following

THEOREM. *Let Σ_{n-1} be a compact $(n-1)$-dimensional differentiable manifold of class C^n in euclidean space E^n which is contained in a ball of radius r. Then, the inequalities (25) and (26) hold, where F, K and M_m $(m=1,2,\cdots,n-1)$ are respsctively the area, total absolute curvature and total absolute mean curvatures of Σ_{n-1} defined by (12). The equality sign holds for the sphere of radius r.*

For $n=3$, (25) gives

$$F(\Sigma_2)\leqq r^2K(\Sigma_2) \tag{27}$$

which is sharper than the inequality $F\leqq(4/\pi)r^2K$ of Fáry [6].

4. Other inequalities. Let dx_i be the projection of the displacement vector at P on the direction e_i, so that $dP_{n-1}=dx_2\wedge\cdots\wedge dx_n$. According to (5) the differential form $d0_{n-i}\wedge dx_{i+1}\wedge\cdots\wedge dx_n$ is the density for lines E^1 (line which contains e_i) in the linear subspace $(e_i,e_{i+1},\cdots,e_n)$ and therefore we have, by (11),

$$\int N(e_i)\,d0_{n-i}\wedge dx_{i+1}\wedge\cdots\wedge dx_n=\frac{0_{n-i+1}}{\pi}F(\Sigma_{n-i}). \tag{28}$$

It follows that the integral (21) satisfies the inequality

$$I\leqq\frac{0_{n-i+1}}{\pi}r^{i-1}\frac{0_{i-2}}{i-1}\int F(\Sigma_{n-i})\,d0_{n-1}\wedge\cdots\wedge d0_{n-i+1}\wedge d0_{n-i-1}\wedge\cdots\wedge d0_1 \tag{29}$$

where we have replaced the integral of $dx_2\wedge\cdots\wedge dx_i$ by its upper bound $V_{i-1}(r)=r^{i-1}0_{i-2}/(i-1)$. By (6) the inequality (29) can be written

$$I\leqq\frac{0_{n-i+1}0_{i-2}}{(i-1)\pi}r^{i-1}0_{i-2}\cdots0_10_{n-i-1}\cdots0_1\int F(\Sigma_{n-i})\,dE^{i-1}(P). \tag{30}$$

From (22), (30) and (12) we have

$$F(\Sigma_{n-1})\leqq\frac{0_{n-i+1}0_{i-2}}{(i-1)0_n}r^{i-1}M_{i-1}(\Sigma_{n-1}) \tag{31}$$

which holds for $i=2,3,\cdots,n-1$.

These inequalities (31) are not sharp. They can probably be replaced by $F(\Sigma_{n-1}) \leqq r^{i-1} M_{i-1}(\Sigma_{n-1})$. For $i=2$, this last inequality has been proved by Chakerian [4].

FACULTAD DE CIENCIAS, UNIVERSIDAD DE BUENOS AIRES (ARGENTINA).

REFERENCES.

[1] T. Bonnesen and W. Fenchel, *Theorie der konvexen Körper*, Berlin, 1934.
[2] H. Busemann, *Convex surfaces*, New York, 1958.
[3] G. D. Chakerian, "An inequality for closed space curves," *Pacific Journal of Mathematics*, vol. 12 (1962), pp. 53-57.
[4] ———, "On some geometric inequalities," *Proceedings of the American Mathematical Society*, vol. 15 (1964), pp. 886-888.
[5] S. S. Chern, "On the kinematic formula in integral geometry," *Journal of Mathematics and Mechanics*, vol. 16 (1966), pp. 101-118.
[6] I. Fáry, "Sur certaines inegalités géométriques," *Acta Sci. Math.* (*Szeged*), vol. 12 (1950), pp. 117-124.
[7] H. Hadwiger, *Vorlesungen über Inhalt, Oberflache und Isoperimetrie*, Berlin, 1957.
[8] L. A. Santaló, *Geometria integral en espacios de curvatura constante*, Publ. Com. Nacional Energia Atómica, Buenos Aires, 1952.
[9] ———, "Sur la mesure des espaces linéaires qui coupent un corps convexe et problemes qui s'y rattachent," *Colloque sur les questions de réalité en Géométrie*, Liége, 1955, pp. 177-190.

L. A. SANTALO

MEAN VALUES AND CURVATURES

We divide this exposition into two parts. Part 1 refers to the mean value of the Euler—Poincaré characteristic of the intersection of two convex hypersufaces in E_4. Part 11 deals with the definition of q-th total absolute curvatures of a compact n-dimensional variety imbedded in euclidean space of $n+N$ dimensions, extending some results given in [10].

1. On convex bodies in E_4

1. Introduction. Let K be a convex body in 4-dimensional euclidean space E_4 and let W_i ($i=0, 1, 2, 3, 4$) be its Minkowski's *Quermass integrale* (see for instance Bonnesen—Fenchel [1]). Recall that

$$W_0 = V = \text{volume of } K$$
$$4W_1 = F = \text{area of } \partial K \tag{1.1}$$
$$W_4 = \pi^2/2,$$

and, if K has sufficiently smooth boundary, we have also

$$4W_2 = M_1 = \text{first mean curvature} = \frac{1}{3}\int_{\partial K}\left(\frac{1}{R_1}+\frac{1}{R_2}+\frac{1}{R_3}\right)d\sigma$$

$$4W_3 = M_2 = 2\text{ th mean curvature} = \frac{1}{3}\int_{\partial K}\left(\frac{1}{R_1R_2}+\frac{1}{R_1R_3}+\frac{1}{R_2R_3}\right)d\sigma \tag{1.2}$$

where R_i are the principal radii of curvature and $d\sigma$ is the element of area of ∂K.

For instance, if $K=$ sphere of radius r, we have

$$V=\frac{1}{2}\pi^2r^4,\quad F=2\pi^2r^3,\quad M_1=2\pi^2r^2,\quad M_2=2\pi^2r. \tag{1.3}$$

We will use throughout the invariants V, F, M_1, M_2 because they have a more geometrical meaning; however we do not assume smoothness to ∂K, so that as definition of M_1, M_2 we take $M_1=4W_2$, $M_2=4W_3$.

The invariants V, F, M_1, M_2 are not independent. They are related by certain inequalities which may be written in the following symmetrical form (following Hadwiger [6]).

$$W_\alpha^{\beta-\gamma}\, W_\beta^{\gamma-\alpha}\, W_\gamma^{\alpha-\beta} \geqslant 1, \quad 0 \leqslant \alpha \leqslant \beta \leqslant \gamma \leqslant 4. \tag{1.4}$$

In explicit form and using the invariants V, F, M_1, M_2 the inequalities (1.4) give the following non-independent inequalities

$$\begin{array}{lll} F^1 \geqslant 4VM_1, & F^3 \geqslant 16\, V^2 M_2, & F^4 \geqslant 128\, \pi^2 V^3, \\ M_1^3 \geqslant 4\, VM_2, & M_1^2 \geqslant 2\, \pi^2 V, & M_2^4 \geqslant 32\, \pi^6 V, \\ M_1^2 \geqslant FM_2, & M_1^3 \geqslant 2\, \pi^2 F^2, & M_2^3 \geqslant 4\, \pi^4 F, \\ M_2^2 \geqslant 2\, \pi^2 M_1. & & \end{array}$$

We will represent throughout the paper by O_i the volume of the i-dimensional unit sphere, that is

$$O_i = \frac{2\pi^{\frac{i+1}{2}}}{\Gamma\left(\frac{i+1}{2}\right)} \tag{1.6}$$

or instance

$$O_0 = 2,\ O_1 = 2\pi,\ O_2 = 4\pi,\ O_3 = 2\pi^2,\ O_4 = \frac{8}{3}\pi^2,\ O_5 = \pi^3. \tag{1.7}$$

2. Mean value of $\chi\,(\partial K \cap g\partial K)$. Let G be the group of isometries of E_4. For any $g \in G$ we represent by $g\partial K$ the image of ∂K by the isometry g. Let dg denote the invariant volume element of G (=kinematic density for E_4). Assume the convex body K fixed and consider the intersections $\partial K \cap g\partial K$, $g \in G$. Then, Federer [5] and Chern [2] have proved the following integral formula

$$\int_G \chi\,(\partial K \cap g\partial K)\, dg = 64\, \pi^2 F M_2 \tag{2.1}$$

where $\chi\,(\partial K \cap g\partial K)$ denote the Euler—Poincaré characteristic of the surface $\partial K \cap g\partial K$.

On the other side, the so-called fundamental kinematic formula of integral geometry, gives

$$\int_{K \cap gK \neq \emptyset} dg = 8\pi^2 \left(4\pi^2 V + 2\, FM_2 + \frac{3}{2}\, M_1^2 \right) \tag{2.2}$$

Therefore the expected value of $\chi\,(\partial K \cap g\partial K)$ is

$$E\,(\chi\,(\partial K \cap g\partial K)) = \frac{8\, F\, M_2}{4\pi^2 V + 2FM_2 + \frac{3}{2}\, M_1^2}. \tag{2.3}$$

Notice that, being K convex, the intersections $\partial K \cap g\partial K$ are closed orientable surfaces. Thus the possible values of χ are, either $\chi = 2$, 4, 6,$\cdots$ or $\chi = 0, -2, -4, -5, \cdots$ If K is an euclidean sphere, obvionsly we have $E\,(\chi) = 2$.

Conjecture. *For all convex sets K of E_4 the inequality*

$$E(\chi(\partial K \cap g\partial K)) \leqslant 2 \tag{2.4}$$

holds good, equality for the euclidean sphere.

Putting

$$\Delta = 8\pi^2 V + 3M_1^2 - 4FM_2 \tag{2.5}$$

the conjecture is equivalent to prove that $\Delta \geqslant 0$. For the euclidean sphere, accrding to (1.3) we have $\Delta = 0$.

In support of this conjecture we will prove it for rectangular parallelepipeds. Let a, b, c, d be the sides of a rectangular parallelepiped in E_4 and assume

$$a \leqslant b \leqslant c \leqslant d \tag{2.6}$$

It is known that (Hadwiger [6])

$$V = abcd, \; F = 2\,(abc + abd + acd + bcd),$$

$$M_1 = \frac{2}{3}\pi\,(ab + ac + ad + bc + bd + cd), \; M_2 = \frac{4}{3}\pi\,(a + b + c + d).$$

With these values we verify the identity

$$\frac{3}{4\pi}\Delta = (4-\pi)[a^2c^2 + a^2(c-b)^2 + b^2(c-a)^2 + a^2(d-b)^2$$

$$+ c^2(d-a)^2 + b^2(b-c)^2 + c^2(d-b)^2] + (18\pi - 56)\,abcd$$

$$+ (4\pi - 12)(a^2b^2 + a^2c^2 + b^2c^2) + (8 - 2\pi)[(b-a)\,acd +$$

$$+ (c-b)\,abd + (d-c)\,acb] + (4-\pi)\,d^2\,[(2A^2 - B^2)$$

$$(a^2 + b^2) + (Ac - Ba)^2 + (Ac - Bb)^2],$$

where $A^2 = (3\pi - 8)/(8 - 2\pi)$, $B^2 = (8 - 2\pi)/(3\pi - 8)$.

Since all terms are positive, we have $\Delta > 0$.

For an ellipsoid of revolution whose semiaxes are $a, a, a, \lambda a$ we have (Hadwiger [6])

$$V = \frac{\pi}{2}\lambda a^4, \; F = 2\pi^2\lambda^2 a^3\, F\left(\frac{5}{2}, \frac{1}{2}, 2; 1-\lambda^2\right),$$

$$M_1 = 2\pi^2\lambda^3 a^2\, F\left(\frac{5}{2}, 1, 2; 1-\lambda^2\right), \tag{2.7}$$

$$M_2 = 2\pi^2\lambda^4 a\, F\left(\frac{5}{2}, \frac{3}{2}, 2; 1-\lambda^2\right)$$

where F denotes the hypergeometric function. In this case the conjecture writes

$$1 + 3\lambda^5 F_1^2 - 4\lambda^5 F_{1/2} F_{3/2} \geqslant 0 \tag{2.8}$$

where

$$F_{1/2} = F\left(\frac{5}{2}, \frac{1}{2}, 2; 1-\lambda^2\right),$$

$$F_1 = F\left(\frac{5}{2}, 1, 2;\ 1-\lambda^2\right),$$

$$F_{3/2} = F\left(\frac{5}{2}, \frac{3}{2}, 2;\ 1-\lambda^2\right).$$

I do not know if (2.8) holds for all values of λ.

II. Absolute total curvatures of compact manifolds immersed in euclidean space

1. Introduction. In this section we extend and complete the contents of [10]. We shall first state some known formulas which will be used in the sequel.

Let L_h be a h-dimensional linear subspace in the $(n+N)$-dimensional euclidean space E_{n+N}. We will call it, simply, a h-space. Let $L_h(0)$ be a h-space in E_{n+N} through a fixed point 0. The set of all oriented $L_h(0)$ constitute the Grassman manifold $G_{h,\, n+N-h}$. We shall represent by $dL_h(0)$ the element of volume of $G_{h,\, n+N-h}$, which is the same thing as the density for oriented h-spaces through 0. The expression of $dL_h(0)$ is well known, but we will recall it briefly for completeness (see [9], [2]).

Let $(O;\, e_1, e_2, \cdots, e_{n+N})$ be an orthonormal frame in E_{n+N} of origin O. In the space of all orthonormal frames of origin O we define the differential forms

$$\omega_{im} = -\omega_{mi} = e_m de_i\,. \tag{1.1}$$

Assuming $L_h(O)$ spanned by the unit vectors $e_1, e_2, \cdots, e_h$, then

$$dL_h(O) = \Lambda\omega_{im} \tag{1.2}$$

where the right side is the exterior product of the forms ω_{im} over the range of indices

$$i = 1, 2, \cdots, h;\quad m = h+1, h+2, \cdots, n+N.$$

The $(n+N-h)$ – space $L_{n+N-h}(O)$ orthogonal to $L_h(O)$ is spanned by $e_{h+1}, \cdots, e_{n+N}$ and (1.2) gives the duality

$$dL_h(O) = dL_{n+N-h}(O) \tag{1.3}$$

The measure of the set of all oriented $L_h(0)$ (= volume of the Grassman manifold $G_{h,\,n+N-h}$) may be computed directly from (1.2) (see 9]). or applying that it is the quotient space $SO(n+N)/SO(h) \times \times SO(n+N-h)$ (see [2]). The result is

$$\int_{G_{h,\,n+N-h}} dL_h(O) = \frac{O_{n+N-1}\, O_{n+N-2} \cdots O_{n+N-h}}{O_1 O_2 \cdots O_{h-1}} \tag{1.4}$$

$$= \frac{O_h O_{h+1} \cdots O_{n+N-1}}{O_1 O_2 \cdots O_{n+N-h-1}}$$

where O_i is the area of the i-dimensional unit sphere (1, (1.6)).

Another known integral formula which we will use is the following.

Consider the unit sphere $\sum_{n+N-1}$ of dimension $n+N-1$ of center O. Let V^s be a s-dimensional variety in $\sum_{n+N-1}$. Let $\mu_{s+h-n-N}(V^s \cap L_h)$ be the $(s+h-n-N)-$dimensional measure of the variety $V^s \cap L_h(O)$ of dimension $s+h-(n+N)$ and let $\mu_s(V^s)$ be the s-dimensional measure of V^s (all these measures considered as measures of subvarieties of the euclidean space E_{n+N}). Then

$$\int_{G_{h,\, n+N-h}} \mu_{s+h-n-N}(V^s \cap L_h(O))\, dL_h(O) =$$

$$= \frac{O_{n+N-h}\, O_{n+N-h+1} \cdots O_{n+N-1}\, O_{h+s-n-N}}{O_1 O_2 \cdots O_{h-1}\, O_s}\, \mu_s(V^s) \tag{1.5}$$

Note that this formula assumes the h-spaces L_h oriented (see [8]). In particular, if $s=1$ an $h=n+N-1$ that is, for a curve V^1 of ength U we have

$$\int_{G_{n+N-1,\, 1}} \nu\, dL_{n+N-1}(O) = \frac{2\, O_{n+N-1}}{O_1}\, U \tag{1.6}$$

where ν is the number of points of the intersection $V^1 \cap L_{n+N-1}(O)$.

2. **Defintions.** Let X^n be a compact n-dimensional differentiable manifold (without boundary) of class C^∞ in E_{n+N}. To each point $\in X^n$ we attach the ρ-space $T^{(q)}(p)$ spanned by the vectors

$$\frac{\partial}{\partial x_1}, \ldots, \frac{\partial}{\partial x_n};\ \frac{\partial^2}{\partial x_1^2}, \ldots, \frac{\partial^2}{\partial x_n^2}; \ldots;\ \frac{\partial^q}{\partial x_1^q}, \ldots, \frac{\partial^q}{\partial x_n^q} \tag{21}$$

which we will call the $q-$th tangent fibre over p. Its dimension is

$$\rho(n, q) = \sum_{i=1}^{q} \binom{n+i-1}{i} \tag{2.2}$$

Assuming

$$1 \leqslant r \leqslant n+N-1, \quad \rho \leqslant n+N-1 \tag{2.3}$$

we define the r-th total absolute curvature of order q of X^n as follows:

a) *Case* $1 \leqslant r \leqslant \rho$. Let O be a fixed point of E_{n+N} and consider a $(n+N-r)$-space $L_{n+N-r}(O)$. Let Γ_r be the set of all r-spaces L_r of E_{n+N} which are contained in some of the fibres $T^{(q)}(p)$, $p \in X^n$, pass through p, and are orthogonal to $L_{n+N-r}(O)$. The intersection $\Gamma_r \cap L_{n+N-r}(O)$ will be a compact variety in $L_{n+N-r}(O)$ whose dimension δ we shall compute in the next section. Let $\mu(\Gamma_r \cap L_{n+N-r}(O))$ be the measure of this variety as subvariety of the euclidean space

$L_{n+N-r}(O)$; if $\delta = 0$, then μ means the number of intersection points of Γ_r and $L_{n+N-r}(O)$.

Then we define the r-th total absolute curvature of order q of $X^n \subset E_{n+N}$ as the mean value of the measures μ for all $L_{n+N-r}(0)$, that is, according to (1.4)

$$K_r^{(q)}(X^n) = \frac{O_1 O_2 \cdots O_{n+N-r-1}}{O_r O_{r+1} \cdots O_{n+N-1}} \int_{G_{n+N-r,\, r}} \mu(\Gamma_r \cap L_{n+N-r}(O))\, dL_{n+N-r}(O). \tag{2.4}$$

The coefficient of the right side may be substituted by

$$O_1 O_2 \cdots O_{r-1} / O_{n+N-r} \cdots O_{n+N-1}.$$

b) *Case* $\rho \leqslant r \leqslant n+N-1$. Instead of the set of L_r which *are contained* in some $T^{(q)}(p)$ we consider now the set of L_r which *contain* some $T^{(q)}(p)$, $p \in X^n$, and are orthogonal to $L_{n+N-r}(O)$. As before we represent this set by Γ_r and the r-th total absolute curvature of order q of $X^n \subset E_{n+N}$ is defined by the same mean value (2.4).

3. Properties. We proceed now to compute the dimension of $\Gamma_r \cap L_{n+N-r}(0)$.

a) *Case* $1 \leqslant r \leqslant \rho$. The set of all $L_r \subset E_{n+N}$ is the Grassman manifold $G_{r+1,\, n+N-r}$ whose dimension is $(r+1)(n+N-r)$. The set of all L_r which are contained in $T^{(q)}(p)$ and pass through p is the Grassman manifold $G_{r,\, \rho-r}$ of dimension $r(\rho-r)$; therefore the set of all L_r which are contained in some $T^{(q)}(p)$, $p \in X^n$, has dimension $r(\rho-r)+n$. On the other side, the set of all $L_r \subset E_{n+N}$ which are orthogonal tc $L_{n+N-r}(O)$ has dimension $n+N-r$. Consequently, the intersection of both sets, as sets of points of $G_{r+1,\, n+N-r}$, has dimension

$$r(\rho - r) + n + n + N - r - (r+1)(n+N-r) = r\rho + n - r(n+N).$$

Since to each L_r orthogonal to $L_{n+N-r}(0)$ corresponds one and only one intersection point with this linear space, the preceding dimension coincide with the dimension δ of $\Gamma_r \cap L_{n+N-r}$, that is,

$$\delta = \dim(\Gamma_r \cap L_{n+N-r}(0)) = r\rho + n - r(n+N).$$

Hence, in order that $K_r^{(q)}(X^n) \neq 0$, it is necessary and sufficient that

$$r\rho + n \geqslant r(n+N) \tag{3.1}$$

b) *Case* $\rho \leqslant r \leqslant n+N-1$. The set of all $L_r \subset E_{n+N}$ which contain a fixed L_ρ, constitute the Grassman manifold $G_{r+\rho,\, n+N-r}$ and therefore the dimension of the set of all L_r which contain some $T^{(q)}(p)$, $p \in X^n$, is $(r-\rho)(n+N-r)+n$. The remainder dimensions are the same as in the case a), so that the dimension of the set of all L_r which contain some $T^{(q)}(p)$, $p \in X^n$, and are orthogonal to $L_{n+N-r}(O)$ is

$$(r-\rho)(n+N-r) + n + n + N - r - (r+1)(n+N-r) = r\rho + n - \rho(n+N)$$

that is

$$\delta = \dim (\Gamma_r \cap L_{n+N-r}(O)) = \rho r + n - \rho(n+N)$$

In order that $K_r^{(q)}(X^n) \neq 0$, it is necessary and sufficient that

$$\rho r + n \geqslant \rho(n+N). \tag{3.2}$$

Of course, to (3.1) and (3.2) we must add the relations (2.3).

The most interesting cases correspond to $\delta = 0$, for which the measure μ in (2.4) is a positive integer and the total absolute curvature is invariant under similitudes. In this case the set of points $p \in X^n$ for which L_r contains or is contained in $T^{(q)}(p)$ can be divided according to the index of p, and we get different curvatures as those defined by Kuiper for the case $q=1$, $r=n+N-1$ [7]. We will not go into details here.

4. Examples.

4.1. Curves, $n=1$. For $n=1$ the condition (3.1) writes

$$1 \geqslant r + r(N-\rho)$$

and since $\rho \leqslant N$ the only possibility is $\rho = N$, $r=1$, which gives $\delta = 0$. The corresponding curvature $K_1^{(N)}(X^1)$ is

$$K_1^{(N)}(X^1) = \frac{1}{O_N} \int_{G_{N,1}} \nu_1 \, dL_N(0) \tag{4.1}$$

where ν_1 is the number of lines in E_{n+N} orthogonal to $L_N(0)$ which are contained in some N—th tangent fiber of the curve X^1. Notice that $G_{N,1}$ is the unit sphere $\sum_N$ and $dL_N(0)$ is the element of area of this sphere in consequence of the duality (1.3). If $e_1, e_2, \cdots, e_{N+1}$ are the principal normals of X^1 then the formula (1.6) says that the right side of (4.1) is equal to the length of the spherical curve $e_{N+1}(s)$ (s=arc length of X^1) up to the factor $1/\pi$. That is, if $\varkappa_N$ is the N-th curvature of X^1 (see, for instance, Eisenhart [4], p. 107) we have

$$K_1^{(N)}(X^1) = \frac{1}{\pi} \int_{X^1} |\varkappa_N| \, ds. \tag{4.2}$$

For the case of curves in E_3, $N=2$, $\varkappa_N$ is the torsion of the curve and $K_1^{(2)}$ is up to the factor π^{-1}, the *absolute total torsion* of X^1.

The condition (3.2) gives $1 \geqslant \rho + \rho(N-r)$ and since $r \leqslant N$, this condition implies $\rho = 1$, $r = N$. We have the curvature

$$K_N^{(1)}(X^1) = \frac{1}{O_N} \int_{G_{1,N}} \nu_N \, dL_1(O), \tag{4.3}$$

where ν_N is the number of hyperplanes L_N of E_{N+1} orthogonal to $L_1(O)$ which contain some tangent line of X^1. The same formula (1.6) gives now that the right side of (4.3) is equal to the length of the curve $e_1(s)$ (=spherical tangential image of X^1), up to the factor $1/\pi$. Therefore, if $\varkappa_1$ is the first curvature of X^1, (4.3) writes

$$K_N^{(1)}(X^1)=\frac{1}{\pi}\int_{X^1}|\varkappa_1|\,ds. \tag{4.4}$$

Notice that for each direction $L_1(O)$ there are at least two hyperplanes orthogonal to $L_1(O)$ which contain a tangent line of X^1 (the hyperplanes which separate the hyperplanes which have common point with X^1 of those which do not have such common point). Therefore the mean value $K_N^{(1)}$ is $\geqslant 2$ and (4.4) gives the classical Fenchel's inequality

$$\int_{X^1}|\varkappa_1|\,ds\geqslant 2\pi. \tag{4.5}$$

If the curve X^1 has at least 4 hyperplanes orthogonal to an arbitrary direction $L_1(0)$ which contain a tangent line of X^1 (as it happens for instance for knotted curves in E_3), the mean value $K_N^{(1)}(X)$ will be $\geqslant 4$, and we have the Fary's inequality

$$\int_{X^1}|\varkappa_1|\,ds\geqslant 4\pi \tag{4.6}$$

4.2 Surfaces, $n=2$.

1. *Total absolute curvatures of order* 1. We have $n=2$, $\rho=2$ and condition (3.1) writes $2\geqslant rN$. Therefore the possible cases are $r=1$, $N=1$; $r=2$, $N=1$ and $r=1$, $N=2$. For $2\leqslant r\leqslant N+1$, condition (3.2) gives $r\geqslant N+1$ and therefore the only possible case is $r=N+1$.

a) *Case* $r=1$, $N=1$. Surfaces in E_3. Having into account that $G_{2,1}$ is the unit sphere $\sum_2$, the curvature (2.4) writes

$$K_1^{(1)}(X^2)=\frac{1}{4\pi}\int_{\Sigma_2}\lambda\,dL_2(0) \tag{4,7}$$

where λ is the length of the curve in the plane $L_2(O)$ generated by the intersections of $L_2(O)$ with the lines of E_3 which are tangent to X^2 and are orthogonal to $L_2(O)$. If H denotes the mean curvature of X^2 and $d\sigma$ denotes the element of area of X^2, it is known that (4.7) is equivalent to the *total absolute mean curvature*

$$K_1^{(1)}(X^2)=\frac{1}{2}\int_{X^2}|H|\,d\sigma. \tag{4.8}$$

b) *Case* $r=2$, $N=1$. Surface $X^2\subset E_3$. The Grassman manifold $G_{1,2}$ is the unit sphere $\sum_2$ and (2.4) can be written

$$K_2^{(1)}(X^2)=\frac{1}{4\pi}\int_{\Sigma_2}\nu_3\,dL_1(0) \tag{4.9}$$

where ν_3 is the number of planes in E_3 which are tangent to X^2 and are orthogonal to the line $L_1(O)$. If K denotes the Gaussian curvature of X^2, since $dL_1(O)$ is the element of area on $\sum_2$, it is easy to see that (4.9) is equivalent, up to a constant factor, to the *total absolute Gaussian curvature* of X^2, that is

$$K_2^{(1)}(X^2)=\frac{1}{2\pi}\int_{X^2}|K|\, d\sigma. \tag{4.10}$$

c) *Case* $r=1$, $N=2$. Surfaces $X^2\subset E_4$. In this case, writting $\sum_3$ = unit 3-dimensional sphere, instead of $G_{3,1}$, we have

$$K_1^{(1)}(X^2)=\frac{1}{2\pi^2}\int_{\Sigma_3}\nu_1\, dL_3(0) \tag{4.11}$$

where ν_1 is the number of tangent lines to X^2 which are orthogonal to the hyperplane $L_3(0)$. Properties of this total absolute curvature it seems to be not known. A geometrical interpretation was given in [10].

d) *Case* $r=N+1$. Surfaces $X^2\subset E_{N+2}$. According to (2.4) we have the following curvature

$$K_{N+1}^{(1)}(X^2)=\frac{1}{O_{N+1}}\int_{\Sigma N}\nu_{N+1}\, dL_1(O) \tag{4.12}$$

where ν_{N+1} is the number of hyperplanes of E_{N+2} which are tangent to X^2 and are orthogonal to the line $L_1(O)$ and $\sum_N$ denotes the N-dimensional unit sphere. Up to a constant factor this curvature coincides with the *curvature of Chern—Lashof* [3]. Since obviously $\nu_{N+1}\geqslant 2$ we have the inequality $K_{N+1}^{(1)}\geqslant 2$, with the equality sign only if X^2 is a convex surface contained in a linear subspace L_3 of E_4.

For $N=2$, X^2 is a surface imbedded in E_4 and the curvature (4.12) is a kind of dual af the curvature (4.11) (see [10].

2. *Total absolute curvatures of order* $q=2$. We have $n=2$, $\rho=5$ and the inequalities (3.1) and (3.2) say that the only possible cases are: a) $r=1$, $N=4$; b) $r=2$, $N=4$; c) $r=1$, $N=5$.

a) *Case* $r=1$, $N=4$. Surface X^2 in E_6. The Grassman manifold $G_{5,1}$ is the unit sphere $\sum_5$ and (2.4) can be written

$$K_1^{(2)}(X^2)=\frac{1}{O_5}\int_{\Sigma_5}\lambda\, dL_5(O) \tag{4.13}$$

where λ is the length of the curve in $L_5(O)$ generated by the intersections of $L_5(O)$ with the lines of E_6 which are orthogonal to $L_5(O)$ and belong to some of the 2-th tangent fibres of X^2.

b) *Case* $r=2$, $N=4$. Surface X^2 in E_6. We have

$$K_2^{(2)}(X^2)=\frac{O_1}{O_4 O_5}\int_{G_{4,2}}\nu_2\, dL_4(O) \tag{4.14}$$

where ν_2 is the number of 2-spaces of E_6 which are orthogonal to $L_4(O)$ and are contained in some 2-th tangent fibre of X^2.

c) Case $r=1$, $N=5$. Surfaces X^2 in E_7.

We have

$$K_1^{(2)}(X^2)=\frac{1}{O_6}\int_{\Sigma_6}\nu_1\,dL_6(O) \tag{4.13}$$

where ν_1 is the number of lines of E_7 which are contained in some 2-th tangent fibre of X^2 and are orthogonal to $L_6(O)$.

The expression of these absolute total curvatures of order 2 by means of differential invariants of X^2 is not known.

Buenos Aires, Argentina Поступило 5.X.1969

Լ. Ա. ՍԱՆՏԱԼՈ. Միջին արժեքներ և կորություն *(ամփոփում)*

Հոդվածի առաջին մասում առաջ է քաշվում (2.4) անհավասարության իրավացիության հիպոթեզը $K \in E_4$ և gK (g-ն E_4-ում իզոմետրիայի ձևափոխություն է) բազմությունների մակերևույթների հատման էյլերյան խարակտերիստիկի միջին արժեքի համար։ Այդ հիպոթեզը ստուգվում է, երբ K-ն զուգահեռանիստ է։

Երկրորդ մասը նվիրված է E_{n+N}-ում խորասուզված n չափանի կոմպակտ դիֆերենցելի բազմաձևության r-րդ q կարգի լրիվ բացարձակ կորության սահմանմանը։

Л. А. САНТАЛО. *Средние значения и кривизны* (резюме)

В первой части статьи выдвигается гипотеза о справедливости неравенства (2.4) для среднего значения эйлеровой характеристики пересечения поверхностей множеств $K \in E_4$ и gK (g—преобразование изомерии в E_4). Эта гипотеза проверяется, когда K есть параллелепипед.

Вторая часть посвящена определению r-той полной абсолютной кривизны порядка q погруженного в E_{n+N} компактного дифференцируемого n-мерного многообразия.

BIBLIOGRAPHY

1. *T. Bonnesen, W. Fenchel,* Theorie der konvexen Körper, Erg. der Mathematik, Berlin, 1934.
2. *S. S. Chern,* On the kinematic Formula in Integral Geometry, J. of Math. and Mechanic, 16, 1966, 101—118.
3. *S. S. Chern, R. K. Lashof,* On the total curvature of immersed manifolds, Am. J. Math. 79, 1957, 306—318.
4. *L. P. Eisenhart,* Riemannian Geometry, Princeton, 19, 1949.
5. *H. Eederer,* Curvature Measures, Trans. Am. Math. Soc. 93, 1959, 418—491.
6. *H. Hadwiger,* Vorlesungen über Inhalt, Oberflache und Isoperimetrie, Springer, Berlin, 1957.
7. *N. Kuiper,* Der Satz von Gauss—Bonnet fur Abbildungen in E^N und damit verwandte Probleme, Jahr. Deutsch. Math. Ver. 69, 1967, 77—88.
8. *L. A. Santaló,* Geometria integral en espacios de curvatura constante, Publicaciones Com. Energia Atómica, Serie Matem. Buenos Aires, 1952.
9. *L. A. Santaló,* Sur la mesure des éspaces linéaires qui coupent un corps convexe et problemes que s'y rattachent, Colloque sur les questions de réalité en Géométrie, Liege, 1955, 177—190.
10. *L. A. Santaló,* Curvaturas absolutas totales de variedades contenidas en un espacio euclidiano, Coloquio de Geometria Diferencial, Santiago de Compostela (Espana), octubre 1967, 29—38.

Istituto Nazionale di Alta Matematica
Symposia Mathematica
Volume XIV (1974)

TOTAL CURVATURES OF COMPACT MANIFOLDS IMMERSED IN EUCLIDEAN SPACE (*)

L. A. SANTALÓ

1. Introduction.

This paper will be concerned with some kind of total absolute curvatures of compact manifolds X^n of dimension n (without boundary) immersed in euclidean space E^{n+N} of dimension $n+N$ ($N \geqslant 1$). Classical Differential Geometry handled almost exclusively with « local » curvatures for such manifolds X^n (assumed sufficiently smooth) and mainly dealed with the case $N=1$. The Gauss-Bonnet theorem, extended by Allendoerfer-Weil-Chern to the case $n > 2$ [1], [7], has been for years the most important, and almost the unique, result of a « global » character. In the classical theory of convex manifolds (boundaries of convex sets) in euclidean space, play an important role the Minkowski's « Quermassintegrale » which may be defined globally without any assumption of differentiability and also, for sufficiently smooth convex manifolds, as integrals of the symmetric functions of the principal curvatures. This classical case shows that, in order to define total curvatures of a given X^n (not necessarily convex) immersed in E^{n+N}, one can either give directly a global definition and then try to express it as the integral of certain local curvatures, or give first a local definition (curvature at a point $x \in X^n$) and then computing the total curvature by integrating this local curvature over X^n. The last method makes necessary some assumptions of smoothness for X^n. A noteworthy example of such curvatures are those introduced by H. Weyl in a classical paper on the volume of tubes [28]. These Weyl's curvatures has been used by Chern to get a general kinematic formula in integral geometry for compact submanifolds of E^{n+N} [10]. For more general subsets of E^{n+N} an analogous formula was given by H. Federer [14] whose « curvature measures » are an extension of the Weyl's curvatures.

(*) I ris ultati conseguiti in questo lavoro sono stati esposti nella conferenza tenuta il 22 maggio 1973.

In 1957-58 two papers of Chern-Lashof [11], [12] call the attention about « absolute » total curvatures, *i.e.* total curvatures obtained by integrating on X^n the absolute values of certain local curvatures. These papers were followed by a series of papers of several authors, mainly N. H. Kuiper, who related this branch of differential geometry with Morse theory of critical points of real valued functions defined over X^n [17], [18]. A survey and new results about this field is to be found in the lecture notes of D. Ferus [16]. See also T. J. Willmore [29].

In the mark of these studies we have introduced in [24], [25], some total curvatures (absolute) for compact manifolds X^n immersed in E^{n+N}. The main purpose of the present paper is to give a local definition of these curvatures, so that they will appear as the integral over X^n of the absolute value of certain differential forms defined in each point $x \in X^n$. These definitions allow to compare the new curvatures with other curvatures previously introduced in the literature. We will then consider some examples, for instance the case of surfaces X^2 immersed in E^4 which presents some remarkable peculiarities.

NOTE: We will consider throughout that X^n is a compact, C^∞-differentiable manifold of dimension n, without boundary, immersed in some euclidean space. In the non-smooth case, a great deal of difficulties arise. For some questions about total curvature of C^1-manifolds, see W. D. Pepe [20].

By E^r_s, $r < s$, we will indicate a r-plane (linear space of dimension r) in the s-dimensional euclidean space E^s. If the euclidean space E^s in which E^r is immersed is apparent form the context, we will write simply E^r instead of E^r_s.

2. Preliminaries.

We will recall the fundamental equations of the differential geometry of a X^n immersed in E^{n+N} and certain know integral-geometric formulae about such manifolds. We use the method of moving frames of Cartan-Chern. See for instance Chern [9] or Willmore [29].

Let $(x; e_1, e_2, \ldots, e_{n+N})$ be a local field of orthonormal frames, such that, restricted to X^n, the vectors $e_1, e_2, \ldots, e_n$ are tangent to X^n and the remaining vectors $e_{n+1}, \ldots, e_{n+N}$ are normal to X^n. The orientation of the unit vectors $e_1, e_2, \ldots, e_{n+N}$ is assumed coherent with that of E^{n+N}. In this section we agree on the following ranges of indices

$$1 \leqslant i, j, k, h, \ldots \leqslant n, \quad n < \alpha, \beta, \gamma, \ldots \leqslant n+N, \quad 1 \leqslant A, B, C, \ldots \leqslant n+N$$

and the summation convention will be used throughout.

The fundamental equations for the moving frames in E^{n+N} are

$$dx = \omega_A e_A, \quad de_A = \omega_{AB} e_B \tag{2.1}$$

where, because $e_A e_B = \delta_{AB}$,

$$\omega_{AB} + \omega_{BA} = 0 \quad \text{and} \quad \omega_A = e_A \cdot dx, \quad \omega_{AB} = e_B \cdot de_A. \tag{2.2}$$

The exterior derivatives satisfy the equations of structure:

$$d\omega_A = \omega_B \wedge \omega_{BA}, \quad d\omega_{AB} = \omega_{AC} \wedge \omega_{CB}. \tag{2.3}$$

The assumption that $e_1, \ldots, e_n$ are tangent to X^n gives

$$\omega_\alpha = 0 \tag{2.4}$$

and the condition that X^n has dimension n insures that the forms ω_i are linearly independent. From (2.3) and (2.4) we deduce $\omega_i \wedge \omega_{i\alpha} = 0$ and therefore, according to the so called lemma of Cartan, we have

$$\omega_{i\alpha} = A_{\alpha,ij}\omega_j, \quad A_{\alpha,ij} = A_{\alpha,ji} \tag{2.5}$$

where $A_{\alpha,ij}$ are the coefficients of the second fundamental form in the normal direction e_α. Notice that we have represented by ω_A, ω_{AB} the forms in (2.3) corresponding to the space of all frames in E^{n+N} as well as the corresponding forms in the bundle of frames such that e_i are tangent vectors and e_α are normal vectors to X^n at x. We think that this simplification in the notation will not cause confusion.

From (2.3), (2.4) and (2.5) we have

$$d\omega_{ij} = \omega_{ih} \wedge \omega_{hj} + \Omega_{ij} \tag{2.6}$$

where

$$\Omega_{ij} = \omega_{i\alpha} \wedge \omega_{\alpha j} = -A_{\alpha,ih} A_{\alpha,jk} \omega_h \wedge \omega_k = \tfrac{1}{2} R_{ijkh} \omega_k \wedge \omega_h \tag{2.7}$$

with

$$R_{ijkh} = A_{\alpha,ik} A_{\alpha,jh} - A_{\alpha,ih} A_{\alpha,jk}. \tag{2.8}$$

We have also

$$d\omega_{\alpha\beta} = \omega_{\alpha\gamma} \wedge \omega_{\gamma\beta} + \Omega_{\alpha\beta} \tag{2.9}$$

where

$$\Omega_{\alpha\beta} = \omega_{\alpha i} \wedge \omega_{i\beta} = -A_{\alpha,ij} A_{\beta,ih} \omega_j \wedge \omega_h = \tfrac{1}{2} R_{\alpha\beta hj} \omega_h \wedge \omega_j \tag{2.10}$$

with

$$R_{\alpha\beta hj} = A_{\alpha,ih} A_{\beta,ij} - A_{\alpha,ij} A_{\beta,ih}\,. \tag{2.11}$$

Note the relations

$$\begin{aligned} &R_{ijkh} = -R_{ijhk} = -R_{jikh}\,, \qquad R_{ijkh} = R_{khij} \\ &R_{ijkh} + R_{ikhj} + R_{ihjk} = 0 \\ &R_{\alpha\beta kj} = -R_{\beta\alpha kj} = -R_{\alpha\beta jk}\,. \end{aligned} \tag{2.12}$$

The expression

$$(A_{\alpha ij}\omega_i\omega_j)\, e_\alpha \tag{2.13}$$

is called the second fundamental form of $X^n \subset E^{n+N}$, and

$$\frac{1}{n}(A_{\alpha ii})\, e_\alpha\,, \tag{2.14}$$

is called the mean curvature vector. X^n is said to be minimal if $A_{\alpha ii} = 0$ for all α.

R_{ijkh} are essentially the components of the Riemann-Christoffel tensor. However, they are not these components. For instance, the Riemannian curvature for the orientation determined by the vectors ξ^i, η^j takes now the form $K(x; \xi^i, \eta^j) = [(R_{ijkh}\xi^i\eta^j\xi^k\eta^h)/(\delta_{ik}\delta_{jh} - \delta_{ih}\delta_{jk})\cdot \cdot \xi^i\eta^j\xi^k\eta^h]$. For $n = 2$, the Gaussian curvature is given by

$$K(x) = R_{1212} \tag{2.16}$$

instead of the classical $K = R_{1212}/g$ when R_{1212} is the component of the Riemann-Christoffel tensor.

3. Densities for linear subspaces and some integral formulae.

We will state some known formulae which will be used in the sequel.

Let E^h_{n+N} denote a h-dimensional linear subspace in E^{n+N}: we will call it, simply, a h-plane. Let $E^h_{n+N}(O)$ denote a h-plane in E^{n+N} through a fixed point O. The set of all oriented $E^h_{n+N}(O)$, with a suitable topology, constitute the Grassman manifold $G_{h,n+N-h}$. We shall represent by $dE^h_{n+N}(O)$ the element of volume in $G_{h,n+N-h}$, which is called the density for oriented h-planes in E^{n+N} through O. The expression of $dE^h_{n+N}(O)$ is well known, but we recall it briefly for completeness (see [22], [23], [10]).

Let $(O; e_1, \ldots, e_2, \ldots, e_{n+N})$ be an orthonormal frame of origin O. In the space of all orthonormal frames of origin O we define the differential forms

$$\omega_{im} = -\omega_{mi} = e_m de_i = -e_i de_m\,, \qquad (i, m = 1, 2, \ldots, n+N)\,. \tag{3.1}$$

Assuming $E^h_{n+N}(O)$ spanned by the unit vectors $e_1, \ldots, e_h$, then

$$dE^h_{n+N}(O) = \Lambda\omega_{im} \tag{3.2}$$

where the right-hand side is the exterior product of the forms ω_{im} over the ranges of indices

$$i = 1, 2, \ldots, h\,, \qquad m = h+1, h+2, \ldots, n+N\,. \tag{3.3}$$

The $(n+N-h)$-plane $E^{n+N-h}_{n+N}(O)$ orthogonal to $E^h_{n+N}(O)$ is spanned by the unit vectors $e_{h+1}, \ldots, e_{n+N}$ and according to (3.2) we have the « duality » (up to the sign)

$$dE^h_{n+N}(O) = dE^{n+N-h}_{n+N}(O)\,. \tag{3.4}$$

Notice that the differential forms $dE^h_{n+N}(O)$ and $dE^{n+N-h}_{n+N}(O)$ are of degree $h(n+N-h)$, which is equal to the dimension of the grassmannian $G_{h,n+N-h}$, as it should be.

The density for sets of h-planes E^h, not through O, in E^{n+N} is given by

$$dE^h = dE^h_{n+N}(O)\wedge\omega_{h+1}\wedge\omega_{h+2}\wedge\ldots\wedge\omega_{n+N} \tag{3.5}$$

where $\omega_{h+1}\wedge\ldots\wedge\omega_{n+N} = (e_{h+1}dx)\wedge(e_{h+2}dx)\wedge\ldots\wedge(e_{n+N}dx)$ is equal to the element of volume in $E^{n+N-h}_{n+N}(O)$ $(=(n+N-h)$-plane spanned by the vectors $e_{h+1}, \ldots, e_{n+N}$ orthogonal to $E^h)$ at the intersection point $E^h \cap E^{n+N-h}(O)$.

The measure of the set of all the oriented $E^h_{n+N}(O)$ (= volume of the Grassman manifold $G_{h,n+N-h}$) may be computed directly from (3.2) or applying the result that it is the quotient space $SO(n+N)/SO(h)\times$ $\times SO(n+N-h)$ (Chern [10]). The result is

$$\int_{G_{h,n+N-h}} dE^h_{n+N}(O) = \frac{O_{n+N-1}O_{n+N-2}\ldots O_{n+N-h}}{O_1 O_2\ldots O_{h-1}} = \frac{O_h O_{h+1}\ldots O_{n+N-1}}{O_1 O_2\ldots O_{n+N-h-1}}\,, \tag{3.6}$$

where O_i is the area of the i-dimensional unit sphere, *i.e.*

$$(3.7)\qquad O_i = \frac{2\pi^{(i+1)/2}}{\Gamma((i+1)/2)}\,.$$

Notice the relation

$$(3.8)\qquad O_1 O_{i-2} = (i-1) O_i\,.$$

For the case $h=1$, the density of oriented lines through O (assuming that $E^1_{n+N}(O)$ is the line spanned by e_1) writes

$$(3.9)\qquad dE^1_{n+N}(O) = \omega_{12}\wedge\omega_{13}\wedge\ldots\wedge\omega_{1n+N} = (e_2 de_1)\wedge(e_3 de_1)\wedge\ldots\wedge(e_{n+N} de_1)\,,$$

which is equal to the element of volume of the $(n+N-1)$ dimensional unit sphere at the end point of e_1. By the duality (3.4) this density (3.9) is equal to the density $dE^{n+N-1}_{n+N}(O)$ of hyperplanes through O (in this case (3.9) corresponds to the hyperplane spanned by $e_2, e_3, \ldots, e_{n+N}$).

Later on we shall need the following formula. Let $E^{n+N}(O)\subset \subset E^{n+N+p}(O)$. Given a line $E^1_{n+N+p}(O)$, let $E^{p+1}(O)$ be the $(p+1)$-plane which contains $E^1_{n+N+p}(O)$ and is perpendicular to $E^{n+N}(O)$ and let $dE^1_{p+1}(O)$ be the density of $E^1_{n+N+p}(O)$ as a line of $E^{p+1}(O)$. If $E^1_{n+N+p}(O)$ denotes the projection of the line $E^1_{n+N+p}(O)$ on $E^{n+N}(O)$ and θ denotes the angle between $E^1_{n+N+p}(O)$ and its projection $E^1_{n+N}(O)$, an easy calculation shows that

$$(3.9)\qquad dE^1_{n+N+p}(O) = \sin^{n+N-1}\theta\, dE^1_{n+N}(O)\wedge dE^1_{p+1}(O)\,.$$

For instance, if $p=1$, we have $\mathrm{d}E^1_2(O) = d\theta$ and (3.9) writes

$$(3.10)\qquad dE^1_{n+N+1}(O) = \sin^{n+N-1}\theta\, dE^1_{n+N}(O)\wedge d\theta\,.$$

Projection formulae. The differential geometry of hypersurfaces $X^n\subset E^{n+1}$ is well known. Calling $R_1, R_2, \ldots, R_n$ the principal radii of curvature of X^n at the point x and putting $d\sigma_n(x) =$ area element of X^n at x (given by $d\sigma_n(x) = \omega_1\wedge\omega_2\wedge\ldots\wedge\omega_n$ according to the notation (2.1)), the total r-th mean curvature of X^n is defined by

$$(3.11)\qquad M_r(X^n) = \frac{1}{\binom{n}{r}}\int_{X^n}\left\{\frac{1}{R_{i_1}}\frac{1}{R_{i_2}}\cdots\frac{1}{R_{i_r}}\right\} d\sigma_n(x)\,,$$

where $\{\}$ denotes the r-th elementary symmetric function of the principal curvatures $1/R_i$ $(i = 1, 2, \ldots, n)$. For $r = 0$, we have $M_0 =$ area of X^n. For $r = n$, if X^n is the boundary ∂D^{n+1} of a domain D^{n+1} of E^{n+1}, it is known that

$$M_n(X^n) = O_n \chi(D^{n+1}) \tag{3.12}$$

where χ denotes the Euler-Poincare characteristic. If n is even, we have $\chi(X^n) = \chi(\partial D^{n+1}) = 2\chi(D^{n+1})$ and (3.12) writes

$$M_n(X^n) = \tfrac{1}{2} O_n \chi(X^n), \qquad n \text{ even}. \tag{3.13}$$

If X^n is a topological sphere, we have $\chi(X^n) = 1 + (-1)^n$.

For closed convex hypersurfaces X_c^n (boundaries of convex bodies of E^{n+1}) we must recall the following «projection formulae» (see [23] and Hadwiger [15]): let X_c^{n-r} be the boundary of the orthogonal projection of X_c^n into $E_{n+1}^{n+N-r}(O)$ and let $\mu_{n-r}(X_c^{n-r})$ denote the measure of X_c^{n-r} (with respect to the euclidean metric in E_{n+1}^{n+1-r}). Then we have

$$\int_{G_{n+1-r,r}} \mu_{n-r}(X_c^{n-r})\, dE_{n+1}^{n+1-r}(O) = \tag{3.14}$$
$$= \frac{O_{n-1} O_{n-2} \ldots O_{n-r}}{O_{r-1} O_{r-2} \ldots O_1} M_r(X_c^n) = \frac{O_{n-1} O_{n-2} \ldots O_r}{O_{n-r-1} \ldots O_1} M_r(X_c^n).$$

For $r = n$ we have $\mu_0(X_c^0) = 2$ and (3.14) coincides with (3.12). For $r = 1$, (3.14) gives the total first mean curvature M_1 as the mean value of the measure of the boundaries of the orthogonal projections of X_c^n on all hyperplanes. For instance, for $n = 2$, the total mean curvature of a convex closed surface X_c^2 in E^3 is given by

$$M_1(X_c^2) = \frac{1}{2\pi} \int_{O_2} u\, dO_2, \tag{3.15}$$

where $dO_2 = dE_3^1(O)$ denotes the element of surface area on the unit sphere and u denotes the length of the boundary of the projection of X_c^2 into a plane perpendicular to the direction defined by dO_2.

For non-convex hypersurfaces, the formulae (3.14) need to be modified: in the right-hand side appear the total «absolute» mean curvatures which we will consider in a next section.

Intersection formulae. Let X^n be a closed hypersurface of E^{n+1}, not necessarily convex (recall that we always assume that X^n is of class C^∞). Let E_{n+1}^r be a moving r-plane in E^{n+1} and consider the manifold $X^{r-1} = X^n \cap E^r_{n+1}$. Call $M_i^{(r)}(X^{r-1})$ the total i-th mean curvature of X^{r-1} as a manifold of dimension $r-1$ in E^r. Then the following formula holds (see [22], [23])

$$\int_{X^n \cap E^r = \emptyset} M_i^{(r)}(X^n \cap E^r)\, dE^r = \frac{O_{n-1} \dots O_{n-r+1}}{O_1 O_2 \dots O_{r-2}} \frac{O_{n-i+1}}{O_{r-i}} M_i(X^n)\,. \tag{3.16}$$

For $i = r-1$, assuming that $X^{r-1} = X^n \cap E^r$ is the boundary of a domain $D^r \subset E^r$, according to (3.12) we have $M_{r-1}^{(r)}(X^{r-1}) = O_{r-1}\chi(D^r)$ and (3.16) gives

$$\int_{X^n \cap E^r \neq \emptyset} \chi(D^r)\, dE^r = \frac{O_{n-1} \dots O_{n-r+1}}{O_1 \dots O_{r-1}} \frac{O_{n-r+2}}{O_1} M_{r-1}(X^n)\,. \tag{3.17}$$

In particular, if X^n is a closed convex hypersurface X^n_c we have $\chi(D^n) = 1$ and (3.17) gives the total measure of all r-planes which intersect X^n_c,

$$\int_{X^n_c \cap E^r \neq \emptyset} dE^r = \frac{O_{n-1} \dots O_{n-r+1}}{O_1 \dots O_{r-1}} \frac{O_{n-r+2}}{O_1} M_{r-1}(X^n_c) = \tag{3.18}$$

$$= \frac{O_{n-1} \dots O_{n-r}}{(n-r+1)\, O_{r-1} \dots O_1} M_{r-1}(X^n_c)\,.$$

If r is odd we have $\chi(D^r) = (\frac{1}{2})\chi(X^r)$ and (3.17) may be written

$$\int_{X^n \cap E^r \neq \emptyset} \chi(X^n \cap E^r)\, dE^r = \frac{2O_{n-1} \dots O_{n-r+1}}{O_1 \dots O_{r-1}} \frac{O_{n-r+2}}{O_1} M_{r-1}(X^n) \quad (r \text{ odd}). \tag{3.19}$$

In order to illustrate the foregoing ideas we will give a typical application. Let $n = 3$, $r = 3$. Then X^3 is a closed hypersurface in E^4; assume that it bounds a domain $D^4 \subset E^4$. According to (3.18) and (3.19) the mean value of $\chi(X^3 \cap E^3)$ is $E(\chi(X^3 \cap E^3)) = 2M_2(X^3)/M_2^*(X^3)$ where M_2^* denotes the 2-th total mean curvature of the convex hull of X^3. If V^* is the volume of the domain bounded by the convex hull, it is known that $M_2^* \geqslant (32\pi^6 V^*)^{\frac{1}{4}} \geqslant (32\pi^6 V)^{\frac{1}{4}}$ (Hadwiger [15]), where V is the volume of D^4. Thus we have

$$E(\chi(X^3 \cap E^3)) \leqslant \frac{M_2(X^3)}{\pi(2\pi^2 V)^{\frac{1}{4}}}\,.$$

The equality sign holds for euclidean spheres.

b) Another known and useful integral formula is the following:

Let X^n be a compact manifold in E^{n+N}. Let $\mu_{r-N}(X^n \cap E^r)$ denote the $(r-N)$-dimensional measure of $X^n \cap E^r$ $(r \geqslant N)$ according to the euclidean metric on E^r. For $r = N$, μ_0 denotes the number of points of the set $X^n \cap E^r$. Then we have

$$\int\limits_{X^n \cap E^r \neq \emptyset} \mu_{r-N}(X^n \cap E^r)\, dE^r = \frac{O_{n+N} O_{n+N-1} \dots O_{n+N-r} O_{r-N}}{O_1 O_2 \dots O_r O_n} \sigma_n(X^n)\,, \tag{3.20}$$

where $\sigma_n(X^n) =$ volume of X^n.

This integral formula holds good for any space of constant curvature, in particular on the $(n+N)$-dimensional sphere, with a suitable definition of dE^r (see [22]).

In all the preceding formulae and in those which will follow, the linear spaces E^r are assumed « oriented ». Otherwise the right-hand side of (3.16), ..., (3.20) should be divided by a factor 2.

4. The total absolute curvatures $K^*_{r,N}(X^n)$.

Let X^n be a compact n-dimensional manifold (without boundary) of class C^∞ in E^{n+N}. To each point $x \in X^n$ we attach the frame $(x; e_1, e_2, \dots, e_{n+N})$ considered in section 2, such that the vectors $e_1, e_2, \dots, e_n$ are tangent to X^n and spann the tangent n-plane $T(x)$. The remaining vectors $e_{n+1}, \dots, e_{n+N}$ are normal to X^n and spann the normal N-plane $N(x)$.

Assuming

$$1 \leqslant r \leqslant n + N - 1 \tag{4.1}$$

we define the r-th total absolute curvature of X^n as follows (see [24], [25]):

a) Case $1 \leqslant r \leqslant n$. Let O be a fixed point in E^{n+N} and consider a $(n+N-r)$-plane say $E^{n+N-r}(O)$ through O. Let Γ_r be the set of all r-planes E^r in E^{n+N} which are contained in some $T(x)$ $(x \in X^n)$, pass through x, and are perpendicular to $E^{n+N-r}(O)$. The intersection $\Gamma_r \cap E^{n+N-r}(O)$ is a compact variety in $E^{n+N-r}(O)$ of dimension

$$\delta_1 = n - rN\,, \tag{4.2}$$

Let $\mu_{n-rN}(\Gamma_r \cap E^{n+N-r}(O))$ denote the measure of this variety as subvariety of $E^{n+N-r}(O)$; if $\delta_1 = 0$, then μ_0 means the number of intersection points of Γ_r and $E^{n+N-r}(O)$.

We define the r-th total absolute curvature of X^n immersed in E^{n+N}, as the mean value of the measures μ_{n-rN} over all $E^{n+N-r}(O)$, that is, according to (3.6),

$$K^*_{r,N}(X^n) = \frac{O_1 \dots O_{n+N-r-1}}{O_r \dots O_{n+N-1}} \int_{G_{n+N-r,r}} \mu_{n-rN}\big(\Gamma_r \cap E^{n+N-r}(O)\big)\, dE^{n+N-r}(O). \tag{4.3}$$

The coefficient on the right-hand side may be replaced by

$$\frac{O_1 O_2 \dots O_{r-1}}{O_{n+N-r} \dots O_{n+N-1}}.$$

b) Case $n \leqslant r \leqslant n+N-1$. In this case, instead of the set of E^r which « are contained » in some $T(x)$, we consider the set of all $E^r \subset E^{n+N}$ which « contain » some $T(x)$ and are perpendicular to $E^{n+N-r}(O)$. As before, we represent this set by Γ_r. The dimension of the variety $\Gamma_r \cap E^{n+N-r}(O)$ is now

$$\delta_2 = n(r+1-n-N) \tag{4.4}$$

and the r-th total absolute curvature of X^n is defined by the same mean value (4.3) which now writes

$$K^*_{r,N}(X^n) = \tag{4.5}$$
$$= \frac{O_1 \dots O_{n+N-r-1}}{O_r \dots O_{n+N-1}} \int_{G_{n+N-r,r}} \mu_{n(r+1-n-N)}\big(\Gamma_r \cap E^{n+N-r}(O)\big)\, dE^{n+N-r}(O).$$

The dimensions δ_1, δ_2, given by (4.2), (4.4) have been calculated elsewhere ([24], [25]). From their values, and since $r \leqslant n+N-1$, we deduce

i) The curvatures $K^*_{r,N}$ are only defined for

$$n \geqslant rN \quad \text{and} \quad r = n+N-1. \tag{4.6}$$

ii) If $r \leqslant n$ and X^n is immersed in $E^{n+N'}$ with $N' < N$, then $K^*_{r,N}(X^n) = 0$. This result follows from the fact that, if $X^n \subset E^{n+N'}$, all tangent spaces $T(x)$ are also contained in $E^{n+N'}$ and therefore μ_{n-rN} in (4.3) is zero except for the spaces $E^{n+N-r}(O)$ which are perpendicular to $E^{n+N'}$, which form a set of measure zero.

iii) The most interesting cases correspond to $n = rN$ and $r = n + N - 1$, for which the measures μ under the integral signs in (4.3) and (4.5) are non negative integers and therefore the total absolute curvatures are invariant at least under similitudes. We will consider these cases separately in the following sections.

iv) Consider the case $n = N$, $r = 1$. This case has the following geometrical interpretation. Let S^{2n-1} denote the unit $(2n-1)$-dimensional sphere in E^{2n} of center O. Let $E^n(x, O)$ be the n-plane through O parallel to the tangent space $T(x)$. The intersection $S^{2n-1} \cap E^n(x, O)$ is a $(n-1)$-dimensional great circle of S^{2n-1}. If we assume identified the pairs of antipodal points on S^{2n-1} we have the $(2n-1)$-dimensional elliptic space P^{2n-1} and the intersections $S^{2n-1} \cap \cap E^n(x, O)$ define a n-parameter family of $(n-1)$-planes in P^{2n-1}, say C_{n-1}. Let $\nu_{n-1}(y)$ be the number of $(n-1)$-planes of C_{n-1} which contain the point $y \in P^{2n-1}$ and let $\nu_{2n-1}(\eta)$ be the number of $(n-1)$-planes of C_{n-1} which are contained in the hyperplane η in P^{2n-1}. Let $d\sigma_{2n-1}(y)$ denote the volume element in P^{2n-1} at y, and let $dE^{2n-1}(\eta)$ denote the density of the hyperplanes of P^{2n-1} at η. Then, the curvatures (4.3) and (4.5) are clearly equal to

$$K^*_{1,n}(X^n) = \frac{2}{O_{2n-1}} \int_{P^{2n-1}} \nu_{n-1}(y)\, d\sigma_{2n-1}(y)\,, \tag{4.7}$$

$$K^*_{2n-1,n}(X^n) = \frac{2}{O_{2n-1}} \int_{E^{2n-1} \subset P^{2n-1}} \nu_{2n-1}(\eta)\, dE^{2n-1}(\eta)\,. \tag{4.8}$$

For $n = 2$, $N = 2$, $r = 1$ we have a congruence of lines C_1 in P^3 and, in a certain sense, the foregoing curvatures are the mean « order » and the mean « class » of the congruence C_1. This relation between compact surfaces of E^4 and congruences of lines in the elliptic space P^3 seems to deserve further attention.

5. A reproductive formula.

Let $X^n \subset E^{n+N}$. Consider the intersection $X^{s-N} = X^n \cap E^s$, $N < s < n + N$, and assume that X^{s-N} is a compact differentiable manifold of dimension $s - N$. Let $K^{*(s)}_{r,N}(X^{s-N})$, $r \leqslant s - N$, denote the total absolute curvature of X^{s-N} as a manifold immersed in E^s. We wish to prove the following « reproductive formula »

$$\int_{E^s \subset E^{n+N}} K^{*(s)}_{r,N}(X^{s-N})\, dE^s = \frac{O_{n+N-s} \cdots O_{n+N-1}\, O_{n+N-r}\, O_{s-N-rN}}{O_1 \cdots O_{s-1}\, O_{s-r}\, O_{n-rN}}\, K^*_{r,N}(X^n)\,, \tag{5.1}$$

where

$$n \geqslant s \geqslant (r+1)N\,. \tag{5.2}$$

Consider first the orthogonal linear spaces $E^s(O)$, $E^{n+N-r}(O)$ through a fixed point O and the intersection $E^{s-r}(O) = E^s(O) \cap E^{n+N-r}(O)$, $s > r$. Let $(O; e_1, e_2, \ldots, e_{n+N})$ be an orthonormal frame and suppose that $E^s(O)$ is spanned by the unit vectors $\{e_1, \ldots, e_s\}$, $E^{s-r}(O)$ is spanned by $\{e_{r+1}, \ldots, e_s\}$ and $E^{n+N-r}(O)$ is spanned by $\{e_{r+1}, \ldots, e_s, e_{s+1}, \ldots, e_{n+N}\}$. The density of $E^s(O)$ in E^{n+N} is

$$\begin{aligned} dE^s_{n+N}(0) = {} & (\omega_{1,s+1} \wedge \omega_{1,s+2} \wedge \ldots \wedge \omega_{1,n+N}) \\ & \wedge (\omega_{2,s+1} \wedge \omega_{2,s+2} \wedge \ldots \wedge \omega_{2,n+N}) \\ & \ldots\ldots\ldots\ldots\ldots \\ & \wedge (\omega_{r,s+1} \wedge \omega_{r,s+2} \wedge \ldots \wedge \omega_{r,n+N}) \\ & \ldots\ldots\ldots\ldots\ldots \\ & \wedge (\omega_{s,s+1} \wedge \omega_{s,s+2} \wedge \ldots \wedge \omega_{s,n+N})\,. \end{aligned} \tag{5.3}$$

The density of $E^{s-r}(O)$ as subspaces of $E^s(O)$ is

$$\begin{aligned} dE^{s-r}_s(O) = {} & (\omega_{r+1,1} \wedge \omega_{r+1,2} \wedge \ldots \wedge \omega_{r+1,r}) \\ & \wedge (\omega_{r+2,1} \wedge \omega_{r+2,2} \wedge \ldots \wedge \omega_{r+2,r}) \\ & \ldots\ldots\ldots\ldots\ldots \\ & \wedge (\omega_{s,1} \wedge \omega_{s,2} \wedge \ldots \wedge \omega_{s,r}) \end{aligned} \tag{5.4}$$

and as subspace of E^{n+N-r}

$$\begin{aligned} dE^{s-r}_{n+N-r}(O) = {} & (\omega_{r+1,s+1} \wedge \ldots \wedge \omega_{r+1,n+N}) \\ & \wedge (\omega_{r+2,s+1} \wedge \ldots \wedge \omega_{r+2,n+N}) \\ & \ldots\ldots\ldots\ldots\ldots \\ & \wedge (\omega_{s,s+1} \wedge \ldots \wedge \omega_{s,n+N})\,. \end{aligned} \tag{5.5}$$

Finally, the density of $E^{n+N-r}(O)$ in E^{n+N} is

$$\begin{aligned} dE^{n+N-r}_{n+N}(O) = {} & (\omega_{r+1,1} \wedge \omega_{r+1,2} \wedge \ldots \wedge \omega_{r+1,r}) \\ & \wedge (\omega_{r+2,1} \wedge \omega_{r+2,2} \wedge \ldots \wedge \omega_{r+2,r}) \\ & \ldots\ldots\ldots\ldots\ldots \\ & \wedge (\omega_{n+N,1} \wedge \omega_{n+N,2} \wedge \ldots \wedge \omega_{n+N,r})\,. \end{aligned} \tag{5.6}$$

Since we are only interested in the absolute value of the densities, we make no question on the order in the exterior products.

From (5.3) to (5.6) we deduce the identity

$$dE_s^{s-r}(O)\wedge dE_{n+N}^s(O) = dE_{n+N-r}^{s-r}(O)\wedge dE_{n+N}^{n+N-r}(O)\,. \tag{5.7}$$

According to the definition (4.3) we have

$$K_{r,N}^{*(s)}(X^{s-N}) = \frac{O_1\ldots O_{s-r-1}}{O_r\ldots O_{s-1}}\int_{G_{s-r,r}} \mu_{s-N-rN}\,dE_s^{s-r}(O)\,, \tag{5.8}$$

where μ_{s-N-rN} denotes the measure of the $(s-N-rN)$-dimensional variety in $E_s^{s-r}(O)$ generated by the intersection points of $E_s^{s-r}(O)$ with the r-planes in E^s which are perpendicular to $E_s^{s-r}(O)$ and are contained in some tangent space of X^{s-N}. From (3.5) and (5.7) we have

$$\int_{E^s\cap X^n\neq\emptyset} K_{r,N}^{*(s)}(X^{s-N})\,dE_{n+N}^s = \tag{5.9}$$

$$= \frac{O_1\ldots O_{s-r-1}}{O_r\ldots O_{s-1}}\int \mu_{s-N-rN}\,dE_{n+N-r}^{s-r}(O)\wedge dE_{n+N}^{n+N-r}(O)\wedge\omega_{s+1}\wedge\ldots\wedge\omega_{n+N}\,.$$

The form $\omega_{s+1}\wedge\ldots\wedge\omega_{n+N}$ is equal to the element of volume in E^{n+N} orthogonal to E^s, which is also equal to the element of volume in E^{n+N-r} orthogonal to E^{s-r} and therefore we have

$$dE_{n+N-r}^{s-r}(O)\wedge\omega_{s+1}\wedge\ldots\wedge\omega_{n+N} = dE_{n+N-r}^{s-r} \tag{5.10}$$

(= density of $(s-r)$-planes, not necessarily through O, in $E^{n+N-r}(O)$) and (5.9) gives

$$\int_{E_{n+N}^s\cap X^n\neq\emptyset} K_{r,N}^{*(s)}(X^{s-n})\,dE_{n+N}^s = \tag{5.11}$$

$$= \frac{O_1\ldots O_{s-r-1}}{O_r\ldots O_{s-1}}\int \mu_{s-N-rN}\,dE_{n+N-r}^{s-r}\wedge dE_{n+N}^{n+N-r}(O)\,.$$

Applying (3.20) to the $(n-rN)$-dimensional variety Y^{n-rN} in $E_{n+N}^{n+N-r}(O)$ generated by the intersection points of $E_{n+N}^{n+N-r}(O)$ with the linear r-spaces of E^{n+N} which are perpendicular to $E_{n+N}^{n+N-r}(O)$ and are contained in some tangent space of X^n and to the $(s-r)$-planes of

$E_{n+N}^{n+N-r}(O)$ which intersect Y^{n-rN} we have

$$(5.12)\qquad \int \mu_{s-N-rN}\, dE_{n+N-r}^{s-r} = \frac{O_{n+N-r}\dots O_{n+N-s}\, O_{s-N-rN}}{O_1\dots O_{s-r}\, O_{n-rN}}\, \mu_{n-rN}\,,$$

where μ_{n-rN} denotes the measure of Y^{n-rN}.

Thus (5.11) writes

$$\int_{E_{n+N}^s \cap X^n \neq \emptyset} K_{r,N}^{*(s)}(X^{s-N})\, dE_{n+N}^s =$$

$$= \frac{O_1\dots O_{s-r-1}\, O_{n+N-r}\dots O_{n+N-s}\, O_{s-N-rN}}{O_r\dots O_{s-1}\, O_1\dots O_{s-r}\, O_{n-rN}} \int_{G_{n+N-r,r}} \mu_{n-rN}\, dE_{n+N}^{n+N-r}(O)\,.$$

This formula and the definition (4.3), give the desired formula (5.1).

6. The case $K^*_{n+N-1,N}(X^n)$: curvature of Chern-Lashof.

The case $r = n + N - 1$ gives rise to the curvature defined by Chern and Lashof [11], [12]. The identity of both curvatures will be apparent from the analytical expression of K^*_{n+N-1}, which will be given in a subsequent section. For the moment, we wish to show how the geometrical definition above allows to obtain directly some known properties of the Chern-Lashof curvature.

a) Notice that $\mu_0(\Gamma_{n+N-1} \cap E^{n+N-1}(O))$ in (4.5) is equal to the number ν of hyperplanes E^{n+N-1} which are perpendicular to a given line $E^1(O)$ and contain some tangent space $T(x)$ of X^n. This number is surely $\geqslant 2$, since there are at least the two support hyperplanes of X^n which are perpendicular to $E^1(O)$. Therefore we have $K^*_{n+N-1,N}(X^n) \geqslant 2$ (theorem 1 of Chern-Lashof [11]).

For an oriented surface X^2 (compact) the number of hyperplanes of support which are perpendicular to a direction $E^1(O)$ is $\geqslant 2(1+g)$, where g is the genus of X^2, related to the Euler characteristic by $\chi(X^2) = 2(1-g)$. Thus we have

$$(6.1)\qquad K^*_{N+1,N}(X^2) \geqslant 2(1+g) = 4 - \chi(X^2)\,.$$

b) The inequality $K_{n+N-1,N}(X^n) < 3$, means that there exists a set of directions $E^1(O)$ (with positive measure) such that the number of hyperplanes in E^{n+N} which contain some $T(x)$ and are perpendicular to $E^1(O)$ is exactly 2, a condition which suffices for X^n to be homeomorphic to a n-dimensional sphere (theorem 2 of Chern-Lashof [11]).

c) Assume that $X^n \subset E^{n+N}(O) \subset E^{n+N+1}(O)$. To each hyperplane E^{n+N} in $E^{n+N+1}(O)$ which is perpendicular to the line $E^1_{n+N+1}(O)$ and contains some $T(x)$ corresponds the hyperplane $E^{n+N-1} = E^{n+N} \cap E^{n+N}(O)$ of E^{n+N} which is perpendicular to the projection $E^1_{n+N}(O)$ of $E^1_{n+N+1}(O)$ into E^{n+N}. According to (3.10) we have

$$K^*_{n+N,N+1}(X^n) = \frac{1}{O_{n+N}} \int\limits_{G_{1,n+N}} \nu \, dE^1_{n+N+1}(O) =$$

$$= \frac{1}{O_{n+N}} \int\limits_{G_{1,n+N-1}} \nu \sin^{n+N-1}\theta \, dE^1_{n+N}(O) \wedge d\theta = \frac{1}{O_{n+N-1}} \int\limits_{G_{1,n+N-1}} \nu \, dE^1_{n+N}(O) =$$

$$= K^*_{n+N-1,N}(X^n) .$$

By induction on N, we get that the total absolute curvature $K^*_{n+N-1,N}(X^n)$ of $X^n \subset E^{n+N}$ does not change if we consider X^n as an immersed manifold in $E^{n+N+p} \supset E^{n+N}$ (Lemma 1 of Chern-Lashof [12]).

7. The case $n = rN$. Local representation of the curvatures $K^*_{r,N}(X^n)$.

Let x be a point of the manifold X^n immersed in E^{n+N} and consider the frame $(x; e_1, e_2, \ldots, e_{n+N})$ of Sect. 2. The density for r-planes through x is given by (3.2) which we will now write

$$dE^r_{n+N}(x) = \Lambda\omega_{im} \quad (i = 1, 2, \ldots, r;\ m = r+1, r+2, \ldots, n+N) \tag{7.1}$$

where $r \leqslant n$. The density for r-planes $E^r_n(x)$ in the tangent space $T(x)$ spanned by $e_1, e_2, \ldots, e_n$ is

$$dE^r_n(x) = \Lambda\omega_{im} \quad (i = 1, 2, \ldots, r;\ m = r+1, \ldots, n) . \tag{7.2}$$

The densities (7.1), (7.2) refers to the r-space spanned by $(e_1, e_2, \ldots, e_r)$. It is important to note that if $r = n$, $N = 1$, the density (7.2) is not defined. Since we have in this case only one $E^n_n (= T(x))$ its average is the same space, so that in this case we must cancel dE^n_n (and the corresponding integrations) in all the formulae in which it appears. On the other hand, this case corresponds to the well known case of hypersurfaces X^n in E^{n+1} and the curvature here defined is the absolute value of the classical Gauss-Kronecker curvature.

The element of volume of X^n at x is

$$d\sigma_n(x) = \omega_1 \wedge \omega_2 \wedge \ldots \wedge \omega_n . \tag{7.3}$$

Assuming $n = rN$, the differential forms $dE^r_{n+N}(x)$ and $dE^r_n(x) \wedge \wedge d\sigma_n(x)$ have the same degree, so that we can define a function $G(x, E^r_n(x))$ by the equation (as noted in section 2, the differential forms in this equality must be considered as forms in the bundle of frames trangent to X^n; for details, see [11] or [29])

$$(7.4) \qquad dE^r_{n+N}(x) = G(x, E^r_n(x))\, dE^r_n(x) \wedge d\sigma_n(x)\,.$$

Calling $\nu = \nu(E^r_{n+N})$ the number of r-planes E^r_{n+N} which are parallel to $E^r_{n+N}(x)$ and belong to some tangent space $T(x)$ of X^n, (7.4) gives

$$(7.5) \qquad \int_{G_{r,n+N-r}} \nu\, dE^r_{n+N}(x) = \int_{X^n} \Big(\int_{G_{r,n-r}} |G(x, E^r_n(x))|\, dE^r_n(x) \Big) \wedge d\sigma_n(x)\,.$$

Thus, setting

$$(7.6) \qquad K^*_{r,N}(X^n) = \int_{X^n} Q^*_{r,N}(x)\, d\sigma_n$$

according to (4.3) and (7.5), (having into account (3.4)), we have

$$(7.7) \qquad Q^*_{r,N}(x) = \frac{O_1 \ldots O_{n+N-r-1}}{O_r \ldots O_{n+N-1}} \int_{G_{r,n-r}} |G(x, E^r_n(x))|\, dE^r_n(x)\,.$$

From (7.4) and (7.1), (7.2), (7.3) we can obtain the expression for the «local» sectional curvature $G(x, E^r_n)$ corresponding to the point x and the section $E^r_n(x)$ (spanned by the unit vectors $e_1, e_2, \ldots, e_r$). We get

$$(7.8) \qquad G(x, E^r_n)\, d\sigma_n = \Lambda \omega_{im}$$

$$(i = 1, 2, \ldots, r;\ m = n+1, n+2, \ldots, n+N)\,.$$

Using (2.5) we get

$$(7.9) \qquad G(x, E^r_n(x)) = \begin{vmatrix} A_{n+1,11} & A_{n+1,12} & \ldots & A_{n+1,1n} \\ \cdots & \cdots & \cdots & \cdots \\ A_{n+N,11} & A_{n+N,12} & \ldots & A_{n+N,1n} \\ A_{n+1,21} & A_{n+1,22} & \ldots & A_{n+1,2n} \\ \cdots & \cdots & \cdots & \cdots \\ A_{n+N,21} & A_{n+N,22} & \ldots & A_{n+N,2n} \\ \cdots & \cdots & \cdots & \cdots \\ \cdots & \cdots & \cdots & \cdots \\ A_{n+1,r1} & A_{n+1,r2} & \ldots & A_{n+1,rn} \\ \cdots & \cdots & \cdots & \cdots \\ A_{n+N,r1} & A_{n+N,r2} & \ldots & A_{n+Nrn} \end{vmatrix}$$

the determinant being of order n because $n = rN$.

This formula corresponds to the r-plane spanned by $e_1, e_2, \ldots, e_r$. For a general r-plane in $T(x)$ spanned by the set of orthogonal vectors $e_1', e_2', \ldots, e_r'$ of the frame $e_i' = \gamma_{ih} e_h$ $(i, h = 1, 2, \ldots, n)$ defined by the orthogonal matrix (γ_{ih}), the elements $A_{\alpha,ij}$ in (7.9) must be substituted by $A'_{\alpha,ij} = \gamma_{ih}\gamma_{jm}A_{\alpha,hm}$ (h, m summed over the ranges $h, m = 1, 2, \ldots, n$) as it follows easily from (2.5) and (2.2).

In order to evaluate $Q^*_{r,N}(x)$ we must compute the mean value of $|G(x, E^r_n(x))|$ over all $E^r_n(x)$ (*i.e.* over the Grassmann manifold $G_{r,n-r}$). Actual evaluation of this mean value seems to be difficult. We will only consider some particular cases in the following sections. As follows either from the geometrical definition or from (7.8), if $r = n$, $N = 1$, $G(x, E^n_n(x)) = G$ is the classical Gauss-Kroneker curvature of X^n at the point x, and we have

$$Q^*_{n,N} = \frac{1}{O_n} |G| . \tag{7.10}$$

8. Local representation of $K^*_{n+N-1,N}(X^n)$.

The hyperplanes in E^{n+N} which contain some tangent space $T(x)$ of X^n, may be determined by its normal vector $E^1_N(x)$ in the normal space to X^n at x, *i.e.* in the N-space spanned by the vectors e_{n+1}, $e_{n+2}, \ldots, e_{n+N}$. Then, instead of the equation (7.4) we consider

$$dE^1_{n+N}(x) = \overline{G}(x, E^1_N(x))\, dE^1_N(x) \wedge d\sigma](x) \tag{8.1}$$

and $K^*_{n+N-1}(X^n)$ may be written

$$K^*_{n+N-1,N}(X^n) = \int_{X^n} Q^*_{n+N-1,N}(x)\, d\sigma_n(x) \tag{8.2}$$

where

$$Q^*_{n+N-1,N}(x) = \frac{1}{O_{n+N-1}} \int_{G_{1,N-1}} |\overline{G}(x, E^1_N(x))|\, dE^1_N(x) . \tag{8.3}$$

(8.1), (8.2) and (8.3) show that the absolute total curvature $K^*_{n+N-1,N}(X^n)$ coincides with the Chern-Lashof curvature [11], [12] as stated in section 6.

Taking $E^1_N(x)$ to be the line of the unit vector e_{n+N} and writting $\overline{G}(x, e_{n+N})$ instead of $\overline{G}(x, E^1_N(x))$, from (8.1) we deduce

$$\overline{G}(x, e_{n+N})\, d\sigma_n(x) = \omega_{n+N,1} \wedge \omega_{n+N,2} \wedge \cdots \wedge \omega_{n+N,n} \tag{8.4}$$

or, by virtue of (2.5),

$$\bar{G}(x, e_{n+N}) = (-1)^n \det (A_{n+N.ij}) \tag{8.5}$$

with $i, j = 1, 2, \ldots, n$.

If, instead of e_{n+N} we consider the general normal vector $e = \cos\theta_s e_{n+s}$ $(s = 1, 2, \ldots, N)$, we get

$$\bar{G}(x, e) = (-1)^n \det (\cos\theta_s A_{n+s.ij}) \tag{8.6}$$

and to get $Q^*_{n+N-1,N}(x)$ (= absolute curvature at x = Chern-Lashof curvature at x) we must evaluate the mean value of $|\bar{G}(x, e)|$ over the $(N-1)$-dimensional unit sphere (*i.e.* over $\cos^2\theta_1 + \cos^2\theta_2 + \ldots + \cos^2\theta_N = 1$). Only in some simple cases, this calculation has been carried out.

9. Total (no absolute) curvatures $K_{n+N-1,N}(X^n)$.

The total absolute curvatures $K^*_{r,N}(X^n)$ are easily defined geometrically by (4.3) or (4.5), but their actual evaluation seems to be difficult, mainly due to the absolute values under the integral sign in (7.7) and (8.3). From the analytical point of view, it is much more natural to consider the curvatures « defined » by the same formulae (7.7), (8.3) and then (7.6) and (8.2) without the absolute value under the integral sign. We shall denote these no absolute curvatures by $Q_{r,N}(x)$ and $K_{r,N}(X^n)$ (or $Q_{n+N-1,N}(x)$ and $K_{n+N-1,N}(X^n)$) respectively. One can handle analytically with these curvatures more easily than with the absolute curvatures, but for a geometrical interpretation like (4.3) or (4.5) it is necessary to provide an orientation (or a sign) to the manifolds $\Gamma_r \cap E^{n+N-1}(O)$ and some difficulties arise.

We will first consider the case $K_{n+N-1,N}(X^n)$. We define

$$Q_{n+N-1,N}(x) = \frac{1}{O_{n+N-1}} \int_{G_{1,N-1}} \bar{G}(x, E^1_N(x))\, dE^1_N(x) \tag{9.1}$$

where $\bar{G}(x, E^1_N(x))$ is defined by (8.5), (8.4) if $E^1_N(x)$ is the line spanned by the vector e_{n+N} or by (8.6) if $E^1_N(x)$ is the line spanned by the vector e. From (9.1) we define

$$K_{n+N-1,N}(X^n) = \int_{X^n} Q_{n+N-1,N}(x)\, d\sigma_n(x) . \tag{9.2}$$

To calculate the mean value (9.1) we consider the unit vector e on the line $E_N^1(x)$, say $e = \cos\theta_s e_{n+s}$ ($s = 1, 2, \ldots, N$; $\cos^2\theta_1 + \cos^2\theta_2 + \ldots + \cos^2\theta_N = 1$). We have

$$(9.3) \qquad \bar{G}(x, e)\, d\sigma_n(x) = (-1)^n (e\, de_1)\wedge(e\, de_2)\wedge\ldots\wedge(e\, de_n)$$
$$= \Lambda(\cos\theta_1\omega_{n+1.i} + \cos\theta_2\omega_{n+2.i} + \ldots + \cos\theta_N\omega_{n+N.i})$$

where in the exterior product on the right-hand side we have $i = 1, 2, \ldots, n$.

The forms $\omega_{n+s.i}$ do not depend on θ_s. Thus, in order to compute (9.1) we must calculate the mean value of monomials $\cos^{\lambda_1}\theta_1 \cos^{\lambda_2}\theta_2 \ldots \ldots \cos^{\lambda_N}\theta_N$ with $\lambda_1 + \lambda_2 + \ldots + \lambda_N = n$ over the N-sphere $\cos^2\theta_1 + \cos^2\theta + \ldots + \cos^2\theta_N = 1$. These mean values are known: they are zero unless all exponents λ_i are even, and in the later case their values are

$$(9.4) \qquad E(\cos^{\lambda_1}\theta_1 \ldots \cos^{\lambda_N}\theta_N) = \frac{\lambda_1)\,\lambda_2)\ldots\lambda_N)}{N(N+2)\ldots(N+n-2)}$$

where λ_i even, $\lambda_1 + \ldots + \lambda_N = n$ and $\lambda) = 1.3 \ldots (\lambda - 1)$. From these mean values, expanding the exterior product (9.3) and using (2.5) and (2.8), by some invariant-theoretic arguments dues to H. Weyl [28], one can deduce the following explicit form of the curvature $Q_{n+N-1,N}(x)$ (n even)

$$(9.5) \qquad Q_{n+N-1,N}(x) = \frac{1}{2^n(2\pi)^{n/2}(n/2)!}\,\delta^{i_1i_2\ldots i_n}_{j_1j_2\ldots j_n} R_{i_1i_2j_1j_2} R_{i_3i_4j_3j_4} \ldots R_{i_{n-1}i_nj_{n-1}j_n}$$

where $\delta^{i_1i_2\ldots i_n}_{j_1j_2\ldots j_n}$ is equal to $+1$ or -1 according as $(i_1 i_2 \ldots i_n)$ is an even or odd permutation of $(j_1 j_2 \ldots j_n)$ and is otherwise zero and the summation is over all $i_1, i_2, \ldots, i_n$ and $j_1, j_2, \ldots, j_n$ independently from 1 to n. If n is odd, $Q_{n+N-1,N}(x) = 0$. Notice that $Q_{n+N-1,N}$ does not depend upon N.

This curvature (9.5) is called the curvature of Lipschitz-Killing (Chern-Lashof [11], Thorpe [27]). It appears in the work of H. Weyl on the volume of tubes [28] and in several papers of Chern ([7], [9], [10]) and others. The total curvature $K_{n+N-1,N}(X^n)$ (n even) gives the Euler-Poincaré characteristic of X^n, according to the formula of Gauss-Bonnet:

$$(9.6) \qquad K_{n+N-1,N}(X^n) = \chi(X^n)\,.$$

The case $n=2$. For surfaces $X^2 \subset E^{2+N}$, we have $Q_{N+1,N} = (1/2\pi)R_{1212} = K/2\pi$, where K is the Gaussian curvature (2.16). The expression of $\bar{G}(x, e)$ (9.3) is a quadratic form in the variables $\cos\theta_i$. Under the hypothesis that this quadratic form is everywhere positive or negative definite, we have

$$Q^*_{N+1,N} = Q_{N+1,N} = K/2\pi \qquad \text{if } K>0$$

$$Q^*_{N+1,N} = -Q_{N+1,N} = -K/2\pi \qquad \text{if } K<0\,.$$

Hence we have

$$K^*_{N+1,N} = (1/2\pi)\Big(\int_U K\,d\sigma - \int_V K\,d\sigma\Big)$$

where $U = \{x \in X^2;\ K(x) > 0\}$, $V = \{x \in X^2;\ K(x) < 0\}$.

The inequality (6.1) and the Gauss-Bonnet theorem give

$$\int_U K\,d\sigma - \int_V K\,d\sigma \geqslant 4\pi(1+g)\,, \qquad \int_U K\,d\sigma + \int_V K\,d\sigma = 2\pi\chi(X^2) = 4\pi(1-g)\,.$$

Thus, we have: If the quadratic form $\bar{G}(x, e)$ (9.3) is everywhere definite (positive or negative) on the surface $X^2 \subset E^{2+N}$, then the following inequalities hold

$$\int_U K\,d\sigma \geqslant 4\pi\,, \qquad \int_V K\,d\sigma \leqslant -4\pi g\,. \tag{9.7}$$

These inequalities are due to B. Y. Chen [3].

10. The case $n = N$, $r = 1$.

If $r = 1$ and e_1 is the unit vector on the line $E^1_n(x)$, equation (7.8) writes

$$G(x, e_1)\,d\sigma_n(x) = \omega_{1,n+1}\wedge\omega_{1,n+2}\wedge\ldots\wedge\omega_{1,n+N}\,. \tag{10.1}$$

For the general tangent vector $e = \cos\theta_i e_i$ $(i = 1, 2, \ldots, n)$ we have

$$G(x, e_1) = \Lambda(\cos\theta_1\omega_{1,n+s} + \cos\theta_2\omega_{2,n+s} + \ldots + \cos\theta_n\omega_{n,n+s}) \tag{10.2}$$

where $s = 1, 2, \ldots, N$.

According to (7.7) we have now

$$Q_{1,N}(x) = \frac{1}{O_{n+N-1}}\int_{G_{1,n-1}} G(x, e)\,dE^1_n(x) \tag{10.3}$$

i.e. $Q_{1,N}(x)$ is the mean value of $G(x, e)$ over the unit sphere $\cos^2\theta_1 + \cos^2\theta_2 + \ldots + \cos^2\theta_n = 1$, which may be evaluated by the same method of H. Weyl of the preceding section. The result is $Q_{1,N}(x) = 0$ if $n = N$ is odd and

$$(10.4) \qquad Q_{1,N}(x) = \frac{1}{2^n(2\pi)^{n/2}(n/2)\}} \delta^{i_1 i_2 \ldots i_n}_{j_1 j_2 \ldots j_n} R_{\alpha_1\alpha_2 j_1 j_2} R_{\alpha_3\alpha_4 j_3 j_4} \ldots R_{\alpha_{n-1}\alpha_n j_{n-1} j_n}$$

where $\alpha_h = n + i_h$ $(h = 1, 2, \ldots, n)$ if $n = N$ is even.

Notice that $Q_{1,N}(X)$ depends on the immersion.

EXAMPLE: For $n = N = 2$, $r = 1$, having into account the properties of symmetry (2.12) we get

$$(10.5) \qquad Q_{1,2}(x) = \frac{1}{2\pi} R_{3412}\,.$$

11. The cases $n + N \leqslant 6$.

In the following sections we wish to consider some particular cases. For $n + N \leqslant 6$, the conditions $n = rN$ and $r = n + N - 1$ give the following possibilities:

a) $n = 2$, $N = 1$, $r = 2$. Corresponds to the classical case of surfaces $X^2 \subset E^3$. We have $Q_{2,1}(x) = (1/2\pi)K$, $K =$ Gaussian curvature. Consideration of $Q^*_{2,1}$ and $K^*_{2,1}$ gives rise to interesting problems (Kuiper [17], Willmore [29]).

b) $n = 2$, $N = 2$, $r = 1$ and $n = 2$, $N = 2$, $r = 3$. These cases cor respond to $X^2 \subset E^4$ and will be considered with detail in the next section.

c) $n = 3$, $N = 3$, $r = 1$: $X^3 \subset E^6$. Particular case of the case considered in sections 7 and 10. Since $n = N = 3$, is odd, we have $Q_{1,3}(x) = 0$.

d) $n = 3$, $N = 2$, $r = 4$: $X^3 \subset E^5$. Particular case of the case considered in section 9. Since $n = 3$ is odd, we have $Q_{4,2}(x) = 0$.

e) $n = 3$, $N = 1$, $r = 3$: $X^3 \subset E^4$. Hypersurfaces in E^4. $Q_{3,1}(x) = (2\pi^2)^{-1}K$ ($K =$ Gauss-Kronecker curvature).

f) $n = 3$, $N = 3$, $r = 5$: $X^3 \subset E^6$. Particular case of the case considered in section 9. $Q_{5,3}(x) = 0$.

g) $n = 4$, $N = 1$, $r = 4$: $X^4 \subset E^5$. Hypersurfaces in E^5. $Q_{4,1}(x) = (8\pi^2/3)^{-1}K$ ($K =$ Gauss-Kronecker curvature).

h) $n=4, N=2, r=2$: $X^4 \subset E^6$. This is a noteworthy case which will be discussed in section 13.

i) $n=4, N=2, r=5$: $X^4 \subset E^6$. Particular case of the case considered in section 9.

j) $n=5, N=1, r=5$: $X^5 \subset E^6$. Particular case of the case considered in section 9.

12. **Surfaces in E^4.**

We will consider separately the cases *a*) $n=2, N=2, r=3$, and *b*) $n=2, N=2, r=1$.

a) *The case* $n=2, N=2, r=3$. Putting $\theta_1=\theta$, $\theta_2=(\pi/2)-\theta$ into (9.3) we have

$$\text{(12.1)} \qquad \bar{G}(x, e)\,\omega_1\wedge\omega_2 = \cos^2\theta\,\omega_{31}\wedge\omega_{32} + \sin^2\theta\,\omega_{41}\wedge\omega_{42} + \sin\theta\cos\theta(\omega_{31}\wedge\omega_{42}+\omega_{41}\wedge\omega_{32})\,.$$

The density for lines about a point in E^2 is $dE_2^1(x)=d\theta$ and thus

$$\text{(12.2)} \qquad \int_0^{2\pi} \bar{G}(x, e)\,\omega_1\wedge\omega_2\wedge d\theta = \pi(\omega_{31}\wedge\omega_{32}+\omega_{41}\wedge\omega_{42})\,.$$

Therefore we have

$$\text{(12.3)} \qquad Q_{3,2}(x)\,\omega_1\wedge\omega_2 = \frac{1}{2\pi}(\omega_{31}\wedge\omega_{32}+\omega_{41}\wedge\omega_{42})\,.$$

The Gaussian curvature of X^2 at x is defined by $d\omega_{12}=-K(x)\omega_1\wedge\omega_2$. Thus, according to (2.3) and (12.3) we get

$$\text{(12.4)} \qquad Q_{3,2}(x) = \frac{1}{2\pi}K(x)\,.$$

Integration over X^2 and application of the Gauss-Bonnet formula for surfaces, gives $K_{3,2}(X^2)=\chi(X^2)$, in accordance with (9.6).

We will now consider the absolute curvature $Q^*_{3,2}$. To this end it is convenient to introduce the normal curvatures of Otsuki [19]. Notice that the form $\omega_{31}\wedge\omega_{42}+\omega_{41}\wedge\omega_{32}$ remains invariant under rotations $e_1 \to \cos\alpha e_1 + \sin\alpha e_2$, $e_2 \to -\sin\alpha e_1 + \cos\alpha e_2$ on the tangent plane, but it can be annihilated by a suitable rotation on the normal plane e_3, e_4. Hence, choosing a suitable pair e_3, e_4 of normal unit vectors one can get

$$\omega_{31}\wedge\omega_{42}+\omega_{41}\wedge\omega_{32}=0\,. \tag{12.5}$$

Then, assuming that the forms ω_{ij} refer to the new frame, we define the normal curvatures λ_n, μ_n (Otsuki's curvatures) by

$$\omega_{31}\wedge\omega_{32}=\lambda_n\omega_1\wedge\omega_2\,,\qquad \omega_{41}\wedge\omega_{42}=\mu_n\omega_1\wedge\omega_2 \tag{12.6}$$

so that according to (12.3) and (12.4) we have

$$\lambda_n+\mu_n=K=\text{Gauss curvature}\,. \tag{12.7}$$

Having into account (12.5), equation (12.1) writes

$$\bar{G}(x, e)=\cos^2\theta\lambda_n+\sin^2\theta\mu_n \tag{12.8}$$

where we may assume

$$\lambda_n\geqslant\mu_n\,. \tag{12.9}$$

If $\lambda_n\mu_n\geqslant 0$, the absolute curvature at x is

$$Q^*_{3,2}(x)=\frac{1}{2\pi^2}\int_0^{2\pi}|G(x, e)|\,d\theta=\frac{1}{2\pi}|\lambda_n+\mu_n|=\frac{1}{2\pi}|K|\,. \tag{12.10}$$

If $\lambda_n\mu_n<0$ we notice that

$$\lambda_n\cos^2\theta+\mu_n\sin^2\theta>0 \quad\text{if}\quad |\theta|<\arctan\sqrt{-\lambda_n/\mu_n}\,,$$

$$\lambda_n\cos^2\theta+\mu_n\sin^2\theta<0 \quad\text{if}\quad \arctan\sqrt{-\lambda_n/\mu_n}<|\theta|\leqslant\pi/2$$

and

$$\int_0^{2\pi}|G(x, e)|\,d\theta=4\int_0^{\pi/2}|\lambda_n\cos^2\theta+\mu_n\sin^2\theta|\,d\theta=$$

$$=4\{\sqrt{-\lambda_n\mu_n}+(\lambda_n+\mu_n)(\arctan\sqrt{-\lambda_n/\mu_n}-\pi/4)\}.$$

Therefore we have

$$Q^*_{3,2}(x)=\frac{2}{\pi^2}\{\sqrt{-\lambda_n\mu_n}+K(\arctan\sqrt{-\lambda_n/\mu_n}-\pi/4)\}. \tag{12.11}$$

We shall do two simple applications of the preceding results.

i) If X^2 is orientable and $K = \lambda_n + \mu_n = 0$ (flat torus), we have

$$K^*_{3,2}(X^2) = \frac{2}{\pi^2}\int_{X^2} \lambda_n \, d\sigma_2 \, .$$

Applying the inequality (6.1), having into account that $K = 0$ implies $g = 1$, we get the following inequality of Otsuki [19]:

$$\int_{X^2} \lambda_n \, d\sigma_2 \geqslant 2\pi^2 \, .$$

ii) If $\mu_n \geqslant 0$, $\lambda_n > 0$, we have $Q^*_{3,2} = K/2\pi$ and the Gauss-Bonnet theorem gives

$$K^*_{3,2} = \int_{X^2} Q^*_{3,2} \, d\sigma_2 = \frac{1}{2\pi}\int_{X^2} K \, d\sigma_2 = \chi(X^2) \, .$$

Inequality (6.1) gives then $\chi(X^2) \geqslant 2$ and we have the following theorem of Chen [4]: if $\mu_n \geqslant 0$, $\lambda_n > 0$, then X^2 is homeomorphic to a 2-sphere.

b) The case $n = 2$, $N = 2$, $r = 1$. This is a particular case of that considered in section 10. Putting $e = \cos\theta e_1 + \sin\theta_2 e_2$, (10.2) becomes

(12.12) $$G(x, e)\,\omega_1\wedge\omega_2 = (\cos\theta\omega_{13} + \sin\theta\omega_{23}) \wedge (\cos\theta\omega_{14} + \sin\theta\omega_{24}) = \cos^2\theta\omega_{13}\wedge\omega_{14} + \sin^2\theta\omega_{23}\wedge\omega_{24} + \sin\theta\cos\theta(\omega_{13}\wedge\omega_{24} + \omega_{23}\wedge\omega_{14}) \, .$$

The form $\omega_{13}\wedge\omega_{24} + \omega_{23}\wedge\omega_{14}$ remains invariant under changes of frames in the normal plane, but by a suitable rotation $e_1 \to \cos\alpha e_1 + \sin\alpha e_2$, $e_2 \to -\sin\alpha e_1 + \cos\alpha e_2$ in the tangent plane, we may attain that

(12.13) $$\omega_{13}\wedge\omega_{24} + \omega_{23}\wedge\omega_{14} = 0 \, .$$

Assuming the frame $(x; e_1, e_2, e_3, e_4)$ chosen in such a way that (12.13) holds, we put

(12.14) $$\omega_{13}\wedge\omega_{14} = \lambda_t\,\omega_1\wedge\omega_2 \, , \qquad \omega_{23}\wedge\omega_{24} = \mu_t\,\omega_1\wedge\omega_2$$

where λ_t, μ_t are the tangent curvatures of X^2 at x.

The curvature $Q_{1,2}(x)$ is then

$$(12.15) \qquad Q_{1,2}(x) = \frac{1}{2\pi^2}\int_0^{2\pi} (\lambda_t \cos^2\theta + \mu_t \sin^2\theta)\, d\theta = \frac{1}{2\pi}(\lambda_t + \mu_t)$$

and the absolute curvature takes the values

$$(12.16) \qquad Q^*_{1,2}(x) = \frac{1}{2\pi}|\lambda_t + \mu_t| \quad \text{if} \quad \lambda_t\mu_t \geqslant 0$$

and

$$(12.17) \qquad Q^*_{1,2}(x) = \frac{2}{\pi^2}\{\sqrt{-\lambda_t\mu_t} + (\lambda_t + \mu_t)(\arctan\sqrt{-\lambda_t/\mu_t} - \pi/4)\}$$

if $\lambda_t\mu_t < 0$.

If we compare with the preceding case $Q^*_{3,2}(x)$ we observe that, instead of the Gaussian curvature K, we now have the invariant $I = \lambda_t + \mu_t$, such that

$$(12.18) \qquad I\omega_1\wedge\omega_2 = (\lambda_t + \mu_t)\omega_1\wedge\omega_2 = \omega_{13}\wedge\omega_{14} + \omega_{23}\wedge\omega_{24}\,.$$

Notice that $d\omega_{34} = -I\omega_1\wedge\omega_2$ and therefore, since every orientable $X^2 \subset E^4$ has a continuous field of normal vectors (Seifert [26]), from the Stokes theorem follows that

$$(12.19) \qquad \int_{X^2} d\omega_{34} = -\int_{X^2} I\omega_1\wedge\omega_2 = 0$$

i.e. the invariant $I(x)$ does not give any non trivial invariant by integration over X^2.

The curvatures $\lambda_t, \mu_t, \lambda_n, \mu_n$ are not independent. From their definition follows easily that

$$(12.20) \qquad \lambda_n\mu_n = \lambda_t\mu_t\,.$$

The invariant I has been introduced by Blaschke [2] and, from a more topological point of view, it has been considered by Chern-Spanier [13]. It is easy to see that I (like K) remains invariant under changes of frames (e_1, e_2) on the tangent plane, and also under changes of frames (e_3, e_4) on the normal plane. From (12.18), using the equations (2.5) one gets

$$(12.21) \qquad I = \begin{vmatrix} A_{3,11} & A_{3,12} \\ A_{4,11} & A_{4,12} \end{vmatrix} + \begin{vmatrix} A_{3,21} & A_{3,22} \\ A_{4,21} & A_{4,22} \end{vmatrix} = R_{3412}\,.$$

13. **Manifolds of dimension 4 immersed in E^6.**

We will now consider the case

$$n=4, \quad N=2, \quad r=2.$$

According to (7.8), if $E_4^2(x)$ is the 2-plane spanned by e_1, e_2 we have

$$G(x, \{e_1, e_2\})\, d\sigma_4 = \omega_{15} \wedge \omega_{16} \wedge \omega_{25} \wedge \omega_{26}. \tag{13.1}$$

For the general 2-space $E_4^2(x)$ spanned by the vectors $e_1' = \gamma_{1h} e_h$, $e_2' = \gamma_{2h} e_h$ $(h = 1, 2, 3, 4)$, we have

$$G(x, \{e_1', e_2'\})\, d\sigma_4 = \gamma_{1h_1}\gamma_{1h_2}\gamma_{2h_3}\gamma_{2h_4}\omega_{h_15}\wedge\omega_{h_26}\wedge\omega_{h_35}\wedge\omega_{h_46} =$$
$$= \frac{1}{4}\begin{vmatrix} \gamma_{1h_2} & \gamma_{1h_4} \\ \gamma_{2h_2} & \gamma_{2h_4} \end{vmatrix}\begin{vmatrix} \gamma_{1h_1} & \gamma_{1h_3} \\ \gamma_{2h_1} & \gamma_{2h_3} \end{vmatrix}\omega_{h_15}\wedge\omega_{h_26}\wedge\omega_{h_35}\wedge\omega_{h_46}.$$

Instead of evaluating the integral at the right side over $G_{2,2}$ it is easier to observe that for any frame $\{e_1', e_2', e_3', e_4'\}$ the sum

$$S' = \sum_{(i,j)} \omega_{i5}' \wedge \omega_{i6}' \wedge \omega_{j5}' \wedge \omega_{j6}' \tag{13.2}$$

where the summation is over all permutations of i, j from 1 to 4, does not depend on the orthogonal frame $\{e_1', e_2', e_3', e_4'\}$. Indeed, setting $e_i' = \gamma_{ih} e_h$ in (13.2), we have

$$S' = \frac{1}{4}\sum_{(i,j)}\begin{vmatrix} \gamma_{ih_2} & \gamma_{ih_4} \\ \gamma_{jh_2} & \gamma_{jh_4} \end{vmatrix}\begin{vmatrix} \gamma_{ih_1} & \gamma_{ih_3} \\ \gamma_{jh_1} & \gamma_{jh_3} \end{vmatrix}\omega_{h_15}\wedge\omega_{h_26}\wedge\omega_{h_35}\wedge\omega_{h_46}$$

where the dummy indices h_i take the values 1, 2, 3, 4. Having into account a well known theorem on orthogonal matrices which states that any minor is equal to its complementary, and since $\det(\gamma_{ij}) = 1$, we get $S' = S = \sum_{(i,j)} \omega_{i5}\wedge\omega_{i6}\wedge\omega_{j5}\wedge\omega_{j6}$.

Consequently S is equal to its mean value over $G_{2,2}$ and according to (3.6) we have

$$Q_{2,2}(x)\, d\sigma_4 = \frac{O_3}{6 O_4 O_5} S = \frac{1}{8\pi^3} S. \tag{13.3}$$

In terms of the invariants R_{ijkh} an easy calculation gives

$$(13.4)\qquad Q_{2,2}(x)=\frac{1}{8\pi^3}\sum_{(i,j)}\begin{vmatrix} A_{5i1} & A_{5i2} & A_{5i3} & A_{5i4}\\ A_{5j1} & A_{5j2} & A_{5j3} & A_{5j4}\\ A_{6i1} & A_{6i2} & A_{6i3} & A_{6i4}\\ A_{6j1} & A_{6j2} & A_{6j3} & A_{6j4}\end{vmatrix}=$$

$$=\frac{1}{8\pi^3}\sum_{(i,j)}(R_{ij12}R_{ij34}+R_{ij13}R_{ij24}+R_{ij14}R_{ij23})\,.$$

It is noteworthy that this invariant does not depend on the immersion of X^4 into E^6. The total curvature $K_{2,2}(X^4)$ coincides, up to a constant factor, with a topological invariant introduced by Chern [8]. For a topological sphere we have $K_{2,2}(X^4)=0$ (as follows from ii) of section 4). Samelson [21] has given examples of manifolds for which $K_{2,2}(X^4)\neq 0$. It can be seen that the differential form (13.1) defines the Pontrjagin class p_1 of X^4 (see Chern [9]).

Tosto porvenuto il 21 maggio 1973.
Bozze licenziate il 18 giugno 1974.

BIBLIOGRAPHY

[1] C. Allendoerfer - A. Weil, *The Gauss-Bonnet theorem for Riemannian polyhedra*, Trans. Amer. Math. Soc., 53 (1943), 101-129.

[2] W. Blaschke, *Sulla geometria differenziale delle superficie S_2 nello spazio euclideo S_4*, Ann. Mat. Pura Appl. (4), 28 (1949), 205-209.

[3] Bang-Yen Chen, *A remark on minimal imbedding of surfaces in E_4*, Kodai Math. Sem. Rep., 20 (1968), 279-281.

[4] Bang-Yen Chen, *A note on the Gaussian curvature of surfaces in E^{2+N}*, Tamkang J. Math., 1 (1970), 11-13.

[5] Bang-Yen Chen, *G-total curvature of immersed manifolds*, Ph. D. dissertation of the author, University of Notre Dame, 1970.

[6] Bang-Yen Chen, *On the total curvature of immersed manifolds III; surfaces in euclidean 4-space*, to appear in Amer. J. Math.

[7] S. S. Chern, *On the curvature integral in a riemannian manifold*, Ann. of Math., 56 (1945), 674-684.

[8] S. S. Chern, *On riemannian manifolds of four dimensions*, Bull. Amer. Math. Soc., 51 (1945), 964-971.

[9] S. S. Chern, *La géométrie des sous-variétés d'un éspace euclidien à plusieurs dimensions*, L'Enseignement Mathématique, 40 (1955), 26-46.

[10] S. S. Chern, *On the kinematic formula in Integral Geometry*, J. Math. and Mech., 16 (1966), 101-118.

[11] S. S. Chern - R. K. Lashof, *On the total curvature of immersed manifolds I*, Amer. J. Math., 79 (1957), 306-313.

[12] S. S. Chern - R. K. Lashof, *On the total curvature of immersed manifolds II*, Michigan Math. J., 5 (1958), 5-12.

[13] S. S. Chern - E. Spanier, *A theorem on orientable surfaces in four dimensional space*, Comm. Math. Helvetici, 25 (1951), 205-209.

[14] H. Federer, *Curvature measures*, Trans. Amer. Math. Soc., 93 (1959), 418-491.

[15] H. Hadwiger, *Vorlesungen über Inhalt, Oberflache und Isoperimetrie*, Springer, Berlin, 1957.

[16] D. Ferus, *Total Absolutkrümmung in Differentialgeometrie und Topologie*, Lecture Notes in Mathematics, n. 66, Springer, Berlin, 1968.

[17] N. H. Kuiper, *On surfaces in euclidean three-space*, Bull. Soc. Math. Belgique, 12 (1960), 5-22.

[18] N. H. Kuiper, *Minimal total absolute curvature for immersions*, Inventiones Math., 10 (1970), 209-238.

[19] T. Otsuki, *On the total curvature of surfaces in Euclidean space*, J. Math. Soc. Japan, 35 (1966), 61-71.

[20] W. D. Pepe, *On the total curvature of C^1 hypersurfaces in E^{n+1}*, Amer. J. Math., 91 (1969), 984-1002.

[21] H. Samelson, *On Chern's invariant for Riemannian 4-manifolds*, Proc. Amer. Math. Soc., 1 (1950), 415-417.

[22] L. A. Santaló, *Geometria Integral en espacios de curvatura constante*, Publ. de la Com. Nac. Energia Atómica, Serie Mat., vol. 1, n. 1, Buenos Aires, 1952.

[23] L. A. Santaló, *Sur la mesure des éspaces linéaires qui coupent un corps convexe et problèmes qui s'y rattachent*, Colloque sur les questions de réalité en Géométrie, Liège (1955), 177-190.

[24] L. A. Santaló, *Curvaturas absolutas totales de variedades contenidas en un espacio euclidiano*, Acta Cientifica Compostelana, 5 (1968-69), 149-158.

[25] L. A. Santaló, *Mean values and curvatures*, Izv. Akad. Nauk. Armjan SSR Sr. mat., 5 (1970), 286-295.

[26] H. Seifert, *Algebraische Approximation von Mannigfaltigkeiten*, Math. Zeits., 41 (1936), 1-17.

[27] J. A. Thorpe, *On the curvatures of riemannian manifolds*, Illinois J. Math., 10 (1966), 412-417.

[28] H. Weyl, *On the volume of tubes*, Amer. J. Math., 61 (1939), 461-472.

[29] T. J. Willmore, *Tight immersions and total absolute curvatures*, Bull. London Math. Soc., 3 (1971), 129-151.

Part II. Integral Geometry, with comments by R. Langevin

Some names are attached to a famous theorem. Santaló is in particular known for a topic, Integral Geometry, and a book "Integral Geometry and Geometric Probability". I see his results as building blocks which at the end form a monument. This chapter picks up less of a third of the articles Santaló published on the subject.

The topic, Integral Geometry, started with two examples described by Buffon in 1777: the franc-carreau game and the needle problem.

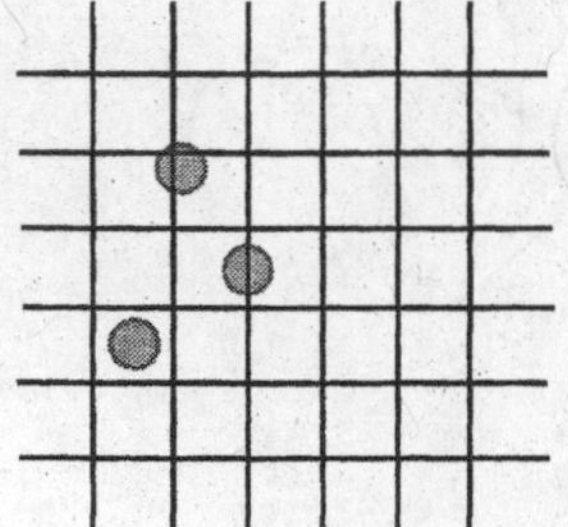

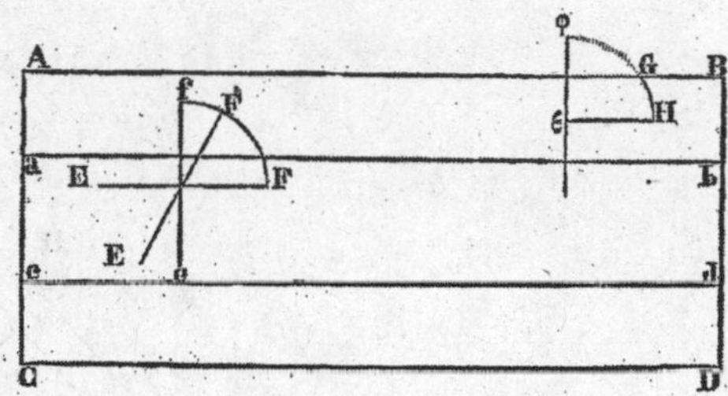

Figure 1: Buffon's franc-carreau game and needle problem

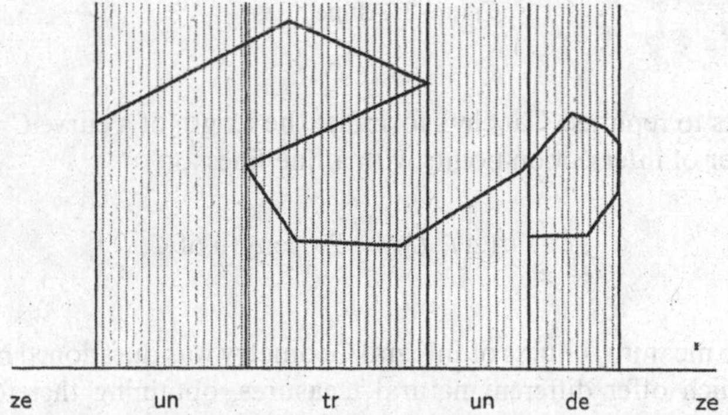

Figure 2: Length of the projection on ℓ counted with multiplicity

Cauchy, in 1832, communicated to the French Academy of Science a formula involving all the projections on lines of a planar curve. Let $m(C, \ell) = \int \#(p^{-1}(y))dy$; $y \in \ell$, where # denotes the number of points of a set. Then we have

$$\int_{-\pi/2}^{\pi/2} m(C, \ell_\theta)\, d\theta = 2 \cdot (\text{length of } C).$$

Crofton was the first to feel the necessity of an explanation to the expression "at random". Let us quote Crofton: *First, since any direction is as likely as any other, as many of the lines are parallel to any direction as to any other. [...] As this infinite system of parallels is drawn at random, they are as thickly disposed along any part of the perpendicular as along any other* ... We recognize the measure $d\theta \otimes dt$ on the set of affine lines (see Figure 3).

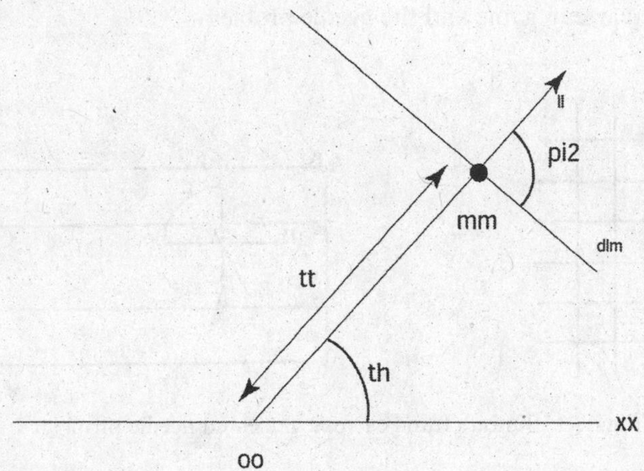

Figure 3: Affine lines

This permits to rephrase Cauchy's formula: the length of a curve C is half of the average number of intersection points with affine lines, or

$$\int_{\mathcal{A}(\in,\infty)} \#(\Delta \cap C) = 2 \cdot \text{ length of } C.$$

A moment, the meaningfullness of Integral Geometry was questioned by Bertrand's paradoxes, which offer different natural measures, obtaining therefore different equalities. Then Poincaré, in the spirit of Klein's Erlangen program recognized the role of the group of affine isometries in the choice of the measure $d\theta \otimes dy$ chosen by Crofton on the set affine lines.

Minkowski introduced curvature in the game. Answering a question of Crofton, he proved that the measure of affine planes intersecting a compact convex body Q of smooth boundary is proportional to the integral of the mean curvature on the boundary of ∂Q.

After important results by Steiner, Hadwiger and Blaschke, who coined the name "Integralgeometrie" in the 1930's, Santaló became the specialist of all results one can obtain mixing classical geometry of submanifolds and some averaging procedure allowed by the homogeneity of the ambient space. In his book, he chosed to systematically look for a density associated to the moving family of objects, linear subspaces of Euclidean space, or orbits of a Lie group action on an homogeneous space. The reader will find many examples in this chapter.

To explain the little music of Santaló's Integral Geometry, let us go back to his article [36.1] *Geometria Integral 7: nuevas aplicaciones del concepto de medida cinemática en el plano y en el espacio.*

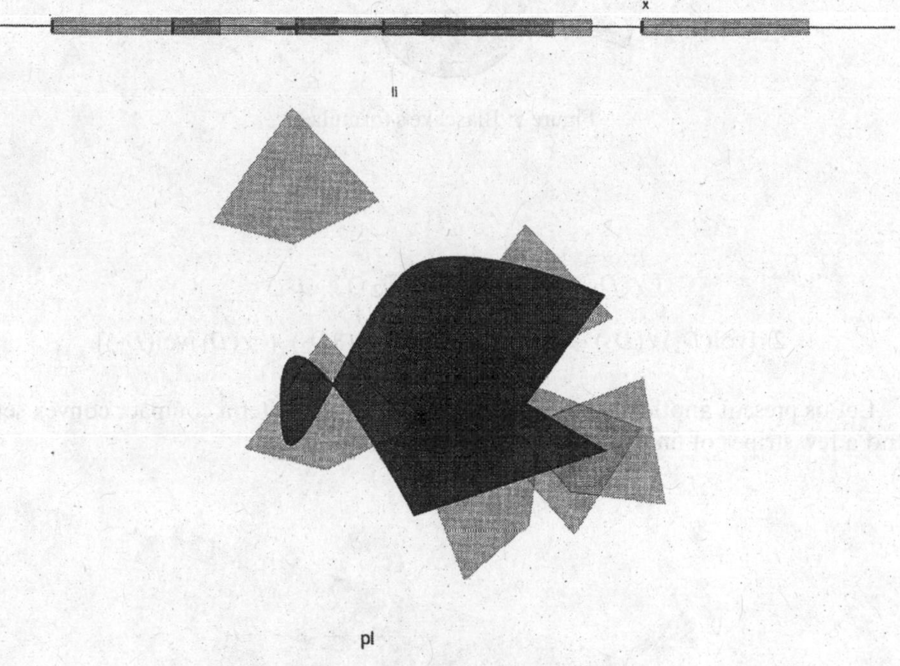

Figure 4: Intersecting intervals, intersecting planar figures

The ambient space is the Euclidean line, the Euclidean plane or the Euclidean space. One object is fixed, another takes all possible positions.

Therefore, to compute an integral, a measure on the set of all possible positions of the second object is needed.

Santaló's starting point is even more elementary than Crofton's: the line and the group of tranlations. Let the first object be a segment of length ℓ, the second a segment of length a; a natural measure of the position of the second segment is dx where x is the position of, say, the left extremity of the segment. Then the measure of the set of positions of the second segment such that it intersects the first is $\ell + a$.

Planar problems involves the density $dm \otimes \theta$ of the group of affine isometries.

A striking result in dimension 2, which announce the more general kinematic formulas of Chern, is Blaschke's formula, which computes averages of Euler characteristics of intersections of a compact domain $D_1 \subset \mathbb{R}^2$ and the image of another, $D_2 \subset \mathbb{R}^2$, by all the affine isometries:

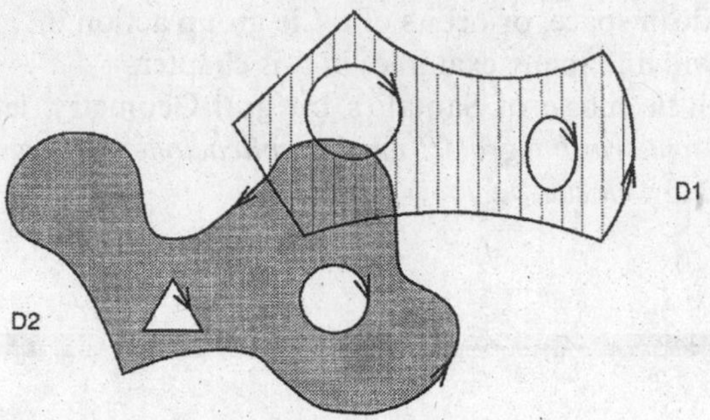

Figure 5: Blaschke's formula

$$\int_{\mathcal{G}} \chi(D_1 \cap g \cdot D_2) = \int_{\mathcal{G}} \chi(D_2 \cap g \cdot D_1) =$$
$$= 2\pi[\mathrm{vol}(D_1)\chi(D_2) + \mathrm{length}(\partial D_1)\mathrm{length}(\partial D_2) + \chi(D_1)\mathrm{vol}(D_2)].$$

Let us present another example developed in this chapter a compact convex set and a few stripes of finite width.

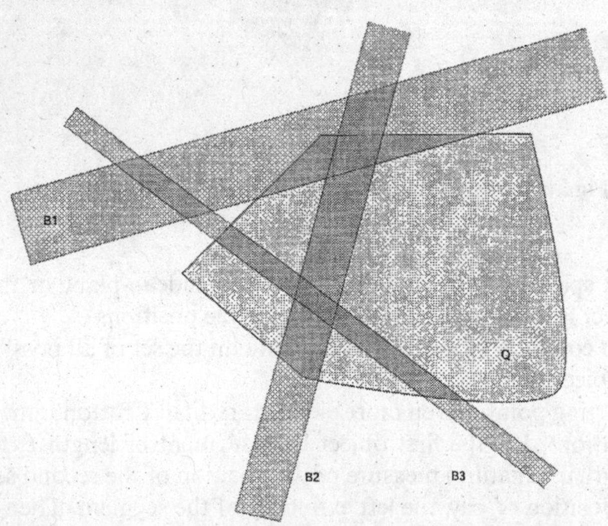

Figure 6: A relative position of Q and a rigid set of stripes

More rigorously, one has to compute the measure of the set of affine isometries $g \in SO(2) \ltimes \mathbb{R}^2$ such that Q and $g(B_1 \cup B_2 \cup B_3)$ have a given configuration like "$B_1 \cap B_2 \cap Q \neq \emptyset$, $B_2 \cap B_3 \cap Q \neq \emptyset$ and $B_1 \cap B_2 \cap B_3 \cap Q = \emptyset$".

Such examples, and analogous periodic or random ones are very usefull in Stereology.

Many of Santaló's papers deal with hyperbolic geometry. The group of hyperbolic isometries, because it is non-compact and of exponential growth, turn averaging methods more cumbersome. This helps to understand why results in hyperbolic geometry often have to link the mobile objects with the fixed one.

The hyperbolic plane can be completed with a circle of "*points at infinity*". This circle does not admit a measure invariant by the group of hyperbolic isometries.

Horocycles are circles tangent to the circle "*at infinity*". It would be natural to look for a measure on the set of horocycles describing them using the tangency point with "*infinity*" and the length of a geodesic going to this point at infinity. Because of previous remark, this does not work. Nevertheless, one can define on the set of horocycle a measure invariant by the action of the group of hyperbolic isometries. These underlying difficulties may help to understand why the topic is so rich. The apparent simplicity of the solution to the problems hide the main difficulty of Integral Geometry: find good questions and compose a large picture with the responses. Definitely Santaló, with his impressionist technique, was one of the great masters of this art.

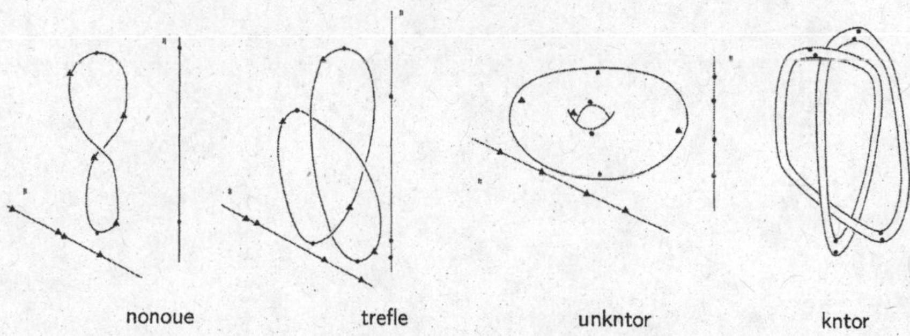

Figure 7: Knotted and unknotted curves and surfaces, and some orthogonal projections on lines

Meanwhile Fenchel, Fary, Milnor, Chern and Kuiper, mixing Integral Geometry and topology, developed another reseach direction (see Figure 7), but this is another story, just lightly touched by Santaló.

Rémi Langevin (Institut de Mathématiques de Bourgogne)

GEOMETRIA INTEGRAL 7

Nuevas aplicaciones del concepto de medida cinemática en el plano y en el espacio

INTRODUCCION

Las probabilidades geométricas, que puede considerarse aparecen en el campo de las matemáticas con el famoso problema de la aguja de BUFFÓN (1707-1788), no alcanzan su verdadero desarrollo en cuanto aplicación de las mismas a la teoría de los cuerpos convexos hasta los trabajos de CROFTON (1826-1915) (2). Este es además el primero que define para conjuntos de puntos y rectas del plano una *densidad* y como integral de la misma una *medida* con la propiedad de ser invariante respecto el grupo de los movimientos. Parece ser CARTAN (5) el primero que en 1896 obtenía densidades análogas para conjuntos de rectas y planos en los espacios euclidiano o no euclidiano tridimensional. Más tarde HERGLOTZ (12) y BLASCHKE (13) obtienen estas densidades para todo subespacio lineal de un espacio euclidiano de cualquier número de dimensiones. Para las aplicaciones ha sido conveniente encontrar ciertas relaciones entre las distintas densidades de estos subespacios, algunas de las cuales utilizaremos en este trabajo, y que han sido obtenidas por VARGA (16) para el espacio ordinario y por PETKANTSCHIN (19) para un espacio euclidiano o elíptico cualquiera.

Sin embargo, en casi su totalidad, estos trabajos se limitan a considerar conjuntos de puntos, rectas o planos ya solos, ya formando distintas combinaciones entre sí. Es primero POINCARÉ (6), en sus lecciones del año 1896, el que introduce la idea de considerar como elementos sistemas de ejes coordenados rectangulares y medir conjuntos de los mismos, lo cual ha de permitir medir conjuntos de figuras cualesquiera. POINCARÉ mismo busca la expresión de esta medida, que luego se llamó *medida cinemática,* para el plano y el espacio ordinarios. BLASCHKE, en el trabajo citado, es quien establece por primera vez la expresión de esta medida cinemática para el espacio *n*-dimensional.

Una vez establecida esta medida cinemática, quedaba la labor de hacer aplicación de la misma, midiendo ciertos conjuntos de cuerpos, superficies o líneas, y ver si de estas medidas se podían obtener, como hizo CROFTON partiendo de las medidas de conjuntos de puntos y rectas, algunas fórmulas integrales u otras aplicaciones a los cuerpos convexos.

Esto es lo que hemos hecho en dos trabajos (17), (18) que preceden al presente. Falta todavía, y no creemos encierre dificultades serias, extender estos nuestros resultados al espacio euclidiano n-dimensional.

Pero no es éste el objeto del presente trabajo, en el cual no nos apartamos todavía de las tres primeras dimensiones. En los trabajos citados nos hemos ocupado exclusivamente de figuras limitadas. Faltaba, pues, ver lo que pasaba cuando se considerasen figuras ilimitadas, como son, por ejemplo, los ángulos completos (planos o diedros), las bandas indefinidas de plano limitadas por dos rectas paralelas y las franjas de espacio limitadas por dos planos paralelos.

Esto constituye el núcleo principal de este trabajo, al cual hacemos preceder algunos resultados, muchos de ellos conocidos (números 1 al 7), pero que nos ha parecido interesante intercalar aquí, porque representan, a nuestro modo de ver, la manera intermediaria por la que se puede pasar de la medida de los elementos punto, recta y plano a la medida cinemática. Al final (núms. 34-37) añadimos también algunos resultados referentes a proyecciones sin gran conexión con todo lo demás, pero que no hemos creído superfluos.

Para la lectura del presente trabajo no son necesarios más que los conocimientos y fórmulas fundamentales de la teoría de Probabilidades Geométricas, que pueden verse, por ejemplo, en DELTHEIL (11) y algunas dependencias entre las densidades de los elementos del espacio que han sido establecidas recientemente por VARGA (16).

PROBLEMAS SOBRE LA RECTA

1. CONJUNTOS DE PUNTOS.—En general, sobre una variedad unidimensional no se pueden dar más elementos que sobre ella puedan moverse sin deformarse que los puntos. Un segmento de una línea, plana o espacial, no puede desplazarse sobre la misma sin sufrir deformación más que si la línea es de curvatura y torsión constantes. En los demás casos y en vistas a problemas de Probabilidades Geométricas no podemos considerar más que puntos aislados o conjuntos de ellos, pero siempre supuestos dados independientemente unos de otros.

Como densidad para conjuntos de puntos de una línea se tiene:

$$dP = ds \qquad [1]$$

siendo ds el elemento de arco.

2. CONJUNTOS DE SEGMENTOS.—Si la línea tiene curvatura y torsión constan-

tes además de conjuntos de puntos se pueden considerar conjuntos de segmentos de la misma tales que, por un movimiento sobre ella puedan llevarse a coincidir. Bastará considerar sólo los casos de la recta (como tipo de línea ilimitada) y la circunferencia.

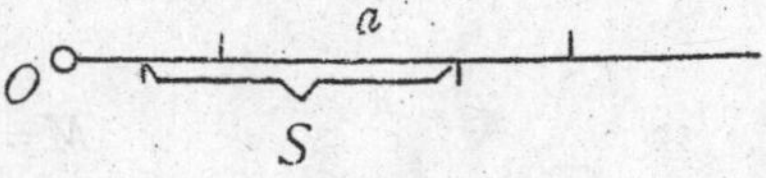

Fig. 1

a) *Línea recta.*—Un segmento de longitud l móvil sobre la recta que lo contiene viene determinado por la distancia de uno cualquiera de sus extremos (siempre el mismo) al origen O (fig. 1). Como densidad para conjuntos de segmentos se tiene, pues, la misma [1], o sea

$$d\mathrm{S} = dx \qquad [2]$$

La medida de los segmentos de longitud l que tienen algún punto común con otro segmento fijo de longitud a es, por tanto,

$$M = \int_{-l}^{a} dx = a + l \qquad [3]$$

En cambio, la medida del conjunto de los segmentos S que están contenidos totalmente en a $(l \leqslant a)$ será:

$$M' = \int_{0}^{a-l} dx = a - l \qquad [4]$$

De estas dos medidas se deduce:

La probabilidad de que un segmento de longitud l *que corta a un segmento de longitud* a, *corte también a otro interior de longitud* b (a $\geqslant$ b) *es*

$$\mathrm{W} = \frac{b + l}{a + l} \qquad [5]$$

Si se impone la condición de *estar contenido* en vez de *cortar* se tiene, por el problema análogo:

$$\mathrm{W} = \frac{b - l}{a - l} \qquad [6]$$

b) *Circunferencia.*—Sea una circunferencia fija de radio r y un arco de la misma de longitud l. Como densidad para un conjunto de tales arcos tomamos también el elemento de arco $d\,s$ y podemos repetir los mismos problemas anteriores,

Sólo hay que observar que aquí el conjunto de todos los arcos de longitud l tiene medida finita e independiente de l, a saber:

$$M = \int ds = 2\pi r \qquad [7]$$

Debido a esta finitud de la medida total aquí se puede resolver el problema: *Probabilidad de que un arco de longitud* l *dado arbitrariamente sobre una circunferencia esté contenido en otro de longitud* a $\geqslant$ l. Será:

$$W = \frac{a - l}{2\pi r} = \frac{\alpha - \lambda}{2\pi} \qquad [8]$$

siendo α y λ los ángulos centrales respectivos.

3. Segmentos de longitud variable.—En lugar de considerar segmentos de longitud constante y, por tanto, dependientes de un solo parámetro que fija su posición sobre la línea de que se trata, se puede también tomar como variable su longitud. Entonces hay que aclarar lo que se entiende por dar arbitrariamente un segmento de longitud arbitraria. Un tal segmento viene determinado por sus extremos, o sea por dos puntos P_1 y P_2, los cuales hay que suponer dados completamente independientes uno del otro. Admitiendo el principio de las probabilidades compuestas, como densidad del par P_1, P_2, hay que tomar el producto de las densidades $dP_1 \,.\, dP_2$. Luego introduciendo en esta expresión la distancia l entre los dos puntos, o sea la longitud variable del segmento, se tiene

$$dP_1 \cdot dP_2 = ds \cdot d(s + l) = ds \cdot dl \qquad [9]$$

Con esta expresión podemos contestar a la pregunta: ¿Cuál es la medida de todos los segmentos de longitud $l \leqslant b$ contenidos en otro de longitud a? Por [4] será

$$M_{\leq b} = \int ds\, dl = \int_0^b (a - l)\, dl = ab - \frac{b^2}{2} \qquad [10]$$

Luego: *la probabilidad de que un segmento dado al azar en el interior de otro de longitud* a *tenga una longitud* $\leqslant$ b *es*

$$W = \frac{ab - \dfrac{b^2}{2}}{\dfrac{a^2}{2}} = \frac{2b}{a} - \frac{b^2}{a^2} \qquad [11]$$

PROBLEMAS SOBRE EL PLANO

4. Conjuntos de puntos.—Una figura dibujada sobre una superficie en general no podrá moverse sobre la misma sin sufrir deformación. Esto es posible, sin embargo, en dos casos: sobre el plano (superficie abierta) y sobre la esfera (superficie cerrada). En los demás casos sólo ha lugar a considerar conjuntos de puntos independientes entre sí para cuya densidad se toma el elemento de área de la superficie

$$d\mathrm{P} = d\sigma \qquad [12]$$

5. Densidad cinemática en el plano.—En el caso del plano y de la esfera toda figura puede moverse sin salirse de la superficie ni deformarse. Se pueden, pues, considerar conjuntos cuyos elementos sean estas figuras e intentar su medida. Esto es lo que vamos a hacer limitándonos al caso del plano.

Toda figura plana viene determinada en su plano por tres parámetros. Estos pueden ser las dos coordenadas x, y que fijan uno de sus puntos y el ángulo φ que una dirección fija en la figura forma con otra invariante del plano. Entonces la densidad cinemática introducida por Poincaré es

$$d\mathrm{K} = d\,x\,d\,y\,d\,\varphi \qquad [13]$$

expresión que es fácil comprobar no varía al cambiar el punto x, y de la figura móvil o la dirección φ de la misma (17) (*).

(*) Esta expresión de la densidad cinemática se puede también obtener sistemáticamente intentando buscar las funciones $f(x, y, \varphi)$ tales que $\int f(x, y, \varphi)\, d\,x\,d\,y\,d\,\varphi$ sea un invariante integral respecto el grupo de los movimientos del plano. En efecto, dicho grupo viene representado por

$$x^* = x + a\cos\varphi - b\,\mathrm{sen}\,\varphi$$
$$y^* = y + a\,\mathrm{sen}\,\varphi + b\cos\varphi$$
$$\varphi^* = \varphi + c$$

y para que $f(x, y, \varphi)$ sea un invariante integral debe ser

$$\frac{\partial}{\partial x}(f\cos\varphi) + \frac{\partial}{\partial y}(f\,\mathrm{sen}\,\varphi) = 0$$
$$\frac{\partial}{\partial x}(-f\,\mathrm{sen}\,\varphi) + \frac{\partial}{\partial y}(f\cos\varphi) = 0$$
$$\frac{\partial}{\partial \varphi} f = 0$$

De la última ecuación se deduce que f es sólo función de x, y y de las dos primeras

$$f_x^2 + f_y^2 = 0$$

luego $f = \mathrm{Cte.}$

Esta densidad cinemática puede también escribirse en otra forma cuando en lugar del punto x, y y el ángulo φ se toman por coordenadas las de una recta invariablemente unida a la figura y la distancia sobre esta recta de un punto fijo en la figura a un punto fijo en la recta (fig. 2). Entonces se tiene también

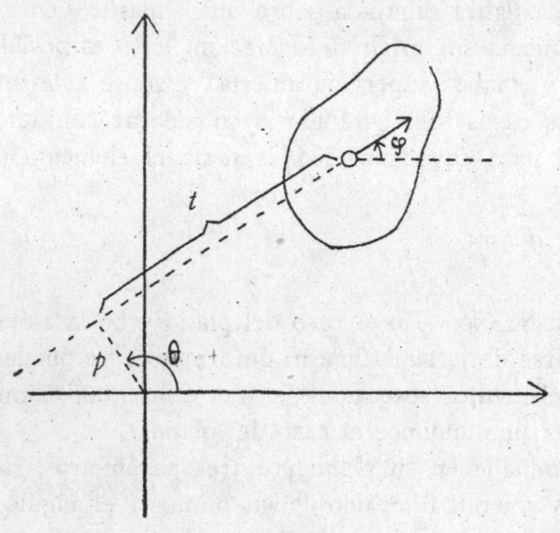

Fig. 2.

$$dK = dp\, d\theta\, dt \qquad [14]$$

siendo p, θ las coordenadas normales de la recta considerada.

6. Segmentos que cortan a una figura convexa. En esta forma [14] de la densidad cinemática es ya fácil el cálculo de la medida del conjunto de los segmentos *orientados* de longitud l que tienen algún punto común con una figura plana convexa K. Se tendrá:

$$M = 2\int dp\, d\theta\, dt = 2\int (\sigma + l)\, dp\, d\theta = 2\pi F + 2\, l L \qquad [15]$$

indicando por σ la longitud de la cuerda que la recta p, θ determina en K, así como por F y L el área y longitud de K respectivamente.

De aquí, recordando la propiedad isoperimétrica del círculo, resulta que: *El círculo es la figura plana que por una área dada tiene menor número de segmentos que le cortan.*

7. Segmentos contenidos en una figura plana convexa.—Si se consideran sólo los segmentos totalmente contenidos en el interior de la figura K anterior y para segmentos orientados se tendrá:

$$M^i = 2\int (\sigma - l)\, dp\, d\theta =$$

$$= 2\int_0^{\pi} A^*\, d\theta - 2\, l \int_0^{\pi} p^*\, d\theta \qquad [16]$$

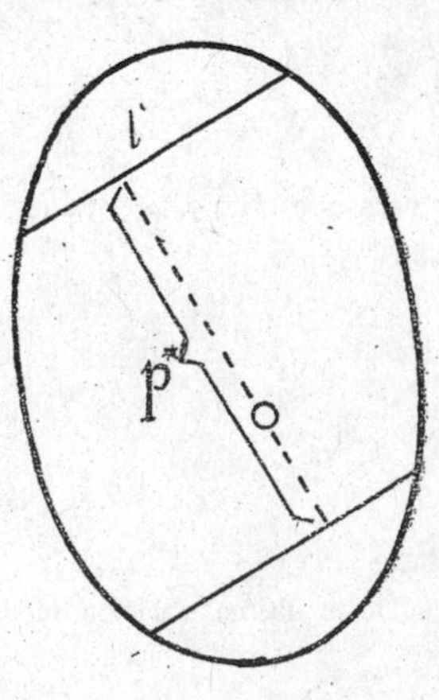

Fig. 3

siendo A* en el área comprendida entre dos cuerdas paralelas de longitud l y p^* la distancia entre ellas (fig. 3). Estas integrales ya no parece posible, en general, expresarlas en forma sencilla. Sin embargo, esto no es difícil en algunos casos particulares. Por ejemplo, los siguientes:

a) *Círculo.*—Llamando r a su radio, es fácil ver que en este caso lo mismo A* que p^* son independientes de θ y valen

$$A^* = \pi r^2 - \left(2 r^2 \operatorname{arc\,sen} \frac{l}{2r} - l \sqrt{r^2 - \frac{l^2}{4}}\right)$$

$$p^* = 2 \sqrt{r^2 - \frac{l^2}{4}}$$

Por tanto:

$$M^i = 2\pi^2 r^2 - 4\pi r^2 \operatorname{arc\,sen} \frac{l}{2r} - 2\pi l \sqrt{r^2 - \frac{l^2}{4}} \qquad [17]$$

b) *Rectángulo.*—También en este caso se pueden terminar las integraciones de [16] en el caso de ser l menor o igual que el menor de los lados del rectángulo. Llamando a_1 y b_1 a las longitudes de estos lados se tiene (fig. 4):

$$A^* = a_1 b_1 - \frac{l^2}{2} \operatorname{sen} 2\theta$$

$$p^* = (a_1 - l \operatorname{sen} \theta) \cos\theta + {} + (b_1 - l \cos\theta) \operatorname{sen}\theta$$

luego

$$M^i = 2\pi a_1 b_1 - 4 l (a_1 + b_1) + 2 l^2 \qquad [18]$$

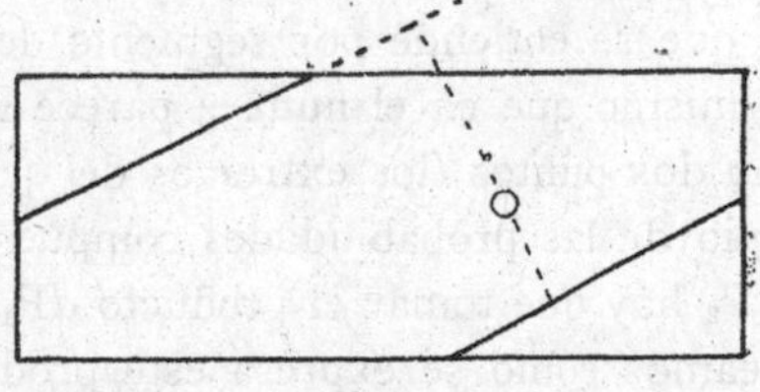

Fig 4.

c) *Triángulo.*—Supongamos que l es menor que cualquiera de las alturas del triángulo para que en toda dirección existan cuerdas de esta longitud. Se observa que $A^* - l p^*$ es precisamente el área del triángulo A P Q rayado cuya área es (fig. 5):

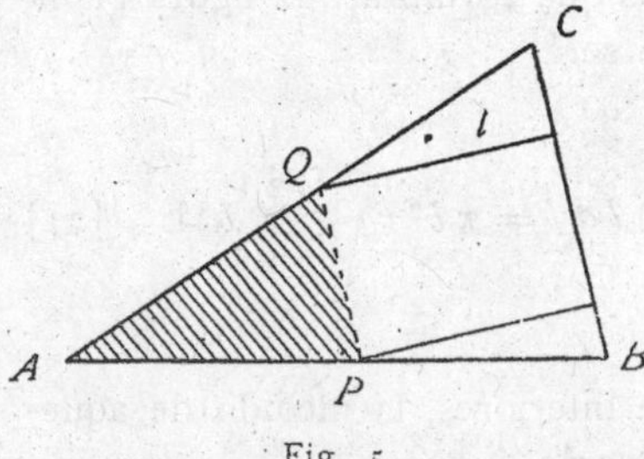

Fig 5.

$$\frac{1}{2} \overline{AP} \cdot \overline{PQ} \operatorname{sen} B =$$

$$= \frac{1}{2} \left[c - \frac{l}{\operatorname{sen} B} \operatorname{sen}(B + \theta) \right] \frac{a}{c} \operatorname{sen} B$$

Para integrar de 0 a π, hagámoslo primero de 0 a A y queda

$$M_A^i = \frac{a}{c} \operatorname{sen} B \int_0^A \left[c - \frac{l}{\operatorname{sen} B} \operatorname{sen}(B + \theta) \right]^2 d\theta =$$

$$= 2 \left[T\,A - a\,l \cos B \cos C + \frac{a\,l^2\,A}{4\,b \operatorname{sen} C} + \frac{a\,l^2}{4\,b} \cos C + \frac{a\,l^2}{4\,c} \cos B \right]$$

donde T es el área del triángulo dado.

Calculando lo análogo entre o, B y o, C y sumando se obtiene:

$$M^i = 2\pi T - 2\,l\,(a \cos B + a \cos C + b \cos A + b \cos C + c \cos A + c \cos B) +$$

$$+ \frac{l^2}{2} \left[\frac{a\,A}{b \operatorname{sen} C} + \frac{b\,B}{c \operatorname{sen} A} + \frac{c\,C}{a \operatorname{sen} B} \right] + \frac{l^2}{2} \left[\frac{a \cos C + c \cos A}{b} + \ldots \right] =$$

$$= 2\pi T - 2\,l\,L + \frac{l^2}{2} \sum_{A,B,C} \left[(\pi - A) \cot A + 1 \right] \qquad [19]$$

7. Segmentos de longitud variable.—Hasta aquí los segmentos, siendo de longitud constante, quedaban determinados por tres parámetros. Pero cuando se supone la longitud también variable, entonces hay que aclarar primeramente lo que se entiende por segmento de longitud arbitraria dado arbitrariamente. Y lo mismo que en el núm. 3 parece natural tomar este concepto equivalente al de dar dos puntos (los extremos del segmento) independientes entre sí. Por el principio de las probabilidades compuestas como densidad de este par de puntos P_1 y P_2 hay que tomar el producto $dP_1 \,.\, dP_2$ de las densidades de cada uno de ellos. Veamos cómo se expresa este producto en función de las coordenadas del segmento y de su longitud. Basta observar que, llamando como antes φ al ángulo que forma con una dirección constante y l a su longitud es

$$d P_1 \cdot d P_2 = l\, d P_1\, d\varphi\, d l \qquad [20]$$

Con esta expresión y la [15] hallada anteriormente podemos resolver el problema: ¿Cuál es la medida de todos los segmentos que cortan a una figura plana convexa y cuya longitud es $\leqslant b$? Evidentemente será

$$M_{\leq b} = \int l\, d P_1\, d\varphi\, d l = \int_0^b (2\pi F + 2\,l\,L)\, l\, d l = \pi b^2 F + \frac{2}{3} b^3 L \qquad [21]$$

Si sólo se considerasen segmentos totalmente interiores, la medida de aquellos que tienen su longitud igual o menor que b será:

$$M^i_{\leq b} = \int_0^b l\, M^i\, d\, l \qquad [22]$$

Por ejemplo, en los casos anteriores del círculo, rectángulo y triángulo se tiene:

Para el círculo

$$M^i_{\leq b} = \pi\, r^2 \left[\pi\, b^2 + 2\, \beta\, (r^2 - b^2) - (2\, r^2 + b^2) \operatorname{sen} \beta \cos \beta\right]$$

siendo

$$\beta = \operatorname{arc\,sen} \frac{b}{2\, r} \qquad [23]$$

Para el rectángulo

$$M^i_{\leq b} = \pi\, a_1\, b_1\, b^2 - \frac{4}{3}\,(a_1 + b_1)\, b^3 + \frac{1}{2}\, b^4 \qquad [24]$$

Para el triángulo

$$M^i_{\leq b} = \left[\pi\, T - \frac{2}{3}\, L\, b + \frac{b^2}{8} \sum_{A,\,B,\,C} \left[(\pi - A) \cot (\pi - A) + 1\right]\right] b^2 \qquad [25]$$

Estas fórmulas nos permiten resolver el problema:

Dado un segmento de longitud arbitraria en el interior de un círculo, rectángulo o triángulo, ¿cuál es en cada caso la probabilidad de que sea igual o menor que otro segmento dado b?

Será

$$W = \frac{M^i_{\leq b}}{F^2}$$

donde el numerador toma los valores anteriores, según el caso de que se trate y F indica el área del recinto considerado.

8. Otros conjuntos de segmentos.—Intentemos medir el conjunto de segmentos de longitud l tales que, cortando a la figura convexa K, tienen los dos extremos fuera de la misma.

Sean p, θ, t las coordenadas del segmento. Llamando σ a la longitud de la cuerda que la recta p, θ determina en K y considerando segmentos orientados es:

$$M = 2 \int_{\sigma \leq} (l - \sigma)\, d\, p\, d\, \theta = 2\, l\, L - 2\, \pi\, F + M^i \qquad [26]$$

siendo M^i la medida de los segmentos de longitud l totalmente contenidos en K [16].

De aquí y de [15] se deduce: *la probabilidad para que un segmento arbitrario de longitud* l *que corta a una figura convexa tenga los dos extremos fuera de la misma es*

$$W = \frac{2\,l\,L - 2\,\pi\,F + M^i}{2\,\pi\,F + 2\,l\,L} \qquad [27]$$

La medida de aquellos segmentos de longitud l que cortando a K tienen uno o los dos extremos interiores será la diferencia entre [15] y [26], a sea

$$M = 4\,\pi\,F - M^i \qquad [28]$$

Las fórmulas [26], [27] y [28] naturalmente se simplifican si l es mayor o igual que el diámetro de K, en cuyo caso es $M^i = 0$.

9. Fórmulas integrales para las figuras convexas planas.—Como densidad para conjuntos de segmentos en lugar de [14] podemos tomar también [13] y con ella vamos a intentar medir de nuevo los segmentos que cortando a K tienen los dos extremos fuera. Fijado el origen P (x, y) del segmento orientado, en la integral

$$\int dx\,dy\,d\varphi \qquad [29]$$

el ángulo φ puede variar hasta llenar el ángulo que forman los radios de centro P y longitud l que cortando a K dejan el otro extremo fuera. Sea ω_2 este ángulo. Por [26] será

$$\boxed{\int \omega_2\,dx\,dy = 2\,(l\,L - \pi\,F) + M^i} \qquad [30]$$

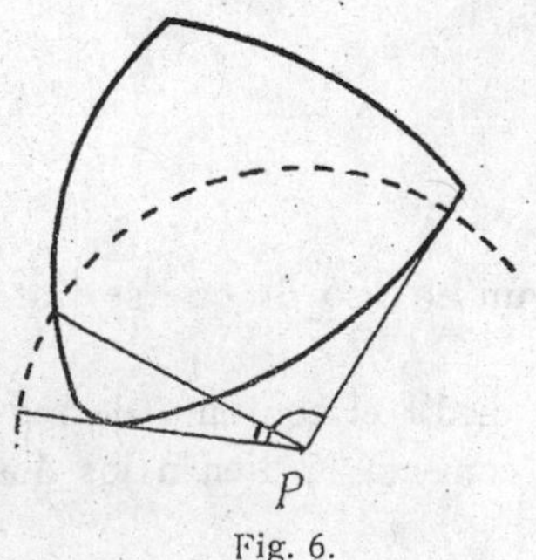

Fig. 6.

En cuanto a los restantes segmentos que cortan a K separando los completamente interiores de medida M^i, para los demás, llamando ω_1 al ángulo que llenan los radios de centro P y longitud l que cortan al contorno K en un solo punto es por [28], y teniendo en cuenta que cada segmento puede tener en P el origen o el extremo

$$\boxed{\int \omega_1\,dx\,dy = 2\,\pi\,F - M^i} \qquad [31]$$

En particular, si l es *mayor o igual que el diámetro de K*, es $M^i = 0$ y queda

$$\int \omega_2 \, dx \, dy = 2 (l \mathrm{L} - \pi \mathrm{F}) \qquad [32]$$

$$\int \omega_1 \, dx \, dy = 2 \pi \mathrm{F} \qquad [33]$$

Sumando [30] y [31] se obtiene la fórmula

$$\int (\omega_1 + \omega_2) \, dx \, dy = 2 l \mathrm{L} \qquad [34]$$

ya establecida por nosotros en otro lugar (17).

10. Otras fórmulas integrales.—Llamando σ a la longitud de la cuerda que la recta $\mathrm{R}(p, \theta)$ determina en la figura convexa K, Crofton mismo se ocupó ya de las integrales

$$\mathrm{I}_n = \int \sigma^n \, d\mathrm{R}$$

para valores naturales de n (12), (13).

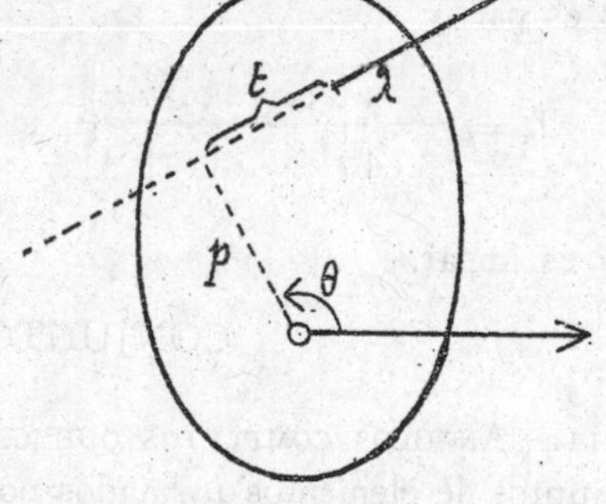

Fig. 7.

Llamando λ a la longitud de la parte del segmento S que en cada posición es interior a K, intentamos calcular nosotros

$$\mathrm{J}_n = \int \lambda^n \, d\mathrm{S} \qquad [35]$$

siendo, como sabemos, $d\mathrm{S} = dp \, d\theta \, dt$.

Con la hipótesis de ser la longitud l del segmento *igual o mayor* que el diámetro de K, no es difícil reducir el cálculo de estas integrales [35] al de las de Crofton. En efecto, se tiene

$$\mathrm{J}_n = \int \left[\int_0^\sigma t^n \, dt + \int_\sigma^l \sigma^n \, dt + \int_l^{l+\sigma} (\sigma + l - t)^n \, dt \right] dp \, d\theta =$$

$$= l \int \sigma^n \, dp \, d\theta - \frac{n-1}{n+1} \int \sigma^{n+1} \, dp \, d\theta$$

o sea

$$J_n = l I_n - \frac{n-1}{n+1} I_{n+1} \qquad [36]$$

Por ejemplo, para $n = 1, 3$ se tiene:

$$J_1 = \int \lambda \, dK = \pi \, l \, F \qquad [37]$$

$$J_3 = \int \lambda^3 \, dK = 3 \, (l - \bar{t}) \, F^2 \qquad [38]$$

siendo $\bar{t}$ la *distancia media* entre dos puntos de K.

Para el caso del círculo de radio unidad será

$$J_n = \frac{2 \cdot 4 \cdot 6 \ldots n}{3 \cdot 5 \cdot 7 \ldots (n+1)} 2^{n+1} l \pi - \frac{1 \cdot 3 \cdot 5 \ldots (n-1)^2}{2 \cdot 4 \cdot 6 \ldots (n+2)} 2^{n+1} \pi^2 \qquad [39]$$

si n es par, y

$$J_n = \frac{1 \cdot 3 \cdot 5 \ldots n}{2 \cdot 4 \cdot 6 \ldots (n+1)} 2^n l \pi^2 - \frac{2 \cdot 4 \cdot 6 \ldots (n-1)^2}{3 \cdot 5 \cdot 7 \ldots (n+2)} 2^{n+2} \pi \qquad [40]$$

si n es impar.

CONJUNTOS DE ANGULOS

11. ANGULOS COMPLETOS QUE CORTAN A UNA FIGURA CONVEXA.—Para medir conjuntos de elementos formados por pares de rectas que se cortan bajo un ángulo constante, tomamos también por densidad la cinemática definida en el número 5. Llamando α al ángulo agudo que forman dos de estas rectas vamos a medir el conjunto de ángulos cuyos lados cortan a una figura convexa K. Tomando por densidad la forma [14] se tiene (fig. 8):

$$M = \int dp \, d\theta \, dt = \int_0^\pi \overline{AB} \,.\, \Delta(\theta) \, d = \theta$$

$$= \frac{1}{\operatorname{sen} \alpha} \int_0^\pi \Delta(\theta) \,.\, \Delta(\theta - \alpha) \,.\, d\theta \qquad [41]$$

siendo $\Delta(\theta)$ la amplitud de K en la dirección perpendicular a θ.

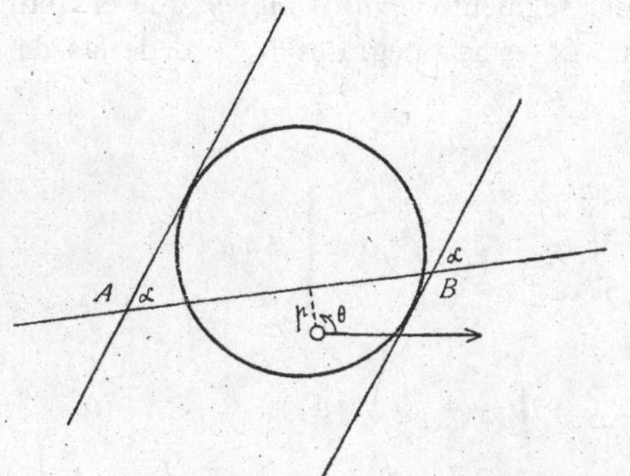

Fig. 8.

No parece fácil hallar el valor de [41] en el caso general, pero sí lo es en los casos siguientes:

a) *Figura de anchura constante.*—Es $\Delta = $ Cte.; luego

$$M = \frac{\pi \Delta^2}{\operatorname{sen} \alpha} \qquad [42]$$

b) *Segmento.*—Cuando K degenera en un segmento de longitud l es

$$\Delta(\theta) = l \cos \theta$$

y como Δ debe ser siempre positivo será

$$M = \frac{l^2}{\operatorname{sen} \alpha} \int_0^{\pi} |\cos(\theta - \alpha)| \, |\cos \theta| \, d\theta$$

y verificando la integración

$$M = \frac{l^2}{2} \left[(\pi - 2\alpha) \cot \alpha + 2 \right] \qquad [43]$$

12. Medida de los segmentos que cortan a un ángulo.—Supongamos ahora el ángulo con sus lados limitados en el vértice y veamos de medir el conjunto de los segmentos de longitud l que cortan a sus dos lados. Se tiene (fig. 9):

$$M = \int dp \, d\theta \, dt = \int_{\alpha}^{\pi} \mathrm{T} \, d\theta \qquad [44]$$

siendo T el área del triángulo O A B, o sea

$$\mathrm{T} = \frac{1}{2} \, l \,.\, \overline{\mathrm{OA}} \operatorname{sen} \theta =$$

$$= \frac{1}{2} \frac{l^2}{\operatorname{sen} \alpha} \operatorname{sen} \theta \operatorname{sen}(\theta - \alpha)$$

luego

$$M = \frac{1}{2} \frac{l^2}{\operatorname{sen} \alpha} \int_{\alpha}^{\pi} \operatorname{sen} \theta \operatorname{sen}(\theta - \alpha) \, d\theta =$$

$$= \frac{l^2}{4} \left[(\pi - \alpha) \cot \alpha + 1 \right] \qquad [45]$$

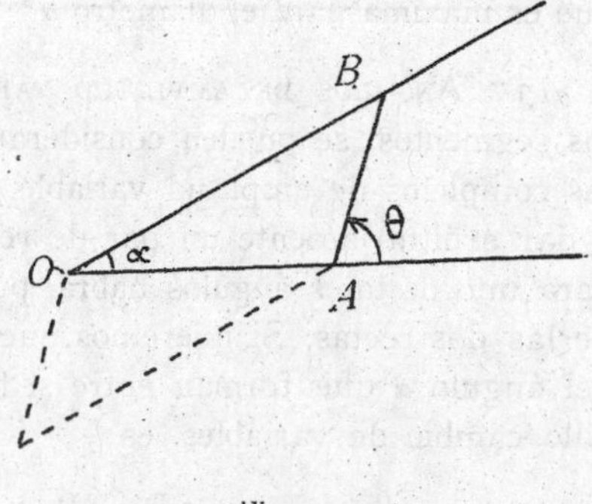

Fig. 9

De aquí se deduce que, recíprocamente, fijando el segmento y tomando como coordenadas del ángulo las x, y de su vértice y el ángulo θ de uno de sus lados con una dirección fija es, con las notaciones de la figura:

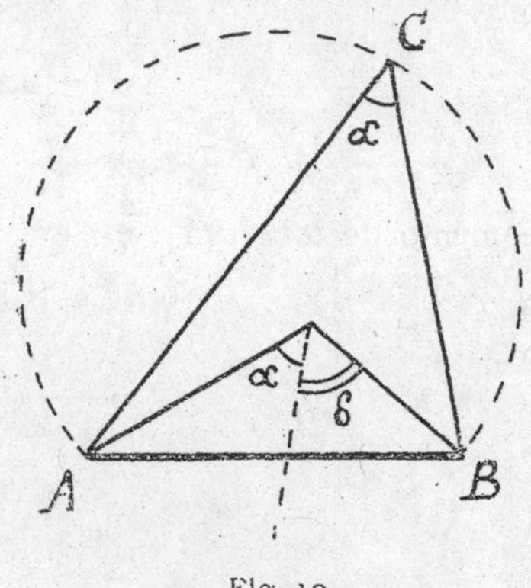

Fig. 10.

$$\int \delta \, d\,x\,d\,y = \frac{l^2}{4}\left[(\pi - \alpha)\cot\alpha + 1\right] \qquad [46]$$

extendida la integración primera al espacio limitado por A B $= l$ y el arco lugar de los puntos desde los cuales se ve l bajo el ángulo α.

Llamando $\Phi = \delta + \alpha$ y teniendo en cuenta el valor del área del segmento circular A C B se tiene

$$\int \Phi \, d\,x\,d\,y = \frac{l^2}{4}\left[\pi\cot\alpha + \frac{(\pi - \alpha)\,\alpha}{\operatorname{sen}^2\alpha} + 1\right] \qquad [47]$$

Poniendo en lugar de α el suplementario y sumando tendremos el valor de la integral anterior extendida a todo el interior del círculo:

$$\int_c \Phi \, d\,x\,d\,y = \frac{l^2}{2}\left[\frac{\alpha\,(\pi - \alpha)}{\operatorname{sen}^2\alpha} + 1\right] \qquad [48]$$

Luego: el valor medio del ángulo bajo el cual se ve una cuerda de un círculo desde sus puntos interiores es

$$\overline{\Phi} = \frac{2}{\pi}\left[\alpha\,(\pi - \alpha) + \operatorname{sen}^2\alpha\right] \qquad [49]$$

que es máxima para el diámetro $\alpha = \frac{\pi}{2}$ como se ve al anular la derivada primera.

13. Angulos de amplitud variable.—Análogamente como en el caso de los segmentos, se pueden considerar aquí conjuntos cuyos elementos sean ángulos completos de amplitud variable. Dar al azar uno de tales ángulos equivaldrá a dar arbitrariamente un par de rectas independientes entre sí. Como densidad para uno de tales ángulos habrá, pues, que tomar el producto de las densidades de las dos rectas. Si queremos que en este producto aparezca el valor variable del ángulo α que forman entre sí basta observar que, como demuestra un sencillo cambio de variables, es:

$$d\,R_1 . d\,R_2 = \operatorname{sen}\alpha \, d\,x\,d\,y\,d\,\theta\,d\,\alpha \qquad [50]$$

siendo x, y las coordenadas del vértice, θ el ángulo que fija la posición de uno de sus lados y α el ángulo de las dos rectas entre sí.

De aquí se deduce que: La medida de todos los ángulos completos de amplitud igual o menor que $\delta \leqslant \frac{\pi}{2}$ cuyos lados cortan a una figura plana convexa es, por [41]

$$M_{\leq \delta} = \int_0^{\delta} \int_0^{\pi} \Delta(\theta) . \Delta(\theta - \alpha)\, d\theta\, d\alpha \qquad [51]$$

Veamos los mismos ejemplos de antes:

a) *Figuras de anchura constante.*—Será

$$M_{\leq \delta} = \pi \Delta^2 \delta \qquad [52]$$

siendo Δ la anchura.

Como además la medida de todos los pares de rectas que cortan a una figura plana convexa es $\frac{L^2}{2}$ (contando cada par una sola vez), tendremos:

La probabilidad de que dos rectas que cortan a una figura plana convexa de anchura constante lo hagan entre sí bajo un ángulo agudo igual o menor que δ es

$$W = \frac{2\pi \Delta^2 \delta}{L^2} = \frac{2\delta}{\pi} \qquad [53]$$

puesto que para tales curvas es $L = \pi \Delta$.

b) *Segmento.*—La medida de los pares de rectas que cortan a un segmento bajo un ángulo agudo igual o menor que δ será por [43] y [51]

$$M_{\leq \delta} = \frac{l^2}{2} \int_0^{\delta} \left[(\pi - 2\alpha) \cos\alpha + 2 \operatorname{sen} \alpha \right] d\alpha =$$

$$= \frac{l^2}{2} \left[(\pi - 2\delta) \operatorname{sen} \delta + 4 (1 - \cos\delta) \right] \qquad [54]$$

y como la medida de todos los pares de rectas que cortan a un segmento de longitud l es $2l^2$, se tiene:

La probabilidad de que dos rectas que cortan a un segmento de longitud l se corten entre sí bajo un ángulo agudo igual o menor que δ es

$$W = \frac{1}{4} \left[(\pi - 2\delta) \operatorname{sen} \delta + 4 (1 - \cos\delta) \right] \qquad [55]$$

RECTAS PARALELAS

14. Densidad para conjuntos de bandas paralelas.—Consideremos ahora conjuntos formados por pares de rectas paralelas a distancia constante Δ. Uno de estos pares vendrá fijado en posición en el plano dando las coordenadas p, θ de una cualquiera de las rectas que lo forman o en general fijando la posición de otra recta cualquiera paralela a las del par e invariablemente unida a él. Por ejemplo, la paralela media. La densidad para conjuntos de tales elementos será, pues, la misma que para conjuntos de rectas, o sea (fig. 11):

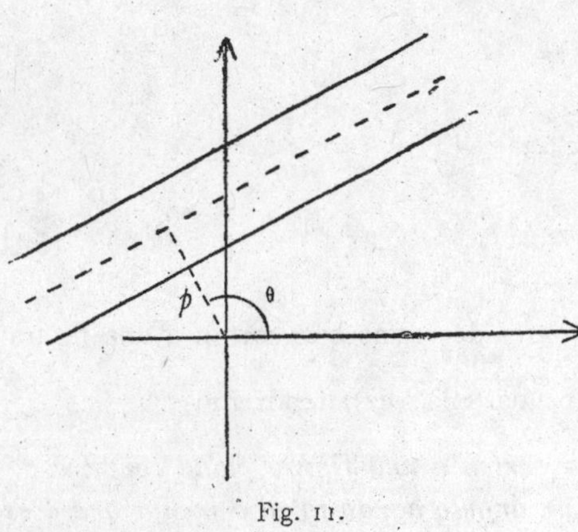

Fig. 11.

$$dP = d\,p\,d\,\theta \text{ (*)} \qquad [56]$$

Con esta expresión se puede ya medir el conjunto de bandas paralelas a distancia Δ tales que tengan algún punto común con una figura convexa de perímetro L. Basta observar, en efecto, que refiriendo [56] a la paralela media la integración deberá extenderse a todas las rectas que cortan a la figura convexa paralela a la dada a distancia $\frac{\Delta}{2}$. Luego:

$$M = \mathrm{L} + \pi\,\Delta \qquad [57]$$

En particular:

La medida del conjunto de bandas paralelas de anchura Δ que contiene a un punto es

$$M = \pi\,\Delta \qquad [58]$$

La medida del conjunto de bandas paralelas de anchura Δ que cortan a un segmento de longitud l *es*

$$M = 2\,l + \pi\,\Delta \qquad [59]$$

15. Fórmulas integrales.—Las dos fórmulas [58] y [59] nos permiten obtener algunas expresiones integrales.

I. Sea una figura plana cualquiera K fija y consideremos la medida de los pares punto-banda paralela (P . P) tales que, estando P contenido en K, pertene-

(*) Rogamos al lector procure, en todo lo que sigue, distinguir las P (representantes de bandas paralelas) de las P (representantes de puntos).

ce también a la banda P. Como densidad para estos pares habrá que tomar el producto de las densidades, luego la integral a calcular es

$$M_{\mathrm{P}.P} = \int_{\mathrm{P}.P \neq 0 \quad \mathrm{P}.K \neq 0} d\mathrm{P} \,.\, dP \qquad [60]$$

Fijando primero P y llamando f al área de la parte de K que pertenece a P es

$$M_{\mathrm{P}.P} = \int f \, dP \qquad [61]$$

Pero fijando primero el punto P y teniendo en cuenta [58] es también:

$$M_{\mathrm{P}.P} = \int \pi \Delta \, d\mathrm{P} = \pi \Delta \mathrm{F} \qquad [62]$$

siendo F el área de K. Igualando estas dos expresiones de una misma medida se tiene:

$$\boxed{\int f \, dP = \pi \Delta \mathrm{F}} \qquad [63]$$

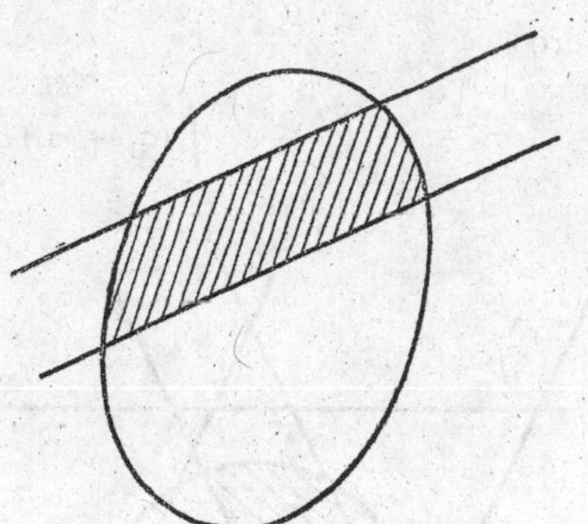

Fig. 12.

II. Si en la misma figura anterior, supuesta ahora convexa, consideramos los pares recta-banda paralela (R . P) tales que, cortando R a K la banda P corta a la cuerda intersección, tendremos por medida

$$M_{\mathrm{R}.P} = \int_{\mathrm{R}.P.K \neq 0} d\mathrm{R} \,.\, dP \qquad [64]$$

Para calcular esta integral, fijando primero P y recordando que la medida de las rectas que cortan a una figura plana convexa es igual a su perímetro, se tiene, llamando l a la longitud de la figura convexa común a K y P:

$$M_{\mathrm{R}.P} = \int l \, dP. \qquad [65]$$

En cambio, fijando primero R y llamando σ a la longitud de la cuerda que ella intercepta en K es, por [59]

$$M_{\mathrm{R}.P} = \int (2\sigma + \pi \Delta) \, d\mathrm{R} = 2\pi \mathrm{F} + \pi \Delta \mathrm{L}$$

luego:

$$\boxed{\int l\, d P = 2 \pi F + \pi \Delta L} \qquad [66]$$

El contorno l se compone de una parte que pertenece al contorno de K y cuya longitud llamaremos s más uno, dos o ningún segmento de las paralelas. Llamando σ_1 y σ_2 a estos últimos (puede ser $\sigma_1 = 0$ ó $\sigma_2 = 0$) es

$$l = \sigma_1 + \sigma_2 + s$$

pero

$$\int (\sigma_1 + \sigma_2)\, d P = 2 \int \sigma\, d R = 2 \pi F$$

luego, por [66]:

$$\int s\, d P = \pi \Delta L \qquad [67]$$

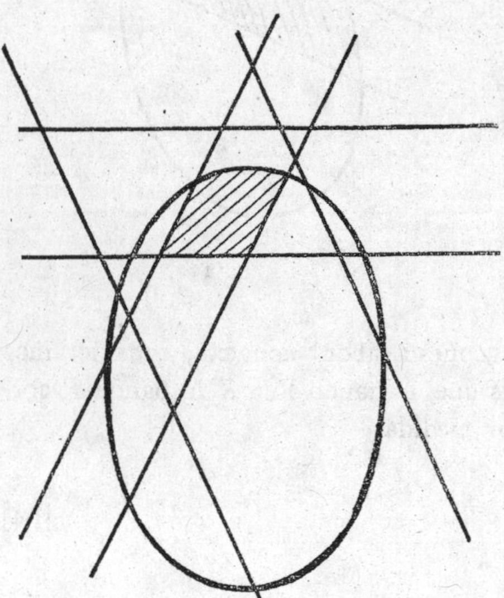

Fig. 13.

16. Generalización de las fórmulas anteriores. — En lugar de considerar un solo par de rectas paralelas podemos considerar n de ellos con anchuras respectivas Δ_i ($i = 1, 2, 3, \ldots, n$). Los problemas del número anterior pueden entonces generalizarse como sigue:

I. Intentemos medir el conjunto cuyos elementos están formados por un punto P más las n bandas P_i de tal manera que P pertenezca a la figura no necesariamente convexa K y a todas las P_i. Se trata, pues, de calcular:

$$M_{P . P_1 \ldots P_n} = \int d P . d P_1\, d P_2 \ldots d P_n \qquad [68]$$

Fijando primero las P_i e integrando respecto dP se tiene

$$M_{P . P_1 \ldots P_n} = \int f_{1 2 3 \ldots n}\, d P_1\, d P_2 \ldots d P_n \qquad [69]$$

llamando $f_{1 3 2 \ldots n}$ al área común a K y a $P_1, P_2, P_3, \ldots, P_n$.

En cambio, fijando primero P, $P_1, P_2, \ldots, P_{n-1}$ e integrando respecto dP_n se tiene por [58]

$$M_{P.P_1.P_2 \ldots P_n} = \int \pi \Delta_n \, dP \, dP_1 \, dP_2 \ldots dP_{n-1} \qquad [70]$$

y procediendo sucesivamente

$$M_{P.P_1.P_2 \ldots P_n} = \pi^n \Delta_1 \Delta_2 \ldots \Delta_n F \qquad [71]$$

luego:

$$\boxed{\int f_{123\ldots n} \, dP_1 \, dP_2 \ldots dP_n = \pi^n \Delta_1 \Delta_2 \ldots \Delta_n F} \qquad [72]$$

El *valor* medio para esta área $f_{123\ldots n}$ será, por tanto:

$$\bar{f}_{123\ldots n} = \frac{\int f_{123\ldots n} \, dP_1 \, dP_2 \ldots dP_n}{\int dP_1 \, dP_2 \ldots dP_n} =$$

$$= \frac{\pi^n \Delta_1 \Delta_2 \ldots \Delta_n F}{(L + \pi \Delta_1)(L + \pi \Delta_2) \ldots (L + \pi \Delta_n)}$$

De esta fórmula [72] se deduce también:

Dado un punto arbitrario en el interior de una figura cualquiera y n *bandas paralelas que cortan a la misma, la probabilidad de que el punto pertenezca a la vez a las* n *bandas es:*

$$W = \frac{\pi^n \Delta_1 \Delta_2 \ldots \Delta_n}{(L + \pi \Delta_1)(L + \pi \Delta_2) \ldots (L + \pi \Delta_n)}$$

II. Sea ahora el problema análogo al anterior (supuesta ahora K convexa), pero en lugar del punto P consideremos la recta R. Es decir, tratemos de calcular la integral:

$$M_{R.P_1.P_2 \ldots P_n} = \int dR \, dP_1 \, dP_2 \ldots dP_n \qquad [73]$$

extendida la integración a todas las posiciones de R y P_i en las cuales la recta R, las bandas P_i y la figura convexa K tienen puntos comunes.

Fijando primero $P_1, P_2, P_3, \ldots, P_n$ es

$$M_{R.P_1.P_2\ldots P_n} = \int l_{123\ldots n}\, dP_1\, dP_2 \ldots dP_n \qquad [74]$$

siendo $l_{123\ldots n}$ el perímetro de la figura convexa común a K y a todas las P_i.

En cambio, fijando primero R y $P_1, P_2, P_3, \ldots, P_{n-1}$ se tiene, llamando $\sigma_{123\ldots n-1}$ a la longitud de la cuerda que R intercepta en la figura intersección de $P_1, P_2, \ldots, P_{n-1}$ y K y por [59]

$$M_{R.P_1.P_2\ldots P_n} = \int (2\sigma_{123\ldots n-1} + \pi\Delta_n)\, dR\, dP_1\, dP_2 \ldots dP_{n-1} =$$

$$= 2\pi \int f_{123\ldots n-1}\, dP_1\, dP_2 \ldots dP_{n-1} + \pi\Delta_n \int dR\, dP_1\, dP_2 \ldots dP_{n-1} \qquad [75]$$

Esta fórmula recurrente, teniendo en cuenta [72], nos da

$$M_{R.P_1.P_2\ldots P} = \pi^n \Delta_1 \Delta_2 \ldots \Delta_n L + 2\pi^n \sum_i \Delta_1 \Delta_2 \ldots \Delta_{i-1} F \Delta_{i+1} \ldots \Delta_n \qquad [76]$$

Igualando ambas expresiones de la medida buscada tenemos:

$$\boxed{\begin{aligned}\int l_{123\ldots n}\, dP_1\, dP_2 \ldots dP_n = \pi^n \Delta_1 \Delta_2 \ldots \Delta_n L + \\ + 2\pi^n \sum_i \Delta_1 \Delta_2 \ldots \Delta_{i-1} F \Delta_{i+1} \ldots \Delta_n\end{aligned}} \qquad [77]$$

El *valor medio* para esta longitud $l_{123\ldots n}$ será, pues:

$$\bar{l}_{123\ldots n} = \frac{\int l_{123\ldots n}\, dP_1\, dP_2 \ldots dP_n}{\int dP_1\, dP_2 \ldots dP_n} =$$

$$= \frac{\pi^n \Delta_1 \Delta_2 \ldots \Delta_n L + 2\pi^n F \sum_i \Delta_1 \Delta_2 \ldots \Delta_{i-1} \Delta_{i+1} \ldots \Delta_n}{(L + \pi\Delta_1)(L + \pi\Delta_2) \ldots (L + \pi\Delta_n)}$$

También:

Dadas n *bandas paralelas que cortan a una figura convexa* K *y una recta arbitraria que corta a la misma, la probabilidad de que la recta y las* n *bandas tengan algún punto común interior a* K *es*:

$$W = \frac{\pi^n \Delta_1 \Delta_2 \ldots \Delta_n L + 2\pi^n F \sum_i \Delta_1 \Delta_2 \Delta_{i-1} \Delta_{i+1} \ldots \Delta_n}{L(L+\pi\Delta_1)(L+\pi\Delta_2)\ldots(L+\pi\Delta_n)}$$

III. Consideremos la integral

$$I_n = \int_{K.P_1.P_2\ldots P_n \neq 0} d P_1 \, d P_2 \ldots d P_n \qquad [78]$$

que representa la medida del conjunto de las posiciones de las n bandas paralelas P_i en las cuales tienen todas algún punto común que también pertenece a K.

Fijando primero $P_1, P_2, P_3, \ldots, P_{n-1}$ y llamando $l_{123\ldots n-1}$ a la longitud de la figura común a estas $n-1$ bandas y a K, se tiene, por [57]

$$I_n = \int (l_{123\ldots n-1} + \pi \Delta_n) \, d P_1 \, d P_2 \ldots d P_{n-1} \qquad [79]$$

o sea por [77]

$$\begin{aligned} I_n = {} & \pi^{n-1} \Delta_1 \Delta_2 \ldots \Delta_{n-1} L + \\ & + 2\pi^{n-1} \sum_i \Delta_1 \Delta_2 \ldots \Delta_{i-1} F \Delta_{i+1} \ldots \Delta_{n-1} + \pi \Delta_n I_{n-1} \end{aligned} \qquad [80]$$

Esta fórmula recurrente nos da:

$$\boxed{\begin{aligned} I_n = {} & \pi^{n-1} L \sum_i \Delta_1 \Delta_2 \ldots \Delta_{i-1} \Delta_{i+1} \ldots \Delta_n + \\ & + 2\pi^{n-1} F \sum_{i,j} \Delta_1 \ldots \Delta_{i-1} \Delta_{i+1} \ldots \Delta_{j-1} \Delta_{j+1} \ldots \Delta_n + \pi^n \Delta_1 \Delta_2 \ldots \Delta_n \end{aligned}} \qquad [81]$$

Fórmula que nos permite resolver el problema:

Dadas n *bandas paralelas que cortan a una figura convexa, ¿cuál es la probabilidad de que todas ellas tengan algún punto común interior a dicha figura?*

Será

$$W = \frac{I_n}{(L+\pi\Delta_1)(L+\pi\Delta_2)\ldots(L+\pi\Delta_n)} \qquad [82]$$

17. Generalización del problema de la aguja de Buffon.—Sean n rectas paralelas entre sí y separadas por las distancias respectivas Δ_i $(i = 1, 2, 3, \ldots, n)$, Considerando el sistema de estas n rectas paralelas como móviles en conjunto, pero manteniendo constante su posición relativa, evidentemente su posición en el plano quedará determinada por las coordenadas p, θ de una de ellas. Por otra parte, recordando que la medida de las rectas que cortan a una figura convexa K es su longitud, se tendrá que la medida de todas las posiciones de estas n paralelas en las que cortan a K (es decir, a su contorno) y en la hipótesis de que esta figura convexa no pueda ser cortada al mismo tiempo por dos paralelas del conjunto, será:

$$M = n\,L \qquad [83]$$

En particular, si K degenera en un segmento de longitud l es

$$M = 2\,n\,l \qquad [84]$$

En cambio, considerando la banda de plano que limitan entre sí la primera y la última de estas paralelas, la medida de aquellas posiciones en que esta banda tiene algún punto común con el segmento es, por [59]

$$2\,l + \pi \sum_1^n \Delta_i \qquad [85]$$

Luego:

Sabiendo que el segmento de longitud l tiene algún punto común con la banda limitada por la primera y la última de las n paralelas, la probabilidad de que corte a alguna de estas paralelas es

$$W = \frac{2\,n\,l}{2\,l + \pi \sum_1^n \Delta_i} \qquad [86]$$

Si el número n tiende a infinito y las bandas limitadas por las paralelas tienden a ocupar todo el plano, es

$$W = \frac{2\,l}{\pi \lim_{n \to \infty} \dfrac{\sum \Delta_i}{n}} \qquad [87]$$

Por ejemplo, si $\Delta_i = \Delta =$ Cte. es

$$W = \frac{2\,l}{\pi\,\Delta} \qquad [88]$$

que es el resultado conocido para rectas paralelas equidistantes.

Si, en cambio, las anchuras Δ_i varían alternativamente de manera que, por ejemplo, sea

$$\Delta_1 = \Delta_1 \ , \quad \Delta_2 = \Delta_3 = 2\,\Delta_1 \ , \quad \Delta_4 = \Delta_5 = \Delta_1 \ , \quad \Delta_6 = \Delta_7 = 2\,\Delta_1 \, , \ldots$$

entonces es

$$\lim_{n\to\infty} \frac{\sum_1^n \Delta_i}{n} = \frac{3}{2}\,\Delta_1$$

y por tanto

$$W = \frac{4\,l}{3\,\pi\,\Delta_1} \qquad [89]$$

18. Línea plana cortada por bandas paralelas.—Recordemos como aplicación de una fórmula de Poincaré (6), (13), (17), que llamando n al número de puntos en que una línea cualquiera (abierta o cerrada) móvil en su plano con densidad cinemática dK corta a un segmento fijo de longitud l se verifica:

$$\int n\,dK = 4\,l\,L \qquad [90]$$

siendo L la longitud de dicha línea.

Sentado esto consideremos un par de rectas paralelas fijas (fig. 14) a distancia Δ. Una recta R de su plano puede venir determinada por el ángulo ψ que forma con estas paralelas y la abscisa t del punto de intersección. Un sencillo cambio de variables da:

$$dR = dp\,d\theta = \operatorname{sen}\psi\,dt\,d\psi \qquad [91]$$

Una línea cualquiera de longitud L viene determinada, salvo una traslación en la dirección de las paralelas por la distancia h de uno de sus puntos a

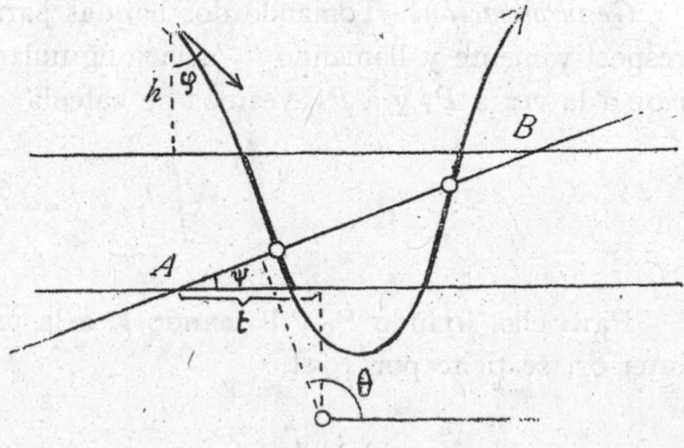

Fig. 14.

una de las paralelas y el ángulo de giro φ. Consideremos

$$I = \int n \, d\,h\,d\,\varphi\,d\,p\,d\,\theta \qquad [92]$$

extendida a todas las posiciones de la línea y de la recta en que se cortan interiormente a la banda P de las paralelas y siendo n el número de estos puntos de intersección. Fijando h, φ y llamando l a la longitud de la parte de línea interior a P queda

$$I = 2 \int l \, d\,h\,d\,\varphi \qquad [93]$$

En cambio, fijando R y observando que lo mismo da suponer una traslación en la dirección de las paralelas por parte de la curva que por parte de la recta queda por [90]

$$I = \int n \operatorname{sen} \psi \, d\,h\,d\,\varphi\,d\,t\,d\,\psi = 4 \int^{\frac{\pi}{2}} \mathrm{L}\,\overline{\mathrm{AB}} \operatorname{sen} \psi \, d\,\psi = 2\,\pi\,\mathrm{L}\,\Delta \qquad [94]$$

Comparando estos dos valores de I y observando que $d\,h\,d\,\varphi$ es también la densidad para conjuntos de paralelas es

$$\boxed{\int l \, d\,P = \pi\,\mathrm{L}\,\Delta} \qquad [95]$$

que comprende a [67] como caso particular.

Generalización.—Tomando dos bandas paralelas P_1 y P_2 de anchuras Δ_1 y Δ_2 respectivamente y llamando l_{12} a la longitud de la parte de la línea que es interior a la vez a P_1 y a P_2, veamos de calcular

$$\int l_{12} \, d\,P_1\,d\,P_2$$

Para ello, fijando P_1 y llamando l_1 a la parte de línea que comprende en su interior, se tiene por [95]

$$\int l_{12} \, d\,P_1\,d\,P_2 = \pi\,\Delta_2 \int l_1 \, d\,P_1$$

de donde

$$\int l_{12}\, dP_1\, dP_2 = \pi^2 \Delta_1 \Delta_2 L$$

El procedimiento es general y nos da:

$$\boxed{\int l_{12\ldots n}\, dP_1\, dP_2 \ldots dP_n = \pi^n \Delta_1 \Delta_2 \ldots \Delta_n L}$$

PROBLEMAS EN EL ESPACIO

19. DENSIDAD CINEMÁTICA.—Toda figura del espacio, línea, superficie o cuerpo viene determinada en posición fijando un sistema de ejes coordenados invariablemente unidos a la misma. Como densidad para un conjunto de tales sistemas de ejes o densidad cinemática se tiene

$$dK = dP\, d\Omega\, d\tau \qquad [96]$$

siendo $dP = dx \,.\, dy \,.\, dz$ la densidad correspondiente al vértice, $d\Omega = \operatorname{sen} \theta\, d\theta\, d\varphi$ el elemento de área correspondiente al punto imagen sobre la esfera unidad de un eje por este origen y $d\tau$ un giro elemental alrededor de este eje. Las propiedades de invariancia de esta densidad cinemática pueden verse en el trabajo de BLASCHKE (14).

CONJUNTOS DE SEGMENTOS

20. CONJUNTOS DE SEGMENTOS DE LONGITUD CONSTANTE.—Llamando dR a la densidad para conjuntos de rectas en el espacio (11), (13) (*), la densidad cine-

(*) Cortando la recta R por un plano perpendicular, llamando $dx \,.\, dy$ al elemento de área del mismo correspondiente al punto de intersección y $d\Omega$ a la imagen esférica de la recta, esta densidad es

$$dR = d\Omega\, dx\, dy$$

Por depender las rectas del espacio de cuatro parámetros, esta forma diferencial es naturalmente de cuarto grado.

mática [96] puede también escribirse

$$dK = dR\, dt\, d\tau \qquad [97]$$

siendo dt la densidad sobre esta recta.

Vamos a aplicar esta fórmula a la medida de todos los segmentos orientados de longitud l que cortan a un cuerpo convexo K de área F y volumen V. En este caso podemos prescindir del factor $d\tau$ y tomar por densidad del segmento $dS = dR\, dt$. Con esto y llamando σ a la longitud de la cuerda que la recta R del segmento determina en K, es

$$M_{\overrightarrow{S}} = 2\int (\sigma + l)\, dR = 4\pi V + \pi l F \qquad [98]$$

luego: *La esfera es el cuerpo que para un volumen dado tiene menor número de segmentos que lo cortan.*

21. Segmentos contenidos en un cuerpo convexo.—Al intentar medir el conjunto de aquellos segmentos de longitud l que están totalmente contenidos en K nos encontramos con la expresión

$$M^i = 2\int_{\sigma > l} (\sigma - l)\, dR = 2\int V^*\, d\Omega - 2l\int A^*\, d\Omega \qquad [99]$$

donde V* es el volumen limitado por las cuerdas de K paralelas a la dirección $d\Omega$ y de longitud l y A* el área de la proyección de este volumen sobre un plano perpendicular a la misma dirección. Las integraciones están extendidas sobre la esfera unidad.

El llevar a cabo las integraciones que figuran en [99] presenta en general dificultades. Para el caso de la esfera es muy fácil observando que V* y A* son independientes de $d\Omega$ y valen

$$V^* = \pi\left(r^2 + \frac{l^2}{4}\right)l + \frac{2}{3}\pi\left(r - \frac{l}{2}\right)^2\left(2r + \frac{l}{2}\right)$$

$$A^* = \pi\left(r^2 - \frac{l^2}{4}\right)$$

siendo r el radio de la esfera. De aquí

$$M^i = \frac{8}{3}\pi^2\left(r - \frac{l}{2}\right)^2\left(2r + \frac{l}{2}\right) \qquad [100]$$

22. CONJUNTOS DE SEGMENTOS DE LONGITUD VARIABLE.—Si del segmento se supone también variable su longitud, su determinación depende ya de seis parámetros. Y como viene fijado por sus dos puntos extremos entenderemos por dar un segmento arbitrario de longitud arbitraria el dar un par de puntos independientemente el uno del otro. Luego la densidad para tales segmentos será $dP_1 \cdot dP_2$ siendo P_1 y P_2 los extremos del segmento. Esta expresión de la densidad se puede transformar de manera que aparezca la longitud l variable del segmento. Queda entonces (16)

$$d\,P_1 \cdot d\,P_2 = l^2\, d\,R\, d\,t\, d\,l \qquad [101]$$

Luego para medir el conjunto de segmentos de longitud igual o menor que b que cortan a un cuerpo convexo K se tiene

$$M_{<b} = \int_0^b l^2 (4\pi V + \pi l F)\, d\,l = \frac{4}{3}\pi V b^3 + \frac{1}{4}\pi b^4 F \qquad [102]$$

Si sólo se consideran segmentos totalmente interiores es

$$M^i_{<b} = \int_0^b l^2 M^i d\,l \qquad [103]$$

Por ejemplo, para una esfera de radio r:

$$M^i_{\leq b} = \left(\frac{4}{3}\pi\right)^2 \left[r^3 - \frac{9}{16} b r^2 + \frac{1}{32} b^3\right] b^3 \qquad [104]$$

23. OTROS CONJUNTOS DE SEGMENTOS.—Consideremos el conjunto de segmentos orientados de longitud l tales que, cortando a un cuerpo convexo K, tienen los dos extremos fuera del mismo. Para medida de tal conjunto se tiene:

$$M_2 = 2\int_{\sigma \leq} (l - \sigma)\, d\,R = \pi F l - 4\pi V + 2 M^i \qquad [105]$$

siendo M^i la medida dada por [99].

Del valor de la medida total [98] y de la anterior se deduce que la medida de aquellos segmentos que cortando a K tienen uno o los dos extremos interiores es

$$M_{1,0} = 8\pi V - 2 M^i \qquad [106]$$

Estas medidas [105] y [106] se simplifican si l es mayor o igual que el diámetro de K, en cuyo caso, naturalmente, es $M^i = 0$.

Las fórmulas [105] y [98] nos permiten además resolver el problema:

Dado un segmento de longitud l *que corta al cuerpo convexo* K, *¿cuál es la probabilidad de que tenga los dos extremos exteriores?*

Será

$$W = \frac{\pi F l - 4\pi V + 2 M^i}{4\pi V + \pi l F}$$

24. Fórmulas integrales para los cuerpos convexos.—Al medir los mismos conjuntos del número anterior, pero tomando por densidad del segmento la expresión $dP \,.\, d\Omega$ equivalente a la $dR \,.\, dt$ usada, resulta, llamando Ω_P^{II} al ángulo sólido que llenan los radios de centro P y longitud l que cortan al contorno de K en dos puntos:

$$M_2 = \int \Omega_P^{II}\, dP = \pi F l - 4\pi V + 2 M^i \qquad [107]$$

Análogamente, si llamamos Ω_P^{I} al ángulo sólido que llenan los segmentos con el origen en P y longitud l que cortan al contorno de K en un solo punto, quitando de [106] la medida de aquéllos totalmente interiores y teniendo en cuenta que cada punto P puede ser origen o extremo, resulta:

$$M_1 = \int \Omega_P^{I}\, dP = 4\pi V - 2 M^i \qquad [108]$$

En el caso de ser l igual o mayor que el diámetro de K, estas fórmulas se simplifican, dando

$$\int \Omega_P^{II}\, dP = \pi F l - 4\pi V \qquad [109]$$

$$\int \Omega_P^{I}\, dP = 4\pi V \qquad [110]$$

extendidas las integraciones a los puntos P exteriores a K.

Sumando [107] y [108] resulta

$$\int (\Omega_P^{I} + \Omega_P^{II})\, dP = \pi l F \qquad [111]$$

fórmula ya obtenida por nosotros en otro lugar (18).

25. Otras fórmulas integrales.—Por el mismo razonamiento que en el número 10, llamando aquí λ a la parte de segmento que es interior a K y siendo la longitud l igual o mayor que el diámetro de este cuerpo convexo, se obtiene

$$\int \lambda^n \, d\mathrm{S} = l \int \sigma^n \, d\mathrm{R} - \frac{n-1}{n+1} \int \sigma^{n+1} \, d\mathrm{R} \qquad [112]$$

siendo $d\mathrm{S} = d\mathrm{R}\, dt$ la densidad para conjuntos de segmentos y σ la longitud de la cuerda que la recta R determina en K.

Por ejemplo, es

$$\int \lambda \, d\mathrm{S} = 2\pi l \mathrm{V}$$

$$\int \lambda^4 \, d\mathrm{S} = 6\,\mathrm{V}^2 (l - \bar{t})$$

siendo $\bar{t}$ la distancia media entre dos puntos de K.

Para la esfera es

$$\int \lambda^n \, d\mathrm{S} = \frac{2^{n+2}}{n+2} l \pi^2 - \frac{n-1}{n+1} \cdot \frac{2^{n+3}}{n+3} \pi^2$$

CONJUNTOS DE ÁNGULOS DIEDROS

26. Angulos diedros completos que cortan a un cuerpo convexo.—Un ángulo diedro completo de abertura $\alpha \leqslant \frac{\pi}{2}$ queda determinado en el espacio por cinco parámetros. Estos pueden ser, por ejemplo, los tres que determinan uno de sus planos y los dos que en este plano determinan a la arista. Representando por $d\mathrm{E}$ a la densidad para conjuntos de planos y $d\mathrm{R}_e$ a la de las rectas de uno de estos planos, se tendrá por densidad de ángulos diedros completos de abertura constante $d\mathrm{E} \,.\, d\mathrm{R}_e$.

Conocida ya la densidad, para obtener la medida del conjunto de aquellos cuyos dos planos cortan a un cuerpo convexo K, se procede como en el caso del plano (N.° 11), resultando:

$$M = \frac{1}{\operatorname{sen} \alpha} \int \Delta(\Omega, \theta^*, \alpha) \,.\, \Delta(\Omega) \,.\, d\theta^* \,.\, d\Omega \qquad [113]$$

donde $\Delta(\Omega, \theta^*, \alpha)$ indica la anchura del cuerpo en la dirección que dista α de un eje de dirección $d\Omega$ y ha girado θ^* alrededor de este eje. Las integraciones están extendidas de o a π respecto $d\theta^*$ y sobre la semiesfera unidad respecto $d\Omega$.

Un caso de integración inmediata es aquel en que K es un cuerpo de anchura constante. Siendo $\Delta(\Omega, \theta^*, \alpha) = \Delta(\Omega) = \text{Cte.}$ es

$$M = \frac{2\pi^2 \Delta^2}{\operatorname{sen} \alpha} \qquad [114]$$

27. Angulos diedros de amplitud variable.—Dar un ángulo diedro completo de magnitud arbitraria equivale a dar un par de planos independientes entre sí. Por densidad de un tal elemento debemos tomar por tanto $dE_1 . dE_2$. Para que aparezca el valor α del ángulo agudo que forman los dos planos observemos la identidad (16)

$$dE_1 . dE_2 = \operatorname{sen}^2 \alpha \, dE_1 \, dR_e \, d\alpha \qquad [115]$$

Con esta expresión podemos ya medir el conjunto de los ángulos diedros completos iguales o menores que $\delta \left(\delta \leqslant \frac{\pi}{2}\right)$ cuyas dos caras cortan a un cuerpo convexo K. En efecto, según [113] y [115], queda

$$M_{\leq \delta} = \int \operatorname{sen} \alpha . \Delta(\Omega, \theta^*, \alpha) . \Delta(\Omega) . d\theta^* \, d\Omega \, d\alpha \qquad [116]$$

Para el caso de los cuerpos de anchura constante es

$$M_{\leq \delta} = \int_0^\delta 2\pi^2 \Delta^2 \operatorname{sen} \alpha \, d\alpha = 2\pi^2 \Delta^2 (1 - \cos \delta) \qquad [117]$$

Recordemos ahora que la medida de los planos que cortan a un cuerpo convexo es la integral de la curvatura media M, luego la medida de los pares de planos que lo cortan contando cada par una sola vez será $\frac{M^2}{2}$. Además, para los cuerpos de anchura constante es $M = 2\pi\Delta$, luego:

La probabilidad de que dos planos que cortan a un cuerpo convexo de anchura constante lo hagan entre sí bajo un ángulo agudo igual o menor que δ es

$$W = 1 - \cos \delta \qquad [118]$$

CONJUNTOS DE PARALELAS Y PLANOS PARALELOS

28. Bandas paralelas planas móviles en el espacio.—La porción de plano limitada por dos rectas paralelas puede fijarse en el espacio bien fijando primero el plano E que la contiene y luego en él una de las paralelas, o bien fijando primero una de éstas y luego el ángulo que ha girado el plano a su alrededor. Según se considere una u otra de estas dos determinaciones, tendremos las dos densidades equivalentes siguientes para conjuntos de bandas de plano en el espacio:

$$dE \,.\, dR_e = dR \,.\, d\theta \qquad [119]$$

Que esta igualdad es cierta, se ve sin más que observar que a los dos miembros les falta sólo la traslación en la dirección de la recta para obtener la densidad cinemática.

Haciendo uso de la primera expresión de [119] y considerando como distintas las bandas que resultan al permutar entre sí las paralelas, obtenemos como medida del conjunto de ellas que cortan a un cuerpo convexo K (*):

$$\int dE\, dR_e = 2 \int (u + \pi \Delta)\, dE = \pi^2 F + 2\pi \Delta M \qquad [120]$$

siendo u la longitud de la sección que E determina en K, Δ la distancia constante entre las dos paralelas y M la medida de los planos que cortan a K (igual por tanto a su curvatura media íntegra cuando la superficie sea de curvatura continua).

Pero tomando la segunda expresión de [119] para medir el mismo conjunto, observamos que, si R corta a K, el ángulo θ puede variar de 0 a 2π, mientras que si R no corta a K, el ángulo θ varía de 0 al valor Ω_R del ángulo diedro que llenan las bandas paralelas de eje R y amplitud Δ que cortan a K, o sea:

$$\int dR\, d\theta = 2\pi \int dR + \int \Omega_R\, dR \qquad [121]$$

y como la medida de las rectas que cortan a K es $\frac{\pi}{2}$ F resulta, igualando esta medida con la [120] antes encontrada:

(*) Recordemos las fórmulas:

$$\int u\, dE = \frac{\pi^2}{2} F \qquad \int dE = M$$

$$\boxed{\int \Omega_R \, dR = 2\pi \Delta M} \qquad [122]$$

extendida la integración a las rectas R que no cortan a K.

29. Conjuntos de planos paralelos.—En un conjunto de pares de planos paralelos a distancia constante Δ, uno de ellos viene determinado dando la posición de otro plano paralelo invariablemente unido al par. Por ejemplo, dando el plano paralelo medio. Por densidad para un tal conjunto de pares de planos paralelos podemos, pues, tomar la forma de tercer grado:

$$dP = dp \,.\, d\Omega \qquad [123]$$

igual a la densidad para conjuntos de planos.

Uno cualquiera de estos pares de planos limita una franja indefinida de espacio de anchura Δ. Para medir el conjunto de estas franjas que tienen algún punto común con un cuerpo convexo K se observa que esta medida será la misma que la de los planos que cortan al cuerpo paralelo exterior al K a distancia $\frac{\Delta}{2}$, o sea:

$$M = \mathrm{M} + 2\pi\Delta \qquad [124]$$

siendo M, como siempre, la medida de los planos que cortan a K o integral de curvatura media.

En particular:

La medida del conjunto de franjas espaciales de anchura Δ que contienen a un punto fijo P es

$$M = 2\pi\Delta \qquad [125]$$

La medida del conjunto de franjas espaciales de anchura Δ que tienen algún punto común con un segmento fijo de longitud l *es*

$$M = \pi l + 2\pi\Delta \qquad [126]$$

La medida del conjunto de franjas espaciales de anchura Δ que tienen algún punto común con una figura plana y convexa de perímetro L es

$$M = \frac{\pi}{2} \mathrm{L} + 2\pi\Delta \qquad [127]$$

30. Fórmulas integrales para los cuerpos convexos.—Estas tres fórmu-

las últimas van a servirnos para hallar unas expresiones integrales análogas a las encontradas en el plano (N.º 15).

I. Sea un cuerpo no necesariamente convexo K y consideremos el conjunto de los elementos formados por un par de planos P paralelos a distancia Δ y un punto P con la condición de estar este punto al mismo tiempo en K y en la franja de espacio que limitan los planos P. Como medida de este conjunto se tiene

$$M_{\mathrm{P}\,.\,P} = \int d\mathrm{P}\,.\,dP \qquad [128]$$

Para calcular esta integral, fijando primero P y llamando v al volumen de la sección de estos planos con K, queda

$$M_{\mathrm{P}\,.\,P} = \int v\,dP \qquad [129]$$

En cambio, fijando P y recordando [125], queda

$$M_{\mathrm{P}\,.\,P} = \int 2\pi\Delta\,d\mathrm{P} = 2\pi\Delta\,\mathrm{V} \qquad [130]$$

siendo V el volumen de K. Luego:

$$\boxed{\int v\,dP = 2\pi\Delta V} \qquad [131]$$

II. Impongamos ahora a K la condición de ser convexo e intentemos medir el conjunto formado por los pares recta-franja de planos paralelos a distancia Δ tales que ambos elementos tengan puntos comunes interiores a K. Esta medida será:

$$M_{\mathrm{R}\,.\,P} = \int d\mathrm{R}\,.\,dP \qquad [132]$$

Fijando primero P y llamando f ai área del cuerpo convexo intersección de K con la franja P una primera integración nos da

$$M_{\mathrm{R}\,.\,P} = \frac{\pi}{2}\int f\,dP \qquad [133]$$

En cambio, fijando primero la recta R, llamando σ a la longitud de la cuerda que ella determina en K y recordando [126], queda:

$$M_{R.P} = \int (\pi\sigma + 2\pi\Delta)\, dR = 2\pi^2 V + \pi^2 \Delta F$$

De aquí:

$$\boxed{\int f\, dP = 2\pi(2V + \Delta F)} \qquad [134]$$

El área f se compone en general de dos áreas planas σ_1 y σ_2 correspondientes a porciones de los planos P y a una parte de área f^* perteneciente a la superficie de K. Se tiene, pues:

$$\int f\, dP = \int (\sigma_1 + \sigma_2 + f^*)\, dP = 2\int \sigma_1\, dP + \int f^*\, dP = 4\pi V + \\ + \int f^*\, dP \qquad [135]$$

que comparado con [134] nos da

$$\int f^*\, dP = 2\pi\Delta F \qquad [136]$$

III. En el mismo cuerpo convexo K anterior consideremos el conjunto de los elementos formados por un plano E y un par de planos paralelos P de manera que su intersección tenga puntos interiores a K. La medida de este conjunto es

$$M_{E.P} = \int dE\,.\,dP \qquad [137]$$

Fijado primero P y llamando m a la medida de los planos que cortan a la parte común a P y a K (*) se tiene:

$$M_{E.P} = \int m\, dP \qquad [138]$$

(*) Como esta intersección en general tendrá líneas de discontinuidad para la curvatura no puede decirse que m sea la integral de la curvatura media, pero puede considerarse como límite de la del cuerpo paralelo exterior a distancia ε cuando ε tiende a cero.

Pero fijando primero E y llamando u a la longitud de su intersección con K y recordando [127] queda

$$M_{E.P} = \int \left(\frac{\pi}{2} u + 2\pi\Delta \right) dE = \frac{\pi^3}{4} F + 2\pi\Delta M \qquad [139]$$

De [138] y [139] se deduce:

$$\boxed{\int m \, dP = \frac{\pi^3}{4} F + 2\pi\Delta M} \qquad [140]$$

31. Generalización de las fórmulas anteriores.—En lugar de considerar un solo par de planos paralelos podemos considerar n de ellos con anchuras respectivas Δ_i ($i = 1, 2, \ldots, n$) con lo cual los problemas del número anterior se generalizan como sigue:

I. Propongámonos medir el conjunto cuyos elementos están formados por un punto más n franjas paralelas P_i de tal manera que P pertenezca al cuerpo, no necesariamente convexo K, y a todas las P_i. Se trata, pues, de calcular

$$M_{P.P_1.P_2\ldots P_n} = \int dP \,.\, dP_1 \,.\, dP_2 \ldots dP_n \qquad [141]$$

Fijando primero $P_1, P_2, \ldots, P_n$ el punto P puede variar por todo el espacio común a K y a las n franjas. Llamando $v_{123\ldots n}$ a este volumen es

$$M_{P.P_1 P_2\ldots P_n} = \int v_{123\ldots n} \, dP_1 \, dP_2 \ldots dP_n \qquad [142]$$

En cambio fijando primero $P, P_1, P_2, \ldots, P_{n-1}$ e integrando respecto dP_n queda, por [125]

$$M_{P.P_1.P_2\ldots P_n} = 2\pi\Delta_n M_{P\,P_1\,P_2\ldots P_{n-1}}$$

Procediendo sucesivamente

$$M_{P.P_1.P_2\ldots P_n} = (2\pi)^n \Delta_1 \Delta_2 \Delta_3 \ldots \Delta_n V \qquad [143]$$

Lo cual, igualado con [142] nos da

$$\boxed{\int v_{123\ldots n} \, dP_1 \, dP_2 \ldots dP_n = (2\pi)^n \Delta_1 \Delta_2 \ldots \Delta_n V} \qquad [144]$$

El *valor medio* para este volumen $v_{1\,2\,3\ldots n}$ será por tanto

$$\bar{v}_{1\,2\,3\ldots n} = \frac{\int v_{1\,2\,3\ldots n}\, d\,P_1\, d\,P_2 \ldots d\,P_n}{\int d\,P_1\, d\,P_2 \ldots d\,P_n} =$$

$$= \frac{(2\,\pi)^n\, \Delta_1\, \Delta_2 \ldots \Delta_n\, V}{(M + 2\,\pi\,\Delta_1)(M + 2\,\pi\,\Delta_2)\ldots(M + 2\,\pi\,\Delta_n)}$$

De esta fórmula [144] se deduce también:

Dado un punto arbitrario en el interior de un cuerpo cualquiera y n *franjas paralelas que cortan al mismo, la probabilidad de que el punto pertenezca a la vez a las* n *franjas será*:

$$W = \frac{(2\,\pi)^n\, \Delta_1\, \Delta_2 \ldots \Delta_n}{(M + 2\,\pi\,\Delta_1)(M + 2\,\pi\,\Delta_2)\ldots(M + 2\,\pi\,\Delta_n)}$$

II. En lugar de un punto sea ahora una recta R y propongámonos calcular

$$M_{R.P_1.P_2\ldots P_n} = \int d\,R\,.\,d\,P_1\,.\,d\,P_2 \ldots d\,P_n \qquad [145]$$

es decir, la medida del conjunto de franjas y rectas tales que, cortando todas a K (ahora sí convexo) tienen puntos interiores comunes.

Fijando R se tiene

$$M_{R.P_1.P_2\ldots P_n} = \frac{\pi}{2}\int f_{1\,2\,3\ldots n}\, d\,P_1\, d\,P_2 \ldots d\,P_n \qquad [146]$$

llamando $f_{1\,2\,3\ldots n}$ al área del cuerpo convexo intersección de K con las P_i $(i = 1, 2, \ldots, n)$.

En cambio, fijando $R, P_1, P_2, \ldots, P_{n-1}$ y llamando $\sigma_{1\,2\,3\ldots n-1}$ a la longitud de la cuerda que R determina en la intersección de K con $P_1, P_2, P_3, \ldots, P_{n-1}$ es:

$$M_{R.P_1.P_2\ldots P_n} = \int (\pi\,\sigma_{1\,2\,3\ldots n-1} + 2\,\pi\,\Delta_n)\, d\,R\, d\,P_1\, d\,P_2 \ldots d\,P_{n-1} =$$

$$= 2\,\pi^2 \int v_{1\,2\,3\ldots n-1}\, d\,P_1\, d\,P_2 \ldots d\,P_{n-1} + 2\,\pi\,\Delta_n\, M_{R.P_1.P_2\ldots P_{n-1}}$$

fórmula recurrente que teniendo en cuenta la [144] antes encontrada nos da

$$M_{R.P_1.P_2\ldots P_n} = \pi (2\pi)^n V \sum_i \Delta_1 \Delta_2 \ldots \Delta_{i-1} \Delta_{i+1} \ldots \Delta_n + \\ + \pi^2 (2\pi)^{n-1} F \Delta_1 \Delta_2 \ldots \Delta_n \qquad [147]$$

De ambas expresiones para la misma medida se deduce

$$\boxed{\int f_{123\ldots n}\, d P_1\, d P_2 \ldots d P_n = 2(2\pi)^n V \sum_i \Delta_1 \Delta_2 \ldots \Delta_{i-1} \Delta_{i+1} \ldots \Delta_n + \\ + (2\pi)^n F \Delta_1 \Delta_2 \ldots \Delta_n} \qquad [148]$$

El *área media* del cuerpo común a K y a las n franjas P_i, será, pues:

$$\bar{f}_{123\ldots n} = \frac{2(2\pi)^n V \sum_i \Delta_1 \Delta_2 \ldots \Delta_{i-1} \Delta_{i+1} \ldots \Delta_n + (2\pi)^n F \Delta_1 \Delta_2 \ldots \Delta_n}{(M + 2\pi \Delta_1)(M + 2\pi \Delta_2) \ldots (M + 2\pi \Delta_n)}$$

Además:

Dadas una recta R *y* n *franjas* P_i *que se sabe cortan al cuerpo convexo* K, *la probabilidad de que la recta corte al eventual cuerpo convexo común a las* P_i *y* K *es*:

$$W = \frac{\int f_{123\ldots n}\, d P_1 \ldots d P_n}{F (M + 2\pi \Delta_1)(M + 2\pi \Delta_2) \ldots (M + 2\pi \Delta_n)}$$

siendo el numerador la expresión [148].

III. Análogamente podemos también considerar la medida del conjunto de plano E más n franjas paralelas $P_1, P_2, \ldots, P_n$ tales que, cortando todos a un cuerpo convexo K también se corten entre sí interiormente a él. Es decir, calcular

$$M_{E\,P_1.P_2\ldots P_n} = \int d E\, d P_1\, d P_2 \ldots d P_n \qquad [149]$$

Para ello, fijando las franjas P_i e integrando respecto d E se tiene:

$$M_{E.P_1.P_2\ldots P_n} = \int m_{123\ldots n}\, d P_1\, d P_2 \ldots d P_n \qquad [150]$$

siendo $m_{1\,2\,3\,\ldots\,n}$ la medida de todos los planos que cortan al cuerpo convexo intersección de K con las P_i.

En cambio, fijando $E, P_1, P_2, \ldots, P_{n-1}$ e integrando respecto dP_n se obtiene por [127]:

$$M_{E.P_1.P_2\ldots P_n} = \int\left(\frac{\pi}{2} u_{1\,2\,3\,\ldots\,n-1} + 2\pi\Delta_n\right) dE\, dP_1\, dP_2 \ldots dP_{n-1} \quad [151]$$

siendo $u_{1\,2\,3\,\ldots\,n-1}$ la longitud de la figura plana convexa sección que E determina en la intersección de K con $P_1, P_2, \ldots, P_{n-1}$.

Recordando las fórmulas de la nota (*) del núm. 28, esto puede escribirse:

$$M_{E\,P_1.P_2\ldots P_n} = \frac{\pi}{2}\cdot\frac{\pi^2}{2}\int f_{1\,2\,3\ldots n-1}\, dP_1\, dP_2 \ldots dP_{n-1} + \\ + 2\pi\Delta_n M_{E.P_1.P_2\ldots P_{n-1}} \quad [152]$$

Fórmula recurrente que nos da

$$M_{E.P_1.P_2\ldots P_n} = \pi\frac{\pi^2}{2}(2\pi)^{n-1}\,V\sum_{i,j}\Delta_1\ldots\Delta_{i-1}\Delta_{i+1}\ldots\Delta_{j-1}\Delta_{j+1}\ldots\Delta_n + \\ + \frac{\pi}{2}\cdot\frac{\pi^2}{2}(2\pi)^{n-1}\,F\sum_i\Delta_1\ldots\Delta_{i-1}\Delta_{i+1}\ldots\Delta_n + (2\pi)^n M\,\Delta_1\Delta_2\ldots\Delta_n \quad [153]$$

Comparando con [150] obtenemos:

$$\int m_{1\,2\,3\ldots n}\, dP_1\, dP_2 \ldots dP_n = \\ = \pi\frac{\pi^2}{2}(2\pi)^{n-1}\,V\sum_{i,j}\Delta_1\ldots\Delta_{i-1}\Delta_{i+1}\ldots\Delta_{j-1}\Delta_{j+1}\ldots\Delta_n + \\ + \frac{\pi}{2}\cdot\frac{\pi^2}{2}(2\pi)^{n-1}\,F\sum_i\Delta_1\ldots\Delta_{i-1}\Delta_{i+1}\ldots\Delta_n + (2\pi)^n M\,\Delta_1\Delta_2\ldots\Delta_n \quad [154]$$

Análogamente como en los casos anteriores esta fórmula nos permite conocer el *valor medio* de $m_{1\,2\,3\,\ldots\,n}$ y resolver el problema de hallar la probabilidad de que un plano y n franjas que cortan a un cuerpo convexo, tengan puntos comunes interiores al mismo.

IV. Sean ahora n franjas paralelas P_i móviles independientemente unas de otras y propongámonos calcular:

$$I_n = \int d P_1 d P_2 \ldots d P_n \qquad [155]$$

extendida la integración al conjunto de posiciones en las que el cuerpo convexo K y las n franjas P_i tienen algún punto común.

Fijando $P_1, P_2, \ldots, P_{n-1}$ y llamando $m_{1\,2\,3\ldots n-1}$ a la medida de los planos que cortan al cuerpo convexo intersección de estos P_i $(i = 1, 2, \ldots, n-1)$ con K es por [124]:

$$I_n = \int (m_{1\,2\,3\ldots n-} + 2\pi \Delta_n)\, d P_1\, d P_2 \ldots d P_{n-1} \qquad [156]$$

o sea:

$$I_n = \int m_{1\,2\,3\ldots n-1}\, d P_1\, d P_2 \ldots d P_{n-1} + 2\pi \Delta_n I_{n-1} \qquad [157]$$

Esta fórmula recurrente nos da:

$$\begin{aligned} I_n = {} & \pi \frac{\pi^2}{2} (2\pi)^{n-2}\, \mathrm{V} \sum_{i,j} \Delta_1 \ldots \Delta_{i-1} \Delta_{i+1} \ldots \Delta_{j-1} \Delta_{j+1} \ldots \Delta_n + \\ & + \frac{\pi}{2} \cdot \frac{\pi^2}{2} (2\pi)^{n-2}\, \mathrm{F} \sum_{i} \Delta_1 \Delta_2 \ldots \Delta_{i-1} \Delta_{i+1} \ldots \Delta_n + \\ & + (2\pi)^{n-1}\, \mathrm{M}\, \Delta_1 \Delta_2 \ldots \Delta_n + (2\pi)^n \Delta_1 \Delta_2 \ldots \Delta_n \end{aligned} \qquad [158]$$

Esta expresión nos permite resolver el siguiente problema:

¿Cuál es la probabilidad de que n *franjas que se sabe cortan a un cuerpo convexo* K *tengan algún punto interior al mismo común?*

Será:

$$W = \frac{I_n}{(\mathrm{M} + 2\pi \Delta_1)(\mathrm{M} + 2\pi \Delta_2) \ldots (\mathrm{M} + 2\pi \Delta_n)} \qquad [159]$$

32. Superficie cortada por franjas paralelas.—Para este número y el siguiente necesitamos hacer uso de una fórmula integral que puede considerarse como generalización en el espacio de una fórmula de Poincaré y que hemos obtenido en otro lugar (18). Es la siguiente:

Si una superficie cualquiera S es móvil en el espacio con densidad cinemática dK y llamamos n al número de puntos en que corta en cada posición a una curva fija de longitud L se verifica:

$$\int n\, dK = 4\pi^2 F L \qquad [160]$$

siendo F el área de S.

Sentado esto supongamos un par de planos paralelos P fijos a distancia Δ, una recta R cualquiera y una superficie S cuya posición, salvo una traslación paralela a los planos, vendrá determinada por la distancia h de uno de sus puntos a uno de estos planos más una dirección $d\Omega$ y más un giro $d\tau$ alrededor de esta dirección.

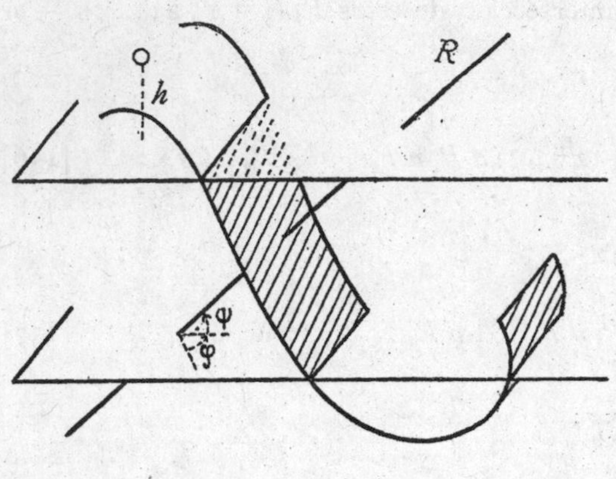

Fig. 15.

La recta R viene determinada por las coordenadas x, y de su intersección con uno de los planos paralelos y los ángulos ψ y φ de su dirección. Su densidad es

$$dR = \operatorname{sen}\psi \cos\psi\, d\psi\, d\varphi\, dx\, dy$$

Consideremos la integral:

$$I_s = \int n \operatorname{sen}\psi \cos\psi\, d\psi\, d\varphi\, dx\, dy\, dh\, d\Omega\, d\tau \qquad [161]$$

siendo n el número de puntos comunes a R y a S interiores a la franja P.

Fijando h, Ω y τ y llamando f al área de la parte de S que en esta posición es interior a la franja P, tenemos:

$$I_s = \pi \int f\, dh\, d\Omega\, d\tau \qquad [162]$$

En cambio, fijando la recta R y observando que la longitud de la parte de la misma comprendida entre los planos P vale $\dfrac{\Delta}{\operatorname{sen}\psi}$ y aplicando [160] resulta (puesto que lo mismo da suponer una traslación paralela a los planos P por parte de la recta que por parte de la superficie):

$$I_s = 4\,\pi^2\,\mathrm{F} \int \Delta \cos\psi\, d\psi\, d\varphi = 8\,\pi^3\,\mathrm{F}\,\Delta \qquad [163]$$

En lugar de suponer los planos paralelos fijos y S móvil podemos suponerlo al revés observando que entonces es $dP = dh\, d\Omega$, con lo cual [162] puede escribirse

$$I_s = \pi \int f\, dP\, d\tau = 2\,\pi^2 \int f\, dP \qquad [164]$$

y considerando cada par de planos contado una sola vez independientemente del orden de los planos entre sí, resulta igualando esta expresión con [163] después de dividirla por 2:

$$\boxed{\int f\, dP = 2\,\pi\,\Delta\,\mathrm{F}} \qquad [165]$$

que comprende a [136] como caso particular.

Generalización.—Tomando dos franjas paralelas P_1 y P_2 de anchuras respectivas Δ_1 y Δ_2 y llamando f_{12} al área de la parte de superficie que es interior a la vez a P_1 y P_2, veamos de calcular

$$\int f_{12}\, dP_1\, dP_2$$

Para ello, fijando P_1 y llamando f_1 a la parte de superficie que comprende en su interior, se tiene por [165]:

$$\int f_{12}\, dP_1\, dP_2 = 2\,\pi\,\Delta_2 \int f_1\, dP_1$$

de donde

$$\int f_{12}\, dP_1\, dP_2 = (2\,\pi)^2\, \Delta_1\, \Delta_2\, \mathrm{F}$$

Este procedimiento general nos da

$$\int f_{123\ldots n}\, dP_1\, dP_2 \ldots dP_n = (2\pi)^n \Delta_1 \Delta_2 \ldots \Delta_n F$$

33. Línea cortada por franjas paralelas.—Sea el mismo par de planos P del número precedente y una línea de longitud L cuya posición, salvo una traslación paralela a dichos planos, vendrá fijada por h, Ω, τ. Análogamente consideremos una superficie cualquiera de área F móvil con densidad cinemática $d\mathrm{K}^* = d\mathrm{P}^*\, d\Omega^*\, d\tau^*$. Consideremos

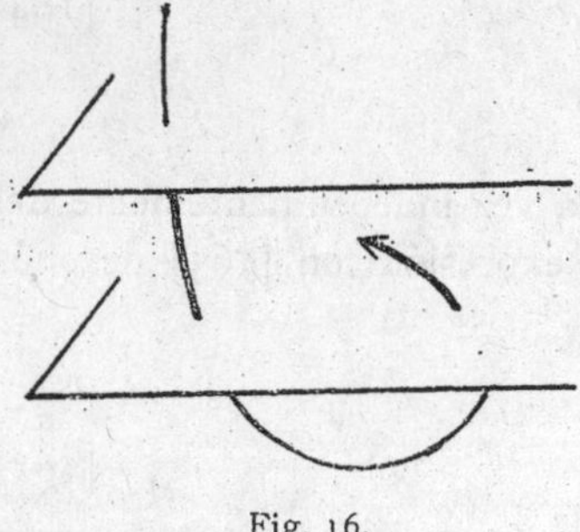

Fig. 16.

$$I_{\mathrm{L}} = \int n\, dh\, d\Omega\, d\tau\, d\mathrm{K}^* \qquad [166]$$

siendo n el número de puntos en que la superficie y la línea se cortan interiormente a la franja P.

Manteniendo fija la línea y teniendo en cuenta [160] queda:

$$I_{\mathrm{L}} = 4\pi^2 \mathrm{F} \int l\, dh\, d\Omega\, d\tau \qquad [167]$$

designando por l la longitud de la parte de la misma comprendida entre los planos paralelos.

Fijemos ahora la superficie y hagamos mover a la línea. Llamando f a la parte de superficie comprendida entre los planos paralelos y teniendo en cuenta que lo mismo da suponer un desplazamiento paralelo a los planos P por parte de la línea que por parte de la superficie, resulta por [160] y [165]:

$$I_{\mathrm{L}} = 8\pi^3 \mathrm{L} \int f\, dP = 8\pi^3 \mathrm{L} \,.\, 2\pi\Delta \mathrm{F} \qquad [168]$$

Igualando con [167] y simplificando, resulta:

$$\int l\, dP = 2\pi\Delta \mathrm{L} \qquad [169]$$

Generalización.—Tomando dos franjas paralelas P_1 y P_2 de anchuras respectivas Δ_1 y Δ_2 y llamando l_{12} a la longitud de la parte de línea que es interior a la vez a P_1 y P_2, intentemos calcular:

$$\int l_{12}\, d P_1\, d P_2$$

Para ello, fijando primero P_1 y llamando l_1 a la parte de línea que comprende en su interior, se tiene por [169]

$$\int l_{12}\, d P_1\, d P_2 = 2\pi \Delta_2 \int l_1\, d P_1$$

de donde

$$\int l_{12}\, d P_1\, d P_2 = (2\pi)^2 \Delta_1 \Delta_2 L$$

Este procedimiento, repetido n veces, nos da en general:

$$\boxed{\int l_{123\ldots n}\, d P_1\, d P_2 \ldots d P_n = (2\pi)^n \Delta_1 \Delta_2 \ldots \Delta_n L.}$$

CONJUNTOS DE RECTAS Y PLANOS DEPENDIENTES DE DOS PARAMETROS Y CUERPOS CONVEXOS

34. Densidades para conjuntos de rectas y planos dependientes de dos parámetros.—Consideremos nada más los haces de rectas paralelas o de planos paralelos a una dirección fija. Las rectas de un tal haz vendrán determinadas por el punto en que cortan a un plano perpendicular a su dirección. Luego la densidad para estas rectas será la misma que para los puntos de uno de estos planos. Es decir,

$$d R^* = d x\, d y \qquad [170]$$

Análogamente, para los planos paralelos a una dirección se puede tomar por densidad la de sus rectas de intersección con otro plano perpendicular, o sea:

$$d E^* = d p\, d \theta \qquad [171]$$

Por densidad de un conjunto de pares de estos elementos, supuestos dados independientemente, se tomará el producto de las densidades de los elementos que lo constituyen.

Llamando σ^* a la longitud de la cuerda que R* determina en un cuerpo K de volumen V, se verifica evidentemente

$$\int \sigma^* \, d\,\mathrm{R}^* = \mathrm{V} \qquad [172]$$

Y llamando f^* al área de la sección que E* determina en el mismo cuerpo es también:

$$\int f^* \, d\,\mathrm{E}^* = \pi\,\mathrm{V} \qquad [173]$$

35. Pares recta-plano (R. E*) que cortan a un cuerpo cualquiera.— Consideremos el conjunto de los pares formados por una recta cualquiera R y un plano E* paralelo a una dirección fija cuyo punto de intersección sea interior al cuerpo no necesariamente convexo K. Como medida de este conjunto se tiene

$$M_{\mathrm{R}\,.\,\mathrm{E}^*} = \int_{\mathrm{R}\,.\,\mathrm{E}^* < \mathrm{K}} d\,\mathrm{R}\,.\,d\,\mathrm{E}^* \qquad [174]$$

Para calcular esta integral, fijando primero R y llamando σ_p a la proyección de la cuerda que ella determina en K sobre un plano perpendicular a la dirección a la que son paralelos los E*, será

$$M_{\mathrm{R}\,.\,\mathrm{E}^*} = 2 \int \sigma_p \, d\,\mathrm{R} \qquad [175]$$

Pero fijando primero E* y llamando f^* al área de la sección que él determina en el cuerpo K es, por [173]

$$M_{\mathrm{R}\,.\,\mathrm{E}^*} = \pi \int f^* \, d\,\mathrm{E}^* = \pi^2\,\mathrm{V} \qquad [176]$$

De estas dos expresiones de la medida se deduce

$$\boxed{\int \sigma_p \, d\,\mathrm{R} = \frac{\pi^2}{2}\,\mathrm{V}} \qquad [177]$$

independiente de la dirección de proyección.

36. Pares plano-recta (E. R*) que cortan a un cuerpo cualquiera.—Sea ahora el conjunto de los pares formados por un plano cualquiera y una recta R* paralela a una dirección fija cuya intersección es interior al cuerpo no necesariamente convexo, K. La medida de tal conjunto es

$$M_{E.R^*} = \int_{R^*.E<K} dE . dR^* \qquad [178]$$

Fijando primero E y llamando f_p al área de la proyección sobre un plano perpendicular a las R* de la sección de E con K, es

$$M_{E.R^*} = \int f_p \, dE \qquad [179]$$

Pero fijando primero R* y llamando σ^* a la longitud de la cuerda que ella intercepta en K es también

$$M_{E.R^*} = \pi \int \sigma^* dR^* = \pi V \qquad [180]$$

De [179] y [180] se deduce

$$\boxed{\int f_p \, dE = \pi V} \qquad [181]$$

37. Conjuntos de rectas paralelas y cuerpos convexos que cortan a otro.—Sean dos cuerpos convexos: uno fijo K_1 y otro móvil K_2, con la densidad cinemática dK_2. Nos proponemos calcular:

$$M_{R^*.K_2} = \int_{R^*.K_1.K_2 \neq 0} dR^* . dK_2 \qquad [182]$$

Fijando primero K_2 y llamando F_p al área de la proyección de la parte común a K_1 y K_2 sobre un plano normal a las rectas R* (fig. 17), se tiene:

$$M_{R^*.K_2} = \int F_p \, dK_2 \qquad [183]$$

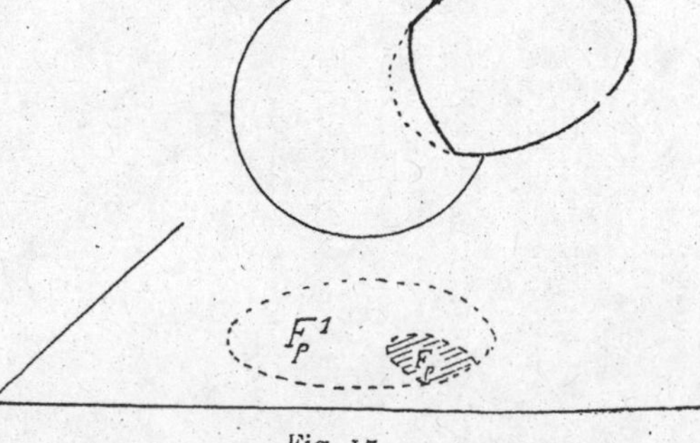

Fig. 17

Pero fijando primero R* y llamando σ^* a la longitud de la cuerda que ella intercepta en K_1 es también por [98]

$$M_{R^*.K_2} = \int (8\pi^2 V_2 + 2\pi^2 F_2 \sigma^*)\, dR^* = 8\pi^2 V_2 F^1_p + 2\pi^2 F_2 V_1 \qquad [184]$$

siendo F^1_p el área de la proyección de K_1 sobre un plano normal a las rectas R^*.

Igualando [183] y [184] resulta

$$\boxed{\int F_p\, dK_2 = 2\pi^2 (4 V_2 F^1_p + F_2 V_1)} \qquad [185]$$

Se observa que el valor de esta integral depende, pues, en general, de la dirección de proyección. Unicamente es independiente de ella cuando K_1 tiene en todas direcciones la misma área de proyección, o sea cuando K_1 es un cuerpo de iluminación constante (*konstanter Helligkeit*).

BIBLIOGRAFIA

(1) E. BARBIER. Note sur le problème de l'aiguille et le jeu du joint convert. Liouv. Jour. (II), 5, 1860.

(2) M. W. CROFTON. On the theory of local probability. Philosophical Transactions of the Royal Soc. of London. 158, 1885.

Además el artículo que bajo la palabra "Probability" escribió en la Encyclop. Britannica. 9 edit., vol, 19 (1885).

(3) E. CZUBER. Geometrische Wahrscheinlichkeiten und Mittelwerte. Leipzig, 1884.

Zur Theorie der geometrischen Wahrscheinlichkeiten. Wiener Berichte, 9.º (1884).

Bericht ueber die Entwicklung der Wahrscheinlichkeitstheorie und ihre Anwendungen. Jahresber. der D .M. V. 7 (1899).

(4) J. J. SYLVESTER. On a funicular solution of Buffon's problem of the needle. Acta Math. 14 (1891).

(5) E. CARTÁN. Le principe de la dualité... Bull. de la Soc. Math. de France, 24 (1896).

(6) H. POINCARÉ. Calcul des probabilités. 2.ª edit. París, 1912.

(7) H. LEBESGUE. Exposition d'un Memoire de M. W. Crofton. Nouvelles Annales. 12 (1912).

(8) G. POLYA. Ueber geometrische Wahrscheinlichkeiten. Wiener Berichte. 126 (1917).

(9) B. HOSTINSKY. Sur les probabilités geométriques. Publications de la fac. des Sc. Université Masaryk. 1925.

(10) E. BOREL. Principes et formules classiques (Traité du calcul des probabilités et de ses applications, Tome I, facs. 1) París, 1924.

(11) R. DELTHEIL. Probabilités geométriques. París, 1926.

(12) G. HERGLOTZ. Geometrische Wahrscheinlichkeiten. Vorlesungen Göttingen, 1933. No publicado.

(13) W. BLASCHKE. Vorlesungen ueber Integralgeometrie. Erstes Heft. Hamburger Math. Einzelschriften. 20, 1935.

Bajo el título de Geometría Integral, anteceden a éste los siguientes trabajos:

(14) W. BLASCHKE. Integralgeometrie 1. Ermittlung der Dichten für lineare Unterräume im E_n. Actualités scientifiques et industrielles 252, París, 1935. Hermann & Cie.

(15) W. BLASCHKE. Integralgeometrie 2. Zu Ergebnissen von M. W. Crofton. Bulletin Mathematique de la Societé Roumaine des Sciences, 37 (1935).

(16) O. VARGA. Integralgeometrie 3. Croftons Formeln für den Raum. Mathematische Zeitschrift. 1935.

(17) L. A. SANTALÓ. Geometría Integral 4. Sobre la medida cinemática en el plano. Abhandlungen aus dem Math. Seminar Hamburg 11 (1936).

(18) L. A. SANTALÓ. Integralgeometrie 5. Über das kinematische Mass im Raum. Actualités Scientifiques et industrielles 357, París, 1936. Hermann & Cie.

(19) B. PETKANTSCHIN. Integralgeometrie 6. Zusammenhänge zwischen den Dichten der linearen Unterräume im n-dimensionalen Raum. Abh. aus dem Math. Seminar Hamburg 11 (1936).

A Generalization of a Theorem of T. Kubota on ovals

L. A. Santaló at Rosario (Argentina).

In[1] [2], T. Kubota has proved the following Theorem:

"*Let the oval K have an everywhere continuous radius of curvature. Let R_M, R_m denote the biggest and smallest value of the curvature radii of K. If one constructs a circle of radius r intersecting K in at least three different or coinciding points, then*

$$R_m \leq r \leq R_M \,."$$

In this note I would like to prove the following general Theorem:

"*Let the oval K have an everywhere continuous radius of curvature. The maximal and the minimal of its values are denoted by R_M and R_m. Let k be a second oval with continuous curvature, and which curvature radius r is either less than R_m or bigger than R_M. Then K and k can intersect in no more than two points.*"

The proof is by the method of Integral Geometry [1].

1. The position of a given oval K in the plane is determined by the coordinates x, y of one of its points and the angle φ of a rotation around this point.

For the measure of a set of positions of the oval k (*kinematic measure*) one uses the integral

$$\int dx\, dy\, d\varphi \tag{1}$$

taken over the set under consideration.

For the number n of intersection points of K with k (clearly n depends on x, y, φ), there is the so-called formula of Poincaré [1], pag. 24.

$$\int n\, dx\, dy\, d\varphi = 4Uu\,, \tag{2}$$

where U, u are the lengths of K, k.

2. Let us suppose that the curvature radius r of k is everywhere smaller than R_m, so $r < R_m$. The same method works for $r > R_M$.

Let us consider a position where k touches K such that both are lying on the same side of the common tangent. Then the oval k is lying completely inside K because $r < R_m$. For this position we choose a point O inside k and K. The support functions $H = H(\theta)$ and $h = h(\theta + \varphi)$ of K and k with respect to O then satisfy $H(\theta) \geq h(\theta + \varphi)$. The curvature radii are given by $R = H + H''$

[1] Translated from German by E. Teufel.

and $r = h + h''$ (the primes denote derivation with respect to θ). Because of the assumption $r < R_m$, we get

$$H(\theta) - h(\theta+\varphi) \geq 0 \quad , \quad (H(\theta) - h(\theta+\varphi)) + (H''(\theta) - h''(\theta+\varphi)) > 0 .$$

Hence $H(\theta) - h(\theta+\varphi)$ is the support function of an oval.

The same is true for $H(\theta) + h(\theta+\varphi+\pi)$.

In order to compute the measure of the set of positions of k intersecting K, we first fix φ, i.e. we only consider translations of k. In this case, the positions of the point x, y (1) fill the area bounded by the ovals with support functions $H(\theta) + h(\theta+\varphi+\pi)$ and $H(\theta) - h(\theta+\varphi)$. This area is [2]

$$\begin{aligned} F(\varphi) &= \frac{1}{2}\int_0^{2\pi} [(H(\theta)+h(\theta+\varphi+\pi))^2 - (H'(\theta)+h'(\theta+\varphi+\pi))^2]\, d\theta \\ &\quad -\frac{1}{2}\int_0^{2\pi} [(H(\theta)-h(\theta+\varphi))^2 - (H'(\theta)-h'(\theta+\varphi))^2]\, d\theta \\ &= \int_0^{2\pi} (H(\theta)h(\theta+\varphi+\pi) - H'(\theta)h'(\theta+\varphi+\pi))\, d\theta \\ &\quad + \int_0^{2\pi} [H(\theta)h(\theta+\varphi) - H'(\theta)h'(\theta+\varphi)]\, d\theta . \end{aligned}$$

In order to compute the kinematic measure, we now have to rotate k, and we find

$$\begin{aligned} M &= \int_0^{2\pi} F(\varphi)\, d\varphi = \int_0^{2\pi}\int_0^{2\pi} H(\theta)h(\theta+\varphi+\pi)\, d\theta\, d\varphi \\ &\quad + \int_0^{2\pi}\int_0^{2\pi} H(\theta)h(\theta+\varphi)\, d\theta\, d\varphi = 2Uu , \end{aligned} \tag{3}$$

because of

$$\int_0^{2\pi} h'(\theta+\varphi)\, d\varphi = \int_0^{2\pi} h'(\theta+\varphi+\pi)\, d\varphi = 0 .$$

Therefore: *The measure of the set of positions of the oval k in which it intersects the oval K, is equal to $2Uu$.*

3. The measure of the sets of positions of k, in which k and K have exactely one point in common, is null. This because in these positions k and K are tangent, and therefore the integral (1) runs over a set of measure zero.

Let us call M_i the measure of the set of positions of k, in which k and K have exactly i intersection points. Formula (2) gives

$$2M_2 + 3M_3 + 4M_4 + 5M_5 + \cdots = 4Uu , \tag{4}$$

[2] As one knows, the area of an oval with support function $p = p(\theta)$ is given by

$$F = \frac{1}{2}\int_0^{2\pi} (p^2 - p'^2)\, d\theta .$$

Cf. [1], pag. 29.

and formula (3) gives

$$M_2 + M_3 + M_4 + \cdots = 2Uu\,. \tag{5}$$

From (4) and (5) we conclude

$$M_3 + 2M_4 + 3M_5 + \cdots = 0\,.$$

Then: "*If the curvature radii of two ovals satisfy the relations* $r < R_m$ *or* $r > R_M$*, then the measure of positions in which they have more than two points in common is null.*"

4. It is left to check whether there are sets of positions of measure zero, in which k and K have more than two points in common.

If k and K have an intersection point without being tangent there, then there exists at least one further intersection point without tangency. Le A and B be two intersection points of k and K without tangency, and let C be a third common point of k and K. In a neighborhood of A there exist two circles with centers on the oval k, such that the first one lies completely inside K and the second one lies completely outside K. Circles with similar behaviour exist in some neighborhood of B. Whenever k has common points with each of these four circles, then k will intersect K in two points nearby A and B. Now, the measure of the set of positions of k, in which k has common points with each of these four circles, is positive. Therefore: If k and K intersect in more than two points without tangency in at least two of them, then there is a set with positive measure of positions, in which k and K have more than two points in common.

It is left to consider the case where k and K have two or more contact points. But this is not possible, because otherwise between two contact points there would be two convex arcs with $r < R_m$ and the same tangent lines at the endpoints, which is a contradiction (Cf. the first part of the cited work of Kubota [2]).

This proves the Theorem.

In case k is a circle, the Theorem is equivalent to the Theorem of T. Kubota noted at the beginning.

References

[1] W. Blaschke, *Vorlesungen über Integralgeometrie*, vol. 20, B. G. Teubner, Leipzig and Berlin, 1936.

[2] T. Kubota, *Ein Satz über Eilinien*, Tôhoku Math. Journal **47** (1940), 96–98.

Rosario, September 20, 1940.

(Submitted November 13, 1940.)

Proof of a theorem of Bottema on ovals

L. A. Santaló at Rosario (Argentina)

Let[1] $H_0(\theta)$ and $H_1(\theta)$ be the support functions of two ovals K_0 and K_1. Following Minkowski, the following terms are called "mixed areas"

$$F_{ij} = F_{ji} = \frac{1}{2}\int_0^{2\pi} (H_i H_j - H_i' H_j')\, d\theta \quad i,j = 0,1\,. \tag{1}$$

or

$$F_{ij} = \frac{1}{2}\int H_i\, ds_j = \frac{1}{2}\int H_j\, ds_i \quad i,j = 0,1\,. \tag{2}$$

where ds_i denotes the arc length element.

As is known, the F_{ij} satisfy the inequality (see [1], pag. 36)

$$F_{01}^2 - F_{00}F_{11} \geq 0\,. \tag{3}$$

We suppose that K_0 and K_1 everywhere have a continuous radius of curvature R_0, R_1. Let r_m be the biggest possible number with

$$rR_1(\theta) \leq R_0(\theta) \tag{4}$$

for all θ, and similarly let r_M be the smallest possible number with

$$rR_1(\theta) \geq R_0(\theta)\,. \tag{5}$$

Then Bottema [4] has proved

$$F_{01}^2 - F_{00}F_{11} \leq \frac{1}{4}F_{11}^2(r_M - r_m)^2\,. \tag{6}$$

In this note I would like to give a different proof of this inequality.

1. Let us consider the oval rK_1 (homothetic to K_1 with dilatation factor r). We suppose that r satisfies the inequality (4). We move the ovals rK_1 and K_0 in a position such that they have a common tangent line in a common point and such that they are lying on the same side of this common tangent line. Then, because of (4), rK_1 lies completely inside K_0. In this situation we choose a point O lying inside rK_1 and inside K_0. Then the support functions rH_1 and H_0 of rK_1 and K_0 with respect to O satisfy $rH_1 \leq H_0$. The curvature radii of rK_1 and K_0 are given by $rR_1 = rH_1 + rH_1''$, $R_0 = H_0 + H_0''$ (the prime indicates derivations). Then because of (4):

[1] Translated from German by E. Teufel.

$$\begin{aligned} H_0 - rH_1 &\geq 0\,, \\ (H_0 - rH_1) + (H_0'' - rH_1'') &\geq 0\,. \end{aligned}$$

Therefore $H_0(\theta) - rH_1(\theta)$ is the support function of an oval.

2. Taking into account all translations of rK_1 leaving rK_1 completely inside K_0, then the point O covers the area bounded by the oval with support function $H_0 - rH_1$. This area, which one may call the measure M_0 of the set of all translations of rK_1 leaving it inside K_0 (see [2]), is given by (1) [2]

$$\begin{aligned} M_0 &= \frac{1}{2}\int_0^{2\pi} \left[(H_0 - rH_1)^2 - (H_0' - rH_1')^2\right] d\theta \\ &= F_{00} - 2rF_{01} + r^2F_{11}\,. \end{aligned}$$

Because of $M_0 \geq 0$, it follows that

$$F_{00} - 2rF_{01} + r^2F_{11} \geq 0 \tag{7}$$

for all r fulfilling (4).

One can prove similarly: If r satisfies the inequality (5), then the measure of all translations of rK_1 leaving K_0 inside, is also given by

$$M_0 = F_{00} - 2rF_{01} + r^2F_{11}\,.$$

Hence, the inequality (7) holds for all r satisfying (4) or (5).

3. The identity

$$F_{00} - 2rF_{01} + r^2F_{11} = F_{11}\left(\frac{F_{01}}{F_{11}} - r\right)^2 - \left(\frac{F_{01}^2}{F_{11}} - F_{00}\right)$$

and (7), taking $r = r_m$ and $r = r_M$, imply

$$\left.\begin{aligned} \tfrac{F_{01}^2}{F_{11}} - F_{00} \leq F_{11}\left(\tfrac{F_{01}}{F_{11}} - r_m\right)^2 \\ \tfrac{F_{01}^2}{F_{11}} - F_{00} \leq F_{11}\left(\tfrac{F_{01}}{F_{11}} - r_M\right)^2 \end{aligned}\right\}. \tag{8}$$

These inequalities already give an upper bound for the deficit $\frac{F_{01}^2}{F_{11}} - F_{00}$.

4. Because of (2), (4) and (5) one has

$$\begin{aligned} F_{01} &= \frac{1}{2}\int H_1\,ds_0 = \frac{1}{2}\int H_1R_0\,d\theta \leq \frac{1}{2}r_M\int H_1R_1\,d\theta \\ &= \frac{1}{2}r_M\int H_1\,ds_1 = r_MF_{11}\,, \end{aligned} \tag{9}$$

[2] As one knows, the area of an oval with support function $H = H(\theta)$ is given by

$$F = \frac{1}{2}\int_0^{2\pi} (H^2 - H'^2)\,d\theta\,.$$

and

$$F_{01} = \frac{1}{2}\int H_1\,ds_0 = \frac{1}{2}\int H_1 R_0\,d\theta \geq \frac{1}{2} r_m \int H_1 R_1\,d\theta =$$
$$= \frac{1}{2} r_m \int H_1\,ds_1 = r_m F_{11}\,. \tag{9'}$$

Hence

$$\frac{F_{01}}{F_{11}} - r_m \geq 0 \quad , \quad r_M - \frac{F_{01}}{F_{11}} \geq 0\,. \tag{10}$$

Because of (3) we can multiply the terms in (8), and using (10) one finds

$$\frac{F_{01}^2}{F_{11}} - F_{00} \leq F_{11}\left(\frac{F_{01}}{F_{11}} - r_m\right)\left(r_M - \frac{F_{01}}{F_{11}}\right). \tag{11}$$

Because of the relation

$$xy \leq \left(\frac{x+y}{2}\right)^2$$

from (11) it follows

$$\boxed{\frac{F_{01}^2}{F_{11}} - F_{00} \leq \frac{1}{4} F_{11}(r_M - r_m)^2\,.} \tag{12}$$

This is the inequality (6) we wanted to prove.

From (8) we know that equality in (12) holds if and only if either there is equality in both formulas in (8) or one of the formulas in (8) is null. In both cases, (8) and (11) imply $r_m = r_M$. Then (4) and (5) imply $rK_1 \equiv K_0$, hence K_1 and K_0 are homothetic.

5. If K_1 is a unit circle, then r_m and r_M are equal to the smallest and biggest curvature radii of K_0, and $2F_{01} = U_0$ (U_0 = perimeter of K_0), $F_{00} = F_0$, $F_{11} = \pi$. The inequality (12) then writes

$$U_0^2 - 4\pi F \leq \pi^2 (r_M - r_m)^2\,.$$

This inequality also goes back to Bottema [4]. See also [3], pag. 83. Equality occurs only for circles.

References

[1] W. Blaschke, *Vorlesungen über Integralgeometrie*, vol. 20, B. G. Teubner, Leipzig and Berlin, 1936.

[2] ———, *Integralgeometrie 21. Über Schiebungen*, Math. Zeits. **42** (1937), 399–410.

[3] T. Bonnesen and W. Fenchel, *Theorie der konvexen Körper*, Ergebnisse der Math. Springer, Berlin, 1934.

[4] O. Bottema, *Eine obere Grenze für das isoperimetrische Defizit einer ebenen Kurve*, Akad. Wetensch. Amsterdam, 1933, Proc. 36.

Rosario (Rep. Argentina) Math. Inst. December, 1940.

Submitted February 3, 1941.

A THEOREM AND AN INEQUALITY REFERRING TO RECTIFIABLE CURVES.*

By L. A. SANTALÓ.

Introduction. The principal purpose of this paper is the demonstration of the following theorem:

Given a rectifiable curve of length L in the euclidean space E_m and calling n the number of common points of this with the surface of a sphere of centre $P(x_1, x_2, \cdots, x_m)$ and radius R, we find

$$\int n\, dP = 2LV_{m-1}(R)$$

where $dP = dx_1\, dx_2 \cdots dx_m$ and $V_{m-1}(R)$ represents the volume of the $(m-1)$-dimensional sphere. The integration is extended to all the points P for which $n \neq 0$.

For example: for the plane $(m = 2)$ if x_1, x_2 are the coordinates of the centre of a circle of radius R we obtain

$$\int n\, dx_1\, dx_2 = 4LR,$$

For space $(m = 3)$

$$\int n\, dx_1\, dx_2\, dx_3 = 2L\pi R^2.$$

In all cases n is an integral-valued function of x_i.

In § 4 we give some applications of this formula. For the case of the plane the formula may be considered as a particular case of the so called formula of Poincaré of integral geometry and it has been demonstrated more generally by Maak [1].[1] When $m = 3$ and under the assumption that the curve possesses a continuously turning tangent (or is composed of a finite number of arcs of this nature) we have obtained this result in another paper [2].

§ 1. Notations.

Let us represent by $V_m(R)$ the volume of the m-dimensional sphere of radius R which is, as is known,

$$V_m(R) = \frac{(\pi R^2)^{m/2}}{\Gamma(m/2 + 1)}. \tag{1}$$

The volume of a spherical segment of this sphere which has a semiangle

* Received August 20, 1940.

[1] Numbers in brackets refer to the bibliography at the end of the paper.

in the center equal to α may be represented by $S_m(\alpha, R)$ and may be calculated in the following form. On intersecting the sphere by an hyperplane at a distance x from the center, the section is an $(m-1)$-dimensional sphere whose volume is $V_{m-1}(\sqrt{R^2-x^2})$ and the volume of the spherical segment of one base distant from the center $h=R\cos\alpha$ will be

$$S_m(\alpha, R) = \int_h^R V_{m-1}(\sqrt{R^2-x^2})\,dx. \tag{2}$$

Placing $x=R\cos\theta$ and using (1)

$$S_m(\alpha, R) = \frac{\pi^{(m-1)/2}R^m}{\Gamma\left(\frac{m+1}{2}\right)}\int_0^\alpha \sin^m\theta\,d\theta = V_{m-1}(R)\cdot R\cdot\int_0^\alpha \sin^m\theta\,d\theta. \tag{3}$$

On evaluating the last integral by integration by parts and writing $\beta=\frac{\pi}{2}-\alpha$ we obtain:

For m even

$$S_m(\alpha, R) = RV_{m-1}\Big[\frac{(m-1)(m-3)\cdots 3.1}{m(m-2)\cdots 4.2}\frac{\pi}{2}-\sin\beta \tag{4}$$
$$\left(\frac{1}{m}\cos^{m-1}\beta+\frac{(m-1)}{m(m-2)}\cos^{m-3}\beta+\cdots+\frac{(m-1)(m-3)\cdots 5.3}{m(m-2)\cdots 4.2}\cos\beta\right)$$
$$-\frac{(m-1)(m-3)\cdots 5.3}{m(m-2)\cdots 4.2}\beta\Big];$$

and for m odd

$$S_m(\alpha, R) = RV_{m-1}\Big[\frac{(m-1)(m-3)\cdots 4.2}{m(m-2)\cdots 3.1}-\sin\beta\Big(\frac{1}{m}\cos^{m-1}\beta \tag{5}$$
$$+\frac{(m-1)}{m(m-2)}\cos^{m-3}\beta+\cdots+\frac{(m-1)(m-3)\cdots 4.2}{m(m-2)\cdots 3.1}\Big)\Big].$$

For $m=2$ the formula given above should be replaced by

$$S_2(\alpha, R) = V_1(R)R\left[\frac{1}{2}\frac{\pi}{2}-\frac{1}{2}(\sin\beta\cos\beta+\beta)\right]. \tag{6}$$

§ 2. Case of a polygonal line.

1. Measure of a set of spheres. Let E_m represent the euclidean space of dimension m. Let us call the measure of a set of spheres of radius R, the measure in E_m of the set of their centres.

Consequences. a). Given a finite number or an enumerable infinity of points, the measure of the spheres, whose surfaces contain some of these, is zero.

b). The measure of the spheres tangent to a segment of a straight line is zero.

2. Spheres which have common points with a segment. For the case of the plane (fig. 1), and the result is completely general for a space of any number of dimensions, the measure M_0 of the set of spheres of radius R which contains totally in its interior a segment of length l $(l \leq 2R)$, is the part indicated by points, or twice the volume of the spherical segment of a sphere of radius R and semiangle $\alpha = \arc\cos \frac{l}{2R}$; in accordance with the notation (2)

(7) $$M_0 = 2S_m(\alpha, R).$$

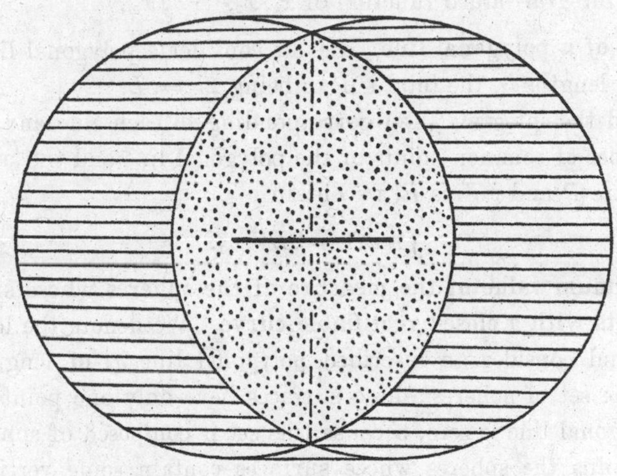

Fig. 1.

The measure M_1 of the spheres of radius R which have with the segment only one point of intersection is composed of two "lunules" (see the shaded area in fig. 1 for E_2), the volume of which will be, in E_m

(8) $$M_1 = 2V_m(R) - 4S_m(\alpha, R).$$

The total measure of the m-dimensional spheres of radius R which have common points with a segment of length l is equal to the volume of a cylinder of height l whose base is an $(m-1)$-dimensional sphere of radius R, plus two hemispheres of radius R. This measure is the sum of the measure of those spheres which contain the segment in their interior (M_0), plus the measure of the spheres which have with the segment one only point of intersection (M_1), plus the measure of the spheres which have with the segment two points of intersection (M_2). Therefore

(9) $$M_0 + M_1 + M_2 = V_m(R) + lV_{m-1}(R).$$

By (7), (8) and (9) we obtain

(10) $$M_2 = lV_{m-1} - V_m + 2S_m(\alpha, R)$$

and by (8)

(11) $$M_1 + 2M_2 = 2lV_{m-1}.$$

If $x_1, x_2, \cdots, x_m$ are the coordinates of the centre of the sphere, we find on putting $dP = dx_1\, dx_2 \cdots dx_m$ and representing by n the number of the points of intersection of the sphere with the fixed segment of length l, that (11) is equivalent to

(12) $$\int n\, dP = 2lV_{m-1}$$

where n is an integral-valued function of $x_1, x_2, \cdots, x_m$.

3. Case of a polygonal line. Let us consider a polygonal line the sides of which have lengths l_i, the total length being $\Sigma l_i = L$.

If we add the integrals (12) corresponding to each side and still denote by n the number of common points of the polygonal line and the surface of the sphere of centre P and radius R, we obtain

(13) $$\int n\, dP = 2LV_{m-1}.$$

4. Maximum value of the measure of the spheres whose surface have common points with a closed rectifiable curve. We denote the length of the curve by L and consider an inscribed polygonal line σ_i of length L_i. The measure of the set of spheres whose surfaces have only one point in common with the polygonal line is zero, because this set is composed of spheres tangent to the sides plus the spheres whose surfaces contain some vertices. Representing by $M_t(\sigma_i)$ the total measure of the spheres which have common points with the polygonal line, we have by (13)

(14) $$M_t(\sigma_i) \leq \tfrac{1}{2}\int n\, dP = L_i V_{m-1} \leq LV_{m-1}$$

because, as we have seen, save for a set of positions of measure zero, n is ≥ 2.

When the number of sides of the polygonal line grows indefinitely, and at the same time their lengths tend towards zero, the volume (14) filled by the centres of the spheres whose surfaces have points in common with the polygonal line, tends towards the volume M_t corresponding to the centres of the spheres whose surfaces have some points in common with the curve. Consequently, by (14) we obtain in the limit

(15) $$M_t \leq LV_{m-1}.$$

5. Lemma. We consider a rectifiable curve of lenth L and denote by l_i the length of the sides of an inscribed polygonal line. Summing Σl_i^r extended on every side, where $r = 1 + a$ and $a > 0$, we have

(16) $$\lim \Sigma l_i^r = 0.$$

This limit is considered for any succession of inscribed polygonal lines which tend towards the curve. In fact if we denote by λ_i the side of greatest length, we have

$$\lim \Sigma l_i^r \leq \lim \ (\lambda_i^a \Sigma l_i) = L \lim \lambda_i^a = 0.$$

As a consequence we observe that, since

$$l_i - \frac{l_i^3}{3!} < \sin l_i < l_i$$

we have

$$\lim \Sigma \sin l_i = L. \tag{17}$$

§ 3. Case of a rectifiable curve.

1. Theorem. If a curve contained in E_m is rectifiable and has length L, and we denote by n the number of common points with the surface of a sphere of radius R and centre P $(x_1, x_2, \cdots, x_m)$ and put $dP = dx_1\, dx_2 \cdots dx_m$, we obtain

$$\int n\, dP = 2LV_{m-1}(R) \tag{18}$$

The radius R may have any value. The integration is extended to all the points for which $n \neq 0$.

Proof. Let us consider two classes of points common to the curve and the surface of the sphere: I. *Intersection points,* when in any neighborhood of these there are points of the curve interior and exterior to the sphere. II. *Contact points,* when in the neighborhood of these the points of the curve are all from the interior or from the exterior of the sphere.

Let us indicate by $n^{(i)}$ the number of the first class for any position of the sphere, and by $n^{(c)}$ the number of the second group. We have $n = n^{(i)} + n^{(c)}$ and both numbers are functions of the coordinates $x_1, x_2, \cdots, x_m$ of the centre of the sphere.

I. We shall first consider the intersection points. Let us take a polygonal line σ_1 inscribed in the curve. For each position of the sphere, we may represent by $n_1^{(i)}$ the number of points of intersection with the polygonal line in which the corresponding side l_i presents one extremity outside and the other inside the sphere, plus the points of intersection which are vertices of the polygonal line and in which this cuts through the spherical surface, the other extremes of these sides which concur in this point remaining in a distinct region of the sphere.

By (8) we obtain

$$\int n_1^{(i)}\, dP = \sum_i (2V_m - 4S_m(\alpha_i, R)) \tag{19}$$

in which $S_m(\alpha_i, R)$ has the value (3) with $\alpha_i = \operatorname{arc}\cos \frac{l_i}{2R}$ and the sum is extended to all the sides of the polygonal line.

On increasing the number of sides of the inscribed polygonal line, in order to unite the extremities of each l_i which has an intersection point or to unite the extremities of two consecutive sides which concur at one point of the surface of the sphere being separate from this, it is always necessary to pass from an interior point to another which is exterior to the sphere. In consequence for a constant position of P, the number $n_1^{(i)}$ does not diminish when the number of sides of the polygonal line increases. That is to say

$$n_1^{(i)} \leq n_2^{(i)} \leq \cdots \leq n_\nu^{(i)} \leq \cdots$$

and when the number of sides increases and their respective lengths tend towards zero, we obtain $\lim_{\nu\to\infty} n_\nu^{(i)} = n^{(i)}$.

In accordance with the fundamental property of the Lebesgue integral, we have

$$\int n^{(i)}\, dP = \lim_{\nu\to\infty} \int n_\nu^{(i)}\, dP = \lim \sum_i (2V_m - 4S_m(\alpha_i, R)). \tag{20}$$

To find this limit we substitute for $S_m(\alpha_i, R)$ its value given by (4), (5) and also set

$$\frac{(m-1)(m-3)\cdots 3.1}{m(m-2)\cdots 4.2}\frac{\pi}{2} R V_{m-1} = \tfrac{1}{2} V_m \qquad \text{for } m \text{ even}$$

and

$$\frac{(m-1)(m-3)\cdots 4.2}{m(m-2)\cdots 3.1} R V_{m-1} = \tfrac{1}{2} V_m \qquad \text{for } m \text{ odd.}$$

We obtain, if m is an even number,

$$\begin{aligned} 2V_m - 4S_m(\alpha_i, R) = 4RV_{m-1} \Big[\sin\beta_i \Big(& \frac{1}{m}\cos^{m-1}\beta_i \\ & + \frac{(m-1)}{m(m-2)}\cos^{m-3}\beta_i + \cdots + \frac{(m-1)(m-3)\cdots 5.3}{m(m-2)\cdots 4.2}\cos\beta_i \Big) \\ & + \frac{(m-1)(m-3)\cdots 5.3}{m(m-2)\cdots 4.2}\beta_i \Big] \end{aligned}$$

and if m is an odd number

$$\begin{aligned} 2V_m - 4S_m(\alpha_i, R) = 4RV_{m-1} \sin\beta_i \Big(& \frac{1}{m}\cos^{m-1}\beta_i \\ & + \frac{(m-1)}{m(m-2)}\cos^{m-3}\beta_i + \cdots + \frac{(m-1)(m-3)\cdots 4.2}{m(m-2)\cdots 3.1} \Big) \end{aligned}$$

where

$$\beta_i = \frac{\pi}{2} - \alpha_i = \operatorname{arc}\sin \frac{l_i}{2R}.$$

On adding these expressions for all sides l_i of the inscribed polygonal line and approximating this polygonal line towards the curve, we have in the limit (considering (16), (17)):

For m even

$$(21)\qquad \lim \sum_i \left(2V_m - 4S_m(\alpha_i, R)\right)$$

$$= 2LV_{m-1}\left(\frac{1}{m} + \frac{(m-1)}{m(m-2)} + \cdots + 2\frac{(m-1)(m-3)\cdots 5.3}{m(m-2)\cdots 4.2}\right)$$

and for m odd

$$(22)\qquad \lim \sum_i \left(2V_m - 4S_m(\alpha_i, R)\right)$$

$$= 2LV_{m-1}\left(\frac{1}{m} + \frac{(m-1)}{m(m-2)} + \cdots + \frac{(m-1)(m-3)\cdots 4.2}{m(m-2)\cdots 3.1}\right).$$

But the sums which are in the parentheses are equal to unity in both cases, and so

$$(23)\qquad \int n^{(i)}\, dP = 2LV_{m-1}.$$

II. We must now consider the contact points. Denoting by $n^{(c)}$ the number of these corresponding to a position of the sphere we shall show that

$$(24)\qquad \int n^{(c)}\, dP = 0.$$

Let us consider a side of length l_i of an inscribed polygonal line and the arc of the curve comprehended between its extremes. This arc (of length u_i) plus the side of the polygonal line forms a closed curve of length $l_i + u_i$. Accordingly by (15), the measure M_t of the spheres whose surface has common points with the curve is $M_t \leq (l_i + u_i)V_{m-1}$. Subtracting from this value the measure of the spheres whose surfaces cut the side of the polygonal line at only one point (given by (8)), there results

$$(25)\qquad (l_i + u_i)V_{m-1} - [2V_m - 4S_m(\alpha_i, R)].$$

On summing these values for all sides of the polygonal line we obtain an upper bound to the measure of the spheres whose surface cuts the curve without cutting any side of the polygonal line in one point only (it might occur that the sphere cuts in two points some of the sides). Consider now a sphere with $n^{(c)}$ contact points. From an inscribed polygonal line of sides sufficiently small, this sphere for each point of the $n^{(c)}$ is contained in the sum of the expressions (25). Consequently

$$(26)\qquad \int n^{(c)}\, dP \leq \lim \sum_i \left[(l_i + u_i)V_{m-1} - 2V_m + 4S_m(\alpha_i, R)\right]$$

As before, the limit is considered for a succession of inscribed polygonal lines which tends towards the curve. But $\lim \sum_i (l_i + u_i) V_{m-1} = 2LV_{m-1}$ and having in mind (21) and (22) we obtain

$$\lim \sum_i (2V_m - 4S_m(\alpha_i, R)) = 2LV_{m-1}.$$

(24) is thus demonstrated.

On summing (23) and (24) and taking into account the fact that $n = n^{(i)} + n^{(c)}$ we obtain the formula (18).

2. Reciprocal theorem. *Calling n the number of common points of a Jordan curve with the surface of a sphere of radius R and centre* $P(x_1, x_2, \cdots, x_m)$ *the curve is rectifiable and its length is equal to* $J : 2V_{m-1}$ *if the integral*

$$J = \int n \, dP \tag{27}$$

is finite.

Proof. For each position of the sphere we decompose n into $n = n^{(i)} + n^{(c)}$, $n^{(i)}$ being the number of intersection points and $n^{(c)}$ the contact points defined previously. Fixing a sphere of centre $P(x_1, x_2, \cdots, x_m)$ and inscribing a polygonal line in the given curve, and representing by $n_1{}^{(i)}$ the number of sides of this polygonal line which cut the surface of the sphere in one only point, we have when we inscribe successive polygonal lines, for a constant position of the sphere,

$$n_1{}^{(i)} \leq n_2{}^{(i)} \leq \cdots \leq n_\nu{}^{(i)} \leq \cdots \leq n^{(i)}.$$

Integrating this succession and applying the formula (8) for polygonal lines we obtain

$$\int n_\nu{}^{(i)} \, dP = \sum_i (2V_m - 4S_m(\alpha_i, R)) \leq \int n^{(i)} \, dP \leq J.$$

Consequently the first sum is bounded as l_i tends towards zero, which shows, on taking into account the values (4) (5), that Σl_i is also bounded. The curve is thus rectifiable and by the direct theorem its length is equal to $J : 2V_{m-1}$.

§ 4. Applications.

1. A definition of length of a curve. The theorem of the previous section permits the following definition of length of a set of points of E_m:

Let us consider a set of points of E_m. *Calling n the number of these which belong to the surface of the sphere of radius R and centre* $P(x_1, x_2, x_3, \cdots, x_m)$, *the length of the set of points is*

$$(28)\qquad \frac{1}{2V_{m-1}(R)}\int n\,dP$$

where $dP = dx_1, dx_2 \cdots dx_m$ *and the integration is extended over all space or, what is the same, over the positions of* P *for which* $n \neq 0$.

In accordance with the reciprocal theorem of the last section this present definition coincides with the usual definition if the set of points is a Jordan curve.

2. A sufficient condition in order that a curve should be rectifiable. If a curve is situated at finite distance and the maximum number of common points that the curve could have with the surface of a sphere of any radius is limited, the curve is rectifiable.

This is a consequence of the reciprocal theorem. This property is due to Marchaud [3].

3. An inequality for rectifiable curves. THEOREM: *If* C *is a rectifiable curve of length* L *in* E_m *and* V *is the volume filled by the points in space whose distance from any point of the curve is* $\leq R$, *then*

$$(29)\qquad V \leq LV_{m-1}(R) + V_m(R).$$

This inequality, for the case of the plane ($m = 2$), has been obtained by Hornich [4]. The procedure we propose to follow will permit us also to give an interpretation of the "deficit," that is to say, of the difference of the two members of the inequality (29). Let M_i be the measure of the spheres of radius R which have i common points with the curve. In accordance with (18), we have

$$(30)\qquad M_1 + 2M_2 + 3M_3 + \cdots = 2LV_{m-1}.$$

If M_0 indicates the measure of the spheres which contain the curve C in their interior, in accordance with the definition of V we obtain also

$$(31)\qquad M_0 + M_1 + M_2 + M_3 + \cdots = V.$$

From (30) and (31) we can deduce that

$$(32)\qquad M_3 + 2M_4 + \cdots - (2M_0 + M_1) = 2LV_{m-1} - 2V.$$

We consider the segment of length D which unites the extremities of the given curve (if the curve C is closed, $D = 0$). Let us represent by M^*_0 the measure of the spheres of radius R which contain this segment totally in their interior, and by M^*_i ($i = 1, 2$) the measure of the spheres whose surfaces have i common points with the same segment. According to (9) and (11) we have

$$M^*_0 + M^*_1 + M^*_2 = V_m + DV_{m-1}; \qquad M^*_1 + 2M^*_2 = 2DV_{m-1}$$

so that

$$2M^*_0 + M^*_1 = 2V_m.$$

If the sphere contains the curve C in its interior it will also contain the segment that unites its extremities, and in consequence $M_0 \leq M^*_0$. Also, if the surface of the sphere cuts the curve in only one point, this curve will have one of its extremities in the interior, and the other exterior, and so the segment will cut the sphere also at only one point. That is to say, $M_1 \leq M^*_1$. It follows that

$$2M_0 + M_1 \leq 2M^*_0 + M^*_1 = 2V_m.$$

Applying this inequality to (32), we obtain

$$\tfrac{1}{2}(M_3 + 2M_4 + \cdots) + V \leq LV_{m-1} + V_m \tag{33}$$

which implies (29).

The equality in (33) will be verified, if $M_i = 0$, only by $i \geq 3$ and moreover $M^*_0 = M_0$, $M^*_1 = M_1$. The condition $M_i = 0$ $(i \geq 3)$ carries with it $M^*_1 = M_1$ and in consequence the conditions for equality are:

1°. $M_i = 0$ $(i \geq 3)$. The curve can not be cut by the surface of the sphere in more than two points (with the exception perhaps of a set of positions of zero measure).

2°. $M^*_0 = M_0$. That is to say, every position of the sphere in which it contains in its interior the two extremities of the curve, contains also all the curve.

In particular, if the given curve C is closed, the equality only is valid in the case of reduction to a point.

ROSARIO (RÉP. ARGENTINA).

BIBLIOGRAPHY.

[1]. W. Maak, "Ueber stetige Kurven (Integralgeometrie 27)." *Abhandlungen aus dem Mathematischen Seminar der Hansischen Universität*, vol. 12 (1938), pp. 163-178.

[2]. L. A. Santaló, "Integralgeometrie. Ueber das kinematische Mass im Raum," *Actualités Scientifiques et industrielles No. 357*, Hermann et Cie, editeurs: Paris (1936).

[3]. A. Marchaud, "Sur les continus d'ordre borné," *Acta Mathematica*, vol. 55 (1930), pp. 67-115.

[4]. H. Hornich, "Eine allgemeine Ungleichung für Kurven," *Monatshefte für Mathematik und Physik*, vol. 47 (1939), pp. 432-438.

Curves of extremal total torsion and D-curves

L. A. Santaló

Abstract

D-curves are the curves on a surface at each of whose points the osculating sphere is tangent to the surface: these curves were studied by Darboux. In this paper, I wish to establish that, under certain conditions, the D-curves coincide with the extremal curves of the total torsion; that is, with the curves whose first variation of the integral $\int \tau \, ds$ is zero. In Section IV we perform an application of these results in order to find the D-curves on the surface of the torus; between these curves, there are some closed curves which have positive torsion at each point.

Total torsion is the name given a straight line connecting two points A and B to the integral $\mathcal{T} = \int_A^B \tau \, ds$, where $\tau(s)$ is the torsion at each point and s the arc.

The problem we wish to solve consists in finding the curves on a surface that are extremal to this torsion; that is, those which vanish at the first variation of the integral. To this end, we apply Bonnet's formula (see, for example, [1], p. 296-298, or [6], t.I, p. 622).

$$\tau = \frac{d\sigma}{ds} + \tau_g,$$

which relates the torsion τ with the *geodesic torsion* τ_g and the derivative with respect to the arc s of the angle σ that the plane osculating to the curve and the plane tangent to the surface form at each point. Thus, it is first necessary to study the *extremal curves of the total geodesic torsion*; that is, of $\mathcal{T}_g = \int_A^B \tau_g \, ds$.

We find the following result: *The extremal curves of the total geodesic torsion coincide with the D-curves of the surface.* We give the name D-curves to those curves whose osculating sphere is tangent to the surface at each point; these curves were first studied by Darboux [4].[1]

In order to adapt the notation of the D-curve equations to those suitable for the calculations below, in Section I we obtain the differential equation of the D-curves with respect to the system of curvilinear coordinates formed by the lines of curvature. This system has the advantage of greatly simplifying the calculations, especially for the problem of variational calculus solved in Section II.

In Section II, on applying the classical formulae of variational calculus, we obtain the equation for the extremal curves of the total geodesic torsion and see how this equation coincides with the previous one for D-curves. We then find the restrictions that must be imposed on the variational curves in order for the extremal curves of the total geodesic torsion also to be those of the total torsion.

[1]Bibliography on these cruves can be found in *Encyklopädie der Math. Wiss.* III_3 D 3, p. 181-182.

In Section III, a further direct proof of the preceding result is given.

In Section IV, we apply the foregoing results to the particular case when the surface is a torus.

I. D-CURVES

As already stated, we call D-curves of a surface those curves whose osculating sphere is tangent to the surface at every point. We now find the equation for such curves.

1. Let $X = X(u, v)$ be the vector equation of the surface. A line belonging to this surface is given by two functions $u = u(t), v = v(t)$ of a parameter t. If the point $Z(t)$ is the centre of the sphere osculating to the curve $X = X(u(t), v(t))$ and r the radius, then the following vector condition is fulfilled:

$$(X - Z)^2 = r^2,$$

and by the definition of an osculating sphere, the conditions obtained must also be fulfilled by deriving the foregoing equation three times, assuming Z and r to be constant. In other words,

$$\begin{aligned} (X - Z)X' &= 0 \\ (X - Z)X'' + X'^2 &= 0 \\ 3X'X'' + (X - Z)X''' &= 0, \end{aligned} \tag{1}$$

where the products are scalar products of vectors, and the primes denote derivatives with respect to the parameter t; for example, $X' = X_u u' + X_v v'$.

If the curve $X = X(u(t), v(t))$ is a D-curve, the centre Z should be on the normal to the surface; that is, by denoting the unit vector normal to the surface by V, we have $Z = X - \lambda V$, where λ is a parameter. On substituting this expression in (1), the first equation is fulfilled for every value of λ, since $VX' = 0$, and the last two equations give

$$\begin{aligned} X'^2 + \lambda V X'' &= 0 \\ 3X'X'' + \lambda V X''' &= 0. \end{aligned} \tag{2}$$

The differential equation of the D-curves is obtained by removing λ from these two equations. It appears that the resulting equation is a third-order equation; however, one observes that by deriving twice we obtain

$$\begin{aligned} V'X' + VX'' &= 0 \\ V''X' + 2V'X'' + VX''' &= 0. \end{aligned}$$

From this latter equation we deduce VX''', and then by removing λ from equations (2) we have

$$\frac{3X'X''}{X''} = \frac{2V'X'' + V''X'}{V'X'}. \tag{3}$$

This is the differential equation for the D-curves in the form given by Darboux [4].

For our purpose, it is convenient to develop this equation further by expressing the elements it contains in terms of the coefficient of the two fundamental quadratic forms of surface theory.

2. Let us take as the parametric lines $u =$ constant, $v =$ constant, the lines of curvature of the surface. Then the two fundamental forms of surface theory give

$$\begin{aligned} X'^2 &= Eu'^2 + Gv'^2 \\ -V'X' &= Lu'^2 + Nv'^2, \end{aligned} \tag{4}$$

where, as is known,

$$E = X_u^2, \quad G = X_v^2, \quad L = -V_u X_u, \quad N = -V_v X_v$$

where the sub-indices denote partial derivatives.

Deriving the first of the expressions (4), we obtain

$$\begin{aligned} 2X'X'' &= E_u u'^3 + E_v u'^2 v' + G_u u' v'^2 \\ &\quad + G_v v'^3 + 2Eu'v'' + 2Gv'v'', \end{aligned} \tag{5}$$

and deriving the second

$$\begin{aligned} -(V'X'' + V''X') &= L_u u'^3 + L_v u'^2 v' + N_u u' v'^2 \\ &\quad N_v v'^3 + 2Lu'u'' + 2Nv'v''. \end{aligned} \tag{6}$$

If $u =$ constant and $v =$ constant are the lines of curvature (which requires that $X_u X_v = 0, V_u X_v = V_v X_u = 0$), expressing the vectors X_{uu}, X_{uv} and X_{vv} by their components with respect to the right-angle trihedron formed by the three vectors X_u, X_v, V, we obtain the Gauss formulae

$$\begin{aligned} X_{uu} &= \frac{E_u}{2E} X_u - \frac{E_v}{2G} X_v + LV \\ X_{uv} &= \frac{E_v}{2E} X_u + \frac{G_u}{2G} X_v + MV \\ X_{vv} &= -\frac{G_u}{2E} X_u + \frac{G_v}{2G} X_v + NV, \end{aligned}$$

and hence,

$$\begin{aligned} V_u X_{uu} &= -\frac{E_u}{2E} L, \quad V_u X_{uv} = -\frac{E_v}{2E} L, \quad V_u X_{uv} = \frac{G_u}{2E} L \\ V_v X_{uu} &= \frac{E_v}{2G} N, \quad V_v X_{uv} = -\frac{G_u}{2G} N, \quad V_v X_{vv} = -\frac{G_v}{2G} N, \end{aligned} \tag{7}$$

whereby we have

$$\begin{aligned} V'X'' &= (V_u u' + V_v v') \cdot \\ &\quad \cdot (X_{uu} u'^2 + 2X_{uv} u'^2 + 2X_{uv} u'v' + X_{vv} v'^2 + X_u u'' + X_v v'') \\ &= -\frac{1}{2EG}[GLE_u u'^3 + (2GLE_v - NEE_v)u'^2 v' \\ &\quad + (2ENG_u - LGG_u)u'v'^2 + ENG_v v'^3] - Lu'u'' - Nv'v''. \end{aligned}$$

Adding this expression to (6), and taking into account the Mainardi-Codazzi relations (see, for example, [2], p. 117 or [1], p. 175) which in our case are reduced to

$$\begin{aligned} 2EGL_v &= (EN + GL)E_v \\ 2EGN_u &= (EN + GL)G_u, \end{aligned} \tag{8}$$

we arrive at

$$\begin{aligned} &2V'X'' + V''X' = \\ &-\frac{1}{2EG}[(2EL_u + LE_u)Gu'^3 + 3GLE_v u'^2 v' + 3ENG_u u'v'^2 \\ &+(2GN_v + NG_v)Ev'^3] - 3Lu'u'' - 3Nv'v''. \end{aligned} \tag{9}$$

With the obtained expressions (9), (5) and (4), and substituting in equation (3), we obtain the equation sought for the D-curves, which can be written as follows:

$$a(u'v'' - v'u'')u'v' + bu'^5 + cu'^3 v'^2 + du'^2 v'^3 + ev'^5 = 0, \tag{10}$$

where

$$\begin{aligned} a &= 6(GL - EN)EG \\ b &= 2EG(LE_u - EL_u) \\ c &= 3(EG_u - GE_u)(GL - EN) + 2G^2(LE_u - EL_u) \\ d &= 3(EG_v - GE_v)(GL - EN) + 2E^2(NG_v - GN_v) \\ e &= 2EG(NG_v - GN_v). \end{aligned} \tag{11}$$

II. Extremal curves of the total torsion

1. *Expression for total torsion.* 1. Let $\tau(s)$ be the torsion of a line on a surface, and $\sigma(s)$ the angle formed by its osculating plane with the plane tangent to the surface. Denoting the principal radii of curvature by R_1 and R_2, and by ω the angle formed by the tangent to the line considered with the first principal direction (that according to which the radius of curvature is R_1), the following Bonnet formula (see, for example, [1], p. 296-298, or [6], t. I, p. 622.) holds

$$\tau = \frac{d\sigma}{ds} + (\frac{1}{R_1} - \frac{1}{R_2}) \sin\omega \cos\omega. \tag{12}$$

The second term of the sum of this expression is the so-called *geodesic torsion.*

As before, let us consider the surface $X(u,v)$ in terms of its lines of curvature as parametric lines. Thus, it is known (see, for example, [2], p. 95) that the values of the principal radii of curvature in terms of the coefficients of the two fundamental quadratic forms (4) are

$$\frac{1}{R_1} = \frac{L}{E}, \quad \frac{1}{R_2} = \frac{N}{G}. \tag{13}$$

Furthermore, the direction of the tangent to the line considered is that of the vector $X_u u' + X_v v'$, so the angle ω that it forms with the tangents X_u and X_v to the principal directions is determined by

$$\sin\omega = \frac{(X_u u' + X_v v')X_v}{[X_v^2(X_u u' + X_v v')^2]^{1/2}} = \frac{Gv'}{[G(Eu'^2 + Gv'^2)]^{1/2}}$$

$$\cos\omega = \frac{(X_u u' + X_v v')X_u}{[X_u^2(X_u u' + X_v v')^2]^{1/2}} = \frac{Eu'}{[E(Eu'^2 + Gv'^2)]^{1/2}}.$$

With these values and with (13), the value of the torsion τ takes the form

$$\tau = \frac{d\sigma}{ds} + \frac{GL - EN}{\sqrt{EG}} \cdot \frac{u'v'}{Eu'^2 + Gv'^2}. \tag{14}$$

2. Consider a curve of the surface connecting two points A and B of the surface. Denoting by σ_A and σ_B the angles formed by the plane osculating to the curve and the plane tangent to the surface, the *total torsion* of the arc of the curve (since $ds^2 = Eu'^2 + Gv'^2$) is

$$T = \int_A^B \tau\, ds = \sigma_B - \sigma_A + \int_A^B \frac{GL - EN}{\sqrt{EG}} \cdot \frac{u'v'}{Eu'^2 + Gv'^2}\, dt. \tag{15}$$

The angle $\sigma_B - \sigma_A$ is the increment of the angle σ on going from A to B, and may therefore be greater than 2π.

2. Extremal curves of the total geodesic torsion: 1. Let us first consider the last term of (15), that is, the total geodesic torsion between A and B. We wish to find the curves for which this total geodesic torsion

$$T_g = \int_A^B \frac{GL - EN}{\sqrt{EG}} \cdot \frac{u'v'}{[Eu'^2 + Gv'^2]^{1/2}}\, dt \tag{16}$$

is an extremal.

Here we are dealing with a problem of variational calculus in a parametric form. In this case, if $f(u,v;u',v')$ is an abbreviated representation of the sub-integral function, the form given by Weierstrass for the differential equation of extremal curves is (see, for example, [9], p. 96 or [6] t. III, p. 630)

$$(u'v'' - v'u'')f_1 + f_{uv'} - f_{vu'} = 0, \tag{17}$$

where the sub-indices denote partial derivatives, and where

$$f_1 = \frac{f_{u'v'}}{v'^2} = -\frac{f_{u'v'}}{u'v'} = \frac{f_{v'v'}}{u'^2}.$$

In our case, this is

$$\begin{aligned} f_1 &= -3\sqrt{EG}(GL-EN)(Eu'^2+Gv'^2)^{-5/2}u'v' \\ f_{uv'} &= -\frac{1}{2}(EG)^{-3/2}(Eu'^2+Gv'^2)^{-5/2}\{(Eu'^2+Gv'^2)[(EN \\ &\quad +GL)(GE_u-EG_u)+2EG(EN_u-GL_u)]Eu'^3 \\ &\quad +(EN-GL)[(2GE_u-3EG_u)u'^3v'^2-EE_uu'^5]EG\} \\ f_{vu'} &= \frac{1}{2}(EG)^{-3/2}(Eu'^2+Gv'^2)^{-5/2}\{(Eu'^2+Gv'^2)[(EN \\ &\quad +GL)(EG_v-GE_v)+2EG(GL_v-EN_v)]Gv'^3 \\ &\quad +(GL-EN)[(2EG_v-3GE_v)u'^2v'^3-GG_vv'^5]EG\}. \end{aligned}$$

Substituting these values in (17) and simplifying (bearing in mind the Mainardi-Codazzi relations (8)), we obtain equation (10)

$$a(u'v''-v'u'')u'v'+bu'^5+cu'^3v'^2+du'^2v'^3+ev'^5=0$$

where the values of the coefficients are the same as those in (11).

We therefore obtain that:

The extremal curves of the total geodesic torsion coincide with the D-curves of the surface.

The study of the extremal curves of the total geodesic torsion is thus reduced to that of the D-curves. Since (10) is a second order equation, the extremal curve giving the initial and final points or giving a point and the tangent at that point, within a certain region of the surface, is thereby determined. In this latter case, we may point out that there is certainly an extremal curve in any direction of the tangent, except perhaps for the principal directions $u'=0, v'=0$. For these directions, a discussion for each surface and for each particular point is required.

2. We now check to see if these extremal curves of $\mathcal{T}_g$, given by (10), correspond to a maximum or a minimum. To that end, let us form Weierstrass' function $\mathcal{E}$. Denoting the sub-integral of (16) by f(u, v; u',v'), this function $\mathcal{E}$ is (see, for example, [3], p. 243):

$$\begin{aligned} \mathcal{E}(u,v;u',v';\bar{u'},\bar{v'}) &= \bar{u'}[f_{u'}(u,v;\bar{u'},\bar{v'})-f_{u'}(u,v;u',v')] \\ &\quad +\bar{v'}[f_{v'}(u,v;\bar{u'},\bar{v'})-f_{v'}(u,v;u',v')] \end{aligned}$$

In our case, this is

$$\mathcal{E}(u,v;u',v';\bar{u'},\bar{v'}) = \frac{GL-EN}{\sqrt{EG}}\left\{\bar{u'}\left[\frac{G\overline{v'^3}}{\overline{W^3}}-\frac{Gv'^3}{W^3}\right]+\bar{v'}\left[\frac{E\overline{u'^3}}{\overline{W^3}}-\frac{Eu'^3}{W^3}\right]\right\}, \tag{18}$$

where $W = \sqrt{Eu'^2 + Gv'^2}$ and $\overline{W}$ the same expression after replacing u', v' by $\bar{u}', \bar{v}'$.

A necessary condition (see [3], p. 244) for a *strong minimum* (maximum) is that $\mathcal{E} \geq 0$ ($\mathcal{E} \leq 0$) for values of u, v, u', v' corresponding to the extremal, and for any values of $\bar{u}', \bar{v}'$. This condition is not satisfied for the previous value of $\mathcal{E}$, since for one same point u, v, u', v', taking $\bar{u}' = 0, \bar{v}' > 0$, the function $\mathcal{E}$ has the sign of the product $(GL - EN)u'$, and for $\bar{u}' = 0, \bar{v}' < 0$ the opposite sign. Therefore:

The extremals are neither strong minima nor strong maxima. Among the curves connecting two points on a surface, there exist none that provide a strong minimum or a strong maximum of the total geodesic torsion.[2]

3. Nevertheless, we now prove that for a sufficiently limited region, and in which the difference $GL - EN$ is different from zero (assuming it to be positive), *the extremals of the total geodesic torsion correspond to a weak maximum (minimum), providing* $u'v' > 0$ ($u'v' < 0$) *is fulfilled.*

For this proof, it is preferable to employ the non-parametric form of the integral (16). Taking u as the independent variable, we obtain

$$\mathcal{T}_g = \int_A^B \frac{GL - EN}{\sqrt{EG}} \cdot \frac{v'}{[E + Gv'^2]^{1/2}} \, du.$$

Legendre's necessary condition for the maximum is (see, for example, [7], p. 325) $f_{v'v'} < 0$ along all the extremal (where f is the sub-integral function). In our case, this is

$$f_{v'v'} = -3\sqrt{EG}(GL - EN)\frac{v'}{(E + Gv'^2)^{5/2}}.$$

Therefore, assuming that $GL - EN > 0$, the condition $f_{v'v'} < 0$ is fulfilled providing that $v' = \frac{dv}{du} > 0$.

The origin A of the extremals (10) having been fixed, there is a sufficiently limited region in a neighbourhood of any extremal passing through this point such that for every point in the region only one extremal with origin A will pass. For this region, the above-mentioned Legendre condition $f_{v'v'} < 0$ is sufficient ([7], p. 397) to ensure that the extremal curve corresponds to a weak maximum.

3. Extremals of the total torsion. We now see how to pass from the extremals of the total geodesic torsion to the curves that make the total torsion either maximum or minimum (15).

---

[2]Recall that an extremal $v_0 = v_0(u)$ referring to non-parametric coordinates is said to be a *strong minimum* (maximum) when the corresponding value of the integral is equal to or less (greater) than that taking for any other curve $v = v(u)$, such that

$$|v(u) - v_0(u)| < \epsilon_1,$$

where ϵ_1 is independent of u.

However, $v_0 = v_0(u)$ is said to be a *weak minimum* (maximum) when we impose on the variational curves the second condition

$$|v'(u) - v_0'(u)| < \epsilon_2,$$

where ϵ_2 is also independent of u. See, for example, [3], p. 91.

1. Let us first assume that the surface has only elliptic points; in other words, it has positive curvature at all its points. Let A and B be two points on the surface. The total torsion and the total geodesic torsion of a line connecting A with B are related by (15)

$$\mathcal{T} = \mathcal{T}_g + \sigma_B - \sigma_A.$$

If we consider only curves with a radius of curvature ρ different from zero, even at the initial and final points A and B, its osculating plane cannot coincide at any point with the tangent to the surface (since in that case, by Meusnier's theorem, the radius of curvature would be zero), and therefore the maximum of $\sigma_B - \sigma_A$ is π. This maximum value cannot be reached, since it would only be so for $\sigma_A = 0, \sigma_B = \pi$, in which case the radius of curvature at the intial and final points would be zero, and the curve would no longer be among those considered. With this observation it is easy to prove that among the curves with $\rho \neq 0$ there are none corresponding to a maximum or minimum of the total torsion.[3]

This is indeed the case; assume, for instance, that $GL - EN > 0$. We have already seen that the extremal curves of $\mathcal{T}_g$ are weak maxima (assuming that along the extremal $\frac{dv}{du} > 0$; in the opposite case it would be minimum and the reasoning would be similar, observing that the minimum value of the total torsion would be $\mathcal{T}_g^0 - \pi$). Let C_0 be the extremal curve of $\mathcal{T}_g$ connecting A with B, and $\mathcal{T}_g^0$ the corresponding value of $\mathcal{T}_g$. The maximum value of $\mathcal{T}$ will be $\mathcal{T}_g^0 + \pi$. This value cannot be reached, since $\mathcal{T}_g$ cannot be greater than $\mathcal{T}_g^0$, and for curves with $\rho \neq 0$ is $\sigma_B - \sigma_A < \pi$.

Nevertheless, it is possible to find curves whose total torsion differs very little from $\mathcal{T}_g^0 + \pi$. In fact, given ϵ_1, let us take on the curve C_0, extremal of $\mathcal{T}_g$, two points A_1 and B_1 sufficiently close to A and B so that the value $\mathcal{T}_g^1$ of integral (16) from A_1 to B_1 along the same curve C_0 fulfils the condition

$$\mathcal{T}_g^1 > \mathcal{T}_g^0 - \frac{\epsilon_1}{3}. \tag{19}$$

At A and B we take two planes whose angles with the tangents to the surface satisfy the condition $\sigma_B - \sigma_A = \pi - \frac{\epsilon_1}{3}$. Let us then connect A with A_1 and B with B_1 by two arcs having at A and B these two planes as osculating planes, and which at A_1 and B_1 connect the curve C_0 with the osculating planes. Points A_1 and B_1 can be taken sufficiently close to A and B such that, in addition to fulfilling (19), the sum of the two integrals (16) of the geodesic torsion of A to A_1 and B to B_1 is less than $\frac{\epsilon_1}{3}$ in absolute value. The total torsion $\mathcal{T}$ of the curve AA_1BB_1 thus constructed is

$$\mathcal{T} > -\frac{\epsilon_1}{3} + \mathcal{T}_g^0 - \frac{\epsilon_1}{3} + \pi - \frac{\epsilon_1}{3} = \mathcal{T}_g^0 + \pi - \epsilon_1.$$

By taking a sequence $\epsilon_1, \epsilon_2, \epsilon_3, \cdots$, that tends to zero, we have a sequence of curves whose total torsion will tend to the maximum value $\mathcal{T}_g^0$. However, this value is not reached, since the curve limit of this sequence has at its extremes

[3] Here we refer to weak maximum or minimum. With greater reason, it is not possible for them to correspond to a strong maximum or minimum.

osculating planes tangent to the surface, and therefore radii of curvature equal to zero.

2. If the points A and B that we wish to connect by an extremal of the total torsion are hyperbolic or parabolic points of the surface, then the osculating plane at these points may be tangent to the surface without the radius of curvature ρ of the curve being zero. This occurs only when the tangents at A and B correspond to asymptotic directions. In this case, the extremals (10) may correspond to a maximum of the total torsion. However, even if the surface has hyperbolic points, and A and B are elliptic or the tangent at A and B does not have an asymptotic direction, the reasoning used in n° 1 can be repeated and then it is clear that the curve cannot correspond to either a maximum or a minimum. In other words, in order for a curve C_0, with $\rho \neq 0$, connecting A with B to correspond to a maximum (see the footnote on page 240) of the total torsion, it is necessary that at these points C_0 should have asymptotic directions as tangents and tangents to the surface as osculating planes.

4. *Variations with osculating planes fixed at the initial and final points and closed extremal curves*: I. As in n° 3, let C_0 be the extremal curve of the total geodesic torsion connecting two points A and B on a surface with *positive and continuous curvature.* If only variational curves having at A and B the same osculating planes as C_0 are considered, the difference $\sigma_B - \sigma_A$ (since it cannot be greater than π) will be the same for all these curves, and therefore C_0 will also be an extremal curve of the total torsion. In other words, for surfaces with positive curvature at all their points, the following holds: *D-curves are extremal curves of the total torsion among all the curves connecting two points A and B and having at these points the same osculating planes as the D-curve considered.*

Furthermore, according to (2, 3): "For two sufficiently close points A and B, the total torsion of the D-curve ($v_0 = v_0(u)$) connecting these points is maximum (assuming $(GL-EN)\cdot v_0' > 0$ between A and B; otherwise it would be minimum) compared with the other curves $v = v(u)$ connecting the same points, with the same osculating planes at these points, and fulfil the conditions

$$|v(u) - v_0(u)| < \epsilon_1, \quad |v'(u) - v_0'(u)| < \epsilon_2 \tag{20}$$

for all the points on the arc AB and for ϵ_1, ϵ_2 sufficiently small".

2. If on a surface of positive and continuous curvature there is a closed D-curve with continuous torsion, we have $\sigma_B - \sigma_A = 0$ (because $A \equiv B$), and since for any other closed curve with continuous torsion we also have $\sigma_B - \sigma_A = 0$, we obtain: *Closed D-curves are extremal curves of the total torsion.*

Observing that all the curves on a spherical surface are D-curves, it follows that the total torsion of the closed curves on a spherical surface has a stationary value. Indeed, it is known (see [5])[4] that the total torsion of the closed spherical curves is zero. Moreover, it is easy to prove this property directly. In fact, if we denote the radius of curvature of the curve by ρ, the radius of the osculating

[4]Furthermore, this property characterizes the spherical surface, in the sense that: "Any surface with a continuous Gauss curvature at all its points, such that the total torsion of all the closed lines of the surface is equal to zero, is a sphere or a part of a spherical surface". See [8].

sphere is equal to (see [2], p. 33)

$$r^2 = \rho^2 + (\frac{\rho'}{\tau})^2,$$

where $\rho' = \frac{d\rho}{ds}$. Hence,

$$\int_c \tau\, ds = \int_c \frac{d\rho}{\sqrt{r^2 - \rho^2}} = \left[\arcsin \frac{\rho}{r}\right]_c,$$

and if the curve is closed and at all its points $\rho \neq 0$, the value of this expression is zero.

3. If the surface has regions with negative curvature, there is no assurance that the extremal curves of the total geodesic torsion will also be extremal curves of the total torsion, even assuming as the only possible curves, varied curves with the same osculating planes at the initial and final points. Indeed, in this case it may occur that for "neighbouring" curves fulfilling conditions (20), the difference $\sigma_B - \sigma_A$ differs from $\sigma_B^0 - \sigma_A^0$ by a multiple of 2π. Here it is necessary to restrict the possible variational curves to those which, in addition to fulfilling (20), also fulfil the new condition

$$|v''(u) - v_0''(u)| < \epsilon_3. \tag{21}$$

Hence, since the directions of the osculating planes differ only slightly, it is not possible for one of these planes to revolve more times than the other, and if at A and B the osculating planes are the same, then $\sigma_B - \sigma_A = \sigma_B^0 - \sigma_A^0$. So for surfaces with hyperbolic points, we have that: *Within the field of the variational curves fulfilling* (20) *and* (21), *and having the same osculating planes at the extremes or being close curves with continuous torsion, the D-curves are extremal curves of the total torsion. For sufficiently limited arcs, these curves correspond to a maximum* (see the footnote on page 240) *if* $(GL - EN) \cdot v_0' > 0$, *and to a minimum in the opposite case.*

III. Direct proof of the preceding result

We now prove directly the preceding fundamental property that the condition for a curve on a surface to be an extremal curve of the total torsion is that at all its points the osculating sphere is tangent to the surface.

I. Leaving aside the surface for a moment, let us consider a curve $X = X(s)$ in space connecting two points A, B. Let $T(s), N(s), B(s)$ be the tangent, principal normal and binormal vectors of module 1. A variational curve of the foregoing, with the same extremes, will be of the form

$$X_1 = X(s) + \epsilon Y(s), \tag{22}$$

where $Y(s)$ is a vector that is zero for the extreme values of the arc s.

The torsion $\tau_1(s)$ of the variational curve $X_1(s)$ is (see [2], p. 27)

$$\tau_1 = \frac{(X_1' X_1'' X_1''')}{X_1'' X_1''' - (X_1' X_1'')^2}, \tag{23}$$

where the primes denote derivatives with respect to the parameter s, and the numerator denotes the mixed product of the three vectors.

According to (22), we have

$$\begin{aligned}(X_1' X_1'' X_1''') &= (X' X'' X''') + \epsilon[(Y' X' X'') + (X' Y'' X''') \\ &\quad + (X' X'' Y''')] + \epsilon h_1.\end{aligned} \tag{24}$$

$$X_1'^2 X_1''^2 - (X_1' X_1'')^2 = X''^2 + 2\epsilon[X'' Y'' + (X' Y') X''^2] + \epsilon h_2,$$

where $\lim h_i = 0$ $(i = 1, 2)$ for $\epsilon \to 0$.

To obtain the last expression, it suffices to observe that, since the arc of $X(s)$ is the derivation variable, $X'^2 = 1$, and thus $X' X'' = 0$.

By applying Frenet's formulae, it is deduced that

$$X' = T, \quad X'' = \kappa N, \quad X''' = -\kappa^2 T + \kappa' N + \kappa \tau B,$$

where κ is the curvature and τ the torsion of the curve $X(s)$.

Furthermore, since $T \wedge N = B$, $N \wedge B = T$, $B \wedge T = N$, we have

$$\begin{aligned}(Y' X'' X''') &= (\kappa B + \tau T)\kappa^2 Y' \\ (X' Y'' X''') &= (-\kappa B + \kappa \tau N) Y'' \\ (X' X'' Y''') &= \kappa B Y''' \\ X'' Y'' + (X' Y') X''^2 &= \kappa N Y'' + \kappa^2 T Y'.\end{aligned}$$

Substituting these values in (24) and then performing the division in (23), we obtain

$$\begin{aligned}\tau_1 &= \tau + \epsilon\Big[(\kappa B - \tau T) Y' + \frac{1}{\kappa}(B Y'' - \tau N Y'') \\ &\quad - \frac{\kappa'}{\kappa^2} B Y''\Big] + \epsilon h_3 \\ &= \tau + \epsilon\Big[(\kappa B - \tau T) Y' + \frac{d}{ds}\Big(\frac{1}{\kappa} B Y''\Big)\Big] + \epsilon h_3,\end{aligned}$$

with $\lim_{\epsilon \to 0} h_3 = 0$.

The limit of the quotient between $\tau_1 - \tau$ and ϵ for $\epsilon \to 0$, multiplying by ϵ, is the *first variation of the torsion*: denoting this by $\delta\tau$, we have[5]

$$\delta\tau = \epsilon\Big[(\kappa B - \tau T) Y' + \frac{d}{ds}\Big(\frac{1}{\kappa} B Y''\Big)\Big]. \tag{25}$$

[5]See [2], p. 64. Here a work by Hamel is cited, in *Sitzungsber. Akad. Berlin*, 1925, p. 5, which we have not been able to consult. We do not know if in this work, in addition to formula (25), there are also other properties of the extremal curves of total torsion studied in the present paper.

2. We now consider the total torsion $\mathcal{T} = \int_A^B \tau\, ds$ of the curve connecting A with B. In order to find the total torsion of the varied curve $X_1(s)$ (22), it is first necessary to find the arc element of this new curve. If $X' = T$, we have

$$\begin{aligned} ds_1 &= \sqrt{X_1'^2} ds = \sqrt{1 + 2\epsilon T Y' + \epsilon^2 Y'^2} ds \\ &= [1 + \epsilon T Y' + \epsilon h_4] ds. \end{aligned}$$

The total torsion of $X_1(s)$ is therefore

$$\mathcal{T}_1 = \int_A^B \tau_1\, ds_1 = \int_A^B (\tau + \delta\tau + \epsilon h_3)(1 + \epsilon T Y' + \epsilon h_4) ds.$$

Since h_3 and h_4 tend to zero with ϵ, the product of ϵ multiplied by the limit of the quotient between the difference $\mathcal{T}_1 - \mathcal{T}$ and ϵ, for ϵ tending to zero (that is, *the first variation of the total torsion*), is equal to

$$\delta\mathcal{T} = \epsilon \int_A^B \Big(\kappa B Y' + \frac{d}{ds}(\frac{1}{\kappa} B Y'') \Big) ds.$$

By imposing the condition that the variational curves must pass through A and B, and *having at these points the same tangents and the same osculating planes*, we arrive at

$$Y_A = Y_B = Y_A' = Y_B' = Y_A'' = Y_B'' = 0,$$

whereby the integral of the second term of the sub-integral expression is zero. Furthermore, integrating by parts (where $B' = -\tau N$),

$$\int_A^B \kappa B Y'\, ds = \Big[\kappa B Y \Big]_A^B - \int_A^B (\kappa' B - \kappa\tau N) Y\, ds.$$

Thus, with the aforementioned conditions at the initial and final points, we have

$$\delta\mathcal{T} = -\epsilon \int_A^B (\kappa' B - \kappa\tau N) Y\, ds.$$

In order for this first variation to be zero for *any value of the module of the vector* Y, it is necessary for this vector to be perpendicular to the vector $\kappa' B - \kappa\tau N$ at each point.

However, the angle formed by the radius of the osculating sphere to a curve with the principal normal satisfies the condition (see [2], p. 33)

$$\tan\alpha = \frac{\rho'}{\rho\tau} = -\frac{\kappa'}{\kappa\tau};$$

that is, it has the same direction as $\kappa' B - \kappa\tau N$. Therefore, *the vector* Y *must be tangent to the sphere osculating to the curve at each point.*

If the curve belongs to a surface, on considering variations on the curve (that is, the vector Y is always tangent to the curve), for the total torsion to be extremal it is therefore necessary that the osculating sphere at each point on the curve be tangent to the surface. We thus obtain once again the result from Section II.

IV. Example. Extremal curves of the total torsion on the torus

I. As an application of the foregoing results, we now find the extremal curves of the total torsion on the surface of the torus; that is, on the surface engendered by a circumference revolving around a straight line that does not pass through its centre.

Denoting the latitude of the meridian section by θ, and the angle that determines this section by φ (Fig. 1), the parametric equations of the torus, that is, the components of the vector $X(\theta, \varphi)$ defining the surface $(I, 1)$, are as follows:

$$\begin{aligned} x &= (a + r\cos\theta)\sin\varphi \\ y &= (a + r\cos\theta)\cos\varphi \\ z &= r\sin\theta, \end{aligned}$$

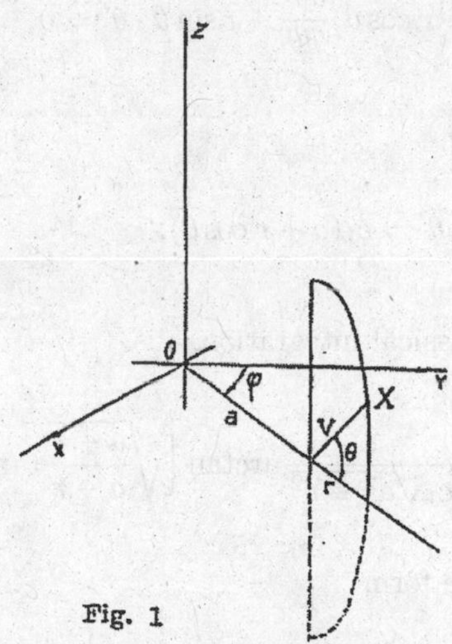

Fig. 1

where r is the radius of the meridian circumference and a the distance of this circumference from the centre to the axis of rotation. The components of the normal unit vector are

$$V(-\cos\theta\sin\varphi, -\cos\theta\cos\varphi, -\sin\theta).$$

The coefficients of the two fundamental quadratic forms are

$$\begin{aligned} E &= X_\theta^2 = r^2, \qquad G = X_\varphi^2 = (a + r\cos\theta)^2 \\ L &= -X_\theta V_\theta = r, \qquad N = -X_\varphi V_\varphi = (a + r\cos\theta)\cos\theta. \end{aligned} \tag{26}$$

With these values, the coefficients of equation (10) for the D-curves or extremal curves of the total geodesic torsion are

$$\begin{aligned} a &= 6ar^3(a + r\cos\theta)^3, \quad b = 0, \quad c = -6ar^4(a + r\cos\theta)^2\sin\theta \\ d &= e = 0. \end{aligned}$$

Hence, equation (10), after simplifying, is reduced to

$$(a + r\cos\theta)(\theta'\varphi'' - \theta''\varphi') - r\sin\theta \cdot \theta'^2\varphi' = 0.$$

Taking φ as the independent variable, this equation is equivalent to

$$(a + r\cos\theta)\frac{d\theta'}{d\varphi} + r\sin\theta \cdot \theta'^2 = 0.$$

By writing $\frac{d\theta'}{d\varphi} = \frac{d\theta'}{d\theta}\theta'$, we obtain

$$(a + r\cos\theta)\frac{d\theta'}{d\theta} + r\sin\theta \cdot \theta' = 0,$$

and a first integration gives

$$\theta' = c_1(a + r\cos\theta). \tag{27}$$

Thus, by means of a classical integration,

$$\varphi = c_2 + \frac{2}{c_1\sqrt{a^2 - r^2}} \arctan\left[\sqrt{\frac{a+r}{a-r}} \tan\frac{\theta}{2}\right].$$

which can be written in the form

$$\theta = 2\arctan\left[\sqrt{\frac{a+r}{a-r}} \tan\frac{1}{2}c_1\sqrt{a^2 - r^2}(\varphi - c_2)\right]. \tag{28}$$

This is the equation for extremal curves of the total geodesic torsion, or, what amounts to the same, for D-curves of the surface of the torus. The constant c_2 is determined by giving the origin A of the extremal curve, and c_1 by the tangent at this point.

2. Assume that $a > r$ so that we have a surface without double lines.

The condition $\frac{dv}{du}$, which as we saw in $(II, 2, 3)$ should be fulfilled along the extremal curve in order for it to be a weak maximum,[6] is equivalent in this case to $\theta' > 0$, and according to (27), and since $a > r$, this condition is fulfilled for any point on the extremal if it is fulfilled at one of its points (that is, if $c_1 > 0$).

[6]In $(II, 2, 3)$, it was assumed that $GL - EN > 0$. This in fact occurs in the present case, since according to (26) $GL - EN = (a + r\cos\theta) \cdot ar > 0$.

Considering all the extremals passing through a fixed point θ_0, φ_0, all their points of contact with the envelope of all these extremals are determined, as is known, by the roots of the Jacobi equations (see [7], p. 108)

$$\frac{\partial(\theta, \theta_0)}{\partial(c_1, c_2)} = 0, \tag{29}$$

where the left-hand side indicates the functional determinant with respect to c_1, c_2 of the function (28) and that resulting from the substitution in this function of θ and φ for θ_0 and φ_0. On performing the operations, we find that this equation (29) is reduced to $\varphi - \varphi_0 = 0$; that is, its only solution is $\varphi = \varphi_0$. Therefore: *The extremals lack conjugate focus; for any arc, they thus correspond to a weak maximum of the total geodesic torsion.*

Let us for greater simplicity take the point $\theta = 0, \varphi = 0$ as the origin of the extremals. In this case, equation (28) is reduced to

$$\theta = 2 \arctan \left[\sqrt{\frac{a+r}{a-r}} \tan \frac{1}{2} c_1 \sqrt{a^2 - r^2} \varphi \right], \tag{30}$$

c_1 being determined by (27), which in this case is $\theta_0' = c_1(a+r)$.

By studying (30), one observes the following: The necessary and sufficient condition for the extremal curve to be closed is that $\frac{1}{2} c_1 \sqrt{a^2 - r^2}$ is rational. Indeed, in this case there will always be an integer k such that $\frac{1}{2} c_1 \sqrt{a^2 - r^2} 2\pi k$ is a multiple of π, and therefore θ will have the same value θ_0 as for $\varphi = 0$. If $c_1 \sqrt{a^2 - r^2} = \frac{p}{q}$, where p and q are prime integers, the curve will closed after wrapping p times around the torus in the direction of its meridians (that is, according to θ), and q times according to its parallels (that is, according to φ). One also observes that, since θ' has constant sign, there are no closed extremal curves that do not wrap around the torus.

As examples, Fig. 2 (a) shows the case $a = 5, r = 3, c_1 = \frac{1}{2}, \theta_0' = 4$; in this case, $p = 2, q = 1$. In Fig. 2 (b), with the same values of a and r, we have $c_1 = \frac{1}{8}, \theta_0' = 1$ and $p = 1, q = 2$.

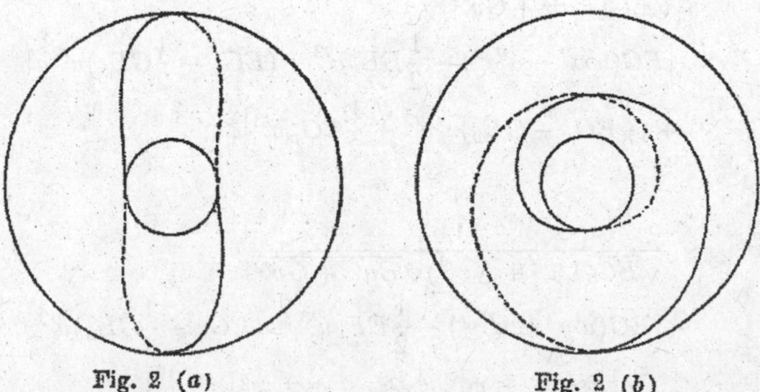

Fig. 2 (a) **Fig. 2 (b)**

3. If, instead of the total geodesic torsion, we consider the total torsion, the foregoing conditions hold providing we consider as possible variational curves solely those in which not only θ and θ' have values close to those corresponding to the extremal, but also θ'' (condition (21)).

One thereby prevents the variation of the angle σ formed by the plane osculating to the curve and the normal to the surface from differing by a multiple of π between the curve and its variational.

4. It would be interesting to find the increment of the angle σ along a closed extremal, from those that exist in 2.

One way of calculating the angle σ consists in observing that the radius of curvature ρ of the curve is equal on the one hand to $R \sin \sigma$ (Meusnier's theorem), and by the definition of radius of geodesic curvature it is also equal on the other hand to $\rho_g \cos \sigma$ (Fig. 3). Hence,

$$\cot \sigma = \frac{R}{\rho_g}.$$

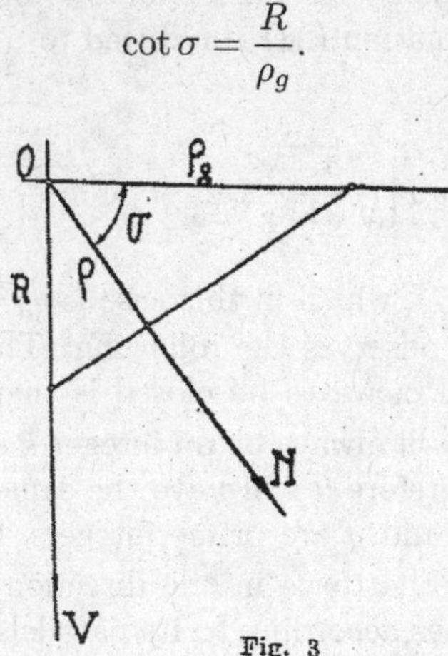

Fig. 3

When the coordinated lines are lines of curvature, **R** and ρ_g are equal to (see [2], pp. 91 and 175)

$$R = \frac{Eu'^2 + Gv'^2}{Lu'^2 + Nv'^2}$$

$$\begin{aligned} \frac{1}{\rho_g} &= \frac{1}{\sqrt{EG}(Eu'^2 + Gv'^2)^{3/2}} \cdot \\ &\Big[EG(u'v'' - u''v') - \frac{1}{2}EE_v u'^3 + (EG_u - \frac{1}{2}GE_u)u'^2 v' \\ &+ (\frac{1}{2}EG_v - GE_v)u'v'^2 + \frac{1}{2}GG_u v'^3\Big], \end{aligned}$$

and therefore,

$$\begin{aligned} \cot \sigma &= \frac{1}{\sqrt{EG}(Lu'^2 + Nv'^2)\sqrt{Eu'^2 + Gv'^2}} \cdot \\ &\Big[EG(u'v'' - u''v') - \frac{1}{2}EE_v u'^3 + (EG_u - \frac{1}{2}GE_u)u'^2 v' \\ &+ (\frac{1}{2}EG_v - GE_v)u'v'^2 + \frac{1}{2}GG_u v'^3\Big]. \end{aligned}$$

In the case of the torus studied here, bearing in mind that we take θ instead of u, and that φ (corresponding to v) is the independent variable, by substituting the values found in $IV, 1$ in the previous expression, as well as the values of θ' and θ'' in terms of θ, we obtain

$$\cot\sigma = \frac{(3r^2c_1^2 - 1)r\sin\theta}{a\sqrt{1+r^2c_1^2}(arc_1^2 + (1+r^2c_1^2)\cos\theta)}. \tag{31}$$

This formula contains some interesting consequences:

a) If $c_1 = \frac{1}{\sqrt{3}r}$, then $\sigma = \frac{\pi}{2}$, and the curve is geodesic. Furthermore, if $c_1\sqrt{a^2-r^2} = \frac{\sqrt{a^2-r^2}}{\sqrt{3}r}$ is rational, we have a closed geodesic line that at the same time is a D-curve; that is, an extremal curve of the total geodesic torsion.

b) If $arc_1^2 - 1 - r^2c_1^2 > 0$, that is, $c_1 > \frac{1}{\sqrt{r(a-r)}}$, $\cot\sigma$ is always finite, and therefore the increment of σ along a closed extremal curve is zero.

c) If $arc_1^2 - 1 - r^2c_1^2 < 0$, that is,

$$c_1 < \frac{1}{\sqrt{r(a-r)}},$$

expression (31) of $\cot\sigma$ becomes infinite for some values of θ. Observing how σ varies with θ, after describing a closed line we find that the increment of σ is equal to the increment of the angle θ.

d) On deriving expression (31), we find that $\frac{d\sigma}{d\theta}$ has constant sign if $arc_1^2 - 1 - r^2c_1^2 < 0$. In this case, the difference $\sigma_B - \sigma_A$ will always increase.

5. Lastly, observe that the geodesic torsion of the extremal curves of the torus is equal to (applying (16), (26) and (27))

$$\mathcal{T}_g = \int_0^\varphi \frac{ac_1}{\sqrt{1+r^2c_1^2}}\,d\varphi = \frac{ac_1}{\sqrt{1+r^2c_1^2}}\varphi.$$

Choosing c_1 such that it satisfies the condition, and also such that $c_1\sqrt{a^2-r^2}$ is rational; that is,

$$c_1\sqrt{a^2-r^2} = \frac{p}{q},$$

where p, q are prime integers, we have a closed curve whose total torsion (according to (15) and the consequence c) from 4) is equal to

$$\mathcal{T} = \frac{ac_1}{\sqrt{1+r^2c_1^2}}2\pi q + 2\pi p.$$

Since it is possible to choose p and q such that (33) is fulfilled, and also that (34) takes values as high as one wishes, we have that: *On the torus there are infinite closed curves with continuous torsion and always with constant sign, and whose total torsion takes values as high as one wishes.*

For example, on the torus engendered by a circle of radius $r = 3$ revolving around an axis located at a distance $a = 5$ from the centre, the curve

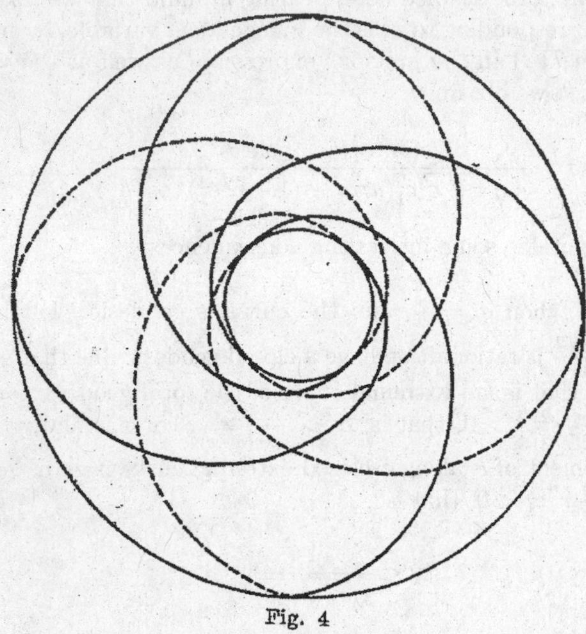

Fig. 4

$$\theta = 2\arctan[2\tan\frac{2}{5}\varphi]$$

corresponding to equation (30) for $c_1\sqrt{a^2-r^2} = \frac{4}{5}$ always has positive torsion, and on closing after wrapping 4 times around the torus according to θ, and according to φ, has a total torsion (according to (34)) equal to $2\pi(\frac{25}{\sqrt{34}}+4)$. This curve is shown in Figure 4.

References

[1] L. Bianchi, *Lezioni di geometria differenziale*, vol. I, N. Zanichelli, Bologna, 1924.

[2] W. Blaschke, *Differentialgeometrie*, vol. I, Springer, Berlin, 1930.

[3] O. Bolza, *Vorlesungen über variationsrechnung*, B. G. Teubner, Leipzig and Berlin, 1909.

[4] G. Darboux, *Les courbes tracées sur une surface, et dont la sphère osculatrice est tangente en chaque point à la surface*, C. R. Acad. Sc. París **LXXIII** (1871), 732–736.

[5] W. Fenchel, *Ueber einen jacobischen satz der kurventheorie*, Tôhoku Math. Journal **39** (1934), 95–97.

[6] E. Goursat, *Cours d'analyse mathemátique*, vol. I,II,III, Gauthier-Villars, Paris, 1902,1905.

[7] J. Hadamard, *Leçons sur le calcul des variations*, Hermann, Paris, 1910.

[8] L. A. Santaló, *Algunas propiedades de las curvas esféricas y una característica de la esfera*, Rev. Mat. Hispano Americana **X** (1935), 9–12.

[9] Tonelli, *Fondamenti di calcolo delle variazioni*, vol. II, Zanichelli, 1923.

Rosario, Instituto de Matemática, January 1941.

INTEGRAL FORMULAS IN CROFTON'S STYLE ON THE SPHERE AND SOME INEQUALITIES REFERRING TO SPHERICAL CURVES

By L. A. Santaló

Introduction. Several integral formulas referring to convex plane curves, notable for their great generality, were obtained by W. Crofton in 1868 and successive years from the theory of geometrical probability [6], [7], [8], [9], [10].

A direct and rigorous exposition of Crofton's principal results, adding some new formulas, was made in 1912 by H. Lebesgue [12]. Another systematic exposition of Crofton's most interesting formulas, together with the generalization of many of them to space, is found in the two volumes on integral geometry by Blaschke [2].

The purpose of the present paper is to give a generalization of Crofton's formulas to the surface of the sphere. This is what we do in part I. We find further integral formulas on the sphere (for instance, (16), (17), (20), (21)) which have no equivalent in the plane. Other formulas, if we consider the plane as the limit of a sphere whose radius increases indefinitely, give integral formulas referring to plane convex curves (c. g., (34), (35)) which we think are new.

In part II, with simple methods of integral geometry [2], we obtain three inequalities referring to spherical curves. Inequality (38) is the generalization to the sphere of an inequality that Hornich [11] obtained for plane curves. (52) and (58) contain the classical isoperimetric inequality on the sphere. Finally, inequality (61) gives a superior limitation for the "isoperimetric deficit" of convex curves on the sphere.

I. FORMULAS IN THE CROFTON STYLE ON THE SPHERE

1. Notation and useful formulas. The element of area on the sphere of unit radius will be represented by $d\Omega$; that is, if θ and φ are the spherical coördinates of the point Ω, we have

$$d\Omega = \sin \theta \, d\theta \, d\varphi. \tag{1}$$

A great non-directed circle C on the same sphere of unit radius can be determined by one of its poles, that is, by either of the extremities of the diameter perpendicular to it. Since $d\Omega$ is the element of area of one of these extremities, the "density" for measuring sets of great circles on the sphere is [2; 61, 80]

$$dC = d\Omega; \tag{2}$$

Received January 9, 1942.

that is, the "measure" of a set of great circles on the sphere is defined as the integral of (2) extended over this set.

It is possible to give the density (2) another form, which will sometimes be useful. We consider a fixed great circle C_0 and a fixed point A on it. The great circle C can be determined for the abscissa t of one of the intersection points from C and C_0 and the angle α between the two circles. If θ and φ are the spherical coördinates of the pole Ω of C with regard to the pole Ω_0 of C_0, $\theta = \alpha$, $\varphi = t$, and (1), (2) give

$$(3) \qquad dC = \sin \alpha \, d\alpha \, dt.$$

Let us consider two great circles C_1, C_2 and one of their intersection points Ω. If α_1 and α_2 are the angles that C_1 and C_2 make with another fixed great circle which also passes through Ω, the following differential formula [2; 78] is known:

$$(4) \qquad dC_1 \, dC_2 = |\sin (\alpha_2 - \alpha_1)| \, d\alpha_1 \, d\alpha_2 \, d\Omega.$$

By (2), formula (4) can be transformed into a "dual" form. Let Ω_1 and Ω_2 be two points on the unit sphere and let C be the great circle determined by them. If β_1 and β_2 are the abscissas of Ω_1 and Ω_2 on C in relation to a fixed origin on this circle, (4) is equivalent to

$$(5) \qquad d\Omega_1 \, d\Omega_2 = |\sin (\beta_1 - \beta_2)| \, d\beta_1 \, d\beta_2 \, dC.$$

2. **First integral formulas. Convex curves on the sphere.** A closed curve on the sphere is said to be *convex* when it cannot be cut by a great circle in more than two points.

A convex curve divides the surface of the sphere into two parts, one of which is always wholly contained in a hemisphere; that is, there is always a great circle which has the whole convex curve on the same side; we only have to consider, for example, a tangent great circle.

When we say a "convex figure", we understand that part of the surface of the sphere which is limited by a convex curve and is smaller than or equal to a hemisphere.

Let us consider a convex figure K on the sphere of unit radius. The radii which are perpendicular to the tangent planes (or, more generally, to the planes of support) to the cone which projects K from the center of the sphere form another cone whose intersection with the sphere is a new convex curve K^*. We shall call K^* the "dual" curve of K. The lengths and areas of K and K^* are connected by the known relations

$$(6) \qquad F^* = 2\pi - L, \qquad L^* = 2\pi - F.$$

All the great circles C that cut K have their poles in the area bounded by K^* and the symmetrical curve of the same K^* with respect to the center of the sphere. This area equals $4\pi - 2F^* = 2L$. Counting each pair of points which are the extremities of a diameter as a single point, and taking into account the value (2) of the density dC, we have

$$\int_{C \cdot K \neq 0} dC = L; \tag{7}$$

this means: on the sphere, the measure of the great circles which cut a convex curve is equal to the length of this curve. This result is given by [2; 81].

3. **Integral of the chords.** Let Ω_1 and Ω_2 be two points inside the convex curve K (always on the unit sphere) and let C be the great circle determined by them. The differential expression (5) can be integrated for all pairs of points within K.

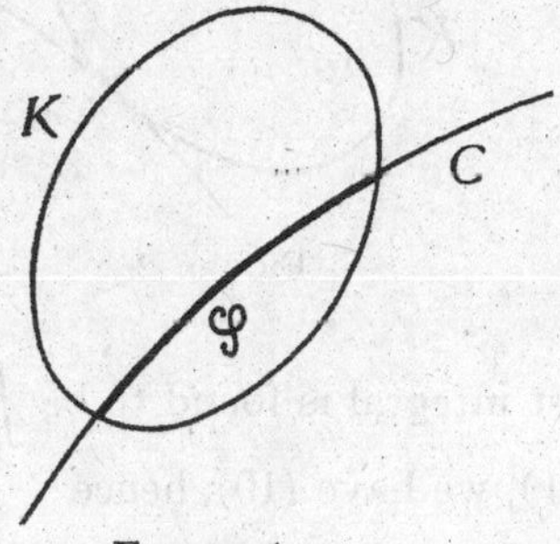

FIGURE 1

The integral of the left side is F^2. By calculating the integral of the right side, if φ represents the length of the arc of C that is contained in K (Fig. 1), we have

$$\int_0^\varphi \int_0^\varphi |\sin(\beta_1 - \beta_2)| \, d\beta_1 \, d\beta_2 = 2(\varphi - \sin\varphi). \tag{8}$$

Hence

$$\int_{C \cdot K \neq 0} (\varphi - \sin\varphi) \, dC = \tfrac{1}{2}F^2. \tag{9}$$

This formula generalizes, as we shall see (§11), Crofton's formula for chords in plane geometry.

4. **Principal Crofton formula.** Let us consider all the pairs of great circles C_1, C_2 that cut K. From (7) we deduce

$$\int_{\substack{C_1 \cdot K \neq 0 \\ C_2 \cdot K \neq 0}} dC_1 \, dC_2 = L^2. \tag{10}$$

Now we can make the integration of formula (4) extend only to the pairs of great circles which cut K. If Ω is fixed inside K, α_1 and α_2 can vary from 0 to π and

$$\int_{\Omega\subset K} \left(\int_0^\pi \int_0^\pi |\sin(\alpha_1 - \alpha_2)| \, d\alpha_1 \, d\alpha_2\right) d\Omega = 2\pi \int_{\Omega\subset K} d\Omega = 2\pi F; \tag{11}$$

if Ω is outside K, α_1 and α_2 can vary from 0 to the angle ω between the great circles which are tangent to K and which pass through Ω (Fig. 2). By applying

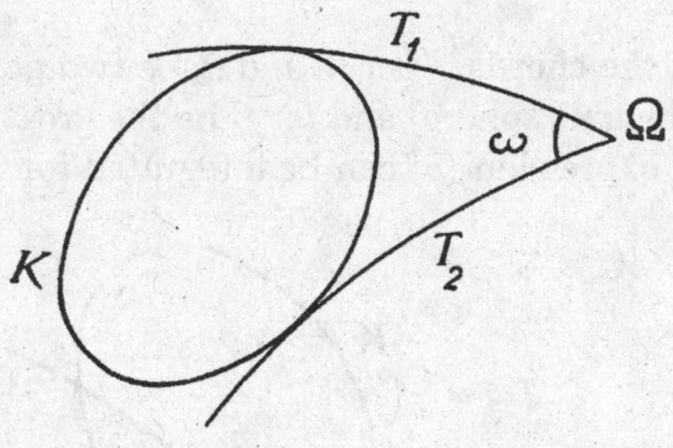

FIGURE 2

(8), the value of this last integral is found to be $\int 2(\omega - \sin\omega)\, d\Omega$ for $\Omega \not\subset K$. Adding this result to (11), we have (10); hence

$$\int (\omega - \sin\omega)\, d\Omega = \tfrac{1}{2}L^2 - \pi F \qquad (\Omega \not\subset K). \tag{12}$$

This formula has the same form as Crofton's fundamental formula of plane geometry. The integration in (12) is extended to all points Ω outside K, each pair of points situated in the extremities of a diameter being considered as a single point.

5. **"Dual" formulas.** From a convex curve K we can deduce the "dual" curve K^* as we have seen in §2. To a great circle C which cuts K corresponds a point Ω^* (the pole of C) which is not inside K^*. The arc φ of C inside K is equal to $\pi - \omega^*$, ω^* being the angle between the two great circles tangent to K^* drawn through Ω^*. Since $F = 2\pi - L^*$ (by (6)), formula (9) can be written

$$\int (\pi - \omega^* - \sin\omega^*)\, d\Omega^* = \tfrac{1}{2}(2\pi - L^*)^2 \qquad (\Omega^* \not\subset K^*). \tag{13}$$

The integration is extended over the outside of K^* (the points which are the extremities of the same diameter being considered as a single point) and consequently

$$\int \pi\, d\Omega^* = \pi(2\pi - F^*) \qquad (\Omega^* \not\subset K^*).$$

Then (13) gives

$$\int (\omega^* + \sin \omega^*) \, d\Omega^* = 2\pi L^* - \pi F^* - \tfrac{1}{2}L^{*2} \qquad (\Omega^* \not\subset K^*). \tag{14}$$

This formula holds for any convex curve K^*; hence it is valid for K:

$$\int (\omega + \sin \omega) \, d\Omega = 2\pi L - \pi F - \tfrac{1}{2}L^2 \qquad (\Omega \not\subset K). \tag{15}$$

From (15) and (12), we deduce

$$\int \omega \, d\Omega = \pi L - \pi F \qquad (\Omega \not\subset K) \tag{16}$$

and

$$\int \sin \omega \, d\Omega = \pi L - \tfrac{1}{2}L^2 \qquad (\Omega \not\subset K). \tag{17}$$

The same procedure shows that formula (12) is equivalent to

$$\int_{C^* \cdot K^* \neq 0} (\pi - \varphi^* - \sin \varphi^*) \, dC^* = \tfrac{1}{2}(2\pi - F^*)^2 - \pi(2\pi - L^*), \tag{18}$$

where the integration is extended over all the great circles C^* which cut K^*. By (7) we have $\pi \int dC^* = \pi L^*$ and by substitution of this value in (18) and writing the formula for K, we have

$$\int_{C \cdot K \neq 0} (\varphi + \sin \varphi) \, dC = 2\pi F - \tfrac{1}{2}F^2, \tag{19}$$

where φ is the length of the arc of C which is inside K.

From (9) and (19) we deduce

$$\int_{C \cdot K \neq 0} \varphi \, dC = \pi F, \tag{20}$$

and

$$\int_{C \cdot K \neq 0} \sin \varphi \, dC = \pi F - \tfrac{1}{2}F^2. \tag{21}$$

We repeat. In (16), (17), ω is the angle between the two great circles tangent to K through Ω; in (20), (21), φ is the length of the arc of the great circle C which is inside K.

The formulas (16), (17), (20), (21) that hold for any convex curve on the unit sphere have no equivalent in the plane.

6. **Formulas for the tangents.** Let K be a convex curve on the unit sphere with continuous radius of geodesic curvature.

If τ is the angle between a variable tangent great circle and a fixed tangent great circle and if s is the length of the arc of K, the radius of geodesic curvature ρ_g is given by

$$\rho_g = \frac{ds}{d\tau}, \tag{22}$$

and the Gauss-Bonnet formula gives

$$\oint_K \frac{ds}{\rho_g} = \int d\tau = 2\pi - F. \tag{23}$$

Let us consider two great circles tangent to K; let Ω be one of the intersection points of these circles. T_1 and T_2 will be the lengths of the arcs of these great circles bounded by Ω and the points of contact (T_1 and $T_2 \leq \pi$), and we represent by ω the angle between the two tangent circles at Ω (Fig. 2).

We wish to express the element of area $d\Omega$ as a function of the angles τ_1, τ_2 which determine the tangent great circles.

For fixed τ_2, as we pass from τ_1 to $\tau_1 + d\tau_1$, the arc T_2 is increased by $dT_2 = (\sin T_1/\sin \omega)d\tau_1$.

In the same way, as we pass from τ_2 to $\tau_2 + d\tau_2$, the arc T_1 is increased by $dT_1 = (\sin T_2/\sin \omega)d\tau_2$.

Since the element of area $d\Omega$ can be expressed in the form $d\Omega = \sin \omega \, dT_1 \, dT_2$, we find the desired expression

$$d\Omega = \frac{\sin T_1 \cdot \sin T_2}{\sin \omega} d\tau_1 \, d\tau_2$$

or

$$\frac{\sin \omega}{\sin T_1 \cdot \sin T_2} d\Omega = d\tau_1 \, d\tau_2 . \tag{24}$$

7. We can make the integration of (24) extend over all pairs of circles tangent to K and, by counting each pair once only (to do this we must divide the integral by 2), we have, by (23),

$$\int \frac{\sin \omega}{\sin T_1 \cdot \sin T_2} d\Omega = \tfrac{1}{2}(2\pi - F)^2 \qquad (\Omega \not\subset K). \tag{25}$$

Likewise, as in the preceding cases, the notation $\Omega \not\subset K$ indicates that the integration must be extended over all points Ω outside K; the points situated in the extremities of a diameter are considered as a single point.

8. Let $\rho_g^{(1)}$, $\rho_g^{(2)}$ be the radii of geodesic curvature of K at the points of contact of the tangent great circles through Ω. By virtue of (22), (24), we have

$$\sin \omega \frac{\rho_g^{(1)} \rho_g^{(2)}}{\sin T_1 \cdot \sin T_2} d\Omega = ds_1 \, ds_2 .$$

By integrating this expression over all pairs of tangent great circles, counting each pair once only, we get

$$\int \sin \omega \frac{\rho_g^{(1)} \rho_g^{(2)}}{\sin T_1 \cdot \sin T_2} d\Omega = \tfrac{1}{2}L^2 \qquad (\Omega \not\subset K). \tag{26}$$

9. By (22) and (24), we have

$$\sin \omega \frac{\rho_g^{(1)}}{\sin T_1 \cdot \sin T_2} d\Omega = ds_1 \, d\tau_2$$

and by integrating over all great circles tangent to K and observing that each point Ω is a common factor of two terms, it follows that

$$\int \sin \omega \frac{\rho_g^{(1)} + \rho_g^{(2)}}{\sin T_1 \cdot \sin T_2} d\Omega = L(2\pi - F) \qquad (\Omega \not\subset K). \tag{27}$$

10. **"Dual" formulas.** According to §5, from formulas (25), (26), (27) we can deduce the respective "dual" formulas.

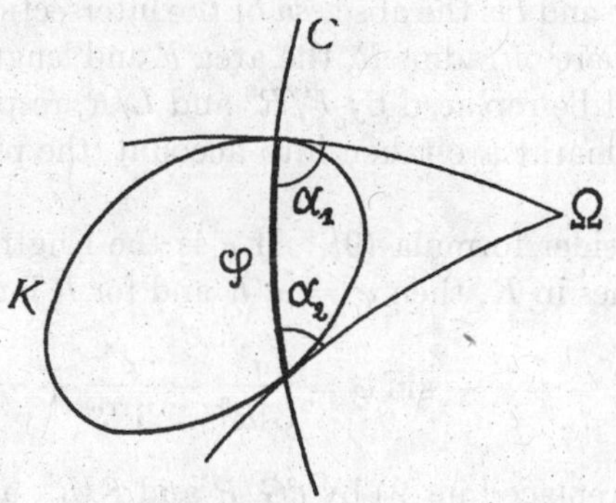

FIGURE 3

If φ is the length of the arc of the great circle C which is inside K and α_1, α_2 are the angles that C makes with the great circles tangent to K at the intersection points of C with K (Fig. 3), formula (25) gives

$$\int_{C \cdot K \neq 0} \frac{\sin \varphi}{\sin \alpha_1 \cdot \sin \alpha_2} dC = \tfrac{1}{2}L^2. \tag{28}$$

We observe that the "dual" element of ds is $d\tau^*$ for the dual curve K^* and reciprocally. Then the dual expression of $\rho_g = ds/d\tau$ will be $d\tau^*/ds^* = 1/\rho_g^*$. Hence, formula (26) gives

$$\int_{C\cdot K\neq 0} \frac{\sin\varphi}{\sin\alpha_1\cdot\sin\alpha_2}\cdot\frac{1}{\rho_g^{(1)}\rho_g^{(2)}}\,dC = \tfrac{1}{2}(2\pi - F)^2. \tag{29}$$

Likewise, formula (27) gives

$$\int_{C\cdot K\neq 0} \frac{\sin\varphi}{\sin\alpha_1\cdot\sin\alpha_2}\left(\frac{1}{\rho_g^{(1)}}+\frac{1}{\rho_g^{(2)}}\right)dC = (2\pi - F)L. \tag{30}$$

11. **Passage to the case of the plane.** The classical Crofton formulas for the plane must result as a special case of the preceding formulas when the radius of the sphere increases indefinitely. Moreover, by this procedure, we shall find some new integral formulas.

We observe the following. (i) The element of area $d\Omega$ on the unit sphere can be replaced by dP/R^2, where dP is the element of area on the sphere of radius R and, as $R \to \infty$, dP will be the element of area in the plane. (ii) Let us consider the form (3) for dC; for the sphere of radius R this expression (3) must be replaced by $dC = \sin\alpha\, d\alpha(dt_R/R)$, where t_R is the length of the arc of the great circle of the sphere of radius R; when R increases to ∞, (3) is $\lim \mathrm{R}\cdot dC = dG$, dG being the "density" of the straight lines of the plane (recall that the "density" dG can be written $dG = \sin\alpha\, d\alpha\, dt$, where α is the angle which G forms with another fixed straight line and t is the abscissa of the intersection point [2; 7]). (iii) When we consider a sphere of radius R, the area F and length L which are in formulas from §§2-10 must be replaced by F/R^2 and L/R, respectively.

When these remarks are taken into account, the preceding formulas give the following results.

(i) Let us consider formula (9). If σ is the length of the arc that the great circle C determines in K, then $\varphi = \sigma/R$ and for R large we have

$$\varphi - \sin\varphi = \frac{\sigma^3}{3!R^3} - \frac{\sigma^5}{5!R^5} + \cdots .$$

If dC and F are replaced in (9) by dG/R and F/R^2, as $R \to \infty$ we have

$$\int_{G\cdot K\neq 0} \sigma^3\, dG = 3F^2. \tag{31}$$

This is the classical chord formula from Crofton [9; 84], [10; 27], [2; 20].

(ii) Formula (12) maintains the same form for the plane. Indeed, ω and $\sin\omega$ do not change; $d\Omega$ becomes dP/R^2, F becomes F/R^2, and L becomes L/R; in the limit as $R \to \infty$, formula (12) does not change. It is the "principal" Crofton formula for the plane [9; 78], [10; 26], [2; 18].

(iii) Formulas (16), (17), (20), (21) have no equivalent in the plane, since, when these formulas are written for the sphere of radius R, as $R \to \infty$ the right side increases indefinitely.

(iv) In formula (25), we must replace $\sin T_1$ and $\sin T_2$ by T_1/R and T_2/R, the element of area $d\Omega$ by dP/R^2, and F by F/R^2. In the limit as $R \to \infty$, we find

$$\int \frac{\sin \omega}{T_1 T_2}\, dP = 2\pi^2 \qquad (P \not\subset K). \tag{32}$$

In this well-known formula ([12]; see also W. Blaschke, *Differentialgeometrie* I, p. 49), T_1 and T_2 are the lengths of the tangents to the convex curve K drawn through P, $dP = dx\, dy$ is the element of area on the plane, and ω is the angle between the tangents at P.

For formulas (26), (27), it is only necessary to observe that the radii of geodesic curvature become the radii of the ordinary curvature of the plane curve. Hence formulas (26) and (27) give the known formulas [12]

$$\int \sin \omega \frac{\rho_1 \rho_2}{T_1 T_2}\, dP = \tfrac{1}{2}L^2, \qquad \int \sin \omega \frac{\rho_1 + \rho_2}{T_1 T_2}\, dP = 2\pi L \tag{33}$$
$$(P \not\subset K).$$

(v) Formula (28), when φ is replaced by σ/R (σ is the length of the arc that the great circle C determines in K and in the limit it is the length of the chord that the straight line G determines in K) and R increases indefinitely, gives

$$\int_{G \cdot K \neq 0} \frac{\sigma}{\sin \alpha_1 \cdot \sin \alpha_2}\, dG = \tfrac{1}{2}L^2. \tag{34}$$

α_1 and α_2 are the angles that the straight line G makes with the tangents to K at the intersection points of G with K. The integration in (34) is extended over all the straight lines G which cut K.

Likewise, (29) and (30) give for the plane

$$\int_{G \cdot K \neq 0} \frac{\sigma}{\rho_1 \rho_2 \sin \alpha_1 \sin \alpha_2}\, dG = 2\pi^2,$$
$$\int_{G \cdot K \neq 0} \frac{(\rho_1 + \rho_2)\sigma}{\rho_1 \rho_2 \sin \alpha_1 \sin \alpha_2}\, dG = 2\pi L, \tag{35}$$

where ρ_1 and ρ_2 are the radii of curvature of the convex curve K at the intersection points of G with K.

II. SOME INEQUALITIES REFERRING TO SPHERICAL CURVES

12. A known formula. Hitherto we have only considered relations on the sphere between a convex curve K and points and great circles. Now we wish to establish some new relations which arise from considering on the sphere sets of variable small circles of constant spherical radius.

Let $\mathfrak{L}$ be a rectifiable curve (not necessarily convex) of length L on the unit sphere. We consider on the same sphere a small circle C_0 of spherical radius ρ $(\rho \leq \frac{1}{2}\pi)$, whose length and area will be

$$L_0 = 2\pi \sin \rho, \qquad F_0 = 2\pi(1 - \cos \rho). \tag{36}$$

Let Ω be the center of the circle C_0 and, as in §1, $d\Omega$ the corresponding element of area of the sphere. If n represents the number of intersection points of the curve $\mathfrak{L}$ with the circle C_0 (n will be a function of Ω), we have the known formula

$$\int n \, d\Omega = \frac{2}{\pi} L L_0 ,$$

or

$$\int n \, d\Omega = 4L \sin \rho; \tag{37}$$

the integration is extended over the whole sphere.

This formula is a particular case of Poincaré's formula of integral geometry [2; 81]. In [2], the formula is established only for spherical curves composed of a finite number of arcs with a continuously turning tangent. More generally, formula (37) is also valid for the case of a curve $\mathfrak{L}$ only supposed to be rectifiable and a circle C_0 . The proof can be copied step by step from that given for Euclidean space of n dimensions in [13].

13. **An inequality referring to rectifiable curves on the sphere.** In this section we generalize for curves on the sphere an inequality that Hornich obtained for Euclidean space [11]. The proof is analogous to that given for Euclidean space in [13].

Let us consider on the sphere of unit radius the rectifiable curve $\mathfrak{L}$ of length L. Let F be the area filled by the points of the sphere whose spherical distance from $\mathfrak{L}$ is $\rho \leq \frac{1}{2}\pi$.

We shall prove that

$$F \leq 2L \sin \rho + 2\pi(1 - \cos \rho) \tag{38}$$

and establish the conditions for the equality in (38).

Let M_i $(i = 0, 1, 2, 3, \cdots)$ be the area covered by the centers of the circles of radius ρ whose distance to $\mathfrak{L}$ is not greater than ρ and which have i points in common with $\mathfrak{L}$.

By (37), we have

$$M_1 + 2M_2 + 3M_3 + 4M_4 + \cdots = 4L \sin \rho, \tag{39}$$

and according to the definition of the area F,

$$M_0 + M_1 + M_2 + M_3 + \cdots = F. \tag{40}$$

From (39) and (40) we deduce

$$2F - 4L \sin \rho = 2M_0 + M_1 - (M_3 + 2M_4 + \cdots). \tag{41}$$

We consider the arc of a great circle of length D ($\leq \pi$) which joins the extremities of the given curve $\mathfrak{L}$ (if this curve is closed, $D = 0$). Let us call M_i^* ($i = 0, 1, 2$) the area covered by the centers of the circles of spherical radius ρ which have i points in common with this arc of length D (for $i = 0$ the arc is interior to the circle).

The area filled by the points whose distance from the arc of length D is not less than $\rho \leq \frac{1}{2}\pi$ is limited by two arcs of circles parallel to this arc at the distance ρ and two semicircles of radius ρ at the ends. The value of this area is $2D \sin \rho + 2\pi(1 - \cos \rho)$ and we can write

$$M_0^* + M_1^* + M_2^* = 2D \sin \rho + 2\pi(1 - \cos \rho). \tag{42}$$

By (37) we have also

$$M_1^* + 2M_2^* = 4D \sin \rho. \tag{43}$$

From (42) and (43) we deduce

$$2M_0^* + M_1^* = 4\pi(1 - \cos \rho). \tag{44}$$

We observe that if the circle C of radius ρ contains in its interior the curve $\mathfrak{L}$, it contains also the arc D. Hence $M_0 \leq M_0^*$. Likewise if C cuts $\mathfrak{L}$ in only one point, it has one of its extremities in the interior and the other in the exterior and so the arc D cuts the circle C also at only one point, that is to say, $M_1 \leq M_1^*$. It follows that, by (41) and (44),

$$\begin{aligned} 2F - 4L \sin \rho &\leq 2M_0^* + M_1^* - (M_3 + 2M_4 + \cdots) \\ &= 4\pi(1 - \cos \rho) - (M_3 + 2M_4 + \cdots); \end{aligned}$$

hence

$$F + \tfrac{1}{2}(M_3 + 2M_4 + \cdots) \leq 2\pi(1 - \cos \rho) + 2L \sin \rho. \tag{45}$$

This inequality implies (38).

The equality in (38) will be verified only if $M_i = 0$ for $i \geq 3$ and moreover $M_0 = M_0^*$, $M_1 = M_1^*$. The condition $M_i = 0$ for $i \geq 3$ carries with it $M_1 = M_1^*$; since in the case when the circle C cuts in only one point the arc of the great circle which joins the extremities of $\mathfrak{L}$, it must cut $\mathfrak{L}$ in an odd number of points. Consequently, the conditions for equality are:

(i) $M_i = 0$ (for $i \geq 3$). The curve $\mathfrak{L}$ cannot be cut by the circle C in more than two points.

(ii) $M_0 = M_0^*$, that is to say, if the circle C contains in its interior the two extremities of the curve $\mathfrak{L}$, it contains also the whole curve.

In particular, if the given curve $\mathfrak{L}$ is closed, the equality in (38) is valid only in the case of reduction to a point.

14. **Isoperimetric inequality on the sphere.** Let K be a convex curve on the sphere of unit radius. We consider the exterior parallel curve to K at the distance $\rho \leq \frac{1}{2}\pi$. This curve cannot have double points and its area is easy to calculate. The area is [3; 81]

$$S = F + L \sin \rho + 2\pi(1 - \cos \rho) - F(1 - \cos \rho), \tag{46}$$

or, with the values (36) of the area and the length of the circle of radius ρ,

$$S = F + F_0 + \frac{1}{2\pi}(LL_0 - FF_0). \tag{47}$$

Let us put, as in the last section, M_i $(i = 0, 2, 4, 6, \cdots)$ for the area covered by the centers of the circles of radius ρ which have i points in common with K (M_0 will be the area covered by the centers of the circles of radius ρ each of which contains K in its interior or which is contained in the interior of K). Since K is a closed curve, i is always even.

The expression (47) is equivalent to

$$M_0 + M_2 + M_4 + \cdots = F + F_0 + \frac{1}{2\pi}(LL_0 - FF_0) \tag{48}$$

and formula (37) gives

$$M_2 + 2M_4 + 3M_6 + \cdots = \frac{1}{\pi} LL_0 . \tag{49}$$

Let us consider a radius ρ such that $M_0 = 0$, that is, such that the circle of radius ρ neither can be totally interior to K nor can contain K in its interior. From (48) and (49) we deduce then

$$M_4 + 2M_6 + \cdots = \frac{1}{2\pi}(LL_0 + FF_0) - (F + F_0). \tag{50}$$

We observe that, by (36), $L_0^2 + F_0^2 - 4\pi F_0 = 0$; hence we can write the identity

$$\begin{aligned} &\frac{1}{2\pi}(LL_0 + FF_0) - (F + F_0) \\ &\qquad = \frac{1}{4\pi}[(L^2 + F^2 - 4\pi F) - (L - L_0)^2 - (F - F_0)^2] \end{aligned} \tag{51}$$

and (50) gives

$$L^2 + F^2 - 4\pi F = (L - L_0)^2 + (F - F_0)^2 + 4\pi(M_4 + 2M_6 + \cdots). \tag{52}$$

Since the second member of this equality always ≥ 0, we obtain the classical isoperimetric inequality on the sphere

$$L^2 + F^2 - 4\pi F \geq 0. \tag{53}$$

This inequality has often been proved. See [1], [3] and [2], and the bibliography in [4; 113]. For proof with methods of integral geometry analogous to those we follow in this paper, see [2; 83].

Equality (52) is valid when F_0 and L_0 are the area and length of any circle which neither contains K in its interior nor is contained in the interior of K. In particular, if C_0^* is the smallest circle which contains K in its interior and C_0 is the greatest circle which is contained in K, by neglecting the non-negative sum $M_4 + 2M_6 + \cdots$, we have

$$L^2 + F^2 - 4\pi F \geq (L - L_0)^2 + (F - F_0)^2, \tag{54}$$

$$L^2 + F^2 - 4\pi F \geq (L_0^* - L)^2 + (F_0^* - F)^2. \tag{55}$$

Taking into account the general inequality

$$x^2 + y^2 \geq \tfrac{1}{2}(x + y)^2, \tag{56}$$

we may combine inequalities (54) and (55) into the inequality

$$L^2 + F^2 - 4\pi F \geq \left(\frac{L_0^* - L_0}{2}\right)^2 + \left(\frac{F_0^* - F_0}{2}\right)^2. \tag{57}$$

This is a better form than (53) for the isoperimetric inequality.

If we substitute for L_0, L_0^*, F_0, F_0^* their values (36), relation (57) gives

$$L^2 + F^2 - 4\pi F \geq 4\pi^2 \sin^2 \frac{r_M - r_m}{2}, \tag{58}$$

where r_M and r_m are the spherical radii of the circles C_0^* and C_0.

T. Bonnesen [3; 82] has obtained the inequality

$$L^2 + F^2 - 4\pi F \geq 4\pi^2 \tan^2 \frac{r_M - r_m}{2},$$

which is better than our (58). His proof is completely different from ours.

For a sphere of radius R, inequality (57) takes the form

$$L^2 - 4\pi F + \left(\frac{F}{R}\right)^2 \geq \left(\frac{L_0^* - L_0}{2}\right)^2 + \left(\frac{F_0^* - F_0}{2R}\right)^2, \tag{59}$$

which as $R \to \infty$ gives the inequality

$$L^2 - 4\pi F \geq \left(\frac{L_0^* - L_0}{2}\right)^2 = \pi^2(r_M - r_m)^2, \tag{60}$$

which is a well-known isoperimetric inequality for plane curves established by Bonnesen [3; 63], [4; 113].

15. An upper limitation for the isoperimetric deficit of convex spherical curves. We now consider only convex curves K with *continuous radius of spherical curvature.* We understand by radius of spherical curvature the limit of the spherical radius of the circle which has three points in common with the curve as these points approach coincidence. This radius ρ ($\rho \leq \frac{1}{2}\pi$) is connected with the radius of geodesic curvature ρ_g by

$$\rho_g = \tan \rho.$$

Let ρ_M be the greatest radius and ρ_m the smallest radius of spherical curvature (both $\leq \frac{1}{2}\pi$). We wish to prove that

$$L^2 + F^2 - 4\pi F \leq \left(\frac{L_0^* - L_0}{2} + \frac{F_0^* - F_0}{2}\right)^2, \tag{61}$$

where L_0, F_0, L_0^*, F_0^* are now the lengths and areas of the circles whose radii are ρ_m and ρ_M respectively.

Likewise, as the area of the *exterior* parallel curve to K at distance ρ was expressed by (46), when we consider the *interior* parallel curve to K at a distance $\rho \leq \rho_m$, this curve will not have double points and its area is equal to

$$- L \sin \rho + F \cos \rho + 2\pi(1 - \cos \rho). \tag{62}$$

If we take $\rho = \rho_m$, area (62) will be the area covered by the centers of the circles of radius ρ_m which are contained in the interior of the convex curve K. If we represent this area by M_0, we can write

$$M_0 = - L \sin \rho_m + F \cos \rho_m + 2\pi(1 - \cos \rho_m). \tag{63}$$

We now wish to find the value of the area covered by the centers of the circles of radius ρ_M each of which contains K entirely in its interior. For this purpose we note that when the circle of radius ρ_M contains K in its interior, by a "dual" transformation (§2) the transformed circle (of radius $\frac{1}{2}\pi - \rho_M$) will be contained in the interior of the transformed curve K^* (whose length and area are $2\pi - F$ and $2\pi - L$ respectively). The area covered by the centers of the circles of radius ρ_M each of which contains K in its interior will then be given by (62) if we substitute ρ for $\frac{1}{2}\pi - \rho_M$, F for $2\pi - L$, and L for $2\pi - F$.

It follows that this area is given by

$$M_0^* = -L \sin \rho_M + F \cos \rho_M + 2\pi(1 - \cos \rho_M). \tag{64}$$

This has the same form as (63).

Let L_0, F_0 and L_0^*, F_0^* be the lengths and areas of the circles of radius ρ_m and ρ_M respectively, given by (36). Formulas (63) and (64) take the form

$$M_0 = F + F_0 - \frac{1}{2\pi}(LL_0 + FF_0) \tag{65}$$

and

$$M_0^* = F + F_0^* - \frac{1}{2\pi}(LL_0^* + FF_0^*). \tag{66}$$

When we take into account identity (51), these equalities give

$$L^2 + F^2 - 4\pi F = (L - L_0)^2 + (F - F_0)^2 - 4\pi M_0\,, \tag{67}$$

$$L^2 + F^2 - 4\pi F = (L_0^* - L)^2 + (F_0^* - F)^2 - 4\pi M_0^*. \tag{68}$$

Since M_0 and M_0^* are non-negative, we have

$$L^2 + F^2 - 4\pi F \le (L - L_0)^2 + (F - F_0)^2, \tag{69}$$

$$L^2 + F^2 - 4\pi F \le (L_0^* - L)^2 + (F_0^* - F)^2. \tag{70}$$

These inequalities give a first upper limit for the isoperimetric deficit $L^2 + F^2 - 4\pi F$.

From inequalities (69) and (70) we find

$$L^2 + F^2 - 4\pi F \le (L - L_0 + F - F_0)^2. \tag{71}$$

$$L^2 + F^2 - 4\pi F \le (L_0^* - L + F_0^* - F)^2. \tag{72}$$

Since the left sides are non-negative by (53) and since

$$xy \le \left(\frac{x + y}{2}\right)^2,$$

by multiplication of (71) and (72), we find

$$L^2 + F^2 - 4\pi F \le \left(\frac{L_0^* - L_0}{2} + \frac{F_0^* - F_0}{2}\right)^2. \tag{73}$$

For a sphere of radius R we have

$$L^2 - 4\pi F + \frac{F^2}{R^2} \le \left(\frac{L_0^* - L_0}{2} + \frac{F_0^* - F_0}{2R}\right)^2, \tag{74}$$

and as $R \to \infty$,

$$L^2 - 4\pi F \le \tfrac{1}{4}(L_0^* - L_0)^2 = \pi^2(\rho_M - \rho_m)^2, \tag{75}$$

where ρ_M and ρ_m are the greatest and the smallest radii of curvature of the plane convex curve K of length L and area F.

This inequality (75) is a known inequality obtained by Bottema [5]; see also [4; 83].

BIBLIOGRAPHY

1. F. BERNSTEIN, *Über die isoperimetrische Eigenschaft des Kreises auf der Kugeloberfläche und in der Ebene*, Mathematische Annalen, vol. 60(1905), pp. 117–136.
2. W. BLASCHKE, *Vorlesungen über Integralgeometrie*, Hamburger Mathematische Einzelschriften, Leipzig und Berlin, 1935.
3. T. BONNESEN, *Les problèmes des isopérimètres et des isépiphanes*, Gauthier-Villars, Paris, 1929.
4. T. BONNESEN AND W. FENCHEL, *Theorie der konvexen Körper*, Ergebnisse der Mathematik, Berlin, 1934, part I.
5. O. BOTTEMA, *Eine obere Grenze für das isoperimetrische Defizit einer ebenen Kurve*, Proceedings of the Köninklijke Akademie van Wetenschappen, vol. 36(1933), pp. 442–446.
6. W. CROFTON, *On the theory of local probability*, Philosophical Transactions of the Royal Society of London, vol. 158(1868), pp. 181–199.
7. W. CROFTON, *Probability*, Encyclopaedia Britannica, 9th edition, 1885.
8. E. CZUBER, *Geometrische Wahrscheinlichkeiten und Mittelwerte*, Leipzig, 1884.
9. R. DELTHEIL, *Probabilités géométriques* (*Traité du Calcul des Probabilités et de ses applications*, published under the direction of E. Borel), vol. II, fasc. II, Gauthier-Villars, Paris, 1926, pp. 74–76.
10. *Encyclopédie des Sciences Mathématiques*, vol. IV, 1904, pp. 26–27.
11. H. HORNICH, *Eine allgemeine Ungleichung für Kurven*, Monatshefte für Mathematik und Physik, vol. 47(1939), pp. 432–438.
12. H. LEBESGUE, *Exposition d'un mémoire de M. W. Crofton*, Nouvelles Annales de Mathématiques, (4), vol. 12(1912), pp. 481–502.
13. L. A. SANTALÓ, *A theorem and an inequality referring to rectifiable curves*, American Journal of Mathematics, vol. 63(1941), pp. 635–644.

INSTITUTO DE MATEMÁTICAS, ROSARIO, ARGENTINE REPUBLIC.

On the isoperimetric inequality for surfaces of constant negative curvature

L. A. Santaló
Instituto de Matemática, Universidad Nacional del Litoral, Rosario

1 Introduction

In the plane, the "isoperimetric problem" consists in the determination, among all the curves of given length, of the curve bounding the greatest area. As is well known, the solution is the circumference.

Indeed, many proofs have been given that the circumference is the curve with a given length enclosing a maximum area. For the background and bibliography to this problem, the reader may refer to [4], [8], [9], [13]. The square brackets refer to the bibliography at the end of the paper. The first of these proofs, some of which are very elementary and well-known, were incomplete for having accepted the existence of the solution without proof. The other proofs can be divided into two types: some based on methods of variational calculus (see [7], [11], page 202), and others of a geometric nature, consisting generally in the proof of certain "isoperimetric inequalities" leading to the solution of the problem with prior proof of the existence of a maximizing figure (see [4], [8], [6], page 26).

The same problem has been addressed and solved for curves situated on the spherical surface; for these, too, the lesser circumferences are the curves that for a given length bound the greatest area (see [2], [8], page 80; [9], page 113; [14]).

In the case of any surface, the problem consists in determining on such a surface, among all the closed curves of given length limiting a simply connected region, those curves for which the limited area is maximal. It is easy to prove (see, for example, [5], page 154; [10] t. III, page 151; [11] page 203) by methods of variational calculus that the curves fulfilling this condition must have *constant geodesic curvature*, but the question of the existence of these maximizing curves remains open.

In this paper we propose to solve the isoperimetric problem on surfaces of negative and constant curvature; that is, on pseudo-spherical surfaces. The method followed is not based on variational calculus, but rather is entirely geometric, and leads us furthermore to the fundamental isoperimetric inequality (22) and to other improved inequalities such as (21), (26) and (27), which avoid all questions of existence and at the same time solve the question of uniqueness.

Another proof for the same question, but employing completely different methods, was given by E. Schmidt [17], of whose fundamental inequality (31) we provide a new proof in Section 8.

In Section 9, sufficient conditions are given for a convex figure C of area F and length L to be contained in, or itself contain, a geodesic circle of radius R.

2 The Gauss-Bonnet formula and geodesic circles

In this Section 2 we summarize some known results which will be useful in what follows.

For the sake of simplicity, in the formulae we assume that the pseudo-spherical surface has a curvature $K = -1$. Then the classical Gauss-Bonnet formula applied to any boundary C limiting a portion of a simply connected surface of area F can be written

$$\int_C \frac{ds}{\rho_g} 2\pi + F, \tag{1}$$

denoting the radius of geodesic curvature at each point of the boundary by ρ_g, the arc element of the neighbourhood by ds, and with the integration extended to the whole boundary C.

If we are dealing with a geodesic polygon; that is, with a closed curve formed arcs of geodesic lines, then denoting its exterior angles by ω_i, the foregoing formula is written as follows

$$\sum \omega_i = 2\pi + F. \tag{2}$$

Let P be a point on the surface. A further arbitrary point can be determined by its geodesic distance to P; that is, by the length r of the geodesic arc connecting it with P[1], plus the angle φ that this geodesic forms with a direction fixed by P; we thus have what is known as a system of geodesic polar coordinates. In this system, the arc element ds of the surface is expressed by (see, for example, [3], Vol. I, p.335.)

$$ds^2 = dr^2 + \sinh^2 R \, d\varphi. \tag{3}$$

The curves that are the geometric locus of the points $r = R =$ constant are called *geodesic circles* of radius R and centre P. The length and the area of these geodesic circles are easily obtained from (3) and are equal to

$$L_0 = 2\pi \sinh R, \quad F_0 = 4\pi \sinh^2 \frac{R}{2} = 2\pi(\cosh R - 1). \tag{4}$$

3 Parallel convex curves

We give the name *convex curve* to any closed curve C limiting a simply connected portion of a pseudo-spherical surface, and which cannot be intersected by any geodesic line at more than two points.

A consequence of this definition is that the geodesics tangent to C, since they already have two common points that are the same, cannot intersect C

[1]Recall that only one geodesic passes through two real points of a surface of negative constant curvature. See, for example, [10], t. III, page 400.

again, and therefore they leave all the curve on the same side. It is thereby deduced that by fixing a direction on C, and considering 3 infinitely close tangent geodesics, the first and last of these tangent geodesics will have their point of contact on the same side of the second. Therefore, the angle between the tangent geodesics always changes in the same direction. Recalling the definition of geodesic curvature as the limit of the quotient between the angle of the two geodesics tangent to the curve and the length of the arc separating their points of contact, it is thereby deduced that *the convex curves have the geodesic curvature of constant sign.* In particular, when dealing with a convex geodesic polygon, the angles ω_i that appear in (2) will all be of the same sign.

Given a convex figure C, we draw through each of its points an arc of normal geodesic with a constant length R towards the exterior of C. The geometric locus of the endpoints of these arcs is known as the *exterior parallel curve* at distance R from C. If C has vertices, the parallel curve is closed by the arcs of geodesic circles of radius R whose centres are these vertices. We wish to find the area of this parallel curve.

Let us first consider the case of a convex curve formed by an infinite number of arcs of geodesic lines; that is, a convex geodesic polygon.

Let A_iA_{i+1} be a side of this polygon. Through the points on this geodesic arc A_iA_{i+1} we draw normal geodesics, and on each one we take the constant length R towards the exterior part of the polygon[2]. We then repeat the same for all the sides of the polygon. In this way, through each vertex A_i we have two geodesic arcs $A_iA'_i$, $A_iA''_i$ of length R, which together form an angle ω_i equal to the exterior angle corresponding to vertex A_i. The endpoints of all these arcs of geodesics normal to A_iA_{i+1} form the arcs $A''_iA'_{i+1}$, which are no longer geodesics but which together with arcs of geodesic circle $A'_iA''_i$, with centre A_i and radius R, form the parallel curve exterior to distance R from the given polygon.

The area of a sector of geodesic circle with radius R and angle ω_i, according to (4), is equal to $(\cosh R - 1)\omega_i$, and therefore the area of all the sectors is $(\cosh R - 1)\sum \omega_i$, or, according to (2)

$$(\cosh R - 1)(2\pi + F). \tag{5}$$

We still have the quadrilaterals $A_iA''_iA'_{i+1}A_{i+1}$. In order to calculate their area, for each side A_iA_{i+1} we take as a system of curvilinear coordinates formed by the geodesics normal to each side the lines $v =$ constant, and as the curves connecting the points of these geodesics that are equidistant from A_iA_{i+1} the lines $u =$ constant. It is known (see, for example, [3], Vol I, page 333) that in a system of this type the arc element is expressed by the form

$$ds^2 = du^2 + \cosh^2 u\, dv^2, \tag{6}$$

and therefore the arc element is equal to $df_i = \cosh u\, du\, dv$. Denoting the length of arc A_iA_{i+1} by l_i, and the area of the quadrilateral $A_iA''_iA'_{i+1}A_{i+1}$ by f_i, we

[2]These geodesic arcs cannot be intersected, since it is known that more than one geodesic normal to another does not pass through a point on a pseudo-spherical surface. This is a consequence of formula (2), since if two normal geodesics passed through such a point, we would have a geodesic triangle with two exterior angles equal to $\frac{\pi}{2}$, and therefore, according to (2), the third exterior angle would have to be greater than π, which is absurd.

then have

$$f_i = \int_0^{l_i} dv \int_0^R \cosh u\, du = l_i \sinh R.\cosh^2 u\, dv^2. \tag{7}$$

Adding this expression for all the sides of the polygon, we obtain $L \sinh R$. The area between the given geodesic polygon and its parallel exterior to distance R is equal to this sum $L \sinh R$ plus expression (5), and therefore, by adding the area F of the original polygon, we have

$$F(R) = L \sinh R + (2\pi + F)(\cosh R - 1) + F. \tag{8}$$

This is the value of the area limited by the parallel curve exterior to distance R from a convex geodesic polygon of perimeter L and area F.

Assuming now the general case of a an arbitrary convex curve C, and approximating by inscribed geodesic polygons that tend to C, the area and the length of these polygons will tend to the area and length of C, and since the curves parallel to the inscribed polygons also tend to the curve parallel to C, formula (8) holds for any convex curve C.

4 An integral formula

Let C be an arbitrary curve of a pseudo-spherical surface. It need not be convex; we assume only that it has a continuous geodesic curvature, or that it is formed by an infinite number of arcs of this type.

Let us also assume each point P on the pseudo-spherical surface to be the centre of a geodesic circle of radius R. Let dP be the area element corresponding to point P, and n the number of points that the geodesic circle of centre P shares in common with the given curve C. We wish to calculate the integral

$$I = \int\int n\, dP, \tag{9}$$

the integration being extended to all the points P for which $n \neq 0$. To this effect, we express dP in a suitable system of curvilinear coordinates.

Consider on the given curve C two points A and B whose distance is the arc element ds. We draw the geodesics forming with C a constant angle θ through the points on ds, and on these geodesics we take the arcs AA_1, BB_1 equal to R. Denoting the arc A_1B_1 by ds_1, we find a relation between ds, ds_1 and the angle θ.

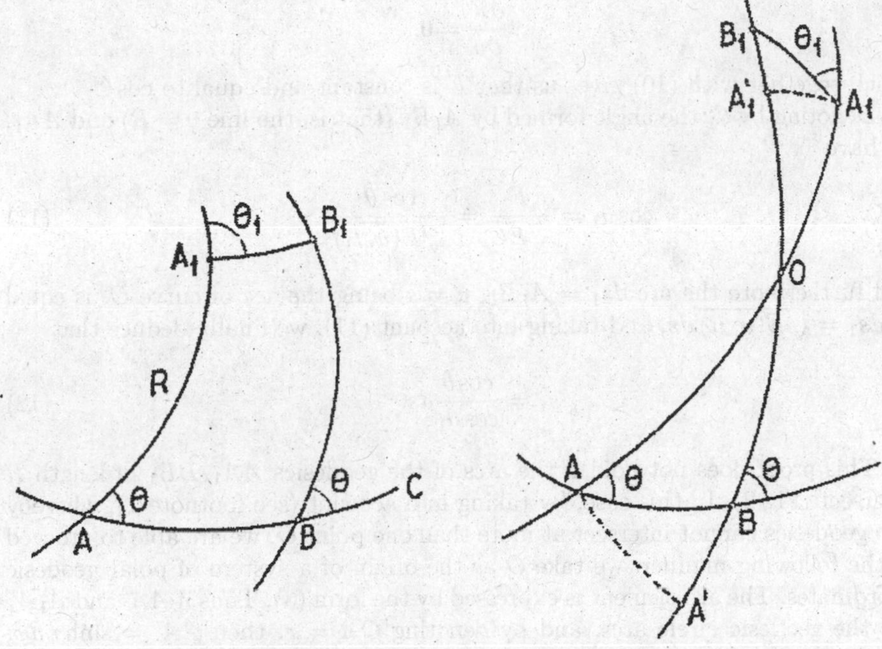

Fig. 1 Fig. 2

Let us first assume that arcs AA_1 and BB_1 of length R are not intersected (Fig.1). For the region ABB_1A_1, by taking the arc AB of curve C sufficiently small, we can define a regular system of curvilinear coordinates formed by the geodesics $AA_1, BB_1, \ldots$ that form the constant angle θ with C as curves $u =$ constant, and with the curves obtained by taking on these geodesics equal arcs from the points on C as curves $v =$ constant. In this system of curvilinear coordinates, the arc element is expressed in the general form

$$ds^2 = E\,du^2 + 2F\,du\,dv + G\,dv^2.$$

Bearing in mind that for $u =$ constant we take $s = v$, we have $G = 1$. Furthermore, for $v =$ constant we have the curve C, and therefore $E(u, 0) = 1$. In accordance with these values E and G, we obtain[3]

$$\cos\theta = F(u, 0) = \text{constant}. \tag{10}$$

By expressing the fact that the lines $u =$ constant are geodesics, we also

[3]Recall that the angle formed by the coordinate lines is given by $\cos\theta = \frac{F}{\sqrt{EG}}$.

obtain the condition[4]

$$\frac{\partial F}{\partial v} = 0,$$

which together with (10) gives us that F is constant and equal to $\cos\theta$.

Denoting by θ_1 the angle formed by A_1B_1 (that is, the line $v = R$) and AA_1, we have

$$\cos\theta_1 = \frac{F}{\sqrt{EG}} = \frac{\cos\theta}{\sqrt{E(u,R)}}, \tag{11}$$

and furthermore the arc $ds_1 = A_1B_1$, $u = s$ being the arc of curve C, is equal to $ds_1 = \sqrt{E(u,R)}\,ds$, and taking into account (11), we finally deduce that

$$ds_1 = \frac{\cos\theta}{\cos\theta_1} ds. \tag{12}$$

This proof does not hold if the arcs of the geodesics AA_1, BB_1 of length R intersect (Fig.2). In this case, by taking into account (see footnote 1), whereby two geodesics cannot intersect at more than one point O, we are able to proceed in the following manner: we take O as the origin of a system of polar geodesic coordinates. The arc element is expressed by the form (3). Thus if AA' and A_1A_1' are the geodesic circle arcs, and by denoting $OA = r$, then $AA' = \sinh r\,d\varphi$, $A_1A_1' = \sinh(R-r)\,d\varphi$. Moreover, by denoting as before the arc AB by ds and A_1B_1 by ds_1, we have

$$\sin\theta = \frac{AA'}{AB} = \sinh r \frac{d\varphi}{ds}, \quad \sin\theta_1 = \frac{A_1A_1'}{A_1B_1} = \sinh(R-r)\frac{d\varphi}{ds_1},$$

as well as

$$\tan\theta = \frac{AA'}{A'B} = \frac{\sinh r\,d\varphi}{dr}, \quad \tan\theta_1 = \frac{A_1A_1'}{A_1'B_1} = \frac{\sinh(R-r)\,d\varphi}{dr}.$$

From these equalities, by dividing in an ordered way we once again obtain (12), which is therefore proven in all cases.

Once having established relation (12), we observe that in order to fix on the surface a point P that is the centre of a geodesic circle of radius R intersecting the fixed curve C at point A, we are able to give point A by its curvilinear absciss s on the curve C, and the angle θ formed by the geodesic radius AP with C. Point P is thus determined by the coordinates s, θ. The area element dP is expressed in these coordinates by the product of the element ds_1 of the curve $\theta =$ constant, by the arc element of the circle $s =$ constant, which according to

[4]The differential equation of the geodesic lines is

$$(EG - F^2)(u'v'' - v'u'') \;+\; (Eu' + Fc')\;(F_u - \frac{1}{2}E_v)u'^2 + G_u u'v' + \frac{1}{2}G_v v'^2$$
$$-\;(Fu' + Gv')\;\frac{1}{2}E_u u'^2 + E_v u'v' + (F_v - \frac{1}{2}G_u)v'^2\;.$$

See, for example, [5], page 175. Expressing the fact that the lines $u =$ constant satisfies this equation, and since $G = 1$, we obtain $F_v = 0$.

(3) is equal to $\sinh R\,d\theta$, and by the sinus of the angle formed by both elements; that is, by the cosine of its complementary, which is θ_1. In other words,

$$dP = \sinh R \cos\theta_1\, d\theta\, ds_1,$$

or, according to (12),

$$dP = \sinh R \cos\theta\, d\theta\, ds.$$

Since the area element is always considered to be positive, in this formula, $\cos\theta$ must be taken at its absolute value.

If we integrate this expression for all the values of s along the curve C, and for each value of s we rotate θ from 0 to 2π, then each element dP will be counted as many times as there are points of intersection with C belonging to the geodesic circle with centre P and radius R. If we denote this number by n, then we have

$$\int\!\!\int n\, dP = \sinh R \int_0^L ds \int_0^{2\pi} |\cos\theta|\, d\theta = 4L\sinh R,$$

where L is the length of curve C.

Integral (9) is thereby calculated, its value being

$$\int\!\!\int n\, dP = 4L\sinh R. \tag{13}$$

This formula generalizes on pseudo-spherical surfaces a particular case of the Poincaré Integral Geometry formula. For the plane, the analogous formula was proven in [6], page 23; for the sphere in [6], page 81, and for n-dimensional Euclidean space in [15].

5 Isoperimetric property of geodesic circles

We now apply formulae (8) and (13) to obtain the fundamental isoperimetric inequality on pseudo-spherical surfaces.

Let C be a convex figure of length L and area F. Let R be a value such that no geodesic circle of radius R can be completely contained within C and no such geodesic circle itself can completely contain C. All the geodesic circles of radius R whose centre is contained within the area limited by the parallel curve exterior to C at distance R intersect C, and reciprocally any geodesic circle of radius R intersecting C has its centre in the said area, whose value is given by (8). We denote by M_i the area filled by the points P that are centres of geodesic circles of radius R intersecting C at points i. Expression (8) can thus be written

$$M_2 + M_4 + M_6 + \cdots = L\sinh R + (2\pi + F)(\cosh R - 1) + F, \tag{14}$$

since if C is a closed curve, the points P for which i is odd do not fill any area.

Analogously, (13) can be written as follows:

$$2M_2 + 4M_4 + 6M_6 + \cdots = 4L\sinh R. \tag{15}$$

From (14) and (15) we deduce that

$$M_4 + 2M_6 + 3M_8 + \cdots = L\sinh R - F - (2\pi + F)(\cosh R - 1), \tag{16}$$

and since the left-hand side of this equality, being a sum of areas, is essentially positive, we have that while R fulfils the afore-mentioned condition that no geodesic circle of radius R either contains or is contained by the curve C, the following inequality is verified

$$L\sinh R - F - (2\pi + F)(\cosh R - 1) \geq 0. \tag{17}$$

Taking into account the values L_0, F_0 given in (4) for the length and area of the geodesic circle of radius R, this inequality (17) can be written as follows:

$$\frac{1}{2\pi}(LL_0 - FF_0) - (F + F_0) \geq 0. \tag{18}$$

According to (4), it can immediately be checked that the following identity is fulfilled:

$$L_0^2 - F_0^2 - 4\pi F_0 = 0, \tag{19}$$

which when taken into account also enables the next equality to be checked identically.

$$\begin{aligned}\frac{1}{2\pi}(LL_0 - FF_0) - (F + F_0) &= \frac{1}{4\pi}\left((L^2 - F^2 - 4\pi F)\right. \\ &+ \left.(F - F_0)^2 - (L - L_0)^2\right).\end{aligned} \tag{20}$$

We now take R in such a way that the geodesic circle of radius R has the same area as the given convex figure C; that is, so that $F = F_0$. Obviously, this circle of radius R can neither contain nor be contained by C. Consequently, it is applicable to inequality (18), which, taking into account (20), can be written thus:

$$L^2 - F^2 - 4\pi F \geq (L - L_0)^2, \tag{21}$$

that is, with even more reason,

$$L^2 - F^2 - 4\pi F \geq 0. \tag{22}$$

This is the *fundamental isoperimetric inequality* on surfaces of constant curvature equal to -1, which in fact tells us that, given the area F, the minimum value of L is $\sqrt{F^2 + 4\pi F}$, and which according to (19) is reached by geodesic circles. Inequality (21) is an "improved" form of the isoperimetric inequality.

6 Other isoperimetric inequalities

It still remains to prove that geodesic circles are the *only* figures for which the equals sign in (22) holds. This can be achieved immediately from other inequalities we now intend to obtain, which as in (21) improve the fundamental isoperimetric inequality (22).

For brevity's sake we denote the left-hand side of inequality (22) by $\triangle$; that is, the so-called *isoperimetric deficit.* In other words,

$$\triangle = L^2 - F(4\pi + F). \tag{23}$$

With this notation it is easy to check that the following identity holds:

$$L \sinh R - F - (2\pi + F)(\cosh R - 1) = \frac{\sinh^2 \frac{R}{2}}{F}\left(\triangle - (L - F \coth \frac{R}{2})^2\right), \tag{24}$$

and therefore, according to (17), we have

$$\triangle \geq (L - F \coth \frac{R}{2})^2.$$

As in (17), this inequality holds for all R such that no geodesic circle of radius R can contain or be contained by C. In particular, for the maximum radius R_e of all the geodesic circles that cannot contain C within their interior, and for the minimum radius R_i of all the geodesic circles that cannot be contained within C, we have

$$\triangle \geq \left(L - F \coth \frac{R_e}{2}\right)^2$$

$$\triangle \geq \left(F \coth \frac{R_i}{2} - L\right)^2.$$

Hence, recalling that whatever the value of x and y, the following is always verified:

$$x^2 + y^2 \geq \frac{1}{2}(x + y)^2, \tag{25}$$

and we obtain

$$\triangle \geq \frac{1}{4}F^2 \left(\coth \frac{R_i}{2} - \coth \frac{R_e}{2}\right)^2. \tag{26}$$

Since its right-hand side is always ≥ 0, this new inequality contains the fundamental isoperimetric inequality (22) and is furthermore an improvement on it, because instead of 0 on the right-hand side there is in general a quantity that is greater.

From (26) we deduce that $\triangle = 0$ only when $R_e = R_i$, in which case figure C must coincide with the geodesic circle having this radius. *Uniqueness* is therefore proven.

A further isoperimetric inequality, analogous to (26), can be obtained by considering instead of identity (24) the following analogous identity, which is also easy to check,

$$\begin{aligned} &L \sinh R - F - (2\pi + F)(\cosh R - 1) \\ &= \frac{\cosh^2 \frac{R}{2}}{4\pi + F}\left(\triangle - \left((4\pi + F)\tanh \frac{R}{2} - L\right)^2\right), \end{aligned}$$

from which, by taking (17) into account, we deduce that

$$\triangle \geq \left((4\pi + F)\tanh\frac{R}{2} - L\right)^2.$$

This inequality is verified in the same way as before for all the radii R of the geodesic circles that can neither contain nor be contained by C; in particular, for their extreme values R_e and R_i. Consequently,

$$\triangle \geq \left((4\pi + F)\tanh\frac{R_e}{2} - L\right)^2$$

$$\triangle \geq \left(L - (4\pi + F)\tanh\frac{R_i}{2}\right)^2.$$

Hence, adding and taking (25) into account, the new inequality is deduced thus

$$\triangle \geq \frac{1}{4}(4\pi + F)^2\left(\tanh\frac{R_e}{2} - \tanh\frac{R_i}{2}\right)^2. \tag{27}$$

which, as in (26), implies the uniqueness of the isoperimetric problem; that is, the fact that geodesic circles are the only figures for which the equals sign holds in (22).

7 Passing to the case of the plane

If the curvature of the pseudo-spherical surface considered is $K = -\frac{1}{a^2}$ instead of $K = -1$, the foregoing formulae still hold simply by observing that, by similarity, the lengths are proportional to a and the areas to a^2. After multiplying both sides by a^2, the fundamental isoperimetric inequality (22) is written thus,

$$L^2 - \frac{F^2}{a^2} - 4\pi F \geq 0.$$

The other inequalities (21), (26) and (27) are written analogously. For example, after multiplying both sides by a^2, (27) gives

$$L^2 - \frac{F^2}{a^2} - 4\pi F \geq \frac{1}{4}(4\pi + \frac{F}{a^2})^2\left(a\tanh\frac{R_e}{2a} - a\tanh\frac{R_i}{2a}\right)^2,$$

which for $a \to \infty$ yields

$$L^2 - 4\pi F \geq \pi^2(R_e - R_i)^2,$$

which is a classical isoperimetric inequality due to Bonnesen [8], page 69, for convex plane figures. In this inequality, R_e is the minimum radius of circles containing C, and R_i is the maximum radius of those contained in the same figure.

8 E. Schmidt's isoperimetric inequality

We now see how the isoperimetric inequality in the form given by E. Schmidt [17] can be deduced from (17).

Let Γ be a geodesic of the pseudo-spherical surface. By drawing the geodesics normal to Γ and drawing on each one, on both sides of Γ, arcs of constant length R, we obtain two curves known as *curves of equal distance with respect to* Γ. If the convex curve C is inscribed between two curves of equal distance R of a geodesic Γ, Γ is said to be a *bisecting geodesic*[5] of C, and with respect to this geodesic $2R$ is the *width* of C. All the bisecting geodesics of a geodesic circle of radius R pass through its centre, and its width is always equal to $2R$ with respect to any of these geodesics.

Instead of the width $2R$ as we have defined it above, let us introduce an angle α linked to R by the relation[6]

$$R = -\log\tan(\frac{\pi}{4} - \frac{\alpha}{2}). \tag{28}$$

According to which

$$\sinh R = \frac{1}{2}(e^R - e^{-R}) = \tan\alpha, \quad \cosh R = \frac{1}{2}(e^R + e^{-R}) = \frac{1}{\cos\alpha}, \tag{29}$$

whereby the area F_0 and the length L_0 of a geodesic circle of radius R are expressed by

$$\begin{aligned} L_0 &= 2\pi\sinh R = 2\pi\tan\alpha \\ F_0 &= 2\pi(\cosh R - 1) = 2\pi(\frac{1}{\cos\alpha} - 1). \end{aligned} \tag{30}$$

If $2R$ is a width of C with respect to any bisecting geodesic, obviously no geodesic circle of radius R can contain or be completely contained by C, and therefore inequality (17) holds.

Let α' be the value of x for which the corresponding value of R according to (28) is the radius of the geodesic circle whose length is equal to that of the given convex curve C, and let α'' be the value for which the geodesic circle whose radius is the corresponding R has the same area F as C. Then in (17), according to (30), we can write

$$L = 2\pi\tan\alpha' \quad F = 2\pi(\frac{1}{\cos\alpha''} - 1),$$

---

[5] E. Schmidt in [17] refers to Poincaré's representation of pseudo-spherical surfaces on the semiplane $x > 0$; then instead of geodesics he speaks of *lines* in the Hyperbolic plane, and what we refer to as bisecting geodesics he calls *Mittelgeraden*.

[6] This angle α, which appears here artificially, appears in the cited paper by Schmidt as a consequence of the plane representation employed in pseudo-spherical surfaces and is known as the *angle of width* (Breitenwinkel). In the non-Euclidean interpretation of geometry on surfaces of negative constant curvature (n° 8), this angle α is equal to the complement of the *angle of parallelism* at distance R. In fact, the angle of parallelism β is related to the distance R by the formula

$$R = -\log\tan\frac{\beta}{2}.$$

See, for example, [3] t. I, page 621.

and also by substituting the values (29), after simple transformations we have

$$\cos\alpha'' \geq \frac{\cos\alpha'}{\cos(\alpha-\alpha')}. \tag{31}$$

This is the form given by E. Schmidt ([17], page 206) to the isoperimetric inequality, from which it can in fact be deduced that $\alpha'' \leq \alpha'$; that is, according to (28), $R'' \leq R'$. Therefore, the radius of the geodesic circle whose area is equal to that of C is smaller than or equal to the radius of the geodesic circle of the same length; the area of C is therefore always $\leq$ than the area of the geodesic circle of the same length, which is the isoperimetric property.

9 Sufficient conditions for a convex figure C to be contained within geodesic circle of radius R

Let[7] C_0 be a geodesic circle of radius R, and denote its length by L_0 and its area by F_0, given by (4). We have already proven that if R is such that no geodesic circle of radius R can contain or be completely contained by C, then (18) holds. Consequently, if this inequality does not hold, there must exist some geodesic circle C_0 with radius R that is either completely contained within C or contains C in its interior.

Therefore, a *sufficient* (although not necessary) condition in order for some geodesic circle of radius R that contains or is contained by C to exist is that the inverse inequality of (18) should hold; that is,

$$2\pi(F+F_0) - LL_0 + FF_0 > 0. \tag{32}$$

The disadvantage of this condition is that it does not enable us to determine whether it is the convex figure C that is contained within some C_0, or whether there exists some C_0 that can be contained within C. It is possible to distinguish between the two cases by doubling inequality (32).

Observe that, according to (22) and (19), we can write

$$L^2 \geq F(4\pi+F), \quad L_0^2 = F_0(4\pi+F_0), \tag{33}$$

and hence

$$L^2L_0^2 \geq FF_0(4\pi+F)(4\pi+F_0), \tag{34}$$

an inequality that holds for any convex figure C (with area F and length L), and for any geodesic circle C_0 (with area F_0 and length L_0).

Now consider the equality

$$LL_0 - F(4\pi+F_0) > \sqrt{L^2L_0^2 - FF_0(4\pi+F)(4\pi+F_0)} \tag{35}$$

[7] Analogous questions to these in Section 9 for plane figures have been studied by Hadwiger [12]. For figures on the spherical surface, see [16].

whose sub-radical quantity cannot be negative because of (34).

If (35) is fulfilled, it is easy to check by squaring and simplifying that (32) is also fulfilled, and therefore either of the two figures, C or C_0, can be contained within the interior of the other.

We now prove that if (35) is fulfilled, $F < F_0$ also holds, and consequently it is clear that the only possibility is that figure C is the one that is contained within the interior of the geodesic circle C_0. Indeed, from (32), which is a consequence of (35), we may deduce that

$$LL_0 < 2\pi(F + F_0) + FF_0, \tag{36}$$

and according to this inequality, if $F \geq F_0$, then with even more reason $LL_0 < (4\pi + F_0)F$, and thus (35) cannot hold, since the right-hand side is essentially positive.

Replacing L_0 and F_0 by its values (4), and by recalling that

$$\cosh^2 R - \sinh^2 R = 1, \qquad \frac{1 + \cosh R}{\sinh R} = \coth\frac{R}{2},$$

inequality (35), after simple operations, is written as follows:

$$L - F\coth\frac{R}{2} > \sqrt{L^2 - F(4\pi + F)}. \tag{37}$$

Therefore: *Inequality* (37) *is a sufficient (although not necessary) condition in order for a convex figure C with area* F *and length* L *to be completely contained within a geodesic circle of radius* R.

Analogously, the following inequality can be considered:

$$LL_0 - F_0(4\pi + F) > \sqrt{L^2L_0^2 - FF_0(4\pi + F)(4\pi + F_0)}, \tag{38}$$

which has as a consequence (32), and furthermore requires that $F_0 < F$; indeed, if in (36) $F_0 \geq F$, one would deduce that $LL_0 < (4\pi + F)F_0$, and therefore the left-hand side of (38) would be negative, which is not possible. Condition (38) is thus a *sufficient* condition for stating that C can completely contain a geodesic circle C_0within its interior.

By replacing in (38) the values L_0 and F_0 given in (4), and subsequent to immediate transformations, we obtain

$$L - (4\pi + F)\tanh\frac{R}{2} > \sqrt{L^2 - F(4\pi + F)}. \tag{39}$$

Therefore: *Condition* (39) *is a sufficient (although not necessary) condition in order for a convex figure C with area* F *and length* L *to contain a geodesic circle of radius R within its interior.*

10 Interpretation in Hyperbolic geometry

Ever since Beltrami [1], it has been known that geometry on surfaces of negative constant curvature is equivalent to Hyperbolic non-Euclidean geometry. Consequently, all the foregoing can be interpreted as relations established among convex figures in the Hyperbolic plane. In particular, inequalities (22), (26) and (27) express the isoperimetric property of the circle in Hyperbolic non-Euclidean or Lobachewsky-Bolyai plane geometry.

References

[1] E. Beltrami, *Saggio di interpretazione della geometria non euclidea*, Giornale di Matematica, **6** (1868), 248–312, Also *Op. Mat.* 1, 374-405; Annales Scientifiques de l'École Normale Supérieure, t. 6, 251-286, (1869) Gauthier-Villars, París.

[2] F. Bernstein, *Ueber die isoperimetrische Eigenschaft des Kreises auf der Kugeloberfläche und in der Ebene*, Mathematische Annalen **60** (1905), 117–136.

[3] L. Bianchi, *Lezioni di geometria differenziale*, vol. I, N. Zanichelli, Bologna, 1924.

[4] W. Blaschke, *Kreis und Kugel*, Veit, Leipzig, 1916.

[5] ———, *Differentialgeometrie*, vol. I, Springer, Berlin, 1930.

[6] ———, *Vorlesungen über Integralgeometrie*, vol. 20, B. G. Teubner, Leipzig and Berlin, 1936.

[7] T. Bonnesen, *Quelques problèmes isopérimètriques*, Acta Mathematica **48** (1926), 123–178.

[8] ———, *Les problèmes des isopérimètres et des isépiphanes*, Collection de monographies sur la théorie des fonctions, Gauthier-Villars, Paris, 1929.

[9] T. Bonnesen and W. Fenchel, *Theorie der konvexen Körper*, Ergebnisse der Math. Springer, Berlin, 1934.

[10] G. Darboux, *Leçons sur la théorie générale des surfaces*, vol. III, Gauthier-Villars, Paris, 1894.

[11] J. Hadamard, *Leçons sur le calcul des variations*, Hermann, Paris, 1910.

[12] H. Hadwiger, *Ueberdeckung ebener Bereiche durch Kreise und Quadrate*, Comm. Math. Helvetici **13** (1941), 195–200.

[13] T. I. Porter, *A history of the classical isoperimetric problem*, These submitted to the department of Mathematics of the University of Chicago, The University of Chicago Press, 1931-32.

[14] L. A. Santaló, *Una demostración de la propiedad isoperimétrica del círculo*, Publicaciones del Instituto de Matemáticas de la Universidad Nacional del Litoral, Rosario **II** (1940), 37–46.

[15] ———, *A theorem and an inequality referring to rectifiable curves*, Amer. J. Math **63** (1941), 635–644.

[16] L. A. Santaló, *Algunos valores madios y desigualdades referentes a curvas situadas sobre la superficie esférica*, Rev. Unión Matemática Argentina **8** (1942), 113–125.

[17] E. Schmidt, *Ueber die isoperimetrische Aufgabe im n-dimensionalen Raum konstanter negativer Krümmung*, Mathematische Zeitschrift **46** (1940), 204–230.

INTEGRAL GEOMETRY ON SURFACES OF CONSTANT NEGATIVE CURVATURE

By L. A. Santaló

1. **Introduction.** We use the expression "integral geometry" in the sense given it by Blaschke [4]. In a previous paper [11] we generalized to the sphere many formulas of plane integral geometry and at the same time applied these to the demonstration of certain inequalities referring to spherical curves.

The present paper considers analogous questions for surfaces of constant negative curvature and consequently for hyperbolic geometry [1].

In §§2–7 we define the measure of sets of geodesic lines and the cinematic measure, making application of both in order to obtain various integral formulas such as, for example, (4.6) which generalizes a classic result of Crofton for plane geometry and (7.5) which is the generalization of Blaschke's fundamental formula of plane integral geometry.

In §8 we apply the above results to the proof of the isoperimetric property of geodesic circles (inequality (8.4)). In §9 we obtain a sufficient condition that a convex figure be contained in the interior of another, thus generalizing to surfaces of constant negative curvature a result which H. Hadwiger [8] obtained for the plane.

For what follows we must remember that on the surfaces of curvature $K = -1$ the formulas of hyperbolic trigonometry are applicable [2; 638], that is, for a geodesic triangle of sides a, b, c and angles α, β, γ, we have

$$\begin{gathered} \cosh a = \cosh b \cosh c - \sinh b \sinh c \cos \alpha, \\ \sinh a/\sin \alpha = \sinh b/\sin \beta = \sinh c/\sin \gamma, \\ \sinh a \cos \beta = \cosh b \sinh c - \sinh b \cosh c \cos \alpha. \end{gathered} \tag{1.1}$$

2. **Measure of sets of geodesics.** Let us consider a surface of constant curvature $K = -1$. Let O be a fixed point on that surface and G a geodesic which does not pass through O. We know that through O passes only one geodesic perpendicular to G [6; 410]. Let v be the distance from O to G measured upon this perpendicular. We shall call θ the angle which the perpendicular geodesic makes with a fixed direction at O. We define as the "density" to measure sets of geodesics the differential expression

$$dG = \cosh v \, dv \, d\theta, \tag{2.1}$$

that is, the measure of a set of geodesics will be the integral of the expression (2.1) extended to this set.

Received March 12, 1943.

To admit this definition it is necessary to prove that the measure does not depend on the point O or on the direction origin of the angles θ.

Let A be the point in which the normal geodesic traced through O cuts G. We shall consider another point O_1 and denote by A_1 the analogous point in which the normal traced through O_1 cuts G. Call α the angle formed by the geodesic OO_1 and the direction origin of angles at O; analogously, α_1 will be the angle which the same geodesic OO_1 makes with the direction origin of angles at O_1. For brevity write

$$OA = v, \quad O_1A_1 = v_1, \quad OO_1 = \rho, \quad OA_1 = \mu, \quad O_1A = \nu, \quad AA_1 = \lambda,$$

where the left sides are the arcs of geodesics. We can also state

$$\varphi = \text{angle } OA_1O_1, \quad \psi = \text{angle } OAO_1, \quad \theta - \alpha = \text{angle } AOO_1,$$

$$\pi - (\theta_1 - \alpha_1) = \text{angle } OO_1A_1.$$

With these notations (Fig. 1) the third formula of hyperbolic geometry (1.1) applied to the triangle OO_1A_1 gives

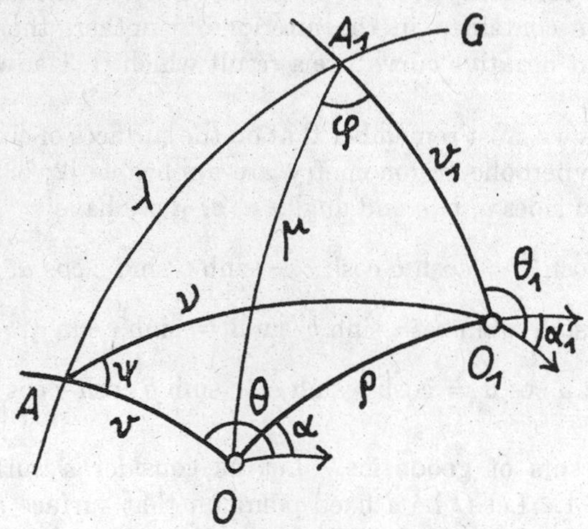

FIGURE 1

$$\sinh \mu \cos \varphi = \cosh \rho \sinh v_1 + \sinh \rho \cosh v_1 \cos (\theta_1 - \alpha_1). \tag{2.2}$$

In the rectangular triangle OAA_1 the second formula (1.1) gives $\sinh \mu = \sinh v/\cos \varphi$ and in consequence (2.2) may be written as

$$\sinh v = \cosh \rho \sinh v_1 + \sinh \rho \cosh v_1 \cos (\theta_1 - \alpha_1). \tag{2.3}$$

Analogously

$$\sinh v_1 = \cosh \rho \sinh v - \sinh \rho \cosh v \cos (\theta - \alpha). \tag{2.4}$$

In order to pass from (2.3) to (2.4) we changed the sign of the second term of the right side because in the quadrilateral OO_1A_1A the internal angle AOO_1 has value $\theta - \alpha$, while angle $OO_1A_1 = \pi - (\theta_1 - \alpha_1)$.

From (2.3) and (2.4) an easy calculation gives

$$\frac{\partial(v, \theta)}{\partial(v_1, \theta_1)} = \frac{\cosh^2 v_1 \sin(\theta_1 - \alpha_1)}{\cosh^2 v \sin(\theta - \alpha)}. \tag{2.5}$$

The third formula (1.1) applied to the triangle OAA_1 gives $\sinh \mu \sin \varphi = \cosh v \sinh \lambda$ and in the triangle OO_1A_1 we also have $\sinh \mu/\sin(\theta_1 - \alpha_1) = \sinh \rho/\sin \varphi$. These two equalities give

$$\cosh v \sinh \lambda = \sinh \rho \sin(\theta_1 - \alpha_1),$$

and by analogy

$$\cosh v_1 \sinh \lambda = \sinh \rho \sin(\theta - \alpha).$$

From these last equalities we deduce

$$\cosh v \sin(\theta - \alpha) = \cosh v_1 \sin(\theta_1 - \alpha_1),$$

and if we take into account this equality, the Jacobian (2.5) has the value $\cosh v_1/\cosh v$ and consequently

$$\cosh v \, dv \, d\theta = \cosh v_1 \, dv_1 \, d\theta_1 ,$$

that is, the density (2.1) and also the measure of any set of geodesics are independent of the point O and the direction origin of the angles θ.

3. **Measure of the geodesics which cut a line.** In the preceding section we determined the geodesic G by its coördinates v, θ. If G cuts a fixed curve C in a point P, it can also be determined by the abscissa s of the point P upon C, that is, by the length of the arc of C between P and an origin of arcs and the angle φ formed at the point P by the curve C and the geodesic G. We shall express the density (2.1) as a function of s, φ.

Since through any point there is only one geodesic perpendicular to another geodesic, G is also determined by s, θ. We shall pass first from v, θ to s, θ and afterwards to s, φ. In the system of curvilinear coördinates whose curves $v =$ const. are the geodesics normal to OA (the letters designate the same points as in §2) and the curves $u =$ const. are their orthogonal trajectories, we have [2; 335]

$$ds^2 = du^2 + \cosh^2 u \, dv^2. \tag{3.1}$$

Consequently, if the geodesic G ($v =$ const.) forms an angle φ with the curve C, we have

$$\sin \varphi = \frac{\cosh u \, dv}{ds}$$

from which

$$dv = \frac{\sin \varphi}{\cosh u} ds.$$

Therefore (2.1) may be written as

$$dG = \cosh v \frac{\sin \varphi}{\cosh u} ds\, d\theta. \tag{3.2}$$

In place of θ we wish now to introduce the angle φ. Let us call ρ the arc of geodesic OP, α the angle which this arc OP makes with the direction origin of angles at O, and α_1 the angle which OP makes with the curve C (Fig. 2). In

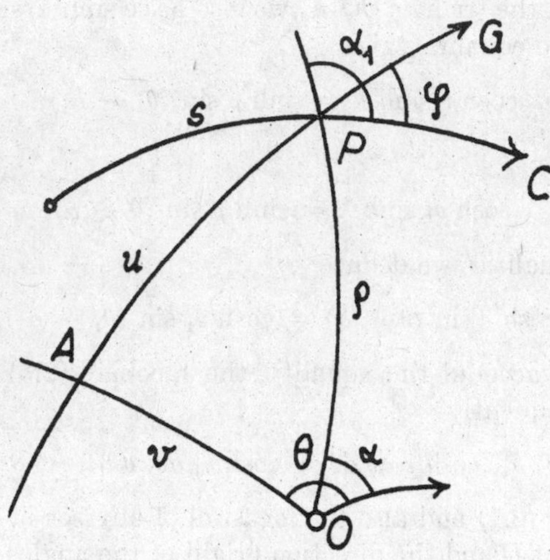

FIGURE 2

the rectangular triangle OAP we have angle $AOP = \theta - \alpha$, angle $APO = \alpha_1 - \varphi$, and from (1.1) we deduce

$$\cot (\alpha_1 - \varphi) = \cosh \rho \tan (\theta - \alpha),$$

from which, by differentiation, we get

$$\frac{d\varphi}{\sin^2 (\alpha_1 - \varphi)} = \frac{\cosh \rho\, d\theta}{\cos^2 (\theta - \alpha)}. \tag{3.3}$$

But, in the same rectangular triangle OAP

$$\cos (\theta - \alpha) = \cosh u \sin (\alpha_1 - \varphi), \qquad \cosh \rho = \cosh u \cosh v$$

and as a consequence

$$\frac{\cosh \rho \sin^2 (\alpha_1 - \varphi)}{\cos^2 (\theta - \alpha)} = \frac{\cosh v}{\cosh u}.$$

Substituting in (3.3), we have

$$d\theta = \frac{\cosh u}{\cosh v} d\varphi, \tag{3.4}$$

and the density (3.2), after the change of the variable θ for the variable φ, will take the form

$$dG = \sin \varphi \, ds \, d\varphi. \tag{3.5}$$

This expression of the density for sets of geodesics which cut a fixed curve C has the same form as for the density for straight lines of the plane [4; 13].

In order to get the measure of all the geodesics which cut a fixed curve C of length L we must integrate (3.5) with respect to s from O to L and with respect to φ from 0 to π; the value of the integral is $2L$. In this way, if the geodesic G cuts the curve C in n points it will be counted n times. As a consequence we have

$$\int_{G \cdot C \neq 0} n \, dG = 2L, \tag{3.6}$$

the integration being extended over all the geodesics which cut the curve C.

We say that a closed curve C is *convex* when it cannot be cut by any geodesic in more than two points. In this case, in (3.6), $n = 2$ always and we find that: *The measure of the geodesics cutting a convex curve is equal to the length of this curve.* This result and formula (3.6) have the same form for the plane [4; 11] and for the sphere [4; 81].

From (3.5) may be obtained also an integral formula which can be considered as the "dual" of (3.6). Multiplying both sides of (3.5) by the angle φ and integrating over all values of s and φ $(0 < \varphi < \pi)$ we find that the integral of the right side has the value

$$\int_0^L ds \int_0^\pi \varphi \sin \varphi \, d\varphi = \pi L$$

and in the left side for every position of G we must add the angles φ_i $(0 < \varphi_i \leq \pi)$ which G makes with C at the n intersections. Hence, we have

$$\int \sum_1^n \varphi_i \, dG = \pi L,$$

the integration being extended over all the geodesics which cut C.

4. Density by pairs of points and integral formula for chords. The density to measure sets of points is equal to the element of area; we shall represent it by dP. Let us consider a pair of points P_1, P_2. In order to measure sets of pairs of points we shall take for density the product of both densities, that is, $dP_1 dP_2$. Through P_1 and P_2 only one geodesic G may pass. Once this geodesic is fixed, the points P_1, P_2 are determined by their abscissa u_1, u_2 measured upon G from an arbitrary origin. We wish to express the product $dP_1 dP_2$ by means of dG and du_1, du_2.

In a system of polar geodesic coördinates the element of arc is expressed [2; 335] by

$$ds^2 = dr^2 + \sinh^2 r \, d\varphi^2 \tag{4.1}$$

and thus the element of area is $\sinh r \, dr \, d\varphi$. When P_1 is fixed, the differential of the area dP_2 expressed in the system of polar geodesic coördinates of origin P_1 is

$$dP_2 = \sinh r \, dr \, d\varphi, \tag{4.2}$$

r being the length of the arc of geodesic G which joins P_1 and P_2, that is, $r = | u_1 - u_2 |$. From any fixed point O we trace the geodesic normal to G and as in §2 and §3 we call A the foot of this perpendicular.

As before, we call v the arc OA and θ the angle made by OA with a fixed direction traced by O. P_1 is supposed to be fixed. In the expression (4.2) we may introduce the angle θ in the place of the angle φ because the relation between them is the same (3.4) already found in what precedes. Then we have

$$dP_2 = \sinh r \frac{\cosh v}{\cosh u_1} dr \, d\theta. \tag{4.3}$$

In the system of rectangular geodesic coördinates in which the curves $v =$ const. are the normal geodesics to OA and the curves $u =$ const. are their orthogonal trajectories, the differential ds has the form (3.1) and the differential of the area for the point P_1 is

$$dP_1 = \cosh u_1 \, du_1 \, dv_1 . \tag{4.4}$$

Taking into account (2.1) and substituting for symmetry du_2 instead of dr, from (4.3) and (4.4) we deduce

$$dP_1 \, dP_2 = \sinh r \, du_1 \, du_2 \, dG, \tag{4.5}$$

G being the geodesic which joins P_1 and P_2 and $r = | u_1 - u_2 |$ being the length of the arc P_1P_2. This differential formula (4.5) permits us to generalize to Crofton's formula for chords surfaces of constant negative curvature. Let C be a convex curve of area F; P_1, P_2 two interior points to C; and G the geodesic which joins them. We desire to integrate (4.5) over all pairs of points contained in C. The integral of the left side is evidently F^2. To calculate the integral of the right side we observe that, if σ represents the length of the arc of the geodesic G which is interior to C, we have

$$\int_0^\sigma \int_0^\sigma \sinh | u_1 - u_2 | \, du_1 \, du_2 = 2(\sinh \sigma - \sigma)$$

and consequently

$$\int_{G \cdot C \neq 0} (\sinh \sigma - \sigma) \, dG = \tfrac{1}{2} F^2. \tag{4.6}$$

This formula (4.6) is the generalization of Crofton's formula for chords.

If K is $-1/R^2$ instead of -1, we have

$$\int_{G \cdot C \neq 0} \left(\sinh \frac{\sigma}{R} - \frac{\sigma}{R}\right) \frac{dG}{R} = \frac{1}{2}\left(\frac{F}{R^2}\right)^2. \tag{4.7}$$

Multiplying both sides by R^4 and letting $R \to \infty$, we find

$$\int_{G \cdot C \neq 0} \sigma^3 \, dG = 3F^2, \tag{4.8}$$

dG being the density for straight lines on the plane and σ the length of the chord which the straight line G determines in the figure plane convex C of area F. This formula (4.8) is Crofton's integral for chords in plane geometry [4; 20], [7; 84].

5. **Density for pairs of geodesics which intersect.** If two geodesics G_1 and G_2 cut each other at a point P and if v_i and θ_i are the coördinates of G_i ($i = 1, 2$) with respect to the origin O (§2) in accordance with (2.1) we have $dG_i = \cosh v_i dv_i d\theta_i$. To measure a set of pairs of geodesics, we take the integral of the expression

$$dG_1 \, dG_2 = \cosh v_1 \cosh v_2 \, dv_1 \, d\theta_1 \, dv_2 \, d\theta_2 \, . \tag{5.1}$$

The geodesics G_1, G_2 may also be determined by their point of intersection P and the angles φ_1, φ_2 which they respectively make with a fixed direction at P. The angle $\varphi = |\varphi_1 - \varphi_2|$ is that formed by G_1 and G_2. If we take into account (3.1) when the geodesic v_1, θ_1 becomes the geodesic $v_1 + dv_1$, θ_1, the arc $u_1 =$ const. described by the point P has the length $\cosh u_1 \, dv_1$. On the other hand, if ds_2 is the arc described upon G_2 by the intersection of G_1 and G_2, the same arc is also equivalent to $\sin \varphi \, ds_2$. Consequently,

$$\cosh u_1 \, dv_1 = \sin \varphi \, ds_2 \, , \tag{5.2}$$

and by analogy

$$\cosh u_2 \, dv_2 = \sin \varphi \, ds_1 \, . \tag{5.3}$$

Also, if we suppose P fixed, the relation between the angles θ_i and φ_i is given by (3.4), that is,

$$\cosh v_i \, d\theta_i = \cosh u_i \, d\varphi_i \, . \tag{5.4}$$

From these equalities and from (5.1) we deduce $dG_1 dG_2 = \sin^2\varphi \, ds_1 ds_2 d\varphi_1 d\varphi_2$. But $\sin \varphi \, ds_1 \, ds_2$ is equal to the element of the area dP and consequently

$$dG_1 \, dG_2 = \sin \varphi \, d\varphi_1 \, d\varphi_2 \, dP. \tag{5.5}$$

This formula which expresses the product of the densities of two intersecting geodesics as a function of the density of their intersection point P and the densities of the angles φ_1, φ_2 at P has the same form as on the plane [4; 17] and on the sphere [4; 78].

In the cases of the plane and the sphere, as two geodesics always cut each other (the exception of straight parallel lines in the plane has no importance), integrating both sides of the formula (5.5) over all the pairs of geodesics which cut a convex curve C, we arrive at Crofton's fundamental formula [4; 18], [7; 78], [11]. For surfaces of constant negative curvature this reasoning cannot be applied because we may find sets of pairs of geodesics of finite measure which cut C without intersecting each other in any point P. Nevertheless formula (5.5) is of use in obtaining the following integral formula. Let us integrate the two sides of (5.5) over all the pairs of geodesics which intersect each other in the interior of a convex curve C of area F. The integral of the right side is

$$\int_{P<C} dP \int_0^\pi \int_0^\pi \sin |\varphi_1 - \varphi_2| \, d\varphi_1 \, d\varphi_2 = 2\pi F. \tag{5.6}$$

To calculate the integral of the left side we first fix G_1. If we call σ_1 the length of the arc of G_1 which is inside C, in accordance with (3.6) the integral of dG_2 extended over all the G_2 which cut σ_1 has value $2\sigma_1$. Thus the integral of the left side of (5.5) is equivalent to $2 \int \sigma_1 dG_1$. Equating to (5.6) and writing σ and G in place of σ_1 and G_1, we get the integral formula

$$\int_{G \cdot C \neq 0} \sigma \, dG = \pi F. \tag{5.7}$$

From this formula and from (4.6) we deduce

$$\int_{G \cdot C \neq 0} \sinh \sigma \, dG = \pi F + \tfrac{1}{2} F^2. \tag{5.8}$$

In (5.8) and (5.7) as in (4.6) σ is the length of the arc of the geodesic G which is inside C.

6. **Cinematic measure.** Hitherto we have only considered sets of points and geodesics. Now we wish to consider sets of elements each of which is formed by a point P and a direction φ at P. To measure a set of such elements we take the integral of the differential form

$$dC = dP \, d\varphi, \tag{6.1}$$

which is called cinematic density. For its definition on the plane and on the sphere, see [4; 20, 81].

On the surfaces of constant negative curvature two figures are called "congruent" if they can be superposed by a motion of the surface into itself [2; 333], [6; 409]. The position of a figure C is determined by fixing an element P, φ invariably bound to C. Consequently, the cinematic density serves also to measure any set of congruent figures.

Let C_0 be a fixed curve of length L_0 and C a mobile curve of length L. Suppose that both curves are formed by a finite number of arcs of continuous geodesic curvature. Calling n the number of intersection points of C and C_0, a number which depends on the position of C, we wish to calculate

$$I = \int_{C \cdot C_0 \neq 0} n \, dC, \tag{6.2}$$

where dC is the cinematic density (6.1) referred to the mobile curve C, and the integration is extended over all the positions of C.

We shall require a preliminary formula. Let C_0 be a fixed curve. For a point A of C_0 consider the geodesic which makes with C_0 an angle θ and upon this geodesic take an arc $AA' = r$. If the point A describes upon C_0 an arc $AB = ds_0$, θ and r remaining constant, the end A' will describe $A'B' = ds_0'$. Let θ' be the angle made by AA' with $A'B'$. The elements ds_0 and ds_0' may be considered to be in first approximation arcs of geodesics and accordingly we can apply formulas (1.1) of hyperbolic trigonometry. Considering only a first approximation, we have $\cosh AB = \cosh A'B' = 1$, $\sinh AB = ds_0$, $\sinh A'B' = ds_0'$. Thus the first formula (1.1) applied to the triangle $AB'A'$ gives

$$\cosh AB' = \cosh r + \sinh r \cos \theta' \, ds_0',$$

and applied to the triangle ABB',

$$\cosh AB' = \cosh r + \sinh r \cos \theta \, ds_0 .$$

From these equalities it follows that

$$\cos \theta' \, ds_0' = \cos \theta \, ds_0 . \tag{6.3}$$

This is the preliminary formula sought and it is verified whether the arcs of geodesic AA', BB' intersect or not.

We return now to the calculation of the integral (6.2). Let C, C_0 intersect in point A at angle α (Fig. 3). We fix at C and C_0 an origin of arcs, s being the

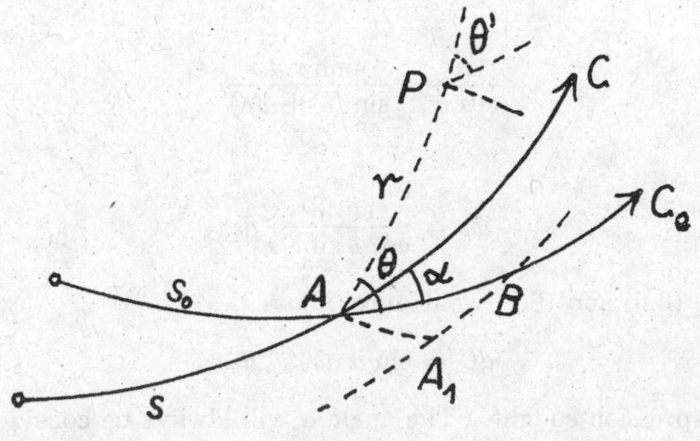

FIGURE 3

curvilinear abscissa of A upon C and s_0 the abscissa of A upon C_0. In order to determine the position of C, in place of P, φ which figures in (6.1), one may substitute s, s_0, α. We wish to express the cinematic density (6.1) with these new variables s, s_0, α. For this we must observe the following.

Let P be the point invariably bound to C which figures in (6.1), PA the geodesic arc which unites P with the intersection point A, and θ the angle which PA makes with the fixed curve C_0. If s and α are fixed and s_0 passes from s_0 to $s_0 + ds_0$, the angle θ will not vary and P will describe an element of arc ds_0' with value, according to (6.3),

$$ds_0' = \frac{\cos\theta}{\cos\theta'}\, ds_0, \tag{6.4}$$

where θ' is the angle formed by the prolongation of AP with the direction of ds_0'. Also, s and s_0 being fixed, when α passes to $\alpha + d\alpha$, the point P describes an arc ds_0'' normal to AP with value

$$ds_0'' = \sinh r\, d\alpha, \tag{6.5}$$

r being the length of the arc of geodesic PA. This value (6.5) is obtained from the expression of the element of the arc in polar geodesic coördinates (4.1). The angle formed by the elements ds_0' and ds_0'' is $\frac{1}{2}\pi - \theta'$ and as a consequence the element of area dP expressed by the coördinates s_0, α has the value $dP = \sin(\frac{1}{2}\pi - \theta')\, ds_0'\, ds_0''$, that is, according to (6.5) and (6.4),

$$dP = \cos\theta \sinh r\, ds_0\, d\alpha. \tag{6.6}$$

We shall suppose now that, having fixed P, we make PA rotate, and with it all the curve C, through an angle $d\varphi$. The point A will describe an arc AA_1 of a geodesic circle of center P, where $AA_1 = \sinh r\, d\varphi$. After turning through the angle $d\varphi$ the curve C will cut C_0 at the point B and the arc A_1B is the arc ds which has increased in passing from φ to $\varphi + d\varphi$. The infinitesimal triangle AA_1B may be considered a geodesic triangle and accordingly the second formula (1.1) gives

$$\frac{ds}{\cos\theta} = \frac{\sinh r\, d\varphi}{\sin(\alpha + d\alpha)},$$

that is,

$$d\varphi = \frac{\sin\alpha}{\cos\theta \sinh r}\, ds. \tag{6.7}$$

From (6.7), (6.6) and (6.1) we deduce

$$dC = \sin\alpha\, ds\, ds_0\, d\alpha. \tag{6.8}$$

This is the expression sought. The angle α will always be considered between 0 and π.

This expression (6.8) of the cinematic density has the same form as that for the plane [4; 23] and permits the immediate calculation of (6.2). Integrating (6.8) over all the values of s, s_0, α, we shall have the integral of dC extended over all the positions in which C cuts C_0, but if in some position C and C_0 intersect in n points, this position will have been counted n times. Consequently,

$$I = \int_{C \cdot C_0 \neq 0} n\, dC = \int_0^L ds \int_0^{L_0} ds_0 \int_{-\pi}^{\pi} \sin |\alpha|\, d\alpha = 4LL_0 ,$$

that is,

$$\int_{C \cdot C_0 \neq 0} n\, dC = 4LL_0 . \tag{6.9}$$

This formula expresses the generalization to the surfaces of constant negative curvature of Poincaré's formula and it has the same form as in the case of the plane [4; 24] and the sphere [4; 81].

The expression (6.8) also permits us to obtain an integral formula which, in a certain form, is the dual formula of (6.9). If we multiply both sides of (6.8) by α and integrate over all the values of s, s_0, α $(0 < \alpha < \pi)$, the sum $\sum_1^n \alpha_i$ of the angles at which the curves C and C_0 intersect will appear on the left side and the integral of the right side will have the value

$$\int_0^L ds \int_0^{L_0} ds_0 \int_{-\pi}^{\pi} | \alpha \sin \alpha |\, d\alpha = 2\pi LL_0 .$$

Consequently,

$$\int_{C \cdot C_0 \neq 0} \sum_1^n \alpha_i\, dC = 2\pi LL_0 . \tag{6.10}$$

It should be noted that this formula also has the same form as in the case of the plane [9; 101] and of the sphere [4; 82].

7. Fundamental formula of cinematic measure. On the surface of constant negative curvature $K = -1$, let us consider a closed curve C_1 of length L_1 without double points and formed by a finite number of arcs of continuous geodesic curvature. Let F_1 be the area bounded by C_1. The total geodesic curvature K_1 of C_1 is composed of the sum of the integrals $\int \kappa_g^1\, ds_1$ of the geodesic curvature along the arcs which form C_1 plus the sum of the exterior angles at the angular points if these appear. Then the Gauss-Bonnet formula gives

$$K_1 = 2\pi + F_1 . \tag{7.1}$$

If C_0 is another closed curve of area F_0 with length L_0 and total geodesic curvature K_0, then $K_0 = 2\pi + F_0$ also. Suppose C_0 fixed and C_1 of variable position.

In each position of C_1 the intersection of the domains bounded by C_0 and C_1 will be composed of a certain number of partial domains whose boundaries are formed by arcs of C_0 and C_1. We represent by F_{01} the area, by L_{01} the length and by K_{01} the total geodesic curvature of the domain C_{01} intersection of the domains bounded by C_0 and C_1. C_{01} may be multiply connected (Fig. 4).

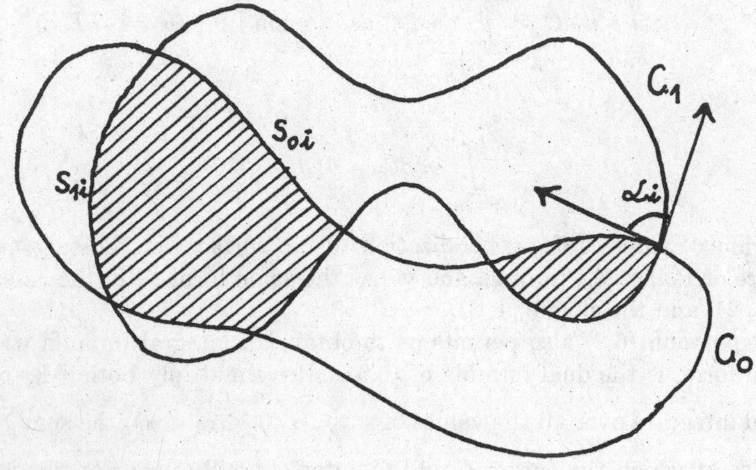

FIGURE 4

We wish to demonstrate the integral formula

$$\int_{C_0 \cdot C_1 \neq 0} K_{01} \, dC_1 = 2\pi(K_0 F_1 + K_1 F_0 + L_0 L_1), \tag{7.2}$$

where dC_1 is the cinematic density (6.1) with reference to the mobile figure C_1, the integration being extended over all the positions of C_1 in which the domain bounded by this curve has any common point with that bounded by C_0.

Formula (7.2) is the generalization on the surfaces of curvature $K = -1$ (that is, the generalization to hyperbolic geometry) of Blaschke's fundamental formula of integral plane geometry. The proof we shall give is analogous to that given for the plane by Maak [9] and Blaschke [4; 37].

Calling s_{0i}, s_{1i} the lengths of the arcs of C_0 and C_1 which contribute to form the boundary of C_{01} and α_i the angles in which C_0 and C_1 intersect, by definition of K_{01} we have

$$K_{01} = \sum_i \int_{s_{0i}} \kappa_g^0 \, ds_0 + \sum_i \int_{s_{1i}} \kappa_g^1 \, ds_1 + \sum_i \alpha_i , \tag{7.3}$$

κ_g^0 and κ_g^1 being the geodesic curvatures of C_0 and C_1. Let us consider the integral $I_0 = \int \kappa_g^0 \, ds_0 \, dC_1$ extended over all the positions in which the point s_0 belongs to the boundary of C_0 and is contained in the interior of C_1. This integral I_0 may be calculated in two ways. Having first fixed the point s_0, we must integrate dC_1 over all the positions of C_1 for which this fixed point is interior to C_1 ; according to (6.1) this integral has the value $2\pi F_1$; there

$$\int_{C_0} \kappa_g^0 \, ds_0$$

of value K_0 remains. Consequently, $I_0 = 2\pi F_1 K_0$. The same integral may be calculated in another way. If we first fix C_1 , the integral $\int \kappa_g^0 \, ds_0$ extended over all the values of s_0 which are interior to C_1 is the sum

$$\sum_i \int_{s_{0i}} \kappa_g^0 \, ds_0$$

which appears in (7.3). We must now integrate the product of this sum by dC_1 . Equating this value of I_0 to that found before, we have

$$\int_{C_0 \cdot C_1 \neq 0} \sum_i \int_{s_{0i}} \kappa_g^0 \, ds_0 \, dC_1 = 2\pi F_1 K_0 ,$$

and analogously, for symmetry, must be

$$\int_{C_0 \cdot C_1 \neq 0} \sum_i \int_{s_{1i}} \kappa_g^1 \, ds_1 \, dC_1 = 2\pi F_0 K_1 .$$

Taking into account these values and (6.10), we have formula (7.2), which we wished to prove.

Formula (7.2) may be written in a more convenient form. For this we must calculate the integral $\int F_{01} \, dC_1$ in which F_{01} is the area of the intersection of C_0 and C_1 and the integration is extended over all the positions for which $C_{01} = C_0 \cdot C_1 \neq 0$. Let us consider the integral $I_{01} = \int dP_0 \, dC_1$, in which dP_0 is the element of area, extended over all the positions in which C_1 contains the point P_0 interior to C_0 . Having fixed P_0 , we find that the integral of dC_1 has a value $2\pi F_1$ and when P_0 is varied over all the interior of C_0 we obtain $I_{01} = 2\pi F_0 F_1$. Also if we fix C_1 first, the point P_0 can vary over all the points of the intersection of C_0 and C_1 . The integral of dP_0 will then be F_{01} and consequently $I_{01} = \int F_{01} \, dC_1$. Equating the two values obtained for I_{01} , we have

$$\int_{C_0 \cdot C_1 \neq 0} F_{01} \, dC_1 = 2\pi F_0 F_1 . \tag{7.4}$$

By the Gauss-Bonnet theorem, if the intersection of C_0 and C_1 is composed of ν simply connected pieces (for example, in Fig. 4, $\nu = 2$), we have $K_{01} = 2\pi\nu + F_{01}$ and moreover $K_0 = 2\pi + F_0$, $K_1 = 2\pi + F_1$. Substituting these values in (7.2) and taking into account (7.4), we find

$$\int_{C_0 \cdot C_1 \neq 0} \nu \, dC_1 = 2\pi(F_0 + F_1) + F_0 F_1 + L_0 L_1 . \tag{7.5}$$

In particular, if C_0 and C_1 are convex, their intersection is always simply connected, that is, composed of only one piece. As a consequence $\nu = 1$ and we have the result:

The measure of the positions of a convex figure C_1 in which it has some common point with another convex figure C_0 has the value

$$\int_{C_0 \cdot C_1 \neq 0} dC_1 = 2\pi(F_0 + F_1) + F_0F_1 + L_0L_1 . \tag{7.6}$$

In the following sections we shall apply this formula and (6.9).

8. **Isoperimetric propriety of geodesic circles.** On the surface of constant negative curvature $K = -1$, let us consider a closed curve C which has no double points, and which has length L and area F. We consider the set of curves congruent to C which have points in common with C. Calling M_i the cinematic measure of the set of these curves which have i points in common with C, we can write formula (7.6) as

$$M_2 + M_4 + M_6 + \cdots = 4\pi F + F^2 + L^2, \tag{8.1}$$

since now $C_0 = C_1 = C$. Analogously, formula (6.9) gives

$$2M_2 + 4M_4 + 6M_6 + \cdots = 4L^2. \tag{8.2}$$

From these two equalities we deduce

$$L^2 - F^2 - 4\pi F = M_4 + 2M_6 + 3M_8 + \cdots \tag{8.3}$$

and, as the M_i, which are the measure of certain sets, are always non-negative, we have

$$L^2 - F^2 - 4\pi F \geq 0. \tag{8.4}$$

This is the isoperimetric inequality on surfaces of constant negative curvature $K = -1$. In fact, from (8.4) can be deduced that for all the curves which limit an area F the minimum value of the length is $(F^2 + 4\pi F)^{\frac{1}{2}}$. This minimum value L_0 is reached by the geodesic circles. Hence if C_0 is a geodesic circle of radius ρ_0 we have [6; 404]

$$L_0 = 2\pi \sinh \rho_0 , \qquad F_0 = 2\pi(\cosh \rho_0 - 1) \tag{8.5}$$

and therefore $L_0^2 = F_0^2 + 4\pi F_0$.

This proof of the isoperimetric inequality (8.4) does not permit the assertion that the geodesic circles are the only figures for which the equality in (8.4) is valid. For this we shall give another proof leading to an inequality stronger than (8.4).

Let ρ_0 have such a value that no geodesic circle of radius ρ_0 is contained in the interior of C nor contains C in its own interior. Also let C_0 be the geodesic circle of radius ρ_0. Calling M_i the measure of the set of circles C_0 which intersect C in i points, in accordance with (7.6) and (6.9) we have

$$M_2 + M_4 + M_6 + \cdots = 2\pi(F + F_0) + FF_0 + LL_0, \tag{8.6}$$

and

$$2M_2 + 4M_4 + 6M_6 + \cdots = 4LL_0. \tag{8.7}$$

From these equalities we deduce

$$LL_0 - FF_0 - 2\pi(F + F_0) = M_4 + 2M_6 + 3M_8 + \cdots \geq 0. \tag{8.8}$$

To abbreviate, we put

$$\Delta = L^2 - F^2 - 4\pi F, \tag{8.9}$$

where Δ is the "isoperimetric deficit". F_0 and L_0 given by (8.5) thus satisfy

$$L_0^2 - F_0^2 - 4\pi F_0 = 0, \tag{8.10}$$

with which we easily prove the identity

$$\frac{1}{2FF_0}[\Delta F_0^2 - (LF_0 - FL_0)^2] = LL_0 - FF_0 - 2\pi(F + F_0). \tag{8.11}$$

Taking into account (8.8), we deduce from (8.11) that

$$\Delta \geq \frac{1}{F_0^2}(LF_0 - FL_0)^2,$$

or by substituting for F_0, L_0 their values (8.5), we get

$$\Delta \geq (L - F \coth \tfrac{1}{2}\rho_0)^2. \tag{8.12}$$

This inequality is verified for any ρ_0 so that no circle of radius ρ_0 could contain C or itself be contained in C. In particular, if ρ_e is the minimum radius of the geodesic circles which enclose C and ρ_i the maximum of those contained in the interior of C, we have

$$\Delta \geq (L - F \coth \tfrac{1}{2}\rho_e)^2, \qquad \Delta \geq (F \coth \tfrac{1}{2}\rho_i - L)^2, \tag{8.13}$$

and taking into account the inequality

$$x^2 + y^2 \geq \tfrac{1}{2}(x + y)^2, \tag{8.14}$$

from (8.13) we deduce

$$\Delta \geq \tfrac{1}{4}F^2(\coth \tfrac{1}{2}\rho_i - \coth \tfrac{1}{2}\rho_e)^2. \tag{8.15}$$

Analogously, taking into account (8.10), we easily prove the identity

$$\frac{4\pi + F_0}{2L_0^2(4\pi + F)}[\Delta L_0^2 - ((4\pi + F)F_0 - LL_0)^2] = LL_0 - FF_0 - 2\pi(F + F_0), \tag{8.16}$$

and consequently, in accordance with (8.8), we have

$$\Delta \geq \frac{1}{L_0^2}((4\pi + F)F_0 - LL_0)^2. \tag{8.17}$$

If we substitute the values (8.5), we may write this inequality

$$\Delta \geq ((4\pi + F)\tanh \tfrac{1}{2}\rho_0 - L)^2. \tag{8.18}$$

Writing this inequality for ρ_e and ρ_i and taking into account (8.14), we deduce

$$\Delta \geq \tfrac{1}{4}(4\pi + F)^2(\tanh \tfrac{1}{2}\rho_e - \tanh \tfrac{1}{2}\rho_i)^2. \tag{8.19}$$

The isoperimetric inequalities (8.15) and (8.19) are stronger than (8.4). They make clear that the equality $\Delta = 0$ can be verified only when $\rho_i = \rho_e$, that is, when C is a geodesic circle. It is thus completely proved that on surfaces of constant negative curvature the geodesic circles are the only curves which for a specified length enclose maximum area. A direct proof of the inequalities (8.15) and (8.19) was given by us in [12]. The isoperimetric problem on the surfaces of constant negative curvature has also been solved in a completely distinct manner in [13].

For a surface of constant curvature $K = -1/R^2$, the inequalities (8.15) and (8.19) are written respectively

$$\left(\frac{L}{R}\right)^2 - \left(\frac{F}{R^2}\right)^2 - 4\pi\frac{F}{R^2} \geq \frac{1}{4}\frac{F^2}{R^2}\left(\frac{1}{R}\coth\frac{\rho_i}{2R} - \frac{1}{R}\coth\frac{\rho_e}{2R}\right)^2,$$

$$\left(\frac{L}{R}\right)^2 - \left(\frac{F}{R^2}\right)^2 - 4\pi\frac{F}{R^2} \geq \frac{1}{4}\left(4\pi + \frac{F}{R^2}\right)^2\left(\tanh\frac{\rho_e}{2R} - \tanh\frac{\rho_i}{2R}\right)^2.$$

Multiplying by R^2 and making $R \to \infty$, we obtain

$$L^2 - 4\pi F \geq F^2\left(\frac{1}{\rho_i} - \frac{1}{\rho_e}\right)^2 \tag{8.20}$$

and

$$L^2 - 4\pi F \geq \pi^2(\rho_e - \rho_i)^2, \tag{8.21}$$

which are isoperimetric inequalities for plane figures. In these, ρ_i is the maximum radius of those circles which are contained in C and ρ_e the minimum of these which contain C. Inequality (8.21) is a classic inequality due to Bonnesen [5; 63].

9. **A sufficient condition that a convex curve congruent to C be contained in the interior of another convex curve C_0.** Let C_1 be a convex curve of length L_1 which limits a domain of area F_1. C_0 is another convex curve of length L_0 and area F_0. We wish to find a sufficient condition that a curve congruent to C_1 be contained in the interior of C_0. As in §8, let M_i be the measure of the set of curves congruent to C_1 which intersect C_0 in i points. M_0 is the measure of the set of curves congruent to C_1 which are in the interior of C_0 or which contain C_0.

According to (7.6) and (6.9) we have

$$M_0 + M_2 + M_4 + M_6 + \cdots = 2\pi(F_0 + F_1) + F_0F_1 + L_0L_1 \tag{9.1}$$

and

$$2M_2 + 4M_4 + 6M_6 + \cdots = 4L_0L_1 . \tag{9.2}$$

From these equalities

$$2\pi(F_0 + F_1) + F_0F_1 - L_0L_1 = M_0 - M_4 - 2M_6 - \cdots . \tag{9.3}$$

If $M_0 = 0$, the left side of this equality is non-positive. Hence a sufficient condition that a curve congruent to C_1 contain C_0 or be contained in C_0 is

$$2\pi(F_0 + F_1) + F_0F_1 - L_0L_1 > 0. \tag{9.4}$$

In order to sharpen this result, we first observe that by (8.4) for any C_0 and C_1

$$L_0^2 \geq F_0^2 + 4\pi F_0 , \qquad L_1^2 \geq F_1^2 + 4\pi F_1$$

and hence

$$L_0^2L_1^2 \geq F_0F_1(4\pi + F_0)(4\pi + F_1). \tag{9.5}$$

Consider the inequality

$$L_0L_1 - F_1(4\pi + F_0) > [L_0^2L_1^2 - F_0F_1(4\pi + F_0)(4\pi + F_1)]^{\frac{1}{2}}, \tag{9.6}$$

whose right side is always real by (9.5). If we square and simplify (9.6), we see that (9.4) is also verified. Consequently, one of the two curves C_0 or C_1 can be contained in the interior of the other. We shall prove that if (9.6) is verified, $F_1 < F_0$ and consequently C_1 can be contained in C_0. In fact, if $F_1 \geq F_0$, in accordance with (9.4), which is a consequence of (9.6), we have $L_0L_1 < 4\pi F_1 + F_0F_1$ and the inequality (9.6) is not verified since the left side must be positive. Consequently,

C_0 and C_1 being two convex curves on the surface of curvature $K = -1$, the inequality (9.6) *is a sufficient (but not necessary) condition that a curve congruent to C_1 be contained in the interior of C_0.*

In particular, if C_1 is a geodesic circle of radius ρ_1, taking into account the values (8.5), we can write the inequality (9.6) as

$$L_0 - (4\pi + F_0) \tanh \tfrac{1}{2}\rho_1 > (L_0^2 - F_0(4\pi + F_0))^{\frac{1}{2}} \tag{9.7}$$

and this inequality is a sufficient condition that C_0 contain in its interior a geodesic circle of radius ρ_1.

Analogously,

$$L_1 - F_1 \coth \tfrac{1}{2}\rho_0 > (L_1^2 - F_1(4\pi + F_1))^{\frac{1}{2}} \tag{9.8}$$

is a sufficient condition that C_1 be contained in the interior of a geodesic circle of radius ρ_0.

If the curvature of the surface is $K = -1/R^2$, the corresponding condition (9.6) can be written without difficulty. Multiplying both sides by R^2 and making $R \to \infty$, we obtain

$$L_0L_1 - 4\pi F_1 > (L_0^2L_1^2 - 16\pi^2F_0F_1)^{\frac{1}{2}}, \tag{9.9}$$

which is a sufficient condition that a plane convex curve congruent to C_1 of area F_1 and length L_1 be contained in the interior of C_0 whose area and length are F_0 and L_0 respectively.

Condition (9.9) for the plane has been obtained by H. Hadwiger [8]. For the analogous condition for the curves on the sphere see [10].

BIBLIOGRAPHY

1. E. Beltrami, *Saggio di interpretazione della Geometria non-Euclidea*, Giornale di Matematiche, vol. VI(1868), pp. 285–315; Opere matematiche de E. Beltrami, tomo I, p. 374.
2. L. Bianchi, *Lezione di Geometria Differenziale*, vol. I, Terza edizione, Bologna, 1927.
3. W. Blaschke, *Vorlesungen über Differentialgeometrie*, vol. I, J. Springer, Berlin, 1929.
4. W. Blaschke, *Vorlesungen über Integralgeometrie*, Hamburger Mathematische Einzelschriften, Heften 20–22, 1935–1937.
5. T. Bonnesen, *Les problèmes des isopérimètres et des isépiphanes*, Paris, 1929.
6. G. Darboux, *Leçons sur la théorie générale des surfaces*, Troisième partie, Paris, 1894.
7. R. Deltheil, *Probabilités géométriques*, Traité du calcul des probabilités et de ses applications publié sous la direction de E. Borel, tome II, fasc. II, Gauthier-Villars, Paris, 1926.
8. H. Hadwiger, *Ueberdeckung ebener Bereiche durch Kreise und Quadrate*, Commentarii Mathematici Helvetici, vol. 13(1941), pp. 195–200.
9. W. Maak, *Integralgeometrie 18, Grundlagen der ebenen Integralgeometrie*, Abhandlungen aus dem Mathematischen Seminar der Hansischen Universität, vol. 12(1937), pp. 83–110.
10. L. A. Santaló, *Algunos valores medios y desigualdades referentes a curves situades sobre la superficie de la esfera*, Revista de la Union Matematica Argentina, vol. VIII(1942).
11. L. A. Santaló, *Integral formulas in Crofton's style on the sphere and some inequalities referring to spherical curves*, this Journal, vol. 9(1942), pp. 707–722.
12. L. A. Santaló, *La desigualdad isoperimetrica sobre las superficies de curvatura constante negativa*, Revista de Matematicas y Fisica teorica, Tucuman, vol. 3(1942), pp. 243–259.
13. E. Schmidt, *Ueber die isoperimetrische Aufgabe im n-dimensionalen Raum konstanter negativer Krümmung*, Mathematische Zeitschrift, vol. 46(1940), pp. 204–230.

Mathematical Institute, Rosario, Argentina.

NOTE ON CONVEX SPHERICAL CURVES

L. A. SANTALÓ

1. Introduction. The formula

$$L = \int_0^\pi \alpha d\tau \tag{1}$$

for plane convex curves in which L is the length and α the breadth according to the direction τ is well known [2, p. 65].[1]

The principal object of the present note is to obtain the formula (8) which generalizes (1) to convex curves on the sphere of unit radius and to deduce from this some consequences.

2. Principal formula. Let us consider the sphere of unit radius. A closed curve on the sphere is said to be *convex* when it cannot be cut by a great circle in more than two points. It is well known that a convex curve divides the surface of the sphere into two parts, one of which is always wholly contained in a hemisphere; that is, there is always a great circle which has the whole convex curve on the same side. When we say the area of a convex curve K we understand the area of that part of the surface of the sphere which is bounded by K and is smaller than or equal to a hemisphere.

Let K be a convex curve on the sphere of unit radius of length L and area F ($L \leqq 2\pi$, $F \leqq 2\pi$). The great circles which have only one common point or include a complete segment common with the curve

Received by the editors January 21, 1944.

[1] Numbers in brackets refer to the references cited at the end of the paper.

K are called "great circles of support" of K. In each point of K for which there exists a tangent great circle the great circle of support coincides with this.

Let τ be the angle between a fixed and a variable great circle of support and s the arc of K. If ρ_g is the radius of geodesic curvature of K the following formulas are known [4, p. 712],

$$(2) \qquad \rho_g = ds/d\tau, \qquad \int_L ds/\rho_g = \int_K d\tau = 2\pi - F.$$

Let C be a great circle which cuts K. We suppose C "oriented" and to each orientation we make correspond one of the points P_c of the two in which the diameter perpendicular to C cuts the sphere. If we consider all the great circles which cut K (each one of these counted twice to correspond with the two orientations) the area covered by the points P_c has the value $2L$ [4, p. 709]; that is to say, if dP_c represents the element of area of the sphere of unit radius corresponding to the point P_c,

$$(3) \qquad \int_{C\cdot K\neq 0} dP_c = 2L.$$

The convex curve K being fixed, the oriented great circle C or, what is the same, the point P_c, can be determined by the point A of K in which the great circle orthogonal to a circle of support of K is also orthogonal to C, and by the distance a from A to C (Fig. 1). If τ is

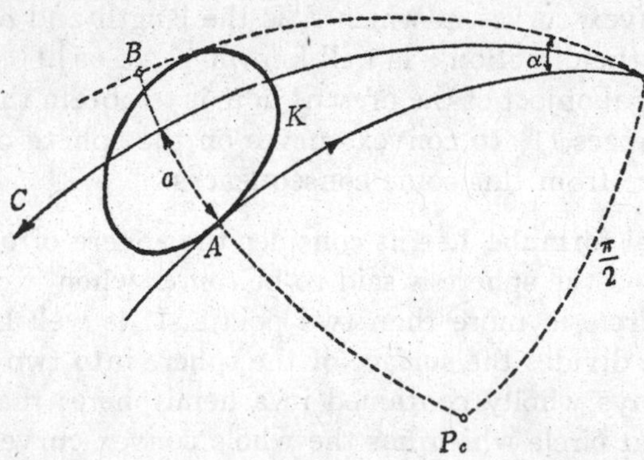

FIG. 1

the angle which determines the great circle of support of K in the point A, the coordinates to determine P_c will be a, τ. We must then express the element of area P_c as a function of the coordinates a, τ.

Suppose first that a remains fixed and τ passes to $\tau+d\tau$. The point P_c will describe an arc $P_cP_c'=d\sigma$, the value of which is (Fig. 2)

$$d\sigma = \sin(\pi/2 - a + \phi)d\psi = \cos(a-\phi)d\psi, \tag{4}$$

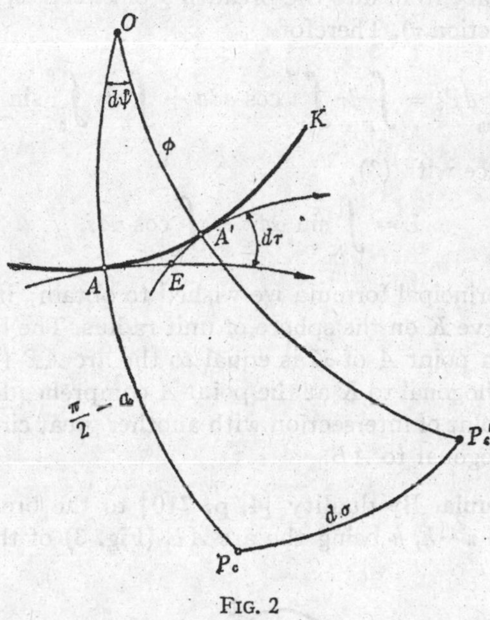

FIG. 2

$\phi=AO$ being the radius of spherical curvature of K at the point A. As a function of the radius of geodesic curvature it is

$$\tan\phi = \rho_g = ds/d\tau. \tag{5}$$

The element $d\psi=AOA'$ is the angle between the great circles orthogonal to K in A and A'. The area of the quadrilateral $AOA'E$ has the value $(\pi-d\tau)+d\psi+\pi-2\pi=d\psi-d\tau$. But save for infinitesimals of secondary order the same area is also equal to the area of the circular sector AOA' which has the value $(1-\cos\phi)d\psi$. In consequence

$$d\tau = \cos\phi d\psi. \tag{6}$$

If τ remains fixed and a passes to $a+da$, the point P_c describes an arc da upon the great circle P_cA. Therefore, taking into account (4) and (6) the element of area dP_c will be expressed

$$dP_c = d\sigma da = (\cos(a-\phi)/\cos\phi)da d\tau,$$

that is to say, in accordance with (5),

$$dP_c = \cos a da d\tau + \sin a da ds. \tag{7}$$

Let us substitute this expression in (3). For each value of s (or τ) the arc a can vary from 0 to the breadth α of K corresponding to the point s (or direction τ). Therefore

$$\int_{C \cdot K \neq 0} dP_c = \int_K d\tau \int_0^\alpha \cos a da + \int_L ds \int_0^\alpha \sin a da,$$

or, in accordance with (3),

$$L = \int_K \sin \alpha d\tau - \int_L \cos \alpha ds. \tag{8}$$

This is the principal formula we wished to obtain; it is valid for any convex curve K on the sphere of unit radius. The breadth α corresponding to a point A of K is equal to the arc AB (Fig. 1) of the great circle orthogonal to K at the point A comprehended between A and the next point of intersection with another great circle of support of K also orthogonal to AB.

3. Dual formula. By duality [4, p. 710] to the breadth α corresponds the arc $\pi - h$, h being the arc AE (Fig. 3) of the great circle

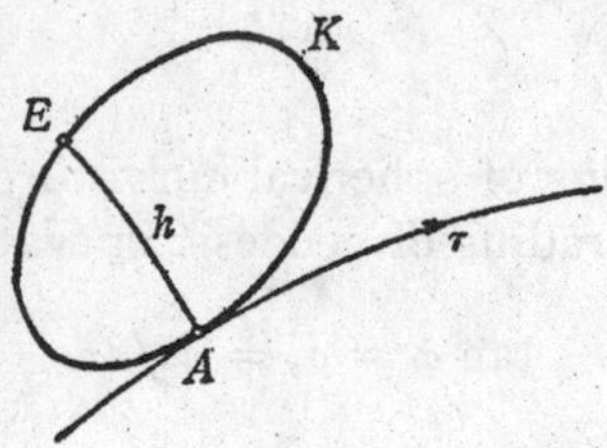

FIG. 3

orthogonal to K at the point A which is interior to K. Moreover the length L must be replaced by $2\pi - F$, ds by $d\tau$ and $d\tau$ by ds. Therefore transforming by duality the formula (8) we shall obtain

$$2\pi - F = \int_L \sin h ds + \int_K \cos h d\tau. \tag{9}$$

4. Consequences. (a) Let β be the minimum breadth of K, that is to say the minimum value of α, and δ the maximum value of α; δ is equal to the "diameter" of K. Let us suppose that $\beta \leqq \alpha \leqq \delta \leqq \pi/2$. Then $\sin \delta \geqq \sin \alpha$, $\cos \delta \leqq \cos \alpha$, and from (8) and (2) we deduce

$$L \leqq (2\pi - F) \sin \delta - L \cos \delta$$

from which we have

$$(10) \qquad L/(2\pi - F) \leqq \tan (\delta/2).$$

Similarly, $\sin \alpha \geqq \sin \beta$, $\cos \alpha \leqq \cos \beta$, and consequently

$$(11) \qquad L/(2\pi - F) \geqq \tan (\beta/2).$$

Therefore: *On the sphere of unit radius for any convex curve K of minimum breadth β and diameter $\delta \leqq \pi/2$, the inequalities* (10) *and* (11) *are verified.*

(b). In (10) and (11) there is equality only when $\alpha = \beta = \delta = \text{const.}$, that is to say for the cases of *convex curves of constant breadth.* These curves of constant breadth on the sphere have been studied by Blaschke [1]. For such curves it is deduced from (8) that

$$(12) \qquad L = (2\pi - F) \tan (\alpha/2).$$

This formula, obtained also by Blaschke [1], appears here as a particular case of our general formula (8).

(c). It is well known that isoperimetric inequality $L^2 + (2\pi - F)^2 \geqq 4\pi^2$ holds for any convex curve K on the unit sphere and that the equality is only true for the circles [4, p. 718]. This isoperimetric inequality and (10), (11) give us

$$(13) \qquad F \leqq 2\pi(1 - \cos (\delta/2)), \qquad L \geqq 2\pi \sin (\beta/2).$$

Consequently: *upon the sphere, given the diameter $\delta \leqq \pi/2$ the convex curve of maximum area is the circle of radius $\delta/2$; given the minimum breadth β the curve of minimum length is the circle of radius $\beta/2$.*

(d). If the convex curve K has continuous geodesic curvature, from (9) and (2) we deduce

$$(14) \qquad \int_L \sin (h/2)(\rho_g^{-1} \sin (h/2) - \cos (h/2)) ds = 0$$

and we therefore have the theorem: *In any convex curve on the sphere of unit radius with continuous radius of geodesic curvature there are at least two points for which*

$$\rho_g = \tan (h/2),$$

h being the chord of K orthogonal to K at the point considered.

Analogously from (8) and (2) we deduce

$$(15) \qquad \int_L \cos (\alpha/2)(\rho_g^{-1} \sin (\alpha/2) - \cos (\alpha/2)) ds = 0$$

and therefore: *in any convex curve on the sphere of unit radius with continuous radius of geodesic curvature there are at least two points for which*

$$\rho_g = \tan(\alpha/2),$$

α being the breadth of K corresponding to the point considered.

5. **Passage to the case of the plane.** Considering a sphere of radius R and making $R\to\infty$, the anterior formulas transform themselves into valid formulas for plane convex curves.

The principal formula (8) if K belongs to a sphere of radius R is written

$$\frac{L}{R} = \int_K \sin\frac{\alpha}{R}\,d\tau - \int_L \cos\frac{\alpha}{R}\frac{ds}{R};$$

multiplying both sides by R and making $R\to\infty$ we obtain the well known formula (1).

The formula (9) taking into account (2) may be written

$$\int_K \sin^2\frac{h}{2}\,d\tau = \int_L \sin\frac{h}{2}\cos\frac{h}{2}\,ds.$$

Writing this formula for a sphere of radius R, multiplying afterwards both sides by R^2 and making $R\to\infty$ we obtain

$$J = \int_L h ds = 2^{-1}\int_0^{2\pi} h^2 d\tau. \tag{16}$$

In this formula h is the chord of the plane convex curve K normal to this at the point s (or τ).

The element hds which appears in the first expression of J in (16) is the area of a strip of K with the height h and base ds. If a point $P(x, y)$ interior to K belongs to ν of these strips, from P there will be ν normals to K and the element of area $dxdy$ will be counted ν times in (16). Hence, if ν has a finite value for any P interior to K, the integral J is also equal to

$$J = \iint_F \nu dxdy, \tag{17}$$

ν being the number of normals to K from the point $P(x, y)$ interior to K.

If there are points with $\nu=\infty$ (for instance the case of the circle) the integral (17) must be considered as the limit of the similar ones referring to convex curves K_n (with ν always finite) which tend to K.

It is interesting to study the limits between which the integral J

is bounded. From any point $P(x, y)$ interior to K may be traced always at least two normals to K, joining in straight lines P with the points of K of minimum and maximum distance to P. Consequently

$$J \geqq 2F. \tag{18}$$

From another aspect, the second expression for J in (16) shows that J is equal to the area that is obtained by extending from a fixed point of the plane segments of length h parallel to the direction τ; as h is less than or equal to the length of the maximum chord of K which also has direction perpendicular to the line of support of direction τ, a known theorem of Rademacher about vector regions of convex curves [**3**; **2**, p. 105] tells us that

$$J \leqq 6F. \tag{19}$$

(18) cannot be bettered, it being sufficient to consider a triangle with an angle approximating to π in order that J comes near $2F$ in such degree as we should wish.

The inequality (19), on the contrary, is probably excessive. For the convex curves of constant breadth α we have

$$J = L\alpha = \pi\alpha^2. \tag{20}$$

If we recall that the triangle of Reuleaux [**2**, p. 132] is the figure which, given the constant breadth α, has minimum area of value $(\pi - 3^{1/2})\alpha^2/2$, we see that $F \geqq (\pi - 3^{1/2})\alpha^2/2$ and consequently from (20) we deduce

$$J \leqq 2\pi F/(\pi - 3^{1/2}) = 4{,}4576 \cdots F. \tag{21}$$

It is probable that this superior bound is valid not only for convex curves of constant breadth but also for any convex curve of the plane. Nevertheless I have not been able to prove this last assertion.

References

1. Blaschke, W. *Einige Bemerkungen über Kurven und Flächen konstanter Breite.* Bericht über die Verhandlungen der Sächsischen Akademie der Wissenschaften vol. 67 (1915).

2. Bonnesen, T., and Fenchel, W., *Theorie der konvexen Körper*, Ergebnisse der Mathematik und ihrer Grenzgebiete, Berlin, 1934.

3. Rademacher, H., *Ueber den Vektorenbereich eines konvexen ebenen Bereiches*, Jber. Deutschen Math. Verein. vol. 34 (1925).

4. Santaló, L. A., *Integral formulas in Crofton's style on the sphere and some inequalities referring to spherical curves*, Duke Math. J. vol. 9 (1942) pp. 707–722.

Universidad Nacional del Litoral, Rosario

[illegible]

[illegible]

[illegible] point of S [illegible] distance to P. [illegible]

[illegible]

[illegible] of the [illegible] the [illegible] which [illegible]

[illegible]

[illegible] sufficient to [illegible] a single [illegible] than [illegible]

[illegible]

[illegible]

[illegible]

It is probable that this equation [illegible] valid [illegible] plane [illegible]

References

[illegible]

Mean value of the number of regions in which a body in space is divided by n arbitray planes

L. A. Santaló

In a previous paper [5], the present author found the mean value of the number of parts into which a convex figure in the plane is divided by n arbitrary straight lines.

We now wish to generalize that result to the case of a body in space intersected by n arbitrary planes. We assume the body to have the connection of the sphere, but without necessarily being convex.

1. Let us first recall some known formulae. A plane E is determined in space by its distance p from a fixed point and by the direction of its normal; if a radius parallel to the said normal direction is drawn through the centre of a unit sphere, this direction will be defined by the point Ω at which the said radius intersects the surface of the sphere. If $d\Omega$ is the area element of the unit sphere that corresponds to the point Ω, then to measure sets of planes in space we take the integral, extended to the set, with the differential form

$$dE = dp\, d\Omega, \tag{1}$$

which is called the *plane density* (see, for example, [4], pag. 92 and also [2], pag. 66).

Let K be the body under consideration, whose volume we denote by V and its area by F. We also require the invariant M, which we refer to as the *mean curvature integral* and whose meaning we recall in the different cases. If the surface limiting K has a continuous curvature, M can be defined as the integral of its mean curvature:

$$M = \frac{1}{2}\int (\frac{1}{R_1} + \frac{1}{R_2}) d\sigma,$$

where R_1 and R_2 are the main radii of curvature corresponding to the surface element $d\sigma$, and taking the integration as extended to the whole surface of K. If K is a polyhedron, by calling the length of its edges l_i and the corresponding internal dihedral angle α_i, we have

$$M = \frac{1}{2}\sum l_i \alpha_i, \tag{2}$$

the sum being extended to all the edges of K (in this case, M is the curvature on the edges, or Steiner's *Kantenkrümmung*, [2], pag. 94). If K is a convex body, M is the measure of the set of planes having some common point with this convex body; if K is reduced to a plane convex figure of perimeter u, we have $M = \frac{\pi}{2}u$, and by extension to the convex bodies in space, Cartan proposed calling the perimeter of the surface limiting K as the product $\frac{2}{\pi}M$ (see [4], pag. 94, or [3], pag. 176). In many cases, in order to find M it is convenient to find the value $M(\epsilon)$ of the same invariant corresponding to the parallel body exterior to the distance ϵ, as well as determining the limit of $M(\epsilon)$ for $\epsilon \to 0$. Let us assume that the surface of K is such that some of the foregoing definitions are applicable to M, and therefore the invariants V, F and M are perfectly defined.

Let E be an arbitrary plane in space. Let f be the area of the section of E with K, u the length of the section of E with the surface of K, and v the number of closed boundaries of which the same intersection of E with the surface of K is composed. If dE has the value (1), and the integrations being extended to all the planes intersecting K, then the following formulae hold (see [2], pages 70-73-74-100, or also [4], pages 93-94.)

$$\int f dE = 2\pi V, \quad \int u dE = \frac{\pi^2}{2} F, \quad \int \nu dE = M. \tag{3}$$

Furthermore, a curve of length L being fixed in space, and n being the number of points at which the curve is intersected by the plane E at each variable position of the plane, the following formula is also verified

$$\int n dE = \pi L, \tag{4}$$

the integration being extended to all the planes in space.

2. Let us now consider the problem as stated. Let $E_1, E_2, E_3, \ldots, E_n$, be n planes intersecting K. If we denote by R the number of regions in which K is divided by these n planes, the mean value desired is by definition

$$R = \frac{\int R dE_1\, dE_2\, dE_3 \ldots dE_n}{\int dE_1\, dE_2\, dE_3 \ldots dE_n}, \tag{5}$$

where dE_h has the values corresponding to the form (1), and the integrations are extended to all the planes intersecting K.

The denominator integral is then immediately calculated. Let M_0 be the integral of the mean curvature of a minimum convex body containing K whose surface as called the *minimum convex envelope* of K. Since the integral of dE_h is equal to the measure of the planes intersecting K, and therefore equal to the measure of the planes intersecting its minimum convex envelope, we have

$$\int dE_1\, dE_2\, dE_3 \ldots dE_n = M_0^n. \tag{6}$$

If K is convex, then $M_0 = M$.

The positions of the planes at which there are more than 3 passing through the same point, or more than 2 passing through the same straight line, are exceptional positions, of zero measure, and therefore they need not be taken into account in the calculation of the numerator integral of (5).

Thus let $E_1, E_2, E_3, \ldots, E_n$ be a position at which no more than 3 pass through the same point, and no more than 2 pass through the same straight line. Let R be the number of regions in which the n planes, for this particular position, divide K. Each of these regions will be a body of the connection of the sphere. The connection points interior to K of 3 planes E_h, or of two planes and of the surface of K, will be the *vertices* V of the decomposition of K into regions; the intersecting line segments of two E_h, or the arcs intersecting one E_h with the surface of K, which join two vertices, will be the *edges* A, and the parts of planes E_h, or of the surface of K limited by the edges, will be the *faces* C of the said decomposition of K into regions by the planes E_h.

We then know that since the regions R are formed by an open manifold, Euler's generalized formula (see, for example [6], pag. 82, or [1], pag. 250)[1] holds

$$V - A + C - R = 1,$$

whereby

$$R = V - A + C - 1. \tag{7}$$

In order to apply this formula, all the faces C must be topologically equivalent to a circle. There is no doubt this will always be so for the interior faces of K, since they come from the intersection of planes E_h. However, the faces formed on the surface of K may not be so; then we add an appropriate number of *auxiliary vertices* V_a and *auxiliary edges* A_a on the surface of K in order to decompose the faces having no connection of the circle into others that do. This can always be done; it suffices to take points interior to the faces to be decomposed and join them appropriately to points of the edges situated on the surface. In this way some edges are decomposed into two, of which we consider only one as being an auxiliary edge.

Thus the vertices V are classified into three types: the V_i interior to K, the V_s situated on the surface and arising from the intersection of the surface of K with the intersecting straight line of two planes E_h, and the auxiliary vertices V_a. Denoting the number of vertices of each type by the same letters, we have

$$V = V_i + V_s + V_a. \tag{8}$$

The edges A can also be of three types: the interior edges A_i, those contained on the surface of K and coming from the intersections of this surface with the planes E_h, which we call A_0, and finally the auxiliaries A_a. Observe that 6 interior edges pass through each interior vertex, and only one interior edge passes through each non-auxiliary vertex of the surface. Thus each edge is counted twice and we have

$$A_i = \frac{1}{2}(6V_i + V_s). \tag{9}$$

The faces C can also be interior faces C_i, or faces situated on the surface, which we denoted by C_s. To calculate the number of interior faces, we consider

[1] In these works one may find only the formula for closed tri-dimensional manifolds, in whose case $V - A + C - R = 0$. In the case we require, being the surface limiting K of the connection of the sphere, in order to convert the open manifold into a closed manifold it suffices to add a region having the same surface by boundary. Thus V, A, C do not change, and only R increases by one; if then the formula for closed manifolds is applied, we immediately obtain the desired formula corresponding to open manifolds.

a particular plane E_h. Let ν_h be the number of closed boundaries of which the intersection of K with E_h is composed; let v_{hi}, v_{hs} and v_{ha} be the numbers of vertices (internal, surface and auxiliary, respectively) contained in this section, and a_{hi}, a_{hs} and a_{ha} the analogous numbers for the edges. Furthermore, let c_h be the number of faces contained in the same section of E_h with K. Then by Euler's theorem mentioned above, now applied to the plane section of E_h with K, we have ([6], pag. 54)

$$(v_{hi} + v_{hs} + v_{ha}) - (a_{hi} + a_{hs} + a_{ha}) + c_h = \nu_h. \tag{10}$$

Since here we are dealing with points on closed boundaries and the arcs they divide, we have $v_{hs}+v_{ha} = a_{hs}+a_{ha}$ and therefore the previous formula becomes

$$v_{hi} - a_{hi} + c_h = \nu_h. \tag{11}$$

Adding together all the equalities analogous to this one for the different planes $h = 1, 2, 3, \ldots, n$ we have: each internal vertex V_i, since it belongs to 3 planes, will have been counted 3 times, and each edge A_i, since it belongs to two planes, will have been counted twice, and therefore

$$3V_i - 2A_i + C_i = \sum_1^n \nu_h. \tag{12}$$

As regards the faces C_s of the surface, by Euler's theorem on surfaces of the connection of the sphere, we have

$$(V_s + V_a) - (A_s + A_a) + C_s = 2. \tag{13}$$

From these equalities (8), (9), (12), (13) and from (7)

$$R = (V_i + V_s + V_a) - (A_i + A_s + A_a) + (C_i + C_s) - 1,$$

we deduce that

$$R = V_i + \frac{1}{2}V_s + \sum_1^n \nu_h + 1. \tag{14}$$

3. We are now ready to calculate the numerator integral of (5), and thereby the mean value we wish to find. Denoting by N_{hjk} a function that takes the value 1 if the planes E_h, E_j, E_k intersect inside K, and the value 0 in the opposite case; denoting by L_{hj} the length of that part of the intersecting line of E_h, E_j which is interior to K, and by F_h the area of the section of E_h with K, then according to (3) and (4), and with the meaning of M_0 as explained when obtaining (6), we have

$$\begin{aligned}\int V_i dE_1\, dE_2 \ldots dE_n &= \binom{n}{3} M_0^{n-3} \int N_{hjk} dE_h\, dE_j\, dE_k \\ &= \binom{n}{3} M_0^{n-3} \pi \int L_{hj} dE_h\, dE_j = \binom{n}{3} M_0^{n-3} \pi \frac{\pi^2}{2} \int F_h dE_h \\ &= \binom{n}{3} M_0^{n-3} \pi^4 V. \end{aligned} \tag{15}$$

Hence: *Given n arbitrary planes intersecting a body K topologically equivalent to a sphere, the mean value of the number of points of intersection of three planes, which are interior to K, is*

$$\bar{V}_i = \binom{n}{3} \pi^4 \frac{V}{M_0^3}, \tag{16}$$

where V is the volume of K, and M_0 is the integral of the mean curvature of its minimum convex envelope. If K is a convex body, then $M_0 = M$.

4. Analogously, let N_{hk} be a function that takes the value 0 if the intersecting line of the planes E_h, E_k does not intersect K and is equal to the number of points of intersection of this straight line with the surface of K in the opposite case. Furthermore, let L_h be the length of the section of the plane E_h with K. Then we have

$$\int V_s dE_1\, dE_2 \ldots dE_n = \binom{n}{2} M_0^{n-2} \int N_{hk} dE_h\, dE_k \tag{17}$$

$$= \binom{n}{2} M_0^{n-2} \pi \int L_h dE_h = \binom{n}{2} M_0^{n-2} \frac{\pi^3}{2} F. \tag{18}$$

Hence: *Given n planes intersecting a body K topologically equivalent to a sphere, the mean value of the number of points at which their line of intersection meet on the surface of K is*

$$\bar{V}_s = \binom{n}{2} \frac{\pi^3 F}{2M_0^2}. \tag{19}$$

5. On applying the third equality of (8), we also have

$$\int \sum_1^n dE_1\, dE_2 \ldots dE_n = M_0^{n-1} \sum_1^n \int \nu_h dE_h = M_0^{n-1} nM. \tag{20}$$

With this equality, and with (17), (15) and (14), we arrive at the mean value desired

$$\bar{R} = \binom{n}{3} \frac{\pi^4 V}{M_0^3} + \binom{n}{2} \frac{\pi^3 F}{4M_0^2} + n\frac{M}{M_0} + 1. \tag{21}$$

Hence: *Given n arbitrary planes intersecting a body K topologically equivalent to a sphere, the mean value of the number of regions in which K is divided by the n planes is given by* (21). V is the volume, F the area, and M the integral of the mean curvature of K. M_0 is the integral of the mean curvature of the minimum convex envelope of K, and therefore if K is convex, $M = M_0$.

If K degenerates into a plane figure of area f, limited by a simple curve of length u, we have

$$V = 0, \quad F = 2f, \quad M = \frac{\pi}{2} u, \quad M_0 = \frac{\pi}{2} u_0,$$

where u_0 is the length of the minimum convex envelope of the plane curve K. Thus

$$\bar{R} = \binom{n}{2} \frac{2\pi f}{u_0^2} + n\frac{u}{u_0} + 1,$$

which is a more general formula than that obtained in the cited work$^{(1)}$, and which agrees with that formula when K is a convex plane figure.

References

[1] P. Alexandroff and H. Hopf, *Topologie*, vol. I, Gauthier-Villars, 1925, Berlin.

[2] W. Blaschke, *Vorlesungen über Integralgeometrie*, vol. 20, B. G. Teubner, Leipzig and Berlin, 1936.

[3] E. Cartan, *Le principe de dualité et certaines integrales multiples de l'espace tangentiel et de l'espace réglé*, Bull. Soc. Math. France **24** (1896), 140–177.

[4] R. Deltheil, *Probabilités Géométriques*, Gauthier-Villars, 1926, París.

[5] L. A. Santaló, *Mean value of the number of parts in which a convex figure is divided by n arbitrary straight lines*, Rev. Un. Mat. Argentina **VII** (1940), 32–37.

[6] O. Veblen, *Analysis Situs*, AMS, 1931, Vol. V, Part II, 2nd edition.

Instituto de Matemáticas de la Facultad de Ciencias
Matemáticas de la Universidad N. del Litoral
Rosario, Setember 1944.

D-curves on cones

L. A. Santaló

1 Introduction

Let us recall that those curves whose osculating sphere at each point is tangent to the surface are known as D-curves of Darboux curves.[1]

In this work, D-curves on conical surfaces are studied. We shall see that in this case the differential equation for such curves can be integrated by quadratures to obtain a relatively simple result that enables a complete discussion.

2 The differential equation for D-curves

The differential equation for D-curves was given for the first time by Darboux [2]. We proceed to obtaining it quickly in a form that will be useful for the purpose we wish to achieve.

Let

$$X = X(u, v) \tag{2.1}$$

be the vector equation of the surface; that is, the three parametric equations $X_i = X_i(u, v)$ $(i = 1, 2, 3)$ are denoted in abbreviated form by (2.1).

A curve of the surface is given by two functions $u = u(t), v = v(t)$ of a parameter t. If the point $Z(t)$ is the centre of the sphere osculating to the curve $X = X(u(t), v(t))$ and r is the radius of the sphere, the following vector relation holds:

$$(X - Z)^2 = r^2. \tag{2.2}$$

By the definition of an osculating sphere, the relations obtained by deriving the foregoing equation three times, assuming Z and r to be constant, should also be fulfilled. In other words,

$$\begin{aligned} (X - Z)X' &= 0 \\ (X - Z)X'' + X'^2 &= 0 \\ 3X'X'' + (X - Z)X''' &= 0 \end{aligned} \tag{2.3}$$

where the products are scalar products of vectors, and the primes denote derivatives with respect to the parameter t.

If the curve $X = X(u(t), v(t))$ is a D-curve, the centre Z is the point of the normal to the surface at distance r from point X; that is, denoting by ξ the unit vector normal to the surface, we have

$$X - Z = r\xi.$$

[1] Bibliography on these curves can be found in the *Encyclopädie der Mathematischen Wissenschaft III_3*, D3, pp. 181-182. See also [4], p. 225.

On substituting this expresssion in (2.3), the first equation is satisfied identically, since $\xi X' = 0$ (because X' is tangent to the surface) and the two remaining equations give

$$\begin{aligned} X'^2 + r\xi X'' &= 0 \\ 3X'X'' + r\xi X''' &= 0. \end{aligned} \tag{2.4}$$

The differential equation of the D-curves is obtained by removing r from these two equationss. It would appear that the resulting equation is of third order, but by deriving $\xi X' = 0$ twice, we obtain

$$\begin{aligned} \xi' X' + \xi X'' &= 0 \\ \xi'' X' + 2\xi' X'' + \xi X''' &= 0, \end{aligned} \tag{2.5}$$

and from this last equation $\xi X''' = 0$ can be obtained, a value which when substituted in the second equation (2.4), and r being removed from the first, enables us to arrive at the final result:

$$\frac{3X'X''}{X'^2} = \frac{2\xi' X'' + \xi'' X'}{\xi' X'}, \tag{2.6}$$

which is the differential equation of the D-curves in vector form.

3 D-curves on cones

In order to define a cone, let us consider as the directrix curve of the generatices the intersection curve of the cone with a unit sphere whose centre is the vertex of the cone. This curve will be a spherical curve whose vector equation is of the form

$$E = E(s), \tag{3.1}$$

where s is the arc of the curve and the following relation being fulfilled:

$$E^2 = 1. \tag{3.2}$$

The equation of the cone can then be written in the form

$$X = \lambda E(s), \tag{3.3}$$

where λ is the parameter that fixes the points on each generatrix. In other words, the parameters u, v from Section 1 are now s, λ; the former fixes the point on the directrix and therefore the generatrix of the cone, while the latter fixes the point on this generatrix. A curve on the cone will be given by two functions $\lambda = \lambda(t), s = s(t)$. For the sake of simplicity, we take $t = s$, whereby every curve of the cone is determined by the single function $\lambda = \lambda(s)$.

In order to apply equation (2.6), we require the first two derivatives of $E(s)$; recall in this regard that $E'(s) = T(s)$, where T denotes the unit tangent vector to the curve (3.1), and furthermore that the following Frenet formulae hold,

$$T' = \kappa N, \quad N' = -\kappa T + \tau B, \quad B' = -\tau N, \tag{3.4}$$

where N and B are the normal principal and binormal unit vectors, respectively, and κ, τ, the curvature and torsion of $E(s)$.

Thereby, deriving (3.3), we have

$$\begin{aligned} X' &= \lambda' E + \lambda T \\ X'' &= \lambda'' E + 2\lambda' T + \lambda\kappa N. \end{aligned} \tag{3.5}$$

Furthermore, the normal vector ξ is expressed by

$$\xi = E \wedge T, \tag{3.6}$$

where the vector product is denoted by $\wedge$, and hence,

$$\begin{aligned} \xi' &= \kappa(E \wedge N) \\ \xi'' &= \kappa'(E \wedge N) + \kappa B - \kappa^2(E \wedge T) + \kappa\tau(E \wedge B) \end{aligned} \tag{3.7}$$

since $B = T \wedge N$.

From the foregoing formulae we may deduce that

$$\begin{aligned} X'^2 &= \lambda^2 + \lambda'^2 \\ X'X'' &= \lambda'\lambda'' + 2\lambda\lambda' + \lambda\lambda'\kappa(EN) \\ \xi' X' &= \lambda\kappa T(E \wedge N) = -\lambda\kappa(T \wedge N)E = -\lambda\kappa(EB) \\ \xi' X'' &= 2\kappa\lambda'(E \wedge N)T = 2\kappa\lambda' E(N \wedge T) = -2\lambda'\kappa(EB) \\ \xi'' X' &= \kappa\lambda'(EB) - \lambda\kappa'(EB) + \lambda\kappa\tau(EN). \end{aligned} \tag{3.8}$$

Let α be the angle formed by the normal principle N with the radius of the sphere containing $E(s)$ that goes to the point of contact. Since the vector E has the direction of the radius that goes towards the exterior of the sphere, then $EN = \cos(\pi - \alpha) = -\cos\alpha$. Moreover, by applying Meusnier's formula to the sphere (of radius 1) containing $E(s)$, we have $\cos\alpha = \rho$, where ρ is the radius of the curvature of the curve $E(s)$, that is, $\rho = 1/\kappa$. Therefore,

$$EN = -\frac{1}{\kappa}. \tag{3.9}$$

Since E, N, B are perpendicular to T they are on the same plane, and we then have

$$EB = \sin\alpha = \sqrt{1 - \rho^2} = \frac{\sqrt{\kappa^2 - 1}}{\kappa}. \tag{3.10}$$

On substituting these values in (3.8) and by replacing the values obtained in equation (2.6) of the D-curves, after simple transformations we obtain

$$\frac{3(\lambda'\lambda''+\lambda\lambda')}{\lambda^2+\lambda'^2}-2\frac{\lambda'}{\lambda}=\frac{\kappa'}{\kappa}+\frac{\tau}{\sqrt{\kappa^2-1}}. \tag{3.11}$$

This equation can be simplified further. Indeed, the curve $E(s)$ is a curve situated on a unit sphere; therefore, the radius of its osculating sphere at any point is equal to 1, and thus the following relation is fulfilled (see, for exemple, [1], p. 34; or [3], p. 73):

$$\frac{1}{\kappa^2}+\frac{\kappa'^2}{\kappa^4\tau^2}=1,$$

from which it is deduced that

$$\tau=\frac{\kappa'}{\kappa\sqrt{\kappa^2-1}},$$

and hence,

$$\frac{\kappa'}{\kappa}+\frac{\tau}{\sqrt{\kappa^2-1}}=\frac{\kappa\kappa'}{\kappa^2-1}. \tag{3.12}$$

By substituting in (3.11) we obtain that the definitive equation for D-curves on the conical surface (3.3) is

$$\frac{3(\lambda'\lambda''+\lambda\lambda')}{\lambda^2+\lambda'^2}-3\frac{\lambda'}{\lambda}=\frac{\kappa\kappa'}{\kappa^2-1}. \tag{3.13}$$

A first integration gives

$$\frac{3}{2}\log(\lambda^2+\lambda'^2)-3\log\lambda=\frac{1}{2}\log(\kappa^2-1)+a,$$

where a is a constant of integration.

Hence

$$(\lambda^2+\lambda'^2)^{3/2}=b\lambda^3\sqrt{\kappa^2-1},$$

where b is the constant e^a. From this equation we may deduce that

$$\frac{\lambda'}{\lambda}=\pm\sqrt{C(\kappa^2-1)^{1/3}-1},$$

where C is the new constant $b^{2/3}$.

A final integration gives as the *equation of the D-curves on conical surfaces*

$$\lambda=A\exp(\pm\int_0^s\sqrt{C(\kappa^2-1)^{1/3}-1\,ds}), \tag{3.14}$$

where A is a new integration constant and $\exp x$ denotes the function e^x.

4 Discussion

Two arbitrary constants A and C appear in equation (3.14). The constant A is determined by giving the initial point of the D-curve, since by making $s = 0$ we have $A = \lambda(0)$. Constant C is determined by the angle formed by the D-curve with the generatrix of the cone corresponding to the initial point. In fact, denoting this angle by ω, we have that

$$\tan\omega = \frac{\lambda}{\lambda'} = \frac{1}{\pm\sqrt{C(\kappa^2-1)^{1/3}-1}}, \tag{4.1}$$

from which C is isolated, taking into account that κ has the value corresponding to the generatrix that passes through the initial point. Raising to the power of two in order to isolate C, the double sign of the radical disappears; that is, given ω, a unique value is determined for C.

On the other hand, two D-curves correspond to each pair of values A and C, these curves corresponding to the double sign that appears in (3.14). According to (4.1), these two curves intersect at the initial point, in accordance with the symmetric directions with respect to the generatrix.

The following notable consequence can be deduced from (4.1): *any D-curve of a conical surface intersects all the generatrices that correspond to a given value of κ of the directrix curve making the same angle.*

Therefore: 1°. *If a D-curve intersects the same generatrix of the cone several times, it always does so making the same angle.* 2°. *The necessary and sufficient condition for a conical surface to have a D-curve that intersects all the generatrices making the same angle (that is, it is a loxodromic curve), is that this surface is a cone of revolution (that is, $\kappa =$ constant). In this case, all the D-curves possess the same property.*

According to the values of C, the D-curves of a cone can show two different forms. Indeed, it is necessary to distinguish two cases:

1. *The D-curve goes all the way round the cone.* Let κ_m be the minimum value of the curvature of the directrix curve $E(s)$. Since this curve is situated on the unit sphere, we have $\kappa_m \geq 1$.

In order for a D-curve to go completely around the cone; that is, for it to intersect all the generatrices, the radical appearing in (3.14) must always be real, and thus must fulfill the following condition:

$$C \geq \frac{1}{(\kappa_m^2-1)^{1/3}}. \tag{4.2}$$

In this case, assuming a closed conical surface, the D-curve will perform infinite revolutions, on one side moving infinitely further away from the vertex, and on the other moving asymptotically closer to the vertex. We denote by $\lambda_0.\lambda_1, \lambda_2, \cdots$ the values of λ corresponding to the points at which a D-curve intersects the same generatrix. Consider, for example, the D-curve corresponding to the + sign in (3.14), and write

$$K = \exp\int_0^L \sqrt{C(\kappa^2-1)^{1/3}-1}\,ds \tag{4.3}$$

where L is the total length of the closed curve $E(s)$. According to equation (3.14), the D-curve will be

$$\lambda_0 = A, \quad \lambda_1 = AK, \quad \lambda_2 = AK^2, \quad \cdots$$

and therefore,

$$\lambda_n = \lambda_0 K^n \tag{4.4}$$

and, analogously, towards the side of the vertex,

$$\lambda_{-n} = \lambda_0 K^{-n}. \tag{4.5}$$

These expressions tell us that *the points at which a D-curve intersects the same generatrix recede from or approach the vertex of the cone according to the potential law* (4.4) *or* (4.5), *respectively.*

Furthermore, since $\lambda_{i+1}/\lambda_i = K =$ constant, the equality that holds for any generatrix, is verified, we have that *the arcs of the same D-curve between consecutive pairs of intersection points belonging to one generatrix are homothetic with respect to the vertex of the cone.*

If, instead of considering the D-curve corresponding to the $+$ sign in (3.14), we consider the D-curve corresponding to the $-$ sign with the same values of A and C, and at the same time we invert the sign of the integration variable s, all the foregoing is equally valid. Moreover, after the initial point, the two curves intersect each other again at a point on the generatrix corresponding to the value of s, for which we have

$$\int_0^s = \frac{1}{2}\int_0^L \tag{4.6}$$

where it should be understood that below the integration sign we have the same function that appears beneath the integral sign in (3.14).

Furthermore, if from a point of intersection two curves λ and $\bar{\lambda}$ are considered, corresponding to the two signs of (3.14), we have $\lambda\bar{\lambda} = A^2$, where the constant A is the value of λ corresponding to the point of intersection considered. This tells us that the two curves are inverse with respect to the sphere whose centre is the vertex of the cone passing through the point of intersection.

It is therefore possible to state that:

The two D-curves corresponding to the two signs of (3.14), each curve being prolonged in the two directions of the arc s, intersect each other infinitely, the points of intersection on the two generatrices being: that corresponding to the initial point and the other corresponding to the value of s defined by the equation (4.6). Both curves are inverse with respect to any sphere whose centre is the vertex of the cone passing through one of the points of intersection.

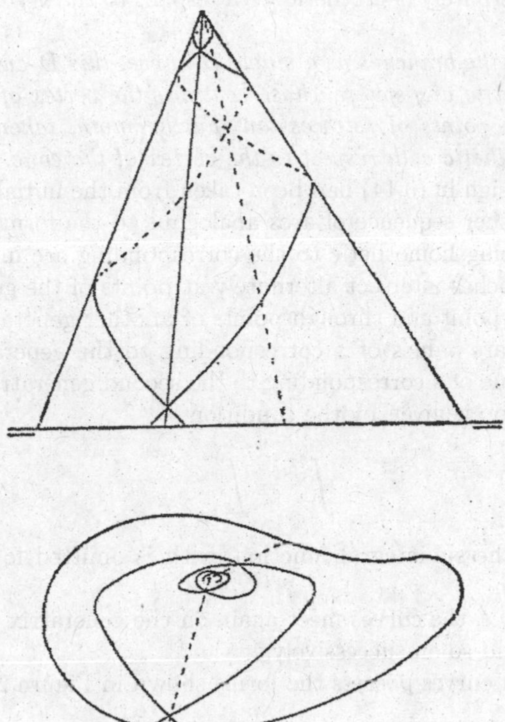

Fig. 1

In Figure 1, one may observe two D-curves corresponding to this case.

2. The D-curve does not revolve completely around the cone. If condition (4.2) is not fulfilled; that is,

$$C < \frac{1}{(\kappa_m^2 - 1)^{1/3}}, \tag{4.7}$$

the D-curve does not intersect all the generatrices and is interrupted at the generatrix corresponding to the value of $\kappa (\kappa > \kappa_m)$ such that

$$C = \frac{1}{(\kappa^2 - 1)^{1/3}}, \tag{4.8}$$

and according to (4.1) the D-curve intersects this detaining generatrix orthogonally.

It is possible for the D-curve to move backwards from this detaining point, changing the sign in (3.14) and at the same time the direction of the integration variable s. In this way we obtain the prolongation of the D-curve according to a new branch that is the inverse of the former with respect to the sphere whose centre is the vertex of the cone, and which passes through the point of retrocession. A sequence of branches is thereby obtained between the first two generatrices for which (4.8) holds and which include the initial point, these branches being alternately homothetic with respect to the vertex of the cone. In other words:

Considering all the branches as a single D-curve, this D-curve is the inverse of itself with respect to any sphere whose centre is the vertex of the cone passing through one of the points of retrocession. Furthermore, taken alternately, the branches are homothetic with respect to the vertex of the cone.

If the opposite sign in (3.14) had been taken from the initial point, we would have obtained another sequence of arcs analogous to the former, with each arc of this sequence being homothetic to the corresponding arc in the previous sequence. These branches intersect alternately at points of the generatrix passing through the initial point and through points of another generatrix.

If s_1 and $-s_2$ are values of s corresponding to the generatrices for which (4.8) holds, the value of s corresponding to the second generatrix containing the points of intersection is given by the condition

$$\int_{s}^{s_1} = \int_{-s_2}^{0}$$

where, as before, the subintegral function, with is omitted for brevity, is that appearing in (3.14).

From this point s, the curves meet again on the generatrix corresponding to the initial point, and so on successively.

Therefore, both curves possess the forms shown in Figure 2.

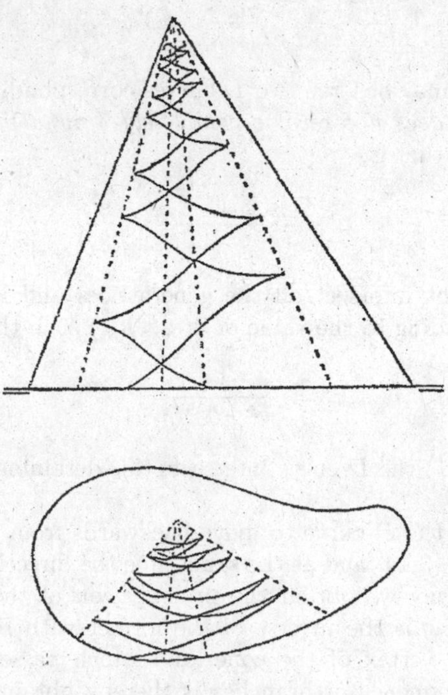

Fig. 2

5 The case of the cone of revolution

Note that the function $\lambda(s)$ is precisely the equation in polar coordinates of the plane curve into which the D-curve of the cone is transformed as the cone develops on the plane. The arc s becomes the polar angle φ.

For the case of a cone of revolution, the spherical directrix $E(s)$ is a smaller circle, and thus $\kappa =$ constant. In this case, equation (3.14) yields $\lambda = A\exp(Bs)$ ($B =$ constant); that is: *the development of the D-curves of the cones of revolution are logarithmic spirals.*

It should be pointed out that, since the development of the cone only fills one angle of the plane, in order to have the complete logarithmic spiral it is necessary to assume that the cone is formed by infinite superimposed leaves which are developed one after another.

Instead of considering the development, one may consider the projection of the D-curve onto the plane containing the base of the cone of revolution. Since the vector radius R of the projection is related to λ by $R = \lambda \sin\theta$, where θ is the angle formed by the generatrices with the axis of the cone, then this projection is also a logarithmic spiral, and reciprocally; that is:

The orthogonal projection of the D-curves of a cone of revolution onto the plane containing the base are logarithmic spirals, and, reciprocally, any logarithmic spiral of the plane at the base of a cone of revolution whose asymptotic point is the centre of the base is projected onto the cone according to a D-curve.

References

[1] W. Blaschke, *Differentialgeometrie*, vol. I, Springer, Berlin, 1930.

[2] G. Darboux, *Les courbes tracées sur une surface, et dont la sphère osculatrice est tangente en chaque point à la surface*, C. R. Acad. Sc. París **LXXIII** (1871), 732–736.

[3] W. Graustein, *Differential geometry*, Macmillan, New York, 1935.

[4] G. Loria, *Curve sghembe speciali algebriche e trascendenti*, vol. 2, N. Zanichelli, Bologna, 1925.

Rosario, Instituto de Matemática

Integral Geometry in three-dimensional spaces of constant curvature

L. A. Santaló

1 Introduction

In its most general form, the core problem of the so-called Integral Geometry is as follows: let S_n be a space of n dimensions in which a continuous and transitive group of transformations G acts. Assume, furthermore, that G possesses a left-invariant measure or volume due to the operations of the group. In other words, if we denote the invariant volume element by dg, and s is any transformation belonging to the group, the last condition can be written symbolically as $d(sg) = dg$. Let C_p be a manifold of dimension p contained in S_n, and C_q another manifold, also contained in S_n, of dimension q. We denote by gC_q the transformed manifold of C_q by the transformation g of G, and by $C_q.gC_q$ the intersection of C_p with gC_q, which depends on g. Let $F(C_p.gC_q)$ be a function of this intersection. We are concerned with the study of integrals of the type

$$I = \int_G F(C_p.gC_q)dg, \tag{1.1}$$

where the integration is extended to the whole volume of the group G.

The most studied case is that in which S_n is the Euclidean space and G the group of movements in the space; dg is then known as the kinematic density. In this case, the integral I can be calculated for various functions F, and from the results obtained it has sometimes been possible to deduce geometric propositions concerning the manifolds C_p and C_q. For example, if C_p and C_q are manifolds of p-dimensional volume $V(C_p)$ and q-dimensional volume $V(C_q)$, respectively, and F is the $(p+q-n)$-dimensional volume of the intersection, we have

$$\int_G V(C_p.gC_q)dg = \alpha V(C_p)V(C_q), \tag{1.2}$$

where α is a constant independent of C_p and C_q (dependent only on p, q, n).

Under certain regularity conditions for C_p and C_q, if F is the Euler characteristic $N(C_p.gC_q)$ of the intersection $C_p.gC_q$ and always in the case of movements in Euclidean space, integral (1.1) can also be calculated, giving the so-called Blaschke fundamental formula for $n = 3$, which was generalized to n dimensions by Chern and Yien [6]. For $n = 3$, see [4]. In fact, in these works only the case $p = q = n$ is considered, but by passing to the limit it is possible to pass easily to the general case.

In this work we consider the case where S_n is a space of three dimensions of constant curvature K, and G the group of movements in this space. More precisely, we might say that we consider the case of the Elliptic ($K>0$) and the Hyperbolic ($K>0$) spaces whose projective representation is well-known; the group G is then that of the projectivities that leave an imaginary ($K>0$) or real non-ruled quadric ($K<0$) invariant. The case $K>0$ has already been studied by Ta-Jen Wu ([13], p. 495-521), but the approach we follow here is more geometric and enables us to deal simultaneously with the aforesaid case and that of $K\leq 0$. As the principle problem, we consider the case where C_p and C_q are two 3-dimensional manifolds of finite volume ($p = q = 3$); that is, two bounded spatial bodies where the function F is the Euler characteristic of the intersection of both bodies. Let us assume furthermore that both bodies are limited by surfaces of finite area, and with finite and continuous curvature at each point, a condition which, once the result has been obtained, can be greatly reduced.

Later we will see some applications, including the generalization to spaces of constant curvature from a result obtained by O. Varga for Euclidean space. Lastly, we consider the case in which C_q is a plane in space (that is, a totally geodesic surface); then, since for each plane e there exists a subgroup E of G that leaves it invariant, it is necessary to consider the homogeneous space G/E, and as invariant measure in this space we have the so-called density for sets of planes. With this density it is possible to generalize many known integral formulae for the Euclidean case, the majority of which, however, are identical for spaces of constant curvature. We thus confine ourselves to pointing out (10.3), which depends on K and contains the inequality (10.5), which is valid for convex bodies in Hyperbolic space.

In all that follows, when we refer to space we mean space of 3 dimensions. Moreover, in the interests of brevity, in order to denote the curvature of the space we use K or k wherever they are most appropriate, which are related by

$$k^2 = K. \tag{1.3}$$

2 Kinematic density in spaces of constant curvature. Densities for planes, lines and points

Assume the three-dimensional space of constant curvature K and the group G (dependent on 6 parameters) of the movements of the space. In order to find the invariant volume element of G, it is convenient to use the method of the "*moving frame*" by E. Cartan ([5], Chapters IX and XII). For this Section 2 it may be useful to consult the fundamental report by S. S. Chern [7].

Let P be a point in the space and consider the trirectangular trihedron formed by three vectors $\overline{e}_1, \overline{e}_2, \overline{e}_3$ of unit module and origin P. Therefore, the following conditions are fulfilled

$$\overline{e}_i.\overline{e}_j = \begin{cases} 1\; if \;\; i=j \\ 0\; if \;\; i\neq j, \end{cases} \tag{2.1}$$

where the scalar products of the first term on the right-hand side are understood as being taken according to the metric of the ambient space.

The vector PP' determined by P and the infinitely close point P' will have as regards the said trihedron certain components that we denote by ω^i; that is, by writing $d_1P = PP'$, we have

$$d_1P = \omega^i.\bar{e}_i, \tag{2.2}$$

where in the second term on the left-hand side it is implicitly understood a sum of terms from $i = 1$ to $i = 3$, whose symbol we remove, as is usual in tensorial calculus; the same convention holds for all that follows in this Section 2.

Furthermore, the absolute differences $D\bar{e}_i$ of each of the vectors $\bar{e}_i$ can also be expressed in the form

$$D\bar{e}_i = \omega_i^k \bar{e}_k, \tag{2.3}$$

where ω_i^k are new linear differential forms; that is, both ω^i and ω_i^k are linear forms in the differentials dx^i of the coordinates x^i of the space, plus the other three coordinates that fix the trihedron formed by the vectors $\bar{e}_i$. Its value can easily be deduced from (2.2) and (2.3) by multiplying the terms on both sides in a scalar fashion by the convinient vector $\bar{e}_i$. We thus obtain

$$\omega_i = \bar{e}_i.dP, \quad \omega_i^k = -\omega_k^i = \bar{e}_k\,.D\bar{e}_i. \tag{2.4}$$

The exterior differentials of the forms ω_i , ω_i^k satisfy the following *structure equations* of the space ([5], p. 322)

$$d\omega_i = [\omega^k \omega_k^i], \tag{2.5}$$

$$d\omega_i^j = [\omega_i^k \omega_k^j] + \frac{1}{2} R_{ikh}^j [\omega^k \omega^h], \tag{2.6}$$

where the square brackets [ ] denote the exterior product and R_{ikh}^j is the curvature tensor of the space.

We now consider the group G of the movements of the space. Assuming a trirectangular trihedron of vertex P_0 formed by the vectors $\bar{e}_{i0}$ (denoted by (P_0, $\bar{e}_{i0}$)) as fixed, each movement g of G can be determined by the trihedron (P, $\bar{e}_i$) transformed from (P_0, $\bar{e}_{i0}$), by g, and reciprocally each trihedron (P, $\bar{e}_i$) determines a movement g: the one that transforms (P_0, $\bar{e}_{i0}$) into (P, $\bar{e}_i$). The family of trihedra P, $\bar{e}_i$ therefore constitutes a family of trihedra adapted to group G. The forms ω^i, ω_i^k are then the relative components of G, and (2.5) and (2.6) are the equations of structure of the group G. The relative components ω^i, ω_i^k are invariant by G, and thus so is the differential form of sixth degree

$$dg = [\omega^1 \omega^2 \omega^3 \omega_1^2 \omega_1^3 \omega_2^3]. \tag{2.7}$$

The group G being transitive, every other invariant differential form of sixth degree should differ from the preceding dg by a constant factor. Thus (2.7) is the invariant volume element of the space of G: this is known as the kinematic density of the space of constant curvature considered.

The foregoing expression of dg can be given a simple geometric interpretation. First observe that according to (2.2) the product $[\omega^1\omega^2\omega^3]$ is equal to the product of the elementary displacements of P, according to three rectangular axes; that is, it is equal to the volume element of the space, which we denote by dP. Thus we have

$$dP = [\omega^1\omega^2\omega^3]. \tag{2.8}$$

Furthermore, each of the ω_i^k, according to (2.4), represents an elementary rotation about the axis $\bar{e}_h (h \neq i \neq k)$. Consequently, the product $[\omega_1^3\omega_2^3]$, for example, represents the area element $d\Omega$ of the unit sphere and centre P (Euclidean sphere, situated in the three-dimensional space tangent to that given in P) corresponding to the representative point of the direction of $\bar{e}_3$. If finally we write $d\varphi$ instead of the form ω_1^2 in order to represent an elementary rotation around $\bar{e}_3$, we have for dg the highly intuitive expression

$$dg = [dP\, d\Omega\, d\varphi], \tag{2.9}$$

which has the same form as for the Euclidean case. Here, and henceforth, *the densities are considered in their absolute value.*

Let us assume, for example, a point P fixed in space and any body Q of finite volume V, moving in the space. The measure of the set of positions of Q in which P is contained (that is, the measure of the set of movements that transform Q to a position in which P is contained), is as follows

$$\int_{P\subset Q} dg = 8\pi^2 V. \tag{2.10}$$

Indeed, it suffices to observe that this measure must be equal to that of the positions of P (assumed to be moving) in which it is interior to Q (assumed to be fixed), and therefore by integrating (2.9), Ω changes over the area 4π of the unit sphere, φ between 0 and 2π and P in the volume of Q; thus result (2.10) is obtained.

Now let us consider the subgroup E of the movements that leave invariant a plane e in the space[1]. Taking as the plane e that determined by the vectors $\bar{e}_1, \bar{e}_2$, and considering displacements of the trihedron $P, \bar{e}_i$, in which $\bar{e}_3$ remains normal to e, equations (2.2) and (2.3) give

$$\omega^3 = 0, \quad \omega_1^3 = 0, \quad \omega_2^3 = 0. \tag{2.11}$$

This system of equations by Pfaff, by the manner in which it is obtained, must be completely integrable, and its integral manifolds represent, in the representative space of the group G, the subgroup E and its transforms by operations of G [2]. The product

$$de = [\omega^3\, \omega_1^3\, \omega_2^3] \tag{2.12}$$

[1] Recall that here and in all that follows, as usual, we refer to the totally geodesic surfaces in space as "planes".

[2] For greater understanding, we consider as the clarifying model the group G of movements in the Euclidean plane referring to an origin O and two rectangular axes x and y. Each movement, being reducible to a translation and a rotation, will be determined by the coordinates x and y of the transformed point of O and by a rotation φ : (x, y, φ) will be the coordinates in the representative three-dimensional space of group G of the movements in the plane (it suffices to

is invariant by G and can be taken as the volume element of the homogeneous space G/E, always providing that its value depends exclusively on the plane e, not on the point P and not on the tangent vectors $\overline{e}_1, \overline{e}_2$ chosen in it. This independence can be deduced by general methods, taking into account the equations of structure (2.5) and (2.6), and also that the space has constant curvature, although in this particular case it is simpler to proceed geometrically. To this effect, observe that (2.12) is equivalent to

$$de = [d\Omega\, dt], \tag{2.13}$$

where $d\Omega = [\omega_1^3\, \omega_2^3]$ is the area element on the Euclidean unit sphere of centre P corresponding to the direction $\overline{e}_3$, and $dt = \omega^3$ is an elementary displacement in the same direction; that is, normal to e. Displacing by parallelism the trihedron $(P, \overline{e}_i)$ along e, the vector $\overline{e}_3$ remains normal, and by also conserving the angle between vectors and their lengths, $d\Omega$ and dt will also be conserved; that is, (2.13) does not depend on the point P chosen in e, nor on the particular position of the tangent vectors $\overline{e}_1, \overline{e}_2$.

Expression (2.12) or its equivalent (2.13) can therefore be used to measure sets of planes (or, what amounts to the same, sets of points in the homogeneous space G/E), thereby obtaining an invariant measure with respect to G, which except for a constant factor is unique, since G is transitive with regard to the planes e; either of these expressions (2.12), (2.13) is known as the *density for sets of planes.*

It is convenient to give this density a more adapted form for the calculation. We thus consider a fixed point O in space and determine each plane e by its distance ρ to O, plus the spherical coordinates θ, φ of the direction of the normal to e from O. The arc element of the space, in these geodesic coordinates ρ, θ, φ (assuming $x^1 = \rho, x^2 = \theta, x^3 = \varphi$), is as follows:

$$ds^2 = d\rho^2 + K^{-1} \sin^2 k\rho d\theta^2 + K^{-1} \sin^2 k\rho \sin^2 \theta\, d\varphi^2. \tag{2.14}$$

---

observe that φ can only change between 0 and 2π; that is, the planes $\varphi = 0$ and $\varphi = 2\pi$ must be identified, and thereby the space of G can be said to be a cylinder of three dimensions). Consider now a line r of the plane considered, which for the sake of simplicity we may assume to be the axis x. The subgroup R of G that leaves r invariant is represented by the line $y = 0, \varphi = 0$. Moreover, observe that each point x, y, φ is transformed by the operation (x_0, y_0, φ_0) of G according to the formulae (group of the parameters),

$$\begin{aligned} x' &= x\cos\varphi_0 - y\sin\varphi_0 + x_0 \\ y' &= x\sin\varphi_0 + y\cos\varphi_0 + y_0 \\ \varphi' &= \varphi + \varphi_0 \end{aligned}$$

Therefore, the line $y = 0, \varphi = 0$, is transformed by the operation (x_0, y_0, φ_0) on the line.

$$x' = x\cos\varphi_0 + x_0, \quad y' = x\sin\varphi_0 + y_0, \quad \varphi' = \varphi_0 \tag{*}$$

where x is the variable parameter. Through each point of the representative space of G pass one and only one line given by (*): they are the integral manifolds of the system analogous to (2.11) for the case of the plane we will consider. To each one corresponds a line in the plane considered (the line passing through (x_0, y_0) with direction φ_0). Considering each line (*) as a point of the space G we have a space of two dimensions: it is the homogeneous space G/R and the invariant volume element of this space is the density for lines in the plane.

We choose as point P from e the origin of the normal from O, and as vectors $\overline{e}_1$, $\overline{e}_2$ the tangents to the directions according to which only φ or θ change, respectively.

The components of $\overline{e}_1, \overline{e}_2$ and $\overline{e}_3$ are therefore

$$\overline{e}_1(0, 0, \frac{k}{\sin k\rho \sin\theta}), \quad \overline{e}_2(0, \frac{k}{\sin k\rho}, 0), \quad \overline{e}_3(1, 0, 0).$$

According to (2.4), and since the components of dP_1 are $d\rho, 0, 0$, we have

$$\begin{aligned}
\omega^3 &= \overline{e}_3.d_1P = d\rho \\
\omega_1^3 &= \overline{e}_3.De_1 \\
&= (\Gamma_{31}^1 d\rho + \Gamma_{32}^1 d\theta + \Gamma_{33}^1 d\varphi)\frac{k}{\sin k\rho \sin\theta} = -\cos k\rho \sin\theta d\varphi \\
\omega_2^3 &= \overline{e}_3.De_2 = \\
&= (\Gamma_{21}^1 d\rho + \Gamma_{22}^1 d\theta + \Gamma_{23}^1 d\varphi)\frac{k}{\sin k\rho} = -\cos k\rho d\varphi,
\end{aligned}$$

and therefore, referring to the area element on the Euclidean unit sphere with centre O corresponding to the normal direction to e as $d\Omega = \sin\theta \, d\theta \, d\varphi$, we obtain

$$de = \cos^2 k\rho \, [d\Omega \, d\rho]. \tag{2.15}$$

For example, the measure of the planes intersecting a sphere of radius ρ, integrating the previous expression from 0 to ρ with respect to ρ, and on the unit sphere $d\Omega$, is as follows:

$$\pi(\rho + k^{-1} \cos k\rho \sin k\rho). \tag{2.16}$$

In particular, if $K > 0$ (that is, k is real), the whole space is equivalent to a sphere of radius $\pi/2k$, and therefore we arrive at the known result (2) that *the measure of all the planes in the Elliptic space of curvature $K = k^2$ is equal to π^2/k* . If we wish to express this result from the point of view of the groups, we have: in the case of $K>0$, the homogeneous space G/E has a finite volume equal to π^2/k. Analogously to the foregoing, the density for sets of lines (that is, geodesics) in the space can also be obtained. Let r be a straight line and consider the subgroup R of the movements that leave it invariant. By taking the trihedron $(P, \overline{e}_i)$ in such a way that $\overline{e}_3$ is tangent to r, the differential equations of the manifolds that in the space of group G represent the subgroup R and its transforms by the operations of G, will be

$$\omega^1 = 0, \quad \omega^2 = 0, \quad \omega_3^1 = 0, \quad \omega_3^2 = 0.$$

The *density for sets of lines*, that is, the invariant elementary measure of the homogeneous space G/R is thus

$$dr = [\omega^1 \, \omega^2 \, \omega_3^1 \, \omega_3^2] \tag{2.17}$$

always providing that this expression does not depend on the particular point P chosen in r. Geometrically, we may observe that $[\omega^1\,\omega^2]$ represents the area element $d\sigma$ normal to r at point P, and $[\,\omega_3^1\,\omega_3^2] = [\omega_1^3\,\omega_2^3]$ is, as stated above, equal to the area element in the unit Euclidean sphere with centre P corresponding to the direction of r. We may therefore write

$$dr = [d\sigma\, d\Omega]. \tag{2.18}$$

In this form, one sees quite clearly that the expression does not depend on the point P, since by parallel translation along r neither $d\sigma$ nor $d\Omega$ change. As G is transitive with regard to the lines in the space, the above-mentioned density is unique except for a constant, factor. It is interesting to point out that the form dr is independent of P for the geodesics of any space, even though they do not have constant curvature, which is the case under consideration in this work. Then, of course, neither group G nor R exist, but from the density dr sets of geodesics can also be measured to obtain properties of the space, as was performed for surfaces ($n = 2$) in a previous work [9], and which will subsequently be done for a general Riemannian space of dimension n in a later work.

As regards the calculation, it may be useful to refer the lines R to a fixed system of coordinates, as was done in the case of the planes. Taking a fixed point O, each line r determines a plane normal to the line from O. Now let $d\Omega$ be the area element in the Euclidean unit sphere with centre O corresponding to the direction normal to the said plane, and $d\sigma$ the area element in the same plane corresponding to its point of intersection with r. Then if ρ is the distance from O to r, by a method analogous to the case of the planes or directly by a change of variables in (2.18), we obtain

$$dr = \cos k\rho [d\sigma\, d\Omega]. \tag{2.19}$$

Let us calculate, for example, the measure of all the straight lines intersecting a sphere of radius ρ. As on a plane in the space, in polar coordinates ρ, θ, we have $d\sigma = k^{-1} \sin k\rho\, d\rho\, d\theta$, and we arrive at

$$\int_{\frac{1}{2}\Omega} d\Omega \int_0^{2\pi} d\theta \int_0^{\rho} k^{-1} \sin k\rho \,\cos k\rho\, d\rho = 2\pi^2 k^{-1} \sin^2 k\rho.$$

In particular, if $K>0$, that is, k is real, the whole space is equivalent to a sphere of radius $\pi/2k$, and thus we have the known result (2): *the measure of all the lines in the Elliptic space of curvature* $K = k^2$ is equal to $2\pi^2/k$. If we wish this to be expressed from the point of view of the groups, the result is as follows: the volume of the homogeneous space G/R corresponding to the case $K > 0$ is equal to $2\pi^2/k$.

Even though it is trivial, we conclude by observing that the density for sets of points P, in a way analogous to the foregoing for planes and lines, is

$$dP = [\omega^1 \omega^2 \omega^3], \tag{2.20}$$

which as we have already seen in (2.8) is equivalent to the volume element of the space. For example, the measure of the set of points contained in a sphere of radius ρ is equal to the volume of the said sphere, which is $2\pi k^{-3}(k\rho - \sin k\rho \cos k\rho)$. For the case $K>0$, on making $\rho = \pi/2k$, the value of the total volume of the space turns out to be π^2/k^3.

3 On the total curvature of a closed surface

Let S be a closed surface contained in the space of constant curvature K. At each point of S two curvatures can be distinguished: a) *The intrinsic or absolute curvature* K^α, which is the Riemannian curvature of S considered as a Riemannian space of two dimensions; b) *The relative curvature* K^r, which is equal to the product of the principal curvatures $1/R_1$, $1/R_2$ of S as a surface contained in the considered space of constant curvature. In other words

$$K^r = \frac{1}{R_1 R_2}. \tag{3.1}$$

We also employ *the mean curvature*, which is always understood to be taken with the radii R_1, R_2 ; that is

$$H = \frac{1}{2}(\frac{1}{R_1} + \frac{1}{R_2}). \tag{3.2}$$

Since the principal radii of curvature R_1 and R_2 play a key role in what follows, it is important to recall their geometric meaning. The study of the geometric differential of surfaces contained in a space of constant curvature (including in many cases those contained in a general Riemannian space) can be carried out in a way analogous to that of surfaces in ordinary space, simply by replacing the straight lines by geodesics and the ordinary parallelism by that of Levi-Civita ([5], p. 96-99). It is thereby proven that through each point P in the surface there pass two normal curves c_i , $(i = 1,2)$ whose property is that the normal to the surface at P, the tangent to c_i and the direction resulting from translating by parallelism the curve normal to the surface at the infinitely close point P' of c_i to point P, are in the same plane. These curves are the lines of curvature of the surface. Two geodesics in space, normal to S at two infinitely close points P, P' of the line of curvature c_i intersect[3] at a point O_i and the distance $O_i P$ $(i = 1,2)$ (measured on the geodesic in space) linked to the principal radius of curvature R_i by the relation (see, for instance [8], p. 214)

$$R_i = k^{-1} \tan kr_i, \tag{3.3}$$

which can be taken as the definition of the principal radii of curvature R_i. Equality (3.3) is general for any curve in space; that is, the relative radius of curvature R of a curve in space is linked to the radius r of the osculating circle of the curve by the same relation

$$R = k^{-1} \tan kr. \tag{3.4}$$

[3] In this paper, we employ the language usually employed in differential geometry, which may not appear rigorous but which in fact is, if the meaning that should be given to the words is well understood. Saying that two infinitely close curves in space intersect means that they belong to a pencil with an envelope, and that the point at which each curve meets its infinitely close neighbour, is its point of contact with the envelope.

Geometrically, the definition of osculating circle is very clear, since it is the same as for the Euclidean case; from this definition, relation (3.4) then defines the relative radius of curvature of the curve. With this definition it is proven that for the curves of a surface the ordinary Euclidean geometry theorems of Meusnier and Euler ([5], p. 201) hold without modification. In other words, if θ is the angle between the principal normal of a curve of the surface and the normal to the surface, then between the radius of curvature R of the curve and R_n of the normal section the following relation holds

$$R = R_n \cos\theta, \tag{3.5}$$

and if the normal section forms an angle φ with one of the principal directions, we have

$$\frac{1}{R_n} = \frac{\cos^2\varphi}{R_1} + \frac{\sin^2\varphi}{R_2}. \tag{3.6}$$

Let us now return to the absolute total curvature K^α and the relative curvature K^r of the surface S. If K is the total curvature of the ambient space, it is known that between both there exists the relation ([5], p. 194)

$$K^r = K^\alpha - K. \tag{3.7}$$

Let $d\sigma$ be the area element of S. If N is the Euler characteristic of S, then we have the formula ([2], p. 166)[4]

$$C^\alpha = \int_S K^\alpha \, d\sigma = -2\pi N, \tag{3.8}$$

and therefore, denoting the area of S by F, from (3.7) we deduce that

$$C^r = \int_S K^r \, d\sigma = -2\pi N - KF. \tag{3.9}$$

The integrals C^α and C^r are known respectively as t*he total absolute and relative curvature* of the surface S. If the surface S is closed and orientable, it limits a body Q and the Euler characteristic N' of Q can be introduced into (3.9). It suffices to recall that between N and N' there is the following simple relation

$$N = 2N', \tag{3.10}$$

and thereby (3.9) can be written

$$C^r = -4\pi N' - KF. \tag{3.11}$$

Finally, it is worth recalling that in the space of constant curvature K, the arc element ds of a circle of radius r corresponding to the central angle of $d\alpha$ is given by the formula

$$ds = k^{-1} \sin kr \, d\alpha. \tag{3.12}$$

[4]For the definition of the Euler characteristic, see [10], p. 87, and for relation (3.10), p. 223.

4 An expression by Blaschke for kinematic density

It is useful for many purposes to give to expression (2.9) for kinematic density another more convenient form; this is the form given by Blaschke for the case of Euclidean space [3], which as we will see holds for the case of spaces of constant curvature.

Let S_0 and S_1 be two surfaces of continuous curvature in the space of constant curvature K. Assume that S_0 is fixed and consider the different positions gS_1 of S_1 obtained by subjecting this fixed surface to operations g of the group of movements in the space. We have seen that g is determined by the elements P, Ω, φ mentioned in Section 2, and thereby the kinematic density takes the form (2.9). Observe furthermore that dg can depend neither on point P nor on the directions serving as reference or origin for expressing Ω and φ, which are chosen in order to determine g.

Let us assume a position of S_1 that intersects S_0. Let P be a point on the intersection curve, and denote the tangent of this curve at P by t. Moreover, let θ be the angle formed by the normals to both surfaces at P, and ds_0, ds_1 be the arc elements of the curves normal to t contained by S_0 and S_1, respectively. If ds is the arc element of the intersection curve corresponding to point P, the volume element of the space can be expressed as follows:

$$dP = \sin\theta ds\, ds_0\, ds_1. \tag{4.1}$$

If the area elements of S_0 and S_1 at point P are denoted by $d\sigma_0$ and $d\sigma_1$, we have

$$d\sigma_0 = ds\, ds_0, \quad d\sigma_1 = ds\, ds_1, \tag{4.2}$$

and therefore from (4.1) we deduce that

$$[dP\, ds] = \sin\theta [d\sigma_0 d\sigma_1]. \tag{4.3}$$

Furthermore, taking the direction of t as the direction Ω that appears in (2.9), and denoting by φ_0, φ_1, the elementary rotations around the normals to S_0 and S_1 at P, we have

$$d\Omega = \sin\theta\, d\varphi_0\, d\varphi_1, \tag{4.4}$$

and furthermore, $d\varphi = d\theta$. Therefore, by exterior multiplication of the latter equalities, we have

$$[ds\, dg] = \sin^2\theta [d\sigma_0\, d\varphi_0\, d\sigma_1\, d\varphi_1], \tag{4.5}$$

which is the formula we wished to obtain. Recall that, although it is not explicitly stated on every occasion, both the densities and the exterior products analogous to those appearing on both sides of (4.5) are always considered in their absolute value.

5 Total curvature of the surface of the intersection of two bodies

Now let Q_0, Q_1 be two bodies in the space of constant curvature that we assume to be bounded and limited by the surfaces S_0, S_1 with *continuous curvatures.* Assume Q_0 to be fixed and consider the different positions gQ_1 of Q_1; that is, the different bodies obtained by applying the movements g in the space to Q_1. The body of intersection

$$Q_{01} = Q_0 \cdot gQ_1 \tag{5.1}$$

is limited by a closed surface S_{01} (which can be considered as being of several disjoint parts) formed by parts of S_0 and S_1 that are connected in the intersection curve H of S_0 and S_1, which is a discontinuity curve for the curvature of S_{01}. The total relative curvature $C^r(Q_0.gQ_1)$ of S_{01}, defined by the integral (3.9), consists of the integral C^r_{01} of the relative curvature of S_0 extended to the part of S_0 interior to gQ_1, plus the integral C^r_{10} of the relative curvature of S_1 extended to the part of gS_1 that is interior to S_0, and the part of C^r_i corresponding to the intersection curve of S_0 with gS_1. In other words,

$$C^r = C^r_{1o} + C^r_{01} + C^r_i. \tag{5.2}$$

It would be useful to have an expression for C^r_i; that is, the part that should be attributed to the intersection curve H when the total relative curvature of S_{01} is formed. One method for obtaining such an expression consists in replacing S_{01} by a nearby parallel surface having continuous relative curvature at all its points, then calculating the total relative curvature of this surface and passing to the limit when the distance between the two parallel surfaces tends to zero. Then the curve H corresponds to the band of the surface generated in the following manner: at each point P of H we assume the normals to n_0, n_1, to S_0, gS_1, and we consider the arc of circle radius ϵ and centre P limited by these normals; when P moves along H, this circle arc will describe a surface band, denoted by$B$, whose total relative curvature for $\epsilon \to 0$ is C^r_i. In order to calculate this total curvature, we begin by calculating the area element $d\sigma$ on B. On the considered circle arcs of radius ϵ, according to (3.12), we have

$$ds_1 = k^{-1} \sin k\epsilon \, d\alpha, \tag{5.3}$$

where α is the angle between the corresponding radius and the principal normal of H at P. These circle arcs are lines of curvature of band B, since they are normal in the plane; the other lines of curvature are the sections with B of the normals to H tangent to the evolutes of this curve. Therefore, if $r + \epsilon$ is the distance from P to the point of contact with the evolute (a distance that depends on α), then according to (3.12), between the arc element ds of H and ds_2 of B normal to ds_1, there will exist the relation

$$\frac{ds}{ds_2} = \frac{\sin k(r+\epsilon)}{\sin kr}.$$

Hence,

$$ds_2 = \frac{\sin kr}{\sin k(r+\epsilon)} ds. \tag{5.4}$$

Therefore, according to (5.3) and (5.4), we have

$$d\sigma = ds_1 ds_2 = \frac{\sin k\epsilon \sin kr}{k \sin k(r+\epsilon)} d\alpha ds. \tag{5.5}$$

Furthermore, the principal radii of curvature of B are $k^{-1} \tan k\epsilon$ and $k^{-1} \tan kr$; consequently, by writing the total relative curvature of B, and making $\epsilon \to 0$, we obtain

$$C_i^r = \int \frac{k}{\tan kr} \, d\alpha \, ds, \tag{5.6}$$

where for each value of s (that is, each point P), α changes between the angles α_0, α_1, that the normals to S_0, gS_1 form with the principal normal of H; s subsequently changes along the whole curve H. To check how r depends on α, observe that the distance $r + \epsilon$ from point P to the point of contact with the evolute is the length of the hypotenuse of a right triangle, one of whose catheti is equal to the radius of the osculating circle ρ and the angle spanned is equal to α. Then by a simple trigonometric formula of right-angled geodesic triangles in spaces of constant curvature k, we have

$$\tan k\rho = \cos\alpha \tan k(r+\epsilon).$$

Therefore, by making $\epsilon \to 0$ and substituting this in (5.6), we obtain

$$C_i^r = \int_H \int_{\alpha_0}^{\alpha_1} \frac{k \cos\alpha}{\tan k\rho} \, d\alpha \, ds;$$

that is,

$$C_i^r = \int_H \frac{\sin\alpha_1 - \sin\alpha_0}{\tan k\rho} ds. \tag{5.7}$$

6 Integral of the relative total curvature of the intersection of two bodies

We have seen that the relative total curvature $C^r(Q_0.gQ_1)$ of the intersection of the bodies mentioned in the previous Section decomposes according to (5.2), where the last part is given by (5.7). The aim now is to calculate the integral

$$\int_G C^r(Q_0.gQ_1) dg, \tag{6.1}$$

where dg is the kinematic density (2.9), and the integration is extended to all the group G of movements in the space of constant curvature.

The integral of C_{01}^r is calculated immediately by a customary trick in integral geometry. If K_0^r denotes the relative curvature of S_0 at the point whose area element is$d\sigma$, then we consider the integral

$$\int K_0^r d\sigma_0 dg$$

as extended, with respect to σ_0, to all the points of S_0 interior to gQ_1, and with respect to g, to the whole of group G. If we first fix g and integrate with respect to $d\sigma_0$, we obtain precisely the integral of C_{01}^r. However, if first σ_0 is fixed, according to (2.10), we have

$$\int K_0^r d\sigma_0 dg = 8\pi^2 V_1 \int_{S_0} K_0^r d\sigma_0 = 8\pi^2 V_1 C_0^r.$$

Thus, we have

$$\int C_{01}^r dg = 8\pi^2 V_1 C_0^r. \tag{6.2}$$

Analogously, by symmetry, we have

$$\int_G C_{10}^r dg = 8\pi^2 V_0 C_1^r, \tag{6.3}$$

where C_0^r and C_1^r are the relative total curves of the surfaces S_0 and S_1 which limit the bodies Q_0 and Q_1.

Now the integral C_i^r remains to be calculated. We proceed from expression (5.7); if α_i $(i = 0, 1)$ are the angles formed by the principal normal of H at P with the normal to S_i, taking into account (3.3), Meusnier's theorem gives

$$\tan k\rho = \tan kr_{in} \cos \alpha_i,$$

from which

$$\cos\alpha_0 + \cos\alpha_1 = \tan k\rho(\tan^{-1} kr_{on} + \tan^{-1} kr_{1n}).$$

Thus, employing the relation

$$\frac{\sin\alpha_1 - \sin\alpha_0}{\cos\alpha_1 + \cos\alpha_0} = \tan\frac{1}{2}(\alpha_1 - \alpha_0),$$

the value (5.7) of C_i^r can be written as follows:

$$C_i^r = k\int_H (\tan^{-1} kr_{on} + \tan^{-1} kr_{1n}) \tan\frac{1}{2}(\alpha_1 - \alpha_0) ds,$$

or by introducing the relative radii of curvature R_{0n}, R_{1n}, and by writing $\alpha_1 - \alpha_0 = \alpha$,

$$C_i^r = \int_H \left(\frac{1}{R_{0n}} + \frac{1}{R_{1n}}\right) \tan\frac{1}{2}\alpha\, ds. \tag{6.4}$$

Euler's theorem (3.6) now enables us to write

$$(6.5) \qquad C_i^r = \int (\frac{\cos^2 \varphi_0}{R_{01}} + \frac{\sin^2 \varphi_0}{R_{02}} + \frac{\cos^2 \varphi_1}{R_{11}} + \frac{\sin^2 \varphi_0}{R_{12}}) \tan \frac{1}{2} \alpha \, ds,$$

where the principal radii of curvature of S_i are denoted by R_{i1} and R_{i2}. Hence, taking into account (4.5), and that the angle between the normals to the surfaces is now α instead of θ, we have

$$(6.6) \qquad \int_G C_i^r dg = \int_G (\frac{\cos^2 \varphi_0}{R_{01}} + \frac{\sin^2 \varphi_0}{R_{o2}} + \\ + \frac{\cos^2 \varphi_1}{R_{11}} + \frac{\sin^2 \varphi_1}{R_{12}}) \tan \frac{1}{2} \alpha \sin^2 \alpha d\sigma_0 d\varphi_0 d\sigma_1 d\varphi_1,$$

bearing in mind all the positions of Q_1; that is, all the movements of g, angle α changes between $-\pi$ and $+\pi$ for each value of $\sigma_0, \varphi_0, \sigma_1, \varphi_1$, and therefore, since they must be considered as absolute values, we obtain

$$\int_{-\pi}^{\pi} | \tan \frac{1}{2} \alpha | \sin^2 \alpha d\alpha = 4.$$

The angles φ_0, φ_1 may change independently between 0 and 2π, when the integrals of $\sin^2 \varphi$ and $\cos^2 \varphi$ appear, which are equal to π. Then it is necessary to divide by 2, since if for any value of φ_0, φ_1 both angles are increased by π, and we have the same position for Q_1; that is, the same g. Finally, by writing

$$(6.7) \qquad \frac{1}{2} \int_{S_i} (\frac{1}{R_{i1}} + \frac{1}{R_{i2}}) d\sigma_i = M_i, \quad i = 0, 1,$$

which are the *total average curvatures* of S_i, we have

$$\int_G C_i^r dg = 8\pi^2 (M_0 F_1 + M_1 F_0),$$

and together with (6.2) and (6.3), we arrive at the final formula

$$(6.8) \qquad \int_G C^r (Q_0 . g Q_1) dg = 8\pi^2 (V_0 C_1^r + V_1 C_0^r + M_0 F_1 + M_1 F_0).$$

It is curious to observe that the spatial curvature does not appear in this result. Formula (6.8) thus has an identical form for any space of constant curvature, in particular for Euclidean space, in which case it coincides with Blaschke's fundamental formula, which is well known (see notes (1) and (2)).

7 The integral of the Euler characteristic

From (6.8) and (3.11) we immediately obtain the integral on G of Euler's characteristic $N'(Q_0.gQ_1)$ for the body Q_{01} obtained by intersection of Q_0 with gQ_1. To this end, it suffices to find the integral of the area F_{01} of the surface limiting Q_{01}. By following exactly the same method as before for finding the integral of C^r_{01} and C^r_{1o}, we obtain

$$\int_G F_{01} dg = 8\pi^2(F_1V_0 + F_0V_1). \tag{7.1}$$

Hence, taking into account (6.8) and (3.11), we have

$$\int_G N'(Q_0.gQ_1)dg = 8\pi^2(N'_1V_0 + N'_0V_1) - 2\pi(M_0F_1 + M_1F_0), \tag{7.2}$$

where N'_0 and N'_1 are the Euler characteristics of Q_0 and Q_1. This is the *fundamental formula* sought. Once again one may observe that it does not depend on the constant curvature of the space: this formula is the same as that in the Euclidean case.

Remarks: *a*) The fact that the spatial curvature does not appear in (7.2) is not general, but rather is related to the number of dimensions of the space. In the case of two dimensions this does not occur. The formula analogous to (7.2) for two dimensions is

$$\int_G N'(Q_0.gQ_1)dg = 2\pi(N'_1F_0 + N'_0F_1) + KF_0F_1, \tag{7.3}$$

where F_i are the areas, L_i the boundary lengths, and N'_i the Euler characteristics ($N' = -$vertices + edges − faces) of the two considered bi-dimensional domains in the plane of constant curvature K. *b)* In the proof of (7.2), we employed (6.8), which was established by assuming that S_0 and S_1 possessed continuous curvature, which is a highly restrictive condition. However, once (7.2) is established, it is straightforward to demonstrate by continuity its validity for much more general cases. Let us assume, for example, that S_0 and S_1 are formed by an infinite number of surface patches (or "faces"), each one having continuous curvature, but joined by edges that are discontinuity curves for the curvature. Then by rounding the edges to change S_0 and S_1 into surfaces with continuous curvature, and by applying (7.2) to these new surfaces, on passing to the limit (7.2) also holds in this case. It merely suffices to clarify the value of the mean curvature M for bodies with edges, which can be deduced from the method followed in Section 5. Indeed, the mean curvature at a point on the band B considered therein was

$$\frac{1}{2}k(\tan^{-1} k\epsilon + \tan^{-1} kr),$$

and thus, multiplying by (5.5), integrating, and making $\epsilon \to 0$, it turns out that to the total mean curvature of the faces it is necessary to add, as a contribution to the edges, the value

$$M = \frac{1}{2}\int \alpha\, ds, \tag{7.4}$$

where α is the angle formed by the normals to the faces along the edges, and s is the arc length of these edges; α is a function of s and the integration is extended to all the edges of the surface.

8 Application: Steiner's formula for spaces of constant curvature

Let us consider a particular case where Q_0 and Q_1 are two *convex bodies;* this means that the geodesic arc connecting any two points on these bodies is completely contained within them. Then $N_0' = N_1' = N' = -1$, and the fundamental formula (7.2) is

$$\int dg = 8\pi^2(V_0 + V_1) + 2\pi(M_0F_1 + M_1F_0), \tag{8.1}$$

where the integration is extended to all the movements g for which $Q_0.gQ_1 \neq 0$. Assume in particular that Q_1 is a sphere of radius ρ. Then by taking the point P that appears in expression (2.9) of dg as the centre of the sphere, for each position of P the integral of $d\Omega d\varphi$ gives π^2, and the integral of dP extended to all the positions in which $Q_0.gQ_1 \neq 0$ is precisely the *volume of the parallel body exterior* to Q_0 at distance ρ. Denoting this volume by V_0, and where

$$\begin{aligned} M_1 &= 4\pi k^{-1}\sin k\rho\cos k\rho \\ F_1 &= 4\pi k^{-2}\sin^2 k\rho \\ V_1 &= 2\pi k^{-3}(k\rho - \sin k\rho\cos k\rho), \end{aligned}$$

we obtain

$$\begin{aligned} V_\rho = V_0 + F_0k^{-1}\sin k\rho\cos k\rho &+ M_0k^{-2}\sin^2 k\rho \\ &+ 2\pi k^{-3}(k\rho - \sin k\rho\cos k\rho), \end{aligned} \tag{8.2}$$

which generalizes to spaces of constant curvature Steiner's formula on the volume of the parallel body exterior to a convex body. For $k = 0$ we have the classical Steiner formula. A direct proof of (8.2), valid for a space of n dimensions, was given by C. B. Allendoerfer [1]. For the bidimensional case, see also [12]. The case $K>0$ (Elliptic space) can also be found in [13]; in this case, the value of ρ must be less than $(\pi/2)k^{-1}$, since the geodesics are the finite length equal to πk^{-1}.

9 Generalization of some results by O. Varga to spaces of constant curvature

Starting with formula (4.5), which has the same form as for Euclidean space, O. Varga obtained different mean values of functions defined on the intersection curve of two surfaces in Euclidean space, when one of the surfaces is assumed to be fixed and the other moving without deformation to all the possible positions [11].

All the formulae obtained by Varga hold equally for spaces of constant curvature, merely by replacing the absolute by relative curvatures with respect to the ambient space. We may mention as an example only one of them, which leads to the generalization to spaces of constant curvature (or, if one prefers, to non-Euclidean geometries) of a theorem given by Varga for Euclidean space. As usual, let S_0 and S_1 be two surfaces with continuous curvature in a space of constant curvature. It is not necessary now to assume that they limit a spatial body; they can be non-orientable open or closed surfaces. As before, let H be the intersection curve of S_0 with gS_1. Let τ be the torsion of H at point P (understood as relative torsion; that is, with respect to the metric of the ambient space), and s be the arc of H. Then denoting by α_0 the angle of the principal normal of H with the normal to S_0, and by φ_0 the angle of the tangent to H with the first principal direction of S_0 (all at point P), it is known that the following formula (18) holds ([5], p. 99)

$$\tau = -\frac{d\alpha_0}{ds} + (\frac{1}{R_{o2}} - \frac{1}{R_{01}}) \sin\varphi_0 \cos\varphi_0, \tag{9.1}$$

in which R_{o2}, R_{01} are the principal radii of curvature of S_0 at P. This formula has an identical form as for ordinary space. Analogously, for $S1$ we have

$$\tau = -\frac{d\alpha_1}{ds} + (\frac{1}{R_{12}} - \frac{1}{R_{11}}) \sin\varphi_1 \cos\varphi_1, \tag{9.2}$$

and thus, by subtracting both expressions, and denoting the angle of both surfaces at P by $\alpha = \alpha_1 - \alpha_0$, we obtain

$$\frac{d\alpha}{ds} = (\frac{1}{R_{12}} - \frac{1}{R_{11}}) \sin\varphi_1 \cos\varphi_1 - (\frac{1}{R_{o2}} - \frac{1}{R_{01}}) \sin\varphi_0 \cos\varphi_0.$$

Squaring and writing in an abbreviated form

$$\Delta_1 = \frac{1}{R_{12}} - \frac{1}{R_{11}}, \quad \Delta_0 = \frac{1}{R_{02}} - \frac{1}{R_{01}}$$

we obtain

$$\begin{aligned}(\frac{d\alpha}{ds})^2 = {} & \Delta_1^2 \sin^2\varphi_1 \cos^2\varphi_1 + \Delta_0^2 \sin^2\varphi_0 \cos^2\varphi_0 - \\ & -2\Delta_1\Delta_0 \sin\varphi_1 \cos\varphi_1 \sin\varphi_0 \cos\varphi_0\end{aligned} \tag{9.3}$$

Let us multiply this equality side by side with (4.5) and integrate the whole group G. In the term on the left-hand side, it is necessary to integrate $\alpha, \varphi_1, \varphi_0$, from 2π and then divide by 2, since each pair of values $\varphi_0, \varphi_1, \varphi_0 + \pi$ and $\varphi_1 + \pi$ corresponds to the same gS_1. Let

$$\int_0^{2\pi} \sin^2\varphi \cos^2\varphi d\varphi = \frac{1}{4}\pi,$$

and observing that the integral of the negative term of expression (9.3) is zero, we arrive at

$$(9.4) \qquad \int_G \left(\int_H \left(\frac{d\alpha}{ds}\right)^2 ds\right) dg = \frac{\pi^3}{4}\left(F_0 \int_{S_1} \Delta_1^2 F d\sigma_1 + F_1 \int_{S_0} \Delta_0^2 d\sigma_0\right).$$

This is the formula obtained by Varga for Euclidean space and it holds equally for spaces of constant curvature. From this formula we deduce that if $\alpha = a$ constant, then $\Delta_1 = 0$, $\Delta_2 = 0$ and therefore the surfaces have all their umbilical points; that is, they are spheres or parts of spheres of spherical surface. It is now possible to state the following: *If the angle in which two intersecting surfaces in a space of constant curvature is constant along the intersection line for all positions of the surfaces, then they are spheres or parts of spheres (including the planes as spheres of infinite radius).*

10 Fundamental formula for planes

The density for sets of planes has already been defined in Section 2; that is, the invariant volume element in the homogeneous space G/E. Let us now see how formula (7.2) is generalized for the case where, instead of Q_1, there is a moving plane e. First, it is necessary to obtain a formula analogous to (4.5). To that effect, let us consider a position of e intersected by a fixed surface S_0. Let P be a point on the intersection curve H and n_0, n_1 the normals to S_0 and to e at P. We take expression (2.13) for de. If θ i is the angle between n_0 and n_1, and φ_0 denotes the rotation about n_0 that determines the position of the plane n_0, n_1, we have $d\Omega = \sin\theta d\theta d\varphi$. Furthermore, if ds_0 is the arc element on S_0 normal to H, we have $dt = \sin\theta ds_0$. Thus $de = \sin^2\theta[d\theta d\varphi ds_0]$ By introducing the arc element ds of H into the terms on both sides, and denoting by $d\sigma_0$ the area element of S_0 corresponding to point P, we obtain the formula we were seeking:

$$(10.1) \qquad [ds\, de] = \sin^2\theta\, [d\theta\, d\sigma_0\, d\varphi_0].$$

Now let Q_0 be a fixed body limited by the closed surface S_0 of continuous curvature. Let e be the plane intersecting Q_0, and consider a point P of the intersection curve H. Multiply the terms on both sides of (10.1) by the geodesic curvature $\varkappa_g$ of H, considered as a curve of e, at P. Since e is a plane, this geodesic curvature coincides with the ordinary curvature (understood as relative to the space of constant curvature under consideration). By Meusnier's theorem, if R_n is the radius of curvature of the normal section of S_0 tangent to H, we have

$$(10.2) \qquad \frac{1}{\varkappa_g} = R_n \sin\theta,$$

and therefore, recalling Euler's formula and the definition of mean curvature, we have

$$\begin{aligned} \int_{G/E} \left(\int_H \varkappa_g ds\right) de &= \frac{1}{2}\int_0^{2\pi} |\sin\theta|\, d\theta \int_{S_0} d\sigma_\rho \int_0^{\pi} \left(\frac{\cos^2\varphi_0}{R_{01}} + \frac{\sin^2\varphi_0}{R_{o2}}\right) d\varphi_0 \\ &= 2\pi M_0. \end{aligned}$$

Furthermore, the Gauss-Bonnet formula applied to the plane e and the intersection curve H of the plane with S_0 yields

$$\int_H \varkappa_g ds = -2\pi N' - KF_{01},$$

where F_{01} is the area of intersection of e with Q_0 (the area limited by H and which may consist of several disjoint parts) and N' is the Euler characteristic of this area. Moreover, as is easy to calculate by analogy with the Euclidean case,

$$\int F_{1o} de = 2\pi V_0,$$

where V_0 is the volume of Q_0, we arrive at the integral formula

$$\int_{G/E} N' de = -M_0 - KV_0. \tag{10.3}$$

This formula is known for the cases $K = 0$ (see (1)) and $K = 1$ (see (2)). If Q_0 is convex, such that the intersection curve H consists always of only one part, we have $N' = -1$ if e intersects Q_0, and $N' = 0$ in the opposite case. Therefore, in this case we have

$$\int_{e \cdot Q_0 \neq \emptyset} de = M_0 + KV_0, \tag{10.4}$$

which gives the *measure of the set of planes intersecting a convex body in the space of constant curvature* K. Since this measure cannot be negative, we may deduce that: *Between the mean curvature* M_0 *and the volume* V_0 *of the whole convex body in the space of constant negative curvature* $K = -1/\alpha^2$, there exists the relation

$$V_0 < \alpha^2 M_0. \tag{10.5}$$

11 An interpretation of the fundamental formula for convex bodies

Let Q_0 be a convex body in the space of constant curvature K, and V_0 its volume, F_0 its area and M_0 its integral of mean curvature. The measure of the set of planes having a common point with Q_0 is given by (10.4); this is now denoted by e_0; that is,

$$e_0 = M_0 + KV_0. \tag{11.1}$$

The measure of the set of planes contained in Q_0 is its volume; denoting this volume by p_0, we have

$$p_0 = V_0. \tag{11.2}$$

The measure of the set of lines having a common point with Q_0 is immediately deduced from (2.18) in exactly the same way as for the Euclidean or Elliptic case (notes (1) and (2)), thus yielding

$$r_0 = \frac{1}{2}\pi F_0. \tag{11.3}$$

Hence, assuming now another moving convex body Q_1, *the measure of the set of positions of Q_1 having a common point with Q_0*, which as we have seen is given by (8.1), can be written as follows:

$$\int_{Q_0 \cdot gQ_1 \neq \emptyset} dg = 8\pi^2(p_0 + p_1) + 4(e_0 r_1 + e_1 r_0)) - 4K(p_0 r_1 + p_1 r_0), \tag{11.4}$$

which does not give the said measure when we take as invariants of the convex bodies the measures of the points they contain, and of the planes and lines intersecting the bodies.

Facultad de Ciencias Físicomatemáticas
La Plata.

References

[1] C. B. Allendoerfer, *Steiner's formulae on a general S^{n+1}*, Bull. Am. Math. Soc. **54** (1948), 128–135.

[2] W. Blaschke, *Differentialgeometrie*, vol. I, Springer, Berlin, 1930.

[3] W. Blaschke, *Integralgeometrie 17. Über Kinematik.*, Deltion, Athens, 1936.

[4] W. Blaschke, *Vorlesungen über Integralgeometrie*, vol. 20, B. G. Teubner, Leipzig and Berlin, 1937.

[5] E. Cartan, *Leçons sur la géométrie des spaces de Riemann*, 2d. edition ed., Gauthier- Villars, Paris, 1946.

[6] S. S. Chern and T. Yien, *Sulla formula principale cinematica dello spazio ad n dimensioni*, Boll. della Unione Matematica Italiana **2** (1940), 432–437.

[7] Shiing-shen Chern, *On integral geometry in Klein spaces*, Ann. of Math. **43** (1942), 178–189.

[8] E. P. Eisenhart, *Riemannian Geometry*, Princeton, 1926.

[9] L. A. Santaló, *Integral geometry on surfaces*, Duke Math. J. **16** (1949), 361–375.

[10] H. Seifert and W. Threlfall, *Lehrbuch der Topologie.*, Leipzig und Berlin: B. G. Teubner. VII, 353 S., 132 Fig. , 1934.

[11] O. Varga, *Integralgeometrie. XIX: Mittelwerte an dem Durchschnitt bewegter Flächen.*, Math. Z. **41** (1936), 768–784.

[12] E. Vidal-Abascal, *A generalization of Steiner's formulae*, Bull. A. Math. Soc. **53** (1947), 841–844.

[13] Ta-Jen Wu, *Ueber elliptische Geometrie*, Math.Zeits. **43** (1938), 495–521.

INTEGRAL GEOMETRY ON SURFACES

By L. A. Santaló

1. Introduction. The Integral Geometry on surfaces was initiated by Blaschke and Haimovici [2], [6]. The particular cases of the elliptic and hyperbolic plane were studied by the author in two preceding papers [12], [13].

Starting from the concepts introduced by Blaschke and Haimovici, it is the purpose of this paper to continue the study of the Integral Geometry on a more general class of surfaces.

In §2 we recall the definition of density for sets of geodesics and obtain the expression (2.7) which plays a fundamental role in all subsequent discussions and is used in §3 to obtain some integral formulas related to the set of geodesics which intersect a fixed curve of finite length on the given surface. In §4 we give a definition of convex domains on the surface and apply it to obtain some mean value theorems. In §5 we consider sets whose elements are pairs of geodesics and obtain some integral formulas and a theorem referring to geodesic lines. In §§6 and 7 we are concerned with the definition of cinematic density on surfaces and with the generalization of the Poincaré formula of Integral Geometry [3; 23]. Finally, in §8 we give a very general inequality (8.7) for convex curves on surfaces which contains as a particular case the isoperimetric inequality (8.15) for surfaces of constant gaussian curvature.

We shall restrict ourselves to *complete analytic Riemannian surfaces* in the sense of Hopf and Rinow [9]. Consequently the following two properties are satisfied: (a) every geodesic ray can be continued to infinite length; (b) any pair of points can be joined by a curve of shortest length which is a geodesic. Furthermore we assume throughout the whole paper that our surface satisfies Condition A stated at the end of the Introduction and in §§3, 4 and 5 we add the stronger Condition B stated in §3.

We shall represent the surface by S, and in general it will be useful to use on it geodesic polar coordinates with pole (in each case chosen in a suitable position) represented by O. The first fundamental formula becomes

$$ds^2 = d\rho^2 + g(\rho, \theta)^2 \, d\theta^2. \tag{1.1}$$

The pole O together with the function $g(\rho, \theta)$ is called an *element* of S and will be represented by $E(0, g)$ or simply by E. We suppose that each element E is continuable to the complete surface S in the sense of Rinow [11] and Myers [10]. Then $g(\rho, \theta)$ must be an analytic function of ρ for $\theta =$ constant and all $\rho > 0$, which satisfies the conditions

$$g(0, \theta) = 0, \qquad g_\rho(0, \theta) = 1. \tag{1.2}$$

Received October 1, 1948; in revised form, January 14, 1949.

According to the property (b) of the complete surfaces, every point on S has at least one pair (ρ, θ) associated with it. The polar coordinate system $(0; \rho, \theta)$ defined by the element E will be called *regular* on a domain D when every point $P \varepsilon D$ has only one coordinate pair ρ, θ. The maximal domain D with this property will be called the *domain of regularity* of the system $(0; \rho, \theta)$. Throughout the whole paper we shall assume that our surface satisfies the following

Condition A. The surface S can be covered by a finite number of domains of regularity with respect to suitable polar coordinate systems.

2. **Density and measure for sets of geodesics.** Let (x_1, x_2) be a system of curvilinear coordinates on S and

$$ds^2 = g_{11}\, dx_1^2 + 2g_{12}\, dx_1\, dx_2 + g_{22}\, dx_2^2 \tag{2.1}$$

its first fundamental differential quadratic form.

It is customary to introduce the notations

$$p_1 = \partial F/\partial x_1', \qquad p_2 = \partial F/\partial x_2', \tag{2.2}$$

where $F(x_1, x_2, x_1', x_2') = (g_{11}x_1'^2 + 2g_{12}x_1'x_2' + g_{22}x_2'^2)^{\frac{1}{2}}$.

It is well known that, given a point $P(x_1, x_2)$ on S and two values p_1, p_2 compatible with (2.2), there is determined a geodesic G which passes through P and has at this point a direction (x_1', x_2') such that the values x_i, x_i', p_i $(i = 1, 2)$ satisfy the equations (2.2).

Given a two dimensional field of geodesics $G(\alpha, \beta)$ depending upon two parameters α, β, that is, given four functions $x_i(\alpha, \beta)$, $p_i(\alpha, \beta)$ $(i = 1, 2)$, let us consider the exterior differential form

$$dG = [dx_1\, dp_1] + [dx_2\, dp_2], \tag{2.3}$$

where the square brackets denote that the multiplication is exterior (see, for instance, E. Cartan [4]). It is known that this differential form has the following two fundamental properties of invariance:

1. It is invariant with respect to a change of coordinates on the surface.
2. It is invariant under displacements of the elements (x_1, x_2, p_1, p_2) on the respective geodesic.

The proof of these properties of invariance can be found in the cited papers of Blaschke and Haimovici.

The differential form dG, *taken always in absolute value*, is called the *density* for sets of geodesics, and the *measure* of a set of geodesics is defined as the integral of dG extended over the set.

In order to give a geometrical interpretation of the density dG let us take on our surface S a system of geodesic polar coordinates (ρ, θ) with the pole at a fixed point O. Then the first fundamental form takes the form (1.1) and $F = (\rho'^2 + g^2\theta'^2)^{\frac{1}{2}}$.

If $\rho = \rho(t)$, $\theta = \theta(t)$ are the equations of a geodesic G, the angle V formed at the point $P(\rho, \theta)$ between G and the geodesic radius OP is given by

$$\tan V = g\theta'/\rho'. \tag{2.4}$$

Consequently we have

$$\begin{aligned} p_1 &= \frac{\partial F}{\partial \rho'} = \rho'(\rho'^2 + g^2\theta'^2)^{-\frac{1}{2}} = \cos V, \\ p_2 &= \frac{\partial F}{\partial \theta'} = g\theta'(\rho'^2 + g^2\theta'^2)^{-\frac{1}{2}} = g \sin V, \end{aligned} \tag{2.5}$$

$$\begin{aligned} dp_1 &= -\sin V\, dV, \\ dp_2 &= \frac{\partial g}{\partial \rho} \sin V\, d\rho + \frac{\partial g}{\partial \theta} \sin V\, d\theta + g \cos V\, dV. \end{aligned} \tag{2.6}$$

Since $x_1 = \rho$, $x_2 = \theta$, the density for geodesics assumes the form

$$dG = -\sin V\, [d\rho\, dV] + \frac{\partial g}{\partial \rho} \sin V\, [d\theta\, d\rho] + g \cos V\, [d\theta\, dV]. \tag{2.7}$$

This expression for dG is general for any surface S and for any set of geodesics which have a common point with a regular domain with respect to the geodesic polar coordinate system defined by the element $E(0, g)$ (see Introduction). However this condition does not mean a loss of generality. Indeed any set of geodesics can be divided into partial subsets each of which have a common point with the regular domains of certain polar coordinate systems. In this case we shall apply (2.7) to each particular subset.

The expression (2.7) takes a remarkably simple form if from the pole O it is possible to draw a geodesic normal to each geodesic of the considered set. In this case, according to property 2 of invariance, we can choose on each geodesic the point in which it is cut by the normal geodesic through O; that is, the point for which it is $V = \pi/2$. Then (2.7) takes the simple form

$$dG = \left| \frac{\partial g}{\partial \rho} \right| [d\theta\, d\rho]. \tag{2.8}$$

For instance, in the case of the plane, $g = \rho$ and the density for straight lines becomes

$$dG = [d\rho\, d\theta] \tag{2.9}$$

as is well known [3; 7], [5; 59]. For surfaces of constant positive gaussian curvature $k = 1$, $g = \sin \rho$ and consequently

$$dG = |\cos \rho\, | [d\theta\, d\rho]. \tag{2.10}$$

For surfaces of constant negative curvature $k = -1$, $g = \sinh \rho$ and we obtain

$$(2.11) \qquad dG = |\cosh \rho| \, [d\theta \, d\rho]$$

as may be obtained directly [13].

3. **Geodesics which intersect a fixed curve.** Before continuing, let us observe that the foregoing definition of measure for sets of geodesics is not applicable to surfaces on which all or almost all geodesics are *transitive*. (For the definition of transitivity see Hedlund [8].) In this case it is obvious that, according to the given definition, the measure of any set of geodesics is zero. In order to avoid this case we shall assume in this section and in the successive §§4 and 5 that the surface S satisfies the following additional condition:

Condition B. There exists a constant $r > 0$ such that any geodesic arc of length $\leq r$ is intersected by almost all the geodesics in only a finite number (zero included) of points.

Here the words "almost all" must be understood in the sense of Hedlund [8].

Let C be a rectifiable curve on S composed of a finite number of arcs with continuously turning tangent (we shall say that C is of class D^1). Let L be the length of C. We want to consider the set of geodesics which intersect C.

Let P be an intersection point of the geodesic G with C. We choose a polar coordinate system $(0; \rho, \theta)$ the domain of regularity of which contains P. If s is the arc length on C and ζ the angle at P between C and the geodesic radius OP we have

$$(3.1) \qquad d\rho = \cos \zeta \, ds, \qquad g \, d\theta = \sin \zeta \, ds.$$

In order to determine G, according to property 2 of invariance, we can choose on each G one point, say P, of its intersection points with C. That means that ρ and θ in (2.7) are related by the equation $\rho = \rho(\theta)$ of C; in other words, $d\rho$ and $d\theta$ are those deduced from (3.1).

Consequently (2.7) takes the form

$$(3.2) \qquad \begin{aligned} dG &= -\sin V \cos \zeta \, [ds \, dV] + \cos V \sin \zeta \, [ds \, dV] \\ &= \sin (\zeta - V) \, [ds \, dV]. \end{aligned}$$

If φ denotes the angle between C and G at P, then $\varphi = \zeta - V$ and since ζ is a function of s only, we have $dV = \zeta' \, ds - d\varphi$ and thus (3.2) may be written $dG = \sin (\zeta - V) \, [ds \, d\varphi]$, or since dG will be considered always in absolute value,

$$(3.3) \qquad dG = |\sin \varphi| \, [d\varphi \, ds].$$

This expression of dG, independent of any coordinate system (depending only upon the fixed curve C), yields a remarkable and very general integral formula.

Let $f(s, \varphi)$ be an integrable function defined on C depending upon the point $P(s)$ and the direction φ at it. Multiplying both sides of (3.3) by $f(s, \varphi)$ we get

$$f(s, \varphi)\, dG = f(s, \varphi) \mid \sin \varphi \mid [d\varphi\, ds]. \tag{3.4}$$

Performing the integration of these equivalent differential forms over the field $0 \leq \varphi \leq \pi$, $0 \leq s \leq L$ (L = length of C), in the left side each geodesic G appears as a common factor of the sum $\sum_1^n f(s_i, \varphi_i)$ of the values of $f(s, \varphi)$ at the n intersection points of G with C. Consequently we have

$$\int_{G \cdot C \neq 0} \sum_1^n f(s_i, \varphi_i)\, dG = \int_0^L \int_0^\pi f(s, \varphi) \sin \varphi\, d\varphi\, ds. \tag{3.5}$$

We want to consider as examples some particular cases of this integral formula.

(a) If $f(s, \varphi) = 1$ we get

$$\int_{G \cdot C \neq 0} n(C \cdot G)\, dG = 2L, \tag{3.6}$$

where $n(C \cdot G)$ denotes the number of intersection points of C and G and the integral is extended over all geodesics which intersect C. This formula (3.6), which generalizes to surfaces a well-known formula for the plane, was previously obtained by Blaschke and Haimovici in their cited papers.

(b) If $f(s, \varphi) = f(\varphi)$ we get

$$\int_{G \cdot C \neq 0} \sum_1^n f(\varphi_i)\, dG = L \int_0^\pi f(\varphi) \sin \varphi\, d\varphi. \tag{3.7}$$

For instance, if $f = \varphi$, we have

$$\int_{G \cdot C \neq 0} \sum_1^n \varphi_i\, dG = \pi L, \tag{3.8}$$

where φ_i ($i = 1, 2, \cdots, n$) are the angles under which G cuts C at the n intersection points.

Other examples, obtained by easy calculation, are

$$\int_{G \cdot C \neq 0} \sum_1^n \mid \sin^m \varphi_i \mid dG = \frac{\Gamma((m/2) + 1)}{\Gamma((m + 3)/2)} \pi^{\frac{1}{2}} L, \tag{3.9}$$

$$\int_{G \cdot C \neq 0} \sum_1^n \frac{1}{\mid \sin \varphi_i \mid} dG = \pi L,$$

where m is any integer >0.

(c) If $f(s, \varphi) = f(s)$ we obtain

$$(3.10) \qquad \int_{G \cdot C \neq 0} \sum_1^n f(s_i)\, dG = 2 \int_0^L f(s)\, ds.$$

4. **Convex curves and convex domains.** We shall say that a simple closed curve C of class D^1 on S is *convex* when any geodesic which intersects C has either two points or a whole arc in common with C. In this case, if C has a finite length L, the measure of the geodesics which have a common arc with C is zero and consequently (3.6) gives: *The measure of the geodesics cutting a convex curve is equal to the length of this curve.*

A domain K on S will be said to be *convex* when the following three properties are satisfied: (1) It is bounded by a closed convex curve. (2) It is homeomorphic to a circle. (3) Every geodesic with a point P interior to K can be prolonged from P in both senses to points outside K.

Let C be a curve of class D^1 and length L contained in the interior of a convex domain K. Let L_0 be the length of the boundary curve C_0 of K. According to the foregoing condition (3) every geodesic which intersects C will also intersect C_0. Consequently (3.6) gives the following mean value for the number of intersection points of C and all geodesics which cut C_0:

$$(4.1) \qquad n^* = \int_{G \cdot C_0 \neq 0} n(C \cdot G)\, dG \Big/ \int_{G \cdot C_0 \neq 0} dG = 2L/L_0 .$$

As an immediate consequence we get the theorem:

Given on S a curve C of class D^1 and length L contained inside a convex domain bounded by a convex curve of length L_0, there exist geodesic lines which intersect C in a number of points $\geq 2L/L_0$.

Exactly the same method, applied to (3.5), yields the more general theorem:

Given a curve C of class D^1 contained inside a convex domain bounded by a convex curve of length L_0 and an integrable function $f(s, \varphi)$ depending upon the points $P(s)$ of the given curve and upon the angles around P, there exist geodesic lines G for whose intersection points $P(s_i)$ $(i = 1, 2, 3, \cdots, n)$ with C the relation

$$\sum_1^n f(s_i, \varphi_i) \geq \frac{1}{L_0} \int_0^L \int_0^\pi f(s, \varphi) \sin \varphi \, d\varphi\, ds$$

holds, where φ_i is the angle at P_i under which G cuts C.

5. **Pairs of geodesics which intersect.** Let G_1 and G_2 be two geodesics which intersect at the point P. We choose a polar coordinate system $(0; \rho, \theta)$, the domain of regularity of which contains P. According to property 2 of in-

variance of the density for geodesics (§2), in order to evaluate the densities dG_1, dG_2 we can choose as point (ρ, θ) the same point $P(\rho, \theta)$ for both geodesics. Then (2.7) gives

$$dG_1 = -\sin V_1 [d\rho\, dV_1] + \frac{\partial g}{\partial \rho} \sin V_1 [d\theta\, d\rho] + g \cos V_1 [d\theta\, dV_1],$$

$$dG_2 = -\sin V_2 [d\rho\, dV_2] + \frac{\partial g}{\partial \rho} \sin V_2 [d\theta\, d\rho] + g \cos V_2 [d\theta\, dV_2],$$

where V_1 and V_2 are the angles which the geodesic OP form at P with G_1 and G_2 respectively.

By exterior multiplication of the last equalities we get

$$[dG_1\, dG_2] = g \sin (V_1 - V_2) [d\rho\, d\theta\, dV_1\, dV_2]. \tag{5.1}$$

On the other hand, the element of area dP of S at P is

$$dP = g\, [d\rho\, d\theta]. \tag{5.2}$$

Consequently we may write

$$[dG_1\, dG_2] = | \sin (V_1 - V_2) | [dP\, dV_1\, dV_2], \tag{5.3}$$

a formula which has the same form as for pairs of straight lines in the plane [3; 17].

We want to make three applications of the formula (5.3).

(a) Let D be a domain on S of area F, let $f(P)$ be an integrable function defined over D and let us consider the integral

$$I = \int_{G_1 \cdot G_2 \cdot D \neq 0} \sum_1^m f(P_i)\, dG_1\, dG_2\,, \tag{5.4}$$

where the sum is extended over all intersection points P_i of G_1 and G_2 which are contained in D, and the integral is extended over all pairs of geodesics which intersect among themselves inside D.

According to (5.3) the integral I may be written

$$I = \int_D \int_0^\pi \int_0^\pi f(P) \,|\sin (V_1 - V_2)|\, dP\, dV_1\, dV_2 = 2\pi \int_D f(P)\, dP. \tag{5.5}$$

On the other hand, according to (3.10), the integral I has also the value

$$I = 2 \int_{G_1 \cdot C \neq 0} \left(\int_\sigma f(P)\, ds \right) dG_1 \tag{5.6}$$

where s means the arc length on G_1 and σ is the arc of G_1 which lies within D.

Consequently we have the integral formula

$$(5.7) \qquad \int_{G_1 \cdot C \neq 0} \left(\int_\sigma f(P)\, ds \right) dG = \pi \int_D f(P)\, dP.$$

For instance, if $f(P) = 1$, we get

$$(5.8) \qquad \int_{G_1 \cdot C \neq 0} \sigma\, dG = \pi F$$

which generalizes to surfaces a well-known formula for the plane [3; 19], [5; 75].

(b) If G_1 and G_2 intersect in more than one point, it may be observed that (5.3) holds for every intersection point. That is, if G_1 and G_2 intersect also at the point P', and V_1', V_2' are the corresponding angles between G_1, G_2 and OP', then

$$(5.9) \qquad |\sin (V_1 - V_2)|\, [dP\, dV_1\, dV_2] = |\sin (V_1' - V_2')|\, [dP\, dV_1'\, dV_2'].$$

As an immediate consequence we get the theorem:

If D and D' are two domains on S such that every pair of geodesics which intersect in D intersect also with the same number of intersection points (assumed finite) in D' and conversely, then D and D' have the same area.

Indeed the integral (5.5) for $f(P) = 1$ equals $2\pi F$; according to the assumptions of the theorem and to (5.9) the integral must have the same value when D is substituted by D', consequently $F = F'$ (F' = area of D').

(c) Let us assume that S is such that any pair of geodesics always has one and only one common point. Let C be a convex curve boundary of a convex domain K (§4). We then have

$$(5.10) \qquad J = \int_{G_1 \cdot K \neq 0,\, G_2 \cdot K \neq 0} dG_1\, dG_2 = L^2.$$

On the other hand, applying (5.3), we have

$$(5.11) \qquad \begin{aligned} J = &\int_{P \epsilon K} dP \int_0^\pi \int_0^\pi |\sin (V_1 - V_2)|\, dV_1\, dV_2 \\ &+ \int_{P \epsilon' K} \int_\omega \int_\omega |\sin (V_1 - V_2)|\, dV_1\, dV_2\, dP, \end{aligned}$$

where ω means the angle at P covered by the geodesics through P which meet K. From (5.10) and (5.11) we deduce

$$(5.12) \qquad \int_{P \epsilon' K} (\omega - \sin \omega)\, dP = \tfrac{1}{2} L^2 - \pi F.$$

This is a formula obtained by Crofton for the case of the plane [3; 18], [5; 78]. We see that it holds without modification for surfaces such that any pair of geodesics always intersects in one and only one point (for instance for the elliptic plane).

6. **Cinematic density.** Let $K = PQ$ be now an arc of length l of a geodesic G; we shall say that K is an oriented segment of origin P and end point Q. In order to measure a set of oriented segments we shall take the integral extended over the set of the differential form

$$dK = [dG^{\rightarrow} ds] \tag{6.1}$$

where $dG^{\rightarrow} = 2\,dG$ is the density for oriented geodesics and s means the abscissa measured on G of the origin P, that is, the distance measured on G between P and a fixed point on G. Obviously (6.1) does not depend on this fixed point on G.

The expression (6.1) will be called the *cinematic density* on S (see Blaschke [2]).

If ρ, θ are the geodesic polar coordinates of P and V is the angle between G and OP (see §2) we have the relation

$$ds = d\rho/\cos V. \tag{6.2}$$

Consequently by exterior multiplication of (2.7) and (6.2), we get

$$dK = g[d\theta\, dV\, d\rho], \tag{6.3}$$

or, if dP is the element of area at P, according to (5.2) we can write the following new form for the cinematic density

$$dK = [dP\, dV]. \tag{6.4}$$

Let us make an application of these two equivalent forms obtained for the cinematic density.

Let C be a convex curve on S which bounds a domain D of area F. We want to measure the set of segments K of length l which meet the domain D. According to (6.1), and calling σ the length of the arc of G which lies within D, we obtain by use of (5.8)

$$m(K\cdot D \neq 0) = \int_{K\cdot D\neq 0} dG^{\rightarrow}\, ds = \int_{G\cdot D\neq 0} (\sigma + l)\, dG^{\rightarrow} = 2\pi F + 2lL, \tag{6.5}$$

where L is the length of C.

On the other hand if we take the form (6.4) for dK we get

$$\begin{aligned} m(K\cdot D \neq 0) &= \int_{K\cdot D\neq 0} dP\, dV \\ &= \int_{P\epsilon D} dP\, dV + \int_{P\epsilon' D} dP\, dV = 2\pi F + \int_{P\epsilon' D} \omega\, dP, \end{aligned} \tag{6.6}$$

where ω means the angle covered by the segments K of origin $P \not\varepsilon D$ which meet D. From (6.5) and (6.6) we get the integral formula

$$\int_{P\varepsilon' D} \omega \, dP = 2lL. \tag{6.7}$$

Another application is obtained if we observe that $[dP \, dV]$, being equivalent to (6.1), is invariant by displacements of P on G. For instance, we have

$$[dP \, dV] = [dQ \, dV'] \tag{6.8}$$

where Q is the end point of the segment K. Let us perform the integration of (6.8) over all segments with $Q \varepsilon D$. The right side gives $\int_{Q \varepsilon D} dQ \, dV' = 2\pi F$ and the left side gives $\int_{Q\varepsilon D} dP \, dV = \int_{Q \varepsilon D} \omega_1 \, dP$.

Consequently we obtain the integral formula

$$\int_{Q \varepsilon D} \omega_1 \, dP = 2\pi F \tag{6.9}$$

where ω_1 means the angle covered by the segments K of length l and origin P which have their end point $Q \varepsilon D$, and the integral is extended over all points P for which $\omega_1 \neq 0$. As we see from the proof, (6.9) holds for any domain D, not necessarily convex.

7. The Poincaré formula for surfaces. As in the preceding number, let K be a geodesic segment of length l and let C be a fixed curve on S of class C^1. If $n(K \cdot C)$ is the number of intersection points of C and K and $n(G \cdot C)$ the number of intersection points of G and C, from (3.6) we deduce

$$\int_{K \cdot C \neq 0} n(K \cdot C) \, dK = \int_{K \cdot C \neq 0} n(K \cdot C) \, dG^{\rightarrow} \, ds = l \int_{G \cdot C \neq 0} n(G \cdot C) \, dG^{\rightarrow} = 4lL. \tag{7.1}$$

Now let K be a polygonal line $P_1 P_2 P_3 \cdots P_m$ on S formed by m geodesic segments $K_i = P_i P_{i+1}$ of length l_i and with given angles $P_{i-1} P_i P_{i+1} = \alpha_i$ at the vertices. Given the position of an arbitrary side $K_i = P_i P_{i+1}$ of the polygonal, the position of the whole polygonal is determined.

The cinematic density for K_i will be $dK_i = [dP_i \, dV_i]$. Since this density is invariant under displacements on geodesics, we have $[dP_i \, dV_i] = [dP_{i+1} \, dV_i']$, where V_i' is the angle between OP_{i+1} and the geodesic $P_i P_{i+1}$. Consequently $V_i' = \pi - (\alpha_{i+1} + V_{i+1})$ and since $\alpha_{i+1} = \text{constant}$ we get (in absolute value) $[dP_i \, dV_i] = [dP_{i+1} \, dV_{i+1}]$ and thus

$$dK_1 = dK_2 = dK_3 = \cdots = dK_m = dK. \tag{7.2}$$

Any one of the densities dK_i can be consequently taken as density for the polygonal line K. If there does not exist a coordinate system $(0; \rho, \theta)$ with domain of regularity containing P_i and P_{i+1}, we divide the side $P_i P_{i+1}$ into the necessary number of parts in order that each partial arc be contained in the domain of regularity of a suitable coordinate system; by this method we see that (7.2) also holds in this case.

Let us apply (7.1) to each side of K. We obtain

$$\int_{K_i \cdot C \neq 0} n(K_i \cdot C)\, dK_i = 4 l_i L. \tag{7.3}$$

Adding for $i = 1, 2, 3, \cdots, m$ and taking into account (7.2), we get

$$\int_{K \cdot C \neq 0} n(K \cdot C)\, dK = 4 L_0 L \tag{7.4}$$

where $L_0 = \sum l_i$ denotes the total length of K.

This formula (7.4) can be included as a particular case of a more general one in which instead of a polygonal line we have a curve composed of a finite number of arcs with continuous geodesic curvature (we will say a curve of class D^2).

Now let K be a curve of such a class and let $\kappa_g = \kappa_g(\sigma)$ be its natural equation (κ_g = geodesic curvature, σ = arc length). Let $P_0 = P(\sigma_0)$ be a point of K. Given an arbitrary point P on S and a direction V at P, by means of the equation $\kappa_g = \kappa_g(\sigma)$, we can construct a unique curve which passes through P and is in this point tangent to the direction V. We shall say that this curve is *congruent* to K. Two congruent curves on S have the same length and the same κ_g at corresponding points, however, their form can obviously be very different; one of them, for instance, may be closed or have double points and the other be open and simple.

In order to measure a set of congruent curves K we shall take the integral over the set of the cinematic density $dK = [dP\, dV]$ (see (6.4)). To admit this definition we must prove that the measure does not depend upon the point $P(\sigma_0)$ chosen on K. That is, however, an immediate consequence of (7.2) if we consider a polygonal line inscribed in K with the sides sufficiently small.

Now let C be a fixed curve of class D^1 and length L. If C and K have a common point we can choose an intersection point as point $P(\sigma_0)$. Then the element of area dP may be written in the form

$$dP = |\sin\theta|\, [d\sigma\, ds] \tag{7.5}$$

where s means the arc length on C, σ the arc length on K, and θ the angle between C and K at P. The cinematic density takes the form (since we consider it always non-negative)

$$dK = |\sin\theta|\, [d\sigma\, ds\, dV]. \tag{7.6}$$

If V_1 is the angle between OP and C at P, then $V = V_1 - \theta$, and because V_1 depends only upon s (not upon σ, θ) we obtain

$$dK = |\sin\theta|\, [ds\, d\sigma\, d\theta]. \tag{7.7}$$

This expression has the same form as in the case of the plane [3; 23]. Consequently by integration over the field $0 \leq \theta \leq 2\pi$, $0 \leq s \leq L$, $0 \leq \sigma \leq L_0$ (L_0 = length of K) we obtain the same result

$$\int_{C \cdot K \neq 0} n(C \cdot K)\, dK = 4LL_0 . \tag{7.8}$$

This formula, for the plane, is called the Poincaré formula [3; 24]. Consequently (7.8) is the generalized Poincaré formula for surfaces. For the hyperbolic plane, see [13].

8. **An inequality for curves on surfaces.** In this section, K will represent a geodesic circle of variable center Q and fixed radius λ.

Let C be a curve of length L and class D^1 and let P be an intersection point of K and C.

According to §6 the expression $dK = [dQ\, dV]$ is invariant if we displace it along the geodesic radius QP, that is, $[dQ\, dV] = [dP\, dV_1]$ (V = angle between OQ and QP; V_1 = angle between OP and QP). Further if s and σ are the arc lengths of C and K respectively and θ is the angle which these curves form at P, then $dP = |\sin\theta|\, [d\sigma\, ds]$. Consequently we have

$$dK = [dQ\, dV] = |\sin\theta|\, [d\sigma\, ds\, d\theta]. \tag{8.1}$$

This formula is not the same as (7.7). Now K has a variable length. However by integration of (8.1) over the field $0 \leq \theta \leq 2\pi$, $0 \leq s \leq L$, $0 \leq \sigma \leq L_0(s, \theta)$, where $L_0(s, \theta)$ means the length of the geodesic circle of radius λ and center $Q(s, \theta)$, we obtain

$$\int_{K \cdot C \neq 0} n(K \cdot C)\, dK = \int_0^L \int_0^{2\pi} |\sin\theta|\, L_0(s, \theta)\, ds\, d\theta, \tag{8.2}$$

where $n(K \cdot C)$ means the number of intersection points of K with C, the integral being extended over all geodesic circles of radius λ which intersect C.

By definition we call *mean value* of the length of all geodesic circles of radius λ with a common point with C the value L_0^* such that

$$\int_0^L \int_0^{2\pi} |\sin\theta|\, L_0(s, \theta)\, ds\, d\theta = L_0^* \int_0^L \int_0^{2\pi} |\sin\theta|\, ds\, d\theta = 4LL_0^* . \tag{8.3}$$

With this definition (8.2) may be written

$$\int_{K \cdot C \neq 0} n(K \cdot C)\, dK = 4LL_0^* . \tag{8.4}$$

This is the formula we shall need in what follows. However we may observe that the given proof is general and that by introduction of the mean value of the length of a curve (defined by (8.3)) the Poincaré formula (7.8) holds also for curves K of variable length in all cases in which these curves are determined by a point P and a direction at it.

Now let C be the boundary curve of a convex domain D of finite area F. Let λ_i be the maximum radius of the geodesic circles contained in C and λ_e the minimum of those which enclose C. Let λ be a value such that

$$\lambda_i \leq \lambda \leq \lambda_e . \tag{8.5}$$

The area covered by the centers of the geodesic circles of radius λ with a common point with C will be $F + A_\lambda$, where A_λ means the area between C and its parallel exterior curve at distance λ. Hence we have

$$\int_{K \cdot C \neq 0} dK = \int_{Q \epsilon F + A_\lambda} dQ \int_0^{2\pi} dV = 2\pi(F + A_\lambda). \tag{8.6}$$

Since C and K are closed curves, we have always $n \geq 2$ (except for positions of contact of zero measure) and consequently, from (8.4) and (8.6) we deduce

$$LL_0^* - \pi(F + A_\lambda) \geq 0. \tag{8.7}$$

This is our fundamental inequality, which can be stated as follows: *Given on the surface S a convex domain of area F bounded by a curve C of length L, if L_0^* is the mean length of the geodesic circles of radius λ which intersect C and $F + A_\lambda$ is the area bounded by the exterior parallel curve to C at distance λ, where λ satisfies* (8.5), *then the inequality* (8.7) *holds.*

The inequality (8.7) may be considered as an isoperimetric inequality. Let us see its form for the case of surfaces of constant gaussian curvature k.

In this case the length L_0 and area F_0 of the geodesic circles of radius λ are given by

$$L_0 = 2\pi k^{-\frac{1}{2}} \sin k^{\frac{1}{2}}\lambda, \qquad F_0 = 2\pi k^{-1}(1 - \cos k^{\frac{1}{2}}\lambda) \tag{8.8}$$

and the area $F + A_\lambda$ bounded by the exterior parallel curve to C at distance λ has the value (see for instance Allendoerfer [1])

$$F + F_0 + (LL_0 - kFF_0)(2\pi)^{-1}. \tag{8.9}$$

Consequently the inequality (8.7) becomes

$$LL_0 - 2\pi(F + F_0) + kFF_0 \geq 0. \tag{8.10}$$

If we introduce the "isoperimetric deficit"

$$\Delta = L^2 + kF^2 - 4\pi F \tag{8.11}$$

and take into account that it is zero for the geodesic circles, we can write the identity

$$LL_0 - 2\pi(F + F_0) + kFF_0 = \frac{1}{2FF_0}\left[\Delta F_0^2 - (LF_0 - FL_0)^2\right]. \tag{8.12}$$

From (8.10) and (8.12) we deduce

$$\Delta \geq \frac{1}{F_0^2} (LF_0 - FL_0)^2. \tag{8.13}$$

This inequality contains already the isoperimetric inequality $\Delta \geq 0$. In order to see that the equality $\Delta = 0$ holds only for the geodesic circles, let us introduce in (8.13) the values (8.8). We obtain

$$\Delta \geq (L - Fk^{\frac{1}{2}} \cot \tfrac{1}{2}k^{\frac{1}{2}}\lambda)^2. \tag{8.14}$$

Since this inequality holds for any λ between λ_i and λ_e we have $\Delta \geq (L - Fk^{\frac{1}{2}} \cot \frac{1}{2}k^{\frac{1}{2}}\lambda_i)^2$, $\Delta \geq (L - Fk^{\frac{1}{2}} \cot \frac{1}{2}k^{\frac{1}{2}}\lambda_e)^2$, and taking into account the inequality $x^2 + y^2 \geq \frac{1}{2}(x - y)^2$, we get

$$\Delta \geq \tfrac{1}{4}kF^2(\cot \tfrac{1}{2}k^{\frac{1}{2}}\lambda_e - \cot \tfrac{1}{2}k^{\frac{1}{2}}\lambda_i)^2. \tag{8.15}$$

In order to see that the right side is always positive we observe that for $k < 0$, (8.15) may be written

$$\Delta = \tfrac{1}{4} \mid k \mid F^2(\coth \tfrac{1}{2} \mid k \mid^{\frac{1}{2}}\lambda_e - \coth \tfrac{1}{2} \mid k \mid^{\frac{1}{2}}\lambda_i)^2. \tag{8.16}$$

The inequality (8.15) is then stronger than the inequality

$$L^2 + kF^2 - 4\pi F \geq 0 \tag{8.17}$$

which is the "isoperimetric inequality" for surfaces of constant curvature k. It proves that the equality $\Delta = 0$ holds only if $\lambda_i = \lambda_e$, that is, if C is a geodesic circle.

Finally, let us observe that for the plane ($k = 0$), the inequality (8.15) takes the form

$$L^2 - 4\pi F \geq F^2\left(\frac{1}{\lambda_i} - \frac{1}{\lambda_e}\right)^2 \tag{8.18}$$

which is an improvement of the classical inequality $L^2 - 4\pi F \geq 0$.

Bibliography

1. Carl B. Allendoerfer, *Steiner's formulae on a general S^{n+1}*, Bulletin of the American Mathematical Society, vol. 54(1948), pp. 128–135.
2. W. Blaschke, *Integralgeometrie XI: Zur Variationsrechnung*, Abhandlungen aus dem Mathematischen Seminar der Hansischen Universität, vol. 11(1936), pp. 359–366.
3. W. Blaschke, *Vorlesungen über Integralgeometrie. I*, Hamburger mathematische Einzelschriften, Heft 20, 1935.
4. Élie Cartan, *Les Systèmes Différentiels Extérieurs et Leurs Applications Géométriques*, Actualités Scientifiques et Industrielles, No. 994, Paris, 1945.
5. R. Deltheil, *Probabilités Géométriques*, Paris, 1926.
6. M. Haimovici, *Géométrie intégrale sur les surfaces courbes*, Annales Scientifiques de l'Université de Jassy, vol. 23(1936), pp. 57–74.
7. M. Haimovici, *Géométrie intégrale sur les surfaces courbes*, Comptes Rendus de l'Académie des Sciences, Paris, vol. 203(1936), pp. 230–232.

8. GUSTAV A. HEDLUND, *The dynamics of geodesic flows*, Bulletin of the American Mathematical Society, vol. 45(1939), pp. 241–260.

9. H. HOPF AND W. RINOW, *Ueber den Begriff der vollständigen differentialgeometrischen Fläche*, Commentarii Mathematici Helvetici, vol. 3(1931), pp. 209–225.

10. SUMNER BYRON MYERS, *Connections between differential geometry and topology*, this Journal, vol. 1(1935), pp. 376–391.

11. W. RINOW, *Über Zusammenhänge zwischen der Differentialgeometrie im Grossen und im Kleinen*, Mathematische Zeitschrift, vol. 35(1932), pp. 512–528.

12. L. A. SANTALÓ, *Integral formulas in Crofton's style on the sphere and some inequalities referring to spherical curves*, this Journal, vol. 9(1942), pp. 707–722.

13. L. A. SANTALÓ, *Integral geometry on surfaces of constant negative curvature*, this Journal, vol. 10(1943), pp. 687–704.

THE INSTITUTE FOR ADVANCED STUDY
AND
FACULTAD DE CIENCIAS EXACTAS.

Reprinted from Vol. I, PROCEEDINGS OF THE INTERNATIONAL CONGRESS OF MATHEMATICIANS, 1950
Printed in U.S.A.

INTEGRAL GEOMETRY IN GENERAL SPACES

L. A. SANTALÓ

Let E be a space of points in which a locally compact group of transformations G operates transitively. Let dx be the left invariant element of volume in G. Let H_0 and K_0 be two sets of points in E and denote by xH_0 the transformed set of H_0 by x ($x \in G$). Let us first assume that the identity is the only one transformation of G which leaves H_0 invariant. If $F(K_0 \cap xH_0)$ is a function of the intersection $K_0 \cap xH_0$, the main purpose of the so-called Integral Geometry (in the sense of Blaschke) is the evaluation of integrals of the type

$$I = \int_G F(K_0 \cap xH_0)\, dx \tag{1}$$

and to deduce from the result some geometrical consequences for the sets K_0 and H_0.

Let us now suppose that there is a proper closed subgroup g of G which leaves H_0 invariant. The elements $H = xH_0$ will then be in one to one correspondence with the points of the homogeneous space G/g. If there exists in G/g an invariant measure and dH denotes the corresponding element of volume, the Integral Geometry also deals with integrals of the type

$$I = \int_{G/g} F(K_0 \cap H)\, dH \tag{2}$$

from which it tries to deduce geometrical consequences for K_0.

In what follows we shall give some examples and applications of the method.

1. Immediate examples. Let us assume G compact and therefore of finite measure which we may suppose equal 1. In order to define a measure $m(K_0)$ of a set of points K_0, invariant with respect to G, we choose a fixed point P_0 in E and set

$$m(K_0) = \int_G \varphi(x)\, dx \tag{3}$$

where $\varphi(x) = 1$ if $xP_0 \in K_0$ and $\varphi(x) = 0$ otherwise.

If the measures $m(K_0)$, $m(H_0)$, and $m(K_0 \cap xH_0)$ exist, it is then known and easy to prove that

$$\int_G m(K_0 \cap xH_0)\, dx = m(K_0)m(H_0), \tag{4}$$

and since $\int_G dx = 1$, the mean value of $m(K_0 \cap xH_0)$ will be $m(K_0)m(H_0)$. Therefore we have: *Given in E two sets K_0, H_0, there exists a transformation x of G such that $m(K_0 \cap xH_0)$ is equal to or greater than $m(K_0)m(H_0)$.*

If K_0 consists of N points P_i $(i = 1, 2, \cdots, N)$ and call $\nu(K_0 \cap xH_0)$ the number of points P_i which belong to xH_0, we want to evaluate $\int_G \nu(K_0 \cap xH_0)\, dx$. We set $\varphi_i(x) = 1$ if $xP_i \in H_0$ and $\varphi_i(x) = 0$ otherwise. According to (3) and the invariance of dx we have

$$m(H_0) = \int_G \varphi_i(x)\, dx = \int_G \varphi_i(x^{-1})\, dx$$

where $\varphi_i(x^{-1}) = 1$ if $P_i \in xH_0$ and $\varphi_i(x^{-1}) = 0$ otherwise. Consequently we have

$$\int_G \nu(K_0 \cap xH_0)\, dx = \sum_1^N \int_G \varphi_i(x^{-1})\, dx = Nm(H_0). \tag{5}$$

Thus the mean value of ν is equal to $Nm(H_0)$ and we have: *Given N points P_i in E and a set H_0 of measure $m(H_0)$, there exists a transformation x of G such that xH_0 contains at least $Nm(H_0)$ of the given points; it contains certainly a number greater than $Nm(H_0)$ if H_0 is closed.*

2. An application to convex bodies. Let E be now the euclidean 3-space and G the group of the unimodular affine transformations which leave invariant a fixed point O. Let H be the planes of E. The subgroup g will consist of all affinities of G which leave invariant a fixed plane H_0. Each plane H can be determined by its distance p to O and the element of area $d\omega_2$ on the unit 2-sphere corresponding to the point which gives the direction normal to H. The invariant element of volume in G/g is then given by

$$dH = p^{-4}\, dp\, d\omega_2\,. \tag{6}$$

Let K_0 be a convex body which contains O in its interior, and let $p(\omega_2)$ be the support function of K_0 with respect to O. If we set $F(K_0 \cap H) = 0$ if $K_0 \cap H \neq 0$ and $F(K_0 \cap H) = 1$ if $K_0 \cap H = 0$, (2) reduces to

$$I(O) = \int_{K_0 \cap H = 0} dH = \frac{1}{3} \int p^{-3}\, d\omega_2 \tag{7}$$

where the last integral is extended over the whole 2-sphere. If O is an affine invariant point of K_0 (for instance, its center of gravity), (7) gives an affine invariant for convex bodies (with respect to unimodular affinities). The minimum of I with respect to O is also an affine invariant which we shall represent by I_m.

By comparing I_m with the volume V and the affine area F_a of K_0 the following theorem can be shown: *Between the unimodular affine invariants I_m, F_a, and V of a convex body the inequalities*

$$I_m V \leqq (4\pi/3)^2, \qquad I_m F_a^2 \leqq (2^6/3)\pi^3 \tag{8}$$

hold, where the equalities hold only if K is an ellipsoid.

For the analogous relations for the plane see [3]. I do not know if in (8) I_m can be replaced by the invariant $I(O)$ corresponding to the center of gravity of K_0.

3. The group of motions in a space of constant curvature. The best known case is that in which $E \equiv S_n$ is an n-dimensional space of constant curvature k and G is the group of motions in it. In this case the invariant element of volume dx in G is well known. If (P_0, e_0^i) $(i = 1, 2, 3, \cdots, n)$ denotes a fixed n-frame (i.e., a point P_0 and n unit mutually orthogonal vectors with the origin at P_0), any motion x can be determined by the n-frame $(P = xP_0, e^i = xe_0^i)$. Let dP be the element of volume in S_n at P and let $d\omega_{n-1}$ be the element of area on the unit euclidean $(n-1)$-sphere corresponding to the direction of e^1; let $d\omega_{n-2}$ be the element of area on the unit euclidean $(n-2)$-sphere orthogonal to e^1 corresponding to the direction of e^2 and so forth. Then dx can be written

$$dx = [dP\, d\omega_{n-1}\, d\omega_{n-2} \cdots d\omega_1]. \tag{9}$$

Let C_p be a p-dimensional variety $(p \leqq n)$ of finite p-dimensional area A_p and C_q a q-dimensional variety $(q \leqq n)$ of q-dimensional area A_q. Assuming $p + q \geqq n$, let $A_{p+q-n}(C_p \cap xC_q)$ be the $(p + q - n)$-dimensional area of $C_p \cap xC_q$. Then the formula

$$\int_G A_{p+q-n}(C_p \cap xC_q)\, dx = \frac{\omega_{p+q-n}}{\omega_p\, \omega_q}\, \omega_1 \omega_2 \cdots \omega_n A_p A_q \tag{10}$$

holds, where ω_i denotes the area of the euclidean unit i-sphere, that is,

$$\omega_i = \frac{2\pi^{(i+1)/2}}{\Gamma((i+1)/2)}. \tag{11}$$

If $p + q = n$, $A_{p+q-n}(C_p \cap xC_q)$ denotes the number of points of the intersection $C_p \cap xC_q$. Notice that (10) is independent of the curvature k of S_n.

If, instead of C_q, we consider a q-dimensional linear subspace L_q^0 of S_n, and denote by $\bar{A}_{p+q-n}(C_p \cap L_q)$ the area of the intersection of C_p with $L_q = xL_q^0$, we obtain

$$\int_{G/g} A_{p+q-n}(C_p \cap L_q)\, dL_q = \frac{\omega_{p+q-n}\, \omega_{n-q}}{\omega_p} A_p \tag{12}$$

where g is the group of motions which leaves invariant L_q^0 and dL_q is the invariant element of volume in the homogeneous space G/g normalized in such a way that the measure of the L_q which cut the unit $(n-q)$-sphere in S_n be its area (depending upon the curvature k of the space).

We shall give two applications:

a) Let S_n be the euclidean n-sphere; G is then compact and according to (9) its total volume will be $\int_G dx = \omega_1\omega_2 \cdots \omega_n$. In this case the mean value of $A_{p+q-n}(C_p \cap xC_q)$ will be $\omega_{p+q-n}\omega_p^{-1}\omega_q^{-1}A_pA_q$. Consequently we have: *Given on the euclidean n-sphere two varieties C_p, C_q of dimensions p, q $(p + q \geqq n)$ and finite areas A_p, A_q, there exists a motion x such that the area of the intersection $C_p \cap xC_q$ is equal to or greater than $\omega_{p+q-n}\omega_p^{-1}\omega_q^{-1}A_pA_q$.*

b) We shall now give an application to the elementary non-euclidean geometry.

Let T be a tetrahedron in the 3-dimensional space of constant curvature k and let L_2 be the planes of this space. One can show that

$$\text{(13)} \qquad \int_{T \cap L_2 \neq 0} dL_2 = \frac{1}{\pi}\left(\sum_1^6 (\pi - \alpha_i) l_i + 2kV\right)$$

where l_i are the lengths of the edges of T and α_i the corresponding dihedral angles; V is the volume of T.

On the other hand (12) applied to the edges of T gives

$$\text{(14)} \qquad \int_{G/0} N(T \cap L_2)\, dL_2 = 2 \sum_1^6 l_i$$

where $N(T \cap L_2)$ denotes the number of edges which are intersected by L_2 and therefore is either $N = 3$ or $N = 4$. From (13) and (14) we can evaluate the measures of the sets of planes L_2 corresponding to $N = 3$ and $N = 4$. These measures being non-negative we get the inequalities (for the euclidean case see [5])

$$2 \sum_1^6 \alpha_i l_i - 4kV \leqq \pi \sum_1^6 l_i \leqq 3 \sum_1^6 \alpha_i l_i - 6kV$$

which for $k = 1$, $k = -1$ gives the following inequalities for the volume V of a tetrahedron in non-euclidean geometry

$$\frac{1}{4} \sum_1^6 (2\alpha_i - \pi) l_i \leqq V \leqq \frac{1}{6} \sum_1^6 (3\alpha_i - \pi) l_i \quad \text{for the elliptic space,}$$

$$\frac{1}{6} \sum_1^6 (\pi - 3\alpha_i) l_i \leqq V \leqq \frac{1}{4} \sum_1^6 (\pi - 2\alpha_i) l_i \quad \text{for the hyperbolic space.}$$

These inequalities may have some interest because, as is known, V cannot be expressed in terms of elementary functions of l_i and α_i.

4. A definition of p-dimensional measure of a set of points in euclidean n-space. Let E be now the euclidean n-dimensional space E_n. The methods of Integral Geometry can be used in order to give a definition of area for p-dimensional surfaces (see Maak [3], Federer [2], and for a comparative analysis Nöbeling [4]). The idea of the method is as follows. The formulas (10) and (12) hold for varieties which have a well-defined p- and q-dimensional area in the classical sense. For more general varieties the same formulas (10), (12) can be taken as a definition for A_p (taking for C_q a variety with A_q well-defined), *provided the integrals on the left-hand sides exist.* The problem is therefore to find the conditions of regularity which C_p must satisfy in order that the integrals (10) or (12) exist.

We want to give an example.

Let C be a set of points in E_n and let dP_i be the element of volume in E_n at the point P_i. Let N_s be the number of common points of C with s unit $(n - 1)$-

spheres whose centers are the points $P_1, P_2, P_3, \cdots, P_s$. Let us consider the following integrals (in the sense of Lebesgue)

$$I_s = \int N_s \, dP_1 \, dP_2 \, dP_3 \cdots dP_s \qquad (s = 1, 2, 3, \cdots) \tag{15}$$

extended with respect to each P_i to the whole E_n.

The q-dimensional measure of C can be defined by the formula

$$m_q(C) = \frac{\omega_q}{2\omega_n^q} I_q. \tag{16}$$

Notice that if I_r is the first integral in the sequence $I_1, I_2, I_3, \cdots$ which has a finite value (may be zero), then $m_q(C) = \infty$ for $q < r$ and $m_q(C) = 0$ for $q > r$. The number r can be taken as the definition for the *dimension* of C.

If C is a q-dimensional variety with tangent q-plane at every point, (16) gives the ordinary q-dimensional area of C (as may be deduced from (10)). The definition (16) may be applied whenever the integrals (15) exist. Following a method used by Nöbeling [4] in similar cases, it is not difficult to prove that the integrals I_s exist if C is an analytic set (or Suslin set).

5. Application to Hermitian spaces. Let $E \equiv P_n$ be now the n-dimensional complex projective space with the homogeneous coordinates $\xi_0, \xi_1, \cdots, \xi_n$ and let G be the group of linear transformations which leaves invariant the Hermitian form $(\xi\bar{\xi}) = \sum \xi_i \bar{\xi}_i$.

If we normalize the coordinates ξ_i such that $(\xi\bar{\xi}) = 1$, every variety C_p of complex dimension p posseses an invariant integral of degree $2p$, namely $\Omega^p = (\sum [d\xi_i d\bar{\xi}_i])^p$ (see Cartan [1]). Let us put

$$J_p(C_p) = \frac{p!}{(2\pi i)^p} \int_{C_p} \Omega^p. \tag{17}$$

It is well known that if C_p is an algebraic variety of dimension p, $J_p(C_p)$ coincides with its order.

If C_p is an analytic variety ("synectic" according to Study, i.e., defined by complex analytic relations) the methods of Integral Geometry give a simple interpretation of the invariant J_p. Let L_{n-p}^0 be a linear subspace of dimension $n - p$ and put $L_{n-p} = xL_{n-p}^0$. If g is the subgroup of G which leaves L_{n-p}^0 invariant, and dL_{n-p} means the invariant element of volume in the homogeneous space G/g normalized in such a way that the total volume of G/g is equal 1, the formula

$$\int_{G/g} N(C_p \cap L_{n-p}) \, dL_{n-p} = J_p(C_p) \tag{18}$$

holds, where $N(C_p \cap L_{n-p})$ denotes the number of points of intersection of C_p with L_{n-p}.

A more general formula, assuming $p + q \geqq n$, is the following

(19) $$\int_{G/g} J_{p+q-n}(C_p \cap L_q)\, dL_q = J_p(C_p)$$

which coincides with (18) for $q = n - p$.

If dx is the element of volume in G normalized in such a way that the total volume of G is equal 1, given two analytic varieties C_p, C_q we also have

(20) $$\int_G J_{p+q-n}(C_p \cap xC_q)\, dx = J_p(C_p) J_q(C_q)$$

which may be considered as the generalization to analytic varieties of the theorem of Bezout.

For $p + q = n$, the foregoing formula (20) is a particular case of a much more general result of de Rham [6].

6. Integral geometry in Riemannian spaces. The Integral Geometry in an n-dimensional Riemannian space R_n presents a different aspect. Here we do not have, in general, a group of transformations G. However, if we take as geometrical elements the geodesic curves Γ of R_n, it is possible to consider integrals analogous to (2), though conceptually different, and to deduce from them geometrical consequences.

Let $ds^2 = g_{ij}\, du^i\, du^j$ be the metric in R_n and let us set $\varphi = (g_{ij}\dot{u}^i\dot{u}^j)^{1/2}$ and $p_i = \partial\varphi/\partial\dot{u}^i$. The exterior differential form $d\Gamma = (\sum[dp_i\, du^i])^{n-1}$ of degree $2(n-1)$ is invariant under displacements of the elements u^i, p_i on the respective geodesic. Therefore we can define the "measure" of a set of geodesic curves as the integral of $d\Gamma$ extended over the set.

Let us consider a bounded region D_0 in R_n and let different arcs of geodesic contained in D_0 be taken as different geodesic lines. Let us consider a geodesic Γ which intersects an $(n-1)$-dimensional variety C_{n-1} contained in D_0 at the point P. Let $d\sigma$ denote the element of $(n-1)$-dimensional area on C_{n-1} at P. If $d\omega_{n-1}$ denotes the element of area on the unit euclidean $(n-1)$-sphere corresponding to the direction of the tangent to Γ at P and θ denotes the angle between Γ and the normal to C_{n-1} at P, the differential form $d\Gamma$ may be written in the form $d\Gamma = |\cos\theta|\,[d\omega_{n-1} d\sigma]$. If C_{n-1} has a finite $(n-1)$-dimensional area A_{n-1} and $N(C_{n-1} \cap \Gamma)$ denotes the number of intersection points of C_{n-1} and Γ, from the last form for $d\Gamma$ it follows that

(21) $$\int_{D_0} N(C_{n-1} \cap \Gamma)\, d\Gamma = \frac{\omega_{n-2}}{n-1} A_{n-1}$$

where the integral is extended over all geodesics of D_0.

If dt denotes the element of arc on Γ and dP is the element of volume in R_n at P, clearly we have $[d\Gamma\, dt] = [dP\, d\omega_{n-1}]$. From this relation if we consider all arc elements (Γ, t) with the origin within a given region D (contained in D_0)

of finite volume $V(D)$ and call $L(D \cap \Gamma)$ the length of the arc of Γ which lies within D, we obtain

$$\int_{D_0} L(D \cap \Gamma)\, d\Gamma = \frac{1}{2}\, \omega_{n-1} V(D). \tag{22}$$

Some consequences of the formulas (21) and (22) for the case $n = 2$ have been given in [8]. They have particular interest for the Riemannian spaces of finite volume whose geodesic lines are all closed curves of finite length (for $n \geqq 3$ it seems, however, not to be known if such spaces, other than spheres, exist). For instance, one can easily show: *If the geodesic lines of a Riemannian space R_n of finite volume V are all closed curves of constant length L and there exists in R_n an $(n - 1)$-dimensional variety of area A_{n-1} which intersects all the geodesic curves, the inequality*

$$LA_{n-1} \geqq \frac{(n-1)\omega_{n-1}}{2\omega_{n-2}} V$$

holds (equality for the elliptic space).

References

1. E. Cartan, *Sur les invariants intégraux de certains espaces homogènes clos et les propriétés topologiques de ces espaces*, Annales de la Société Polonaise de Mathematiques vol. 8 (1929) pp. 181–225.
2. H. Federer, *Coincidence functions and their integrals*, Trans. Amer. Math. Soc. vol. 59 (1946) pp. 441–466.
3. W. Maak, *Oberflächenintegral und Stokes Formel im gewöhnlichen Raume*, Math. Ann. vol. 116 (1939) pp. 574–597.
4. G. Nöbeling, *Ueber den Flächeninhalt dehnungsbeschränkter Flächen*, Math. Zeit. vol. 48 (1943) pp. 748–771.
5. G. Pólya and G. Szegö, *Aufgaben und Lehrsätze aus der Analysis*, Berlin, 1925, vol. 2, p. 166.
6. G. de Rham, *Sur un procède de formation d'invariants intégraux*, Jber. Deutschen Math. Verein. vol. 49 (1939) pp. 156–161.
7. L. A. Santaló, *Un invariante afin para las curvas planas*, Mathematicae Notae vol. 8 (1949) pp. 103–111.
8. ———, *Integral geometry on surfaces*, Duke Math. J. vol. 16 (1949) pp. 361–375.

University of Rosario,
Rosario, Argentina.

MEASURE OF SETS OF GEODESICS IN A RIEMANNIAN SPACE AND APPLICATIONS TO INTEGRAL FORMULAS IN ELLIPTIC AND HYPERBOLIC SPACES (*)

By L. A. SANTALÓ

1. Introduction

The Integral Geometry on surfaces has been considered by BLASCHKE [2], HAIMOVICI [5], VIDAL ABASCAL [9] and the author [7]. In § 2, 3, 4, 5 of the present paper we consider some points of the Integral Geometry in a Riemannian n-dimensional space. We start with the definition of density for sets of geodesics and obtain some integral formulas (for instance (3.2) and (5.3)) which generalize to Riemannian spaces well known results of the Euclidean space.

In § 6 we consider, as particular cases, the elliptic and hyperbolic spaces and we give some integral formulas referring to convex bodies in these spaces.

In § 7 the elliptic space is considered in more detail. The "duality" which holds in this space permits the obtention of some more integral formulas.

In what follows ω_i shall represent the area of the Euclidean i-dimensional unit sphere and χ_i the volume bounded by it, that is,

$$(1.1) \qquad \omega_i = \frac{2\,\pi^{(i+1)/2}}{\Gamma((i+1)/2)}\;, \qquad \chi_i = \frac{2\,\pi^{(i+1)/2}}{(i+1)\Gamma((i+1)/2)}$$

2. Density for sets of geodesics in a n-dimensional Riemannian space

Let R_n be a n-dimensional Riemannian space defined by

(*) Manuscrito recebido a 2 de julho de 1951.

$$ds^2 = g_{ij}\, dx^i dx^j \tag{2.1}$$

and let us introduce the following notations

$$F = (g_{ij}\dot{x}^i\dot{x}^j)^{1/2} \quad , \quad p_i = \partial F/\partial \dot{x}^i. \tag{2.2}$$

The density for sets of geodesics is the following exterior differential form, *taken always in absolute value,*

$$dG = \sum_{i=1}^{n} [dp_1 dx^1 \ldots dp_{i-1} dx^{i-1} dp_{i+1} dx^{i+1} \ldots dp_n dx^n]. \tag{2.3}$$

The measure of a set of geodesics will be the integral of dG extended over the set.

The density (2.3) is the $(n-1)$th power of the exterior diffetential invariant form $\Sigma\,[dp_i\, dx^i]$ which integral constitutes the invariant integral of Poincaré of the dynamics [3, p. 19,78]. Therefore it possesses the following two fundamental properties of invariance: *a*) It is invariant with respect to a change of coordinates in the space; *b*) It is invariant under displacements of the elements (x^i, p_i) on the respective geodesic.

In order to give a geometrical interpretation of the density dG let us consider a fixed hypersurface S^{n-1} and a set of geodesics which intersect S^{n-1}. Let G be such a geodesic and P its intersection point with S^{n-1}. In a neighborhood of P we may assume that the equation of S^{n-1} is $x^n = 0$ and that the coordinate system is orthogonal, that is

$$ds^2 = g_{11}(dx^1)^2 + g_{22}(dx^2)^2 + \ldots + g_{nn}(dx^n)^2$$

and thus

$$p_i = g_{ii}\,\frac{dx_i}{ds}\,.$$

If α^i represents the cosine of the angle between G and the x^i-coordinate curve at P, we have

$$\alpha^i = \sqrt{g_{ii}}\,\frac{dx^i}{ds}$$

and

$$(2.4) \qquad p_i = \sqrt{g_{ii}}\alpha^i \quad , \quad dp_i = \sqrt{g_{ii}}d\alpha^i + \frac{\partial\sqrt{g_{ii}}}{\partial x^h}\alpha^i\, dx^h .$$

In order to determine G, acording to the property $b)$ of invariance of dG, we may choose its intersection point P with S^{n-1}. At the point P it is $x^n = 0$, $dx^n = 0$ and consequently the density (2.3) takes the form

$$dG = [dp_1\, dx^1 \ldots dp_{n-1}\, dx^{n-1}]$$

or, according to (2.4),

$$(2.5) \qquad dG = (g_{11}g_{22} \ldots g_{n-1,\,n-1})^{1/2}\,[d\alpha^1\, d\alpha^2 \ldots d\alpha^{n-1} dx^1 \ldots dx^{n-1}\;].,$$

If $d\sigma$ represents the element of $(n-1)$-dimensional area on S^{n-1} we have

$$d\sigma = (g_{11}\, g_{22} \;\ldots g_{n-1,\,n-1})^{1/2}\, dx^1\, dx^2 \ldots dx^{n-1}.$$

On the other hand the element of area on the $(n-1)$-dimensional unit sphere of center P corresponding to the direction of the tangent to G at P has the value

$$(2.6) \qquad d\omega_{n-1} = \frac{[d\alpha^1 \ldots d\alpha^{n-1}\,]}{|\,\alpha^n\,|} .$$

Hence we have

$$(2.7) \qquad dG = |\alpha^n| \quad [d\omega_{n-1} d\sigma\,] = \left|\,\cos\varphi\,\right| \quad [d\omega_{n-1} d\sigma\,]$$

where $\alpha^n = \cos\varphi$ is the cosine of the angle φ between the tangent to G and the normal to S^{n-1} at the point P.

3. Geodesics which intersect a fixed hypersurface

The expression (2.7) of dG gives immediately a very general integral formula. Let $f(\sigma, \varphi)$ be an integrable function defined on S^{n-1} depending upon the point $P(\sigma)$ und upon the direction φ at it. Multiplying both sides of (2.7) by $f(\sigma, \varphi)$ and performing the integration over the hypersurface S^{n-1} and the half of the $(n-1)$-dimensional unit sphere (in order to consider non-oriented geodesics), in the left side each geodesic G appears as common factor of the sum $\& f(\sigma_i, \varphi_i)$ of the values of $f(\sigma, \varphi)$ at the m intersection points of G with S^{n-1}. Consequently we have

$$(3.1) \qquad \int \sum_1^m f(\sigma_i, \varphi_i)\, dG = \int_{S^{n-1}} \int_{\frac{1}{2}\omega_{n-1}} f(\sigma, \varphi) \mid \cos \varphi \mid d\sigma\, d\omega_{n-1}.$$

For instance, if $f(\sigma, \varphi) = 1$, the integral of $| \cos \varphi |\, d\omega_{n-1}$ gives a half of the projection of the $(n-1)$-dimensional unit sphere upon a diametral plane; consequently we get

$$(3.2) \qquad \int m\, dG = \chi_{n-2} F$$

where χ_{n-2} is given by (1.1) and m denotes the number of intersection points of G and S^{n-1}. The integral is extended over all geodesics which intersect S^{n-1} and F represents the area of S^{n-1}.

4. Convex domains

We shall say that a simple closed hypersurface S^{n-1} is *convex* when any geodesic which intersects it has either two points or a whole arc in common with S^{n-1}. In this case, if S^{n-1} has a finite area F, the measure of the geodesics which have a common arc with S^{n-1} is zero and consequently (3.2) gives: *the measure of the geodesics cutting a convex hypersurface of area F is equal to $\frac{1}{2}\chi_{n-2}F$.*

A domain Q in our Riemannian space will be said to be convex when the following three properties are satisfied: 1. It is bounded by a closed convex hypersurface; 2. It is homeomorphic to a $(n-1)$-dimensional sphere; 3. Every geodesic with a point P interior to Q can be prolonged from P in both senses to points outside Q.

Let S^{n-1} be a hypersurface of area F contained in the interior of a convex domain Q; let F_o be the area of the boundary S_o^{n-1} of Q. According to the foregoing condition 3 every geodesic which intersects S^{n-1} will also intersect S_o^{n-1}. Consequently (3.2) gives the following mean value for the number of intersection points of S^{n-1} and all geodesics which cut S^{n-1}

$$m^* = \int m\, dG / \int dG = 2\, F/F_o. \tag{4.1}$$

As an immediate consequence we have the theorem

Given a hypersurface of area F contained inside a convex domain bounded by a hypersurface of area F_o, there exist geodesic lines which intersect S^{n-1} in a number of points $\geq 2\, F/F_o$.

Exactly the same method applied to (3.1), yields the more general theorem:

Given a hypersurface S^{n-1} contained inside a convex domain bounded by a hypersurface of area F_o and an integrable function $f(\sigma, \varphi)$ depending upon the points $P(\sigma)$ of S^{n-1} und upon the angles φ around the normal to S^{n-1} at P, there exist geodesic lines G for whose intersection points $P_i = P(\sigma_i)$ $(i = 1, 2, \ldots, m)$ with S^{n-1} the relation

$$\sum_1^n f(\sigma_i, \varphi_i) \geq \frac{2}{\chi_{n-2} F_o} \int_{S^{n-1}} \int_{\frac{1}{2}\omega^{n-1}} f(\sigma, \varphi) |\cos\varphi|\, d\sigma\, d\omega_{n-1} \tag{4.2}$$

holds, where φ_i is the angle at P_i between G and the normal to S^{n-1}.

5. Sets of geodesic segments

Let t be the arc length on the geodesic G. From (2.7) we deduce

$$[dG\, dt] = |\cos\varphi|\, [d\omega_{n-1} d\sigma\, dt]. \tag{5.1}$$

The product $|\cos\varphi|\, dt$ equals the projection of the arc element dt upon the normal to the hypersurface S^{n-1} at the point P. Consequently $|\cos\varphi|\, d\sigma\, dt$ represents the element of volume dP of the given Riemannian space at P. Consequently (5.1) may be written in the form

$$[dG\, dt] = [dP\, d\omega_{n-1}]. \tag{5.2}$$

An oriented segment S of geodesic can be determined either by G, t (G = geodesic which contains S, t = abscissa on G of the origin of S) or by P, ω_{n-1} (P = origin of S, ω_{n-1} = point on the unit sphere which gives the direction of S). The two equivalent differential forms (5.2) may therefore be taken as density for sets of segments of geodesic lines.

Let us consider the measure of the set of oriented segments S with the origin inside a fixed domain D. The integral of the left hand side of (5.2) gives $2\int \sigma dG$ where σ denotes the lenght of the arc of G which lies inside D (the factor 2 appears as a consequence that dG means the density for non-oriented geodesic lines). The integral of the right side of (5.2) is equal to $\omega_{n-1}V$, where V is the volume of D. Consequently we have the following integral formula

$$\int \sigma dG = \tfrac{1}{2}\, \omega_{n-1} V \tag{5.3}$$

which for $n = 2$, 3 generalizes well known results of the integral geometry of the Euclidean spaces.

6. An integral formula for convex bodies in spaces of constant curvature

Let R_n be now a Riemannian space of constant curvature K. With respect to a system of polar coordinates it is known that the element of length can be written in the form [4, p. 240],

$$ds^2 = d\rho^2 + \frac{\operatorname{sen}^2\sqrt{K}\rho}{K}\, d\lambda^2_{n-1} \tag{6.1}$$

where ρ denotes the geodesic distance from a fixed point (origin of coordinates) and $d\lambda_{n-1}$ represents the element of length of the $(n-1)$-dimensional unit sphere. The element of volume will take the form

$$dP = \frac{\operatorname{sen}^{n-1}\sqrt{K}\rho}{K^{(n-1)/2}}\,[d\rho\; d\omega_{n-1}] \tag{6.2}$$

where $d\omega_{n-1}$ denotes the element of area on the $(n-1)$-dimensional unit sphere.

Let P_1, P_2 be two points in R_n and let G be the geodesic whicn unites them. Let ρ_1, ρ_2 be the abscissas on G of P_1 and P_2. With respect to a system of geodesic polar coordinates with the origin at P_1, the element of volume dP_2 has the form

$$(6.3) \qquad dP_2 = \frac{\operatorname{sen}^{n-1}\sqrt{K}\,|\rho_2-\rho_1|}{K^{(n-1)/2}}\,[d\rho_2\, d\omega_{n-1}]\,.$$

By exterior multiplication by dP_1 we have in consequence of (5.2)

$$(6.4) \qquad [dP_1\, dP_2] = \frac{\sin^{n-1}\sqrt{K}\,|\rho_2-\rho_1|}{K^{(n-1)/2}}\,[d\rho_1\, d\rho_2\, dG].$$

This formula was given following different way by HAIMOVICI [6].

Let us consider the case $n = 3$. If Q is a convex domain of volume V and we consider all the pairs of points P_1, P_2 inside Q, the integral of the left side of (6.4) is equal to V^2. If σ denotes the length of the arc of G which lies inside Q, by calculating the integral of the right side, we have

$$\int_0^\sigma\!\!\int_0^\sigma \sin^2\sqrt{K}\,|\rho_2-\rho_1|\,d\rho_2 d\rho_1 = \tfrac{1}{2}\left(\sigma^2 - \frac{1}{K}\sin^2\sqrt{K}\sigma\right)$$

Hence we have the integral formula

$$(6.5) \qquad \frac{1}{K}\int\left(\sigma^2 - \frac{1}{K}\sin^2\sqrt{K}\,\sigma\right)dG = 2\,V^2$$

where the integral is extended over all the geodesics which intersect Q.

For the elliptic space ($K = 1$) this formula reduces to

$$(6.6) \qquad \int(\sigma^2 - \sin^2\sigma)\,dG = 2\,V^2\,,$$

and for the hyperbolic space ($K = -1$)

(6.7) $$\int (\text{sh}^2\sigma - \sigma^2)\, dG = 2\, V^2 .$$

For the Euclidean space ($K = 0$) we observe that

$$\lim_{K\to 0} \frac{1}{K} \int \left(\sigma^2 - \frac{1}{K} \left(\sqrt{K}\sigma - \frac{K\sqrt{K}}{3!}\sigma^3 + \dots \right)^2 \right) dG = \frac{1}{3} \int \sigma^4 dG$$

and consequently we have

(6.8) $$\int \sigma^4 dG = 6\, V^2 ,$$

which is a well known formula [1, p. 77].

7. Integral formulas for convex bodies in the elliptic space

In the elliptic n-dimensional space all geodesics are closed and have the finite length π. The hyperplanes have finite area $\frac{1}{2}\omega_{n-1}$. Since any geodesic intersects a fixed hyperplane in one and only one point, formula (3.2) gives the measure of the set of all geodesics of the n-dimensional elliptic space:

(7.1) $$\int dG = \tfrac{1}{2}\, \chi_{n-2}\, \omega_{n-1}.$$

Let Q be a convex body of area F and volume V and let us consider the set of geodesic segments of length π which intersect Q. The integral of the left side of (5.2) extended over this set is

(7.2) $$\int dG\, dt = \pi \int dG = \tfrac{1}{2}\, \pi\, \chi_{n-2}\, F$$

and the integral of the right side of (5.2) is

(7.3) $$\int dP d\omega_{n-1} = \tfrac{1}{2}\, \omega_{n-1}\, V + \int \Omega dP$$

where Ω denotes the angle under which Q is seen from P (P exterior to Q). From (7.2) and (7.3) we deduce

(7.4) $$\int \Omega\, dP = \tfrac{1}{2}\, \pi\, \chi_{n-2}\, F - \tfrac{1}{2}\, \omega_{n-1}\, V.$$

For instance for $n = 2$ we get the known formula [8],

$$\int \Omega \, dP = \pi L - \pi F \tag{7.5}$$

where L denotes the length of Q and F its area.

For $n = 3$ we have

$$\int \Omega \, dP = \tfrac{1}{2} \pi^2 F - 2 \pi V . \tag{7.6}$$

In the elliptic space to each integral formula referring to convex bodies corresponds another one by "duality". For the sake of simplicity we shall consider the case $n = 3$; the case $n = 2$ was already considered in [8].

Let M, F, V be the integrated mean curvature, area and volume of a given convex body Q. For the dual conex body Q^* it is known that we have

$$F^* = 4\pi - F, \quad M^* = M \quad , \quad V^* = \pi^2 - M - V . \tag{7.7}$$

By duality to each straight line G corresponds another straight line G^* and hence, having into account (7.7), the formula (5.3) writes

$$\int (\pi - \varphi^*) \, dG^* = 2 \pi (\pi^2 - M^* - V^*)$$

where φ^* denotes the angle between the two support planes to Q through G^* and the integral is extended over all G^* exterior to Q. Having into account (7.1) and (3.2), and by replacing G^* by G, we have the integral formula

$$\int \varphi \, dG = 2 \pi (M + V) - \tfrac{1}{2} \pi^2 F \tag{7.8}$$

which has no analogous in the Euclidean geometry.

Let us now consider the formula (6.6). Applied to the dua convex body Q^* we have

$$\int [(\pi - \varphi^*)^2 - \sin^2 \varphi^*] \, dG^* = 2 (\pi^2 - M^* - V^*)^2$$

from which and (7.8), (7.1), (3.2) it follows that

$$\int (\varphi^2 - \sin^2 \varphi)\, dG = 2\,(M + V)^2 - \tfrac{1}{2}\,\pi^3 F \tag{7.9}$$

where, as in (7.8), φ denotes the angle between the two planes of support through G and the integral is extended over all G exterior to Q.

For the elliptic space of curvature $K = 1/R^2$ the formula (7.9) becomes

$$\int (\varphi^2 - \sin^2 \varphi)\, \frac{dG}{R^2} = 2\left(\frac{M}{R} + \frac{V}{R^3}\right)^2 - \tfrac{1}{2}\,\pi^3 \frac{F}{R^2}$$

and by $R \to \infty$, after multiplication by R^2, we get

$$\int (\varphi^2 - \operatorname{sen}^2 \varphi)\, dG = 2\, M^2 - \tfrac{1}{2}\,\pi^3\, F$$

which is the well known formula due to HERGLOTZ which corresponds to (7.9) for the 3-dimensional Euclidean space [1, p. 79].

Bibliography

[1] W. Blashcke, *Vorlesungen uber Integralgeometrie*, Hamburger Mathematische Einzelschriften, vol. 20, 1936.

[2] W. Blashcke, *Integralgeometrie* 11: *zur Variationsrechnung*, Abhandlungen aus dem Math. Sem. Hamburg, 11, 1936.

[3] E. Cartan, *Leçons sur les invariants integraux*, Paris, 1922

[4] E. Cartan, *Leçons sur la géométrie des espaces de Riemann*, Paris, 1946.

[5] M. Haimovici, *Géométrie intégrale sur les surfaces courbes*, Ann. Scien. Univ Jassy, vol. 23, 1936.

[6] M. Haimovici, *Generalisation d'une formule de Crofton dans un espace de Riemann a n dimensions*, C. R. Acad. Sc. de Roumanie, vol. 1, 1936.

[7] L. A. Santaló, *Integral Geometry on surfaces*, Duke Mathematical Journal, vol. 16, 1949.

[8] L. A. Santaló, *Integral formuals in Crofton's style on the sphere*, Duke Mathematical Journal, vol. 9, 1942.

[9] E. Vidal Abascal, *Geometria Integral sobre superficies curvas*, Publ. del Consejo de Inv. Cient. Madrid, 1949.

Universidad Nacional de la Plata
Argentina

INTEGRAL GEOMETRY IN HERMITIAN SPACES.*

By L. A. SANTALÓ.

1. Introduction. Let P_n be the n-dimensional complex projective space with the homogeneous coordinates x^i $(i=0,1,2,\cdots,n)$. Let x^{*i} denote the complex conjugate of x^i.

We consider the group U (unitary group) of linear transformations

$$x'^i=\sum_{k=0}^{n}\xi_k{}^i x^k \tag{1.1}$$

which leaves the form

$$(xx^*)=\sum_{i=0}^{n}x^i x^{*i} \tag{1.2}$$

invariant. Here and in the sequel, a^* denotes the complex conjugate $(=\bar{a})$ of a. The coefficients of $\xi_k{}^i$ satisfy, then, the relations

$$(\xi_k\xi^*{}_l)=\sum_{i=0}^{n}\xi_k{}^i\xi^*{}_l{}^i=\delta_{kl}, \tag{1.3}$$

where $\delta_{kl}=0$ if $k\neq l$ and $=1$ if $k=l$.

Since the coefficients $\xi_k{}^i$ are complex numbers, the group U depends upon $(n+1)^2$ real parameters.

The geometry of P_n with the fundamental group U of transformations is called the Hermitian geometry (more precisely the elliptic hermitian geometry) and the space P_n itself is called an Hermitian space. The Integral Geometry in these spaces was initiated by Blaschke [2] who gave the densities for linear subspaces, without making applications to integral formulas. The case $n=2$ was first considered by Varga [8] and later, in a more complete form, by Rohde [6].

In the present paper we generalize to the n-dimensional case some of the results which Varga and Rohde obtained for the plane. The main results we obtain are the following:

We first determine the explicit form of the left invariant element of volume du of the group U. If $L_r{}^0$ is a fixed linear subspace of dimension r (throughout the paper we shall mean by *dimension* the complex dimension)

* Received August 27, 1951.

and Γ_r is the subgroup of U which leaves L_r^0 invariant, we determine also the explicit form of the invariant element of volume dL_r of the homogeneous space U/Γ_r. Then, an easy calculation gives the total volumes (6.3) and (6.4) of U and U/Γ_r.

If the coordinates x^i are normalized such that $(xx^*) = 1$ it is well known that every variety C_p of dimension p $(p < n)$ has an invariant integral of degree $2p$, namely $J_p(C_p)$ defined in section **4**. Let C_h and C_r $(r + h \geq n)$ be two analytic varieties. Let uC_r denote the transform of C_r by the transformation u of U, and let $C_h \cap uC_r$ denote the $(h + r - n)$-dimensional variety intersection of C_h and uC_r. If the invariant element of volume du of U is normalized in such a way that the total volume of U is equal 1, we prove the following integral formula:

$$\text{(1.4)} \qquad \int_U J_{h+r-n}(C_h \cap uC_r)\, du = J_h(C_h) J_r(C_r).$$

If instead of C_r we consider a linear subspace L_r, and the invariant element of volume dL_r in U/Γ_r is normalized such that the total volume of U/Γ_r is equal 1, we also have

$$\text{(1.5)} \qquad \int_{U/\Gamma_r} J_{h+r-n}(C_h \cap L_r)\, dL_r = J_h(C_h).$$

We do not consider the case in which C_h is non-analytic. For $n = 2$ this more general case has been considered by Rohde [6], and it remains an interesting open question to extend the results of Rohde to the n-dimensional case.

2. Relative components and equations of structure of the unitary group. The coefficients ξ_k^i of the unitary transformation (1.1) can be interpreted as coordinates of the transformed points ξ_k $(k = 0, 1, \cdots, n)$ of the $n + 1$ vertices of the n-simplex of reference; according to (1.3) they will be the vertices of an autoconjugate n-simplex with respect to the fundamental quadric $(\xi\xi^*) = 0$. These autoconjugate n-simplexes may be considered as the "frames" for the unitary group U (according to the theory of Cartan [3]) and the relative components ω_{jk} will then be defined by the equations

$$\text{(2.1)} \qquad d\xi_k = \sum_{j=0}^{n} \omega_{kj} \xi_j.$$

From (2.1) and (1.3) de deduce

$$\text{(2.2)} \qquad \omega_{kj} = (\xi^*{}_j d\xi_k),$$

and the equations of structure, deduced from (2. 2) by remembering that $d(d\xi_k)=0$, are

$$d\omega_{ih}=\sum_{k=0}^{n}[\omega_{ik}\omega_{kh}], \tag{2. 3}$$

where here and throughout the paper square brackets denote exterior multiplication [3].

Notice that if ω^*_{kj} denotes the complex conjugate of the pfaffian form ω_{kj}, from (1. 3) and (2. 2) we deduce

$$\omega_{kj}+\omega^*_{jk}=0. \tag{2. 4}$$

3. Density for linear subspaces and cinematic density. We wish to define a measure for sets of r-dimensional linear subspaces L_r invariant with respect to U. If $L_r{}^0$ is a fixed L_r, and Γ_r denotes the subgroup of U which leaves $L_r{}^0$ invariant, then the problem is equivalent to the determination of an invariant element of volume in the homogeneous space U/Γ_r.

We follow the general method [5], [7]. Let $L_r{}^0$ be defined by the points $\xi_0, \xi_1, \cdots, \xi_r$. If it is fixed, in the equations (2. 1) we will have

$$\omega_{jk}=0 \quad \text{for} \quad 0\leq j\leq r, \quad r+1\leq k\leq n. \tag{3. 1}$$

Since ω_{jk} are complex pfaffian forms, from $\omega_{jk}=0$ we deduce $\omega^*_{jk}=0$. Consequently the number of forms ω_{jk}, ω^*_{jk} $(0\leq j\leq r,\ r+1\leq k\leq n)$ is $2(r+1)(n-r)$. The density for sets of L_r, that is, the invariant element of volume in the space U/Γ_r will be, up to a constant factor,

$$dL_r=[\prod\omega_{jk}\omega^*_{jk}], \tag{3. 2}$$

where the exterior product is taken always in absolute value and the indices range between the limits

$$0\leq j\leq r, \qquad r+1\leq k\leq n. \tag{3. 3}$$

According to the equations of structure (2. 3), it is easy to verify that $d(dL_r)=0$, which is a sufficient condition in order that dL_r be effectively a density [7].

The invariant element of volume du in the space of the group U ("cinematic density" in the nomenclature of Blaschke) will be, up to a constant factor,

$$du=[\prod\omega_{jk}\omega^*_{jk}\prod\omega_{hh}] \tag{3. 4}$$

where $j<k$, $0\leq j, k, h\leq n$, that is, according to (2. 4), du is equal to the absolute value of the exterior product of all the relative components.

11

4. The invariant integral $J_r(C_r)$ of an analytic variety C_r. Let C_r be an analytic variety of complex dimension r, that is, a variety defined by a set of $n+1$ analytic functions $x^i = x^i(t_1, t_2, t_3, \cdots, t_r)$ $(i = 0, 1, 2, \cdots, n)$ of r complex variables $t_1, t_2, \cdots, t_r$ in a domain D.

Assuming the homogeneous coordinates x^i normalized such that

$$(4.1) \qquad (xx^*) = \sum_{i=0}^{n} x^i x^{*i} = 1,$$

let us consider the following differential form of degree $2r$:

$$(4.2) \qquad \Omega^r = \sum [dx^{i_1} dx^{*i_1} \cdots dx^{i_r} dx^{*i_r}],$$

the summation being extended over all the combinations of $i_1, i_2, \cdots, i_r$ from 1 to n.

It is well known [4] that Ω^r is the only differential form of degree $2r$ which is invariant with respect to the group U.

The integral of Ω^r over an r-dimensional linear subspace L_r has the value [4],

$$(4.3) \qquad \int_{L_r} \Omega^r = (2\pi i)^r / r!.$$

For a general analytic variety C_r we introduce the invariant integral $J_r(C_r)$ defined by

$$(4.4) \qquad J_r(C_r) = r!/(2\pi i)^r \int_{C_r} \Omega^r.$$

If C_r is an algebraic variety, $J_r(C_r)$ coincides with its order (Cartan [4]).

For some purpose it is convenient to write Ω^r in another form. Let T_r be the r-dimensional linear subspace tangent to C_r at the point x, and take in T_r r points α_p $(p = 1, 2, \cdots, r)$ such that

$$(4.5) \qquad (x\alpha^*_p) = 0, \qquad (\alpha_p \alpha^*_q) = \delta_{pq}.$$

We will have

$$(4.6) \qquad dx^i = a x^i + \sum_{p=1}^{r} b_p \alpha_p{}^i, \qquad dx^{*i} = a^* x^{*i} + \sum_{p=1}^{r} b^*{}_p \alpha^*{}_p{}^i$$

where a and b_p are pfaffian forms given by

$$(4.7) \qquad a = (x^* dx), \qquad b_p = (\alpha^*{}_p dx).$$

From (4.6) and (4.5) we deduce

$$(4.8) \qquad \sum_{i=0}^{n} [dx^i dx^{*i}] = [aa^*] + \sum_{p=1}^{r} [b_p b^*{}_p].$$

Since $a = -a^*$, we have $[aa^*] = 0$, and taking the r-th power of both sides of (4.8) we get $\Omega^r = [b_1 b^*_1 b_2 b^*_2 \cdots b_r b^*_r]$, or, according to (4.7),

$$(4.9) \qquad \Omega^r = [\prod_{p=1}^{r} (\alpha_p{}^* dx)(\alpha_p dx^*)],$$

a formula which will be useful in the following sections.

5. The invariant integral J_r for the variety generated by the $(r-1)$-osculating spaces of a given analytic curve. As an example, we wish to evaluate J_r for the variety generated by the $(r-1)$-osculating linear spaces of a given analytic curve of the n-dimensional complex projective space.

Let C be an analytic curve defined by the $n+1$ parametric equations $y^i = y^i(t)$ $(i = 0, 1, 2, \cdots, n)$, where $y^i(t)$ are analytic functions of the complex variable t. In order to evaluate $J_1(C)$ we must normalize the coordinates y^i as indicated in (4.1). We set $x^i = y^i/(yy^*)^{\frac{1}{2}}$; then

$$dx^i = \{2(yy^*)dy^i - (dyy^*)y^i - (ydy^*)y^i\}/2(yy^*)^{3/2},$$
$$dx^{*i} = \{2(yy^*)dy^{*i} - (dy^*y)y^{*i} - (y^*dy)y^{*i}\}/2(yy^*)^{3/2},$$

and by exterior multiplication and addition,

$$\sum_{i=0}^{n} [dx^i dx^{*i}] = \{(yy^*)\Sigma [dy^i dy^{*i}] - [(y^*dy)(ydy^*)]\}/(yy^*)^2.$$

Notice that

$$(yy^*)\Sigma[dy^i dy^{*i}] - [(y^*dy)(ydy^*)]$$
$$= \{(yy^*)(y'y^{*\prime}) - (y^*y')(yy^{*\prime})\}[dtdt^*] = ((y\wedge y')(y^*\wedge y^{*\prime}))[dtdt^*],$$

where the notation $a \wedge b$ denotes the bivector with the components

$$a^{ij} = a^i b^j - a^j b^i.$$

Consequently we have

$$(5.1) \qquad \Omega^1 = \sum_{i=0}^{n} [dx^i dx^{*i}] = |\, y\wedge y' \,|^2/|\, y \,|^4 [dtdt^*]$$

and finally, according to the definition (4.4),

$$(5.2) \qquad J_1(C) = 1/2\pi i \int_C |\, y\wedge y' \,|^2/|\, y \,|^4 [dtdt^*].$$

If C is an algebraic curve, then (5.2) gives the order of C (except the sign which depends upon the orientation assumed for C).

Let us now consider the variety C_r generated by the $(r-1)$-osculating linear spaces of C. In order to evaluate $\int_{C_r} \Omega^r$ we consider two consecutive $(r-1)$-osculating spaces, $L_{r-1}(t)$, defined by the points $y(t), y'(t), \cdots, y^{(r-1)}(t)$, and $L_{r-1}(t+dt)$, defined by $y(t+dt), y'(t+dt), \cdots, y^{(r-1)}(t+dt)$, which are contained in the linear space $L_r(t)$ defined by the points $y(t), y'(t), \cdots, y^{(r)}(t)$. The angle $d\tau$ between $L_{r-1}(t)$ and $L_{r-1}(t+dt)$ is equal to the distance Ω^1 between the poles of $L_{r-1}(t)$ and $L_{r-1}(t+dt)$ considered as linear subspaces of L_r. Choosing the coordinate system so that the equations of L_r are $y^{r+1}=y^{r+2}=\cdots=y^n=0$, the pole of $L_{r-1}(t)$ in $L_r(t)$ will be the point Y_r whose coordinates are the determinants of order r in the r by $r+1$ matrix $(y^{k(i)})$, where $i=0,1,\cdots,r-1$, and $k=0,1,\cdots,r$.

We now apply (5.1). The coordinates of the point Y'_r are the determinants of order r of the r by $r+1$ matrix $(y^{k(i)})$, where $i=0,1,\cdots, r-2, r$, and $k=0,1,\cdots,r$, and following a device due to Ahlfors and H. Weyl ([1], [9], p. 144), one can prove that

$$|Y_r \wedge Y'_r| = |Y_{r-1}| \cdot |Y_{r+1}|, \tag{5.2}$$

where Y_{r-1} and Y_{r+1} have a meaning analogous to that of Y_r, i. e. they are the multivectors defined by $(y, y', \cdots, y^{(r-2)})$ and $(y, y', \cdots, y^{(r)})$ respectively.

The device consists of choosing a coordinate system in which $y, y', \cdots, y^{(r)}$ have the coordinates

$$y(y^0, 0, \cdots, 0), y'(y^{0\prime}, y^{1\prime}, 0, \cdots, 0), \cdots, y^{(r)}(y^{0(r)}, y^{1(r)}, \cdots, y^{r(r)}).$$

In this system of coordinates the relation (5.2) becomes trivial.

Therefore we have

$$d\tau = |Y_{r-1}|^2 \cdot |Y_{r+1}|^2 / |Y_r|^4 [dt dt^*].$$

Since according to (4.3) $\int d\tau = \int \Omega^1 = 2\pi i$ we have the relation

$$d\tau / 2\pi i = \Omega^r / \int_{L_r} \Omega^r$$

and consequently we get

$$\begin{aligned} J_r(C_r) &= r!/(2\pi i)^r \int_{C_r} \Omega^r = 1/2\pi i \int_C d\tau \\ &= 1/2\pi i \int_C |Y_{r-1}|^2 \cdot |Y_{r+1}|^2 / |Y_r|^4 [dt dt^*]. \end{aligned} \tag{5.3}$$

This expression is due to Ahlfors and H. Weyl [1], [9]. If C is an algebraic curve (5.3) give the classes of different orders.

For instance, for a plane algebraic curve $y^0 = y^0(t)$, $y^1 = y^1(t)$, $y^2 = y^2(t)$, the class is given (except for the sign) by the integral

$$J_2(C) = 1/2\pi i \int_C |y|^2 \cdot |yy'y''|^2 / |y \wedge y'|^4 [dtdt^*],$$

where $|yy'y''|$ denotes the absolute value of the determinant formed by the components of y, y', y''.

6. Total volume of the unitary group and of the homogeneous spaces U/Γ_r. In order to calculate the volume $\int_U du$ of the unitary group U with the invariant element of volume du given by (3.4), we put $\xi_0{}^k = \rho_k e^{i\theta_k}$ and consequently

$$\omega_{00} = (\xi^*{}_0 d\xi_0) = \sum_{k=0}^{n} (\rho_k d\rho_k + i\rho_k{}^2 d\theta_k).$$

Since $(\xi^*{}_0\xi_0) = \sum_0^n \rho_k{}^2 = 1$, we have $\sum_0^n \rho_k d\rho_k = 0$, and therefore the integral of ω_{00} over all possible values of the variables has the value

$$\int \omega_{00} = i \sum_{k=0}^{n} \int \rho_k{}^2 d\theta_k = 2\pi i.$$

The same proof gives

$$\int \omega_{jj} = 2\pi i. \tag{6.1}$$

On the other hand we observe that, according to (4.9), the exterior product

$$[\prod_{h=1}^{n-r} \omega_{r,r+h}\omega^*{}_{r,r+h}] = [\prod_{h=1}^{n-r} (\xi^*{}_{r+h} d\xi_r)(\xi_{r+h} d\xi^*{}_r)]$$

denotes the element Ω^{n-r} relative to the linear $(n-r)$-dimensional space defined by the points $\xi_r, \xi_{r+1}, \cdots, \xi_n$. Consequently (4.3) gives

$$\int [\prod_{h=1}^{n-r} \omega_{r,r+h}\omega^*{}_{r,r+h}] = (2\pi i)^{n-r}/(n-r)!. \tag{6.2}$$

From (3.4), (6.1) and (6.2) we get

$$\int_U du = (2\pi i)^{n+1} \prod_{r=0}^{n} (2\pi i)^{n-r}/(n-r)!$$

which can be written

$$\int_U du = \prod_{h=1}^{n+1} (2\pi i)^h/(h-1)!. \tag{6.3}$$

This formula gives the total volume of the unitary group U.

We wish now to calculate the total volume of the homogeneous space U/Γ_r, or, what is the same, the total measure of all the linear r-dimensional spaces contained in the hermitian n-dimensional space.

Let us write, for the moment, $du = du_n$, in order to exhibit clearly the dimension n of the space. According to (3.2) and (3.4), we have the following relation between the elements of volume du_n, dL_r, and the elements du_r, du_{n-r}, of the r- and $(n-r)$-dimensional unitary groups:

$$du_n = [du_r\, du_{n-r-1}\, dL_r].$$

Integration of both sides of this equality over all possible values of the variables, taking (6.3) into account, gives

$$\prod_{h=1}^{n+1} (2\pi i)^h/(h-1)! = \prod_{h=1}^{r+1} (2\pi i)^h/(h-1)! \prod_{h=1}^{n-r} (2\pi i)^h/(h-1)! \int_{U/\Gamma_r} dL_r,$$

and consequently

$$\int_{U/\Gamma_r} dL_r = (2\pi i)^{(n-r)(r+1)} 1!2!\cdots r!/[n!(n-1)!\cdots(n-r)!]. \tag{6.4}$$

This is the measure of all the linear r-dimensional spaces of the hermitian n-dimensional space, i. e. the total volume of the homogeneous space U/Γ_r.

The finite values (6.3) and (6.4) induce a normalization of the elements of volume du and dL_r in such a way that the total volumes of the corresponding spaces are equal to 1. These normalized elements will be

$$du' = \prod_{h=1}^{n+1} (h-1)!/(2\pi i)^h du \tag{6.5}$$

$$dL_r' = (2\pi i)^{-(n-r)(r+1)} n!(n-1)!\cdots(n-r)!/1!2!\cdots r![dL_r]. \tag{6.6}$$

7. Linear subspaces which intersect an analytic variety. Let C_h be a fixed analytic variety of complex dimension h. Let L_r^0 be a fixed r-dimensional linear subspace and $L_r = uL_r^0$ be the transform of L_r^0 by $u \epsilon U$. We assume $r + h - n \geqq 0$, and let $C_h \cap L_r$ be the $(r+h-n)$-dimensional variety intersection of C_h with L_r. We wish to evaluate the integral

$$I = \int_{U/\Gamma_r} J_{h+r-n}(C_h \cap L_r)\, dL_r'. \tag{7.1}$$

Let ξ_0 be a point of $C_h \cap L_r$. In order to define L_r, we may take the points $\xi_0, \xi_1, \cdots, \xi_{h+r-n}$ on the linear $(h+r-n)$-space tangent to the intersection $C_h \cap L_r$ at ξ_0, and the points $\alpha_{h+r-n+1}, \cdots, \alpha_r$ such that

$$(7.2) \qquad (\xi_i \xi^*{}_j) = \delta_{ij}, \qquad (\xi^*{}_i \alpha_l) = 0, \qquad (\alpha^*{}_l \alpha_m) = \delta_{lm}.$$

Let $\gamma_{r+1}, \cdots \gamma_n$ be $n-r$ points such that

$$(7.3) \qquad (\gamma^*{}_p \xi_i) = 0, \qquad (\gamma^*{}_p \alpha_l) = 0, \qquad (\gamma^*{}_p \gamma_q) = \delta_{pq},$$

and let $\beta_{h+r-n+1}, \cdots, \beta_h$ be $n-r$ points on the tangent space to C_h at ξ_0 such that

$$(7.4) \qquad (\beta^*{}_a \beta_b) = \delta_{ab}, \qquad (\beta^*{}_a \xi_i) = 0.$$

In (7.2), (7.3), and (7.4) we agree on the range of indices

$$0 \leq i, j \leq h+r-n, \quad h+r-n+1 \leq l, m \leq r, \quad r+1 \leq p, q \leq n$$
$$h+r-n+1 \leq a, b \leq h.$$

According to (3.2) and (2.2) we have

$$(7.5) \qquad dL_r = [\prod (\gamma^*{}_p d\xi_i)(\gamma_p d\xi^*{}_i) \prod (\gamma^*{}_p d\alpha_l)(\gamma_p d\alpha^*{}_l)],$$

the product being extended over all γ, ξ, α. Since $d\xi_0$ is on the tangent h-space to C_h, we have

$$d\xi_0 = \sum_i A_i \xi_i + \sum_a B_a \beta_a,$$

where, according to (7.2) and (7.4), $A_i = (\xi^*{}_i d\xi_0)$, $B_a = (\beta^*{}_a d\xi_0)$.

Consequently

$$(\gamma^*{}_p d\xi_0) = \sum_a (\gamma^*{}_p \beta_a)(\beta^*{}_a d\xi_0), \qquad (\gamma_p d\xi^*{}_0) = \sum_a (\gamma_p \beta^*{}_a)(\beta_a d\xi^*{}_0),$$

and

$$(7.6) \qquad [\prod_p (\gamma^*{}_p d\xi_0)(\gamma_p d\xi^*{}_0)] = \|(\gamma^*{}_p \beta_a)\| \cdot \|(\gamma_p \beta^*{}_a)\| [\prod_a (\beta^*{}_a d\xi_0)(\beta_a d\xi^*{}_0)].$$

According to (4.9), the differential invariant Ω^h referred to C_h and the differential invariant Ω^{r+h-n} referred to $C_h \cap L_r$ may be written

$$(7.7) \qquad \Omega^h = [\prod_i (\xi^*{}_i d\xi_0)(\xi_i d\xi^*{}_0) \prod_a (\beta^*{}_a d\xi_0)(\beta_a d\xi^*{}_0)],$$

$$(7.8) \qquad \Omega^{h+r-n} = [\prod_i (\xi^*{}_i d\xi_0)(\xi_i d\xi^*{}_0)].$$

Consequently, from (7.5) and (7.6) we have

$$(7.9) \qquad [\Omega^{h+r-n} dL_r] = [\theta \Omega^h],$$

where θ is a differential form which does not depend upon ξ_0. According to (6.6) the same formula holds, up to a constant factor, for dL_r' instead of dL_r. The integral of θ over all possible values of the variables gives a constant value c which does not depend upon C_h. Consequently, integration of (7.9) gives $I = cJ_h(C_h)$. The value of c can be found by considering the case in which C_h is a linear subspace L_h. In this case we have $J_{h+r-n}(L_h \cap L_r) = J_{h+r-n}(L_{h+r-n}) = 1$, $J_h(L_h) = 1$, and $\int dL_r' = 1$. Consequently $c = 1$.

We get the final result

$$\text{(7.10)} \qquad \int_{U/\Gamma_r} J_{r+h-n}(C_h \cap L_r)\, dL_r' = J_h(C_h).$$

If $r + h - n = 0$, $J_0(C_h \cap L_r)$ denotes the number of intersection points of L_r with C_h.

8. Analytic varieties which intersect each other. Let C_h, C_r $(h + r - n \geqq 0)$ be two analytic varieties of dimension h, r respectively. Let uC_r be the transform of C_r by $u \varepsilon U$. We consider the integral

$$\text{(8.1)} \qquad I = \int_U J_{r+h-n}(C_h \cap uC_r)\, du'.$$

The frame which determines u may be chosen in the following way. Let ξ_0 be a point of the intersection $C_h \cap uC_r$. We choose the points $\xi_1, \xi_2, \cdots, \xi_{h+r-n}$ on the linear $(h + r - n)$-dimensional space tangent to the variety $C_h \cap uC_r$ at ξ_0, and the points $\alpha_{h+r-n+1}, \cdots, \alpha_r$ on the tangent r-space to C_r at ξ_0, in such a way that

$$\text{(8.2)} \qquad (\xi^*_i \xi_j) = \delta_{ij}, \qquad (\xi^*_i \alpha_l) = 0, \qquad (\alpha^*_l \alpha_m) = \delta_{lm}.$$

Let $\gamma_{r+1}, \cdots, \gamma_n$ be $n - r$ points such that

$$\text{(8.3)} \qquad (\gamma^*_p \xi_i) = 0, \qquad (\gamma^*_p \alpha_l) = 0, \qquad (\gamma^*_p \gamma_q) = \delta_{pq}.$$

In (8.2), (8.3), and in the remainder of this section we agree on the ranges of indices $0 \leq i, j \leq h + r - n$, $h + r - n + 1 \leq l, m \leq r$, $r + 1 \leq p, q \leq n$.

The frame which determines u is the n-simplex ξ_i, α_l, γ_p. According to (3.4) we have

$$\text{(8.4)} \qquad du = [\Pi(\xi^*_i d\xi_i)\Pi(\alpha^*_l d\alpha_l)\Pi(\gamma^*_p d\gamma_p)\Pi(\xi^*_i d\xi_j)(\xi_i d\xi^*_j)\Pi(\alpha^*_l d\alpha_m)(\alpha_l d\alpha^*$$
$$\Pi(\alpha^*_l d\xi_i)(\alpha_l d\xi^*_i)\Pi(\gamma^*_p d\gamma_q)(\gamma_p d\gamma^*_q)\Pi(\gamma^*_p d\xi_i)(\gamma_p d\xi^*_i)\Pi(\gamma^*_p d\alpha_l)(\gamma_p d\alpha^*_l)],$$

where $i \neq j$, $l \neq m$, $p \neq q$.

Let $\beta_{h+r-n+1}, \cdots, \beta_h$ be $n-r$ points on the tangent h-space to C_h at ξ_0, such that

$$(8.5) \qquad (\beta^*_a\beta_b) = \delta_{ab}, \qquad (\beta^*_a\xi_i) = 0,$$

where $h+r-n+1 \leq a, b \leq h$.

Since we always take ξ_0 on C_h, we have

$$d\xi_0 = \sum_i A_i\xi_i + \sum_a B_a\beta_a$$

where, according to (8.2) and (8.3), $A_i = (\xi^*_i d\xi_0)$, $B_a = (\beta^*_a d\xi_0)$.

Consequently we have

$$(\gamma^*_p d\xi_0) = \sum_a (\gamma^*_p\beta_a)(\beta^*_a d\xi_0), \qquad (\gamma_p d\xi^*_0) = \sum_a (\gamma_p\beta^*_a)(\beta_a d\xi^*_0),$$

and by exterior multiplication,

$$(8.6) \qquad [\prod_p(\gamma^*_p d\xi_0)(\gamma_p d\xi^*_0)] = \|(\gamma^*_p\beta_a)\|\cdot\|(\gamma_p\beta^*_a)\|[\prod_a(\beta^*_a d\xi_0)(\beta_a d\xi^*_0)].$$

The differential invariant Ω^r referred to C_r, according to (4.9), is

$$(8.7) \qquad \Omega^r = [\prod_i(\xi^*_i d\xi_0)(\xi_i d\xi^*_0)\prod_l(\alpha^*_l d\xi_0)(\alpha_l d\xi^*_0)].$$

From (7.7), (7.8), (8.4), (8.6), and (8.7) we deduce

$$(8.8) \qquad [\Omega^{h+r-n}du] = [\psi\Omega^r\Omega^h],$$

where ψ is a differential form which does not depend upon ξ_0. Up to a constant factor, according to (6.5), the same formula holds for du' instead of du. The integral of ψ over all possible values of the variables gives a constant value c. Consequently from (8.1) and (8.8) we deduce $I = cJ_h(C_h)J_r(C_r)$.

In order to determine the value of the constant c we consider the case in which C_h and C_r are linear spaces L_h and L_r. In this case we have $J_{h+r-n}(L_h \cap uL_r) = 1$, $\int du' = 1$, $J_h(C_h) = J_r(C_r) = 1$. Consequently we have proved the integral formula

$$(8.9) \qquad \int_U J_{r+h-n}(C_h \cap uC_r)\,du' = J_h(C_h)J_r(C_r).$$

If $r+h-n=0$, $J_0(C_h \cap uC_r)$ denotes the number of intersection points of C_h with uC_r.

For algebraic varieties, since the invariant J coincides with the order of the variety, (8.9) is an integrated form of the theorem of Bezout. Therefore (8.9) may be considered as a generalization of the theorem of Bezout to analytic varieties.

If we have $p+1$ analytic varieties $C_{h_0}, C_{h_1}, \cdots, C_{h_p}$ such that $h_0+h_1+\cdots+h_p \geqq np$, from (8. 9) we get immediately by recurrence

$$\int_U J_{h_0+h_1+\ldots+h_p-np}(C_h \cap u_1 C_{h_1} \cap u_2 C_{h_2} \cap \cdots \cap u_p C_{h_p}) du_1' du_2' \cdots du_p'$$
$$= J_{h_0}(C_{h_0}) J_{h_1}(C_{h_1}) \cdots J_{h_p}(C_{h_p}).$$

THE UNIVERSITY OF LA PLATA, ARGENTINA.

BIBLIOGRAPHY.

[1] L. V. Ahlfors, "The theory of meromorphic curves," *Acta Societatis Scientiarum Fennicae*, A, vol. III, 1941.

[2] W. Blaschke, "Densita negli spazi di Hermite," *Rendiconti dell' Academia dei Lincei*, vol. 29, 6, pp. 105-108, 1939.

[3] E. Cartan, *La théorie des groupes finis et continus et la géométrie differentielle traités par la méthode du repere mobile*, Paris, 1937.

[4] E. Cartan, "Sur les invariants intégraux de certains espaces homogènes clos et les proprietés topologiques de ces espaces," *Annales de la Société Polonaise des Mathématiques*, vol. 8, pp. 181-225, 1929.

[5] S. S. Chern, "On integral geometry in Klein spaces," *Annals of Mathematics*, vol. 43, pp. 178-189, 1942.

[6] H. Rohde, "Unitäre Integralgeometrie," *Hamburger Abhandlungen*, vol. 13, pp. 295-318, 1940.

[7] L. A. Santaló, "Integral Geometry in projective and affine spaces," *Annals of Mathematics*, vol. 51, pp. 739-755, 1950.

[8] O. Varga, "Ueber die Integralinvarianten die zu einer Kurve in der Hermitischen Geometrie gehören," *Acta Litt. Scientarum Szeged*, vol. 9, pp. 88-102, 1939.

[9] H. Weyl, "Meromorphic functions and analytic curves," *Annals of Mathematics Studies*, No. 12, Princeton, 1943.

L. A. Santaló

CUESTIONES DE GEOMETRIA DIFERENCIAL E INTEGRAL EN ESPACIOS DE CURVATURA CONSTANTE

Introducción.

Uno de los resultados más notables de la Geometría Diferencial obtenidos en los últimos años ha sido la generalización de la fórmula de Gauss-Bonnet a variedades multidimensionales hecha por Allendoerfer-Weil [2], de la cual una demostración simple y elegante utilizando los métodos de E. Cartan fué dada poco después por Chern [5]. Para el caso particular de hipersuperficies en espacios de curvatura constante los elementos que intervienen en la fórmula toman significado geométrico bien preciso y el resultado fué obtenido, casi en la misma fecha pero independientemente de los autores anteriores y por camino muy diferente, por Herglotz [9].

Lo demostración de Chern es intrínseca, es decir, no supone a la variedad sumergida en un espacio de mayor número de dimensiones. Sin embargo, para el caso particular de una hipersuperficie de un espacio de curvatura constante, el mismo camino de Chern conduce al resultado de Herglotz de manera simple y natural, apareciendo claramente el significado geométrico de ciertos invariantes, significado que en el caso general queda un poco obscuro. Es por esto que creemos puede ser útil insistir sobre dicha demostración, adaptándola al caso particular mencionado de las hipersuperficies en espacios de curvatura constante. Esto es lo que hacemos como primera parte de este trabajo.

Como segunda parte hacemos aplicación de la fórmula obtenida al cálculo de ciertas expresiones duales (nº 3) y a algunas cuestiones de geometría integral. Así, en el nº 4 obtenemos las fórmulas integrales (4.10) y (4.11) que generalizan a n dimensiones unos resultados de BLASCHKE para $n=2,3$. En los nº 6 y 7 se obtienen las fórmulas (6.7), (7.2), (7.7) y (7.9) que análogamente generalizan a espacios n dimensionales fórmulas conocidas para $n=2$ y $n=3$.

1. FÓRMULAS FUNDAMENTALES.

Como espacio de curvatura constante K entendemos el espacio no-euclidiano, elíptico si $K>0$ e hiperbólico si $K<0$. Es decir, suponemos en el espacio proyectivo n-dimensional la hipercuádrica fundamental (no reglada)

$$\Phi(x_i) \equiv \Sigma a_{ij} x_i x_j = 0 \qquad (i, j = 0, 1, 1, \ldots, n) \tag{1.1}$$

con los coeficientes a_{ij} reales. Si Φ es imaginaria tomamos el signo de los a_{ij} de manera que para todo punto real sea $\Phi(x_i)>0$ y el conjunto de todos los puntos no pertenecientes a $\Phi(x_i)=0$ constituye el espacio elíptico; si Φ es real, el espacio hiperbólico es el conjunto de puntos para los cuales es $\Phi(x_i)<0$. En ambos casos, los movimientos del espacio son las proyectividades que dejan invariante la hipercuádrica fundamental.

Las coordenadas homogéneas x_i se pueden normalizar de manera que para todo punto no perteneciente a la hipercuádrica sea

$$\Phi(x_i) = \frac{1}{K} \tag{1.2}$$

siendo K una constante (curvatura del espacio), positiva en el caso elíptico y negativa en el hiperbólico.

Dados dos puntos $A(x_0, x_1, \ldots, x_n)$, $B(y_0, y_1, \ldots, y_n)$ se define su «producto escalar» por la expresión

$$(A, B) = (B, A) = \frac{1}{2}\, \Sigma y_i \frac{\partial \Phi}{\partial x_i} = \frac{1}{2}\, \Sigma x_i \frac{\partial \Phi}{\partial y_i} \tag{1.3}$$

con lo cual la condición (1.2) de normalización se escribe

$$(A, A) = \frac{1}{K}\,. \tag{1.4}$$

La condición $(A, B)=0$ expresa que A, B son conjugados respecto de Φ. Sean $A_0, A_1, \ldots, A_n$ n puntos que sean vértices de un simplex autoconjugado, es decir

$$(A_i, A_i) = \frac{1}{K}, \quad (A_i, A_j) = 0 \qquad (i \neq j) \tag{1.5}$$

de donde, por diferenciación,

$$(A_i, dA_i) = 0, \quad (A_i, dA_j) + (A_j, dA_i) = 0. \tag{1.6}$$

Consideremos un movimiento elemental que lleve los puntos A_i a los $A_i + dA_i$. Los desplazamientos dA_i pueden expresarse en la forma

$$dA_i = \sum_{h=0}^{n} \omega_i^h A_h \tag{1.7}$$

siendo los coeficientes ω_i^h formas diferenciales lineales (componentes relativas del movimiento según E. Cartan) cuyo valor se obtiene multiplicando escalarmente (1.7) por el A_h correspondiente, resultando según (1.5)

$$\omega_i^h = K(A_h, dA_i) \tag{1.8}$$

y por tanto, según (1.6),

$$\omega_i^i = 0, \quad \omega_i^h + \omega_h^i = 0. \tag{1.9}$$

Estas componentes relativas no son independientes. Escribiendo las condiciones de integrabilidad de (1.7), o sea, anulando la diferencial exterior del segundo miembro teniendo en cuenta las ecuaciones mismas, resultan las «ecuaciones de estructura»,

$$d\omega_i^h = \sum_{j=0}^{n} [\omega_i^j \omega_j^n] \tag{1.10}$$

donde los paréntesis cuadrados indican multiplicación exterior.

La distancia s entre dos puntos A, B se define por la relación

$$(A, B) = \frac{\cos \sqrt{K}\, s}{K} \tag{1.11}$$

de manera que si B está sobre la recta que une los puntos conjugados A_0 y A_i y s_i es su distancia a A_0, será

$$B = \cos(\sqrt{K}\, s_i)\, A_0 + \operatorname{sen}(\sqrt{K}\, s_i)\, A_i \,. \tag{1.12}$$

De aquí, manteniendo fijos A_0 y A_i y haciendo variar B sobre la recta que los une,

$$dB = \sqrt{K}\,(-\operatorname{sen}(\sqrt{K}\, s_i)\, A_0 + \cos(\sqrt{K}\, s_i)\, A_i)\, ds_i \,. \tag{1.13}$$

En particular, para $s_i = 0$, o sea $B \equiv A_0$, resulta $dA_0 = \sqrt{K} A_i ds_i$ y por tanto *el elemento de arco sobre* $A_0 A_i$ *a partir de* A_0 *vale*

$$ds_i = \sqrt{K}\,(A_i, dA_0) = \frac{\omega_0{}^i}{\sqrt{K}} \tag{1.14}$$

y el elemento de volumen del espacio, correspondiente al punto A_0 se escribirá

$$dV = \frac{1}{K^{n/2}}\,[\omega_0{}^1 \omega_0{}^2 \dots \omega_0{}^n] \,. \tag{1.15}$$

Para definir el ángulo φ entre dos rectas que pasan por A_0, se toman los puntos A_i, B_i conjugados de A_0 sobre estas rectas y entonces φ se define por

$$\cos\varphi = K\,(A_i, B_i) \,. \tag{1.16}$$

De aquí se deduce, de manera análoga a la anterior, que *el elemento de ángulo sobre el plano* $A_0 A_i A_j$ *a partir de la recta* $A_0 A_i$ *está dado por*

$$d\varphi_{ij} = K\,(A_j, dA_i) = \omega_i^j \,. \tag{1.17}$$

Por tanto, el elemento de ángulo sólido correspondiente a la dirección $A_0 A_i$ será

$$d\Omega_i = [\omega_i^1 \omega_i^2 \dots \omega_i^{i-1} \omega_i^{i+1} \dots \omega_i^n] \,. \tag{1.18}$$

2. La fórmula de Gauss-Bonnet.

En el espacio n-dimensional de curvatura constante K consideremos una hipersuperficie cerrada, orientable S de clase ≥ 3, que sea el contorno de un cuerpo Q.

En cada punto A_0 de S consideremos el simplex autoconjugado $A_0, A_1, \ldots, A_n$ tal que el hiperplano determinado por $A_0, A_1, \ldots, A_{n-1}$ sea el hiperplano tangente a S y por tanto A_0A_n la normal a S y, además, las direcciones A_0A_i $(i=1, 2, \ldots, n-1)$ sean las direcciones principales de S en el punto A_0. Según las notaciones (1.14), (1.17) y la definición de los radios de curvatura principales, será

$$\omega_n^{\ i} = d\varphi_{ni} = -\frac{ds_i}{R_i} \tag{2.1}$$

siendo R_i el radio de curvatura principal según la dirección A_0A_i (fórmula de O. Rodrigues para espacios de curvatura constante). Según (1.18) es también

$$d\Omega_n = (-1)^{n-1}\frac{[ds_1 ds_2 \ldots ds_{n-1}]}{R_1 R_2 \ldots R_{n-1}} = \frac{[-1]^{n-1}\, dF}{R_1 R_2 \ldots R_{n-1}} \tag{2.2}$$

siendo dF el elemento de área de S en el punto A_0.

Siguiendo a Chern [5] consideremos ahora las formas diferenciales, definidas para valores de h tales que $0 \leq 2h \leq n-1$,

$$\Phi_h \equiv \Sigma \varepsilon_{\alpha_1\alpha_2\ldots\alpha_{n-1}} [\omega_{\alpha_1}^{\ 0}\, \omega_0^{\ \alpha_2}]\, [\omega_{\alpha_3}^{\ 0}\, \omega_0^{\ \alpha_4}] \ldots [\omega_{\alpha_{2h-1}}^{\ 0} \omega_0^{\ \alpha_{2h}}] \\ [\omega_{\alpha_{2h+1}}^{\ n}\, \omega_{\alpha_{2h+2}}^{\ n} \ldots \omega_{\alpha_{n-1}}^{\ n}] \tag{2.3}$$

donde $\varepsilon_{\alpha_1\alpha_2\ldots\alpha_{n-1}}$ vale cero si hay algún índice α_i repetido y $+1$ o -1 según que la permutación $(\alpha_1\alpha_2 \ldots \alpha_{n-1})$ sea par o impar respecto el orden natural $1, 2, \ldots, n-1$. La sumatoria está extendida a todas las permutaciones de los índices $1, 2, \ldots, n-1$.

Para ver la interpretación geométrica de estas formas diferenciales, observemos que según (2.1), (1.14) y (1.9) se tiene

$$\Phi_h = (-1)^h K^h\, \Sigma \varepsilon_{\alpha_1 \ldots \alpha_{n-1}} \frac{[ds_{\alpha_1} ds_{\alpha_2} \ldots ds_{\alpha_{n-1}}]}{R_{\alpha_{2h+1}} R_{\alpha_{2h+2}} \ldots R_{\alpha_{n-1}}}\,. \tag{2.4}$$

Introduciendo las curvaturas medias

$$m_i = \frac{1}{\binom{n-1}{i}} \left\{\frac{1}{R_{\alpha_1}}\ \frac{1}{R_{\alpha_2}} \ldots \frac{1}{R_{\alpha_i}}\right\}, \qquad m_0 = 1 \tag{2.5}$$

donde el paréntesis indica la función simétrica elemental de orden i formada con los $1/R_{\alpha_i}$ $(i=0, 1, 2, ..., n-1)$ y teniendo en cuenta que $dF=[ds_1\, ds_2 \dots ds_{n-1}]$, se tiene también

$$(2.6)\qquad \Phi_h = (-1)^h (n-1)!\, K^h m_{n-2h-1}\, dF\,.$$

Por otra parte, consideremos también las formas diferenciales (para valores de h tales que $0 \leq 2h \leq n-2$),

$$(2.7)\qquad \Psi_h = 2(h+1)\, \Sigma \varepsilon_{\alpha_1 \dots \alpha_{n-1}} [\omega_{\alpha_1}^{0}\, \omega_0^{\alpha_2}]\, [\omega_{\alpha_3}^{0}\, \omega_0^{\alpha_4}] \dots [\omega_{\alpha_{2h+1}}^{0}\, \omega_0^{n}]$$
$$[\omega_{\alpha_{2h+2}}^{n} \dots \omega_{\alpha_{n-1}}^{n}]$$

cuya significación geométrica, procediendo igual que antes resulta ser

$$(2.8)\qquad \Psi_h = 2(-1)^{h+1}(h+1)\, K^{h+1}(n-1)!\; m_{n-2h-2}\, dV$$

siendo dV el elemento de volumen del espacio, o sea, $dV = [ds_1\, ds_2 \dots ds_n]$.

Teniendo en cuenta las ecuaciones de estructura (1.10) y las reglas de diferenciación exterior, de (2.3) y (2.7) se deduce la relación fundamental siguiente (debida a CHERN [5])

$$(2.9)\qquad d\Phi_h = \Psi_{h-1} + \frac{n-2h-1}{2(h+1)}\, \Psi_h$$

o sea

$$(2.10)\qquad \Psi_h = \frac{2(h+1)}{n-2h-1}\, (d\Phi_h - \Psi_{h-1})$$

fórmula recurrente que permite escribir

$$(2.11)\qquad \Psi_h = d\theta_h$$

siendo

$$(2.12)\qquad \theta_h = \sum_{\lambda=0}^{h} (-1)^{h-\lambda}\, \frac{(2h+2) \dots (2\lambda+2)}{(n-2\lambda-1) \dots (n-2h-1)}\, \Phi_\lambda$$

para valores de h tales que $0 \leq 2h \leq n-2$.

Distingamos ahora dos casos según la paridad de n.

1º *n par*. - Pongamos $n=2p$. Queremos aplicar la fórmula de STOKES

$$\int_S \theta_{p-1} = \int_Q d\theta_{p-1} = \int_Q \Psi_{p-1} \tag{2.13}$$

a la hipersuperficie S y al cuerpo Q que ella limita. Para ello observemos que el campo de las normales A_0A_n a S se puede prolongar por continuidad al interior de S excepto para un número finito de puntos cuyo número es precisamente igual a la característica de EULER-POINCARÉ $\chi(Q)$ del cuerpo Q[1].

Rodeando estos puntos por esferas de radio ε y haciendo luego $\varepsilon \to 0$, la parte correspondiente a estas esferas en la integral del primer miembro de (2.13), según (2.12) y (2.6) se reduce a

$$-(-1)^{p-1} \frac{2p\,(2p-2)\ldots 2}{(n-1)\,(n-3)\ldots 1}\,(n-1)!\; O_{n-1}\,\chi(Q) \tag{2.14}$$

siendo O_{n-1} el área de la esfera euclidiana $(n-1)$-dimensional de radio unidad, o sea, en general

$$O_i = \frac{2\pi^{(i+1)/2}}{\Gamma\left(\frac{i+1}{2}\right)}\,. \tag{2.15}$$

El signo menos en (2.14) es debido a que estas esferas de radio ε limitan a Q por su parte exterior.

Introduciendo las integrales de curvatura media

$$M_i = \int_S m_i\,dF \tag{2.16}$$

---

[1] Recordemos que si Q está descompuesto en símplices y a_i es el número de ellos de dimensión i, es

$$\chi(Q) = a_0 - a_1 + a_2 - \ldots + (-1)^n a_n\,.$$

Para el contorno S de Q es

$$\chi(S) = 0 \quad (n \text{ par})\,, \qquad \chi(S) = 2\chi(Q) \quad (n \text{ impar})$$

según (2.12), (2.6), (2.8) la fórmula (2.13) se escribe por tanto

$$\sum_{\lambda=0}^{p-1} (-1)^{p-1} \frac{2p(2p-2)\dots(2\lambda+2)}{(n-2\lambda-1)\dots 3.1} K^{\lambda}(n-1)!\, M_{n-2\lambda-1}$$

$$-(-1)^{p-1} \frac{2p(2p-2)\dots 2}{(n-1)(n-3)\dots 3.1}(n-1)!\, O_{n-1}\chi(Q) =$$

$$= 2(-1)^{p} p K^{p}(n-1)!\, V,$$

siendo V el volumen de Q.

Introduciendo las áreas O_i definidas por (2.15), simples transformaciones de los coeficientes permiten escribir esta fórmula en la forma

$$(2.17) \qquad \sum_{\lambda=0}^{p-1} \binom{n-1}{2\lambda} \frac{O_n}{O_{n-1-2\lambda}\, O_{2\lambda}} K^{\lambda} M_{n-2\lambda-1} + K^{p} V = \frac{1}{2} O_n \chi(Q)$$

que es la fórmula generalizada de GAUSS-BONNET para n par.

2º *n impar*. - Pongamos $n=2p+1$. En este caso de (2.9) se deduce $d\Phi_p = \Psi_{p-1}$ y por tanto, de (2.11),

$$(2.18) \qquad d(\theta_{p-1} - \Phi_p) = 0.$$

La fórmula de STOKES aplicada a la hipersuperficie S y al cuerpo Q que ella limita, nos da ahora

$$(2.19) \qquad \int_S \theta_{p-1} - \int_S \Phi_p = 0$$

o bien, teniendo en cuenta, como antes, que para llenar el cuerpo Q con un campo de normales continuo, debemos aislar un número de puntos igual a $\chi(Q)$ y rodearlos por esferas de radio ε que luego puede tender a cero, resulta

$$\sum_{\lambda=0}^{p-1} (-1)^{p-1} \frac{2p(2p-2)\dots(2\lambda+2)}{(n-2\lambda-1)\dots 4.2} K^{\lambda}(n-1)!\, M_{n-2\lambda-1}$$

$$-(-1)^{p} K^{p}(n-1)!\, F - (-1)^{p-1} \frac{2p(2p-2)\dots 2}{(n-1)(n-3)\dots 2}(n-1)!\, O_{n-1}\chi(Q) = 0$$

donde F es el área de S.

Simplificando y transformando los coeficientes de manera que aparezcan las áreas O_i, resulta

$$(2.20) \qquad \sum_{\lambda=0}^{p-1} \binom{n-1}{2\lambda} \frac{O_n}{O_{n-1-2\lambda}\, O_{2\lambda}}\, K^{\lambda} M_{n-2\lambda-1} = \frac{1}{2}\, O_n \chi(Q)$$

donde debe entenderse que es $M_0 = F$, de acuerdo con (2.5) y (2.16). En resumen:

La fórmula generalizada de GAUSS-BONNET *para espacios de curvatura constante K se escribe*

$$(2.21) \qquad c_{n-1} M_{n-1} + c_{n-3} M_{n-3} + \ldots + c_1 M_1 + K^{n/2}\, V = \frac{1}{2}\, O_n \chi(Q)$$

para n par, y

$$(2.22) \qquad c_{n-1} M_{n-1} + c_{n-3} M_{n-3} + \ldots + c_2 M_2 + c_0 F = \frac{1}{2}\, O_n \chi(Q)$$

para n impar. En ambos casos M_i son las integrales de curvatura media definidas por (2.5) *y* (2.16) *y*

$$(2.23) \qquad c_i = \binom{n-1}{i} \frac{O_n}{O_i\, O_{n-1-i}}\, K^{(n-1-i)/2}\,.$$

Para n impar puede sustituirse $\chi(Q) = \frac{1}{2}\chi(S)$.

3. Fórmulas duales en el espacio elíptico.

Consideremos el espacio elíptico ($K > 0$) y, por simplicidad, el caso $K = 1$.

Dada la hipersuperficie orientable y cerrada S, la paralela a distancia $\pi/2$ se llama la « dual » o « polar » de S; la representaremos por S^*. Si Q es el cuerpo limitado por S, el cuerpo Q^* limitado por S^* por el lado que no contiene a Q será el dual polar de Q.

Entre las integrales de curvatura media M_i de S y las M_i^* de S^* existe la relación [11]

$$(3.1) \qquad M_i^* = M_{n-1-i}\,. \qquad (i = 0, 1, 2, \ldots, n-1)$$

Por tanto, escribiendo (2.21) para S^*, teniendo en cuenta

(3.1) y que $\chi(Q^*)=\chi(Q)$, resulta (poniendo $M_0=F$),

$$c_{n-1}F+c_{n-3}M_2+\ldots+c_1M_{n-2}+V^*=\frac{1}{2}O_n\chi(Q) \tag{3.2}$$

que permite calcular el volumen V^* del cuerpo polar de Q, para *n par*.

La misma (2.21), teniendo en cuenta (3.1) y que $M_0{}^*=F^*$ (área de S^*) nos da

$$c_{n-1}F^*+c_{n-3}M_{n-3}+\ldots+c_1M_1+V=\frac{1}{2}O_n\chi(Q) \tag{3.3}$$

de la cual se deduce el área F^* de la superficie polar S^* para *n par*.

Para *n impar*, la fórmula dual de (3.22) es

$$c_{n-1}F+c_{n-3}M_2+\ldots+c_2M_{n-3}+c_0F^*=\frac{1}{4}O_n\chi(S) \tag{3.4}$$

donde se ha aplicado la relación $\chi(Q)=\frac{1}{2}\chi(S)$.

Para *n impar*, el volumen V^* del cuerpo polar no se puede obtener por dualidad de la fórmula de Gauss-Bonnet. En este caso V^* se puede calcular restando al doble del volumen total del espacio elíptico (o sea O_n) el volumen del cuerpo paralelo exterior a Q a distancia $\pi/2$, dado por la fórmula generalizada de Steiner y aplicando luego (2.22) para simplificar el resultado (ver Allendoerfer [1]). Se obtiene así, para *n impar*

$$c_{n-2}M_{n-2}+c_{n-4}M_{n-4}+\ldots+c_1M_1+V+V^*=\left(1-\frac{1}{2}\chi(Q)\right)O_n \tag{3.5}$$

donde también se puede sustituir $\chi(Q)$ por su igual $\frac{1}{2}\chi(S)$.

Ejemplos. - Para $n=2$, (3.2) y (3.3) nos dan las fórmulas elementales $L+F^*=2\pi\chi(Q)$, $L^*+F=2\pi\chi(Q)$, habiendo llamado en este caso L a la longitud del contorno de Q y F al área.

Para $n=3$, (3.4) da

$$F+F^*=2\pi\chi(S)$$

relación debida a Blaschke [3] y (3.5) da

$$M_1+V+V^*=2\pi^2-\pi^2\chi(Q)$$

Para $n=4$, (3.2) y (3.3) dan respectivamente

$$2F + 3M_2 + 3V^* = 4\pi^2\chi(Q)$$
$$2F^* + 3M_1 + 3V = 4\pi^2\chi(Q)$$

Para $n=5$, según (3.4) se tiene

$$F + 2M_2 + F^* = \frac{4}{3}\pi^2\chi(S)$$

y según (3.5)

$$M_3 + M_1 + V + V^* = \left(1 - \frac{1}{2}\chi(Q)\right)\pi^3 .$$

4. Una aplicación.

Consideremos el espacio elíptico n-dimensional ($K=1$) y en él el cuerpo Q limitado por la hipersuperficie S del número anterior.

Recordemos que si dos normales a S en puntos consecutivos de una misma línea de curvatura (por ejemplo la tangente a la dirección A_0A_i, nº 2) se cortan en el punto C_i tal que $C_iA=\varrho_i$ y el elemento de arco sobre la línea de curvatura es ds_i, se tiene $ds_i = \text{sen}\,\varrho_i d\varphi_i$, siendo $d\varphi_i$ el ángulo entre las dos normales. Por tanto el elemento de área dF de S se puede expresar

$$dF = [\prod_1^{n-1} \text{sen}\,\varrho_i\, d\varphi_i] . \tag{4.1}$$

Por otra parte, entre los ϱ_i y los radios de curvatura principales R_i introducidos en el nº 2 existe la relación [6, pág. 214],

$$R_i = \text{tg}\,\varrho_i . \tag{4.2}$$

Tomemos ahora a partir de los puntos A_0 de S y sobre las normales, segmentos de longitud $\lambda\,(-\pi/2 \leq \lambda \leq \pi/2)$. Si $\lambda > 0$ tomaremos el segmento en la dirección positiva C_iA_0 y si $\lambda < 0$ en la dirección negativa A_0C_i. El lugar geométrico de los extremos de estos segmentos constituye la hipersuperficie S_λ paralela a S a distancia λ. El elemento de área dF_λ de S_λ, según (4.1), será

$$(4.3) \qquad dF_\lambda = [\prod_1^{n-1} \operatorname{sen}(\varrho_i + \lambda)\, d\varphi_i]$$

o bien, según (4.1) y (4.2)

$$(4.4) \qquad dF_\lambda = \prod_1^{n-1} \left(\cos\lambda + \frac{\operatorname{sen}\lambda}{R_i}\right) dF$$

y el elemento de volumen del espacio correspondiente a un punto P de S_λ, será

$$(4.5) \qquad dP = [dF_\lambda\, d\lambda] \; .$$

Veamos el signo de este elemento de volumen. Cada factor sen $(\varrho_i + \lambda) d\varphi_i$ de (4.3) es positivo para $-\varrho_i < \lambda \leq \pi/2$ y negativo para $-\pi/2 \leq \lambda < -\varrho_i$; en el primer caso la distancia PA_0 es un *mínimo* de las distancias de P a los puntos de la línea de curvatura tangente a la dirección A_0A_i, y en el segundo caso un *máximo*. Para cada punto P de la normal A_0A_n a S y cada línea de curvatura A_0A_i definimos el número δ_i igual a $+1$ si la distancia PA_0 es un mínimo e igual a -1 si es un máximo; si no es ni máximo ni mínimo pondremos $\delta_i = 0$. El signo del dP correspondiente a P será el signo del producto $\nu(P, A_0) = \delta_1 \delta_2 \dots \delta_{n-1}$. Por tanto, integrando (4.5) a toda la superficie F de S y a todo λ entre $-\pi/2 \leq \lambda \leq \pi/2$, en el primer miembro el elemento de volumen dP aparece factor común de la suma de los coeficientes ν correspondientes a las distintas normales trazadas desde P a S. Es decir, si estas normales son $PA_{01}, PA_{02}, \dots$ poniendo $N = \sum_h \nu(P, A_{0h})$, resulta

$$(4.6) \qquad \int N dP = \int_S \int_{-\pi/2}^{\pi/2} \prod_1^{n-1} \left(\cos\lambda + \frac{\operatorname{sen}\lambda}{R_i}\right) dF d\lambda \; .$$

Teniendo en cuenta que

$$(4.7) \qquad \int_{-\pi/2}^{\pi/2} \cos^{n-i}\lambda \operatorname{sen}^{i-1}\lambda\, d\lambda = \begin{cases} 0 & \text{si } i \text{ es } \textit{par} \\ \dfrac{2 O_n}{O_{n-i} O_{i-1}} & \text{si } i \text{ es } \textit{impar}, \end{cases}$$

según la definición (2.16), (2.5) de las integrales de curvatura media y la notación (2.23), resulta

$$\int NdP = 2(c_0 M_0 + c_2 M_2 + \ldots + c_{n-1} M_{n-1}) \tag{4.8}$$

si n es *impar*, y

$$\int NdP = 2(c_0 M_0 + c_2 M_2 + \ldots + c_{n-2} M_{n-2}) \tag{4.9}$$

si n es *par*.

Teniendo en cuenta (2.22) y (3.2) estas fórmulas toman la forma simple (puesto que para $K=1$ es $c_i = c_{n-1-i}$),

$$\int NdP = O_n \chi(Q) \qquad \text{para } n \text{ impar} \tag{4.10}$$

$$\int NdP = O_n \chi(Q) - 2V^* \qquad \text{para } n \text{ par}\ . \tag{4.11}$$

En el primer miembro las integrales están extendidas a todo el espacio elíptico. Para $n=2,3$ estas fórmulas son debidas a Blaschke [4].

5. Densidad para conjuntos de rectas.

Para determinar una recta G en el espacio de curvatura constante K (o sea, una geodésica del espacio), supongamos una hipersuperficie S fija que corte a G; sea dF el elemento de área de S correspondiente al punto P de intersección, dO_{n-1} el elemento de área sobre la esfera euclidiana unidad de centro P correspondiente a la dirección de la recta y φ el ángulo entre G y la normal a S en P. Entonces la densidad para medir conjuntos de rectas es [12]

$$dG = |\cos\varphi| [dO_{n-1} dF]\ . \tag{5.1}$$

De aqui se deduce inmediatamente, por ejemplo, que la medida del conjunto de rectas que cortan a una hipersuperficie convexa y cerrada es igual a $\dfrac{1}{4\pi} O_n F$ [12].

Sea t el arco sobre G. Si dt es el elemento de este arco en el punto P de G, de (5.1) se deduce

$$[dG\, dt] = [dO_{n-1}\, dP] \tag{5.2}$$

siendo $dP = |\cos\varphi|\,[dF\, dt]$ el elemento de volumen del espacio correspondiente al punto P. Integrando ambos miembros de (5.2) a todos los puntos P interiores al cuerpo Q limitado por S, en el primer miembro el arco t podrá variar dentro de la cuerda λ que G determina en Q y el segundo miembro da $\frac{1}{2}O_{n-1}V$ (teniendo en cuenta que a direcciones opuestas corresponde la misma geodésica). Se tiene así la fórmula integral

$$\int \lambda\, dG = \frac{1}{2}\, O_{n-1} V \tag{5.3}$$

que es independiente de K y vale aun para espacios de RIEMANN cualesquiera [12]. En (5.3) la integral está extendida a todas las rectas G que cortan a Q.

6. Fórmula integral de las cuerdas.

En el espacio euclidiano ($K=0$) de n dimensiones, generalizando una conocida fórmula de CROFTON para $n=2$, HADWIGER ha obtenido la fórmula integral [7],

$$\int \lambda^{n+1}\, dG = \frac{1}{2}\, n(n+1)\, V^2 \tag{6.1}$$

donde, como en el número anterior, λ significa la longitud de la cuerda que la recta G determina en el cuerpo Q, *supuesto ahora convexo*, de volumen V. La integración está extendida a todas las rectas del espacio.

Nuestro objeto es generalizar esta fórmula al caso de los espacios de curvatura constante.

Sean P_1, P_2 dos puntos de la recta G y sean t_1, t_2 los valores de t correspondientes a los mismos. En un sistema de coordenadas polares de origen P_1 el elemento de volumen correspondiente a P_2 vale como se sabe

$$dP_2 = \frac{\operatorname{sen}^{n-1} \sqrt{K}\, |t_2 - t_1|}{K^{(n-1)/2}}\, [dt_2\, dO_{n-1}] \tag{6.2}$$

Multiplicando exteriormente por dP_1 y teniendo en cuenta (5.2) resulta

$$[dP_1 dP_2] = \frac{\operatorname{sen}^{n-1} \sqrt{K}\,|t_2 - t_1|}{K^{(n-1)/2}}\,[dt_1 dt_2 dG] \tag{6 3}$$

fórmula obtenida por otro camino por HAIMOVICI [8].

Sea ahora Q un cuerpo convexo e integremos (6.3) a todos los pares de puntos P_1, P_2 interiores a Q. El primer miembro vale V^2, siendo V el volumen de Q. En el segundo miembro, si λ es la longitud de la cuerda que G determina en Q aparece la integral

$$\Phi_{n-1}(\lambda, K) = \int_0^\lambda\int_0^\lambda \frac{\operatorname{sen}^{n-1} \sqrt{K}\,|t_2 - t_1|}{K^{(n-1)/2}}\,dt_1 dt_2 \tag{6.4}$$

que vale

$$\Phi_{n-1}(\lambda, K) = -\frac{2}{(n-1)\,K^{(n+1)/2}}\left[\frac{1}{n-1}\operatorname{sen}^{n-1}\sqrt{K}\lambda\right. \tag{6.5}$$

$$\left.+\sum_{i=1}^{(n-3)/2}\frac{(n-2)\ldots(n-2i)}{(n-3)\ldots(n-1-2i)^2}\operatorname{sen}^{n-1-2i}\sqrt{K}\lambda\right]+\frac{(n-2)\ldots 3.1}{(n-1)\ldots 4.2}\,K\lambda^2,$$

para *n impar*, y

$$\Phi_{n-1}(\lambda, K) = -\frac{2}{(n-1)\,K^{(n+1)/2}}\left[\frac{1}{n-1}\operatorname{sen}^{n-1}\sqrt{K}\lambda\right. \tag{6.6}$$

$$\left.+\sum_{i=1}^{(n-2)/2}\frac{(n-2)\ldots(n-2i)}{(n-3)\ldots(n-1-2i)^2}\operatorname{sen}^{n-1-2i}\sqrt{K}\lambda-\frac{(n-2)\ldots 4.2}{(n-3)\ldots 3.1}\sqrt{K}\lambda\right]$$

para *n par*. Para $n=2$ resulta $\Phi_1(\lambda, K) = 2(\sqrt{K}\lambda - \operatorname{sen}\sqrt{K}\lambda)\,K^{-\frac{3}{2}}$.

La integración de ambos miembros de (6.3) nos da por tanto

$$\int \Phi_{n-1}(\lambda, K)\,dG = V^2 \tag{6.7}$$

donde en el primer miembro la integración está extendida a todas las rectas G que cortan a Q.

Por ejemplo, para $n=2$ se tiene

$$\frac{1}{K}\int\left(\lambda - \frac{\operatorname{sen}\sqrt{K}\lambda}{\sqrt{K}}\right)dG = \frac{1}{2}\,F^2$$

siendo F el área del dominio convexo Q. Para $n=3$,

$$\frac{1}{K}\int\left(\lambda^2-\frac{1}{K}\operatorname{sen}^2\sqrt{K}\lambda\right)dG=2V^2$$

y para $n=4$,

$$\frac{1}{K^2}\int\left(\lambda-\frac{1}{6\sqrt{K}}\operatorname{sen}^3\sqrt{K}\lambda-\frac{1}{\sqrt{K}}\operatorname{sen}\sqrt{K}\lambda\right)dG=\frac{3}{4}V^2 .$$

Obsérvese que para $K\to 0$ resulta siempre el resultado de HADWIGER (6.1).

7. FÓRMULAS DUALES DE LAS INTEGRALES DE LAS CUERDAS.

En el espacio elíptico, $K=1$, a las fórmulas del número anterior corresponden otras fórmulas duales.

a) Consideremos primero la fórmula (5.3). Por dualidad, a una recta G corresponde un espacio lineal L_{n-2} de dimensión $n-2$. La densidad para conjuntos de estos espacios la indicaremos por dL_{n-2} y su valor es igual a la densidad dG de su recta dual. Supongamos Q convexo. A la longitud λ de la cuerda que G determina en Q corresponde el ángulo $\pi-\varphi$, siendo φ el ángulo formado por los hiperplanos tangentes a Q trazados por L_{n-2}. Por tanto, la fórmula dual de (5.3) es

$$\int(\pi-\varphi)\,dL_{n-2}=\frac{1}{2}\,O_{n-1}V^* . \tag{7.1}$$

Como la medida total de los L_{n-2} exteriores a Q es igual a la medida de las rectas que cortan a su dual Q^* y por tanto igual a $\frac{1}{4\pi}O_nF^*$, (7.1) se puede escribir

$$\int\varphi\,dL_{n-2}=\frac{1}{4}\,O_n\,F^*-\frac{1}{2}\,O_{n-1}V^* \tag{7.2}$$

donde la integración está extendida a todos los L_{n-2} exteriores a Q, φ es el ángulo formado por los hiperplanos tangentes a Q trazados por L_{n-2}, F^* es el área de la hipersuperficie S^* del cuerpo dual de Q y V^* es el volumen de este cuerpo dual.

Los valores de F^* y V^* pueden sustituirse por invariantes cuerpo Q aplicando las fórmulas del nº 3.

Por ejemplo, para $n=2$, (7.2) se escribe del

$$\int \varphi dP = \pi (L - F) \tag{7.3}$$

siendo φ el ángulo de las tangentes a la figura convexa Q trazadas desde el punto P, L la longitud y F el área de Q; la integración extendida a todo el plano elíptico exterior a Q.

Para $n=3$, resulta

$$\int \varphi dG = 2\pi (M_1 + V) - \frac{1}{2}\, \pi^2 F \; . \tag{7.4}$$

De aquí se deduce, por ejemplo, que para todo cuerpo convexo del espacio elíptico se cumple la desigualdad

$$F \leq \frac{4}{\pi}\,(M_1 + V)\; . \tag{7.5}$$

Para los casos particulares (7.3) y (7.4) ver [10], [12].

b) Consideremos ahora la fórmula dual de (6.7). Observemos que para *n par* es, según (6.6),

$$\Phi_{n-1}(\pi - \varphi, 1) = \Phi_{n-1}(\varphi, 1) + \frac{2(n-2)\dots 4.2}{(n-1)(n-3)\dots 3.1}\,\pi - \frac{4(n-2)\dots 4.2}{(n-1)\dots 3.1}\,\varphi \tag{7.6}$$

y por tanto, teniendo en cuenta la medida total de las rectas que cortan a Q y la fórmula (7.2), resulta que la fórmula dual de (6.7) para *n par* es

$$\int \Phi_{n-1}(\varphi, 1)\, dL_{n-2} = V^{*2} + \frac{(n-2)(n-4)\dots 2}{2(n-1)(n-3)\dots 1}\,(O_n F^* - 4 O_{n-1} V^*) \tag{7.7}$$

donde la integración está extendida a todos los L_{n-2} exteriores a Q, φ representa el ángulo formado por los hiperplanos tangentes a Q trazados por L_{n-2}, $\Phi_{n-1}(\varphi, 1)$ es la función (6.6),

F^* y V^* son el área y el volumen del cuerpo Q^* polar de Q, que pueden calcularse a partir de los invariantes de Q por medio de las fórmulas del nº 3.

Por ejemplo, para $n=2$ resulta

$$\int (\varphi - \operatorname{sen} \varphi)\, dP = \frac{1}{2} L^2 - \pi F$$

que es la clásica fórmula de CROFTON para el plano elíptico.

Si n es *impar*, según (6.5) resulta

$$\Phi_{n-1}(\pi - \varphi, 1) = \Phi_{n-1}(\varphi, 1) + \frac{(n-2)\ldots 3.1}{(n-1)(n-3)\ldots 4.2}\pi^2 - \frac{2(n-2)\ldots 3.1}{(n-1)(n-3)\ldots 4.2}\pi\varphi \tag{7.8}$$

y la fórmula dual de (6.7), aplicando (7.2), resulta

$$\int \Phi_{n-1}(\varphi, 1)\, dL_{n-2} = V^{*2} + \frac{(n-2)(n-4)\ldots 3.1}{(n-1)(n-3)\ldots 4.2}\frac{\pi}{4}(O_n F^* - 4 O_{n-1} V^*) \tag{7.9}$$

donde los distintos términos tienen el mismo significado de antes.

Por ejemplo, para $n=3$, resulta

$$\frac{1}{2}\int (\varphi^2 - \operatorname{sen}^2 \varphi)\, dG = (M_1 + V)^2 - \frac{\pi^3}{4} F$$

como ya obtuvimos en otro lugar [12].

BIBLIOGRAFIA

[1] C. B. ALLENDOERFER, *Steiner's formulae on a general* S^{n+1}, Bull. of the Am. Math. Soc. 54, 1948, 128-135.

[2] C. B. ALLENDOERFER-A. WEIL, *The Gauss-Bonnet theorem for Riemannian polyhedron*, Transactions of the Am. Math. Soc. 53, 1943, 101-129.

[3] W. BLASCHKE, *Integralgeometrie* 22: *Zur elliptischen Geometrie*, Math. Zeits. 41, 1936, 785-786.

[4] — — *Ueber geschlossene Kurven und Flächen in der elliptischen Geometrie*, Hamburg Abh. 12, 1938, 111-113.

[5] S. S. CHERN, *On the curvatura integra in a riemannian manifold*, Annals of Math. 46, 1945, 674-684.

[6] E. P. EISENHART, *Riemannian Geometry*, Princeton, 1926.

[7] H. HADWIGER, *Neue Integralrelationen für Eikörperpaare*, Acta Scientiarum Math. XIII, 1950, 252-257.

[8] M. HAIMOVICI, *Généralisation d'une formule de Crofton dans un espace de Riemann à n dimensions*, C. R. Acad. Sc. de Roumanie, 7, 1936.

[9] G. HERGLOTZ, *Ueber die Steinersche Formel für Parallelflächen*, Hamburg Abh. XV, 1943, 165-177.

[10] L. A. SANTALÓ, *Integral Formulas in Crofton's style on the sphere*, Duke Math. J. 9, 1942, 707-722.

[11] — — *On parallel hypersurfaces in the elliptic and hyperbolic n-dimensional space*, Proc. Am. Math. Soc. 1, 1950, 325-330.

[12] — — *Measure of sets of geodesics in a riemannian space and applications to integral formulas in elliptic and hyperbolic spaces*, Summa Brasiliensis Math. 3, 1952, 1-11.

ON THE MEAN CURVATURES OF A FLATTENED CONVEX BODY

L. A. SANTALÓ

Let E^n be the n-dimensional euclidean space and let L^r be a linear subspace of E^n. A convex body K^r contained in L^r can be considered as a *flattened convex body* of E^n. As a convex body of L^r, K^r possesses the mean curvatures

$$M_q^r \; (0 \leq q \leq r-1)$$

defined by (1). As a flattened convex body of E^n, K^r possesses the mean curvatures

$$M_q^n \; (0 \leq q \leq n-1)$$

defined as the limit of the mean curvatures $M_q^n(\varepsilon)$ of the convex body $K^r(\varepsilon)$ parallel exterior to K^r at the distance ε, as $\varepsilon \rightarrow 0$.

The purpose of the present note is to prove the formulae (13), (14), (15) which relate the mean curvatures M_q^r and M_q^n. As a consequence, we complete an integral-geometric result of HERGLOTZ and PETKANTSCHIN [formula (16)].

1. Let K^r be a convex body contained in a linear subspace L^r of E^n. The boundary ∂K^r is an $(r-1)$-dimensional variety of L^r which is assumed to be twice differentiable.

If ϱ_i $(i = 1, 2, \ldots, r-1)$ denote the principal radii of curvature of ∂K^r at a point P', the q-th mean curvature of K^r (as a convex body of L^r) is defined by

$$M_q^r = \frac{1}{\binom{r-1}{q}} \int_{\partial K^r} \left\{ \frac{1}{\varrho_1}, \ldots, \frac{1}{\varrho_q} \right\} d\sigma_{r-1} \tag{1}$$

where the brackets $\{\}$ denote the q-th elementary symmetric function formed by the principal curvatures $1/\varrho_i$ and $d\sigma_{r-1}$ is the element of area of ∂K^r at P'.

As particular cases we have:

$M_0^r = \sigma_{r-1} =$ area of ∂K^r,

$M_{r-1}^r = O_{r-1} =$ area of the $(r-1)$-dimensional unit sphere, *i. e.*

$$O_{r-1} = \frac{2\pi^{r/2}}{\Gamma(r/2)} \tag{2}$$

For instance, if $r=2$, K^2 is a plane convex figure and we have: $M_0^2=\sigma_1=$ length of the boundary of K^2; $M_1^2=2\pi$. For $r=3$, if K^3 is a convex body in ordinary space, we have: $M_0^3=\sigma_2=$ surface area of ∂K^3; $M_1^3=$ integrated mean curvature of ∂K^3; $M_2^3=4\pi$. For $r=1$, M_0^1 is meaningless; however in this case we have:

$$M_{r-1}^r=O_{r-1}=O_0=2$$

and consequently we will allways take $M_0^1=2$.

We now consider K^r as a flattened convex body of E^n. In order to define its mean curvatures

$$M_q^n \quad (q=1, 2, \ldots, n-1)$$

we consider first the mean curvatures of the convex body $K^r(\varepsilon)$ parallel to K^r at a distance ε (*i. e.* the set of points of E^n whose distance to K^r is $\leqq \varepsilon$) and then pass to the limit as $\varepsilon \to 0$.

The boundary $\partial K^r(\varepsilon)$ is a twice differentiable hypersurface of E^n with a well defined normal at each point P; let P' be the intersection point of K^r (considered as a convex body of L^r) we will say that P belongs to the region (A) of $\partial K^r(\varepsilon)$; if P' is a point of ∂K^r we will say that P belongs to the region (B) of $\partial K^r(\varepsilon)$.

At the points of the region (A), the element of area of $\partial K^r(\varepsilon)$ is equal to $\varepsilon^{n-r-1}\,dO_{n-r-1}\,d\sigma_r$ where $dO_{n-r-1}=$ area element of the unit $(n-r-1)$-dimensional sphere and $d\sigma_r=$ volume element of K^r. At the points of the region (B), the element of area of $\partial K^r(\varepsilon)$ is equal to $\varepsilon^{n-r}\,dO_{n-r}\,d\sigma_{r-1}$ where $d\sigma_{r-1}=$ element of area of ∂K^r. Consequently, the q-th mean curvature of $K^r(\varepsilon)$ is given by

$$(3) \qquad M_q^n(\varepsilon)=\frac{1}{\binom{n-1}{q}}\left[\int_{K^r}\left\{\frac{1}{R_1},\ldots,\frac{1}{R_q}\right\}\varepsilon^{n-r-1}\,dO_{n-r-1}\,d\sigma_r + \int_{\partial K^r}\left\{\frac{1}{R_1},\ldots,\frac{1}{R_q}\right\}\varepsilon^{n-r}\,dO_{n-r}\,d\sigma_{r-1}\right]$$

where the principal radii of curvature $R_h=R_h(\varepsilon)$ have the following values:

a) For the points of the region (A) it is clear that

$$(4) \qquad \begin{aligned} R_h&=\varepsilon \quad &&\text{for} \quad h=1,2,3,\ldots,n-r-1\\ R_h&=\infty \quad &&\text{for} \quad h=n-r,\, n-r+1,\ldots,n-1. \end{aligned}$$

b) In order to find the values of $R_h=R_h(\varepsilon)$ at the points of the region (B), let us consider at each point x of ∂K^r a frame of n orthogonal unit vectors $\mathbf{e}_1, \mathbf{e}_2, \ldots, \mathbf{e}_n$ such that $\mathbf{e}_1, \mathbf{e}_2, \ldots, \mathbf{e}_{n-r}$ be constant (independent of x) and orthogonal to L^r; $\mathbf{e}_{n-r+1}, \ldots, \mathbf{e}_{n-1}$ be the principal tangents to ∂K^r (as a variety of L^r) at the point x, and $\mathbf{e}_n$ be the normal to ∂K^r contained in L^r. The vector equation of $\partial K^r(\varepsilon)$ will be

$$\mathbf{X} = \mathbf{x} - \varepsilon \mathbf{N} \tag{5}$$

where

$$\mathbf{N} = \cos\vartheta\, \mathbf{e}_n + \sum_{h=1}^{n-r} \cos\vartheta_h\, \mathbf{e}_h . \tag{6}$$

For each fixed $\mathbf{x}$, $\mathbf{X}$ will describe a $(n-r)$-sphere and consequently we have

$$R_h = \varepsilon \quad \text{for} \quad h = 1, 2, 3, \ldots, n-r. \tag{7}$$

For $h = n-r+1, \ldots, n-1$, by the equations of Olinde Rodrigues we have

$$d\mathbf{N} \cdot \mathbf{e}_h = -\frac{1}{R_h}\, d S_h \tag{8}$$

where $d S_h$ denotes the arc element on $\partial K^r(\varepsilon)$ the tangent vector of this arc being parallel to $\mathbf{e}_h$, *i. e.*

$$d S_h = d\mathbf{X} \cdot \mathbf{e}_h = ds_h - \varepsilon\, d\mathbf{N} \cdot \mathbf{e}_h \tag{9}$$

where ds_h is the arc element on ∂K^r tangent to $\mathbf{e}_h$.

From (6), taking into account that the vectors $\mathbf{e}_h$, for $h = 1, 2, \ldots, n-r$, are constant, we have

$$d\mathbf{N} \cdot \mathbf{e}_h = \cos\vartheta\, d\,\mathbf{e}_n \cdot \mathbf{e}_h = -\frac{\cos\vartheta}{\varrho_{h-n+r}}\, ds_h \quad (h = n-r+1, \ldots, n-1) \tag{10}$$

where $\varrho_1, \varrho_2, \ldots, \varrho_{r-1}$ are the principal radii of curvature of ∂K^r. From (8), (9) and (10) we have

$$R_h = \frac{\varrho_{h-n+r}}{\cos\vartheta} + \varepsilon \quad \text{for} \quad h = n-r+1, \ldots, n-1. \tag{11}$$

With the values (4) and (7), (11) we can calculate $M_q^n(\varepsilon)$ and pass to the limit as $\varepsilon \to 0$.

There are three possible cases:

(1) $q \geq n-r$. The first integral in (3) vanishes as $\varepsilon \to 0$ and the second integral reduces to

$$M_q^n = \frac{1}{\binom{n-1}{q}} \int_{\partial K^r} \left\{ \frac{1}{\varrho_1}, \ldots, \frac{1}{\varrho_{q-n+r}} \right\} \cos^{q-n+r}\vartheta\, dO_{n-r}\, d\sigma_{r-1} \tag{12}$$

$$= \frac{\binom{r-1}{q-n+r}}{\binom{n-1}{q}} M_{q-n+r}^r \int \cos^{q-n+r}\vartheta\, dO_{n-r} .$$

The area element of the $(n-r)$-dimensional unit sphere may be written

$$dO_{n-r} = \sin^{n-r-1}\vartheta \sin^{n-r-2}\vartheta_1 \ldots \sin\vartheta_{n-r-2}\, d\vartheta\, d\vartheta_1 \ldots d\vartheta_{n-r-1}$$

and the integral in (12) must be extended over the half sphere whose pole is the end point of the normal to ∂K^r (contained in L^r). The limits of integration are then

$$0 \leqq \vartheta \leqq \frac{\pi}{2}, \quad 0 \leqq \vartheta_i \leqq \pi \ (i = 1, 2, \ldots, n-r-2), \quad 0 \leqq \vartheta_{n-r-1} \leqq 2\pi$$

and we have

$$(13) \qquad M_q^n = \frac{\binom{r-1}{q-n+r}}{\binom{n-1}{q}} M^r_{q-n+r}\, O_{n-r-1} \int_0^{\pi/2} \cos^{q-n+r}\vartheta \sin^{n-r-1}\vartheta\, d\vartheta$$

$$= \frac{\binom{r-1}{q-n+r}}{\binom{n-1}{q}} \frac{O_q}{O_{q-n+r}} M^r_{q-n+r}.$$

(2) $q = n-r-1$. The second integral in (3) vanishes as $\varepsilon \to 0$ and the first tends to $O_{n-r-1}\,\sigma_r$. Consequently

$$(14) \qquad M_q^n = \frac{1}{\binom{n-1}{q}} O_{n-r-1}\,\sigma_r(K^r)$$

where $\sigma_r(K^r)$ denotes the volume of K^r.

(3) $q < n-r-1$. Both integrals in (3) vanish as $\varepsilon \to 0$, and consequently

$$(15) \qquad M_q^n = 0.$$

2. Examples. For the ordinary space, $n = 3$, we have the following possibilities:

a) $r = 1$. K^r reduces to a segment of length s. The mean curvatures are

$$M_0^3 = 0, \quad M_1^3 = \pi s, \quad M_2^3 = 2\pi M_0^4 = 4\pi.$$

b) $r = 2$. K^r is a plane convex figure; let s be its perimeter and σ its area. We have

$$M_0^3 = 2\sigma, \quad M_1^3 = \frac{\pi}{2} M_0^2 = \frac{\pi}{2} s, \quad M_2^3 = 2 M_1^2 = 4\pi.$$

3. An integral-geometric application. The mean curvatures M_q^n are related with certain invariants H_q^n of K^r which appear in integral geometry. H_q^n $(q=0, 1, \ldots, n-1)$ denotes the measure of the set of q-dimensional linear spaces of E^n which have a common point with K^r. Analogously, if K^r is contained in L^r, then H_q^r $(q=0, 1, 2, \ldots, r-1)$ denotes the measure of the set of q-dimensional linear spaces of L^r which intersect K^r. The invariants M_q^n and H_q^n are related by (see [2], p. 183).

$$H_q^n = \frac{O_{n-2}\, O_{n-3} \ldots O_{n-q-1}}{2\,(n-q)\, O_{q-1}\, O_{q-2} \ldots O_1} M_{q-1}^n \qquad (1 \leqq q \leqq n-1)$$

and

$$H_q^r = \frac{O_{r-2}\, O_{r-3} \ldots O_{r-q-1}}{2\,(r-q)\, O_{q-1}\, O_{q-2} \ldots O_1} M_{q-1}^r \qquad (1 \leqq q \leqq r-1)\,.$$

Consequently, in terms of H_q^n, H_q^r the formulae (13), (14) and (15) may be written

$$H_q^n = c_{rnq}\, H_{q+r-n}^r \tag{16}$$

where c_{rqn} are the following constants:

$$c_{rqn} = \frac{\binom{r-1}{n-q}}{\binom{n-1}{q-1}} \frac{O_{n-2} \ldots O_{r-1}}{O_{q-1} \ldots O_{q+r-n}} \frac{O_{q-1}}{O_{q-n+r-1}}, \qquad \text{for} \quad q \geqq n-r+1$$

$$c_{rqn} = \frac{O_{n-2}\, O_{n-3} \ldots O_{n-q-1}}{2\,(n-q) \binom{n-1}{q-1} O_{q-1}\, O_{q-2} \ldots O_1} O_{q-1} \qquad \text{for} \quad q = n-r$$

$$c_{rqn} = 0 \qquad \text{for} \quad q < n-r\,.$$

The formula (16) has been given by HERGLOTZ and PETKANTSCHIN (see [1], p. 292); however they do not give the explicit values of the constants c_{rqn}.

Examples.

$$c_{213} = \frac{O_1\, O_0}{2 \cdot 2} = \frac{2 \cdot 2\pi}{4} = \pi, \qquad c_{223} = \frac{O_1\, O_1}{2 O_1\, 2} = \frac{\pi}{2}\,.$$

FACULTY OF SCIENCES
UNIVERSITY OF BUENOS AIRES
BUENOS AIRES, ARGENTINA

(Manuscript received January 8, 1957)

BIBLIOGRAPHY

[1] B. PETKANTSCHIN : *Zusammenhaenge zwischen den Dichten der Linearen Unterraeume im n-dimensionalen Raum*, (Integralgeometrie 6), Abhandlungen aus dem Math. Sem. Hansischen Univ., **11**, 249–310, (1936).

[2] L. A. SANTALÒ : *Sur la mesure des espaces linéaires qui coupent un corps convexe et problèmes qui s'y rattachent*, Colloque sur les questions de réalité en Géométrie, 177–190, Liège, (1955).

ÖZET

E^n, n boyutlu öklidyen uzay ve L^r de bunun lineer bir alt-uzayı olsun. L^r nin ihtiva ettiği K^r gibi bir konveks cisim, E^n uzayında, *yassı* bir konveks cisim gibi düşünülebilir. K^r cismi, L^r içinde tetkik edilecek olursa, (1) formülü ile tarif edilen

$$M_r^n \, (0 \leqq q \leqq r-1)$$

ortalama eğriliklerini; E^n nin yassı bir konveks cismi gibi telâkki olunursa,

$$M_q^n \, (0 \leqq q \leqq n-1)$$

ortalama eğriliklerini haizdir. Bu ortalama eğrilikler, K^r cismine, ε uzaklığında çizilen $K^r(\varepsilon)$ dış paralel konveks cisminin $M_q^n(\varepsilon)$ ortalama eğriliklerinin, ε u sıfıra yaklaştırmak suretiyle bulunan, limitleri olarak tarif edilirler.

Bu makalenin gayesi, bahis konusu M_q^r ve M_q^n ortalama eğrilikleri arasındaki bağıntıyı ifade eden (13), (14), (15) formüllerini ispat etmektir. Bunların neticesi olarak da HERGLOTZ ve PETKANTSCHIN in bir formülü tamamlanmaktadır –formül (16)–.

Two applications of the integral geometry in affine and projective spaces.

To Prof. O. Varga on his 50. anniversary with cordial friendship.

By L. A. SANTALÓ (Buenos Aires).

Introduction.

The integral geometry in projective space was initiated by VARGA [9] and continuated, together with the integral geometry in affine space, by the present author [4].

In this paper we give two applications of these concepts. First we consider the density for sets of pairs of parallel hyperplanes invariant with respect to the unimodular affine group. Then we evaluate the measure of all pairs of parallel hyperplanes which contain a given convex body K: the result is the integral (3. 1) where $\Delta(\sigma)$ is the width of K corresponding to the direction σ. Consequently, J (3. 2) is an unimodular affine invariant of K and we obtain the inequalities (3. 9) which relate J with the volume V of K.

The second application concerns the density for sets of hyperquadrics invariant with respect to the projective group. We give the explicit forms (5. 14), (5. 16) and (5. 19) of this density. For $n=2$ (conics on the plane) the formula (5. 19) was given by STOKA [8].

§. 1. The unimodular affine group.

We consider the n-dimensional affine space and in it the group A of affine transformations modulo 1, which in matrix notation is written

$$(1.1) \qquad x' = Ax + B, \quad \det A = |A| = 1$$

where $A = (a_{ij})$ and $B = (b_i)$ are $n \times n$ and $n \times 1$ matrices respectively; x and x' denote $n \times 1$ matrices whose elements are the n coordinates $x_1, x_2, \ldots, x_n$ of the point x and those $x'_1, x'_2, \ldots, x'_n$ of the transformed point x'.

A transformation of the group A is determined by a pair of matrices (A, B). The identity corresponds to the pair (E, O) where E is the unit matrix and O the $n\times 1$ matrix with all the elements equal to zero. The inverse of (A, B) is

$$(A, B)^{-1} = (A^{-1}, -A^{-1}B)$$

and the law for the product can be written

$$(A_2, B_2)(A_1, B_1) = (A_2A_1, A_2B_1 + B_2).$$

According to the theory of E. CARTAN (see for instance [5]) the *relative components* of the group A will be the elements (which are pfaffian forms) of the matrices Ω_1 (of type $n\times n$) and Ω_2 (of type $n\times 1$) defined by the equation

$$(A, B)^{-1}(A + dA, B + dB) = (E + \Omega_1, \Omega_2).$$

Thus we have

(1.2) $$\Omega_1 = A^{-1}dA, \quad \Omega_2 = A^{-1}dB.$$

By exterior differentiation and taking into account that

(1.3) $$dA^{-1} = -A^{-1}dAA^{-1}$$

we have

(1.4) $$\begin{aligned} d\Omega_1 &= dA^{-1}\wedge dA = -A^{-1}dAA^{-1}\wedge dA = -\Omega_1\wedge\Omega_1 \\ d\Omega_2 &= dA^{-1}\wedge dB = -A^{-1}dAA^{-1}\wedge dB = -\Omega_1\wedge\Omega_2 \end{aligned}$$

which are the equations of structure of MAURER-CARTAN for the unimodular affine group A.

In explicit form, if we set

(1.5) $$A = (a_{ij}), \quad B = (b_i), \quad A^{-1} = (\alpha^{ij}), \Omega_1 = (\omega_{ij}), \Omega_2 = (\omega_i),$$

(1.2) and (1.4) can be written in the form

(1.6) $$\begin{aligned} \omega_{ij} &= \sum_{h=1}^{n} \alpha^{ih}da_{hj} = -\sum_{h=1}^{n} a_{hj}d\alpha^{ih}, \quad \omega_i = \sum_{h=1}^{n}\alpha^{ih}db_h \\ d\omega_{ij} &= -\sum_{h=1}^{n}\omega_{ih}\wedge\omega_{hj}, \quad d\omega_i = -\sum_{h=1}^{n}\omega_{ih}\wedge\omega_h. \end{aligned}$$

By differentiation of the condition $\det A = 1$, we get

(1.7) $$\omega_{11} + \omega_{22} + \cdots + \omega_{nn} = 0.$$

§ 2. Density for sets of parallel hyperplanes.

It is known that there does not exist a density for sets of hyperplanes invariant with respect to A [4]. However we are going now to prove that such an invariant density exists for pairs of parallel hyperplanes.

Let us consider two parallel hyperplanes

(2.1) $$ux = h_1, \quad ux = h_2 \qquad (h_1 \neq h_2)$$

where h_1, h_2 are scalars and u a $1 \times n$ matrix $u = (u_1, u_2, \ldots, u_n)$.

By the unimodular affine transformation (A, B) these hyperplanes transform to

(2.2) $$uA^{-1}(x-B) = h_1, \quad uA^{-1}(x-B) = h_2.$$

In order that the varied transformation $(A+dA,\ B+dB)$ may give rise to the same hyperplanes (2.2), the relations

(2.3) $$d\left(\frac{uA^{-1}}{uA^{-1}B+h_i}\right) = 0 \qquad (i=1,2)$$

must hold (observe that the denominators are scalars) and we have

$$u\,dA^{-1}(uA^{-1}B+h_i) - uA^{-1}(u\,dA^{-1}B + uA^{-1}dB) = 0.$$

Since dA and dB are independent, a first condition is $uA^{-1}dB = 0$, i. e.

(2.4) $$u\Omega_2 = 0.$$

The remaining terms, give by application of (1.3),

$$u\Omega_1 A^{-1}(h_i + uA^{-1}B) = uA^{-1}(u\Omega_1 A^{-1}B).$$

Since the terms inside the parenthesis are scalars, we can set aside A^{-1} and we have

$$u\Omega_1(h_1 + uA^{-1}B) = u(u\Omega_1 A^{-1}B)$$
$$u\Omega_1(h_2 + uA^{-1}B) = u(u\Omega_1 A^{-1}B)$$

and by substraction

$$u\Omega_1(h_1 - h_2) = 0$$

i. e.

(2.5) $$u\Omega_1 = 0.$$

According to the general theory (see for instance [5]) if we have a set of relative components (or a set of linear combinations of relative components), say $\omega^{(1)}, \omega^{(2)}, \ldots, \omega^{(n+1)}$ such that the system

(2.6) $$\omega^{(1)} = 0, \omega^{(2)} = 0, \ldots, \omega^{(n+1)} = 0$$

is equivalent to the conditions (2.3), then the density for sets of pairs of parallel hyperplanes (if it exists) will be the exterior product $\omega^{(1)} \wedge \omega^{(2)} \wedge \wedge \cdots \wedge \omega^{(n+1)}$.

In our case the set (2.6) is given by (2.4) (one condition) and (2.5) (n conditions). Since the pairs of parallel hyperplanes are transformed transitively by the group A, without loss of generality we can take

$$u = (0, 0, \ldots, 0, 1), \quad h_1 = 0, \quad h_2 = 1. \tag{2.7}$$

The transformed hyperplanes (2.2) take then the form

$$\sum_{i=1}^{n} \alpha^{ni} x_i = \sum_{i=1}^{n} \alpha^{ni} b_i, \quad \sum_{i=1}^{n} \alpha^{ni} x_i = \sum_{i=1}^{n} \alpha^{ni} b_i + 1 \tag{2.8}$$

and the system (2.6) will be

$$\omega_n = 0, \quad \omega_{n1} = 0, \quad \omega_{n2} = 0, \ldots, \omega_{nn} = 0. \tag{2.9}$$

Therefore the density for sets of parallel hyperplanes, when they are written in the form (2.8), reads

$$d\mathfrak{P} = \omega_{n1} \wedge \omega_{n2} \wedge \cdots \wedge \omega_{nn} \wedge \omega_n. \tag{2.10}$$

The condition for this differential form to be a density is $d(d\mathfrak{P}) = 0$, which is easily verified if we take into account the equations of structure (1.4).

By (1.6) and since $\det A = 1$, we obtain (we always take the densities in absolute value)

$$d\mathfrak{P} = \sum_{i=1}^{n} \alpha^{ni} d\alpha^{n1} \wedge d\alpha^{n2} \wedge \cdots \wedge d\alpha^{nn} \wedge db_i. \tag{2.11}$$

This formula, together with (2.8) gives the following

Theorem 1. *The density, invariant with respect to the unimodular affine group A, for sets of pairs $\mathfrak{P}$ of parallel hyperplanes written in the form*

$$\sum_{i=1}^{n} l^i x_i = m, \quad \sum_{i=1}^{n} l^i x_i = m + 1 \tag{2.12}$$

is

$$d\mathfrak{P} = dl^1 \wedge dl^2 \wedge \cdots \wedge dl^n \wedge dm. \tag{2.13}$$

We want now to find a geometrical interpretation of the density [2.13]. Let λ^i be a unit vector. The element of area on the unit hypersphere corresponding to the direction of λ^i is expressed by

$$d\sigma = \frac{d\lambda^2 \wedge d\lambda^3 \wedge \cdots \wedge d\lambda^n}{\lambda^1} \tag{2.14}$$

and also by

$$d\sigma = \sum_{i=1}^{n} (-1)^{i-1} \lambda^i d\lambda^1 \wedge \cdots \wedge d\lambda^{i-1} \wedge d\lambda^{i+1} \wedge \cdots \wedge d\lambda^n. \tag{2.15}$$

If λ^i is the unit vector normal to the hyperplanes (2.12) and we put

$$\varrho^2 = \sum_{i=1}^{n} (l^i)^2, \qquad d\varrho = \frac{\sum_{i=1}^{n} l^i dl^i}{\varrho}$$

we have

$$\lambda^i = \frac{l^i}{\varrho}, \quad d\lambda^i = \frac{dl^i}{\varrho} - \frac{l^i}{\varrho^2} d\varrho$$

and

$$d\lambda^2 \wedge d\lambda^3 \wedge \cdots \wedge d\lambda^n = \frac{dl^2 \wedge \cdots \wedge dl^n}{\varrho^{n-1}} - \sum_{i=2}^{n} \frac{l^i dl^2 \wedge \cdots \wedge dl^{i-1} \wedge d\varrho \wedge dl^{i+1} \wedge \cdots \wedge dl^n}{\varrho^n} = \sum_{i=1}^{n} (-1)^{i-1} \frac{l^1 l^i dl^1 \wedge \cdots \wedge dl^{i-1} \wedge dl^{i+1} \wedge \cdots \wedge dl^n}{\varrho^{n+1}}.$$

Consequently, by (2.14) we obtain

$$d\sigma = \frac{1}{\varrho^n} \sum_{i=1}^{n} (-1)^{i-1} l^i dl^1 \wedge \cdots \wedge dl^{i-1} \wedge dl^{i+1} \wedge \cdots \wedge dl^n. \tag{2.16}$$

Next consider the distances

$$p_1 = \frac{m}{\varrho}, \quad p_2 = \frac{m+1}{\varrho}$$

from the origin to the hyperplanes (2.12). We have

$$dp_1 = \frac{dm}{\varrho} - \frac{m}{\varrho^2} d\varrho, \quad dp_2 = \frac{dm}{\varrho} - \frac{(m+1)}{\varrho^2} d\varrho$$

and

$$dp_1 \wedge dp_2 = \frac{1}{\varrho^3} d\varrho \wedge dm = \frac{1}{\varrho^4} \sum_{i=1}^{n} l^i dl^i \wedge dm.$$

Hence

$$d\sigma \wedge dp_1 \wedge dp_2 = \frac{1}{\varrho^{n+2}} dl^1 \wedge dl^2 \wedge \cdots \wedge dl^n \wedge dm = \frac{d\mathcal{G}}{\varrho^{n+2}}.$$

Remember that we consider always the densities in absolute value; therefore we make no question of te sign.

In order to introduce the distances p_1, p_2 from the origin, we observe that $p_2 - p_1 = 1/\varrho$ and thus

$$d\mathcal{G} = \frac{d\sigma \wedge dp_1 \wedge dp_2}{|p_2 - p_1|^{n+2}} \tag{2.17}$$

which is the desired geometrical interpretation for $d\mathcal{G}$.

§. 3. Measure of sets of parallel hyperplanes which contain a given convex body.

Let K be a given convex body in the n dimensional space and let $\Delta=\Delta(\sigma)$ be the width of K corresponding to the direction σ. The measure of all pairs of parallel hyperplanes which contain K will be

$$(3.1) \qquad M=\int\frac{d\sigma\wedge dp_1\wedge dp_2}{|p_2-p_1|^{n+2}}=\frac{1}{n(n+1)}\int_{\frac{1}{2}E}\frac{d\sigma}{\Delta^n}.$$

This measures gives, together with its geometrical interpretation, the following affine invariant of K

$$(3.2) \qquad J=\int_{\frac{1}{2}E}\frac{d\sigma}{\Delta^n}$$

the integral extended over the half of the n dimensional unit sphere.

Elsewhere [6] we gave an analogous affine invariant I defined by the following integral

$$(3.3) \qquad I=\frac{1}{n}\int_E\frac{d\sigma}{p^n}$$

where $p=p(\sigma)$ is the support function of K with respect to an interior point also affine invariant with respect to K. In [6] we proved that between I and the volume V of K the inequality

$$(3.4) \qquad IV\leqq\frac{4\pi^n}{n^2(\Gamma(n/2))^2}$$

holds, where equality occurs only for ellipsoids.

If K possesses a centre of symmetry, we obviously have $nI=2^{n+1}J$ and the invariant J coincides up to a constant factor with I. If K does not possess a centre of symmetry, J and I are not trivially related. Let us consider the inequalities

$$\frac{1}{x^n}+\frac{1}{y^n}\geqq\frac{2}{(xy)^{n/2}}, \qquad (xy)^{n/2}\leqq\left(\frac{x+y}{2}\right)^n$$

from which we deduce

$$(3.5) \qquad \frac{1}{x^n}+\frac{1}{y^n}-\frac{2^{n+1}}{(x+y)^n}\geqq 0$$

valid for $x>0$, $y>0$ and where equality occurs only for $x=y$. Denoting by p_1, p_2 the values of p at opposite points and applying (3.5) we have

$$2J=\int_E \frac{d\sigma}{\Delta^n}=\int_E \frac{d\sigma}{(p_1+p_2)^n} \leqq \frac{1}{2^{n+1}}\int_E \left(\frac{d\sigma}{p_1^n}+\frac{d\sigma}{p_2^n}\right)=\frac{n}{2^n}I$$

and therefore

$$J \leqq \frac{n}{2^{n+1}}I \tag{3.6}$$

where equality occurs if and only if K is centrally symmetric. From (3.6) and (3.4) we obtain

$$JV \leqq \frac{4\pi^n}{2^{n+1}n(\Gamma(n/2))^2} \tag{3.7}$$

with equality only for ellipsoids.

In order to obtain a lower bound for the product JV we remind that between the volume P of the least parallelepiped which contains K and the volume V the inequality

$$P \leqq n!\,V$$

holds (see MACBEATH [2]) and that the value of J for a parallelepiped of volume P is

$$J_P=\frac{2^{n-1}}{(n-1)!\,P}$$

as can be obtained by a direct calculation (since J is invariant with respect to affinities it suffices to consider the case of an n-dimensional cube; see, BAMBAH [1]). Hence we have

$$JV \geqq J_P V \geqq \frac{1}{n!}J_P P=\frac{2^{n-1}}{n!\,(n-1)!}. \tag{3.8}$$

Since equalities cannot hold simultaneously in (3.8), we always have $JV>2^{n-1}/n!\,(n-1)!$. We may summarize the obtained results as follows

Theorem 2. *The measure of the set of pairs of parallel hyperplanes which contain a given convex body K, invariant with respect to the group of unimodular affine transformations, is given by the integral* (3.1). *This measure gives rise to the invariant J* (3.2) *which is related with the volume V of J by the inequalities*

$$\frac{2^{n-1}}{n!\,(n-1)!}<JV \leqq \frac{4\pi^n}{2^{n+1}n(\Gamma(n/2))^2} \tag{3.9}$$

where the upper bound is attained if and only if K is an ellipsoid.

The exact value of the lower bound is not known. Probably it is attained when K is a simplex, but I have not the proof.

A direct proof of the affine invariance of J together with some generalizations for the cases $n=2,3$ was given elsewhere [7].

§ 4. The real projective group.

Let us now consider the n-dimensional projective space and in it the group of projective transformations

$$x' = Ax, \quad \det A = 1 \tag{4.1}$$

where A is an $(n+1)\times(n+1)$ matrix and x, x' denote $(n+1)\times 1$ matrices whose elements are the homogeneous coordinates $x_0, x_1, \ldots, x_n$ and $x'_0, x'_1, \ldots, x'_n$ of the points x and x' respectively.

Similarly as in the case of the affine group, the relative components ω_{ij} of the projective group are the elements of the matrix

$$\Omega = A^{-1} dA \tag{4.2}$$

and satisfy the equations of structure

$$d\Omega = -\Omega \wedge \Omega. \tag{4.3}$$

If we set

$$A = (a_{ij}), \quad A^{-1} = (\alpha^{ij}), \quad \Omega = (\omega_{ij})$$

the explicit forms of (4. 2) and (4. 3) are

$$\omega_{ij} = \sum_{h=0}^{n} \alpha^{ih} da_{hj} = -\sum_{h=0}^{n} a_{hj} d\alpha^{ih}, \quad d\omega_{ij} = -\sum_{h=0}^{n} \omega_{ih} \wedge \omega_{hj}. \tag{4.4}$$

By differentiation of the relation $\det A = 1$, we also obtain

$$\omega_{00} + \omega_{11} + \omega_{22} + \cdots + \omega_{nn} = 0. \tag{4.5}$$

5. Measure of sets of hyperquadrics.

Let us consider the hyperquadric

$$x^t \Phi x = 0 \tag{5.1}$$

where x^t denotes the $1\times(n+1)$ transposed matrix of x and Φ is a $(n+1)\times(n+1)$ diagonal matrix

$$\Phi = \begin{pmatrix} \varepsilon_0 & & & \\ & \varepsilon_1 & & \\ & & \ddots & \\ & & & \varepsilon_n \end{pmatrix}$$

with the elements $\varepsilon_i = \pm 1$. It is well known that every hyperquadric is projectively equivalent to one of the type (5.1).

By the projectivity (4.1) the hyperquadric (5.1) transforms to

$$x^t (A^{-1})^t \Phi A^{-1} x = 0. \tag{5.2}$$

In order that the varied projectivity $A + dA$ may conduce to the same hyperquadric, we must have

$$d((A^{-1})^t \Phi A^{-1}) = 0$$

that is, because of (4.2) and (1.3),

$$(A^{-1})^t (\Omega^t \Phi + \Phi \Omega) A^{-1} = 0$$

and thus

$$\Omega^* = \Omega^t \Phi + \Phi \Omega = 0. \tag{5.3}$$

The density for sets of hyperquadrics whose equation has the form (5.2) will be the exterior product of the independent elements of the symmetric matrix Ω^*. These elements are

$$\omega_{ij}^* = \varepsilon_j \omega_{ji} + \varepsilon_i \omega_{ij}$$

and the relation (4.5) gives

$$\sum_{i=0}^{n} \frac{1}{\varepsilon_j} \omega_{ii}^* = 0. \tag{5.4}$$

Therefore the projective invariant density for sets of hyperquadrics may be written in any one of the following equivalent forms (for $i = 0, 1, 2, \ldots, n$)

$$d C_i = \omega_{00}^* \wedge \omega_{01}^* \wedge \cdots \wedge \widehat{\omega_{ii}^*} \wedge \cdots \wedge \omega_{nn}^* \tag{5.5}$$

where the hat $\wedge$ means that the covered element must be omitted.

These forms (5.5) are differential forms of degree $\frac{1}{2} n(n+3)$ as it should indeed be. An easy calculation, using the equations of structure (4.4), shows that $d(dC_i) = 0$, and consequently (5.5) is really a projective invariant density for sets of hyperquadrics.

The densities (5.5) refer to the hyperquadric (5.2). It is our purpose now to introduce explicitly the coefficients of this hyperquadric, that is, the elements of the symmetric matrix

$$Q = (A^{-1})^t \Phi A^{-1} \tag{5.6}$$

which are

$$q_{hl} = \sum_{i=0}^{n} \varepsilon_i \alpha^{ih} \alpha^{il}. \tag{5.7}$$

From (5.6) and (5.3) we obtain

(5.8) $$dQ = -(A^{-1})^t \Omega^* A^{-1}$$

and hence

(5.9) $$\Omega^* = -A^t dQ A$$

or, explicitly

(5.10) $$\omega_{ij}^* = -\sum_{h,l=0}^{n} a_{hi} a_{lj} dq_{hl}.$$

In virtue of the symmetry $q_{ij} = q_{ji}$, the matrix of the system (5.10) (for $i \leqq j$, $h \leqq l$) is the second power matrix of A, denoted by $P_2(A)$ (see, for instance [3 p. 85]) and since $[P_2(A)]^{-1} = P_2(A^{-1})$ (as it follows from the known property $P_2(AB) = P_2(A) P_2(B)$ when we choose $B = A^{-1}$ and remember that $\det A = 1$, $P_2(E) = E$), substituting (5.10) in (5.5) we get

(5.11) $$dC_i = \sum_{\substack{h,l=0 \\ h \leqq l}}^{n} (-1)^{\nu(i,i)+\nu(h,l)} \alpha^{ih} \alpha^{il} dq_{00} \wedge dq_{01} \wedge \cdots \wedge \widehat{dq_{hl}} \wedge \cdots \wedge dq_{nn}$$

where

(5.12) $$\nu(h,l) = \frac{(2n+1-h)h}{2} + l + 1$$

is the order of the element (h, l) in the sequence $(0,0)$, $(0,1)$, $(0,2), \ldots$, $(h,l), \ldots, (n-1,n)$, (n,n).

Because of (5.4) and (5.5) we observe that $(-1)^{\nu(i,i)} \varepsilon_i dC_i$ does not depend on i; hence we can also take as density for sets of hyperquadrics

(5.13) $$dC = (-1)^{\nu(i,i)} \varepsilon_i dC_i$$

and then, from (5.11) and (5.7) we deduce

(5.14) $$dC = \frac{1}{n+1} \sum_{\substack{h,l=0 \\ h \leqq l}}^{n} (-1)^{\nu(h,l)} q_{hl} dq_{00} \wedge dq_{01} \wedge \cdots \wedge \widehat{dq_{hl}} \wedge \cdots \wedge dq_{nn}.$$

This is a first form for dC. A second form is obtained if we observe that by differentiation of $\det Q = \pm 1$. We get

(5.15) $$\sum_{h,l=0}^{n} q^{hl} dq_{hl} = 0, \quad dq_{nn} = -\sum_{h,l=0}^{n}{}' \frac{q^{hl}}{q^{nn}} dq_{hl}$$

where the accent denotes that the term $h = l = n$ is excluded.

Substituting in (5.14) we have, up to the sign which is inessential since we consider always the densities in absolute value,

(5.16) $$dC = \frac{1}{q^{nn}} dq_{00} \wedge dq_{01} \wedge \cdots \wedge dq_{n-1,n}$$

This is a second form for dC.

The densities (5.14) and (5.16) refer to the hyperquadric (5.2) which satisfies the condition $\det Q = \pm 1$. That is, the forms (5.14) and (5.16) apply when we have normalized the equation of the hyperquadric in such a way that $\det Q = \pm 1$ holds, a normalization which is always possible for non degenerate hyperquadrics.

Another normalization could be to take the equation of the hyperquadric in the form

$$\sum_{i,j=0}^{n} q_{ij}^{*} x_i x_j = 0 \quad \text{with} \quad q_{nn}^{*} = 1. \tag{5.17}$$

In order to apply to this case the above result, it is enough to set

$$q_{ij}^{*} = \frac{q_{ij}}{q_{nn}}, \qquad \Delta = \det(q_{ij}^{*}) = \frac{\det(q_{ij})}{q_{nn}^{n+1}} = \frac{1}{q_{nn}^{n+1}}.$$

We shall have

$$q_{ij} = q_{ij}^{*} \Delta^{-\frac{1}{n+1}}, \quad q^{*nn} = \frac{q^{nn}}{q_{nn}^{n} \Delta} = \frac{q^{nn}}{\Delta^{1/(n+1)}} \tag{5.18}$$

and, since

$$d\Delta = \Delta \sum_{h,l=0}^{n} q^{*hl} dq_{hl}^{*}$$

we get

$$dq_{ij} = \Delta^{-\frac{1}{n+1}} dq_{ij}^{*} - \frac{1}{n+1} \Delta^{-\frac{1}{n+1}} q_{ij}^{*} \sum_{h,l=0}^{n} q^{*hl} dq_{hl}^{*}$$

and

$$dq_{00} \wedge dq_{01} \wedge \cdots \wedge dq_{n-1,n} = \Delta^{-\frac{n(n+3)}{2(n+1)}} dq_{00}^{*} \wedge dq_{01}^{*} \wedge \cdots \wedge dq_{n-1,n}^{*}$$
$$- \frac{1}{n+1} \Delta^{-\frac{n(n+3)}{2(n+1)}} \sum_{i,j=0}^{n}{}' q_{ij}^{*} q^{*ij} dq_{00}^{*} \wedge dq_{01}^{*} \wedge \cdots \wedge dq_{n-1,n}^{*}.$$

Since in the last sum the values $i = j = n$ are excluded and $q^{*}{}_{n} = 1$, we have

$$\sum_{i,j=0}^{n}{}' q_{ij}^{*} q^{*ij} = (n+1) - q^{*nn}$$

and consequently

$$dq_{00} \wedge dq_{01} \wedge \cdots \wedge dq_{n-1,n} = \frac{1}{n+1} \Delta^{-\frac{n(n+3)}{2(n+1)}} q^{*nn} dq_{00}^{*} \wedge \cdots \wedge dq_{n-1,n}^{*}.$$

Taking into account (5.16) and (5.18) we finally get, up to the sign,

$$dC = \frac{dq_{00}^{*} \wedge dq_{01}^{*} \wedge \cdots \wedge dq_{n-1,n}^{*}}{(n+1)\Delta^{\frac{n+2}{2}}}. \tag{5.19}$$

For $n=2$, the density for sets of conics written in the form $q_{00}x_0^2+$ $+2q_{01}x_0x_1+2q_{02}x_0x_2+q_{11}x_1^2+2q_{12}x_1x_2+x_2^2=0$ becomes

$$dC=\frac{dq_{00}\wedge dq_{01}\wedge dq_{02}\wedge dq_{11}\wedge dq_{12}}{3\Delta^2}.$$

This expression, up to the factor 1/3 which is inessential, was given by Stoka [8].

We can summarize the above results in the following

Theorem 3. *The projective invariant density for sets of non degenerate hyperquadrics* $x^tQx=0$ *is given by any one of the equivalent forms* (5.14), (5.16) *when* $\det Q=\pm 1$. *If the equation of the hyperquadrics satisfies the condition* $q_{nn}=1$ *and we set* $\det Q=\Delta$, *then the density takes the form* (5.19).

Bibliography.

[1] R. P. Bambah, Polar reciprocal convex bodies, *Proc. Cambridge Philos. Soc.* **51** (1955) 377—378.

[2] A. M. Macbeath, A compactness theorem for affine equivalence-classes of convex regions, *Canad. J. Math.* **3** (1951), 54—61.

[3] C. C. MacDuffee, The theory of matrices, *(Ergebnisse der Mathematik und ihre Grenzgebiete*, Chelsea Ed.) 1946.

[4] L. A. Santaló, Integral geometry in projective and affine spaces, *Ann. of Math.* **51** (1950), 739—755.

[5] L. A. Santaló, Introduction to the integral geometry, *Paris*, 1952.

[6] L. A. Santaló, Un invariante afin para los cuerpos convexos del espacio de n dimensiones, *Portugal. Math.* **8** (1949), 155—161.

[7] L. A. Santaló, Un nuevo invariante afin para las curvas convexas del plano y del espacio, *Math. Notae*, **17** (1958), 78—91.

[8] M. Stoka, Masura unei multimi de varietati dintr-un spatiu R_n, *Bul. Stiintific*, **7** (1955), 903—937.

[9] O. Varga, Ueber Masse von Paaren linearer Mannigfaltigkeiten im projektiven Raum P_n, *Rev. Mat. Hisp.-Amer.* **10** (1935), 241—264.

(Received January 27, 1959.)

On the Kinematic fundamental formula of Integral Geometry for spaces of constant curvature

L. A. Santaló

Abstract

This paper provides a proof of the kinematic fundamental formula of Integral Geometry for spaces of constant curvature and n dimensions. The result is formula (1.2) and it was given in [7]. The proof is a more detailed and direct version of that given by the author in [9]; some misprints occurring in that paper have been corrected here. Some applications are given at the end of the paper.

1 Introduction

Let Q_0 and Q_1 be two bodies of n-dimensional space with constant curvature K. Let us assume that their boundaries ∂Q_0, ∂Q_1 are two hypersurfaces of class $\mathcal{C}^2$, so that the integrals (2.1) enable us to define the so-called integrals of mean curvature of order r, which we denote by $M_r(Q_0)$ and $M_r(Q_1)$, respectively. If the boundaries display singularities, the integrals $M_r(Q_i)$ can often be defined as their limit value for the parallel body exterior to the distance ϵ when $\epsilon \to 0$.

Let us assume that Q_0 is fixed and Q_1 is variable. Let dQ_1 be the kinematic density for Q_1 in the sense of integral geometry. This is defined in the following way: in order to fix the position of Q_1, it is necessary to fix one of the points, x, and an n-hedron formed by n orthogonal unitary vectors, $e_1, \ldots, e_n$ of vertex x. Once x is fixed, in order to fix the n-hedron it is necessary to fix e_n by its endpoint on the sphere O_{n-1} of dimension $n-1$ and unit radius; then e_{n-1} is fixed by its endpoint on the sphere O_{n-1} of unit radius and dimension $n-2$ orthogonal to e_n; then e_{n-2} on O_{n-3} orthogonal to e_n and e_{n-1}, and so on successively until e_2 is fixed by its endpoint on the circumference O_1 orthogonal to $e_n, e_{n-1}, \ldots, e_3$. Denoting by dO_i the area element on the unit sphere O_i corresponding to the endpoint of e_{i+1} and dx to the volume element of the space at point x, the kinematic density is

$$dQ_1 = dx \wedge dO_{n-1} \wedge dO_{n-2} \wedge \cdots \wedge dO_1 \tag{1.1}$$

As expected, this is a differential form of degree $1+2+3+\cdots+n = n(n+1)/2$, since the position of a body Q_1 in a space of n dimensions depends on this number of parameters.

Let $\chi(Q)$ be the Euler-Poincaré characteristic of the body Q. As previously stated, assuming Q_0 to be fixed and Q_1 to be variable, the kinematic fundamental formula of integral geometry in spaces of constant curvature K is as follows

$$\begin{aligned}(1.2)\qquad & \int \chi(Q_0 \cap Q_1) dQ_1 = \\ & O_1 O_2 \dots O_{n-1}\left(-\frac{2\epsilon_n}{O_n} K^{n/2} V_0 V_1 + V_1 \chi_0 + V_0 \chi_1\right) \\ & + O_1 O_2 \dots O_{n-2} \frac{1}{n} \sum_{h=0}^{n-2} \left(\frac{n}{h+1}\right) M_h^0 M_{n-2-h}^1 \\ & 2 O_1 O_2 \dots O_{n-2} \sum_{s=0}^{n-4} \frac{O_{n-s}}{O_s} M_s^0 \left\{ \sum_{t=[(s-\epsilon_n)/2]+1}^{(n-3-\epsilon_n)/2} c_{stn} M_{2t+\epsilon_n-s-1}^1, \right.\end{aligned}$$

where V_0 and V_1 are the volumes of Q_0 and Q_1, and where

$$(1.3)\qquad O_i = \frac{2\pi^{\frac{i+1}{2}}}{\Gamma(\frac{i+1}{2})}, \quad \epsilon_n = \frac{1}{2}(1+(-1)^n)$$

$$c_{stn} = \begin{pmatrix} n-1 \\ 2t+\epsilon_n \end{pmatrix} \begin{pmatrix} 2t+\epsilon_n \\ s \end{pmatrix} \cdot$$

$$\cdot \frac{O_{n+s-2t-\epsilon_n+1}}{O_{2t+1+\epsilon_n-s} O_{n-2t-\epsilon_n} O'_{n-2t-1-\epsilon_n}} K^{(n-2t-1-\epsilon_n)/2}.$$

Integration on the left-hand side of the equation is extended to all the positions of Q_1 intersecting Q_0.

Furthermore, in the interests of brevity

$$M_h^0 = M_h(Q_0),\ M_h^1 = M_h(Q_1),\ \chi_0 = \chi(Q_0),\ \chi_1 = \chi(Q_1).$$

For the Euclidean space ($K = 0$) and $n = 2, 3$, this formula is found for example in Blaschke [2]. For $K = 0$, it was given in 1940 by Chern-Yien [5], and with a more detailed proof by Chern [4] in 1952.

For the general case of the space of constant curvature K, the previous formula was proved by the present author in [9]; see also [11]. For $n = 3$, a direct proof can be found in [8], and also for the elliptic space $n = 3$, $K > 0$ in Ta-Jeu-Wu [15]. For $n = 2$, with various applications, see [10].

The object of this paper is to provide a more direct and detailed exposition of the proof given in [9], while at the same time correcting some inadvertent typographical errors as well as providing a more compact final result. Some applications will also be provided at the end.

2 Known definitions and formulae

We now recall some required definitions and integral formulae that have been obtained elsewhere.

We indicate the Euclidean space for E_n and the n-dimensional space of constant curvature K for $S_n(K)$.

Let Q be a convex body of E_n. Assuming that the hypersurface ∂Q that limits this body is of class $\mathcal{C}^2$, and by calling R_i $(i = 1, 2, \ldots, n-1)$ the principal radii of curvature at the point whose area element is $d\sigma$, the integral

$$M_r = \frac{1}{\binom{n-1}{r}} \int_{\partial Q} \left\{ \frac{1}{R_1}, \frac{1}{R_2}, \ldots, \frac{1}{R_r} \right\} d\sigma \tag{2.1}$$

is called the *mean curvature integral of order r* of ∂Q, where the parenthesis in the integrand indicate the elementary symmetric function of order r formed by the principle curvatures $1/R_i$. In particular, we have

$$\begin{aligned} M_0 &= F = \text{ area of } \partial Q \\ M_{n-1} &= O_{n-1}, \end{aligned} \tag{2.2}$$

where O_{n-1} indicates the volume of the $(n-1)$-dimensional Euclidean unit sphere. So in general we may write

$$O_i = \frac{2\pi^{\frac{i+1}{2}}}{\Gamma(\frac{i+1}{2})} \tag{2.3}$$

for $i = 0, 1, 2, \ldots,$

Definition (2.1) for the invariants $M_r(Q)$ assumes that ∂Q is of class $\mathcal{C}^2$. The advantage of this definition is that it holds even when Q is not convex, providing that ∂Q is of class $\mathcal{C}^2$. For the cases in which this condition is not fulfilled, if Q is convex, the $M_r(Q)$ can be substituted by other invariants, which are equivalent except for one factor and are defined for any convex body. They are as follows:

a) The *section integrals*[1] $W_r(Q)$ (known in German as "Quermassintegrale", and in French as "travers exterieurs" related to the $M_r(Q)$ by (see Bonnesen-Fenchel [[3], p.63])

$$M_{r-1} = nW_r. \tag{2.4}$$

b) The measures H_r of the sets of linear subspaces L_r of E_n intersecting Q, measures taken in the integral geometry sense (see [13]). Their relation with the M_r is

$$M_{r-1} = \frac{2(n-r)O_1 O_2 \cdots O_{r-1}}{O_{n-r-1} O_{n-r} \cdots O_{n-2}} H_r. \tag{2.5}$$

It is interesting to consider the case of the "flattened" convex body Q; that is, contained within a linear subspace L_q of E_n. This we will denote by Q^q. If Q^q is regarded as a convex body of the Euclidean space E_q, it has mean curvature integrals that we may represent by M_r^q $(0 \le r \le q-1)$. However, if it is regarded as a flattened convex body of E_n, its mean curvature integrals will be $M_r = M_r^n$, which can be calculated by any of the formulae (2.4) and (2.5).

The relations between both mean curvature integrals are as follows (see [12], and for $q = n-1$ also Hadwiger [[6], p.215]):

[1]Note of the Editors: Cross-sectional measures.

1. If $r \geq n-q$, then

$$M_r^n = \frac{\binom{q-1}{r-n+q}}{\binom{n-1}{r}} \frac{O_r}{O_{r-n+q}} M_{r-n+q}^q. \tag{2.6}$$

2. If $r = n-q-1$, then

$$M_r^n = \frac{1}{\binom{n-1}{r}} O_{n-q-1} \sigma_q(Q^q), \tag{2.7}$$

where $\sigma_q(Q^q)$ indicates the volume of Q^q.

3. If $r < n-q-1$, then

$$M_r^n = 0. \tag{2.8}$$

Now let Q be a convex body of E_n, and let us assume that it is intersected by the linear subspaces L_q. The intersections $Q \cap L_q$ will be flattened convex bodies of dimension q. Considering all the L_q intersecting Q, and by denoting the density for sets of L_q in E_n by dL_q, we have the integral formula

$$\int_{Q \cap L_q \neq \emptyset} M_r^q(Q \cap L_q) dL_q = \frac{O_{n-2} \dots O_{n-q}}{2O_1O_2 \dots O_{q-2}} \frac{O_{n-r}}{O_{q-r}} M_r^n(Q). \tag{2.9}$$

The proof may be seen in [9].

For $r = q-1$, according to (2.2) this is $M_{q-1}^q = O_{q-1}$, and therefore (2.9) gives the measure of all the L_q intersecting Q. Thus

$$\int_{Q \cap L_q \neq \emptyset} dL_q = \frac{O_{n-2} \dots O_{n-q} O_{n-q-1}}{2O_1O_2 \dots O_{q-1}(n-q)} M_{q-1}^n(Q), \tag{2.10}$$

in which the following relation is employed:

$$2\pi O_{n-q-1} = (n-q) O_{n-q+1}. \tag{2.11}$$

If we denote the q-dimensional volume of the intersection $Q \cap L_q$ by $\sigma_q(Q \cap L_q)$, then the following formula also holds,

$$\int_{Q \cap L_q \neq \emptyset} \sigma_q(Q \cap L_q) dL_q = \frac{O_n O_{n-1} \dots O_{n-q}}{2O_1O_2 \dots O_q} \frac{O_q}{O_n} V(Q), \tag{2.12}$$

where $V(Q)$ is the volume of Q.

3 The kinematic fundamental formula for the Euclidean space E_n and applications

Let Q be a not necessarily convex body of E_n whose boundary ∂Q is of class $\mathcal{C}^2$. In this case, the mean curvature integrals are defined by the formulae (2.1).

Let $\chi(Q)$ be the Euler-Poincaré characteristic of Q. Let us recall that for n even this is always

$$\chi(\partial Q) = 0, \quad (n \text{ even}) \tag{3.1}$$

and for n odd

$$\chi(\partial Q) = 2\chi(Q), \quad (n \text{ odd}), \tag{3.2}$$

also holding for any n the Gauss-Bonnet formula

$$M_{n-1}(Q) = O_{n-1}\chi(Q), \tag{3.3}$$

which generalizes the second formula (2.2), since

$$\chi(Q) = 1 \quad \text{si } Q = \text{the topological sphere}. \tag{3.4}$$

Now let Q_0 and Q_1 be two bodies of E_n whose boundaries are of class C^3. Let us assume that Q_0 is fixed and that Q_1 is variable, with the kinematic density dQ_1 given by (1.1). The kinematic fundamental formula for E_n is then

$$\begin{aligned}\int_{Q_0\cap Q_1\neq\emptyset} \chi(Q_0\cap Q_1)dQ_1 \;&=\; O_1\ldots O_{n-2}\Big\{O_{n-1}(\chi^0 V_1 + \chi^1 V_0) + \\ &\frac{1}{n}\sum_{h=0}^{n-2}\binom{n}{h+1} M_h^0 M_{n-2-h}^1\Big\},\end{aligned} \tag{3.5}$$

where V_0 and V_1 are the volumes of Q_0 and Q_1, respectively, and where $M_h^0 = M_h(Q_0), M_{n-2-h}^1 = M_{n-2-h}(Q_1)$ is written in the interest of brevity.

The proof of this formula (3.5) can be seen in Chern [4].

Let us now consider the particular case in which $Q_0 = Q_0^q$ is a flattened body contained within an L_q of E_n, and Q_1 any convex body of E_n. In this case we have

$$\chi(Q_0^q \cap Q_1) = \chi(Q_0^q) = \chi(Q_1) = 1, \tag{3.6}$$

and taking into account (2.6), (2.7) and (2.8), we obtain

$$\begin{aligned}\int dQ_1 \;&=\; O_1\ldots O_{n-2}\Big\{O_{n-1}V_1 + \frac{\binom{n}{n-q}}{n\binom{n-1}{q}}O_{n-q-1}\sigma_q(Q_0^q)M_{q-1}^1 \\ &\frac{1}{n}\sum_{h=n-q}^{n-2}\binom{n}{h+1}\frac{\binom{q-1}{h+q-n}}{\binom{n-1}{h}}\frac{O_h}{O_{h+q-n}}M_{h+q-n}^q M_{n-2-h}^1\Big\},\end{aligned} \tag{3.7}$$

where the integral is extended to all the positions of Q_1 that intersect with Q_0^q.

This formula enables us to calculate the mean curvature integrals $M_q(Q_0 \cap Q_1)$ extended to all the positions of Q_1. To this effect, let us consider the integral

$$I = \int dQ_1 dL_q \tag{3.8}$$

extended to all the positions of Q_1 and of L_q for which we have $Q_0 \cap Q_1 \cap L_q \neq \emptyset$.

By first fixing Q_1, and according to (2.10), we have

$$(3.9) \qquad I = \frac{O_{n-2} \ldots O_{n-q-1}}{2(n-q)O_1 \ldots O_{q-1}} \int M_{q-1}(Q_0 \cap Q_1) dQ_1.$$

However, by fixing L_q, and according to (3.7), we have

$$(3.10) \qquad I = O_1 \ldots O_{n-2} \int \Big\{ O_{n-1} V_1 + \frac{\binom{n}{n-q}}{n\binom{n-1}{q}} O_{n-q-1} \sigma_q(Q_0^q) M_{q-1}^1$$
$$\frac{1}{n} \sum_{h=n-q}^{n-2} \binom{n}{h+1} \frac{\binom{q-1}{h+q-n}}{\binom{n-1}{h}} \frac{O_h}{O_{h+q-n}} M_{h+q-n}^q M_{n-2-h}^1 \Big\} dL_q$$

Whereby, on applying (2.10), (2.12) and (2.9), we obtain

$$(3.11) \qquad I = O_1 \ldots O_{n-2} \Big\{ O_{n-1} V_1 \frac{O_{n-2} \ldots O_{n-q-1}}{2(n-q)O_1 \ldots O_{q-1}} M_{q-1}^0$$
$$+ \frac{\binom{n}{q}}{n\binom{n-1}{q}} O_{n-q-1} \frac{O_{n-1} \ldots O_{n-q}}{2O_1 \ldots O_{q-1}} M_{q-1}^1 V_0$$
$$+ \frac{1}{n} \sum_{h=n-q}^{n-2} \frac{\binom{n}{h+1}\binom{q-1}{h+q-n}}{\binom{n-1}{h}} \frac{O_{n-2} \ldots O_{n-q}}{2O_1 \ldots O_{q-2}} \frac{O_{2n-h-q}}{O_{n-h}}$$
$$\cdot \frac{O_h}{O_{h+q-n}} M_{h+q-n}^0 M_{n-2-h}^1 \Big\}.$$

Equalizing with (3.9), and simplifying, we have

$$(3.12) \qquad \int M_{q-1}(Q_0 \cap Q_1) dQ_1 = O_1 \ldots O_{n-2} \Big\{ O_{n-1}(V_1 M_{q-1}^0 + V_0 M_{q-1}^1)$$
$$+ \frac{(n-q)O_{q-1}}{O_{n-q-1}} \frac{O_h}{O_{h+q-n}} \sum_{h=n-q}^{n-2} \frac{\binom{q-1}{q+h-n} O_{2n-h-q}}{(h+1)O_{n-h}} M_{n-2-h}^1 M_{h+q-n}^0 \Big\}$$

This formula holds for $0 < q \leq n-1$. For $q = n$, this must be substituted by the fundamental (3.5), taking into account (3.3).

4 Validity of (3.12) for non-convex bodies and spaces of constant curvature

Now let Q_0, Q_1 be two not necessarily convex bodies of E_n whose boundaries are always of class $\mathcal{C}^2$. Let x be a point at the intersection $\partial Q_0 \cap \partial Q_1$, and let us call $d\sigma_0 =$ the area element of ∂Q_0 at x, $d\sigma_1 =$ the area element of ∂Q_1 at x, and $d\sigma_{01} =$ the area element of $\partial Q_0 \cap \partial Q_1$ at x. Furthermore, let ϕ be the angle formed by the normals to ∂Q_0 and ∂Q_1 at x. If we denote the kinematic density of Q_1 "around x" by $dQ_1[x]$; that is, by assuming that x is fixed, we have a relation of the form

$$d\sigma_{01} \wedge dQ_1 = \Phi(\phi)\, d\sigma_0 \wedge d\sigma_1 \wedge dQ_1[x]. \tag{4.1}$$

The existence of this relation is evident if we observe that both members determine both the position of Q_1 and that of the point x. The explicit form of the function $\Phi(\phi)$ is somewhat complicated, although it is not necessary to obtain it. In fact, it can easily be deduced from the equivalent formula given by Chern in [4], formula (24).

In order now to calculate the integral (3.12), let us observe that $M_{q-1}(Q_0 \cap Q_1)$ is composed of three parts: a) that corresponding to the boundary of Q_0, which is interior to Q_1; b) that corresponding to the boundary of Q_1, which is interior to Q_0; c) that corresponding to $\partial Q_0 \cap \partial Q_1$.

The integrals of the first two parts can be calculated immediately by the usual method in integral geometry by taking a fixed point on the boundary of one of the bodies and integrating the other to all the positions containing this point, which then moves over all the boundary of the first body. Thus, regardless of the convexity of Q_0 and Q_1, we obtain the first two terms of (3.12); that is, the expression.

The contribution of the intersection $\partial Q_0 \cap \partial Q_1$ to the mean curvature $M_{q-1}(Q_0 \cap Q_1)$ can be obtained by taking the parallel body exterior to the distance ϵ, then by calculating the integral of the $(q-1)$-nth mean curvature corresponding to the part parallel to $\partial Q_0 \cap \partial Q_1$, and finding the limit value of the same for $\epsilon \to 0$. This limit value will be of the form $F(x, \Theta_h, R_i^0, R_i^1)\, d\sigma_{01}$, where F is a function (whose explicit form is immaterial) of the point x of the intersection $\partial Q_0 \cap \partial Q_1$, of the angles Θ_h $(h = 1, 2, \ldots, 1/2\, n(n-1))$that fix the position of Q_1 around x, and of the principle radii of curvature R_i^0 and R_i^1 of ∂Q_0 and ∂Q_1, respectively, at the point x. For the explicit calculation of F, it suffices to apply the generalized theorems of Euler and Meusnier to the intersection of the $(n-1)$-dimensional manifolds, and the Olinde Rodrigues formulae to the same intersection. On multiplying the last integral by dQ_1, by applying (4.1) and integrating with regard to $dQ_1[x]$ to all the positions of Q_1 around x, we arrive at a function $F_1(x, R_i^0, R_i^1)$ depending only on x, R_i^0, R_i^1. The explicit form of this function is immaterial to our purpose; it suffices to observe that, having arrived at this form exclusively by "local" considerations around the point x, it does not matter whether Q_0 and Q_1 are convex or not.

As a result of subsequently integrating the product of this function F_1 by $d\sigma_0 \wedge d\sigma_1$, we know that in the case where Q_0 and Q_1 are convex it gives the sum that appears on the right-hand side of (3.12). Therefore, since $F_1(x, R_i^0, R_i^1)$ is independent of the convexity, and even of the connection of Q_0, Q_1, the final result (3.12) will always be the same in every case.

Let us now observe that all the local reasonings we have just performed, with the same functions F and F_1 appearing in them, are valid for bodies Q_0, Q_1 of a space of constant curvature K. Indeed, both formula (4.1) and those of Euler, Meusnier and Olinde Rodrigues, which it has been necessary to employ, hold equally in all cases. Thus we arrive at the key result: *formulae* (3.12) *for*

$1 \leq q < n-1$ *hold for any pair of not necessarily convex bodies* Q_0, Q_1, *and for any space of constant curvature* K.

As previously stated, for $q = n$, (3.12) must be substituted by the fundamental formula (3.5), in which $\chi(Q_0 \cap Q_1)$ is replaced by $(1/O_{n-1})M_{n-1}(Q_0 \cap Q_1)$, in accordance with (3.3). Thus the complement to formulae (3.12) is as follows:

$$(4.2) \qquad \int M_{n-1}(Q_0 \cap Q_1)\, dQ_1 = O_1 \ldots O_{n-1}\Big\{M^0_{n-1}V_1 + M^1_{n-1}V_0 + \frac{1}{n}\sum_{h=0}^{n-2}\binom{n}{h+1}M^0_h M^1_{n-2-h}\Big\}$$

which for the reasons explained above is valid for all bodies (with ∂Q_0 and ∂Q_1 of class $\mathcal{C}^2$), as well as for any space of constant curvature.

5 The kinematic fundamental formula in spaces of constant curvature

We now have all the elements required in order to write the kinematic fundamental formula for spaces of constant curvature K. It only suffices to recall the Gauss-Bonnet formula generalized by Allendoerfer-Weil-Chern, which relates the mean curvature integrals of the boundary of a body Q_1 of constant curvature K (see, for example, [7]). Then, for n even this is:

$$(5.1) \qquad \binom{n-1}{n-1}c_{n-1}M_{n-1} + \binom{n-1}{n-3}c_{n-3}M_{n-3} + \ldots \binom{n-1}{1}c_1 M_1 + K^{n/2}V = \frac{1}{2}O_n\chi(Q),$$

and for n odd

$$(5.2) \qquad \binom{n-1}{n-1}c_{n-1}M_{n-1} + \binom{n-1}{n-3}c_{n-3}M_{n-3} + \cdots + c_0 M_0 = \frac{1}{2}O_n\chi(Q),$$

where the constants c_i have the value

$$(5.3) \qquad c_i = \frac{O_n}{O_i O_{n-1-i}}K^{(n-1-i)/2},$$

where V is the volume of Q.

According to the parity of n, let us apply (5.1) or (5.2) to the intersection $Q_0 \cap Q_1$, and then multiply both sides by dQ_1 and integrate to all the positions of Q_1 having a common point with Q_0. On the righ-hand side there appears the integral of $\chi(Q_0 \cap Q_1)$, whose value is precisely that sought by the kinematic fundamental formula. On the left-hand side there appear the known integrals (3.12) and (4.2), as well as the integral of the volume $V(Q_0 \cap Q_1)$ of the intersection of Q_0 and Q_1.

This integral can be calculated immediately by the usual method, considering the integral of $dx \wedge dQ_1$ being extended to all the points x contained within $Q_0 \cap Q_1$, and to all the positions of Q_1 intersecting Q_0. On first fixing Q_1, we have the integral desired. However, if the point x is fixed first, the integral of dQ_1 yields $O_1 O_2 \dots O_{n-1} V_1$. Since x may subsequently sweep over all the interior of Q_0, we obtain

$$\int V(Q_0 \cap Q_1)\, dQ_1 = O_1 O_2 \dots O_{n-1} V_0 V_1 \tag{5.4}$$

Together with (3.12) and (4.2), this now gives the integrals of all those resulting from the addition of the left-hand sides of (5.1) and (5.2). Isolating the right-hand side integral, we arrive at the kinematic fundamental formula we have been seeking.

If, in order to bring together in a single formula the cases of n even and n odd, we introduce the abbreviations (1.3), we have the formula (1.2) we wished to prove.

6 Applications

Most of the integral geometry results in spaces of constant curvature are contained as particular cases in the general formula (1.2). As an example, we now refer to two applications:

a) *The Steiner formula in spaces of constant curvature.*

In the interest of brevity we put

$$K = k^2. \tag{6.1}$$

Let Q_1 be a sphere of radius ρ. It will be $\chi(Q_1) = 1$, and moreover

$$\begin{aligned} M_i(Q_1) &= O_{n-1} k^{(i+1-n)} \sin^{n-1-i} k\rho \cos^i k\rho \\ V(Q_1) &= O_{n-1} k^{-(n-1)} \int_0^\rho \sin^{n-1} k\rho \, d\rho. \end{aligned} \tag{6.2}$$

In expression (1.1) of dQ_1, by taking the centre of Q_1 as point x, for all positions of Q_1 one may integrate the expression $dO_{n-1} \wedge \cdots \wedge dO_1$, which yields the product $O_{n-1} O_{n-2} \dots O_1$. On dividing both sides of (1.2) by this factor, the result will be the volume of the domain covered by the points x whose distance to Q_0 is equal to or less than ρ; that is, *the volume of the parallel body exterior to Q_0 at distance ρ.*

The expression of this volume in terms of V_0, M_i $(i = 0, 1, \dots, n-1)$ constitutes precisely the generalization to spaces of constant curvature of the Steiner formula in the Euclidean case. The general expression for dimension n takes a complicated form that requires no description here, but which arises from (1.2) with the substitutions (6.2). We confine ourselves to writing the first cases $n = 2, 3, 4$.

$n = 2$. In this case, by writing

$$V_0 \to F_0, \quad V_1 \to F_1, \quad M_0^0 \leftrightarrow L_0, \quad M_0^1 \to L_1,$$

where F_0, F_1 are the areas, and L_0, L_1 are the lengths of the boundaries of the bi-dimensional figures Q_0, Q_1, then formula (1.2) can be written thus

$$\int \chi(Q_0 \cap Q_1)\, dQ_1 = 2\pi(F_1\chi_0 + F_0\chi_1 - \frac{k^2}{2\pi}F_0F_1 + \frac{1}{2\pi}L_0L_1). \tag{6.3}$$

If Q_1 is a sphere of radius ρ, we have

$$F_1 = 2\pi k^{-2}(1 - \cos k\rho), \quad L_1 = 2\pi k^{-1}\sin k\rho, \tag{6.4}$$

whereby, substituting in (6.3) and dividing both sides by 2π, the left-hand side represents the area F_ρ of the parallel domain exterior to Q_0 at distance ρ, we obtain

$$F_\rho = F_0 \cos k\rho + L_0 k^{-1} \sin k\rho + 2\pi k^{-2}(1 - \cos k\rho)\chi_0, \tag{6.5}$$

which is in agreement with the value obtained by Vidal-Abascal [14].

$n = 3$. The general formula (1.2) takes the form

$$\int \chi(Q_0 \cap Q_1) dQ_1 = 8\pi^2(V_1\chi_0 + V_0\chi_1) + 2\pi(F_0M_1 + F_1M_0). \tag{6.6}$$

If Q_1 is a sphere of radius ρ, we have

$$\begin{aligned} V_1 &= 2\pi k^{-3}(k\rho - \sin k\rho\, \cos k\rho), \\ F_1 &= M_0^1 = 4\pi k^{-2}\sin^2 k\rho, \\ M_1 &= M_1^1 = 4\pi k^{-1} \sin k\rho\, \cos k\rho \\ \chi_1 &= 1 \end{aligned} \tag{6.7}$$

Substituting in (6.6), and by dividing both sides by $8\pi^2$, it transpires that the volume of the parallel body exterior to Q_0 at distance ρ is as follows

$$\begin{aligned} V_\rho = V_0 \;&+\; F_0\, k^{-1} \sin k\rho\, \cos k\rho + M_1\, k^{-2}\sin^2 k\rho \\ &+\; 2\pi k^{-3}(k\rho - \sin k\rho\, \cos k\rho)\chi_0, \end{aligned} \tag{6.8}$$

which is in agreement with the result obtained by Allendoerfer [1].

$n = 4$. The general formula (1.2) takes the form

$$\begin{aligned} \int \chi(Q_0 \cap Q_1) dQ_1 &= 16\pi^4(-\frac{3}{4\pi^2}k^4V_0V_1 + V_1\chi_0 + V_0\chi_1) \\ &+ 4\pi^2(2F_0M_2^1 + 3M_1^0M_1^1 + 2F_1M_2^0 + \frac{4}{3}k^2F_0F_1). \end{aligned} \tag{6.9}$$

If Q_1 is a sphere of radius ρ, we have

$$\begin{aligned} V_1 &= \frac{2}{3}\pi^2k^{-4}(-\sin^2 k\rho\, \cos k\rho + 2(1 - \cos k\rho)), \\ F_1 &= 2\pi^2k^{-3}\sin^3 k\rho, \\ M_1^1 &= 2\pi^2k^{-2}\sin^2 k\rho\, \cos k\rho \\ M_2^1 &= 2\pi^2k^{-1}\sin k\rho\, \cos^2 k\rho \\ \chi_1 &= 1 \end{aligned} \tag{6.10}$$

Substituting in (6.9) and dividing both sides by $O_1O_2O_3 = 16\pi^4$, it follows that the volume of the exterior parallel body at distance ρ of Q_0 is

$$
\begin{aligned}
V_\rho &= V_0\,(\frac{1}{2}\sin^2 k\rho\,\cos k\rho + \cos k\rho)\\
&+ F_0\,(\sin k\rho\,\cos^2 k\rho + \frac{2}{3}\sin^3 k\rho)k^{-1}\\
&+ M_1^0\,\frac{3}{2}\,k^{-2}\sin^2 k\rho\,\cos k\rho\\
&+ M_2^0\,k^{-3}\sin^3 k\rho\\
&+ \frac{2}{3}\pi^2 k^{-4}(2 - 2\cos k\rho - \sin^2 k\rho\,\cos k\rho)\chi_0.
\end{aligned} \tag{6.11}
$$

In all cases, if we wish to find the surface area of the exterior parallel body at distance ρ, it suffices to derive the volume with regard to ρ.

b) An integral formula for convex bodies of the Elliptic space.

Let us consider the Elliptic space of n dimensions. This is $K = 1$. The geodesics of this space are closed lines of length π. They may be regarded as a body Q_1 for which

$$
\begin{gathered}
\chi_1 = 0, \quad V_1 = 0, \quad M_0 = M_1 = \dots M_{n-3} = M_{n-1} = 0\\
M_{n-2} = \frac{\pi}{n-1} O_{n-2}.
\end{gathered} \tag{6.12}
$$

In order to see this, it suffices to consider the exterior parallel surface at distance ϵ, and then effect $\epsilon \to 0$.

Let Q_0 be a convex body of the Elliptic space. In expression (1.1) of dQ_1, let x be a point of Q_1. If we apply (1.2) in the left-hand side, we have $\chi(Q_0 \cap Q_1) = 1$. Furthermore, if x is interior to Q_0, on integrating dQ_1 we obtain $O_1O_2 \dots O_{n-1}V_0$.

If x is exterior to Q_0, on integrating dQ_1 we obtain

$$2O_1O_2 \dots O_{n-2}\int \omega\, dx,$$

where ω is the solid angle formed by all the geodesics stemming from x and intersecting Q_0. A factor 2 is introduced, since each point x is a vertex of two opposite angles ω. Substituting the values (6.12) in the right-hand side (1.2) and equalizing, we have

$$
\begin{aligned}
O_1O_2 \dots O_{n-1}V_0 + 2O_1O_2 \dots O_{n-2}\int \omega\, dx\\
= O_1O_2 \dots O_{n-2}\frac{\pi}{n-1}O_{n-2}M_0^1,
\end{aligned}
$$

and since $M_0^1 = F_0 =$ the area of Q_0, we have the following integral formula

$$\int \omega\, dx = \frac{\pi}{2(n-1)}O_{n-2}F_0 - \frac{O_{n-1}}{2}V_0, \tag{6.13}$$

where the integration is extended to all the Elliptic space exterior to Q_0.

Examples. For $n = 2$, we have

$$\int \omega \, dx = \pi L_0 - \pi F_0,$$

where L_0 is the length and F_0 the area of Q_0.

For $n = 3$, we have

$$\int \omega \, dx = \frac{1}{2}\pi^2 F_0 - 2\pi V_0.$$

It does not appear an easy task to obtain these formulae directly.

References

[1] C. B. Allendoerfer, *Steiner's formulae on a general* S^{n+1}, Bull. Am. Math. Soc. **54** (1948), 128–135.

[2] W. Blaschke, *Vorlesungen über Integralgeometrie*, Deutscher Verlag der Wissenschaften, Berlin, 1955, 3te Aufl.

[3] T. Bonnesen and W. Fenchel, *Theorie der konvexen Körper*, Ergebnisse der Math. Springer, Berlin, 1934.

[4] S. S. Chern, *On the kinematic formula in the euclidean space of n dimensions*, American J. of Math. **74** (1952), 227–236.

[5] S. S. Chern and T. Yien, *Sulla formula principale cinematica dello spazio ad n dimensioni*, Boll. della Unione Matematica Italiana **2** (1940), 432–437.

[6] H. Hadwiger, *Vorlesungen über Inhalt, Oberfläche und Isoperimetrie*, Springer-Verlag, Berlin, 1957.

[7] L. A. Santaló, *Questions of differential and integral geometry in spaces of constant curvatura, (Spanish)*, Rendiconti del Sem. Mat. di Torino **14** (1954-55), 277–295.

[8] L. A. Santaló, *Integral Geometry in three-dimensional spaces of constant curvature (Spanish)*, Math. Notae **IX** (1949-50), 1–28.

[9] ———, *Geometría Integral en espacios de curvatura constante*, Publicaciones de la Com. Nac. de Energía Atómica **1** (1952), 1–68.

[10] ———, *Introduction to integral geometry*, Actualités Sci. Ind., no. 1198, Publ. Inst. Math. Univ. Nancago II. Herman et Cie, Paris, 1953.

[11] ———, *On the kinematic formula in spaces of constant curvature*, Proceedings of the International Congress (1954), 251–252, Amsterdam.

[12] ———, *On the mean curvatures of a flattened convex body*, Revue de la Fac. Sciences Univ. Istanbul **21** (1956), 189–194.

[13] ———, *Sur la mesure des espaces linéaires qui coupent un corps convexe et problèmes qui s'y rattachent*, Colloque sur les questions de réalité en géométrie, Liège, 1955, Georges Thone, Liège, 1956, pp. 177–190.

[14] E. Vidal-Abascal, *A generalization of Steiner's formulae*, Bull. A. Math. Soc. **53** (1947), 841–844.

[15] Ta-Jen Wu, *Ueber elliptische Geometrie*, Math.Zeits. **43** (1938), 495–521.

FACULTAD DE CIENCIAS EXACTAS Y NATURALES
Universidad de Buenos Aires.

Groups in the plane with respect to which the sets of points and lines admit an invariant measure

L. A. Santaló

1 Introduction

Let P_n be the group of collineations in real n-dimensional projective space on itself. Given a subgroup G of P_n, there exists a general method for ascertaining whether or not the sets of linear subspaces $L_r (0 \leq r \leq n-1)$ admit an invariant measure with respect to G. More generally, it is easy to calculate whether or not sets of figures formed by linear subspaces transitively transformed by G admit an invariant measure with respect to G; see Chern [2], Santaló [9], Varga [14], Luccioni [4]. Let us consider some examples:

a) The sets of hyperplanes L_{n-1} do not possess invariant measure with respect to P_n, but they do possess invariant measure with respect to the group of unimodular centre-affinities [5], [8] [4], and also with respect to the group of movements, whether Euclidean or not (Blaschke [1]);

b) The sets of "pairs of parallel hyperplanes" do not possess invariant measure with respect to P_n, but they do possess it with respect to the group of unimodular affinities [10];

c) The sets of "triads of parallel lines" in three-dimensional projective space do not possess invariant measure with respect to P_3, but do possess it with respect to the group of unimodular affinities [6];

d) The sets of "pairs of lines" in the projective plane do not possess invariant measure with respect to P_2, but they do possess it with respect to the centre-affine group as well as to the unimodular affine group (Stoka [11]).

Similar examples exist in abundance. However, the inverse problem is more complicated; viz: given L_r (or a configuration formed by linear subspaces), find all the subgroups G of P_n with respect to which the sets of L_r (or the sets of given configurations) admit an invariant measure with respect to G.

We confine ourselves to subgroups G of P_n, since the linear subspaces for whose sets we wish to ascertain whether or not there exists an invariant measure must be transitively transformed by G. If, instead of linear subspaces or configurations of these, other figures are considered, and instead of P_n a more general group of transformations is considered, the direct problem is also easy to solve, while the inverse problem is still posed, but its solution is apparently not so simple.

Obviously, a way of solving the stated inverse problem consists in successively checking all the subgroups G of P_n. Given that since Sophus Lie a method has been known for calculating these subgroups, which are infinite in number, the problem can be solved by checking them one by one. However, for $n = 3$ the procedure is extremely lengthy, and for $n > 3$ the method is impractical given the large number of subgroups of P_n. It is not difficult to provide some criteria in order to simplify the task by eliminating families of subgroups. Nevertheless, a direct and more convenient solution would be desirable.

In this work, we confine ourselves to the case of the projective plane ($n = 2$) for which the only subspaces are the points L_0 and the lines L_1. The case of the points has been studied from another perspective (Stoka [12]), but here we give the explicit form of the invariant measure in those cases where it exists.

In short, the problem to be solved is as follows: *to ascertain all the subgroups of the group of collineations of the real projective plane with respect to which the sets of points and the sets of lines admit an invariant measure, and to give the explicit form of this measure in each case.*

The groups for which there exists an invariant measure for sets of points and for sets of lines are summarized in the last section. They are the only ones for which it makes sense to study an integral geometry in the usual meaning of this phrase. By measuring particular sets of points and lines, and taking into account that the projective transformations conserve the projective convexity of the sets of points, it is possible to arrive at properties of the invariant convex sets with respect to a certain group. For the groups that do not admit an invariant measure for sets of points and lines, it is necessary to consider sets of pairs of these elements, from whose measures interesting properties for the theory of convex sets can also be found, as they were obtained in [8], [10], [6]. We leave this latter study for a future occasion.

Instead of "measure", we speak in general of invariant "density", it being fully understood that the density for a set with respect to a certain group is the differential form whose integral gives an invariant measure. Furthermore, since we here are concerned with transitive groups, it is well known that the density or the invariant measure, should it exist, is always unique except for a constant factor.

2 Projective groups in he plane

The projective groups in the plane (subgroups of the group of collineations) were obtained by Sophus Lie. Employing the usual notation, these groups are as follows (see, for example, G. Kowalewski [3], page 384).

I. *Groups depending on only one parameter.*

These do not concern us here, since they cannot be transitive with respect either to the points or to the lines, which are dependent upon two parameters.

II. *Groups depending on 2 parameters.*

1. p, q; 2. $xp,\ yq$; 3. $xp,\ q$; 4. $p+xq,\ q$;

5. $q,\ xp+yq$; 6. $q,\ \gamma\, xp+yq\ \ (\gamma \neq 0, 1)$;

7. $q,\ p+yq$; 8. $q,\ xp+(x+y)q$;

9. $q\text{-}2yp,\ 2xp+yq$; 10. $q,\ xq$; 11. $q,\ yq$.

III. *Groups depending on* 3 *parameters.*

1. $p,\ q,\ xp+yq$;
2. $p,\ 2xp+yq,\ x(xp+yq)$;
3. $p+yq,\ q,\ xq$;
4. $p,\ q,\ (\alpha+1)xp+(\alpha-1)yq$;
5. $q,\ xp,\ yq$;
6. $p,\ q,\ xp+(x+y)q$;
7. $q,\ xq,\ xp+\gamma\ yq$;
8. $p,\ q,\ xq$;
9. $p-xq,\ q,\ xp+2yq$;
10. $p+y(xp+yq),\ q+x(xp+yq),\ xp-yq$;
11. $q,\ xq,\ yq$.

IV. *Groups depending on* 4 *parameters.*

1. $p,\ xp,\ q,\ yq$;
2. $p,\ q,\ xq,\ yq$;
3. $p,\ q,\ xq,\ yq$;
4. $q,\ xp,\ xq,\ yq$;
5. $p,\ xp,\ yq,\ x(xp+yq)$;

V. *Groups depending on* 5 *parameters.*

1. $p,\ q,\ xq,\ 2xp+yq,\ x(xp+yq)$;
2. $p,\ q,\ xp,\ yq,\ xq$;
3. $p,\ q,\ yp,\ xq,\ xp-yq$;

VI. *Groups depending on* 6 *parameters.*

1. $p,\ q,\ xp,\ yq,\ xq,\ x(xp+yq)$;
2. $p,\ q,\ xp,\ yp,\ xq,\ yq$;

VII. *Groups depending on* 7 *parameters.* Do not exist.

VIII. *Projective group, or group of all the collineations, depending on* 8 *parameters.*

1. $p,\ q,\ xp,\ xq,\ yp,\ yq,\ x(xp+yq),\ y(xp+yq);$

For our purpose, it is convenient for these groups to be expressed explicitly as transformation groups in the projective plane on itself. This is what we do in each case; the result may consist of one of the following two types:

1. *Affine type.* In matrix notation, it will be of the form

$$x' = Ax + B \tag{2.1}$$

where A is a non-singular 2×2 matrix and B a 2×1 matrix. The coordinates are non-homogeneous. In this case, the Maurer-Cartan forms or relative components of the group, which are always denoted by ω_i and are all defined except for one linear combination with constant coefficients, are the elements of the matrices (see [10]):

$$\Omega_1 = A^{-1}\,dA, \quad \Omega_2 = A^{-1}\,dB \tag{2.2}$$

and the equations of structure, obtained by differentiation, exterior to these matrices, take the form

$$d\Omega_1 = -\Omega_1 \wedge \Omega_1, \quad d\Omega_2 = -\Omega_1 \wedge \Omega_2 \tag{2.3}$$

2. *Projective type.* In this case, it is necessary to use homogeneous coordinates, and the matrix form of the transformation formulae is

$$x' = Ax \tag{2.4}$$

where A is now a 3 x 3 matrix, which since we are dealing with homogeneous coordinates can always be normalized such that $\det A = 1$. The Maurer-Cartan forms are now elements of the matrix

$$\Omega = A^{-1}\,dA \tag{2.5}$$

and the equations of structure are expressed as follows:

$$d\Omega = -\Omega \wedge \Omega. \tag{2.6}$$

Another convenient way of finding the Maurer-Cartan forms in this projective case consists in considering the rows of the matrix A as homogeneous coordinates of three points A_0, A_1, A_2; taking these points as the base of a system of homogeneous coordinates, the Maurer-Cartan forms are determined by the relations

$$dA_i = \sum_{k=0}^{2} \omega_{ik} A_k, \quad i = 0, 1, 2, \tag{2.7}$$

from which one may deduce, for example,

$$\omega_{00} = |dA_0\, A_1\, A_2|,\ \omega_{01} = |A_0\, dA_0\, A_2|,\ \omega_{10} = |dA_1\, A_1\, A_2|, \text{ etc.} \tag{2.8}$$

where the right-hand sides indicate the determinants whose rows are the coordinates of the elements indicated (see [10], [5]).

3 Projective groups depending on 2 parameters

In this case, the measure for sets of points or lines should coincide with the invariant measure of the group, known in integral geometry as the kinematic measure of the group, and which is known very well to exist always (it is the Haar measure for these particular groups). It only remains to check whether or not the group is transitive with respect to the points or lines in the plane; if it is transitive, the measure of the group gives us the corresponding measure, and it is only necessary to express it in terms of the coordinates of the point or line according to each case.

Throughout this work, when it is said that a group is transitive with respect to the points or lines, it should be understood that it is transitive "in general"; that is, transitive except for exceptional positions. For example, we say that the group $x' = ax, y' = by$ is transitive with respect to the lines, although exception is made of the lines passing through the origin, since they are always transformed into lines that also pass through the origin. Neither do we distinguish between the local and the global triviality, since this is unnecessary for our purpose; for example, with respect to the group of collineations that leave a real conic section invariant (Cayley group), we say that the points are transformed transitively, although it is obvious that the interior points cannot be transformed into exterior points.

Let us now examine, case by case, the projective groups that depend on 2 parameters.

1. p, q.

This is the group of the translations

$$x' = x + a, \quad y' = y + b. \tag{3.1}$$

A measure exists for sets of points, which is the ordinary area (integral of the density $dP = dx \wedge dy$), but not for sets of lines, since the group is not transitive with respect to these lines.

2. xp, yq.

This is the group

$$x' = ax, \quad y' = by \tag{3.2}$$

which is transitive for points and for lines. Therefore, invariant density exists for sets of points and for sets of lines. To find the explicit form for these densities, one may observe that, according to (2.2), the Maurer-Cartan forms are

$$\omega_1 = \frac{da}{a}, \quad \omega_2 = \frac{db}{b} \tag{3.3}$$

and the kinematic density is $\omega_1 \wedge \omega_2$. In order to interpret this kinematic density in each case, note that the point of coordinates $(1, 1)$ by general transformation (3, 2) becomes the point of coordinates $(a,\ b)$, and therefore, denoting the coordinates of this general point by x, y, as the density of sets of points we have

$$dP = \frac{dx \wedge dy}{xy}. \tag{3.4}$$

For lines, the line that does not pass through the origin of equation $x+y-1=0$ is transformed into $x'/a+y'/b-1=0$; moreover, the normal equation for this general line is

$$x'\cos\phi+y'\sin\phi-p=0, \tag{3.5}$$

where p is the distance to the origin and ϕ the angle of the normal with the axis x. Thus, we have $a=p/\cos\phi, b=p/\sin\phi$. Substituting in (3.3) and making the exterior product, we obtain that the density for sets of lines, except the sign, which is inessential, since the invariant densities are always defined save for a constant factor, is as follows:

$$dG=\frac{dp\wedge d\phi}{p\sin\phi\cos\phi} \tag{3.6}$$

3. xp, q.

This is the group

$$x'=ax, \quad y'=y+h, \tag{3.7}$$

which is transitive for points and for lines, and therefore it admits invariant density for both sets of elements. According to (2.2), the Maurer-Cartan forms are

$$\omega_1=\frac{da}{a}, \quad \omega_2=dh \tag{3.8}$$

and the kinematic density is $\omega_1\wedge\omega_2$. The point $(1,0)$ becomes the general point (a, h), and thus the density for points $P(x, y)$ is

$$dP=\frac{dx\wedge dy}{x} \tag{3.9}$$

The line $x+y-1=0$ becomes $x'/a+y'-h-1=0$, which compared with the normal form (3.5) tells us that $a=\tan\phi, h=p/\sin\phi-1$. Hence, by substituting in (3.8) and making the exterior product, we have that the density for sets of lines $G(p,\phi)$is

$$dG=\frac{dp\wedge d\phi}{\sin^2\phi\cos\phi}. \tag{3.10}$$

4. $p+xq$, q.

This is the group

$$x'=x+c, \quad y'=cx+y+h, \tag{3.11}$$

which is transitive for points and lines; there thus exists invariant density for both sets of elements. To find the explicit form for these densities, one may observe that the Maurer-Cartan forms are

$$\omega_1=dc, \quad \omega_2=-c\,dc+dh \tag{3.12}$$

and therefore the kinematic density is $\omega_1 \wedge \omega_2 = dc \wedge dh$.

The point $(0,0)$ becomes (c,h) and hence, the density for sets of points is

$$dP = dx \wedge dy. \tag{3.13}$$

The line $y = 0$ becomes $cx' - y' + c^2 + h = 0$, whereby, comparing with (3.5), we have that $c = -\cot\phi, c^2 + h = p/\sin\phi$. Hence, the kinematic density in terms of the parameters p, ϕ of the line G is expressed as follows:

$$dG = \frac{dp \wedge d\phi}{\sin^3\phi}, \tag{3.14}$$

which is the density for sets of lines.

5. q , $xp + yq$.

This is the group

$$x' = ax, \quad y' = ay + h. \tag{3.15}$$

Points are transformed transitively, but not the lines, since a line and its transformed are parallel. Hence: invariant density exists for sets of points, but not for sets of lines.

It can be immediately seen that the expression of the density for sets of points is

$$dP = \frac{dx \wedge dy}{x^2}. \tag{3.16}$$

6. q, $\gamma\ xp+yq\ \ (\gamma \neq 0,1)$.

This is the group

$$x' = a^\gamma x, \quad y' = ay + h, \tag{3.17}$$

which is transitive for points and lines; that is, invariant density exists for sets of points and for sets of lines. The Maurer-Cartan forms are

$$\omega_1 = \gamma\frac{da}{a}, \quad \omega_2 = \frac{dh}{a}. \tag{3.18}$$

Point $(1,0)$ becomes the general point (a^γ, h), and thus by writing $x = a^\gamma, y = h$, the density for sets of points is

$$dP = \frac{dx \wedge dy}{x^{\frac{\gamma+1}{\gamma}}}. \tag{3.19}$$

For lines, note that the line $y = x$ becomes the general line $y' = a^{-\gamma+1}x' + h$. Therefore, comparing with (3.5), we obtain $a^{-\gamma+1} = -\cot\phi, h = p/\sin\phi$, and the density for lines is

$$dP = (-\tan\phi)^{\frac{\gamma-2}{\gamma-1}}\frac{dp \wedge d\phi}{\sin^3\phi}. \tag{3.20}$$

7. q , $p + yq$.

This is the group

$$x' = x + \log a, \quad y' = ay + h, \tag{3.21}$$

which is transitive for points and lines; that is, invariant density exists for sets of points and for sets of lines.
The Maurer-Cartan forms are

$$\omega_1 = \frac{da}{a}, \quad \omega_2 = \frac{dh}{a} \tag{3.22}$$

and the kinematic measure is $\omega_1 \wedge \omega_2$. The expression of this density for sets of points is obtained by observing that the point (0,0) is transformed into the general point (log a, h), and therefore, denoting its coordinates by x, y, we have

$$dP = e^{-x} dx \wedge dy. \tag{3.23}$$

For lines, note that the line $y = x$ passes to $y' = ax' - a \log a + h$, and therefore we have $a = -\cot\phi, h - a\log a = p/\sin\phi$, whereby, substituting in (3.22) and making the exterior product, we obtain the following expression of the density for sets of lines:

$$dG = \frac{dp \wedge d\phi}{\cos^2\phi \, \sin\phi}. \tag{3.24}$$

8. q , $xp + (x + y)\, q$.

This is the group

$$x' = ax, \quad y' = a \log ax + ay + b, \tag{3.25}$$

which is transitive for points and lines; that is, invariant density exists for sets of points and sets of lines.

On applying (2.2), we obtain

$$\omega_1 = \frac{da}{a}, \quad \omega_2 = \frac{db}{a}. \tag{3.26}$$

The point $P(1,0)$ becomes P (a, a log a + b); that is, by writing $a = x$, log $a + b = y$, then substituting in (3.26) and making the exterior product, we have that the expression of the density for points P (x, y) is

$$dP = \frac{dx \wedge dy}{x^2}. \tag{3.27}$$

For lines, by taking $y = 0$, which is transformed into the general line $y' = x' \log a + b$, and comparing as usual with the normal equation (3.5), we obtain $a = -\cot\phi, b = p/\sin\phi$, whereby, differentiating, then substituting in (3.26) and making the exterior product, we have

$$dG = e^{\cot\phi} \frac{dp \wedge d\phi}{\sin^3\phi}. \tag{3.28}$$

9. q - $2yp$, $2xp + yq$.

This is the group

$$x' = a^2 x - 2ahy - h^2, \quad y' = ay + h \tag{3.29}$$

of the affinities that leave the conic section $y^2 + x = 0$ invariant. Since both the points and the lines are transformed transitively by this group, we have that invariant measure exists for both elements. By applying (3.2), we obtain

$$\omega_1 = \frac{da}{a}, \quad \omega_2 = \frac{dh}{a}.$$

Point $(1, 0)$ becomes the general point $(a^2 - h^2, h)$; thus, the general coordinates of a point are $a^2 - h^2 = x, h = y$, whereby the kinematic density $\omega_1 \wedge \omega_2$ takes the form

$$dP = \frac{dx \wedge dy}{(y^2 + x)^{3/2}}, \tag{3.30}$$

which is the expression of the density for points.

For lines, note that the line $x = 1$ is transformed into the general line $x' + 2hy' - a^2 - h^2 = 0$, whereby, comparing with (3.5), we obtain $2h = \tan\phi, a^2 + h^2 = p/\cos\phi$. Hence, an immediate calculation yields this new expression for the kinematic density,

$$dG = \frac{dp \wedge d\phi}{(4p\cos\phi - \sin^2\phi)^{3/2}}, \tag{3.31}$$

which is the expression of the density for sets of lines (always excepting a constant factor).

10. q , xq.

This is the group

$$x' = x, \quad y' = ax + y + b, \tag{3.32}$$

which is not transitive for points, but is for lines. Therefore, there exists no invariant density for sets of points, but it does exist for sets of lines.

In order to find this density, we proceed as usual, observing that the Maurer-Cartan forms are $\omega_1 = da, \omega_2 = db$, and that the line $y = 0$ is transformed into the general line $ax' - y' + b = 0$, whereby one may write $a = \cot\phi, b = p/\sin\phi$, and the density for sets of lines (equal to the kinematic density $da \wedge db$) is thus

$$dG = \frac{dp \wedge d\phi}{\sin^3\phi}. \tag{3.33}$$

11. q , yq.

This is the group

$$x' = x, \quad y' = ay + b, \tag{3.34}$$

which is not transitive for points, but is for lines; in other words, no invariant density exists for sets of points, but it does exist for sets of lines. To find this latter density, note that by applying (2.2), we obtain

$$\omega_1 = \frac{da}{a}, \quad \omega_2 = \frac{db}{a} \tag{3.35}$$

and that the line $y = x$ becomes the general line $y' = ax' + b$, whereby one may write $a = -\cot\phi, b = p/\sin\phi$, and consequently the density for sets of lines (equal to the kinematic density $\omega_1 \wedge \omega_2$) is as follows:

$$dG = \frac{dp \wedge d\phi}{\cos^2\phi \sin\phi}. \tag{3.36}$$

4 Projective groups depending on 3 parameters

We analyze these groups one by one.

1. p, q , $xp + yq$.

This is the group

$$x' = ax + c, \quad y' = ay + h. \tag{4.1}$$

It is not transitive for lines, since every line is transformed into a parallel line; however, it is transitive for points. Let us check to see if invariant density exists for points.

According to (2.2), the Maurer-Cartan forms are

$$\omega_1 = \frac{da}{a}, \quad \omega_2 = \frac{dc}{a}, \quad \omega_3 = \frac{dh}{a}, \tag{4.2}$$

and the equations of structure, according to (2.3), are

$$d\omega_1 = 0, \quad d\omega_2 = -\omega_1 \wedge \omega_2, \quad d\omega_3 = -\omega_1 \wedge \omega_3. \tag{4.3}$$

Point $(0, 0)$ is transformed into the general points (c, h); the system defining the points is therefore $\omega_2 = 0, \omega_3 = 0$. If invariant density exists, it is $\omega_2 \wedge \omega_3$, and in order for this form to be a density, according to the general theory [5], [9], its exterior differential must be zero. Taking into account the equations of structure (4.3), we have

$$d(\omega_2 \wedge \omega_3) = -2\omega_1 \wedge \omega_2 \wedge \omega_3 \neq 0.$$

Thus, invariant density does not exist either for sets of points or for sets of lines.

2. p, $2xp + yq$, $(xp + yq)$.

This is the group

$$x' = \frac{ax + b}{cx + h}, \quad y' = \frac{y}{cx + h}, \quad ah - bc = 1. \tag{4.4}$$

The matrices that appear in (2.5) are in this case

$$A = \begin{pmatrix} a & 0 & b \\ 0 & 1 & 0 \\ c & 0 & h \end{pmatrix}, \quad A^{-1} = \begin{pmatrix} h & 0 & -b \\ 0 & 1 & 0 \\ -c & 0 & a \end{pmatrix}$$

and therefore the Maurer-Cartan forms are

(4.5) $$\omega_1 = c\,db - a\,dh, \quad \omega_2 = h\,db - b\,dh, \quad \omega_3 = -c\,da + a\,dc,$$

and the equations of structure, according to (2.6), are

(4.6) $$d\omega_1 = \omega_2 \wedge \omega_3, \quad d\omega_2 = 2\omega_1 \wedge \omega_2, \quad d\omega_3 = 2\omega_3 \wedge \omega_1.$$

Point $(0,1)$ is transformed into the general point $(b/h, 1/h)$. Therefore, the equations defining the points are $\omega_2 = 0, \omega_1 = 0$. Since $d(\omega_1 \wedge \omega_2) = 0$, we have that density exists for sets of points. To find the explicit form of this density, it suffices to write $x = b/h$, $y = 1/h$, whereby we have that (taking into account the relation $ah - bc = 1$), $\omega_1 \wedge \omega_2 = -dh \wedge db = h^3 dx \wedge dy = y^{-3} dx \wedge dy$; that is, the density for sets of points is as follows:

(4.7) $$dP = \frac{dx \wedge dy}{y^3}.$$

For lines, note that the line $y = x$ is transformed into the general line $y' = hx' - b$. The equations defining the lines are therefore $\omega_1 = 0, \omega_2 = 0$. Since $d(\omega_1 \wedge \omega_2) = 0$, we obtain that the lines also have invariant density. To find the explicit form of this density, comparing $y' = hx' - b$ with the normal form (3.5), we have that $h = -\cot\phi, b = -p/\sin\phi$, and therefore the product $\omega_1 \wedge \omega_2 = dh \wedge db$ is written as follows:

(4.8) $$dG = \frac{dp \wedge d\phi}{\sin^3\phi}.$$

Observe that in homogeneous coordinates, the group (4.4) is written as

(4.9) $$x' = ax + bt, \quad y' = y, \quad t' = cx + ht,$$

with the condition $ah - bc = 1$, which since we are dealing with homogeneous coordinates can always be imposed. By the change of coordinates $x \to x$, $y \to t$, $t \to y$, the group (4.9) is written as follows $x' = ax + by$, $y' = cx + hy, t' = t$; or becoming once again non-homogeneous coordinates,

(4.10) $$x' = ax + by, \quad y' = cx + hy, \quad ah - bc = 1,$$

which is the group of unimodular centre-affinities. The existence of density for points and lines, invariant with respect to this group, is known and is of great importance [8].

3. $p + yq,\ q\ ,\ xq$.

This is the group

$$x' = x + \log c, \quad y' = ax + cy + h. \tag{4.11}$$

By applying (2.2), we obtain that the Maurer-Cartan forms are

$$\omega_1 = \frac{da}{c}, \quad \omega_2 = \frac{dc}{c}, \quad \omega_3 = -\frac{a\,dc}{c^2} + \frac{dh}{c}, \tag{4.12}$$

and the equations of structure are

$$d\omega_1 = \omega_1 \wedge \omega_2, \quad d\omega_2 = 0, \quad d\omega_3 = -\omega_2 \wedge \omega_3 - \omega_1 \wedge \omega_2. \tag{4.13}$$

Point $(0,0)$ becomes the general point $(\log c, h)$; therefore, the system defining the points is $\omega_2 = \omega_3 = 0$; since $d(\omega_2 \wedge \omega_3) = 0$ which is the result of applying the equations of structure, then density exists for sets of points. By writing $\log c = x, h = y$, we have that the explicit form of this density is

$$dP = e^{-x}\, dx \wedge dy. \tag{4.14}$$

For sets of lines, note that the line $y = 0$ is transformed into the general line $y' = ax' + h - a \log c$. In order for this line to remain fixed, it is necessary that $da = 0, d(h - a \log c) = 0$; that is, $\omega_1 = 0, \omega_3 = 0$. According to the equations of structure, we have $d(\omega_1 \wedge \omega_3) = 2\omega_1 \wedge \omega_2 \wedge \omega_3$. Therefore, invariant density does not exist for sets of lines.

4. p , q, $(\alpha + 1)\, xp + (\alpha - 1)\, yq$.

This is the group

$$x' = a^{\alpha+1}x + c, \quad y' = a^{\alpha-1}y + h, \tag{4.15}$$

for which the Maurer-Cartan forms are

$$\omega_1 = \frac{da}{a}, \quad \omega_2 = \frac{dc}{a^{\alpha+1}}, \quad \omega_3 = \frac{dh}{a^{\alpha-1}}, \tag{4.16}$$

and the equations of structure are

$$d\omega_1 = 0, \quad d\omega_2 = -(\alpha + 1)\omega_1 \wedge \omega_2, \quad d\omega_3 = -(\alpha - 1)\omega_1 \wedge \omega_3. \tag{4.17}$$

Point $(0,0)$ becomes the general point $(c,\ h)$. The equations defining the points are therefore $\omega_2 = 0, \omega_3 = 0$. Since $d(\omega_2 \wedge \omega_3) = -2\alpha\omega_1 \wedge \omega_2 \wedge \omega_3 \neq 0$, then no invariant density exists for sets of points, except for the case $\alpha = 0$.

The case $\alpha = 0$ corresponds to the group

$$x' = ax + c, \quad y' = \frac{1}{a}\, y + h, \tag{4.18}$$

for which the density for the points is

$$dP = dx \wedge dy. \tag{4.19}$$

For lines, note that the line $y = x$ is transformed into the line $y' = a^{-2}x' - ca^{-2} + h$; therefore, the equations defining the lines are $\omega_1 = 0, \omega_2 - \omega_3 = 0$. Since $d(\omega_1 \wedge (\omega_2 - \omega_3) = 0$, then there exists density for sets of lines. The explicit form of this density is obtained from the equalities $a^2 = -\tan\phi, -c/a^2 + h = p/\sin\phi$, whereby

$$\omega_1 \wedge (\omega_2 - \omega_3) = \frac{dp \wedge d\phi}{2a^{\alpha+1}\sin\phi\cos^2\phi}$$

or, disregarding a constant factor,

$$dG = \frac{dp \wedge d\phi}{(-\tan\phi)^{(\alpha+1)/2}\sin\phi\cos^2\phi}. \tag{4.20}$$

For the case $\alpha = 0$ of group (4.18), this density takes the form

$$dG = \frac{dp \wedge d\phi}{(-\sin 2\phi)^{3/2}}. \tag{4.21}$$

The existence of invariant density for points and lines for the group (4.18) was predictable, since this group is projectively equivalent to that of movements in the Euclidean plane. In fact, the group of movements is

$$x' = x\cos\theta - y\sin\theta + a, \quad y' = x\sin\theta + y\cos\theta + b, \tag{4.22}$$

whereby

$$x' + iy' = e^{i\theta}(x + iy) + a + ib,$$
$$x' - iy' = e^{-i\theta}(x - iy) + a - ib,$$

and by the linear change of coordinates $X = x + iy, Y = x - iy$, by writing $e^{i\theta} = A, a + ib = B, a - ib = C$, we have that

$$X' = AX + B, \quad Y' = \frac{1}{A}Y + C,$$

which is the group (4.18).

5. q , xp , yq.

This is the group

$$x' = ax, \quad y' = by + h. \tag{4.23}$$

The formulae (2.2) and (2.3) give

$$\omega_1 = \frac{da}{a}, \quad \omega_2 = \frac{db}{b}, \quad \omega_3 = \frac{dh}{b}, \tag{4.24}$$

$$d\omega_1 = 0, \quad d\omega_2 = 0, \quad d\omega_3 = -\omega_2 \wedge \omega_3. \tag{4.25}$$

Point $(1, 0)$ becomes the general point (a, h), and therefore the equations defining the points are $\omega_1 = 0, \omega_3 = 0$. Since $d(\omega_1 \wedge \omega_3) = \omega_1 \wedge \omega_2 \wedge \omega_3 \neq 0$, then there is no invariant density for sets of points.

The line $y = x$ is transformed into $y' = (b/a)x' + h$. The equations defining these lines are $\omega_2 - \omega_1 = 0, \omega_3 = 0$, and since $d(\omega_3 \wedge (\omega_2 - \omega_1)) = -\omega_1 \wedge \omega_2 \wedge \omega_3 \neq 0$, then there is no invariant density for sets of lines.

6. p , q , $xp + (x + y)\, q$.

This is the group

$$x' = ax + c, \quad y' = (a \log a)x + ay + h, \tag{4.26}$$

whose Maurer-Cartan forms are

$$\omega_1 = \frac{da}{a}, \quad \omega_2 = \frac{dc}{a}, \quad \omega_3 = -\log a \frac{dc}{a} + \frac{dh}{a}, \tag{4.27}$$

and the equations of structure are

$$d\omega_1 = 0, \quad \omega_2 = -\omega_1 \wedge \omega_2, \quad d\omega_3 = -\omega_1 \wedge \omega_2 - \omega_1 \wedge \omega_3. \tag{4.28}$$

Point $(0,0)$ passes to $(c,\, h)$; therefore, the equations defining the points are $\omega_2 = 0, \omega_3 = 0$. Since $d(\omega_2 \wedge \omega_3) = -2\omega_1 \wedge \omega_2 \wedge \omega_3 \neq 0$, then there is no invariant density for sets of points.

Line $y = 0$ becomes the general line $y' = x' \log a - c \log a + h$. Consequently, the equations defining the lines are $\omega_1 = 0, \omega_3 = 0$. Since $d(\omega_1 \wedge \omega_3) = 0$, then invariant density exists for sets of lines. To find the explicit form, we have the equations $\log a = -\cot \phi, -c \log a + h = p/\sin \phi$, therefore the density $\omega_1 \wedge \omega_3$ for sets of lines can be written as

$$dG = e^{\cot \phi} \frac{dp \wedge d\phi}{\sin^3 \phi}. \tag{4.29}$$

7. q , xq , $xp + \gamma yq$.

This is the group

$$x' = ax, \quad y' = bx + a^\gamma y + h, \tag{4.30}$$

for which we obtain

$$\omega_1 = \frac{da}{a}, \quad \omega_2 = -\frac{b\, da}{a^{\gamma+1}} + \frac{db}{a^\gamma}, \quad \omega_3 = \frac{dh}{a^\gamma}, \tag{4.31}$$

$$d\omega_1 = 0, \quad d\omega_2 = (1 - \gamma)\, \omega_1 \wedge \omega_2, \quad d\omega_3 = -\gamma\, \omega_1 \wedge \omega_3. \tag{4.32}$$

Point $(1,0)$ is transformed into the general point $(a,\, b + h)$, and thus the equations defining the points are $\omega_1 = 0, \omega_2 + \omega_3 = 0$. Since $d(\omega_1 \wedge (\omega_2 + \omega_3)) = 0$, then there exists density for points. This density is $\omega_1 \wedge (\omega_2 + \omega_3)$, which by writing $x = a, y = b + h$, gives

$$dP = \frac{dx \wedge dy}{x^{\gamma+1}}. \tag{4.33}$$

To check whether or not density exists for lines, we proceed as usual. Line $y = x$ is transformed into $y' = (a^{\gamma-1} + b/a)x' + h$. The equations defining the lines are therefore $(\gamma - 1)\omega_1 + \omega_2 = 0, \omega_3 = 0$. Since $d[\omega_3 \wedge ((\gamma - 1)\omega_1 + \omega_2)] = (2\gamma - 1)\omega_1 \wedge \omega_2 \wedge \omega_3 \neq 0$, then invariant density does not exist for sets of lines.

The case $\gamma = \frac{1}{2}$ is an exception for which density exists for lines, and its explicit form is easily obtained thus,

$$dG = \frac{dp \wedge d\phi}{\sin^3 \phi}. \tag{4.34}$$

8. p , q , xq.

This is the group

$$x' = x + c, \quad y' = ax + y + b, \tag{4.35}$$

for which we obtain

$$\omega_1 = da, \quad \omega_2 = dc, \quad \omega_3 = -a\,dc + db, \tag{4.36}$$

$$d\omega_1 = 0, \quad d\omega_2 = 0, \quad d\omega_3 = -\omega_1 \wedge \omega_3. \tag{4.37}$$

Density for points obviously exists in this case and is as follows:

$$dP = dx \wedge dy. \tag{4.38}$$

Line $y = x$ is transformed into the general line $y = (a+1)x' - ac + b - c$. The equations defining the lines are $\omega_1 = 0, \omega_3 - \omega_2 = 0$. Since $d(\omega_1 \wedge (\omega_3 - \omega_2)) = 0$, then density exists for lines. Its explicit form is obtained from the equalities $a + 1 = -\cot\phi, -ac + b - c = p/\sin\phi$, whereby the density for sets of lines $\omega_1 \wedge (\omega_3 - \omega_2)$ can be written in the form

$$dG = \frac{dp \wedge d\phi}{\sin^3 \phi}. \tag{4.39}$$

9. $p - xq$, q , $xp + 2yq$.

This is the group

$$x' = ax + \frac{b}{a}, \quad y' = -bx + a^2 y + h. \tag{4.40}$$

According to (2.2) and (2.3), its Maurer-Cartan forms are

$$\omega_1 = \frac{da}{a}, \quad \omega_2 = \frac{b\,da}{a^3} - \frac{db}{a^2} = -\frac{1}{a}\,d(\frac{b}{a}), \tag{4.41}$$

$$\omega_3 = \frac{b}{a^3}\,d(\frac{b}{a}) + \frac{dh}{a^2}, \tag{4.42}$$

and the equations of structure are

$$d\omega_1 = 0, \quad d\omega_2 = \omega_2 \wedge \omega_1, \quad d\omega_3 = 2\omega_3 \wedge \omega_1. \tag{4.43}$$

Point $(0,0)$ is transformed into the general point $(b/a,\ h)$, and thus the equations defining the points are $\omega_2 = 0, \omega_3 = 0$. Since $d(\omega_2 \wedge \omega_3) = -3\omega_1 \wedge \omega_2 \wedge \omega_3 \neq 0$, then no invariant density exists for sets of points.

Line $y = 0$ is transformed into the general line $y' = -(b/a)x' + b^2/a^2 + h$. Consequently, the equations defining the lines are $\omega_2 = 0, \omega_3 = 0$, as above. In other words, there is no invariant density for sets of lines in this case either.

10. $p + y(xp + yq)$, $q + x(xp + yq)$, $xp - yq$.

This is the group of the projectivities that leave the conic section $2xy+1=0$ (Kowalewski [3], page 389) invariant. It is therefore known as the Cayley group, for which it is known that there exists density for points and for lines. The expression and the properties of these densities are very well known ([9], page 110).

11. q , xq , yq.

This is the group

$$x' = x, \quad y' = ax + by + c. \tag{4.44}$$

It is not transitive with respect to the points, but it is transitive with respect to the lines. If

$$\omega_1 = \frac{da}{b}, \quad \omega_2 = \frac{db}{b}, \quad \omega_3 = \frac{dc}{b}, \tag{4.45}$$

$$d\omega_1 = \omega_1 \wedge \omega_2, \quad d\omega_2 = 0, \quad d\omega_3 = -\omega_2 \wedge \omega_3, \tag{4.46}$$

and observing that the line $y = 0$ is transformed into line $y' = ax' + c$, and thus the equations defining the lines are $\omega_1 = 0, \omega_3 = 0$, it is possible to calculate $d(\omega_1 \wedge \omega_3) = 2\omega_1 \wedge \omega_2 \wedge \omega_3 \neq 0$. Consequently, no invariant density for sets of lines exists in this case either.

5 Projective groups depending on 4 parameters

These are as follows:

1. p , xp , q , $y\,q$.

This is the group

$$x' = ax + b, \quad y' = cy + h. \tag{5.1}$$

By the general method, we obtain

$$\omega_1 = \frac{da}{a}, \quad \omega_2 = \frac{dc}{c}, \quad \omega_3 = \frac{db}{a}, \quad \omega_4 = \frac{dh}{c}, \tag{5.2}$$

with the equations of structure

$$d\omega_1 = 0,\ d\omega_2 = 0,\ d\omega_3 = -\omega_1 \wedge \omega_3,\ d\omega_4 = -\omega_2 \wedge \omega_4. \tag{5.3}$$

Point $(0,0)$ becomes $(b,\ h)$; then the equations defining the points are $\omega_3 = 0, \omega_4 = 0$. We have $d(\omega_3 \wedge \omega_4) = -\omega_1 \wedge \omega_3 \wedge \omega_4 + \omega_3 \wedge \omega_2 \wedge \omega_4 \neq 0$; therefore, no invariant density exists for sets of points.

Line $y = x$ passes to $y' = (c/a)x' - cb/a + h$, and therefore the equations defining the lines are $\omega_1 - \omega_2 = 0, \omega_3 - \omega_4 = 0$. Since $d[(\omega_1 - \omega_2) \wedge (\omega_3 - \omega_4)] \neq 0$, no invariant density exists for sets of lines.

This group is isomorphic to the group of similarities in the plane:

$$\begin{aligned} x' &= \rho(x \cos\theta - y \sin\theta) + a \\ y' &= \rho(x \sin\theta + y \cos\theta) + b. \end{aligned}$$

In fact, we have

$$\begin{aligned} x' + iy' &= \rho e^{i\theta}(x + iy) + a + ib, \\ x' - iy' &= \rho e^{-i\theta}(x - iy) + a - ib, \end{aligned}$$

and with the change of coordinates $X = x + iy, Y = x - iy$, and by writing

$$\rho e^{i\theta} = A, \quad \rho e^{-i\theta} = B, \quad a + ib = C, \quad a - ib = H,$$

we obtain

$$X' = AX + C, \quad Y' = BY + H,$$

which is the group (5.1). We thus arrive at the known fact that for the group of similarities in the plane no invariant density exists either for points or for lines. However, it can be seen that invariant density exists for sets of circles (Stoka [11]), which equals $R^{-2} dx \wedge dy$, if R is the radius and x, y are the orthogonal cartesian coordinates of the centre.

2. p , q , xq , $xp + a\ yq$.

This is the group

$$x' = ax + b, \quad y' = cx + a^\alpha y + h. \tag{5.4}$$

Acoording to (2.2) and (2.3), we obtain

$$\omega_1 = \frac{da}{a}, \quad \omega_2 = -\frac{c\,da}{a^{\alpha+1}} + \frac{dc}{a^\alpha}, \quad \omega_3 = \frac{db}{a}, \tag{5.5}$$
$$\omega_4 = -\frac{c\,db}{a^{\alpha+1}} + \frac{dh}{a^\alpha},$$

with the equations of structure

$$d\omega_1 = 0, \quad \omega_2 = (1-a)\omega_1 \wedge \omega_2, \quad d\omega_3 = -\omega_1 \wedge \omega_3, \tag{5.6}$$
$$d\omega_4 = -\omega_2 \wedge \omega_3 - a\,\omega_1 \wedge \omega_4.$$

Point $(0,0)$ is transformed into the general point $(b,\ h)$, and therefore the equations defining the points are $\omega_3 = 0, \omega_4 = 0$. We have $d(\omega_3 \wedge \omega_4) = -(1 + a)\omega_1 \wedge \omega_3 \wedge \omega_4$, thus no invariant density exists for sets of points, except for the case $a = -1$.

In order to see what occurs with the sets of lines, observe that the line $y=0$ is transformed into $y' = (c/a)x' - cb/a + h$; therefore, the equations defining the lines are $\omega_2 = 0, \omega_4 = 0$. Since $d(\omega_2 \wedge \omega_4) = (1-2a)\omega_1 \wedge \omega_2 \wedge \omega_4$, then no invariant density exists for lines, except in the case $a = 1/2$.

Consequently, it is convenient to distinguish between the two following cases:

a) $a =$ - 1. This is the group

$$x' = ax + b, \quad y' = cx + \frac{y}{a} + h, \tag{5.7}$$

which admits density for points (but not for lines), and is equal to

$$dP = dx \wedge dy. \tag{5.8}$$

b) $a = \frac{1}{2}$. This is the group

$$x' = ax + b, \quad y' = cx + a^{1/2}y + h, \tag{5.9}$$

which admits density for lines (but not for points). The explicit form of this density is obtained, as usual, by comparing the line $y' = (c/a)x' - cb/a$+h, transformed from $y = 0$, with its normal equation (3.5), which gives $c/a = -\cot\phi$, $-cb/a + h = p/\sin\phi$. Hence, the expression $\omega_2 \wedge \omega_4$ of the density for sets of lines can be written

$$dG = \frac{dp \wedge d\phi}{\sin^3\phi}. \tag{5.10}$$

3. p , q , xq , yq.

This is the group

$$x' = x + a, \quad y' = bx + cy + h. \tag{5.11}$$

The Maurer-Cartan forms and the equations of structure are

$$\omega_1 = \frac{db}{c}, \quad \omega_2 = -\frac{dc}{c}, \quad \omega_3 = -\frac{b\,da}{c} + \frac{dh}{c}, \quad \omega_4 = da, \tag{5.12}$$

$$d\omega_1 = \omega_1 \wedge \omega_2,\ d\omega_2 = 0,\ d\omega_3 = -\omega_1 \wedge \omega_4 + \omega_3 \wedge \omega_2,\ d\omega_4 = 0. \tag{5.13}$$

Point $(0,0)$ becomes (a, h); therefore, the equations defining the points are $\omega_3 = 0, \omega_4 = 0$, and since $d(\omega_1 \wedge \omega_4) \neq 0$, there exists no invariant density for sets of points.

Line $y = 0$ becomes $y' = bx' - ba + h$; thus, the lines are defined by $\omega_1 = 0$, $\omega_3 = 0$, and since $d(\omega_1 \wedge \omega_3) = \omega_1 \wedge \omega_2 \wedge \omega_3$, no invariant density exists for sets of lines.

4. *q , xp , xq , yq.*

This is the group

$$x' = ax, \quad y' = bx + cy + h, \tag{5.14}$$

whose Maurer-Cartan forms and equations of structure are, according to (2.2) and (2.3),

$$\omega_1 = \frac{da}{a}, \quad \omega_2 = -\frac{b\,da}{ac} + \frac{db}{c}, \quad \omega_3 = \frac{dc}{c}, \quad \omega_4 = \frac{dh}{c} \tag{5.15}$$

$$d\omega_1 = 0,\ d\omega_2 = \omega_1 \wedge \omega_2 + \omega_2 \wedge \omega_3,\ d\omega_3 = 0,\ d\omega_4 = -\omega_3 \wedge \omega_4. \tag{5.16}$$

Point $(1, 0)$ becomes the general point $(a,\ b + h)$; therefore, the equations defining the points are $\omega_1 = 0, \omega_2 + \omega_4 = 0$. Since $d[\omega_1 \wedge (\omega_2 + \omega_4)] \neq 0$, we have that no invariant density exists for sets of points.

Line $y = 0$ becomes the general line $y' = (b/a)x' + h$; thus, the equations defining the lines are $\omega_2 = 0, \omega_4 = 0$. Since $d(\omega_2 \wedge \omega_4) = \omega_1 \wedge \omega_2 \wedge \omega_4 \neq 0$, no invariant density exists for sets of lines.

5. p , xp , yq , $x(xp + yq)$.

This is the group

$$x' = \frac{ax + b}{cx + h}, \quad y' = \frac{ey}{cx + h}. \tag{5.17}$$

No special study is required. Indeed, this group admits the group p, xp, yq (group n° 5 of 3 parameters, with the permutation x for y) as a subgroup. This is the subgroup obtained for $c = 0$. Since this group does not admit invariant density for either points or lines, neither does the present group admit invariant density for sets of points or for sets of lines.

6 Groups depending on 5 parameters

These are as follows:

1. p , q , xq , $2xp + yq$, $x(xp + yq)$.

This is the group

$$x' = \frac{ax + c}{mx + r}, \quad y' = \frac{bx + y + h}{mx + r}, \quad (ar - mc = 1). \tag{6.1}$$

Applying the method referred to in n° 2, we obtain

$$\begin{aligned}
\omega_1 &= r\,da - c\,dm, \quad \omega_2 = r\,dc - c\,dr, \\
\omega_3 &= (mh - br)\,da + db + (bc - ah)\,dm, \\
\omega_4 &= (mh - br)\,dc + dh + (bc - ah)\,dr, \\
\omega_5 &= -m\,da + a\,dm, \quad \omega_5 = -\omega_1 = -m\,dc + a\,dr,
\end{aligned}$$

with the equations of structure

$$\begin{aligned}
d\omega_1 &= \omega_2 \wedge \omega_5, \quad d\omega_2 = 2\omega_1 \wedge \omega_2, \quad d\omega_3 = \omega_3 \wedge \omega_1 + \omega_4 \wedge \omega_5, \\
d\omega_4 &= \omega_3 \wedge \omega_2 + \omega_1 \wedge \omega_4, \quad d\omega_5 = 2\omega_5 \wedge \omega_1, \quad d\omega_6 = \omega_5 \wedge \omega_2.
\end{aligned}$$

Point $(0,0)$ is transformed into $(c/r, h/r)$, and the equations must be $\omega_2 = 0, \omega_4 = 0$ in order for it to remain fixed. Since

$$d(\omega_2 \wedge \omega_4) = 3\omega_1 \wedge \omega_2 \wedge \omega_4 \neq 0,$$

then no density exists for sets of points.

Line $y = 0$ is transformed into the line $y' = (br - hm)x' + ha - bc$, and the equations must be $\omega_3 = 0, \omega_4 = 0$ in order for it to remain fixed. Since $d(\omega_3 \wedge \omega_4) = 0$, then density exists for lines. In order to check the geometric interpretation of this density, note that $br - hm = -\cot\phi, ha - bc = p/\sin\phi$, whereby $r\omega_3 - m\omega_4 = d\phi/\sin^2\phi, -c\omega_3 + a\omega_4 = dp/\sin\phi + (\dots)d\phi$, and hence the density $\omega_3 \wedge \omega_4$ is written as follows:

$$dg = \frac{dp \wedge d\phi}{\sin^3\phi}.$$

2. p , q , xp , yq , xq.

This is the group

$$x' = ax + m, \quad y' = bx + cy + h, \tag{6.2}$$

which has as subgroup n° 4 of 4 parameters (for $m = 0$). Since this latter group does not admit invariant density either for points or for lines, then neither does the present group.

3. p , q yp , xq , $xp - yq$.

This is the group of unimodular affinities

$$x' = ax + by + c, \quad y' = mx + gy + h \quad (ag - bm = 1). \tag{6.3}$$

It is well known that invariant density exists for this group for points ($dP = dx \wedge dy$), but invariant density does not exist for lines (see [9]).

7 Projective groups depending on 6 parameters

These are as follows:

1. p , q , xp , yq , xq , $x(xp + yq)$.

This is the group

$$x' = \frac{ax + by + c}{ny + r}, \quad y' = \frac{my + h}{ny + r}. \tag{7.1}$$

In order to find the Maurer-Cartan forms and the equations of structure, we follow the second method referred to in n° 2. The matrix A that appears there is now

$$A = \begin{pmatrix} a & b & c \\ 0 & m & h \\ 0 & n & r \end{pmatrix}, \tag{7.2}$$

with the condition $\det A = 1$. Points A_0, A_1, A_2 are the points $A_0(a,b,c)$, $A_1(0,m,h)$, $A_2(0,n,r)$ (in homogeneous coordinates). According to the formulae (2.8) and those analogous, the non-vanishing Maurer-Cartan forms are

$$\begin{aligned} \omega_{00} &= (mr-nh)da,\ \omega_{01} = -(br-nc)da + ar\,db - an\,dc, \\ \omega_{02} &= (bh-mc)da - ah\,db + am\,dc,\ \omega_{11} = ar\,dm - an\,dh, \\ \omega_{12} &= -ah\,dm + am\,dh,\ \omega_{21} = ar\,dn - an\,dr,\ \omega_{22} = -ah\,dn + am\,dr \end{aligned} \tag{7.3}$$

observing the relation $\omega_{00} + \omega_{11} + \omega_{22} = 0$, which is obtained by differentiating the determinant $|A_0A_1A_2| = 1$.

The equations of structure are

$$d\,\omega_{ik} = \sum_s \omega_{is} \wedge \omega_{sk}.$$

In order for point A_0 to remain fixed, from the relation $dA_0 = \omega_{00}A_0 + \omega_{01}A_1 + \omega_{02}A_2$ it is deduced that $\omega_{01} = 0, \omega_{02} = 0$. These are the equations defining the points. Applying the equations of structure, we obtain $d(\omega_{01} \wedge \omega_{02}) = 3\omega_{00} \wedge \omega_{01} \wedge \omega_{02} \neq 0$, which proves that no invariant density exists for sets of points.

For sets of lines, note that $\omega_{02} = 0, \omega_{12} = 0$ in order for the line A_0A_1 to remain fixed. Applying the equations of structure, taking into account that the ω_{ij} that do not appear in (7.3) are zero, we obtain $d(\omega_{02} \wedge \omega_{12}) = 3\omega_{02} \wedge \omega_{22} \wedge \omega_{12} \neq 0$. Therefore, invariant density does not exist for sets of lines.

2. $p\ ,\ q\ ,\ xp\ ,\ xq\ ,\ yp\ ,\ yq.$

This is the general affine group

$$x' = ax + by + c, \quad y' = mx + ny + r, \tag{7.4}$$

with $an - bm \neq 0$, for which it is known that no invariant density exists either for sets of points or for sets of lines [9].

Finally, there remains the general projective group

$$p, q, xp, xq, yp, yq, x(xp+yq), y(xp+yq),$$

whose equations are

$$x' = \frac{ax + by + c}{mx + ny + r}, \quad y' = \frac{ex + gy + h}{mx + ny + r}. \tag{7.5}$$

This group has been well studied (Varga [14]), and also in [5], [9], and it is known not to admit invariant density either for sets of points or for sets of lines.

8 Summary of results

From the foregoing, we have that:

1. The projective groups in the plane depending on 2 parameters for which invariant measure exists for sets of points and for sets of lines are as follows:

$$xp, yq;\; xp, q;\; p+xq, q;\; q, \gamma xp+yq;$$
$$q, p+yq;\; q, xp+(x+y)q;\; q-2yp, 2xp+yq.$$

2. The projective groups in the plane depending on 3 parameters for which invariant measure exists for sets of points and for sets of lines in the plane are as follows:

$$p, 2xp+yq, x(xp+yq);\; p, q, xq;\; q, xq, 2xp+yq;$$
$$p, q, xp-yq;\; p+y(xp+yq), q+x(xp+yq), xp-yq.$$

Of these four groups, the latter corresponds to the Cayley group and has been well studied [9]; the fourth is isomorphic to the group of movements in the plane and its integral geometry is also well known. However, the other three groups probably merit a more detailed study from the point of view of integral geometry.

3. No projective groups in the plane depending on more than 3 parameters exist that admit invariant measure for sets of points and for sets of lines. In this case, it is nevertheless necessary to study what occurs with the sets of pairs of elements (two points, two lines, point and line). There are groups for which these sets of pairs admit invariant measure and others which do not. Some particular cases are known (see [10], [6], and several works by Stoka [11], [12], [13] and by Luccioni [4]), but a systematic study exhausting and ordering all the possible cases has still to be performed, although it would appear to involve greater difficulty (see [7]).

References

[1] W. Blaschke, *Integralgeometrie*, vol. 252, Actualités Scient. et Industrielles, Hermann, 1935, Paris.

[2] Shiing-shen Chern, *On integral geometry in Klein spaces*, Ann. of Math. **43** (1942), 178–189.

[3] G. Kowalewski, *Einführung in die Theorie der Kontinuerlichen Gruppen*, Leipzig, 1931.

[4] R. E. Luccioni, *Geometría Integral en espacios proyectivos*, Revista de Matemática y Física Teórica, Universidad N. de Tucuman **XV** (1964), 53–80.

[5] L. A. Santaló, *Integral Geometry in Projective and Affine Spaces*, Annals of Mathematics **51** (1950), 739–755.

[6] ———, *On the measure of sets of parallel linear subspaces in affine space*, Canadian J. of Mathematics **14** (1962), 313–319.

[7] ———, *Integral geometry of the projective groups of the plane depending on more than three parameters*, Annales Scientifiques Univ. Iasi **XI** (1965), 307–335.

[8] L. A. Santaló, *Un invariante afin para los cuerpos convexos del espacio de n dimensiones*, Portugaliae Mathematica **8** (1949), 155–161.

[9] ———, *Introduction to integral geometry*, Actualités Sci. Ind., no. 1198, Publ. Inst. Math. Univ. Nancago II. Herman et Cie, Paris, 1953.

[10] ———, *Two applications of the integral geometry in affin and projective spaces*, Publ. Math. Debrecen **7** (1960), 226–237, Volume in tribute to O. Varga.

[11] M. I. Stoka, *Geometria integrale in uno spazio euclideo E_n*, Bull. Unione Mat. Italiana **XIII** (1958), 470–485.

[12] ———, *Geométrie intégrale dans un space E_n*, Revue de Mathématiques Pures et Appliquées, Acad. R. P. Roumaine **IV** (1959), 123–156.

[13] ———, *Geométrie intégrale dans un space E_n*, Revista de Matemáticas y Física teórica, Univ. N. de Tucuman **XIV** (1962), 25–29.

[14] O. Varga, *Ueber Masse von Paaren linearer Mannigfaltigkeiten in Projektiven Raum*, Rev. Mat. Hispano-Americana **10** (1935), 241–264.

Facultad de Ciencias Exactas y Naturales
Universidad de Buenos Aires

Average values for polygons formed by random lines in the Hyperbolic plane

L. A. Santaló

Abstract

Goudsmit [2] and more recently R. E. Miles [4], [5], P. I. Richards [6] and G. Kendall-P. A. P. Moran [3] have considered the problem of finding the mean number of sides, the mean perimeter, the mean area, the mean area squared and many other averages of the convex polygons into which the Euclidean plane is divided by a system of straight lines distributed at random homogeneously throughout the plane. The first result of Goudsmit was generalized to the 3-dimensional space in [8], and many other interesting results for polytopes determined by random hyperplanes in n-dimensional Euclidean space have been obtained by R. E. Miles (unpublished papers).

In the present paper we generalize the problem of Goudsmit to the Hyperbolic plane. The straight lines are assumed given at random in the sense of Integral Geometry [9], [10]. First, we obtain some mean values for a finite convex set K intersected by a finite number n of random straight lines (the mean number of vertices inside K (2.6), the mean number of regions into which K is divided (2.8) and the mean number of sides of each region (2.10), the mean perimeter (2.12) and the mean area (2.13)). Then, we consider that K is a circle which expands to all the Hyperbolic plane. If the density for lines is normalized such that the mean area is $E(A) = 1$, then we get for the mean values of the number of sides N, the perimeter S and the area squared A^2, the following values

$$E(N) = 4, \quad E(S) = 1 + \frac{1}{\xi} = 3,683..., \quad E(A^2) = 13,02...$$

where $\xi = 0,3726...$ is the positive root of $4\pi\xi^2 - 2\xi - 1 = 0$. The exact value of $E(A^2)$ is given by (4.14).

1 Introduction and results

The problem we wish to solve is as follows:

"Let us consider a Hyperbolic plane divided into polygons by straight lines at random, with a uniform density such that the mean value E (A) of the area of these polygons is equal to the unit E $(A) = 1$. We require the mean values of the number of sides N of each polygon, of the perimeter S of each polygon, and especially the square area of each polygon".

The problem was considered and solved for the Euclidean plane by S. A. Goudsmit [2]. It was subsequently generalized to the space by the present authors [8], and was recently the object of study and generalization in several directions by R. E. Miles [4], [5], P. I. Richards [6] and D. G. Kendall (see M. G. Kendall – P. A. P. Moran [3]). Some new results by Miles have yet to be published, and the present authors wish to express their appreciation to him for having provided them with a draft copy. These generalizations concern new mean values for the case of the plane and their extension to Euclidean space of n dimensions.

Generalization to the Hyperbolic plane is of interest as a fresh example of the different behaviour of this plane as compared with the Euclidean, because of its sub-divisions into polygonal domains.

Let us summarize the results obtained and compare them with the Euclidean case. We denote the number of sides of each polygon by N; the perimeter of the polygons by S; the area of the polygons by A, and the "mean value" by E according to the law of uniform probability that we specify later. Thus we have

Euclidean plane (known results [2]):

$$E(N) = 4, \quad E(S) = 2\sqrt{\pi} = 3,5448..., \quad E(A) = 1,$$

$$E(A^2) = \frac{1}{2}\pi^2 = 4,934...$$

Hyperbolic plane:

$$E(N) = 4, \quad E(S) = 1 + \frac{1}{\psi} = 3,683..., \quad E(A) = 1,$$

$$E(A^2) = \frac{1}{2}\pi^2 = 13,02...$$

where ξ is the positive root of $4\pi\xi^2 - 2\xi - 1 = 0$, and the exact value of $E(A^2)$ is given by (4.14).

2 Mean values for a convex figure K in the Euclidean plane

We now provide a new proof of the known results for the case of the Euclidean plane, which will help to better understand the path we wish to follow for the case of the Hyperbolic plane. Note that the original proof by Goudsmit, as well as the subsequent variants, do not appear to be easily generalized to the Hyperbolic plane.

Let us first recall some results already obtained by the present authors on a former occasion [7]. Let K be a convex figure in the plane, with perimeter L and area F. We consider n straight lines G_i $(i = 1, 2, \ldots, n)$ which intersect K. If p_i, θ_i are the co-ordinates of G_i (p_i = the distance of G_i to the origin; θ_i = the angle of the normal to G_i with a fixed direction in the plane), and we take as the density for measuring sets of lines the differential form $dG_i = (dp_i \, d\theta_i)$, [10], [3]. The mean value of the number of points V of the intersection of n straight lines, which are interior to K, will be

$$E(V) = \frac{\int V \, dG_1 \, dG_2 \ldots dG_n}{\int dG_1 \, dG_2 \ldots dG_n}, \tag{2.1}$$

where the numerator integral is extended to all the lines in the plane and the denominator integral is extended to all those intersecting K, and is therefore

$$\int dG_1\, dG_2 \dots dG_n = L^n, \tag{2.2}$$

since it is known that, for each straight line, the measure of those intersecting K is equal to L; that is

$$\int dG_1 = L, \tag{2.3}$$

(see [10], p. 13; [3], p. 58]).

In order to calculate the numerator integral of (2.1), we call V_{ij} the function of G_i, G_j, which is equal to 1 if G_i and G_j intersect inside K, and is equal to 0 if they intersect outside K (to complete the definition, we write $V_{ii} = 0$). For each position of the n straight lines, $V = \sum V_{ij}$ and the number of V_{ij} is $\frac{1}{2}n(n-1)$. If we denote by s_i the length of the chord that G_i determines in K, we have

$$\int V_{ij} dG_i\, dG_j = 2\int s_i\, dG_i = 2\pi F, \tag{2.4}$$

since the measure of the straight lines intersecting a segment is equal to double their length ([10], p. 13; [3], p. 58]), and moreover the following formula is known and immediate

$$\int s_i dG_i = \pi F. \tag{2.5}$$

Consequently, we have

$$\begin{aligned} \int V\, dG_1\, dG_2 \dots dG_n &= \frac{1}{2}n(n-1)\int V_{ij}\, dG_1\, dG_2 \dots dG_n \\ &= n(n-1)\pi F L^{n-2}. \end{aligned}$$

Hence:

a) *The mean value of the number of points of intersection of n straight lines intersecting K, which are interior to K, is*

$$E_K(V) = n(n-1)\pi F L^{-2}. \tag{2.6}$$

Now let P be the number of regions in which n randomly distributed straight lines divide the convex figure K. The positions of the n straight lines at which more than 2 cross at the same point are exceptional positions of zero measure with regard to the integrals in (2.1). Except for these positions, the following formula holds

$$P = V + n + 1, \tag{2.7}$$

as may be immediately deduced from the Euler relation for networks in the plane. In fact, we now consider the network formed by the n straight lines and the boundary of K. The number of vertices is equal to V plus the $2n$ points determined by the straight lines on the boundary of K. The number of regions is denoted by P. For the number of sides, one may observe that 4 sides pass through each of the interior vertices V, and 3 sides pass through each of the vertices of the boundary; since each side belongs to two vertices, the number of sides will therefore be $\frac{1}{2}(4V+6n) = 2V+3n$. Then according to Euler's theorem for open surfaces (the number of regions, plus that of the vertices, is equal to the number of sides plus one), we have

$$V + P + 2n - 2V - 3n = 1,$$

which yields relation (2.7), which we wished to prove. Consequently

b) *The mean value of the number of regions in which n randomly distributed straight lines divide the convex figure K is*

$$E_K(P) = n(n-1)\pi F L^{-2} + n + 1. \tag{2.8}$$

As previously observed, the total number of sides of the network formed by the boundary of K, plus the chords determined by the straight lines G_i is $2V+3n$. The number of sides of the boundary is $2n$, so the number of interior sides will be $2V+n$. If we call N_i the number of sides in the region C_i on adding the N_i for all the regions P, we observe that each interior side is counted twice, while each side of the boundary of K is counted only once. Therefore

$$\sum_1^P N_i = 2(2V+n) + 2n = 4V + 4n. \tag{2.9}$$

Hence, *defining* the mean value of the number of sides of the regions in which a convex figure K is divided by n randomly distributed straight lines by

$$E_K(N) = \frac{E_K(\sum_1^P N_i)}{E_K(P)},$$

we have:

c) *The mean value of the number of sides of the regions in which a convex figure K is divided by n randomly distributed straight lines is*

$$E_K(N) = 4 - \frac{4}{E_K(V) + n + 1}, \tag{2.10}$$

where $E_K(V)$ is given by (2.6).

Let us consider now the sum $\sum S_i$ $(i = 1, ;P)$ of the perimeters of the regions into which K is divided. Denoting, as before, by s_i the length of the chord that G_i determines in K, we have

$$\sum_1^P S_i = 2\sum_1^n s_i + L, \tag{2.11}$$

and therefore we obtain:

d) *The mean value of the perimeters of the regions in which a convex figure K is divided by n randomly distributed straight lines intersecting K is*[1]

$$E_K(S) = \frac{2nE_K(s_i) + L}{E_K(P)},$$

or, taking into account that, according to (2.5) and (2.3), $E_K(s_i) = \pi F/L$, we have

$$E_K(S) = \frac{2\pi nFL + L^3}{n(n-1)\pi F + (n+1)L^2}. \tag{2.12}$$

Finally, immediately defining the mean area of the regions in which K is divided by $E_K(A) = F/E_K(P)$, we have:

c) *The mean value of the area of the regions in which a convex figure K is divided by n randomly distributed straight lines intersecting K is*

$$E_K(A) = \frac{FL^2}{n(n-1)\pi F + (n+1)L^2}. \tag{2.13}$$

3 Passing to the entire Euclidean plane

On passing from the convex figure K to the whole Euclidean plane, the above-mentioned mean values depend on the form of K and on the way in which this figure expands until it fills all the plane. To be precise, the final result depends on the following two parameters:

a) The limit of the ratio F/L^2 of the area of K and the square of its perimeter when K expands until it occupies all the plane. *We will assume that K is a circle, and therefore*

$$\frac{F}{L^2} = \frac{1}{4\pi}. \tag{3.1}$$

This hypothesis is fundamental. If, instead of a circle, we assume K to be a square, then $F/L^2 = 1/16$, and all subsequent formulae would be modified.

b) Furthermore, it is necessary to give the mean density of the straight lines, which now cover the entire plane. This mean density is obtained by the limit value of the ratio r/n, where r is the radius of the circle and n is the number of straight lines intersecting it. We may adopt two criteria:

1. Let us assume that $E(A) = 1$. Then, according to (2.13) and by writing $F = \pi r^2$, $L = \pi r$, whenever $r \to \infty$, we obtain $E(A) = 1$, the number of straight

[1]Observe that this definition of "mean value" as the quotient

$$E_K(\sum S_i)/E_K(P)$$

is quite natural. Others may be adopted; for example

$$E_K(\sum S_i/P),$$

which is more in agreement with the usual definition of probability calculus, but which is extremely difficult to calculate exactly.

lines n must expand in such a way that

$$\lim \frac{r}{n} = \frac{1}{2\sqrt{\pi}}. \tag{3.2}$$

This is the hypothesis put forward by Goudsmit [2] when he assumes that the straight lines in the plane are such that the mean value of the area of the polygons into which the plane is divided is equal to unity.

2. One may introduce the number k, which denotes the average number of straight lines intersecting a segment of unit length distributed randomly over the plane. In this case, since the probability that a straight line intersecting the circle K may also intersect a segment of unit length situated in its interior is $2/L$ ([3], p. 58), then the mean value of the number of straight lines intersecting the segment will be $2n/L = n/\pi r$. If we wish this value to be k, then by making $r \to \infty$, the number of straight lines n must expand in such a way that

$$\lim \frac{r}{n} = \frac{1}{\pi k}. \tag{3.3}$$

This is the hypothesis put forward by Richards [6]. Parameter k is related to the parameter τ employed by Miles [4], [5], since $\tau = \pi k/2$.

Now we take Goudsmit's convention; that is, assuming condition (3.2) to be fulfilled, by performing $k = 2/\sqrt{\pi}$ it corresponds to Richards' case.

With this convention, from (2.10) and (2.12), by writing $F = \pi r^2$, $L = 2\pi r$ and taking (3.2) into account, we have that the mean values of the number of sides N, and of the perimeter S of the polygons in which the Euclidean plane is divided by randomly distributed straight lines covering the plane with the condition $E(A) = 1$, are $E(N) = 4$, $E(S) = 2\sqrt{\pi}$, in accordance with the known results already mentioned.

Now we find the mean value of the square area, which constitutes the most complicated operation. Let us see how, by means of the method we are following here, we may arrive at Goudsmit's result.

The mean value of the square area: Let K be the circle of radius r. Let P_1, P_2 be two points interior to this circle, and let us denote by $dP_i = [dx_i\, dy_i]$ $(i = 1, 2)$ the area element of the plane corresponding to each of these points. If $G_1, G_2, \ldots, G_n$ are n straight lines intersecting K, let us consider the integral

$$I = \int dP_1\, dP_2\, dG_1 \ldots dG_n \tag{3.4}$$

extended to all the straight lines intersecting K, and for each position of these straight lines it is likewise extended to the positions of P_1, P_2 belonging to the same region of those by which K is divided; that is, to the positions at which P_1 and P_2 are not separated by any of the straight lines n.

By first fixing the n straight lines G_i, we have

$$I = \int \sum_{1}^{P} A_i^2\, dG_1 \ldots dG_n, \tag{3.5}$$

Hence,

$$(3.6) \qquad E\Big(\sum_1^P A_i^2\Big) = \frac{I}{L^n},$$

and therefore the mean value of the square area of the regions in which the straight lines n divide K, according to the definition adopted, is

$$(3.7) \qquad E_K(A^2) = \frac{I}{E_K(P)L^n},$$

where $E_K(P)$ is given by (2.8).

In order to compute I, we now fix the points P_1, P_2. Denoting by a the distance between P_1, P_2, since the integral of each dG_i extended to the positions at which G_i does not separate the points P_1, P_2 is equal to the measure of the straight lines intersecting K (whose value is $L = 2\pi r$) less the measure of those intersecting the segment P_1P_2 (whose value is $2a$), we obtain

$$(3.8) \qquad I = \int (2\pi r - 2a)^n \, dP_1 \, dP_2.$$

P_1 having been fixed, it is $dP_2 = a \, da \, d\theta$, where θ is the angle of the straight line P_1P_2 with a fixed direction in the plane. a, θ having been fixed, the integral of dP_1 is the area of intersection of K with its translation in the direction opposite to θ of a segment a. For the general case of any convex figure K, this area cannot be expressed by elementary functions, but if K is a circle of radius r the area is composed of two circular segments of height equal to $r - a/2$. Furthermore, this area is independent of the direction θ, and thus the integral of $d\theta$ is 2π. This gives

$$(3.9) \quad I = 4\pi \int_0^{2r} (2\pi r - 2a)^n a (r^2 \arccos(a/2r) - \frac{a}{4}\sqrt{4r^2 - a^2}) \, da.$$

Therefore, according to (3.7) and (2.8), we have

$$(3.10) \quad E_K(A^2) = \frac{16\pi \int_0^{2r} (1 - \frac{a}{\pi r})^n a r^2 \left(\arccos \frac{a}{2r} - \frac{a}{4r}\sqrt{4 - \frac{a^2}{r^2}} \right) da}{n(n-1) + 4(n+1)}.$$

Taking the limit for $r \to \infty$ and $n \to \infty$, with the condition (3.2), we obtain

$$(3.11) \qquad E(A^2) = 2\pi \int_0^\infty e^{-2a/\sqrt{\pi}} a \, da = \frac{1}{2}\pi^2,$$

which is in agreement with Goudsmit's result [2].

Observe that if instead of the integral I (3.4), we consider

$$\int f(a) \, dP_1 \, dP_2 \, dG_1 \dots dG_n,$$

where a is always the distance P_1P_2, the same reasoning brings us to the final formula:

$$(3.12) \qquad E(\int f(a)dP_1\,dP_2) = 2\pi \int_0^\infty e^{-2a/\sqrt{\pi}} af(a)\,da = \frac{1}{2}\pi^2$$

where the integral of the left-hand side is extended to the interior of one of the polygons in which the straight lines divide the plane. This formula (3.12), which contains many particular cases according to the value assigned to the function $f(a)$, was obtained by Richards [6].

4 The case of the Hyperbolic plane

Although the results obtained are already known, the foregoing calculations are made for the Euclidean plane in order to better understand the generalization to the Hyperbolic plane. The method remains the same; only the intermediate formulae and the final results are different.

A straight line in the Hyperbolic plane is determined by its distance v to a fixed origin O, and by the angle θ that its normal by O forms with a fixed direction (also by O). In order to measure a set of straight lines, we take the integral of the differential form

$$(4.1) \qquad dG = \cosh v\,dv\,d\theta,$$

which in Integral Geometry is known as the "density" for straight lines in the Hyperbolic plane, and which is the only invariant form by movements (except for a constant factor). See [9], [10].

With this density, the measures of the straight lines intersecting a convex figure is equal to the perimeter of the figure (as in the Euclidean case), and the formula (2.5) for the Euclidean plane also holds ([9], [10]. Consequently, all the foregoing for the case of a convex figure K in the Euclidean plane in Section 2 is equally valid for the Hyperbolic plane. In other words:

The mean values (2.6), (2.8), (2.10), (2.12) and (2.13) are equally valid for the Hyperbolic plane.

The difference arises if we assume that K expands until it fills the entire plane. Let us assume that K is a circle of radius r. We then have (see, for example, Coxeter [1], p.250)

$$(4.2) \qquad L = 2\pi \sinh r, \quad F = 2\pi(\cosh r - 1),$$

and therefore

$$(4.3) \qquad \frac{2\pi F}{L^2} = \frac{1}{\cosh r + 1}.$$

The mean area (2.13) is thus

$$(4.4) \qquad E_K(A) = \frac{4\pi \sinh^2 r}{n(n-1) + 2(n+1)(\cosh r + 1)}.$$

If we normalize the density of the straight lines so that for $r \to \infty$ it is $E(A) = 1$, we obtain

$$4\pi(\frac{\sinh r}{n})^2 = 1 - \frac{1}{n} + \frac{2(n+1)(\cosh r + 1)}{n^2},$$

and since for $r \to \infty$, $\sinh r/\cosh r \to 1$, by writing

$$\lim \frac{\sinh r}{n} = \xi, \tag{4.5}$$

ξ must be the root of the equation

$$4\pi\xi^2 = 1 + 2\xi; \tag{4.6}$$

that is, since ξ is positive

$$\xi = \frac{1+\sqrt{1+4\pi}}{4\pi} = 0,3726... \tag{4.7}$$

For the second case considered in Section 3, if we wish the mean value of the number of straight lines intersecting a segment of unit length situated randomly on the plane to be k, the radius r and the number n must tend towards ∞ linked by the relation

$$\lim \frac{2n}{2\pi \sinh r} = k;$$

that is,

$$\lim \frac{\sinh r}{n} = \frac{1}{\pi k}. \tag{4.8}$$

In this case $E(A) = 4/(\pi k^2 + 2k)$, and $k = 1/\pi\xi$ in order for $E(A) = 1$.

On passing to the entire Hyperbolic plane from a circle whose radius $r \to \infty$, and with a number of straight lines n which also tend towards ∞ conserving relation (4.5), $E(A) = 1$ and relation (2.10) immediately yield $E(N) = 4$. The mean value of the perimeter S of the polygons in which the plane is divided is no longer the same. Starting from (2.12), by writing $F = 2\pi(\cosh r - 1)$, $L = 2\pi \sinh r$ and passing to the limit, by taking into account (4.5) we arrive at

$$E(S) = 1 + \frac{1}{\xi} = 3,683... \tag{4.9}$$

Thus we have the first results stated in Section 1. Let us now proceed to the most complicated calculation for the mean value of the square area.

The mean value of the square area: In order to follow the same path as for the Euclidean plane, it is necessary to calculate the area of intersection of two circumferences of radius r whose distance between their centres is a $(0 \le a \le 2r)$, (Fig. 1). Let φ be the central angle corresponding to the shaded area which constitutes the said intersection. The area of the circular sector of central angle φ is $\varphi(\cosh r - 1)$. The area of the isosceles triangle OUV, whose value is $\pi - (\varphi + 2\alpha)$, must be subtracted (Coxeter [1], p.246), where α is the base angle.

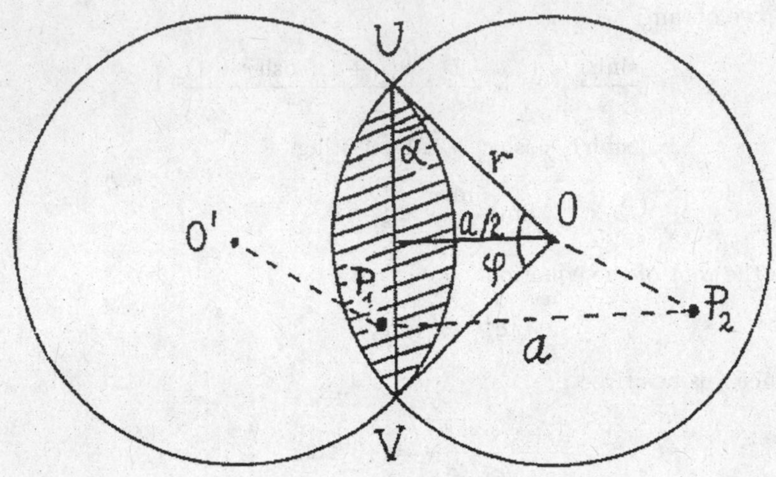

Fig.1

Furthermore, in the right triangle, half of OUV is (Coxeter [1], p.238)

$$\sin\alpha = \frac{\sinh(a/2)}{\sinh r}, \quad \cos\frac{\varphi}{2} = \frac{\tanh(a/2)}{\tanh r}.$$

Therefore, the area of intersection of both circles is

$$\begin{aligned}(4.10)\quad H(a,r) &= 2\varphi(\cosh r - 1) - 2\pi + 2(\varphi + 2\alpha)\\ &= 4\arccos\frac{\tanh(a/2)}{\tanh r}\cosh r - 2\pi + 4\arcsin\frac{\sinh(a/2)}{\sinh r}.\end{aligned}$$

As in the case of the Euclidean plane, we now have the same integral I, which on the one hand has the same value (3.5), and on the other, taking into account that P_1 having been fixed,

$$dP_2 = \sinh a\, da\, d\theta,$$

instead of (3.9) we have

$$(4.11)\qquad I = 2\pi\int_0^{2r}(2\pi\sinh r - 2a)^n H(a,r)\sinh a\, da.$$

In order to find $E(A^2)$, according to (3.7) it is necessary to divide by $E_K(P)L^n$ and then take to the limit for $r, n \to \infty$ linked by relation (4.5). According to the normalization $E_K(A) = 1$, it is $E_K(P) = F = 2\pi(\cosh r - 1)$. Furthermore, if $L = 2\pi\sinh r$, we obtain

$$\begin{aligned}E(A^2) &= \lim_{r,n\to\infty}\frac{2\pi\int_0^{2r}(1 - \frac{a}{\pi\sinh r})^n \sinh a H(a,r)\, da}{2\pi(\cosh r - 1)}\\ &= 4\int_0^{\infty} e^{-a/\pi\xi}\sinh a\arccos(\tanh(a/2))\, da,\end{aligned}$$

or, by writing $a = 2\xi u$,

$$E(A^2) = 8\xi \int_0^\infty e^{-2u/\pi} \sinh(2\xi u) \arccos(\tanh \xi u)\, du. \tag{4.12}$$

Since

$$\int e^{-2u/\pi} \sinh(2\xi u) du = \frac{\pi}{2(\pi^2\xi^2 - 1)} e^{-2u/\pi} (\sinh 2\xi u + \pi\xi \cosh 2\xi u),$$

and also

$$d(\arccos(\tanh \xi u)) = -\frac{\xi}{\cosh \xi u}\, du$$

integrating the integral of (4.12) by parts, thus yielding

$$E(A^2) =$$

$$= 8\xi \left[\frac{\pi}{2(\pi^2\xi^2 - 1)} e^{-2u/\pi} (\sinh(2\xi u) + \pi\xi \cosh(2\xi u)) \arccos(\tanh \xi u) \right]_0^\infty$$

$$+ \frac{8\pi\xi^2}{2(\pi^2\xi^2 - 1)} \int_0^\infty e^{-2u/\pi} (\sinh 2\xi u + \pi\xi \cosh 2\xi u) \frac{du}{\cosh \xi u}$$

$$= -\frac{2\pi^3\xi^2}{\pi^2\xi^2 - 1}$$

$$+ \frac{4\pi\xi^2}{\pi^2\xi^2 - 1} \int_0^\infty e^{-2u/\pi} (\sinh(2\xi u) + \pi\xi \cosh(2\xi u)) \frac{du}{\cosh \xi u},$$

having taken into account that $1/\pi < \xi < 2/\pi$.

The following integrals can now be calculated immediately

$$\int_0^\infty e^{-2u/\pi} \frac{\sinh(2\xi u)}{\cosh \xi u}\, du = 2 \int_0^\infty e^{-2u/\pi} \sinh \xi u\, du = \frac{2\pi^2\xi}{4 - \pi^2\xi^2}$$

$$\int_0^\infty e^{-2u/\pi} \frac{\cosh(2\xi u)}{\cosh \xi u}\, du = 2 \int_0^\infty e^{-2u/\pi} \cosh \xi u\, du$$

$$- \int_0^\infty \frac{e^{-2u/\pi}}{\cosh \xi u}\, du = \frac{4\pi}{4 - \pi^2\xi^2} - \int_0^\infty \frac{e^{-2u/\pi}}{\cosh \xi u}\, du.$$

This last integral is expressed by the logarithmic derived function of $\Gamma(x)$. Then by introducing the functions

$$\Psi(x) = \frac{d}{dx} \log \Gamma(x), \quad \beta(x) = \frac{1}{2}\left(\Psi(\frac{x+1}{2}) - \Psi(\frac{x}{2}) \right), \tag{4.13}$$

we have (see, for instançe, the tables of integrals of I. S. Gradshteyn-I. M. Ryzhik, Moscou, 1962, p. 961)

$$\int_0^\infty \frac{e^{-2u/\pi}}{\cosh \xi u}\, du = \frac{1}{\xi} \beta(\frac{1}{2} + \frac{1}{\pi\xi}).$$

Consequently, on bringing together all the results obtained, we have

$$(4.14) \qquad E(A^2) = \frac{2\pi^3\xi^2}{\pi^2\xi^2 - 1}\left[-1 + \frac{12\xi}{4 - \pi^2\xi^2} - \frac{2}{\pi}\beta(\frac{1}{2} + \frac{1}{\pi\xi})\right].$$

In order to calculate

$$\beta(\frac{1}{2} + \frac{1}{\pi\xi}),$$

we apply (4.13) and employ the Jahnke-Emde tables (p.16). Since $\xi = 0,3726...$, we have

$$x = \frac{1}{2} + \frac{1}{\pi\xi} = 1,354...$$

In order to apply the Jahnke-Emde tables we employ the relation

$$\Psi(x) = \Psi(x-1) + \frac{1}{x},$$

and therefore

$$\Psi(\frac{x+1}{2}) = \Psi(1,177) = \Psi(0,177) + 0,849 = 0,625$$

$$\Psi(x/2) = \Psi(0,677) = 0,190$$

$$\beta(x) = \frac{1}{2}\left(\Psi(\frac{1}{2}(x+1) - \Psi(\frac{1}{2}x))\right) = 0.218...$$

Substituting these values in (4.14), and performing the remaining calculations, we arrive at

$$E(A^2) = 13,02...$$

This is the final result, which is considerably larger than in the Euclidean case.

Analogously to the Euclidean case, if we consider the integral

$$\int f(a)\, dP_1\, dP_2\, dG_1 \ldots dG_n,$$

we arrive at the general result

$$E(\int f(a)\, dP_1\, dP_2) = 4\int_0^\infty e^{-a/\pi\xi} f(a) \sinh a \arccos(\tanh\frac{a}{2})\, da,$$

which generalizes P. I. Richards' formula (3.12) to the Hyperbolic plane.

References

[1] H. S. M. Coxeter, *Non-euclidean Geometry*, University of Toronto Press, 1957, 3a. ed. Toronto.

[2] S. A. Goudsmit, *Random distribution of lines in a plane*, Rev. Mod. Phys. **17** (1945), 321–322.

[3] M. G. Kendall and P. A. P. Moran, *Geometrical probability*, Griffin's statistical Monographs & courses, Hafner Publishing Com., New York, 1963.

[4] R. E. Miles, *Random polygons determined by random lines in a plane*, Proc. Nat. Acad. Sciences **52** (1964), 901–907.

[5] ———, *Random polygons determined by random lines in a plane II*, Proc. Nat. Acad. Sciences **52** (1964), 1157–1160.

[6] P. I. Richards, *Averages for formed by random lines*, Proc. Nat. Acad. Sciences **52** (1964), 1160–1164.

[7] L. A. Santaló, *Valor medio del número de partes en que una figura convexa es dividida por rectas arbitrarias*, Rev. Unión Mat. Argentina **VII** (1941), 33–37.

[8] ———, *Sobre la distribución de planos en el espacio*, Rev. Unión Mat. Argentina **XIII** (1948), 120–124.

[9] L. A. Santaló, *Integral geometry on surfaces of constant negative curvature*, Duke Math. J. **10** (1943), 687–709.

[10] ———, *Introduction to integral geometry*, Actualités Sci. Ind., no. 1198, Publ. Inst. Math. Univ. Nancago II. Herman et Cie, Paris, 1953.

SANTALÓ, L. A.
1975
Annali di Matematica pura ed applicata
(IV), Vol. CIII, pp. 71-79

The Kinematic Formula in Integral Geometry for Cylinders.

L. A. SANTALÓ (Buenos Aires, Argentina)

(Dedicated to Professor BENIAMINO SEGRE on the occasion of his 70-th birthday)

Summary. – *We generalize the kinematic formula of Chern-Federer* (1.2) *to the case in which the moving manifold M^q is a cylinder in E^n. These cylinders and the corresponding kinematic density are suitable defined and some particular cases are considered in detail.*

1. – Introduction.

This paper will be concerned with the so called « kinematic formula » in Integral Geometry, due to FEDERER [2] and CHERN [1]. We shall refer mainly to the work of Chern, which likely assumes some more restrictive conditions than Federer, but remains into the mark of differential geometry. The approach of Federer is more in the mark of measure theory. The formula to which we refer is the following (CHERN [1]).

Let M^p, M^q be a pair of orientable, compact, differentiable manifolds (without boundary) of dimensions p, q immersed in euclidean space E^n. Let dg denote the kinematic density (= Haar measure of the group of motions in E^n) so normalized that the measure of all positions about a point is $O_{n-1}\, O_{n-2} \ldots O_1$ where

$$O_i = \frac{2\pi^{(i+1)/2}}{\Gamma((i+1)/2)} \tag{1.1}$$

is the volume of the i-dimensional unit sphere. Assume M^p fixed and M^q moving with the kinematic density dg. Let $\mu_e(X^k)$ $(0 \leqslant e \leqslant k)$ be the integral invariants (we call them Weyl's curvatures) of the riemannian k-dimensional manifold X^k to be defined below. Then the kinematic formula of Chern-Federer writes

$$\int \mu_e(M^p \cap gM^q)\, dg = \sum_{0 \leqslant i \leqslant e} c_i \mu_i(M^p) \mu_{e-i}(M^q)\,, \qquad i \text{ even} \tag{1.2}$$

where e even and $0 \leqslant e \leqslant p+q-n$. The integral on the left is over the whole

(*) Entrata in Redazione il 22 maggio 1973.

group of euclidean motions in E^n, i.e. over all positions of M^q, and $c_i = c_i(n, p, q, e)$ are numerical constants depending on n, p, q, e which may be calculated as follows. Put

$$c_{e-i} = \frac{O_n \dots O_1 O_{p+q-n+2} O_{q+1-i} O_{p+1-e+i}}{O_p O_{p+1} O_q O_{q+1} O_{p+q-n+2-i} O_{p+q-n+2-e+i}} b_{e,p+q-n+1-i} \tag{1.3}$$

where the b's are given by the following identity (with respect to the indeterminate x) with $m = p + q - n + 2$,

$$\frac{2 O_{m-1} O_{m-2}}{O_{m-e-1}} x^{m-e-1} \sum \frac{(e/2)!}{(2\lambda)!\mu!(e/2 - 2\lambda - \mu)!} 2^{2\lambda} \frac{O_{2\lambda+m-e}}{O_{2\lambda}} x^{2\lambda+2\mu} = b_{e,m-e-1} x^{m-e-1} + \dots + b_{e,m-1} x^{m-1} \tag{1.4}$$

where the sum on the left side is over the following range of indices

$$0 \leqslant 2\lambda + \mu \leqslant e/2\,, \quad 0 \leqslant \lambda, \mu\,.$$

For instance, for $i = 0$, (1.3) gives

$$c_e = \frac{O_n \dots O_1 O_{p+1-e}}{O_{p+1} O_p O_q O_{p+q-n+2-e}} b_{e,p+q-n+1} \tag{1.5}$$

and identifying the coefficients of x^{m-1} of both sides of (1.4) we have (since the relations $0 \leqslant 2\lambda + \mu \leqslant e/2$, $2\lambda + 2\mu = e$ give $\lambda = 0$, $\mu = e/2$)

$$b_{e,m-1} = \frac{O_{m-1} O_{m-2} O_{m-e}}{O_{m-e-1}}\,.$$

Hence

$$b_{e,p+q-n+1} = \frac{O_{p+q-n+1} O_{p+q-n} O_{p+q-n+2-e}}{O_{p+q-n+1-e}}$$

and thus

$$c_e = \frac{O_n \dots O_1 O_{p+1-e} O_{p+q-n+1} O_{p+q-n}}{O_{p+1} O_p O_q O_{p+q-n+1-e}}\,. \tag{1.6}$$

In particular we have

$$c_0(n, p, q, 0) = \frac{O_n \dots O_1 O_{p+q-n}}{O_p O_q}\,. \tag{1.7}$$

In like manner, if we put $i = e$ in (1.3), making use of (1.4), we easily get

$$c_0(n, p, q, e) = \frac{O_n \dots O_1 O_{q+1-e} O_{p+q-n+1} O_{p+q-n}}{O_p O_q O_{q+1} O_{p+q-n+1-e}}\,. \tag{1.8}$$

Our purpose is to extend (1.2) to the case in which M^q is a cylinder $Z_{h,m}$ in E^n. In this case, the kinematic density must be replaced by the density $dZ_{h,m}$ for cylinders, which we will define in section 3. The result is the formula (5.2) which contains as special or limiting cases many formulas in integral geometry in E^n. We consider with detail some of these particular formulas in section 6.

2. – The Weyl's curvatures.

We will define the curvatures $\mu_e(X^k)$ which appear in the kinematic formula (1.2) (see WEYL [6], CHERN [1], FEDERER [2]).

Let X^k be a differentiable riemannian manifold of dimension k and consider the classical differential forms ω_α, $\omega_{\alpha\beta}$ $(1 \leqslant \alpha, \beta, \gamma, \delta \leqslant k)$ of the « moving frames » method, such that

$$\omega_{\alpha\beta} + \omega_{\beta\alpha} = 0 \qquad d\omega_\alpha = \sum_\beta \omega_\beta \wedge \omega_{\beta\alpha} \qquad d\omega_{\alpha\beta} = \sum_\gamma \omega_{\alpha\gamma}\wedge\omega_{\gamma\beta} + \Omega_{\alpha\beta} \tag{2.1}$$

where

$$\Omega_{\alpha\beta} = \tfrac{1}{2}\sum_{\gamma,\delta} S_{\alpha\beta\gamma\delta}\,\omega_\gamma\wedge\omega_\delta \tag{2.2}$$

The coefficients $S_{\alpha\beta\gamma\delta}$ are essentially (though not exactly) the components of the Riemann-Christoffel tensor and have the same well known symmetry properties

$$\begin{cases} S_{\alpha\beta\gamma\delta} = -S_{\alpha\beta\delta\gamma} = -S_{\beta\alpha\gamma\delta} \\ S_{\alpha\beta\gamma\delta} = S_{\gamma\delta\alpha\beta}\,, \qquad S_{\alpha\beta\gamma\delta} + S_{\alpha\gamma\delta\beta} + S_{\alpha\delta\beta\gamma} = 0 \end{cases} \tag{2.3}$$

Put

$$I_e = \frac{(-1)^{e/2}(k-e)!}{2^{e/2}k!}\sum \delta^{\alpha_1\alpha_2\ldots\alpha_e}_{\beta_1\beta_2\ldots\beta_e} S_{\alpha_1\alpha_2\beta_1\beta_2}\ldots S_{\alpha_{e-1}\alpha_e\beta_{e-1}\beta_e} \tag{2.4}$$

where e is an even integer satisfying $0 \leqslant e \leqslant k$ and $\delta^{\alpha_1\ldots\alpha_e}_{\beta_1\ldots\beta_e}$ is equal to $+1$ or -1 according as $\alpha_1, \ldots, \alpha_e$ is an even or odd permutation of $\beta_1, \ldots, \beta_e$ and is otherwise zero, and the summation is taken over all $\alpha_1, \ldots, \alpha_e$ and $\beta_1, \ldots, \beta_e$ independently from 1 to k. When X^k is oriented and compact, we let

$$\mu_e(X^k) = \int_{X^k} I_e\, d\sigma_k \tag{2.5}$$

where $d\sigma_k$ is the volume element. This formula (2.5) defines the Weyl's curvatures (e even, $0 \leqslant e \leqslant k$). In particular we have

$$\mu_0(X^k) = \text{total volume of } X^k \tag{2.6}$$

and, if k is even,

$$\mu_k(X^k) = \tfrac{1}{2}\, O_k\, \chi(X^k) \tag{2.7}$$

where $\chi(X^k)$ denotes the Euler-Poincaré characteristic of X^k. (2.7) is the Gauss-Bonnet formula for compact even dimensional manifolds.

It would be of interest to compare these curvatures $\mu_e(X^k)$ with other curvatures which appear in the literature. For instance, if X^k is the boundary of a bounded convex set of E^{k+1} the volume $V(\varrho)$ of the parallel set to X^k at distance ϱ is (HADWIGER [3])

$$V(\varrho) = \sum_{i=0}^{k+1} \binom{k+1}{i} W_i \varrho^i \tag{2.8}$$

and the volume of the «tube» at distance ϱ is

$$V(\varrho) - V(-\varrho) = 2 \sum_{i=0}^{k+1} \binom{k+1}{i} W_i \varrho^i\,, \qquad i \text{ odd}. \tag{2.9}$$

The invariants $W_i(X^k)$ (quermassintegrale, introduced by Minkowski) may be written

$$W_i = \frac{1}{k+1} M_{i-1} \tag{2.10}$$

where M_i $(i = 0, 1, 2, \ldots, k)$ are the i-th integrated mean curvatures

$$M_i = \frac{1}{\binom{k+1}{i}} \int_{X_k} \left\{ \frac{1}{R_{s_1}} \cdots \frac{1}{R_{s_i}} \right\} d\sigma_k \tag{2.11}$$

where $d\sigma_k$ is the volume element of X^k and $\{1/R_{s_1}, \ldots, 1/R_{s_i}\}$ is the i-th elementary symmetric function of the principal curvatures of X^k. Comparing (2.9) with the Weyl's formula for the volume of tubes [6], we get

$$\mu_e = M_e\,, \qquad e \text{ even}. \tag{2.12}$$

This formula holds for smooth compact hypersurfaces X^k of E^{k+1}, not necessarily convex. We deduce that, for e even, the mean curvatures (2.11) are isometric invariants of X^k which do not depend of its immersion in E^{k+1}, i.e. are intrinsic invariants.

3. – Density for cylinders.

Let M^h be an orientable, compact, differentiable manifold (without boundary) which belongs to a $(n-m)$-dimensional linear space E^{n-m} in E^n. Thus

$$h + m < n\,. \tag{3.1}$$

Through each point $x \in M^h$ we consider the m-dimensional linear space E^m perpendicular to E^{n-m}. The set of all these E^m is called a cylinder $Z_{h,m}$ of dimension $h+m$, whose generators (or generating m-spaces) are the m-spaces E^m and which orthogonal cross section is the manifold M^h.

If we assume $Z_{h,m}$ moving in E^n, its position may determined by a $E^{n-m}(0)$ through a fixed point 0, orthogonal to the geneators E^m, and the position of the cross section M^h in $E^{n-m}(0)$. The density $dE^{n-m}(0)$ (= volume element of the grassmann manifold $G_{n-m,m}$ of all $(n-m)$-planes through 0 in E^n) and dg_{n-m} (kinematic density in E^{n-m}) are well known (see, for instance SANTALÓ [4], [5], CHERN [1], HADWIGER [3]). The density $dZ_{h,m}$ for cylinders $Z_{h,m}$ is then

$$dZ_{h,m} = dE^{n-m}(0) \wedge dg_{n-m} . \tag{3.2}$$

We recall these densities for completeness. If $(x;\, e_1, e_2, \ldots, e_n)$ is an orthogormal frame in E^n and we put

$$\omega_i = (dx \cdot e_i)\,, \qquad \omega_{ih} = de_i \cdot e_h \tag{3.3}$$

then

$$dg_n = \Lambda\omega_i \Lambda\omega_{jk} \tag{3.4}$$

where the exterior products are between the ranges

$$i = 1, 2, \ldots, n\,; \qquad j = 2, 3, \ldots, n\,; \qquad k = 1, 2, \ldots, n-1 \tag{3.5}$$

with

$$j > k\,.$$

The differential form

$$d\sigma_n = \omega_1 \wedge \omega_2 \wedge \ldots \wedge \omega_n \tag{3.6}$$

is the volume element in E^n. If E^{n-m} is spanned by $e_{m+1}, e_{m+2}, \ldots, e_n$ we have

$$dg_{n-m} = \omega_{m+1} \wedge \ldots \wedge \omega_n \Lambda\omega_{jk} \tag{3.7}$$

where

$$j = m+2, \ldots, n\,; \quad k = m+1, \ldots, n-1\,, \quad k < j$$

and if E^m is spanned by $e_1, e_2, \ldots, e_m$ we have

$$dg_m = \omega_1 \wedge \ldots \wedge \omega_m \Lambda\omega_{jk} \tag{3.8}$$

where

$$j = 2, 3, \ldots, m\,; \quad k = 1, 2, \ldots, m-1\,, \quad k < j\,.$$

Finally, assuming $E^{n-m}(0)$ parallel to E^{n-m}, we have

$$dE^{n-m}(0) = \Lambda\omega_{jk} \tag{3.9}$$

where

$$j = m+1, \ldots, n, \quad k = 1, 2, \ldots, m,$$

From (3.4), (3.2), (3.7) and (3.8) we deduce

$$dg = dg_n = dE^{n-m}(0) \wedge dg_{n-m} \wedge dg_m = dZ_{h,m} \wedge dg_m . \tag{3.10}$$

The exterior products in (3.8) and (3.7) have a clear geometrical meaning. Indeed we have

$$\omega_1 \wedge \ldots \wedge \omega_m = d\sigma_m = \text{volume element in } E^m , \tag{3.11}$$

$$\omega_{m+1} \wedge \ldots \wedge \omega_n = d\sigma_{n-m} = \text{volume element in } E^{n-m} . \tag{3.12}$$

$$\Lambda\omega_{jk}(j = 2, \ldots, m;\ k = 1, \ldots, m-1;\ k<j) = dO_{m-1} \wedge dO_{m-2} \wedge \ldots \wedge dO_1 \tag{3.13}$$

where dO_i denotes the area element on the unit i-dimensional sphere in the space spanned by $e_1, e_2, \ldots, e_{i+1}$, and

$$\Lambda\omega_{jk}(j = m+2, \ldots, n;\ k = m+1, \ldots, n-1;\ k<j) = dO_{n-m-1} \wedge \ldots \wedge dO_1 \tag{3.14}$$

where dO_i is now the area element on the unit sphere in the $(i+1)$-space spanned by $e_{m+1}, \ldots, e_{m+i+1}$.

The density dE^m for the generating m-spaces E^m writes

$$dE^m = dE^{n-m}(0) \wedge d\sigma_{n-m} \tag{3.15}$$

where $d\sigma_{n-m}$ is the element of volume in $E^{n-m}(0)$ at the intersection point $E^m \cap E^{n-m}(0)$. Having into account (3.2) and (3.7) we have

$$dZ_{h,m} = dE^m \wedge dO_{n-m-1} \wedge \ldots \wedge dO_1 . \tag{3.16}$$

We always consider the densities in absolute value, so that the order of the exterior products above is immaterial.

4. – The Weyl's curvatures for cylinders.

Choose the frame $(x; e_1, e_2, \ldots, e_n)$ such that $x \in Z_{h,m}$ and $e_1, e_2, \ldots, e_{m+h}$ spann the tangent space to $Z_{h,m}$ in such a way that $e_1, e_2, \ldots, e_m$ spann the generator E_m

through x and $e_{m+1}, \ldots, e_{m+h}$ spann the tangent space to the cross section M^h. The volume elements in E^m, M^h and $Z_{h,m}$ are, respectively

$$(4.1) \qquad d\sigma_m = \omega_1 \wedge \ldots \wedge \omega_m\,, \qquad d\sigma_h = \omega_{m+1} \wedge \ldots \wedge \omega_{m+h}\,, \qquad d\sigma_{m+h} = \omega_1 \wedge \ldots \wedge \omega_{m+h}$$

and we have

$$(4.2) \qquad d\sigma_{m+h} = d\sigma_m \wedge d\sigma_h\,.$$

Since all E^m are perpendicular to E^{n-m}, we have e_α = constant for $\alpha = 1, 2, \ldots, m$ and thus $\omega_{\alpha k} = de_\alpha \cdot e_k = 0$ $(\alpha = 1, \ldots, m;\ k = 1, 2, \ldots, m+h)$. Therefore $d\omega_{\alpha k} = \sum \omega_{\alpha i} \wedge \wedge \omega_{ik} + \Omega_{\alpha k} = 0$ and consequently $\Omega_{\alpha k} = 0$. Therefore, applying (2.2) to the cylinder $Z_{h,m}$ we have $S_{\alpha k i j} = 0$ $(\alpha = 1, 2, \ldots, m;\ k,\ i,\ j = 1, 2, \ldots, m+h)$. According to the symmetry properties (2.3) of $S_{\alpha k i j}$ the equation $S_{\alpha k i j} = 0$ implies that $S_{k \alpha i j} = S_{k i \alpha j} = S_{k i j \alpha} = 0$. The remaining $S_{k i j s}$ with $i, j, k, s = m+1, m+2, \ldots, m+h$ are the functions $S_{k i j s}$ corresponding to the cross section M^h.

Therefore, the sums on the right side of (2.4) are the same for $Z_{h,m}$ and for M^h and an easy calculation gives

$$(4.3) \qquad \begin{aligned} I_e(Z_{h,m}) &= \frac{\binom{h}{e}}{\binom{h+m}{e}}\, I_e(M^h) \qquad &\text{if } e < h\,,\\ I_e(Z_{h,m}) &= 0 &\text{if } e > h\,. \end{aligned}$$

5. – The kinematic formula for cylinders.

Let E^m be a generator of $Z_{h,m}$ and consider a bounded domain $D^m \subset E^m$. Assume that $D^m = D^m(t)$ depends on a parameter t in such a way that $D^m \to E^m$ when $t \to \infty$. Consider the compact manifold $D^m \times M^h$. If $d\sigma_m$, $d\sigma_h$, $d\sigma_{h+m}$ denote respectively the volume elements in $D^m(t)$, M^h and $Z_{h,m}$ we have $d\sigma_{h+m} = d\sigma_m \wedge d\sigma_h$ and from (2.5) and (4.3) we get

$$(5.1) \qquad \begin{aligned} \mu_e(Z_{h,m}) &= \frac{\binom{h}{e}}{\binom{h+m}{e}}\, \mu_e(M^h)\sigma_m \qquad &\text{if } e < h\,,\\ \mu_e(Z_{h,m}) &= 0 &\text{if } e > h\,. \end{aligned}$$

where σ_m denotes the volume of D^m.

We now apply formula (1.2) to M^p and $M^q = D^m \times M^h$. Using (3.10) having into account that $dg_m = d\sigma_m \wedge dO_{m-1} \wedge \ldots \wedge dO_1$ and making $t \to \infty$ (after division of both sides by σ_m), so that $D^m \to E^m$ we get the desired formula

$$(5.2) \qquad \int\limits_{Z_{h,m}\cap M^p\neq\phi} \mu_e(M^p \cap Z_{h,m})\, dZ_{h,m} = \sum_{e-h\leqslant i\leqslant e} \frac{c_i}{O_1 O_2 \dots O_{m-1}} \frac{\binom{h}{e-i}}{\binom{h+m}{e-i}} \mu_i(M^p)\mu_{e-i}(M^h)$$

(e even, $0\leqslant e\leqslant p+h+m-n$, $i\geqslant 0$, i even)

where $c_i = c_i(n, p, h+m, e)$ are the same constants as in Chern's formula (1.2).

6. – Particular cases.

1) Assume that $Z_{h,m}$ reduces to a m-plane E^m ($h=0$). Then, according to (3.16) we have $dZ_{h,m} = dE^m \wedge dO_{n-m-1} \wedge \dots \wedge dO_1$ and $\mu_0(M^0) = 1$. The sum on the right side of (5.2) reduces to the term $i = e$ and according to (1.6) we get (e even, $e \leqslant p+m-n$)

$$(6.1) \qquad \int\limits_{E^m\cap M^p\neq\phi} \mu_e(M^p \cap E^m)\, dE^m = \frac{O_{n-m} \dots O_n\, O_{p+1-e} O_{p+m-n} O_{p+m-n+1}}{O_p O_{p+1} O_{p+m-n+1-e} O_1 \dots O_m} \mu_e(M^p)\,.$$

This formula is due to Chern [1]. For $p = n-1$, see [4], [5].

2) Consider the case $e = p+m-n$. According to (2.7) we have

$$\mu_{p+m-n}(M^p \cap E^m) = \tfrac{1}{2} O_{p+m-n}\, \chi(M^p \cap E^m)$$

and (6.1) gives ($p+m-n$ even)

$$(6.2) \qquad \int\limits_{E^m\cap M^p\neq\phi} \chi(M^p \cap E^m)\, dE^m = \frac{2 O_{n-m} \dots O_n\, O_{n-m+1} O_{p+m-n+1}}{O_1 O_p O_{p+1} O_1 \dots O_m} \mu_{p+m-n}(M^p)\,.$$

3) The case $e = 0$. Applying (1.7), from (5.2) we deduce

$$(6.3) \qquad \int\limits_{Z_{h,m}\cap M^p\neq\phi} \mu_0(M^p \cap Z_{h,m})\, dZ_{h,m} = \frac{O_n \dots O_m\, O_{p+h+m-n}}{O_p O_{h+m}} \mu_0(M^p)\mu_0(M^h)\,.$$

If $p+h+m-n = 0$, then μ_0 is equal to the number ν of intersection points of M^p and $Z_{h,m}$, and (6.3) gives

$$(6.4) \qquad \int \nu(M^p \cap Z_{h,m})\, dZ_{h,m} = \frac{2 O_n \dots O_m}{O_p O_{n-p}} \mu_0(M^p)\mu_0(M^h)$$

and (6.1) gives

$$\int \nu(M^p \cap E^m)\, dE^m = \frac{2O_{n-m} \dots O_n}{O_p O_1 \dots O_m} \mu_0(M^p) . \tag{6.5}$$

The integrals in (6.4) and (6.5) are extended over all positions of $Z_{h,m}$ and E^m, ν being zero if they do not intersect the manifold M^p.

4) As last example, consider the case

$$p + h + m - n = 2 , \quad e = 2 .$$

We have, by (2.7)

$$\mu_2(M^p \cap Z_{h,m}) = \tfrac{1}{2} O_2\, \chi(M^p \cap Z_{h,m})$$

From (1.3), (1.4), having into account that $q = h + m$, $p + q - n = 2$, we have

$$c_2(n, p, h + m, 2) = \frac{O_n \dots O_1\, O_{p-1} O_3 O_2}{O_p O_{p+1} O_{h+m} O_1} . \tag{6.6}$$

Using (6.6), (1.8) and the identity $O_1 O_{i-2} = (i-1)\, O_i$, formula (5.2) gives

$$\int_{M^p \cap Z_{h,m} \neq \phi} \chi(M^p \cap Z_{h,m})\, dZ_{h,m} = \frac{2O_n \dots O_m\, O_3}{O_p O_{h+m} O_1} \cdot \left[\frac{h(h-1)}{2\pi(h+m-1)} \mu_0(M^p) \mu_2(M^h) + \frac{O_{p-1}}{O_{p+1}} \mu_2(M^p) \mu_0(M^h)\right] . \tag{6.7}$$

REFERENCES

[1] S. S. CHERN, *On the kinematic formula in Integral Geometry*, J. Math. and Mech., **16** (1966), pp. 101-118.

[2] H. FEDERER, *Curvature Measures*, Trans. Amer. Math. Soc., **93** (1959), pp. 418-491.

[3] H. HADWIGER, *Vorlesungen über Inhalt, Oberfläche und Isoperimetrie*, Springer, Berlin, 1957.

[4] L. A. SANTALÓ, *Geometria Integral en espacios de curvatura constante*, Publ. Com. Nac. Energia Atòmica, Serie Mat., vol. 1, No. 1, Buenos Aires, 1952.

[5] L. A. SANTALÓ, *Sur la mesure des éspaces linéaires qui coupent un corps convexe et problemes qui s'y rattachent*, Colloque sur les questions de réalité en Géometrie, Liege, 1955, pp. 177-190.

[6] H. WEYL, *On the volume of tubes*, Amer. J. Math., **61** (1939), pp. 461-472.

Sets of segments on surfaces

L. A. Santaló

Abstract

The sets of segments randomly distributed on the plane have been recently investigated by Coleman [1], [2], Parker-Cowan [3], Santaló [9] among others. In the present paper, we consider some questions analogous to that of Parker-Coleman for sets of "geodesic segments" on surfaces, in particular, sets of segments on the sphere and on the Hyperbolic plane. We obtain the mean values of the total length of the part of segments which are interior to a convex set K and the mean value of the intersection points of pairs of segments that are interior to K. For the Hyperbolic plane, we consider also the case of segments distributed on the plane according to a Poisson process.

Introduction

The study of sets of segments on the plane and in Euclidean space has recently given rise to some works (Coleman [1], [2], Parker-Cowan [3], Santaló [9]). The aim of the present paper is to see how some of the results obtained are valid for sets of *geodesic segments* on surfaces, under certain restrictions. Given a convex set K on a surface, to be defined below, and randomly distributed geodesic segments, in the sense of Integral Geometry (probably uniform), certain problems regarding geometric problems are solved and certain mean values are calculated. In particular, given n randomly distributed segments S_i having a common point K, we calculate the mean value of the sum of the lengths of the parts of S_i that are interior to K (1.12), and the mean value of the number of points N of the intersection of segments S_i interior to K (1.14).

As particular surfaces, cases of the sphere and of the Euclidean and Hyperbolic planes are subsequently considered. For these latter, it is assumed that the number of segments is infinite, so that they extend over the entire plane to form a Poisson process. The above-mentined mean values are also studied in this case.

It is interesting to observe certain differences that appear between the Euclidean plane and the Hyperbolic plane, which suggest the need for carrying out a detailed study on Poisson processes for segments in the latter case.

1 Sets of geodesic segments on a surface

1.1 Density for sets of geodesic segments

Let us assume a 2-dimensional Riemannian manifold, which we refer to simply as a surface. It is known that on this surface a *density* for measuring geodesic sets may be defined. We denote this density by dG^* if it refers to oriented geodesics, and by dG where non-oriented geodesics are concerned. Obviously, $dG^* = 2dG$;

this can be seen in [5] and [6]. For the case of the plane, dG is the classical density of lines in the theory of geometric probabilities.

A geodesic segment, denoted by S^* if it is oriented and by S if it is not, is determined by one of its endpoints P (the origin being oriented) and the angle φ formed by the segment with a fixed direction through P (as well as by length s). It is also determined by the geodesic G^* or G containing the segment and the absciss t of the origin P of the same geodesic. It is then known that in order to measure sets of oriented geodesic segments with constant length s (with certain conditions of invariance) any of the two following expressions can be taken:

$$dS_s^* = dP \wedge d\varphi = dG^* \wedge dt, \tag{1.1}$$

where dP denotes the area element of the surface at P. If the segments are not oriented, it suffices to divide the former expression by 2, or to take dG instead of dG^*.

If the length s of the segment is variable, with density function $F(s)$ such that

$$\int_0^\infty F(s)ds = 1, \qquad \int_0^\infty sF(s) = E(s), \tag{1.2}$$

where $E(s)$ denotes the mean length, the density for sets of oriented geodesic segments is written $dS^* = F(s)dS_s^* \wedge ds$, and for non-oriented segments

$$dS = F(s)\, dG \wedge dt \wedge ds. \tag{1.3}$$

Since the measures and densities are always considered at their absolute value, the order of the differences in these expressions is unimportant.

1.2 Geodesic segments intersecting a convex set

A set K on a surface is said to be *convex* if it is bound and limited by a curve ∂K such that any geodesic whose intersection with K is not empty can have only one common point (supporting geodesic), two common points (secant geodesic) with ∂K, unless it has a whole segment as part of ∂K.

The geodesics on the surface can be closed and thus have finite length (for example, in the case of the sphere). In order for the following results also to hold in this case, it is necessary to impose the condition that *only convex sets K and segments S are considered such that the diameter D of K, the maximum length s_m of S and the minimum length L_G of G fulfil the condition*

$$D + s_m \leqq L_G. \tag{1.4}$$

Obviously, if all the geodesics (except a set of zero measure) have infinite length, and we consider segments themselves (that is, $s_m < \infty$), this condition (1.4) is always fulfilled.

With this condition, if σ denotes the length of the chord that the geodesic G determines in K, the following formulae hold (see [5] or [6]),

$$\mu(G; G \cap K \neq \emptyset) = L, \qquad \int \sigma dG = \pi F, \tag{1.5}$$

where μ denotes the measure and with density dG, so that the first equality is the value of the measure of the geodesics having a common point with K, L is the length of ∂K, and F is the area of K. The second integral can be considered as extended to all the geodesics on the surface, since for those that do not intersect K, $\sigma = 0$.

To measure the set of segments S having a common point with K, by applying (1.3), (1.2) and (1.5), we have

$$\mu(S; S \cap K \neq \emptyset) = \int F(s)(s+\sigma)ds \wedge dG = E(s)L + \pi F. \tag{1.6}$$

If K is reduced to a geodesic segment of length α, then considering it as a flattened convex set, $F = 0$, $L = 2\alpha$, and therefore the measure of the set of segments S intersecting a fixed segment S_α of length α is

$$\mu(S; S \cap S_\alpha \neq \emptyset) = 2\alpha E(s). \tag{1.7}$$

Once again, let K be any convex set on the surface, and let S be the intersection $S \cap K$ (the length of α being S_α). Consider the set of pairs (G, S) (geodesic, segment) such that G intersects S_α. The integral

$$I = \int dG\, dS \tag{1.8}$$

extended to this set of pairs, first fixing S and integrating dG, on applying the first formula (1.5) is $I = 2\int \alpha dS$. However, by first fixing G, if σ is the length of the chord $G \cap K$, according to (1.7), we have $I = 2\int \sigma E(s)dG = 2\pi F E(s)$. Therefore, we obtain

$$\int \alpha dS = \pi F E(s). \tag{1.9}$$

Here, too, the integral on the left-hand side can be considered as being extended to all the segments on the surface, since if S does not intersect K we have $\alpha = 0$.

Now let us consider two segments S_1, S_2 intersecting K, and let n_{12} be the function that is equal to 1 if S_1, S_2 intersect inside K, and which is equal to 0 in the opposite case. We wish to calculate the integral

$$J = \int n_{12} dS_1 \wedge dS_2 \tag{1.10}$$

extended to all the segments S_1, S_2 on the surface.

By first fixing S_1 and denoting the length of that part of it interior to α_1 by K, according to (1.7) this integral is

$$J = 2\int \alpha_1 E(s_2) dS_1,$$

and by applying (1.9) we obtain

$$J = \int n_{12} dS_1 \wedge dS_2 = 2\pi F E(s_1) E(s_2), \tag{1.11}$$

where $E(s_i)$ denotes the mean length of the segments S_i.

From these results we may deduce:

a) Given n randomly distributed geodesic segments $S_1, S_2, \ldots, S_n$ intersecting a convex set K, the mean value of the sum of the lengths $\alpha_1, \alpha_2, \ldots, \alpha_n$ and their intersections with K is

$$E\left(\left|\sum \alpha_i\right|\right) = \pi F \sum_1^n \frac{E(s_i)}{\pi F + LE(s_i)}. \tag{1.12}$$

In particular, if all the segments have the same mean length $E(s)$, we obtain

$$E\left(\sum \alpha_i\right) = \frac{\pi F n E(s)}{\pi F + LE(s)}. \tag{1.13}$$

b) Given n randomly distributed geodesic segments $S_1, S_2, \ldots, S_n$ intersecting a convex set K, the mean value of the number N of intersection points of the segments S_i, which are interior to K (for example, in figure 1, $N = 6$), is

$$E(N) = 2\pi F \sum_{i<j} \frac{E(s_i)E(s_j)}{(\pi F + LE(s_i))(\pi F + LE(s))}.$$

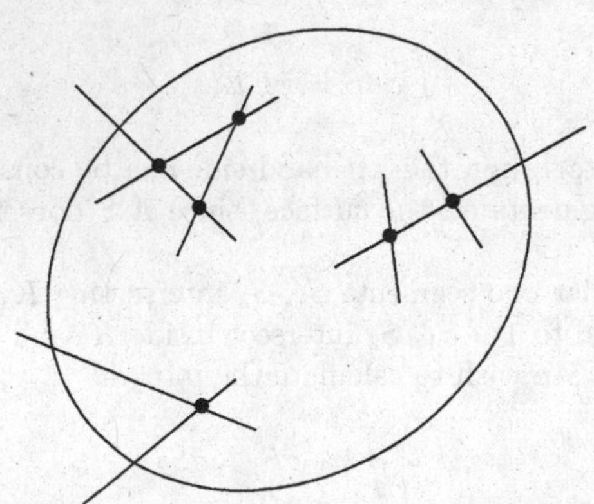

Figure 1

If all the segments have the same mean length E (s), we have

$$E(N) = \frac{n(n-1)\pi F[E(s)]^2}{(\pi F + LE(s))^2}. \tag{1.14}$$

2 Sets of segments on the sphere

2.1 The case of the unit sphere

The foregoing holds for geodesic segments on any surface, providing they fulfil the said conditions for K and s. Let us consider, in particular, the case of the unit sphere.

In this case, the set of all the segments can be considered, and not only those intersecting K. Their measure (non-oriented segments) is

$$\mu(S) = \int F(s)ds \wedge dP \wedge d\varphi = 4\pi^2. \tag{2.1}$$

Hence, and from (1.6), the solution to the following problems of geometric probabilities can be deduced:

a) If K is a fixed convex set on the unit sphere and S a segment given randomly on the unit sphere (always with density (1.3)), the probability of S having a point in common with K is

$$p(S \cap K \neq \emptyset) = \frac{\pi F + LE(s)}{4\pi^2}. \tag{2.2}$$

Condition (1.4) is assumed to be fulfilled, which in this case is written $D + s_m \leq 2\pi$.

From (1.9) and (1.11) one may also deduce that

b) If K is a fixed convex set on the unit sphere and n maximum circle segments are randomly distributed on the unit sphere, the mean value of the sum of the lengths of the parts of these segments that are interior to K is

$$E\left(\sum \alpha_i\right) = \frac{F}{4\pi} \sum_1^n E(s_i). \tag{2.3}$$

If all the segments have the same mean length $E(s)$, we have

$$E\left(\sum \alpha_i\right) = \frac{nE(s)F}{4\pi}. \tag{2.4}$$

c) In the same conditions as those above, the mean value of the number of intersection points of the n S_i segments that are interior to K is

$$E(N) = \frac{F}{8\pi^3} \sum_{i<j} E(s_i)E(s_j). \tag{2.5}$$

In particular, if all the segments have the same mean length $E(s)$, we have

$$E(N) = \frac{n(n-1)F(E(s))^2}{16\pi^3}. \tag{2.6}$$

It is easy to check that this formula holds without restriction (1.4). In particular, it holds when K is extended to the entire sphere ($F = 4\pi$), then the mean value of the number of intersection points of n randomly distributed segments, independently and according to the density (1.3) on the unit sphere, is

$$E(N) = \frac{n(n-1)(E(s))^2}{4\pi^2}. \tag{2.7}$$

3 The case of the Euclidean and Hyperbolic planes

3.1 The case of unlimited surfaces: Poisson processes

Let us assume the case of an unlimited surface on which the convex set K can be extended to cover any point on the surface without losing its condition of convexity. In this case, the geodesics have infinite length and therefore always fulfil (1.4). To fix ideas, we confine ourselves to the case of the Euclidean and Hyperbolic planes.

Let K be the convex set and consider a further convex set K_0 that contains K. Given a randomly distributed segment S intersecting K_0 , the probability (measure quotient) of it also intersecting K, according to (1.6), is

$$p(S \cap K \neq \emptyset) = \frac{\pi F + E(s)L}{\pi F_0 + E(s)L_0}, \tag{3.1}$$

where F, F_0 are the areas of K, K_0, and L, L_0 are the lengths of the bounds ∂K, ∂K_0 respectively.

Given n segments S_i with the same mean length $E(s)$, which are known to intersect S_0, the probability of exactly m of these segments also intersecting K (binomial law) is

$$p_m = \binom{n}{m}\left(\frac{\pi F + E(s)L}{\pi F_0 + E(s)L_0}\right)^m \left(1 - \frac{\pi F + E(s)L}{\pi F_0 + E(s)L_0}\right)^{n-m}. \tag{3.2}$$

Assume that K_0 is extended in such a way that it covers any point on the plane (Euclidean or Hyperbolic), so that $F_0 \longrightarrow \infty$, $L_0 \longrightarrow \infty$. There now appears a fundamental difference between the planes. If K fulfils certain conditions (h-convexity), for example, if it is a circle whose radius tends to the infinite, for the Euclidean plane we have $L_0/F_0 \longrightarrow 0$, while for the Hyperbolic plane we have $L_0/F_0 \longrightarrow 1$ (see [10]). In order to include both cases in a single formula, we introduce the constant ε such that $\varepsilon = 0$ for the Euclidean case, and $\varepsilon = 1$ for the case of the Hyperbolic plane. We then have

$$\frac{L_0}{F_0} \longrightarrow -\varepsilon. \tag{3.3}$$

Now let us assume that at the same time as $F_0 \longrightarrow \infty$ in the way described above (K_0 tending to cover all the plane), the number of segments intersecting

K_0 also tends to the infinite, such that

$$\frac{n}{F_0} \longrightarrow \lambda, \qquad \lambda = \text{ constant.} \tag{3.4}$$

is fulfilled; that is, λ is the number of segments by a unit of area on the plane.

With this condition, by using the classical pass to the limit to change from the binomial law to Poisson's law, from (3.2) we deduce that for $n \to \infty$

$$p_m \longrightarrow \frac{(\lambda H)^m}{m!} \exp(-\lambda H), \qquad H = \frac{\pi F + E(s)L}{\pi - \varepsilon E(s)}. \tag{3.5}$$

A set of segments distributed over the whole plane (either Euclidean or Hyperbolic) under the foregoing conditions is said to constitute a *Poisson process* of intensity λ. If $E(s) = 0$, then $H = F$ and we have a Poisson process of intensity points λ. The mean value of the number of segments having a common point with a convex test set K distributed randomly on the plane is λH; that is, it depends on λ (the number of segments by unit of area), on $E(s)$ (mean value of the lengths of the process segments), which are characteristics of the process, and also on the area F as well as on the perimeter L of K.

According to Parker and Cowan [3], we may observe that if D_r is the distance from the origin to the r-th closest segment, the probability of D_r being greater than x is equal to the probability of a circle of radius x situated randomly on the plane having no common point with more than $r-1$ segments; that is

$$p(D_r > x) = \sum_{m=0}^{r-1} \frac{(\lambda H)^m}{m!} \exp(-\lambda H),$$

with H given by (3.5), and where

$$F = \pi x^2, \qquad L = 2\pi x$$

for the Euclidean case, and

$$F = \pi(\cosh x - 1), \qquad L = 2\pi \operatorname{senh} x$$

in the case of the Hyperbolic plane.

For $r - 1$, there exists the probability that the distance from a randomly given point to the closest segment is x, and therefore the density function for D_1 will be the derivative with the sign changed; that is, it is equal to

$$2\lambda(\pi x + E(s)) \exp(-\lambda(\pi x^2 + 2xE(s))$$

for the Euclidean case (Parker and Cowan [3]), and

$$2\lambda\pi(\pi x + E(s))^{-1}(\pi \operatorname{senh} x + E(s) \cosh x) \exp(-\lambda H))$$

where it is necessary to write

$$H = \frac{2\pi}{\pi E(s)}(\pi(\cosh x - 1 + E(s) \operatorname{senh} x),$$

for the Hyperbolic plane.

3.2 Mean values

We now see what occurs with the mean values studied in 1.2 when changing to a Poisson process.

Observe that if the convex set K is interior to the convex set K_0, and considering that the segments S_i in 1.2 are segments having a common point with K_0 (but not necessarily with K), formulae (1.13) and (1.14) also hold if the denominators (measures of the set of possible cases) are replaced by $\pi F_0 + E(s)L_0$. Thus we have for the infinite case of K and K_0, which contains K.

$$(3.6) \qquad E\left(\sum \alpha_i\right) = \frac{n\pi F E(s)}{\pi F_0 + E(s)L_0}, \qquad E(N) = \frac{n(n-1)\pi F(E(s))^2}{(\pi F_0 + E(s)L_0)^2}.$$

Assuming that K_0 extends to the whole plane at the same time s as n increases according to expression (3.4), we obtain

$$(3.7) \qquad E\left(\sum \alpha_i\right) \longrightarrow \frac{\lambda \pi F E(s)}{\pi - \varepsilon E(s)}, \qquad E(N) = \frac{\lambda^2 \pi F(E(s))^2}{\pi - \varepsilon E(s))^2},$$

which are the mean values of $\sum \alpha_i$ (sum of the lengths of the parts of the segments of the process that are interior to K) and N (the number of intersection points among segments of the process that are interior to K).

For the case of the Euclidean plane, these values are the same as those already obtained through a perhaps more general approach by Parker and Cowan in [3].

References

[1] R. Coleman, *Sampling procedures for the lengths of random straight lines*, Biometrika **59** (1972), 415–426.

[2] ———, *The distance from a given point to the nearest end of one member of a random process of linear segments*, Stochastic geometry (a tribute to the memory of Rollo Davidson), Wiley, London, 1974, pp. 192–201.

[3] P. Parker and R. Cowan, *Some properties of line segment processes*, J. Appl. Probability **13** (1976), no. 1, 96–107.

[4] L. A. Santaló, *Integral geometry on surfaces of constant negative curvature*, Duke Math. J. **10** (1943), 687–709.

[5] ———, *Integral geometry on surfaces*, Duke Math. J. **16** (1949), 361–375.

[6] ———, *Introduction to integral geometry*, Actualités Sci. Ind., no. 1198, Publ. Inst. Math. Univ. Nancago II. Herman et Cie, Paris, 1953.

[7] ———, *On some geometric inequalities in the style of Fáry*, Amer. J. Math. **91** (1969), 25–31, (to appear when the present paper was written).

[8] ———, *Integral Geometry and Geometric Probability*, Addison-Wesley, 1976, Encyclopedia of mathematics and its applications, Vol. 1.

[9] ———, *Sobre segmentos al azar en E^n*, Rev. Mat. y Fis. teórica, Universidad de Tucumán **26** (1976), 229–238.

[10] L. A. Santaló and I. Yañez, *Averages for polygons formed by random lines in Euclidean and hyperbolic planes*, J. Appl. Probability **9** (1972), 140–157.

PORTUGALIAE MATHEMATICA
Vol. 39 — Fasc. 1-4 — 1980

NOTES ON THE INTEGRAL GEOMETRY IN THE HYPERBOLIC PLANE (*)

BY
L. A. SANTALÓ
Faculdade de Ciências Exactas y Naturales
Universidad de Buenos Aires
Ciudade Universitaria, Buenos Aires
Argentina

ABSTRACT. The Integral Geometry in the hyperbolic plane was initiated, many years ago, in [5]. *Later on, it was applied to the geometry of random mosaics in the hyperbolic plane* [7]. *Im the present work we extend to the hyperbolic plane some new results of the euclidean integral geometry which have been given in recent years for several authors, in particular certain results of H. Hadwiger* [3] *and some formulas of H. J. Firey* [2], *R. Schneider* [8], [9] *and W. Weil* [10] *on the kinematic measure for sets of support figures.*

1. Some elementary remarks and two conjectures

Let K (t) be a family of bounded closed convex sets in the hyperbolic plane, depending on the parameter t $(0 \leqslant t)$ and such that $K(t_1) \subset K(t_2)$ for $t_1 < t_2$.

Let F (t) denote the area and L (t) the perimeter of K (t). The isoperimetric inequality

$$L^2 - 4\pi F - F^2 \geqslant 0 \tag{1.1}$$

is well known, where the equality sign holds if and only if K is a circle [5], [6, p. 324].

(*) Dedicado a la memoria de António A. Monteiro, en recuerdo de sincera amistad, admiración y gratitud por su obra matemática en la Argentina. Received December 20, 1982.

Assume that, for any point P of the plane, there is a value t_P of t such that, for all $t > t_P$, we have $P \in K(t)$. We then say that K (t) expands over the whole plane, as $t \to \infty$.

From (1.1) we deduce

$$\lim_{t \to \infty} \frac{F(t)}{L(t)} \leqslant 1. \tag{1.2}$$

The convex domain K (t) is said to be h-convex, or convex with respect to horocycles, when for each pair of points A,B belonging to the domain, the entire segments of the two horocycles AB also belong to the domain. Any h-convex set is convex, but the converse is not true. If the boundary $\partial K(t)$ is smooth, the necessary and suffcient condition for h-convexity is that the curvature of ∂K (geodesic curvature) satisfies the condition $\chi_g \geqslant 1$. For instance, the circles are all h-convex. The Gauss-Bonnet formula

$$\int_{\partial k} \chi_g \, ds = 2\pi + F \tag{1.3}$$

gives then $\lim_{t \to \infty} (F/L) \geqslant 1$ and hence, taking (1.2) into account, we have that for all h-convex sets which expand to the whole hiperbolic plane, the relation

$$\lim_{t \to \infty} \frac{F(t)}{L(t)} = 1 \tag{1.4}$$

holds [7].

We deduce some elementary consequences (the random lines are assumed to be given with the uniform invariant density of the Integral Gometry [6]):

a) If σ denotes the length of the chord that the random line G determines on the h-convex set K, it is well known that the mean value of σ is [6, p. 312]

$$E(\sigma) = \pi F/L. \tag{1.5}$$

Therefore, according to (1.4), we have: *the mean value of the length* σ *of the chord that a random line* G *determines on any h-convex domain* K (t), *tends to* π *as* K (t) *expands to the whole hyperbolic plane.*

b) Given independently at random two lines G_1, G_2 which intersec with a convex set K, the probability that they meet inside K is known to be $p(G_1 \cap G_2 \in K) = 2\pi F/L^2$. Therefore we have: *the probability that two independent lines wich intersect with a given h-convex set* K, *meet inside* K *tends to* 0 *as* K *expands to the whole hyperbolic plane.*

c) The probability that two independent random lines G_1, G_2 which intersect with a given convex domain K be non secant lines in the hyperbolic plane (i.e. $G_1 \cap G_2 = \emptyset$) is

$$p(G_1 \cap G_2 = \emptyset \mid G_1 \cap K \neq \emptyset,\ G_2 \cap K \neq \emptyset) = \tag{1.6}$$

$$= 1 - \frac{2\pi F}{L^2} - \frac{2}{L^2} \int_{P \notin K} (\omega - \sin \omega)\, dP$$

where P is a point exterior to K and ω is the angle between the support lines of K through P. The proof of (1.6) is straightforward by the same method of the euclidean and elliptic planes [6, pp. 51, 319] for which it is $p = 0$.

If K is a circle, by direct computation it is easy to show that the last term of (1.6) tends to 0 as $L \to \infty$. Thus we have: *the probability that two independent random lines which meet a given circle of radius* R, *be non secant lines of the hyperbolic plane, tends to* 1 *as* $R \to \infty$.

We end this section with two conjectures:

1) The equality (1.4) holds good for any convex domain of the hyperbolic plane which expands to the whole plane, i.e. the conditon of h-convexity is superfluous.

2) The relation $\lim p(G_1 \cap G_2 = \emptyset \mid G_1 \cap K \neq \emptyset, G_2 \cap K \neq \emptyset) = 1$, as K expands to the whole hyperbolic plane, which we have proven for circles, holds good for any convex set K of the hyperbolic plane.

The proof of these conjectures will be interesting in order to fill several gaps in the integral geometry of the hyperbolic plane.

2. Generalization of some formulas of Hadwiger to the Hyperbolic plane

2.1. Let K be a convex set of area F and perimeter L in the hyperbolic plane H^2. The lenght L_ρ and area F_ρ of the set K_ρ parallel to K in the distance ρ, are given by [6, p. 322]

$$(2.1)\qquad \begin{aligned} L_\rho &= (2\pi + F)\sinh\rho + L\cosh\rho \\ F_\rho &= (2\pi + F)\cosh\rho + L\,\text{sihn}\,\rho - 2\pi. \end{aligned}$$

Notice that, if K expands to the whole plane, we have

$$(2.2)\qquad \lim (L_\rho/L) = \exp\rho \quad , \quad \lim (F_\rho/F) = \exp\rho.$$

In the enclidean plane, both limits are equal to 1.

Let f (ρ) be an integrable function such that

$$(2.3)\qquad \int_0^\infty f(\rho)\sinh\rho\, d\rho < \infty, \quad \int_0^\infty f(\rho)\cosh\rho\, d\rho < \infty.$$

A point P exterior to K may be determined by the parameter ρ of the exterior parallel set of K which boundary ∂K_ρ contains P (distance from P to K) and the direction Φ of the tangent line to ∂K_ρ at P. The element of area at P is then $dP = ds_\rho\, d\rho$, where ds_ρ means the arc length of ∂K_ρ at P. Therefore, we have, taking (2.1) into account,

$$\int f(\rho)\, dP = \int_0^\infty f(\rho)\, ds_\rho\, d\rho = \int_0^\infty f(\rho)\,[(2\pi + F)\sinh\rho + L\cosh\rho]\, d\rho$$

where the first integral is extended to the exterior of K.

This formula may be written

$$(2.4)\qquad \int_{H_2} f(\rho)\, dP = f(0)\, F + (2\pi + F)\int_0^\infty f(\rho)\sinh\rho\, d\rho + + L\int_0^\infty f(\rho)\cosh\rho\, d\rho$$

where ρ means the distance from P to K and the integral on the left is extended to the whole hyperbolic plane H^2.

This formula (2.4) generalizes to the hyperbolic plane, an analogous formula given by Hadwiger for the euclidean n-space [3].

EXAMPLE. For $f(\rho) = \exp(-a\rho)$, $a > 1$, we have

$$\int_0^\infty \exp(-a\rho) \sinh \rho \, dP = \frac{1}{a^2-1}, \tag{2.5}$$

$$\int_0^\infty \exp(-a\rho) \cosh \rho \, d\rho = \frac{a}{a^2-1}$$

and therefore we have, for any convex curve K of the hyperbolic plane

$$\int_{H^2} \exp(-a\rho) \, dP = \frac{1}{a^2-1} (2\pi + a L + a^2 F) \quad (a > 1). \tag{2.6}$$

The analogous formula in the euclidean plane reads

$$\int_{E^2} \exp(-a\rho) \, dP = a^{-2} (2\pi + a L + a^2 F). \tag{2.7}$$

2.2. The density for lines in H^2 is $dG = \cosh \rho \, d\rho \, d\varphi$, where ρ is the distance from G to the origin and φ is the angle of the normal to G through the origin with a fixed direction [6, p. 306]. The measure of the set of lines whose distance to a convex set K lies between ρ and $\rho + d\rho$ is $L_{\rho+d\rho} - L_\rho = L'_\rho \, d\rho$. Therefore, according to (2.1), if $f(\rho)$ is any function with the conditions (2.3), we have

$$\int_{H^2_*} f(\rho) \, dG = L f(0) + \int_0^\infty f(\rho) L'_\rho \, d\rho = \tag{2.8}$$

$$= L f(0) + (2\pi + F) \int_0^\infty f(\rho) \cosh \rho \, d\rho + L \int_0^\infty f(\rho) \sinh \rho \, d\rho.$$

For instance, for $f(\rho) = \exp(-a\rho)$, $a > 1$, we have

$$\int_{H^2_*} \exp(-a\rho) \, dG = \frac{a}{a^2-1} (2\pi + F + a L), \quad a > 1 \tag{2.9}$$

where H^2_* means the set of all lines of the hyperbolic plane.

The analogous formula for the euclidean plane, reads

$$\int_{E^2_*} \exp(-a\rho)\, dG = a^{-1}(2\pi + aL). \tag{2.10}$$

2.3. Let K, K_1 be two convex sets in H^2. We know that the measure of the set of congruent sets to K_1 which intersect with K_ρ (exterior parallel set to K in the distance ρ) is [6, p. 321]

$$m(K_\rho \cap K_1 \neq \emptyset) = 2\pi (F_\rho + F_1) + F_\rho F_1 + L_\rho L_1 \tag{2.11}$$

here L_ρ and F_ρ are given by (2.1).

The measure of the sets congruent with K_1 whose distance to K lies between ρ and $\rho + d\rho$, using (2.1) will be

$$\begin{aligned} m_{\rho+d\rho} - m_\rho = m'_\rho d\rho = [&(F \sinh\rho + L\cosh\rho + 2\pi\sinh\rho)(2\pi + \\ &+ F) + (F\cosh\rho + L\sinh\rho + 2\pi\cosh\rho) L]\, d\rho. \end{aligned} \tag{2.12}$$

Therefore, for any function $f(\rho)$ which satisfies the conditions (2.3) we have

$$\begin{aligned} \int f(\rho)\, dK_1 = {} & f(0)\,[2\pi (F + F_1) + F F_1 + L L_1] \\ & + [2\pi (L + L_1) + L F_1 + F L_1] \int_0^\infty f(\rho) \cosh\rho\, d\rho \\ & + [(2\pi + F)(2\pi + F_1) + L L_1] \int_0^\infty f(\rho) \sinh\rho\, d\rho \end{aligned} \tag{2.13}$$

where the integral on the left is extended over all sets congruent to K_1 of the hyperbolic plane.

For euclidean n-spaces, this formula was given by HADWIGER [3]. The classical formula (2.11) corresponds to $f(0) = 1$, $f(\rho) = 0$ for $\rho \neq 0$.

3. Kinematic measures for sets of support figures

W. J. FIREY [2], R. SCHNEIDER [8], [9] and W. WEIL [10] have considered sets of compact convex sets congruent to K_1 which touch a fixed compact convex set K, i.e. such that K_1 and K have no interior points in common and $\partial K \cap \partial K_1$ is not empty. We want to extend their results to the hyperbolic plane.

From (2.11) and (2.1) we deduce

$$\text{(3.1)} \qquad \lim_{\rho \to 0} \rho^{-1} [m (K_\rho \cap K_1 \neq \emptyset) - m (K \cap K_1 \neq \emptyset)] =$$

$$= \left[\frac{d}{d\rho} m (K_\rho \cap K \neq \emptyset) \right]_{\rho=0} = (2\pi + F_1) L + (2\pi + F) L_1.$$

This expression is called the kinematic measure of positions of K_1 which support K.

If β, β_1, are subsets of the respective boundaries of K and K_1, their normal images on the unit circle Ω can be written, respectively

$$\text{(3.1)} \qquad \gamma = \int_\beta K_g \, ds, \qquad \gamma_1 = \int_{\beta_1} K_g^1 \, ds_1$$

where K_g, K_g^1 are the geodesic curvatures and s, s_1 the arc elements on ∂K, ∂K_1.

If l, l_1 are the respective lenghts of the arcs β, β_1, the kinematic measure of positions of K_1 which support K in such a way that β_1 supports β is

$$\text{(3.2)} \qquad m (\beta, \beta_1) = \gamma l_1 + \gamma_1 l.$$

This formula follows from (3.1) noting that the classical formula of Gauss-Bonnet, when $\beta = \partial K$, $\beta_1 = \partial K_1$, gives

$$\gamma = 2\pi + F, \qquad \gamma_1 = 2\pi + F_1.$$

4. Some problems of Geometric Probability

4.1. Consider on the hyperbolic plane H^2 a fixed convex set K. We give at random (uniformly, in the sense of the theory of Geometric Probability) a line-segment S of lenght l. Assume that S touches K. We want the probability that the contact point $\partial K \cap S$ be an end point of S.

According to (3.1) the total measure of positions of S, is

$$\text{(4.1)} \qquad m (\partial K, S) = 2 (2\pi + F) l + 2\pi L$$

and the measure of the positions in which S has an end point touching ∂K corresponds to $l_1 = 0$, i.e. $m(\partial K \cap \text{end point of } S \neq \emptyset) = 2\pi L$. Thus, the probability is

$$(4.2) \qquad p(\partial K \cap \text{end point of} \quad S \neq \emptyset) = \frac{\pi L}{(2\pi + F) l + \pi L}.$$

The probability that ∂K and S touches in an interior point of S will be

$$(4.3) \qquad p(\partial K \cap \text{interior poit of} \quad S \neq \emptyset) = \frac{(2\pi + F) l}{(2\pi + F) l + \pi L}.$$

If K is h-convex and expands to the whole hyperbolic plane, the probabilities (4.2) and (4.3) tend to $\pi/(\pi + l)$ and $l/(\pi + l)$ respectively. In the case of the euclidean plane, these limits are 1 and 0 respectively.

4.2. Let K be a fixed convex set in H^2 and K_1 a moving triangle $T_1 = ABC$ which touches ∂K. We want the probability that the contact point be a vertex of the triangle.

According to (3.1) the total measure of contact positions in which the triangle touches K is

$$m_{\text{total}} = (2\pi + F_1) L + (2\pi + F) L_1$$

where L_1 is the perimeter and F_1 the area of the triangle.

The kinematic measure of the set of positions with a vertex touching ∂K, according to (3.2) will be $(2\pi + F_1) L$ and the requested probability results to be

$$p(\partial K \text{ and T touches in a vertex}) = \frac{(2\pi + F_1) L}{(2\pi + F_1) L + (2\pi + F) L_1}.$$

The probability that ∂K and T touches in a side of T, is the complementary.

This kind of problems, in euclidean 3-space, were initiated by P. Mc Mullen (4].

5. Line-segment processes in the hyperbolic plane

5.1. The measure of the set of oriented line-segments $S^1 = 0A$ of fixed lenght l_1 which intersect with a given convex set K, according to (2.11) is

$$m(K \cap S_1 \neq \emptyset) = 2\pi F + 2 l_1 L \tag{5.1}$$

and the measure of the oriented line-segment whose origin is interior to K is

$$m(S_1 \mid 0 \in K) = 2\pi F. \tag{5.2}$$

The formulas (5.1) and (5.2) are the same for the hyperbolic and for the euclidean plane [6, pp. 90, 321]. Therefore, the probability that a random segment S_1 intersecting with K have its origin inside K, is

$$p(0 \in K \mid S_1 \cap K \neq \emptyset) = \frac{\pi F}{\pi F + l_1 L} \tag{5.3}$$

if K is assumed h-convex and expands to the whole hyperbolic plane $(F, L \to \infty)$ this probability tends to 1 for the euclidean plane and to $\pi/(\pi + l_1)$ for the hyperbolic plane. That means that «edge effects» are significant by passing to the limit in the hyperbolic case.

5.2. This «edge effect» is clearly made evident in the following example.

Let K_0 be a convex set contained in K in such a way that any oriented segment of length l_0 which intersects with K_0 has the origin inside K. The probability that a random segment S_1 which has the origin $0 \in K$, intersects with K_0 will be

$$p_1 = \frac{\pi F_0 + l L_0}{\pi F} \tag{5.4}$$

and the probability that a oriented segment S_1 which intersects with K, also intersects with K_0, will be

$$p_2 = \frac{\pi F_0 + l L_0}{\pi F + l L}. \tag{5.5}$$

Given n random segments with the conditions above, the probabilities that m of them intersect with K_0, will be, respectively (binomial law)

$$p_{1m} = \binom{n}{m} p_1^m (1 - p_1)^{n-m}, \qquad p_{2m} = \binom{n}{m} p_2^m (1 - p_2)^m. \tag{5.6}$$

Assume that $n \to \infty$ and K expands to the whole hyperbolic plane in such a way that

$$\frac{n}{F} \to \alpha = \text{constant}. \tag{5.7}$$

The probabilities (5.6) will tend to the limits

$$p_{1m}^* = \frac{(\alpha H_1)^m}{m!} \exp(-\alpha H_1), \tag{5.8}$$

$$p_{2m}^* = \frac{(\alpha H_2)^m}{m!} \exp(-\alpha H_2)$$

where

$$H_1 = F_0 + \frac{l L_0}{\pi}, \qquad H_2 = \frac{\pi F_0 + l L_0}{\pi + l}. \tag{5.9}$$

Therefore we get two kind of oriented line-segment processes in the hyperbolic plane. The first correspond to a Poisson point process of intensity α, each point being the origin of an oriented segment of length l with the orientation uniformly distributed from 0 to 2π. In this case we have $E(m) = \alpha H_1$, and $E(m) \to \infty$ as $l \to \infty$.

The second process gives

$$E(m) = \alpha \frac{\pi F_0 + l L_0}{\pi + l}$$

and therefore, if $l \to \infty$ we have $E(m) \to \alpha L_0$. We can speak of a process of rays which has no analogous in the euclidean plane.

On line-segment processes in the euclidean plane, see R. Cowan [1].

REFERENCES

1. Cowan, Richard, *Homogeneous line-segment processes*, Math. Proc. Camb. Phil. Soc. 86, 1979, 481-489.
2. Firey, W. J., *Kinematic measures for sets of support figures*, Mathematatika, 21, 1974, 270-281.
3. Hadwiger, H., *Eine Erweiterung der kinematische Hauptformel der Integralgeometrie*, Abh. Math. Sem. Hamburg Univ. 44, 1975, 84-90.
4. Mc Mullen, P., *A dice probability problem*, Mathematika, 21, 1974, 193-198.
5. Santaló, L. A., *Integral Geometry on surfaces of constant negative curvature*, Duke Math. J., 10, 1943, 687-704.
6. Santaló, L. A., *Integral Geometry and Geometric Probability*, Encyclopedia of Mathematics, Addison-Wesley, Reading, 1976.
7. Santaló, L. A. and Yanez, I., *Averages for polygons formed by random lines in euclidean and hyperbolic planes*, J. of Applied Probability, 9, 1972, 140-157.
8. Schneider, R., *Kinematische Berührmass für konvexe Körper*, Abh. Math. Sem. Univ. Hanburg, 44, 1975, 12-23.
9. Schneider, R., *Integralgeometrie*, Vorlesungen, Math. Institut Freiburg, 1979.
10. Weil, W., *Kinematicintegral formulas for convex bodies*, Contribution to Geometry (Proc. Geom. Sympos. Siegen, 1978, 60-76) Birkhauser, Basel, 1979.

REND. SEM. MAT.
UNIVERS. POLITECN. TORINO
Vol. 38°, 1 (1980)

Luis A. Santalò

CAUCHY AND KUBOTA's FORMULA FOR CONVEX BODIES IN ELLIPTIC n-SPACE

Summary: *The classical integral formulas de CAUCHY and KUBOTA of the theory of convex bodies are extended to elliptic n-dimensional space. The CAUCHY's formula takes the form (4.7) and that of KUBOTA takes the form (5.5).*

1. Introduction

Let K_n be a compact convex body in elliptic n-dimensional space S^n. Let L_r denote an r-plane (r-dimensional subspace) in S^n which do not intersect K_n and let L^*_{n-r-1} denote the $(n-r-1)$-plane dual of L_r. The $(r+1)$-planes $L_{r+1}[L_r]$ through L_r which meet K_n determine in L^*_{n-r-1} a convex set K^*_{n-r-1} (projection of K_n into L^*_{n-r-1} from L_r). Let V^*_{n-r-1} denote the $(n-r-1)$-dimensional volume of K^*_{n-r-1} and M_i^* $(i=0,1,...,n-r-2)$ the i-th integrals of mean curvature of ∂K^*_{n-r-1}. In this paper we shall obtain some integral formulas referring to V^*_{n-r-1} and M_i^* from which, in particular, one can deduce the mean values of these magnitudes with respect to all r-planes L_r exterior to K_n. These mean values generalize to elliptic space the classical formulas of Cauchy and Kubota for convex bodies in euclidean n-space (see [1], pp. 48-49; [3], pp. 217-218).

For simplicity, we shall assume that ∂K_n is of class C^2 in order that the integrals of mean curvature be well defined. However, using the relation $M_i(\partial K_n) = nW_{i+1}(K_n)$ ([3], p. 224) and applying the theorem that any convex hypersurface can be approximated by a sequence of analytic convex hypersurfaces, it follows that the obtained integral formulas can be expressed

---

Classificazione per soggetto AMS(1979): 52A22.

in terms of the quermassintegrale W_i and therefore they hold for general compact convex bodies in S^n.

2. Some known formulas

For notations and details we refer to [3]. Let dL_r denote the density, in the sense of integral geometry, for r-planes L_r in S^n ([3], p. 305). With this density, the total measure of all L_r in S^n $(0 \leqslant r \leqslant n-1)$ is given by ([3], p. 309),

$$\int_{\text{Total}} dL_r = \frac{0_n 0_{n-1} \ldots 0_{r+1}}{0_{n-r-1} \ldots 0_0} = \frac{0_n \ldots 0_{n-r}}{0_r \ldots 0_0}, \tag{2.1}$$

where $0_i = 2\pi^{(i+1)/2}/\Gamma((i+1)/2)$ denotes the surface area of the i-dimensional unit sphere. For $r=0$ (2.1) gives the volume of S^n, namely $0_n/2$, as is well known.

Let K_n be a compact convex body in S^n. If $M_i = M_i(\partial K_n)$ denotes the i-th integral of mean curvature of ∂K_n (for $i=0$, M_0 is the surface area of ∂K_n) and $M_i(\partial K_n \cap L_{r+1})$ denotes the i-th integral of mean curvature of the intersection of ∂K_n with a moving L_{r+1}, we have ([3], p. 248),

$$\int_{K_n \cap L_{r+1}} \neq \phi M_i(\partial K_n \cap L_{r+1})\, dL_{r+1} = \frac{0_{n-2} \ldots 0_{n-r-1} 0_{n-i}}{0_{r-1} \ldots 0_0 0_{r+1-i}} M_i(\partial K_n) \tag{2.2}$$

This formula holds without change in euclidean and elliptic spaces.

If $\sigma_{r+1}(K_n \cap L_{r+1})$ denotes the volume of the convex set $K_n \cap L_{r+1}$ we have ([3], p. 309)

$$\int_{K_n \cap L_{r+1}} \neq \phi \sigma_{r+1}(K_n \cap L_{r+1})\, dL_{r+1} = \frac{0_{n-1} \ldots 0_{n-r-1}}{0_r \ldots 0_0} V \tag{2.3}$$

where $V = V(K_n)$ denotes the volume of K_n.

Finally, let us recall that the measure of all r-planes L_r intersecting K_n is given by ([3], p. 310)

$$\int_{K_n \cap L_r \neq \phi} dL_r = \frac{0_{n-2} \ldots 0_{n-r}}{0_r \ldots 0_1} \left[0_{n-1} V + \sum_{i=1}^{r'} \binom{r-1}{2i-1} \frac{0_r 0_{r-1} 0_{n-2i+1}}{0_{2i-1} 0_{r-2i} 0_{r-2i+1}} M_{2i-1} \right] \tag{2.4}$$

for r even $(r = 2r')$, and

$$\int_{K_n \cap L_r \neq \phi} dL_r = \frac{0_{n-2} \ldots 0_{n-r}}{0_{r-1} \ldots 0_1} \sum_{i=0}^{r'} \binom{r-1}{2i} \frac{0_{r-1} 0_{n-2i}}{0_{2i} 0_{r-2i-1} 0_{r-2i}} M_{2i} \tag{2.5}$$

for r odd $(r=2r'+1)$.

For $r=1$ $(r'=0)$ (2.5) is not directly applicable. However, using the identity $(0_{n-2}\dots 0_{n-r})(0_{r-1}\dots 0_1)^{-1}=(0_{n-2}\dots 0_r)(0_{n-r-1}\dots 0_1)^{-1}$, we get (putting $M_0=F$),

$$\int_{K_n\cap L_1\neq\phi} dL_1=(0_n/4\pi)\,F \tag{2.6}$$

For $n\geqslant 3$, $r=2$, (2.4) gives

$$\int_{K_n\cap L_2\neq\phi} dL_2=(0_{n-2}\,0_{n-1}/8\pi^2)\,(V+M_1)\,,$$

3. Solid angles

Consider a compact convex body K_n and an r-plane L_r $(0\leqslant r\leqslant n-2)$ in S^n which do not intersect. The set of all $(r+1)$-planes L_{r+1} which contain L_r and meet K_n, can be measured with the invariant density $dL_{r+1}[L_r]$ ([3], p. 202) which has the same explicit form for the elliptic as for the euclidean space and coincides with the invariant element of volume of the Grassmann manifold $G_{1,n-r-1}$. We call this measure the *solid angle* which subtends K_n from L_r and we denote it by

$$\phi_r^{(n)}=\int dL_{r+1}[L_r] \tag{3.1}$$

where the integral is extended over all L_{r+1} which satisfy the following conditions

$$L_r\subset L_{r+1}\,,\qquad L_{r+1}\cap K_n\neq\phi\,.$$

Notice that for $r=0$, $\phi_0^{(n)}$ is the usual solid angle under which K_n is seen from the point L_0. For $r=n-2$, $\phi_{n-2}^{(n)}$ is the angle between the support hyperplanes to K_n through L_{n-2}.

According to the given definition (3.1), the solid angle $\phi_r^{(n)}$ is equal to the $(n-r-1)$-dimensional volume V^*_{n-r-1} of the convex set K^*_{n-r-1} dual of K_n defined in the introduction.

4. Integral formulas

We have denoted by $dL_{r+1}[L_r]$ the density for $(r+1)$-planes L_{r+1} about a fixed r-plane L_r. Calling now $dL_r(L_{r+1})$ the density for r-planes L_r in L_{r+1}, we have the following formula

$$dL_r(L_{r+1}) \wedge dL_{r+1} = dL_r \wedge dL_{r+1}[L_r] \tag{4.1}$$

which is essentially due to B. Petkantschin [2] and has the same form for elliptic as for euclidean space (see also [3], p. 207).

We want to compute the integral of both sides of (4.1) over all r- and $(r+1)$-planes such that

$$L_r \cap K_n = \phi \quad , \quad L_r \subset L_{r+1} \quad , \quad L_{r+1} \cap K_n \neq \phi \; .$$

According to (3.1) the right side member gives

$$\int_{K_n \cap L_r = \phi} \phi_r^{(n)} dL_r \; ,$$

In order to compute the integral of the differential form on the left of (4.1), we first leave L_{r+1} fixed and observe that the integral of all L_r such that $L_r \subset L_{r+1}$ and $L_r \cap K_n = \phi$, is equal to the total measure of the r-planes in L_{r+1} (which, according to (2.1) is O_{r+1}/O_0), less the measure of the r-planes $L_r \subset L_{r+1}$ which meet the convex set $K_n \cap L_{r+1}$. Putting $M_i^{(r+1)} = M_i(\partial K_n \cap L_{r+1})$ and $V^{(r+1)} = V(K_n \cap L_{r+1})$ this measure, according to (2.4) and (2.5) is

a) For $r = 2r'$, $n = r+1 = 2r'+1$,

$$\int_{K_n \cap L_r \neq \phi} dL_r(L_{r+1}) = V^{(r+1)} + \sum_{i=1}^{r'} \binom{r-1}{2i-1} \frac{O_{r-1} O_{r+2-2i}}{O_{2i-1} O_{r-2i} O_{r-2i+1}} M_{2i-1}^{(r+1)} \tag{4.3}$$

b) for $r = 2r'+1$, $n = r+1 = 2(r'+1)$,

$$\int_{K_n \cap L_r \neq \phi} dL_r(L_{r+1}) = \sum_{i=0}^{r'} \binom{r-1}{2i} \frac{O_{r-1} O_{r+1-2i}}{O_{2i} O_{r-2i-1} O_{r-2i}} M_{2i}^{(r+1)} \tag{4.4}$$

Thus, multiplying the difference between O_{r+1}/O_0 and the measures (4.3) or (4.4) by dL_{r+1} and performing the integration over all L_{r+1} which meet K_n, having into account (2.2) and (2.3) and equating with (4.2) we get:

a) For $r = 2r'$ (r even),

$$\int_{K_n \cap L_r \neq \phi} \phi_r^{(n)} dL_r = \frac{O_{r+1} O_{n-2} \dots O_{n-r-1}}{O_0 O_r \dots O_1} \sum_{i=0}^{r'} \binom{r}{2i} \frac{O_r O_{n-2i}}{O_{2i} O_{r-2i} O_{r-2i+1}} M_{2i}$$

$$-\frac{0_{n-1}\dots 0_{n-r-1}}{0_r \dots 0_0} V - \sum_{i=1}^{r'} \binom{r-1}{2i-1} \frac{0_{r-1}0_{r+2-2i}}{0_{2i-1}0_{r-2i}0_{r-2i+1}} \frac{0_{n-2}\dots 0_{n-r-1}0_{n-2i+1}}{0_{r-1}\dots 0_0 0_{r-2i+2}} M_{2i-1} \quad (4.5)$$

b) For $r=2r'+1$ (r odd),

$$\int_{K_n \cap L_r \neq \phi} \phi_r^{(n)} dL_r = \frac{0_{n-2}\dots 0_{n-r-1}}{0_0 0_r \dots 0_1} \Bigg[0_{n-1} V + \sum_{i=1}^{r'+1} \binom{r}{2i-1} \frac{0_{r+1}0_r 0_{n-2i+1}}{0_{2i-1}0_{r+1-2i}0_{r+2-2i}} M_{2i-1} \Bigg] \quad (4.6)$$

$$- \sum_{i=0}^{r'} \binom{r-1}{2i} \frac{0_{r-1}0_{r+1-2i}}{0_{2i}0_{r-2i-1}0_{r-2i}} \frac{0_{n-2}\dots 0_{n-r-1}0_{n-2i}}{0_{r-1}\dots 0_0 0_{r+1-2i}} M_{2i}$$

The case $r=0$. The preceding formulas are not directly applicable for $r=0$. In this case, we can proceed as follows. Let λ denote the length of the chord $K_n \cap L_1$. The integral of the left side of (4.1) gives $\int (\pi-\lambda)\, dL_1$ and since the measure of the set of lines L_1 which meet K_n is $(0_n/4\pi) F$ (according to (2.6)) and it is known that $\int \lambda\, dL_1 = (0_{n-1}/2) V$ ([3], p. 307), we have

$$\int_{L_0 \notin K_n} \phi_0^{(n)} dL_0 = (0_n/4) F - (0_{n-1}/2) V. \quad (4.7)$$

In this formula the solid angle $\phi_0^{(n)}$ is equal to the volume of the projection of K_n into the hyperplane dual of L_0, so that (4.7) is the extension to the elliptic space of the so called formula of Cauchy for convex bodies in euclidean space ([1], p. 48; [3], p. 218).

As a consequence we have that the mean value of the volume of the projection $\phi_0^{(n)}$ is

$$E(\phi_0^{(n)}) = \frac{0_n F - 2 0_{n-1} V}{2 0_n - 4V}$$

where V denotes the volume and F the surface area of K_n.

Some particular cases. The cases $n=2,3$ are known (see [3] p. 318-319 and [4] p. 186). They write

$$\int \phi_0^{(2)} dL_0 = \pi (L-F)$$

$$\int \phi_0^{(3)} dL_0 = (1/2)\pi^2 F - 2\pi V$$

$$\int \phi_1^{(3)} dL_1 = 2\pi (M_1 + V) - (1/2)\, \pi^2 F$$

For $n \geqslant 4$ the results are new. For $n = 4$ we get the following possibilities

$$\int \phi_0^{(4)} dL_0 = (2/3)\, \pi^2 F - \pi^2 V$$

$$\int \phi_1^{(4)} dL_1 = 2\pi^2 (V + M_1) - (4/3)\, \pi^2 F$$

$$\int \phi_2^{(4)} dL_2 = \pi^2 [(2/3) F + M_2 - M_1 - V]$$

In all cases, the integrals are extended over all L_0, L_1, L_2 exterior to the corresponding convex body.

5. More integral formulas

In the elliptic space, the principle of duality allows to assign to each integral formula its dual. Given a compact convex set K_n, the hypersurface parallel to ∂K_n in a distance $\pi/2$, is called the hypersurface "polar" of ∂K_n and it is the boundary of a convex body K_n^p (which does not contain K_n). The integrals of mean curvature of ∂K_n and ∂K_n^p satisfy the relation ([3], p. 304)

$$M_i(\partial K_n^p) = M_{n-1-i}(\partial K_n) \quad , \quad i = 0, 1, \ldots, n-1 \tag{5.1}$$

The formula (2.2) may be written

$$\int_{K_n \cap L_r \neq \phi} M_i^{(r)}(\partial K_n \cap L_r)\, dL_r = \frac{O_{n-2} \ldots O_{n-r} O_{n-i}}{O_{r-2} \ldots O_0 O_{r-i}}\, M_i(\partial K_n) \tag{5.2}$$

which holds for $i = 0, 1, \ldots, r-1$; $r = 2, 3, \ldots, n-1$.

By duality, using (5.1), (5.2) transforms into

$$\int_{K_n \cap L_{n-r-1} = \phi} M_{r-1-i}^{(r)}(\partial K_r^*)\, dL_{n-r-1} = \frac{O_{n-2} \ldots O_{n-r} O_{n-r}}{O_{r-2} \ldots O_0\, O_{r-i}}\, M_{n-1-i}(\partial K_n) \tag{5.3}$$

where K_r^* denotes the projection of K_n from L_{n-r-1} into its dual r-plane.

By the change of indices $n-r-1 \to r$, $r-1-i \to i$ (5.3) takes the form

$$\int_{K_n \cap L_r = \phi} M_i^{(n-1-r)}(\partial K_{n-1-r}^*)\, dL_r = \frac{O_{n-2} \ldots O_{r+1}\, O_{r+i+2}}{O_{n-r-3} \ldots O_0\, O_{i+1}}\, M_{r+i+1}(\partial K_n) \tag{5.4}$$

which holds for $i=0,1,\ldots,n-r-2$; $r=0,1,\ldots,n-3$.

For $r=0$ we have

$$\int_{L_0 \notin K_n} M_i^{(n-1)}(\partial K_{n-1}^*)\, dL_0 = \frac{0_{n-2}\, 0_{i+2}}{2 0_{i+1}} M_{i+1}(\partial K_n) \qquad (5.5)$$

which holds for $i=0,1,\ldots,n-2$. This formula generalizes to elliptic space the so called formula of Kubota for convex bodies in euclidean space ([1], p. 49; [3], p. 217).

For $i=0$, denoting $F^* = M_0(\partial K_{n-1}^*)$ the surface area of K_{n-1}^*, we have

$$\int_{L_0 \notin K_n} F^*\, dL_0 = 0_{n-2} M_1 . \qquad (5.6)$$

For $n=3$, if u denotes the perimeter of the projected set K_2^*, (5.6) writes

$$\int_{L_0 \notin K_3} u\, dL_0 = 2\pi M_1 , \qquad (5.7)$$

The difference between the total measure (1.1) of all L_r in S^n and the measure (2.4) or (2.5) of those which meet K_n gives the measure of all L_r which are exterior to K_n and then (5.4) allows to write down the mean values of the integrals of mean curvature $M_i^{(n-1-r)}(\partial K_{n-1-r}^*)$. For instance, from (5.7) we deduce

$$E(u) = \frac{2\pi M_1}{\pi^2 - V} .$$

BIBLIOGRAPHY

[1] Bonnesen, T. - Fenchel, W., *Theorie der konvexen Körper*, Ergebnisse der Mathematik und ihre Grenzgebiete, Berlin, Springer 1934.

[2] Petkantschin, B., *Zusammenhänge zwischen den Dichten der linearen Unterräume im n-dimensionales Räume*, Abhandlungen Math. Seminar Hansischen Universität, 11, 249-310, (1936).

[3] Santaló, L.A., *Integral Geometry and Geometric Probability*, Encyclopedia of Mathematics and its Applications, Reading: Addison-Wesley 1976.

[4] Santaló, L.A., *Integral Geometry*, Studies in Global Geometry and Analysis (S.S. Chern, editor) Washington: Mathematical Association of America (distributed by Prentice-Hall) 1967.

L.A. SANTALO', Departamento de Matemática. Universidad de Buenos Aires, Ciudad Universitaria (Nuñez), Buenos Aires, Argentina.

Lavoro pervenuto alla redazione il 19-XII-1979

Part III. Convex geometry, with comments by R. Schneider

Contributions to the geometry of convex sets form, in comparison, a smaller part of Santaló's mathematical work, but they concern an appealing diversity of geometric topics and have had considerable influence on later developments. For example, the Blaschke-Santaló inequality is until today, almost sixty years after Santaló gave his proof, at the centre of much interesting research.

The first contribution to consider here is one to combinatorial geometry. The motivation came from Helly's theorem and a question of Vincensini. A *k-transversal* of a family $\mathcal{F}$ of compact convex sets in n-dimensional Euclidean space $\mathbb{R}^n$ is a k-dimensional flat that meets all members of $\mathcal{F}$. Helly's theorem says that $\mathcal{F}$ has a 0-transversal if every subfamily of $n+1$ or fewer sets has a 0-transversal. Vincensini asked whether for $k \in \{1, \ldots, n-1\}$ there is a number j such that $\mathcal{F}$ has a k-transversal whenever each subfamily of j or fewer sets has a k-transversal. Counterexamples showed that this can only hold under strong additional assumptions on the shapes and positions of the members of the family $\mathcal{F}$. In [40.2], Santaló considers the case where $\mathcal{F}$ is a family of parallelotopes with edges parallel to the coordinate axes, and he proves the following results. If every $2^{n-1}(2n-1)$ or fewer elements of $\mathcal{F}$ have a 1-transversal, then $\mathcal{F}$ has a 1-transversal. If every $2^{n-1}(n+1)$ or fewer elements of $\mathcal{F}$ have an $(n-1)$-transversal, then $\mathcal{F}$ has an $(n-1)$-transversal. The proof employs ideas which Radon used in his proof of Helly's theorem. For the case of the plane, where the two results are the same and give the test number 6, Santaló notes in [42.3] that this number is best possible, and he treats similar questions for finite families of segments. Later developments, naturally, concerned more general convex bodies, and today transversal theory is a well established subject, with results contributed by Hadwiger, Danzer, Grünbaum, Eckhoff, Tverberg, Goodman, Pollack, Wenger, and others. Danzer, Grünbaum, and Klee in their famous survey article on Helly type theorems state, and Eckhoff in his article in the Handbook of Convex Geometry agrees, that "the first significant results on common transversals were those of Santaló". Even the special situation studied by Santaló is intricate: it is still not known whether for $n > 2$ the test numbers found by Santaló are optimal, nor what the analogous result is for k-transversals with $1 < k < n-1$.

In contrast to this excursion into combinatorial geometry, most of Santaló's contributions to the theory of convex bodies deal with geometric inequalities. The work [46.1] is devoted to the spherical analogue of Jung's theorem. The latter theorem determines the smallest circumradius (radius of the circumball, the smallest containing ball) of all convex bodies of given diameter $D > 0$ in $\mathbb{R}^n$. The minimum is attained by the regular simplex. Let $\mathbb{S}^n$ denote the unit sphere in $\mathbb{R}^{n+1}$. The geometric notions used in the following refer to the inner geometry of $\mathbb{S}^n$. Santaló notes that in order to determine the minimal circumradius of a convex body in $\mathbb{S}^n$ (contained in a closed hemisphere) of given diameter, one can, as in the Euclidean case, restrict oneself to simplices containing the centre of their circumball. He shows that the minimum is attained by a regular simplex. However then, other than in the Euclidean case, a complication arises: the diameter of a regular spherical simplex is equal to its edge length l only if $l \le \pi/2$. For $l > \pi/2$, the diameter is the distance between two complementary subsimplices, each containing 'half' of the vertices; this requires to distinguish further, between even and odd dimensions. By explicit calculations and estimates, the minimal circumradius is found in each case. Santaló also demonstrates that the spherical analogue of Jung's theorem is more powerful than the Euclidean case. First, Steinhagen's theorem, where diameter and circumradius are replaced by thickness (minimal width) and inradius, is in the spherical case obtained just by dualizing, that is, applying the obtained analogue of Jung's theorem to the polar body. Second, the Euclidean theorems of Jung and Steinhagen can be obtained from their spherical counterparts by a limit procedure. Finally, Santaló deduces a nice result (and its dual) which has no Euclidean counterpart. Let $\tan r = \sqrt{(n+1)/2n}$, and let K be a convex body in $\mathbb{S}^n$. Then there exists a spherical ball of radius r that is either contained in K or contained in the complement of $K \cup (-K)$. The number r is best possible.

The paper [46.2] deals with a prominent class of convex bodies, namely the bodies of constant width. Let K be a convex body of constant width a in $\mathbb{R}^n$. In the plane, by Barbier's classical theorem the perimeter L of K satisfies $L = \pi a$, and in three-dimensional space Blaschke has shown that volume V and surface area F are connected by $2V - aF + 2\pi^3/3 = 0$. (As a consequence, the (unsolved) problem of finding the three-dimensional convex body of constant width a with minimal volume is equivalent to finding such a body with minimal surface area.) Santaló in his paper extends Barbier's and Blaschke's results to higher dimensions, establishing $\lfloor (n+1)/2 \rfloor$ independent linear relations between the quermassintegrals of a body of constant width a. His proof makes use of properties of the radii of curvature of a (smooth) convex body of constant width and the representation of the quermassintegrals in terms of elementary symmetric functions of principal radii of curvature. He was apparently unaware of the fact that, six years before, Dinghas had given a similar proof. A short proof not needing differential geometry was later given by Debrunner, who in turn was unaware of Santaló's paper.

An inequality of lasting importance is established in the work [49.3]. Its origins in two and three dimensions go back to Blaschke's affine differential geometry. Santaló not only extended it to higher dimensions, but gave the proof an important twist, thus avoiding additional assumptions, so that the general inequality with good reason bears the name *Blaschke-Santaló inequality*. For a convex body K in $\mathbb{R}^n$ and an interior point z of K, let $K^z := (K - z)^* + z$, where $(K - z)^*$ is the polar body of $K - z$. Santaló shows that that there is a unique point at which the volume $V(K^z)$ attains its minimum. This point is denoted by s and is nowadays called the *Santaló point* of K. This notion is dual to the *centroid* (centre of gravity for a homogeneous mass distribution): s is the Santaló point of K if and only if s is the centroid of K^s. The product $V(K)V(K^s)$ is invariant under nonsingular affine transformations of K; it is called the *volume product* (or the *Mahler volume*) of K and is probably the simplest and most important affine invariant of convex bodies. Every continuous, affine invariant real functional on the space of convex bodies with interior points attains a maximum and a minimum, and the Blaschke-Santaló inequality settles the case of the maximum. It says that

$$V(K)V^s(K) \leq \kappa_n^2, \tag{1}$$

where κ_n denotes the volume of the n-dimensional unit ball. Santaló's proof proceeds as follows. We can assume $s = 0$. From the minimum property of the Santaló point it follows that the support function $h(K, \cdot)$ of K satisfies

$$\int_{\mathbb{S}^{n-1}} h(K, u)^{-(n+1)} u \, \sigma(\mathrm{d}u) = 0,$$

where σ denotes spherical Lebesgue measure. This means that the indefinite σ-integral of $h(K, \cdot)^{-(n+1)}$ satisfies the assumption for Minkowski's existence theorem. Consequently, there exists a convex body C for which $h(K, u)^{-(n+1)}$ is the curvature function. This has two important consequences, namely that

$$nV(K^s) = \Omega(C),$$

where Ω is the *affine surface area* of C, and

$$V(K^s) = V(K, C, \dots, C),$$

where the right side is a mixed volume. Now one uses *Minkowski's inequality*

$$V(K, C, \dots, C)^n \geq V(K)V(C)^{n-1}$$

and Blaschke's *affine isoperimetric inequality*

$$\Omega(C)^{n+1} \leq n^{n+1}\kappa_n^2 V(C)^{n-1}$$

to deduce (1). If K is an ellipsoid, then equality holds in the latter two inequalities and hence also in the Blaschke-Santaló inequality. At the time when Santaló wrote his paper, the affine isoperimetric inequality and its equality condition were only known under sufficiently strong differentiability assumptions on C, and it was not known whether the convex body C, which is obtained by the use of Minkowski's existence theorem, satisfies this differentiability. Therefore, the proof of the inequality required, strictly speaking, an additional approximation argument, and the equality case could not be settled. That equality in (1) indeed holds only if K is an ellipsoid, was only proved more than three decades later, by Petty, with a rather complicated proof.

Santaló's proof of (1) uses three pearls from classical convex geometry: Minkowski's existence theorem for convex bodies with given curvature function, Minkowski's mixed volume inequality, and Blaschke's affine isoperimetric inequality. The proof of the latter is based on Steiner symmetrization. Only in the 1980s, proofs were found of the Blaschke-Santaló inequality and its equality condition that use Steiner symmetrization and the Brunn-Minkowski theorem in a more direct way, by Saint Raymond, Meyer and Pajor. Ongoing research about the volume product concerns stability versions of the Blaschke-Santaló inequality (Böröczky jr.) and the intricate problem of the counterpart to the Blaschke-Santaló inequality for the minimum of the volume product. It has been conjectured that this minimum is obtained by the simplex, and for centrally symmetric convex bodies by the parallelepiped (and, if this is true, also by the cross-polytope, which is dual to the parallelepiped, but for $n > 3$ also by other polytopes). Positive answers are known in the plane and for some special classes of convex bodies, for example, for zonoids. Non-sharp estimates have been obtained. Contributions to this topic are due to Mahler, Saint Raymond, Reisner, Bourgain, Milman, Pisier, G. Kuperberg, and others.

The Blaschke-Santaló inequality has been used to derive other geometric inequalities (by Lutwak and others), functional versions (Ball, Artstein-Avidan, Milman, and others), a spherical counterpart, and it has found some surprising applications in other fields, like the geometry of Minkowski and Finsler spaces and Stochastic Geometry.

Complete systems of inequalities for convex bodies in the plane are the subject of Santaló's paper [59.1]. Generally, a set of inequalities holding for $k \geq 2$ given functionals $\varphi_1, \ldots, \varphi_k$ of convex bodies in $\mathbb{R}^n$ is said to be a *complete system* if to any k real numbers $a_1, \ldots, a_k$ satisfying these inequalities there exists a convex body K in $\mathbb{R}^n$ with $\varphi_i(K) = a_i$ for $i = 1, \ldots, k$. The first interesting case is $k = 3$, and Santaló investigates inequality systems for three of the following functionals of convex bodies in the plane: the area F, the perimeter L, the diameter D, the thickness E, the circumradius R, and the inradius r. These functionals are continuous, invariant under rigid motions, and homogeneous of degree one or two under dilatations. To study the completenes of inequality systems for a given triplet of these functionals, say $\varphi_1, \varphi_2, \varphi_3$, Santaló takes up an idea used by Blaschke in the case of the quermassintegrals in three-space. With each planar convex body K, he associates the coordinate pair (x, y) defined by $x = c_1\varphi_1^{a_1}(K)\varphi_3^{a_2}(K)$ and $y = c_2\varphi_2^{b_2}(K)\varphi_3^{b_3}(K)$, where $a_1, a_3, b_2, b_3 \in \{\pm 1, \pm 2\}$ are chosen such that (x, y) does not change under similarities of K and the positive constants c_1, c_2 guarantee that $0 \leq x, y \leq 1$. In this way, the set of convex bodies in the plane is mapped to a closed, simply connected set with interior points in the x, y-plane, which is nowadays called a *Santaló diagram*. Each sharp inequality between $\varphi_1, \varphi_2, \varphi_3$ yields with its equality case a part of the boundary of the Santaló diagram. If the curves defined in this way by the inequalities of a system make up the whole boundary of a simply connected domain, then the system of inequalities is complete. In this way, Santaló investigates the triples (F, L, E), (F, D, E), (F, L, R), (F, L, r), (L, D, E) and finds in each case that the known inequalities (due to many authors) make up a complete system. In the case (D, r, R), he proves a new inequality between the three quantities and thus establishes a complete system. In the cases (D, E, R) and (E, R, r) he finds that at least one inequality is still missing. Santaló understood his investigation as an invitation to further research. The topic fascinates by its great variety of extremal bodies that are observed. In later years, complete inequality systems for further seven of the 20 triplets from the considered functionals have been established, mainly by Maria A. Hernández Cifre and her co-workers. It appears that seven cases remain, for which the invitation is still open.

The following three papers of Santaló on different aspects of the geometry of convex bodies are all inspired from Integral Geometry. In [83.1], a convex body K in $\mathbb{R}^n$ is divided by a hyperplane H into two parts K_1, K_2. As proved by Bokowski and Sperner jr., the volumes of these two parts, the $(n-1)$-dimensional volume σ_{n-1} of $H \cap K$ and the diameter D of K satisfy an inequality $V(K_1)V(K_2) \leq c(n)D^{n+1}\sigma_{n-1}$, with an explicit constant $c(n)$. In dimensions two and three, Santaló gives a short integral-geometric proof, which has the advantage of yielding a sharper result in three dimensions and allowing an extension to elliptic and hyperbolic space.

To describe the starting point of [86.1], let μ denote the (suitably normalized) Haar measure on the group of rigid motions of $\mathbb{R}^n$. For convex bodies K and L, the classical principal kinematic formula of Integral Geometry yields explicitly the integral $\int \mathbf{1}\{K \cap gL \neq \emptyset\}\, \mu(\mathrm{d}g)$, that is, the invariant measure of all rigid motions g for which gL has nonempty intersection with K; the result is a bilinear expression in the quermassintegrals of K and L. In contrast to this, there is no explicit formula for the integral $\int \mathbf{1}\{gL \subset K\}\, \mu(\mathrm{d}g)$, giving the invariant measure of all rigid motions g for which gL is contained in K. Thus, these *containment measures* (or *inclusion measures*) define, in a naturtal way, new functionals which deserve further study. Santaló's paper is devoted to the special case where L is a line segment of length l. For a given convex body K, the function $M_0(l) := \int \mathbf{1}\{gL \subset K\}\, \mu(\mathrm{d}g)$ is then the kinematic measure of all line segments of length l that are contained in K. Santaló relates this function to other invariants of convex bodies, like the chord power integrals, an invariant of Enns and Ehlers and, in the plane, an invariant introduced by Pohl. He proves various identities and inequalities, also giving a proof for a conjectured inequality of Enns and Ehlers for the expected chord length of certain random chords of K; this yields an extremal property of the ball. At about the same time, containment measures, for segments or for more general convex bodies, were studied and applied in China by Delin Ren, and later by his former students Gaoyong Zhang and Jiazu Zhou.

Also the motivation for [88.1] comes from Integral Geometry. In Euclidean geometry, the invariant measures of the sets of planes of a fixed dimension that hit a given convex body define important functionals of convex bodies, the quermassintegrals. In equiaffine geometry (the geometry of the group of volume preserving affine transformations), the set of hyperplanes, for example, does not have an invariant measure, but the set of parallel pairs of hyperplanes does. The measure of the pairs of all parallel hyperplanes leaving a convex body K between them defines an equiaffine invariant of K. In Euclidean terms it can (up to a normalizing factor) be expressed as

$$J(K) = \int_{\mathbb{S}^{n-1}} w(K,u)^{-n}\, \sigma(\mathrm{d}u),$$

where $w(K,u)$ is the width of K in direction u. If z is an interior point of K, then $J(K) \leq n2^{-(n+1)}V(K^z)$ (as Santaló shows, e.g., in his book), with equality if K is symmetric with respect to z. It follows from the Blaschke-Santaló inequality (and its equality conditions) that $J(K)V(K)$ attains its maximum precisely if K is an ellipsoid. The problem of the minimum remains open. For convex bodies with a boundary of class C^2 and with positive Gauss-Kronecker curvature κ, Santaló defines a series of equiaffine invariants by

$$J_h(K) = \int_{\mathbb{S}^{d-1}} w(K,u)^h \kappa^{-(h+n)(n+1)} \sigma(\mathrm{d}u),$$

for arbitrary integers h. Thus, $J_{-n} = J$, and J_0 is the affine surface area. The functional $J_{-(n+1)}(K)$ is interpreted as an affine invariant measure of the set of all pairs of parallel hyperplanes that support the body K. Santaló notes some special values and some inequalities satisfied by these invariants. The paper concludes with the sentence, "it should be interesting to find other ineqalities of the same type for the affine invariants J_h". It remains a challenge until today to decide whether, for convex bodies of given volume, the functionals J_h with $h \leq 0$ attain their maximum for ellipsoids, as it holds for $h = 0$ and $h = -n$, and whether J_{-n} attains its minimum for simplices.

ROLF SCHNEIDER (Freiburg i. Br.)

A theorem on sets of parallelepipeds with parallel edges

L. A. Santaló

Abstract

The following theorem is proven: "Given in the plane a set of parallelograms with sides parallel to two fixed directions, if 6 of these parallelograms have points on the same line, there exists a line having a common point with all the parallelograms in the set".
In Section II, this theorem is generalized to ordinary space and some particular cases of n-dimensional space.

The following theorem is well-known:

"Given in the plane a set of convex figures, if every 3 of them have a common point, there exists a point common to all the figures in the set".

This theorem due to Helly can be generalized to sets of convex figures in a space of any number of dimensions. A proof for this general case was given by Radon [2], in 1935, and another by König [1] in 1922.

In 1935, Vincensini [3], posed the following correlative problem: "Given a set of convex figures in the plane, can a number n be given such that if n to n of these figures possess a common secant line, there exists a secant line to all the figures in the set?"

As noted by Schönhardt[1], Vincensini's solution to this problem was erroneous. Indeed, it is possible to check that the question thus posed admits of no solution. Taking as convex figures the segments shown in Fig 1.(formed by the sides, suitably extended, of a regular polygon of 7 sides), one may observe that for each six of such segments there exists a common secant, although there exists no line that intersects them all.

An analogous figure can be constructed for any number of segments, and therefore: for every n, $n+1$ convex figures can always be given such that any n of them has a common secant, although there exists no secant common to them all.

However, the problem has a solution if the convex figures are limited to parallelograms with sides parallel to two fixed directions; in other words, for the case of a set of parallelograms such as those shown in Fig. 2, which can obviously be reduced to segments.

[1] Jahrbuch über die Fortschritte der Mathematik, Bd. 61 (1935), pag. 757

Fig. 1

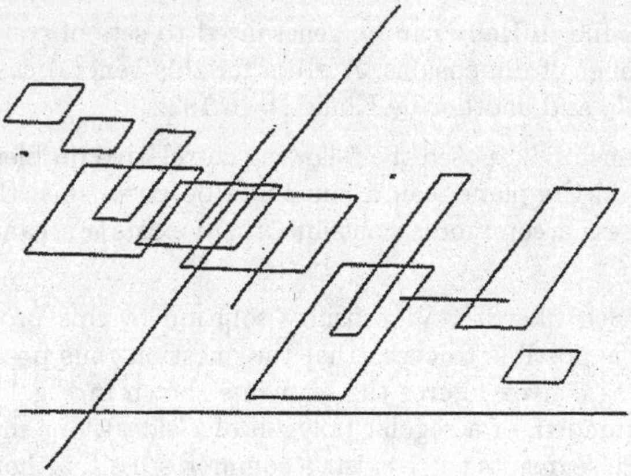

Fig. 2

In this case we prove that the following theorem holds: "*Given a set of parallelograms with sides parallel to two fixed directions, if 6 of these parallelograms have points on the same line, there exists a line having a common point with all the parallelograms in the set*".

The following two theorems also hold for the space:

"*If a set of parallelepipeds is such that for each 16 of these parallelepipeds there is a plane intersecting them, there exists a plane that interests them all*".

"*If a set of parallelepipeds is such that for each 20 of these parallelepipeds there is a plane intersecting them, there exists a line that intersects them all*".

The proof given in this paper for these theorems is based on the idea employed by Radon in his cited work for proving the theorem stated at the beginning.

In Section I, we solve the problem for the plane, and in Section II for the space for finite sets. In Section III, the results are extended to the case of infinite sets. In Section IV, two questions that remain to be solved are stated.

I. The case of the Plane

The theorem to be proven is as follows:

"Given a finite set of parallelograms with sides parallel to two fixed directions, if every 6 of these parallelograms have a common secant, there exists a secant common to them all"[2].

We proceed by induction. While this is clear for a set of 6 parallelograms, it is necessary to prove that if it is true for N, it is also true for N + 1.

Thus let $N+1$ be the parallelograms $P_1, P_2, P_3, \ldots, P_{N+1}$ with sides parallel to two fixed directions such that 6 of these parallelograms possess a common secant. Since for N parallelograms we assume the theorem to be true, any group of N parallelograms taken among the $N+1$ given will admit a common secant.

We take some coordinated axes parallel to the fixed directions of the sides of the parallelograms. In this system of oblique coordinates, represented by

$$ax + b_k y + c_k = 0 \tag{1}$$

to the line having a point common to all the parallelograms, except perhaps to P_k. We thus have $N+1$ lines corresponding to $k = 1, 2, \ldots, N+1$.

One may choose that the first coefficient a is positive and always the same for all the lines. From these lines we choose 4 such that the coefficients b_k are of the same sign. This is certainly possible if $N+1 \geq 7$; that is, $N \geq 6$, which is fulfilled within our hypothesis. Let us assume that these lines are the first 4 and will be either of the type

$$ax + b_k y + c_k = 0 \tag{2}$$

or of the type

$$ax - b_k y + c_k = 0 \tag{3}$$

where a and b_k are positive, and c_k can be of any sign.

Consider the system of homogenous equations

$$\begin{aligned} \sum \lambda_k &= 0 \\ \sum b_k \lambda_k &= 0 \\ \sum c_k \lambda_k &= 0 \end{aligned} \tag{4}$$

with the unknowns λ_k $(k = 1, 2, 3, 4)$.

[2] For the sake of convenience, we call any line having some point common to the parallelogram a *secant*, even though the line does not divide it into two parts.

Taking into account the first equation, $\lambda_k$cannot all be of the same sign. Placing the positive ones on one side of the equation and the negative on the other, we have

$$\begin{aligned} \sum{}'\lambda_k &= \sum{}''\lambda_k \\ \sum{}'b_k\lambda_k &= \sum{}''b_k\lambda_k \\ \sum{}'c_k\lambda_k &= \sum{}''c_k\lambda_k \end{aligned} \tag{5}$$

where with a prime we indicate that the sum of terms is extended to the positive λ_k and with a double prime the negative ones.

If the 4 lines belong to type (2), we consider the line of the equation

$$a(\sum{}'\lambda_k)x + (\sum{}'b_k\lambda_k)y + (\sum{}'c_k\lambda_k) = 0 \tag{6}$$

which is identical to

$$a(\sum{}''\lambda_k)x + (\sum{}''b_k\lambda_k)y + (\sum{}''c_k\lambda_k) = 0 \tag{6'}$$

If they belonged to type (3), we would consider the line that can be represented by either of the following equations

$$\begin{aligned} a(\sum{}'\lambda_k)x - (\sum{}'b_k\lambda_k)y + (\sum{}'c_k\lambda_k) = 0 \\ a(\sum{}''\lambda_k)x + (\sum{}''b_k\lambda_k)y + (\sum{}''c_k\lambda_k) = 0 \end{aligned} \tag{7}$$

The line thus obtained, either (6) or (7), as the case may be, *will have a point common to all the parallelograms.*

Indeed, let P_i be a parallelogram. Let us assume, for example, that we are dealing with the first case, in which 4 lines are of the form (2). Since P_i is intersected by the lines (1), except for that corresponding to $k = i$, it will have N points $x_k^{(i)}, y_k^{(i)}$ that satisfy

$$ax_k^{(i)} + b_k y_k^{(i)} + c_k = 0, \qquad k = 1, 2, 3, \ldots, i-1, i+1, \ldots, N+1 \tag{8}$$

The index i cannot form part of the values of k in the two equations (6) and (6') at the same time; it may not appear in either (if $i \neq 1, 2, 3, 4$), but at the very least it does not appear in one of them. Let (6) be that in which $\sum{}'$ do not include the value $k = i$. Then consider the point

$$\xi^{(i)} = \frac{\sum{}'\lambda_k x_k^{(i)}}{\sum{}'\lambda_k}, \qquad \eta^{(i)} = \frac{\sum{}'b_k\lambda_k y_k^{(i)}}{\sum{}'b_k\lambda_k}$$

This point belongs to P_i and furthermore fulfils equation (6), since when $k \neq i$, (8) are fulfilled. The statement is thus proven.

II. The case of the Space

1. Planes intersecting a set of parallelepipeds. Let us choose some coordinates parallel to the three directions of the edges of the parallelepipeds. The problem is conducted in the same way as for the plane, but now (in order to form the system of λ_k analogous to (4)) 5 equations of the following type are required

$$ax + b_k y + c_k z + d_k = 0$$

such that for all these equations b_k have constant sign and so do c_k. In fact, it is necessary to have 17 equations of planes, then 9 of them can be taken with the b_k of the same sign, and in the remaining 5 with c_k of constant sign. Following the same reasoning as for the plane, we have that $N+1 \geq 17$; that is, $N \geq 16$. The construction of the plane having a point in common with all the parallelepipeds is performed in the same way as for constructing the line in the case of the plane. Thus we have:

"Given in the space a finite set of parallelepipeds with edges parallel to three fixed directions, if 16 of these parallelepipeds possess points situated in the same plane, there exists a plane having a point common to all".

The procedure for the proof is general for any number of dimensions. We obtain:

"*Given in the* n*-dimensional Euclidean space a finite set of parallelepipeds with edges parallel to n fixed directions, if every* $2^{n-1}(n+1)$ *of them has a common secant hyperplane, there exists a hyperplane with a point common to them all".*

2. Lines intersecting a set of parallelepipeds. Let us also take a system of axes parallel to the directions of the three concurrent edges of the parallelepipeds. The result to be proven is as follows:

"Given a finite set of parallelepipeds with edges parallel to 3 fixed directions such that 20 of them possess a common secant line, there exists a secant line common to them all".

As in the case of the plane, and proceeding by induction, the problem boils down to proving if, given $N+1$ parallelepipeds, any N of them have a common secant, there exists a secant common to all.

Let $P_1, P_2, P_3, \ldots, P_{N+1}$ be the parallelepipeds. The line intersecting them all except for P_k can be written thus

$$\begin{aligned} ax + b_k y + c_k &= 0 \\ ax + d_k z + e_k &= 0 \end{aligned} \tag{9}$$

6 of these lines for which b_k and d_k have constant sign are required. In fact, it is enough to have 21 lines, since there will necessarily be 11 having b_k with constant sign, and among these 11 there will certainly be 6 having d_k with constant sign. Thus we have $N+1 \geq 21$; that is, $N \geq 20$, and therefore the statement is fulfilled.

It can be assumed that the 6 lines thus chosen are those corresponding to $k = 1, 2, \ldots, 5, 6$. Let them be, for example, of the form

$$\begin{aligned} ax + b_k y + c_k &= 0 \\ ax - d_k z + e_k &= 0 \quad (k = 1, 2, \ldots, 6). \end{aligned} \tag{10}$$

The reasoning would be followed analogously if the sign of b_k or d_k were different from that indicated.

Let us form the following system

$$\begin{aligned} \sum \lambda_k &= 0 \\ \sum b_k \lambda_k &= 0 \\ \sum c_k \lambda_k &= 0 \\ \sum d_k \lambda_k &= 0 \\ \sum e_k \lambda_k &= 0 \quad (k = 1, 2, \ldots, 6) \end{aligned} \tag{11}$$

with the unknowns λ_k. This system of homogeneous equations has solution, and the positive terms λ_k can be placed on one side and the negative on the other, thus obtaining

$$\begin{aligned} \sum{}' \lambda_k &= \sum{}'' \lambda_k \\ \sum{}' b_k \lambda_k &= \sum{}'' b_k \lambda_k \\ \sum{}' c_k \lambda_k &= \sum{}'' c_k \lambda_k \\ \sum{}' d_k \lambda_k &= \sum{}'' d_k \lambda_k \\ \sum{}' e_k \lambda_k &= \sum{}'' e_k \lambda_k \end{aligned} \tag{12}$$

The straight line

$$\begin{aligned} a(\sum{}' \lambda_k)x + (\sum{}' b_k \lambda_k)y + (\sum{}' c_k \lambda_k) = 0 \\ a(\sum{}' \lambda_k)x - (\sum{}' d_k \lambda_k)y + (\sum{}' e_k \lambda_k) = 0 \end{aligned} \tag{13}$$

which can also be expressed as

$$\begin{aligned} a(\sum{}'' \lambda_k)x + (\sum{}'' b_k \lambda_k)y + (\sum{}'' c_k \lambda_k) = 0 \\ a(\sum{}'' \lambda_k)x - (\sum{}'' d_k \lambda_k)y + (\sum{}'' e_k \lambda_k) = 0 \end{aligned} \tag{13'}$$

intersects all the $N + 1$ parallelepipeds.

Indeed, let P_i be one of these parallelepipeds. Then consider from between (13) and (13') that in which the index k does not take the value i. Let this one be, for example, (13). Since P_i is intersected by all the lines (9), except for that corresponding to $k = i$, then for the values of k appearing in (13) there are points of P_i that fulfil

$$\begin{aligned} ax_k^{(i)} + b_k y_k^{(i)} + c_k &= 0 \\ ax_k^{(i)} - d_k z_k^{(i)} + e_k &= 0 \end{aligned} \tag{14}$$

The point

$$\xi^{(i)} = \frac{\sum{}' \lambda_k x_k^{(i)}}{\sum{}' \lambda_k}, \quad \eta^{(i)} = \frac{\sum{}' b_k \lambda_k y_k^{(i)}}{\sum{}' b_k \lambda_k} \quad \zeta^{(i)} = \frac{\sum{}' d_k \lambda_k z_k^{(i)}}{\sum{}' d_k \lambda_k}$$

belongs to P_i, and taking into account (14), one sees that it satisfies equations (13); that is, it belongs to the said line.

The same procedure can be generalized almost without modification to a Euclidean space of any number of dimensions. If we denote a straight line by $n-1$ equations of the form

$$\begin{aligned} ax_1 + b_2 x_2 + c_2 &= 0 \\ ax_1 + b_3 x_3 + c_3 &= 0 \\ ax_1 + b_4 x_4 + c_4 &= 0 \\ \cdots\cdots\cdots\cdots & \\ ax_1 + b_n x_n + c_n &= 0 \end{aligned}$$

the approach previously adopted can be reproduced without difficulty. We then obtain:

"*Given in the n-dimensional space a finite set of parallelepipeds with edges parallel to n fixed directions, if for every* $2^{n-1}(2n-1)$ *of these parallelepipeds there exists a line that intersects them, there exists a line intersecting them all*".

III. Generalization to infinite sets

The foregoing results also hold for the case of finite sets of parallelograms or parallelepipeds situated at finite distance; that is, all contained within a bounded region in the plane.

Let us first consider the case of the plane.

First, let $P_1, P_2, P_3, \ldots, P_n, \ldots$ be a numerable infinity of parallelograms. Then take any P_0 and consider the straight lines $R_1, R_2, R_3, \ldots, R_n, \ldots$ that intersect $P_0P_1, P_0P_1P_2, P_0P_1P_2P_3, P_0P_1P_2P_3 \ldots P_n, \ldots$. Now let us draw a circumference whose centre is an arbitrary point O containing P_0 in its interior. Each line R_n can be represented by the point Q_n, which is a projection of O on that line. Points Q_n are all interior to the said circumference, and since they are infinite there exists an accumulation point Q. The line corresponding to this point (that is, that perpendicular by Q to the line OQ) intersects all the parallelograms. Indeed, if it did not intersect one of these parallelograms, P_i, there would exist a neighbourhood of Q such that all the lines corresponding to their points would not intersect it either (assuming the parallelograms also to include their sides; that is, considered as closed sets). However, this is not possible, since Q is an accumulation point of points Q_n, and thus there exists in every neighbourhood of Q points whose corresponding lines intersect any P_i, however large i may be in the given numerable sequence.

It may happen that the point of accumulation Q is O itself. Then a sequence of points Q_n tending to O is chosen, and a limit direction of the lines connecting it with O is taken. As before, one observes that the one perpendicular by O to this direction intersects all the parallelograms.

Once proven for numerable sets, the step to arbitrary sets can be performed in the following manner: every parallelogram can be replaced by another that contains it within its interior of parallel sides sufficiently close for any point in the second one to be less than distance ϵ from the first, and such that the vertices are points of rational coordinates. The set thus obtained is numerable, and since for every 6 of its parallelograms there exists a common secant, we have for each ϵ a secant common to the infinite parallelograms belonging to the numerable set. One of these lines that intersects all the parallelograms belonging to the numerable set will have points that are at most at distance ϵ from the parallelograms belonging to the primitive set. By giving ϵ a decreasing sequence of values $\epsilon_1 > \epsilon_2 > \epsilon_3 > \dots$ with $\lim \epsilon_i = 0$, an arbitrary limit line of the system must intersect them all.

For the case of the space, when dealing with planes one may proceed analogously, denoting each plane by the projection on that plane of a fixed point O.

For the case of lines, it does not suffice to determine their intersection; from the lines with the perpendicular drawn by a fixed point, it is also necessary to give the direction of the normal to this plane. One may proceed as follows: first let $P_0, P_1, P_2, P_3, \dots$ be a numerable infinity of parallelepipeds. Consider the lines R_n $(n = 1, 2, 3, \dots)$ intersecting the first parallelepipeds n of the sequence, and a sphere whose centre is an arbitrary point O containing P_0 in its interior. The planes perpendicular to R_n and the lines parallel to them are drawn from O. The intersections of these planes with R_n determine a set of points A_n interior to the sphere, while the parallel lines determine on the surface of the sphere a further set of points B_n. Since A_n are an infinite number, they will have at least one accumulation point A. Let us take a sequence $A_1, A_2, A_3, \dots$ tending to A. The corresponding points B_i of the set of points on the sphere will also have at least one accumulation point B. It is easy to see that the line passing through A and parallel to OB intersects all the parallelepipeds.

The step to non-numerable sets is performed in the same way as for the plane.

IV. Questions to be resolved

The procedure followed thus far leaves the following two questions to be resolved:

1. Are the numbers obtained, 6 for ensuring the existence of a common secant in the plane, 16 for secant planes in the space, and 20 for the case of lines, ... the minimum numbers? In other words, in the case of the plane for example, is it

possible to give 6 parallelograms with sides parallel to two fixed directions such that every five of them has a common secant and without any line intersecting all 6?

2. We have been considering lines and hyperplanes in n-dimensional Euclidean space. What occurs if r-dimensional secant linear manifolds are considered? In other words, for what number ν is it true that if in a set of parallelepipeds with edges parallel to n fixed directions, any ν of them having an r-dimensional linear manifold as a common secant, there exists an r-dimensional manifold secant common to them all?

The cases solved correspond to $r = 1$ and $r = n - 1$.

References

[1] D. König, *Ueber konvexe Körper*, Math. Zeits **14** (1922), 208–210.

[2] J. Radon, *Mengen Konvexer Körper, die cinen gemeinsamen Punkt enthalten*, Math. Ann. **83** (1921), 113–115.

[3] P. Vincensini, *Figures convexes et varietés linéaires de l'espace euclidien a n dimensions*, Bull. Sciences Math. **59** (1935), 163–174.

ROSARIO. INSTITUTO DE MATEMÁTICAS.

Supplement to the note: A theorem on sets of parallelepipeds with parallel edges

L. A. Santaló

Abstract

In this paper, we complete some points of a preceding paper entitled "A Theorem on Sets of Parallelepipeds with Parallel Edges" (these "*Publicaciones*", Vol. II, n° 4), with a proof of the following theorems:

I. Given a set of parallel segments on the plane, if each group of 3 such segments can be intersected by a straight line, there exists a straight line intersecting all the segments of the set.

II. Given a set of $N-1$ parallel segments and 1 segment that is not parallel to these segments, if each group of 5 of these N-segments can be intersected by a straight line, there exists a straight line intersecting all the segments of the set.

In paper n° 4, Vol. II of "*Publicaciones del Instituto de Matemáticas*" entitled "A Theorem on Sets of Parallelepipeds with Parallel Edges", we proved the following theorem:

Given a set of parallelograms with sides parallel to two fixed directions, if each group of 6 of such parallelograms have points on the same straight line, there exists a straight line having a common point with all the parallelograms of the set.

The question of whether or not the number 6 appearing in this statement was minimum remained to be resolved.[1]

This question has been answered in the affirmative by R. Frucht, who gives an example of 6 segments with sides parallel to two fixed directions such that each 5 segments have a common secant, but without the existence of a straight line having a common point with all of them. Obviously, this proves that the number 6 cannot be reduced.

The example provided by Frucht is as follows[2] (Fig. 1):

[1] See section IV of the paper cited, p. 59.

[2] This example was published in the *Revista de la Unión Matemática Argentina,* Vol VII, 1941, p.134.

Segment N°	Cartesian coordinates of the endpoints	Equation of a straight line intersecting only the other 5 segments
I	(1,8) &(4,8)	y=-x+8
II	(1,7) &(2,7)	y=x+4
III	(1,6) &(2,6)	y=-x+9
IV	(1,5) &(4,5)	y=x+5
V	(0,3) &(0,9)	y=-2x+10
VI	(5,0) &(5,10)	y=2x+3

Although this example refers to the limit cases of parallelograms reduced to segments, it is obvious that the example is equally valid for real parallelograms, merely by considering each segment as the base of a rectangle with a sufficiently small height.

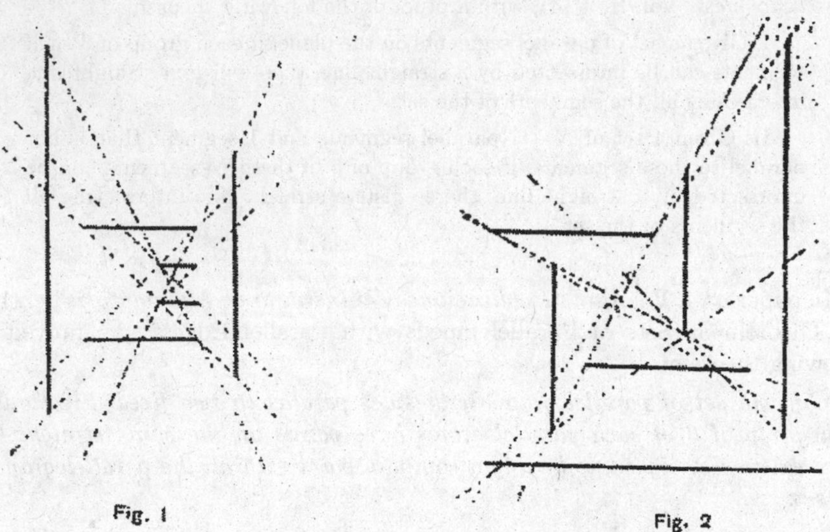

Fig. 1 Fig. 2

A further example, also for the case of segments, has been explained to us verbally by E. Corominas, which is as follows:

Segment N°	Cartesian coordinates of the endpoints	Co-ordinates of the points that determine a line intersecting the other 5 segments
I	(-2,4) & (3,4)	(0,-2) & (4,1)
II	(0,-2) & (0,3)	(4,1) & (7,-2)
III	(1,0) & (6,0)	(7,10) & (3,4)
IV	(4,1) & (4,6)	(-2,4) & (10,-3)
V	(7,-2) & (7,10)	(4,6) & (0,-2)
VI	(-2,-3) & (10,-3)	(-2,4) & (6,0)

This example, considering the segments as bases of very narrow rectangles, could also be used for the general case of parallelograms.

Comparison of the two examples above lead us to the following considerations about the particular case of finite sets of segments in the plane, parallel to two fixed directions.

Frucht's and Corominas' examples are substantially different in that while the first has 2 segments parallel to one direction and 4 parallel to the other, the second has 3 segments parallel to each of the two fixed directions of the problem.

The following question then arises: Is there an example with only one segment parallel to the first direction and 5 parallel to the second, or with 6 segments parallel to one and the same direction?

We now see how this question is answered in the negative in both cases.

1 Finite sets of parallel segments

We prove the theorem:

Given a finite set of parallel segments, if each 3 of such segments have points on the same straight line, there exists a straight line with a point common to all the segments in the set.

We furnish two proofs; the first analogous to that given in the cited work for the general case, and the second direct proof, although based on the same principle.

If two segments are on the same line, all the segments are on the same line and the proposition is obvious. If the segments are on different lines, we consider all the straight lines that intersect at least two segments in the set. If the set is finite, among all these lines there will be one having a maximum inclination with respect to the common direction of the parallel lines. Take a system of oblique co-ordinate axes such that the axis x has the direction of the parallel lines, and the axis y possesses with respect to x a greater inclination than the maximum referred to above. As regards this system of axes, all the straight lines intersecting at least 2 segments in the given set are of the form $x - b_k y + c_k = 0$, with $b_k > 0$.

In order to have 4 of these lines and continue the general reasoning of the work cited, it suffices that the number of segments be equal to or greater than 4, and that each 3 of the segments have points common to a line, in which case the theorem is proven.

If we do not wish to refer to the general proof, for this particular case a further direct proof can be given, based on an analogous reasoning[3]. Proceeding by induction, for the proof of the theorem it suffices to establish that if each $N - 1$ segments of a set N of these segments ($N \geq 4$) situated on parallel lines have points common to a line, there exists a line having a point common to all the N segments. Let

$$x = c_k - b_k y \tag{1}$$

be the line intersecting all the segments, except perhaps the kth segment. If $N \geq 4$, there will be at least 4 of these lines. Let us take any 4 of them to form

[3] I am grateful to Dr. B. Levi for his assistance with this direct proof.

a system of homogeneous equations

$$\sum \lambda_k = 0, \quad \sum \lambda_k c_k = 0, \quad \sum \lambda_k b_k = 0 \tag{2}$$

with the unknowns λ_k ($k = 1, 2, 3, 4$). By the first equation, the λ_k cannot all have the same sign; we thus place the positive on one side and the negative on the other:

$$\begin{aligned} \sum{}' \lambda_k &= \sum{}'' \lambda_k \\ \sum{}' \lambda_k c_k &= \sum{}'' \lambda_k c_k \\ \sum{}' \lambda_k b_k &= \sum{}'' \lambda_k b_k, \end{aligned} \tag{3}$$

where the sum extended to the positive λ_k is denoted by a prime, and the negatives are denoted by two primes.

Let us now consider a straight line that can be represented by either of the following two equations:

$$(\sum{}' \lambda_k)x = \sum{}' \lambda_k c_k - (\sum{}' \lambda_k b_k)y \tag{I}$$

$$(\sum{}'' \lambda_k)x = \sum{}'' \lambda_k c_k - (\sum{}'' \lambda_k b_k)y. \tag{II}$$

We state that this line intersects all the segments. Indeed, let us consider any one of these segments, for example, the mth (situated on the straight line $y = y_m$). Since it is intersected by $N-1$ of the lines (I) (all, except, at most, that corresponding to $k = m$), it will have $N-1$ points of intersection: let $x_m^{(i)}$ ($m \neq i$) be the abscissae of these points; that is,

$$x_m^{(i)} = c_i - b_i y_m \quad (i = 1, 2, \ldots, m-1, m+1, \ldots, N). \tag{4}$$

In the sums (3), it may occur that the value $k = m$ does not appear, but even when it does, it will always appear only either on the left-hand sides or the right-hand sides, so that we may take one of the forms of the line, either (I) or (II), such that in the sums of the coefficients the value $k = m$ does not appear. Let, for example, $\sum'$ be the sum that fulfils this condition. The point of co-ordinates

$$\xi = \frac{\sum' \lambda_k x_m^{(k)}}{\sum' \lambda_k}, \quad \eta = y_m$$

belongs to the mth segment (since λ_k are positive), and it is also on the line (I), as may be seen immediately, taking into account (4).[4]

[4] A geometric way to prove the theorem of this n° 1 was shown to me by Dr. J. Rey Pastor. It is as follows: By a dual transformation, the segments in the statement of the theorem are transformed into angles whose vertices are on a straight line; by passing this line to the infinite by a projectivity, the problem is reduced to proving that: given in the plane a set of parallel strips such that each 3 of them has a common point, there exists a point that is common to all of them. It is not difficult to prove this theorem directly, and furthermore the theorem is no more than a particular case of Helly's theorem proved by Radon, [2] and König [1].

2 Application to a theorem on systems of linear equations

The previous theorem can also be stated in an arithmetical way. Indeed, one may observe that, assuming a system of co-ordinates with the axis x parallel to the segments given and any axis y, if the segments are situated on the lines $y = \gamma_i (i = 1, 2, 3, \ldots, N)$ and the respective endpoints have α_i, β_i for abscissae, then stating that the line $x = \xi y + \eta$ intersects the ith segment is equivalent to writing

$$\alpha_i \leq \xi\gamma_i + \eta \leq \beta_i.$$

Therefore, the theorem from Section n° 1 can be stated as follows: *Given a system of inequalities of the form*

$$\alpha_i \leq x\gamma_i + y \leq \beta_i \qquad (i = 1, 2, 3, \ldots, N)$$

with the unknowns x, y, *if each group of 3 of them are compatible, then so is the total system.*

3 The case of $N - 1$ parallel segments, plus 1 segment not parallel to them

In this case, the number 6 that appears in the statement of the general theorem can be reduced to 5. In fact, we prove that

Given $N - 1$ parallel segments and only one segment not parallel to them (with $N \geq 6$), *if each 5 segments have points common with a straight line, there exists a line having common points with all the segments in the set.*

The proof will also be given by reduction to the general case.

By the theorem of parallel segments, the $N-1$ parallel segments in the statement have a common secant. If this secant intersects the non-parallel segment, the theorem is proven. If it does not intersect this segment, it will always be possible by a projective transformation conserving the parallelism of the first $N - 1$ segments to make this line intersecting them become parallel to the line containing the Nth segment. Consider in this position a system of co-ordinate axes whose axis x has the direction of the first $N - 1$ segments, and axis y the direction of the Nth segment. Let $x - c_N = 0$ be the equation, in this system, of the line intersecting the first $N - 1$ segments. Proceeding by induction, as in the general case of the cited paper, it can be assumed that each set of $N - 1$ segments of the given N segments has a common secant. Consider, then, the $N - 1$ lines intersecting the Nth segment and the different groups of $N - 2$ segments that can be formed with the parallel $N - 1$ segments. These lines are of the form $x + b_k y + c_k = 0$ $(k = 1, 2, 3, \ldots, N - 1)$. Since, by hypothesis, $N \geq 6$ and therefore $N - 1 \geq 5$, then among these lines there are at least 3 having the b_k with the same sign. These three lines, together with the line $x - c_k = 0$, constitute the 4 lines required in order to follow the general reasoning and conclude the proof of the theorem.[5]

[5]In the cited Publication, it was stated that the b_k should be of *the same sign*. It was also

This number 5, of the segments required in this particular case, cannot be reduced. Indeed, the following example (Fig. 3) proves that there can be 5 segments (4 parallel and 1 not parallel to them) such that 4 of these segments have points common to a line, without the existence of any line having a point common to all.

Segment N°	Coordinates of the endpoints			Pairs of points belonging to a line intersecting the other 4		
I	(0,0)	&	(0,10)	(6,0)	&	(6,8)
II	(1,8)	&	(6,8)	(0,0)	&	(6,6)
III	(4,6)	&	(6,6)	(0,10)	&	(6,2)
IV	(4,4)	&	(6,4)	(0,0)	&	(6,8)
V	(1,2)	&	(6,2)	(0,10)	&	(6,4)

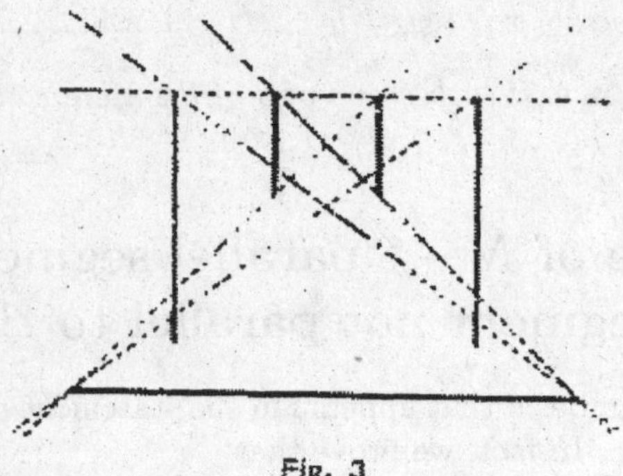

Fig. 3

4 Generalization to infinite sets

The two cases studied here (Nos. 1 and 3) also hold in the case of infinite sets of segments *situated at a finite distance.* In both cases, the proof is the same as that given for the general case in the "Publication" referred to in this paper (see Section III, page 58).

References

[1] D. König, *Ueber konvexe Körper*, Math. Zeits **14** (1922), 208–210.

[2] J. Radon, *Mengen Konvexer Körper, die cinen gemeinsamen Punkt enthalten*, Math. Ann. **83** (1921), 113–115.

ROSARIO, INSTITUTO DE MATEMÁTICA, MARZO, 1942.

necessary to add that some of them, or all, can be *zero*, whereby the reasoning still holds. The fundamental condition was that there should not be two b_k with *opposite signs*.

ANNALS OF MATHEMATICS
Vol. 47, No. 3, July, 1946

CONVEX REGIONS ON THE n-DIMENSIONAL SPHERICAL SURFACE

BY L. A. SANTALÓ

(Received October 29, 1945)

1. Introduction

Let K be any bounded subset of n-dimensional euclidean space. If D is the diameter of K and R the radius of the smallest spherical surface enclosing K, it is known that the following relation holds:

$$R \leqq \left(\frac{n}{2n+2}\right)^{\frac{1}{2}} D. \tag{1.1}$$

This result was obtained by H. W. E. Jung [6], [7]; for bibliography until 1934 see Bonnesen-Fenchel [3, p. 78]. More recent proofs have been given by W. Süss [11] and L. M. Blumenthal-G. E. Wahlin [2].

If K is now a convex set of n-dimensional euclidean space, the "breadth" B of K is defined as the minimum distance of two parallel supporting hyperplanes of K. Let r be the radius of the greatest spherical surface which is contained in K. As a kind of dual of Jung's theorem (1.1) are known the relations

$$r \geqq \begin{cases} \dfrac{(n+2)^{\frac{1}{2}}}{2n+2} B & \text{for } n \text{ even} \\ \frac{1}{2} n^{-\frac{1}{2}} B & \text{for } n \text{ odd.} \end{cases} \tag{1.2}$$

For $n = 2$ this theorem was proved by W. Blaschke [1]; for any n by Steinhagen [10]; for bibliography until 1934 see Bonnesen-Fenchel [3, p. 79]. Another proof was given by H. Gericke [4].

The purpose of the present paper is to give a generalization of inequalities (1.1) and (1.2) to sets on the n-dimensional spherical surface. Whilst in the euclidean case it is necessary to give an independent proof for each inequality (1.1), (1.2) it will be enough to prove in the spherical case the generalization of Jung's inequality (1.1), because the generalization of the inequalities (1.2) will then follow by duality.

As an application of these results we obtain two theorems (Theorem 3 and 4) referring to convex regions on the n-dimensional spherical surface, the last of which generalizes a known theorem [8] of Robinson.

For $n = 2$ a geometrical proof of the results contained in this paper has been given by the author in [9].

2. Spherical Simplexes on $S_{n,1}$

An n-dimensional spherical surface $S_{n,1}$ is the "surface" of an $(n + 1)$-dimensional sphere of unit radius in the $(n + 1)$-dimensional euclidean space.

Let T_n be an equilateral spherical simplex on $S_{n,1}$, that is, the spherical simplex determined by $n + 1$ points $a_1, a_2, a_3, \cdots, a_{n+1}$ of $S_{n,1}$ whose mutual distances measured on $S_{n,1}$ have the constant value l (l = edge of T_n). If a_i also represents the unit vectors with the origin at the center of $S_{n,1}$ and with the end points a_i, we have

$$a_i^2 = 1, \qquad a_i a_j = \cos l. \tag{2.1}$$

The circumscribed sphere of T_n considered as an n-dimensional sphere of $S_{n,1}$ has a spherical center c, which is a point of $S_{n,1}$ and a spherical radius R, so that

$$c^2 = 1, \qquad ca_i = \cos R. \tag{2.2}$$

In order to calculate the value of R as a function of the edge l it suffices to observe that we can put

$$c = \sum_1^{n+1} \lambda_i a_i, \qquad \lambda_i > 0.$$

Since T_n is an equilateral simplex, we have $\lambda_i = \lambda$ = constant, and from (2.1) and (2.2), also

$$c^2 = (n+1)\lambda^2 + n(n+1)\lambda^2 \cos l = 1, \qquad ca_i = \cos R = \lambda + n\lambda \cos l$$

whence

$$\cos R = \left(\frac{1 + n\cos l}{n+1}\right)^{\frac{1}{2}}. \tag{2.3}$$

We thus get the relation $\cos l \geqq -1/n$, which holds for any equilateral spherical simplex of $S_{n,1}$.

We wish now to calculate the spherical diameter of T_n. Let x, y be the end-points of a diameter of T_n. Let $a_1, a_2, \cdots a_\nu$ $(1 \leqq \nu \leqq n+1)$ be the vertices of the simplex $T_{\nu-1}$ of minimal dimension whose vertices are among those of T_n and which contains the point x. The end-point y cannot be a point of $T_{\nu-1}$ because in this case x and y would be points of the boundary of $T_{\nu-1}$ and x would be contained in a simplex of dimension $<\nu$. Consequently y is a point which belongs to the simplex $T_{n-\nu}$ whose vertices are $a_{\nu+1}, a_{\nu+2}, \cdots, a_{n+1}$. Hence we have

$$x = \sum_1^{\nu} \lambda_i a_i, \qquad y = \sum_{\nu+1}^{n+1} \lambda_i a_i, \qquad \lambda_i \geqq 0 \tag{2.4}$$

with

$$\begin{aligned} x^2 &= \sum_1^{\nu} \lambda_i^2 + 2 \sum_{i,j=1}^{\nu}{}' \lambda_i \lambda_j \cos l = 1 \\ y^2 &= \sum_{\nu+1}^{n+1} \lambda_i^2 + 2 \sum_{i,j=\nu+1}^{n+1}{}' \lambda_i \lambda_j \cos l = 1 \end{aligned} \tag{2.5}$$

where $\sum'$ denotes a summation with $i = j$ excluded.

The spherical distance φ from x to y is given by

$$\cos\varphi = xy = \left(\sum_1^{\nu} \lambda_i\right)\left(\sum_{\nu+1}^{n+1} \lambda_i\right) \cos l. \tag{2.6}$$

To find the maximal value of φ we consider two cases:

a) $l \leqq \pi/2$, $\cos l \geqq 0$. By (2.5) we have

$$\begin{aligned} \left(\sum_1^{\nu} \lambda_i\right)^2 &= \sum_1^{\nu} \lambda_i^2 + 2 \sum_{i,j=1}^{\nu}{}' \lambda_i\lambda_j = 1 + 2 \sum_{i,j=1}^{\nu}{}' \lambda_i\lambda_j(1 - \cos l) \\ \left(\sum_{\nu+1}^{n+1} \lambda_i\right)^2 &= \sum_{\nu+1}^{n+1} \lambda_i^2 + 2 \sum_{i,j=\nu+1}^{n+1}{}' \lambda_i\lambda_j = 1 + 2 \sum_{i,j=\nu+1}^{n+1}{}' \lambda_i\lambda_j(1 - \cos l). \end{aligned} \tag{2.7}$$

The maximum value of φ corresponds to the minimum value of $\cos\varphi$, or according to (2.6) and (2.7) to $\lambda_1 = 1$, $\lambda_2 = \lambda_3 = \cdots = \lambda_\nu = 0$, $\lambda_{\nu+1} = 1$, $\lambda_{\nu+2} = \lambda_{\nu+3} = \cdots = \lambda_{n+1} = 0$. Consequently the spherical diameter of T_n equals the edge l.

b) $l > \pi/2$, $\cos l < 0$. By (2.6) the maximum value of φ corresponds to the maximum of the product $(\sum_1^{\nu} \lambda_i)$ $(\sum_{\nu+1}^{n+1} \lambda_i)$ with the relations (2.7). Taking into account the inequality

$$\lambda_i\lambda_j \leqq \tfrac{1}{2}(\lambda_i^2 + \lambda_j^2) \tag{2.8}$$

and $\cos l < 0$, we deduce from (2.5)

$$1 \geqq \sum_1^{\nu} \lambda_i^2 + \sum_{i,j=1}^{\nu}{}' (\lambda_i^2 + \lambda_j^2) \cos l = \sum_1^{\nu} \lambda_i^2(1 + (\nu - 1) \cos l). \tag{2.9}$$

Moreover, since

$$\left(\sum_1^{\nu} \lambda_i\right)^2 \leqq \nu \sum_1^{\nu} \lambda_i^2 \tag{2.10}$$

we deduce from (2.9)

$$\left(\sum_1^{\nu} \lambda_i\right)^2 \leqq \frac{\nu}{1 + (\nu - 1) \cos l}. \tag{2.11}$$

Analogously we get

$$\left(\sum_{\nu+1}^{n+1} \lambda_i\right)^2 \leqq \frac{n - \nu + 1}{1 + (n - \nu) \cos l}. \tag{2.12}$$

From (2.6), (2.11) and (2.12) and $\cos l < 0$ the inequality

$$\cos\varphi \geqq \frac{\nu^{\frac{1}{2}}(n - \nu + 1)^{\frac{1}{2}} \cos l}{(1 + (\nu - 1) \cos l)^{\frac{1}{2}}(1 + (n - \nu) \cos l)^{\frac{1}{2}}} \tag{2.13}$$

follows. There is equality only when $\lambda_1 = \lambda_2 = \cdots = \lambda_\nu$ and $\lambda_{\nu+1} = \lambda_{\nu+2} = \cdots = \lambda_{n+1}$.

By considering the right hand member of (2.13) as a function of the continuous variable ν and equating to 0 its derivative, we find

$$(n - 2\nu + 1)\,(1 + (n - 1) \cos l - n \cos^2 l) = 0.$$

Except the limit cases $\cos l = 1$, $\cos l = -1/n$, the value of ν for which the right hand member of (2.13) is a minimum corresponds to the integral solution of the equation $n - 2\nu + 1 = 0$. Hence we have

$$\cos \varphi \geqq \frac{(n + 1) \cos l}{2 + (n - 1) \cos l} \qquad \text{for } n \text{ odd} \tag{2.14}$$

$$\cos \varphi \geqq \frac{(n(n + 2))^{\frac{1}{2}} \cos l}{(2 + (n - 2) \cos l)^{\frac{1}{2}}(2 + n \cos l)^{\frac{1}{2}}} \qquad \text{for } n \text{ even.} \tag{2.15}$$

The right hand member of (2.14) is the spherical distance between two opposite spherical simplexes $T_{\frac{1}{2}(n-1)}$ of T_n. The right hand member of (2.15) is the spherical distance from a $T_{n/2}$ to the opposite $T_{n/2-1}$ of T_n.

Summing up our conclusions we have

LEMMA 1. *The spherical diameter D_0 of an equilateral spherical simplex T_n of edge l has the following values:*

a) *If $l \leqq \pi/2$, $D_0 = l$, that is, in virtue of* (2.3)

$$\cos D_0 = \frac{1}{n}\,((n + 1) \cos^2 R - 1). \tag{2.16}$$

b) *If $l > \pi/2$ (and always $\cos l \geqq -1/n$), that is, for $\cos R \leqq (n + 1)^{-\frac{1}{2}}$ it is*

$$\cos D_0 = \frac{(n + 1) \cos l}{2 + (n - 1) \cos l} \qquad \textit{for } n \textit{ odd} \tag{2.17}$$

$$\cos D_0 = \frac{(n(n + 2))^{\frac{1}{2}} \cos l}{(2 + (n - 2) \cos l)^{\frac{1}{2}}(2 + n \cos l)^{\frac{1}{2}}} \qquad \textit{for } n \textit{ even} \tag{2.18}$$

or, according to (2.3)

$$\cos D_0 = \frac{(n + 1) \cos^2 R - 1}{1 + (n - 1) \cos^2 R} \qquad \textit{for } n \textit{ odd} \tag{2.19}$$

$$\cos D_0 = \frac{(n + 1) \cos^2 R - 1}{(1 + (n + 1) \cos^2 R)^{\frac{1}{2}}\left(1 + \dfrac{(n + 1)(n - 2)}{n + 2} \cos^2 R\right)^{\frac{1}{2}}} \tag{2.20}$$

for n even.

These formulas (2.19) and (2.20) will be needed in the sequel.

Let us consider now on $S_{n,1}$ a non equilateral simplex T'_n and let $b_1, b_2, b_3, \cdots, b_{n+1}$ be its vertices. Suppose that R is the spherical radius of the circumscribed sphere of T'_n and that the point c of $S_{n,1}$ is the spherical center of this circumscribed sphere. Let us suppose also that c belongs to T'_n. We shall prove the following

LEMMA 2. *Among the n-dimensional spherical simplexes on $S_{n,1}$ whose circumscribed sphere has the spherical radius R and which contain the spherical center of its circumscribed sphere, the equilateral simplex has a minimum diameter.*

PROOF. Let l be the edge of the equilateral spherical simplex T_n inscribed in the sphere of spherical radius R. We consider two cases:

1°. $l \leqq \pi/2$. In this case the spherical diameter of T_n is l. We shall prove that there exist at least one edge of T'_n of length greater than l. Since c is a point of T'_n we can write

$$c = \sum_1^{n+1} \mu_i b_i \quad \text{with} \quad \mu_i \geqq 0 \tag{2.21}$$

where

$$c^2 = \sum_1^{n+1} \mu_i^2 + 2 \sum_{i,j=1}^{n+1}{}' \mu_i \mu_j b_i b_j = 1. \tag{2.22}$$

We have also

$$\cos R = cb_k \quad \text{and} \quad \sum_1^{n+1} \mu_k \cos R = c \sum_1^{n+1} \mu_k b_k = c^2 = 1$$

whence

$$\cos R = \left(\sum_1^{n+1} \mu_k \right)^{-1}. \tag{2.23}$$

If all the edges of T'_n were smaller than or equal to l, we would have $b_i b_k \geqq \cos l$ since $l \leqq \pi/2$ and, since $\mu_i \geqq 0$,

$$\cos R = cb_k = \mu_k + \sum_{i=1}^{n+1}{}' \mu_i b_i b_k \geqq \mu_k + \sum_1^{n+1}{}' \mu_i \cos l.$$

since T'_n is not equilateral we get, by adding for $k = 1, 2, 3, \cdots, n+1$,

$$(n+1) \cos R > (1 + n \cos l) \sum_1^{n+1} \mu_k$$

and from (2.23)

$$\cos^2 R > \frac{1 + n \cos l}{n+1},$$

in contradiction to (2.3). This proves that at least one of the products $b_i b_k$ is smaller than $\cos l$ and consequently at least one of the edges of T'_n is greater than l.

2°. $l > \pi/2$. In this case the diameter D_0 of T_n is given by the formulas (2.19), (2.20). Let us consider two cases:

a) *n odd*. Let us put $m = \frac{1}{2}(n+1)$. We can write (2.21) in the form

$$c = \sum_1^m \mu_i b_i + \sum_{m+1}^{n+1} \mu_i b_i, \qquad \mu_i \geqq 0. \tag{2.24}$$

If δ is the spherical distance between the points $\sum_1^m \mu_i b_i / | \sum_1^m \mu_i b_i |$ and $\sum_{m+1}^{n+1} \mu_i b_i / | \sum_{m+1}^{n+1} \mu_i b_i |$ which belong to $S_{n,1}$, from (2.24) we deduce

$$\begin{aligned} c \sum_1^m \mu_i b_i = \sum_1^m \mu_i \cos R = \left(\sum_1^m \mu_i^2 + 2 \sum_{i,j=1}^{m}{}' \mu_i \mu_j b_i b_j \right) \\ + \left(\sum_1^m \mu_i^2 + 2 \sum_{i,j=1}^{m}{}' \mu_i \mu_j b_i b_j \right)^{\frac{1}{2}} \left(\sum_{m+1}^{n+1} \mu_i^2 + 2 \sum_{i,j=m+1}^{n+1}{}' \mu_i \mu_j b_i b_j \right)^{\frac{1}{2}} \cos \delta. \end{aligned} \tag{2.25}$$

Since $l > \pi/2$, from (2.17) it follows that $D_0 > \pi/2$. If the distance δ were smaller than or equal to D_0, we would have $\cos \delta \geqq \cos D_0$, and from (2.25),

$$\begin{aligned} \sum_1^m \mu_i \cos R - \left(\sum_1^m \mu_i^2 + 2 \sum_{i,j=1}^{m}{}' \mu_i \mu_j b_i b_j \right) - \left(\sum_1^m \mu_i^2 + 2 \sum_{i,j=1}^{m}{}' \mu_i \mu_j b_i b_j \right)^{\frac{1}{2}} \\ \cdot \left(\sum_{m+1}^{n+1} \mu_i^2 + 2 \sum_{i,j=m+1}^{n+1}{}' \mu_i \mu_j b_i b_j \right)^{\frac{1}{2}} \cos D_0 \geqq 0. \end{aligned} \tag{2.26}$$

Since $\cos D_0 < 0$, by applying the theorem of the arithmetic and geometric means, we obtain

$$\begin{aligned} \sum_1^m \mu_i \cos R - \left(\sum_1^m \mu_i^2 + 2 \sum_{i,j=1}^{m}{}' \mu_i \mu_j b_i b_j \right) \\ - \frac{1}{2} \left(\sum_1^{n+1} \mu_i^2 + 2 \sum_{i,j=1}^{m}{}' \mu_i \mu_j b_i b_j + 2 \sum_{i,j=m+1}^{n+1}{}' \mu_i \mu_j b_i b_j \right) \cos D_0 \geqq 0. \end{aligned}$$

Writing this inequality for all the $\binom{n+1}{m}$ possible combinations of $\mu_1, \mu_2, \cdots, \mu_{n+1}$ taken m at a time and summing, we get, since T_n' is not equilateral and $c^2 = 1$,

$$\begin{aligned} \binom{n}{m-1} \sum_1^{n+1} \mu_i \cos R - \left[\binom{n}{m-1} \sum_1^{n+1} \mu_i^2 + \alpha \left(1 - \sum_1^{n+1} \mu_i^2 \right) \right] \\ - \frac{1}{2} \left[\binom{n+1}{m} \sum_1^{n+1} \mu_i^2 + 2\alpha \left(1 - \sum_1^{n+1} \mu_i^2 \right) \right] \cos D_0 > 0 \end{aligned} \tag{2.27}$$

where

$$\alpha = \frac{\binom{m}{2} \binom{n+1}{m}}{\binom{n+1}{2}} = \frac{n-1}{4n} \binom{n+1}{m}. \tag{2.28}$$

Taking into account (2.23) which is also valid for the present case, we have

$$\cos D_0 < \frac{2 \left(\alpha - \binom{n}{m-1} \right) \left(\sum_1^{n+1} \mu_i^2 - 1 \right)}{\left(\binom{n+1}{m} - 2\alpha \right) \sum_1^{n+1} \mu_i^2 + 2\alpha}. \tag{2.29}$$

Furthermore

$$\frac{1}{\cos^2 R} = \sum_1^{n+1} \mu_i^2 + 2 \sum_{i,j=1}^{n+1}{}' \mu_i \mu_j$$

and since $2\mu_i\mu_j \leqq \mu_i^2 + \mu_j^2$, we have

$$\frac{1}{\cos^2 R} \leqq (n+1) \sum_1^{n+1} \mu_i^2 . \tag{2.30}$$

By (2.28), (2.29) and (2.30) we obtain

$$\cos D_0 < \frac{(n+1)\cos^2 R - 1}{1 + (n-1)\cos^2 R}.$$

We have arrived at a contradiction with (2.19). This proves that the assumption that all the distances δ were smaller than or equal to D_0 is false. Consequently the simplex T'_n has not smaller diameter than the equilateral simplex T_n.

b) *n even.* Let us put $m = n/2$. We may proceed as above until the inequality (2.26). Then summing for all the $\binom{n+1}{m}$ possible combinations of μ_1, $\mu_2, \cdots, \mu_{n+1}$ taken m at a time and applying the known Cauchy's inequality (see [5, p. 16])

$$\sum \xi\eta \leqq \left(\sum \xi^2\right)^{\frac{1}{2}}\left(\sum \eta^2\right)^{\frac{1}{2}}$$

we obtain since T'_n is supposed non equilateral and $c^2 = 1$,

$$\begin{aligned} &\binom{n}{m-1} \sum_1^{n+1} \mu_i \cos R - \left[\binom{n}{m-1} \sum_1^{n+1} \mu_i^2 + \alpha\left(1 - \sum_1^{n+1} \mu_i^2\right)\right] \\ &\quad - \left[\binom{n}{m-1} \sum_1^{n+1} \mu_i^2 + \alpha\left(1 - \sum_1^{n+1} \mu_i^2\right)\right]^{\frac{1}{2}} \\ &\quad \cdot \left[\binom{n}{m} \sum_1^{n+1} \mu_i^2 + \alpha'\left(1 - \sum_1^{n+1} \mu_i^2\right)\right]^{\frac{1}{2}} \cos D_0 > 0 \end{aligned} \tag{2.31}$$

where

$$\alpha = \frac{n-2}{2n}\binom{n}{m-1}, \qquad \alpha' = \frac{1}{2}\binom{n}{m}.$$

From (2.31) and (2.23) it follows that

$$\cos D_0 < \frac{1 - \sum_1^{n+1} \mu_i^2}{\left(\frac{n-2}{n+2} + \sum_1^{n+1} \mu_i^2\right)^{\frac{1}{2}}\left(1 + \sum_1^{n+1} \mu_i^2\right)^{\frac{1}{2}}}$$

and taking into account (2.30) we get

$$\cos D_0 < \frac{(n+1)\cos^2 R - 1}{\left(1 + \frac{(n-2)(n+1)}{n+2}\cos^2 R\right)^{\frac{1}{2}}(1 + (n+1)\cos^2 R)^{\frac{1}{2}}}$$

in contradiction to (2.20). Consequently the assumption that all the distances

δ were smaller than D_0 was false. Hence the simplex T'_n does not have a smaller diameter than the equilateral simplex T_n.

This completes the proof of the Lemma 2.

3. Circumscribed Sphere to Sets on the n-Dimensional Sphere

Let us consider sets of points on $S_{n,1}$, that is, on the surface of the sphere of unit radius in the $(n+1)$-dimensional euclidean space. We consider only sets which lie entirely in a hemisphere, hence, its spherical diameter D is always $\leqq \pi$.

Given a set K on $S_{n,1}$, the smallest sphere on $S_{n,1}$ enclosing K is called the "circumscribed sphere" to K; let R be its spherical radius. We have $R \leqq \pi/2$.

Our purpose is to give an inequality between R and D valid for any set K, which will be the generalization to $S_{n,1}$ of the Jung's inequality (1.1). That is to say, given R we wish to find the minimum value of D. For our purpose we can assume without loss of generality that K is a closed set.

Following the same way as in euclidean case (see Bonnesen-Fenchel [3, p. 77]) it is seen that the circumscribed sphere to K contains points of K which form a set K' whose spherical convex cover (*konvexe Hülle*, [3, p. 5]) contains the spherical center c of the circumscribed sphere. Hence we can choose points of K' forming the vertices of a spherical simplex T' whose diameter is not greater than the diameter of K, which contains the center c and has the same circumscribed sphere as T. Consequently to find the minimal value of D it suffices to consider only simplexes with the same circumscribed sphere of spherical radius R which contain the center of this sphere.

It can happen that the dimension of T' be smaller than n, but as the left hand sides of (2.16), (2.19) and (2.20) increase with n, we have, in virtue of Lemmas 1 and 2:

THEOREM 1.—*For any set K on the surface $S_{n,1}$ of the $(n+1)$-dimensional euclidean sphere of unit radius which lie on an hemisphere, the spherical diameter D of K and the spherical radius R of its circumscribed sphere satisfy the following relations*

1°. *If* $\cos R \geqq (n+1)^{-\frac{1}{2}}$ *it is*

$$\cos 2R \leqq \cos D \leqq \frac{(n+1)\cos^2 R - 1}{n} \tag{3.1}$$

2°. *If* $0 \leqq \cos R \leqq (n+1)^{-\frac{1}{2}}$ *it is*

$$\cos 2R \leqq \cos D \leqq \frac{(n+1)\cos^2 R - 1}{1 + (n-1)\cos^2 R} \quad \textit{for } n \textit{ odd} \tag{3.2}$$

$$\cos 2R \leqq \cos D \leqq \frac{(n+1)\cos^2 R - 1}{(1+(n+1)\cos^2 R)^{\frac{1}{2}}\left(1 + \frac{(n+1)(n-2)}{n+2}\cos^2 R\right)^{\frac{1}{2}}} \tag{3.3}$$

for n even.

4. Inscribed Sphere in a Convex Set on the n-Dimensional Sphere

A set K on the n-dimensional spherical surface of unit radius is said to be convex when: 1. It lies in an hemisphere of $S_{n,1}$. 2. Any great circle arc of $S_{n,1}$ whose end points lie in K, lies entirely in K.

A closed convex n-dimensional spherical set K is called a "convex spherical region".

Two great spheres of $S_{n,1}$ (generalization of the great circles of the sphere in ordinary euclidean space) divide $S_{n,1}$ into four "lunes". Let B be the angle of the smallest lune containing the convex spherical region K. We shall call B the "spherical breadth" of K.

The greatest sphere on $S_{n,1}$ which is enclosed in K is called the "inscribed sphere" of K; let r be its spherical radius.

The diametral hyperplanes perpendicular to the radii of $S_{n,1}$ which projects the surface of the convex region K from the center of $S_{n,1}$ envelop a cone whose intersection with $S_{n,1}$ is the surface of a new convex region K^*. We shall call K^* the dual region of K.

The spherical radius r of the inscribed sphere and the spherical breadth B of K are connected with the spherical radius R^* of the circumscribed sphere and the spherical diameter D^* of K^* by the relations

$$D^* + B = \pi, \qquad R^* + r = \pi/2.$$

Consequently, transforming by duality the Theorem 1, we obtain:

THEOREM 2. *For any convex spherical region K on the surface $S_{n,1}$ of the $(n+1)$-dimensional euclidean sphere of unit radius, the spherical breadth B and the spherical radius r of its inscribed sphere satisfy the following relations:*

1°. *If* $\sin r \geqq (n+1)^{-\frac{1}{2}}$ *it is*

$$\cos 2r \geqq \cos B \geqq \frac{1-(n+1)\sin^2 r}{n} \tag{4.1}$$

2°. *If* $0 \leqq \sin r \leqq (n+1)^{-\frac{1}{2}}$ *it is*

$$\cos 2r \geqq \cos B \geqq \frac{1-(n+1)\sin^2 r}{1+(n-1)\sin^2 r} \quad \textit{for } n \textit{ odd} \tag{4.2}$$

$$\cos 2r \geqq \cos B \geqq \frac{1-(n+1)\sin^2 r}{(1+(n+1)\sin^2 r)^{\frac{1}{2}}\left(1+\dfrac{(n+1)(n-2)}{n+2}\sin^2 r\right)^{\frac{1}{2}}} \tag{4.3}$$

for n even.

5. Passage to the Case of the n-Dimensional Euclidean Space

The formulas (1.1) and (1.2) for the euclidean space must result as a limit case of the preceding Theorems 1 and 2 when the radius of the spherical surface $S_{n,1}$ increases indefinitely.

If we now consider the n-dimensional spherical surface $S_{n,a}$ of radius a, the values R, D, r, B which are in the formulas of the Theorems 1 and 2, must be replaced by R/a, D/a, r/a, B/a.

Let us first consider Theorem 1. In order that the convex spherical region K tends to a bounded convex region of the n-dimensional euclidean space as $a \to \infty$, we must take the case $\cos R \geqq (n+1)^{-\frac{1}{2}}$ and we obtain by (3.1)

$$\cos (D/a) \leqq \frac{(n+1)\cos^2 (R/a) - 1}{n}$$

whence, by great values of a

$$1 - \frac{D^2}{2a^2} + \cdots \leqq 1 - \frac{n+1}{n}\frac{R^2}{a^2} + \cdots .$$

Simplifying and multiplying both sides by a^2 and making $a \to \infty$ we obtain the inequality (1.1).

Let us now consider Theorem 2. In order that the convex spherical region K tends to a bounded convex region of the n-dimensional euclidean space as $a \to \infty$, we must take the case $\sin r \leqq (n+1)^{-\frac{1}{2}}$ and we obtain, by (4.2) and (4.3)

$$\cos (B/a) \geqq \frac{1-(n+1)\sin^2 (r/a)}{1+(n-1)\sin^2 (r/a)} \quad \text{for } n \text{ odd}$$

$$\cos (B/a) \geqq \frac{1-(n-1)\sin^2 (r/a)}{(1+(n+1)\sin^2 (r/a))^{\frac{1}{2}}\left(1+\dfrac{(n+1)(n-2)}{n+2}\sin^2 (r/a)\right)^{\frac{1}{2}}} \quad \text{for } n \text{ even}$$

whence, for large values of a,

$$1 - \frac{B^2}{2a^2} + \cdots \geqq 1 - 2n\frac{r^2}{a^2} + \cdots \quad \text{for } n \text{ odd}$$

$$1 - \frac{B^2}{2a^2} + \cdots \geqq 1 - \frac{2(n+1)^2}{n+2}\frac{r^2}{a^2} + \cdots \quad \text{for } n \text{ even.}$$

Simplifying and multiplying both sides by a^2 and making $a \to \infty$ we obtain the inequalities (1.2).

6. Two Theorems on Convex Regions on the n-Dimensional Spherical Surface

Let K be a convex region with spherical diameter D on $S_{n,1}$. Clearly there is always on $S_{n,1}$ an $(n-1)$-dimensional spherical surface of spherical radius $R_1 = \frac{1}{2}(\pi - D)$ which intersect both K and its symmetrical region with respect the center of $S_{n,1}$.

For $R_1 = R$ we shall have the minimum value of R for which, for any K, there exist an $(n-1)$-dimensional sphere of spherical radius R on $S_{n,1}$ which either encloses K in its interior or intersects both K and its symmetrical region with respect the center of $S_{n,1}$.

For $\cos R \leqq (n+1)^{-\frac{1}{2}}$ by (3.2) and (3.3) it is $\cos D \leqq 0$, hence $D \geqq \pi/2$, $R_1 \leqq \pi/4$ and $\cos R_1 \geqq 2^{-\frac{1}{2}}$; consequently $R_1 \neq R$. To be $R = R_1$ we may therefore assume that $\cos R \geqq (n+1)^{-\frac{1}{2}}$ and then introducing in (3.1) the value $D = \pi - 2R$ we have

$$-\cos 2R = \frac{(n+1)\cos^2 R - 1}{n}$$

whence

$$\tan R = \left(\frac{2n}{n+1}\right)^{\frac{1}{2}}. \tag{6.1}$$

For the equilateral spherical simplex on $S_{n,1}$ inscribed in the $(n-1)$-dimensional spherical surface of spherical radius R given by (6.1), it is $D = l$ (l = edge of the simplex) and $\pi - l = 2R$. This proves that the value of R given by (6.1) cannot be diminished.

We have established the following theorem

THEOREM 3. *Let K be a convex spherical region of $S_{n,1}$. There is always an $(n-1)$-dimensional spherical surface of $S_{n,1}$ with spherical radius R given by* (6.1) *such that it either incloses K in its interior or intersects both K and its symmetrical region with respect the center of $S_{n,1}$. The value of R given by* (6.1) *cannot be diminished.*

By duality this theorem can be announced

THEOREM 4. *Let K be a convex region of $S_{n,1}$. There is always an $(n-1)$-dimensional spherical surface of $S_{n,1}$ with spherical radius r given by*

$$\tan r = \left(\frac{n+1}{2n}\right)^{\frac{1}{2}} \tag{6.2}$$

such that it is either enclosed in K or has neither any point in common with K nor with the symmetrical region of K with respect the center of $S_{n,1}$. The value of r given by (6.2) *cannot be increased.*

For $n = 2$ this theorem has been obtained by Robinson [8].

UNIVERSIDAD NACIONAL DEL LITORAL
ROSARIO, ARGENTINA.

BIBLIOGRAPHY

[1] W. BLASCHKE. *Ueber den grössten Kreis in einer konvexen Punktmenge*, Jahresbericht der Deutschen Math. Ver., Vol. 23, 1914, pp. 369–374.

[2] L. M. BLUMENTHAL AND G. E. WAHLIN. *On the spherical surface of smallest radius enclosing a bounded subset of n-dimensional euclidean space*, Bull. of the Am. Math. Soc., Vol. 47, 1941, pp. 771–777.

[3] T. BONNESEN UND W. FENCHEL. *Theorie der konvexen Körper*, Ergebnisse der Mathematik und ihrer Grenzgebiete, Berlin 1934.

[4] H. GERICKE, *Ueber die grösste Kugel in einer konvexen Punktmenge*, Mathematische Zeitschrift, Vol. 40, 1936.

[5] G. H. HARDY, J. E. LITTLEWOOD AND G. POLYA. *Inequalities*, Cambridge University Press, 1934.

[6] H. W. E. JUNG. *Ueber die kleinste Kugel die eine räumliche Figur einschliesst*, Journal für die reine und angewandte Mathematik, Vol. 123, 1901, pp. 241–257.

[7] H. W. E. JUNG. *Ueber den kleinstein Kreis der eine ebene Figur einschliesst*, ibid., vol. 137, 1909, pp. 310–313.

[8] R. M. ROBINSON. *Note on convex regions on the sphere*, Bull. of the Am. Math. Soc., Vol. 44, 1938, pp. 115–116.

[9] L. A. SANTALÓ. *Propiedades de las figuras convexas sobre la esfera*, Mathematicae Notae, Vol. 4, 1944, pp. 11–40.

[10] P. STEINHAGEN. *Ueber die grösste Kugel in einer konvexen Punktmenge*, Abhandlungen aus dem Mathematisches Seminar der Hamburgischen Universität, Vol. 1, 1922, pp. 15–26.

[11] W. SÜSS. *Durchmesser und Umkugel bei mehrdimensionalen Punktmengen*, Mathematische Zeitschrift, Vol. 40, 1936, pp. 315–316.

On convex bodies with constant width in E_n

L. A. Santaló

Abstract

It is known that between the length L and the width a of a plane convex figure of constant width, there exists a relation

$$L = \pi a$$

Analogously, between the volume V, the area F and the integral of the mean curvature M of a convex body in space, with constant width a, there exist the relations

$$2V - aF + 2\pi a^3/3 = 0$$
$$M - 2\pi a = 0$$

The aim of this paper is to generalize these relations, obtaining all the existent relations, to the convex bodies of constant width in n-dimensional Euclidean space. These relations are the equations (4.7) and (4.8) in Section 4.

1 Definitions and notation

Throughout this paper we follow the notations taken from Bonnesen-Fenchel's work "*Theorie der Konvexen Körper*", Berlin, 1934. Let us quickly recall the definitions required.

Let E_n be the n-dimensional Euclidean space. Every convex set of points of E_n that is limited and closed and has interior points is called a *convex body.*

The boundary points of a convex body C constitute a convex hypersurface which, for brevity's sake, we also represent by C.

Every hyperplane that has a common point with C without passing through it is called a *supporting hyperplane* (*Stützebene*) of a convex body C; that is, every hyperplane that has a common point with C but leaves all this body in the same semi-space of the two in which E_n is divided.

Let O be a point interior to C, which we take as the origin of coordinates. A direction $u = (u_1, \ldots, u_n)$ by O will be determined by its director cosines $u_1, \ldots, u_n$, constrained by the relation $u^2 = \sum u_i^2 = 1$. In the direction normal to u there is only one supporting hyperplane; let $H(u) = H(u_1, \ldots, u_n)$ be the distance of this hyperplane from O. The function

$$H = H(u) \tag{1.1}$$

is called the *supporting function* of C.

The convex body $C(h)$ whose supporting function is

$$H_h(u) = H(u) + h \tag{1.2}$$

is called the *parallel exterior* convex body to C at distance h, where h is a positive constant.

It is proven (Bonnesen-Fenchel, p. 49) that the volume $V(h)$ of $C(h)$ is given by the following expression:

$$V(h) = \sum_{\nu=0}^{n} \binom{n}{\nu} h^\nu W_\nu, \tag{1.3}$$

where the W_ν are invariants of the convex body C.

If the hyperspace limiting C has at each point finite principal radii of curvature, the invariants W_ν are expressed as follows (Bonnesen-Fechel, p. 63).

Let $\{R_1, R_2, \ldots, R_\nu\}$ be the elementary symmetrical function of order ν of the principal radii of curvature $R_1, R_2, \ldots, R_{n-1}$, and let us denote by $d\omega$ the area element of the unit sphere corresponding to the direction whose normal supporting plane touches C at a point where the principal radii of curvature are the R_i. With these notations, W_ν is given by

$$W_\nu = \frac{1}{n\binom{n-1}{n-\nu-1}} \int_E H\{R_1, R_2, \ldots, R_{n-\nu-1}\} d\omega, \tag{1.4}$$

where the integration is extended to the whole surface of the unit n-dimensional sphere.

As particular cases of W_ν, we have:

W_0 is equal to the volume V of C; that is

$$W_0 = V; \tag{1.5}$$

nW_1 is equal to the area F of C (or the length L for $n = 2$); that is

$$W_1 = F/n; \tag{1.6}$$

nW_2 is equal to the integral of the mean curvature of C; that is

$$W_2 = M/n; \tag{1.7}$$

where

$$M = \frac{1}{n-1} \int \left(\frac{1}{R_1} + \frac{1}{R_2} + \cdots + \frac{1}{R_{n-1}} \right) ds,$$

and where ds is the element of area of C and the integration is extended to all the surface of C;

W_n is equal to the volume of the unit n-dimensional sphere; that is:

$$W_n = \frac{\pi^{n/2}}{\Gamma(n/2+1)}. \tag{1.8}$$

2 Relations between W_ν for parallel convex bodies

Formula (1.3) tells us that $V(h)$ is a polynomial in h of degree n. Therefore, by applying the Taylor development for polynomials, the following formula holds.

$$V(h) = \sum_{\nu=0}^{n} \frac{h^\nu}{\nu!} \left(\frac{d^\nu V(h)}{dh^\nu} \right)_{h=0} . \tag{2.1}$$

By comparison with (1·3), we deduce that

$$W_\nu = \frac{1}{\nu! \begin{pmatrix} n \\ \nu \end{pmatrix}} \left(\frac{d^\nu V(h)}{dh^\nu} \right)_{h=0} . \tag{2.2}$$

This new expression for W_ν is preferable to (1.4) in so far as it does not require the existence or finiteness of the principal radii of curvature R_i.

Instead of the convex body C, we now consider the exterior parallel $C(h)$: the same formula (2.2) should also hold for this. Therefore, if we call the value of the invariant W_ν for the convex body $W_\nu(h)$, then $C(h)$ will be

$$W_\nu(h) = \frac{1}{\nu! \begin{pmatrix} n \\ \nu \end{pmatrix}} \left(\frac{d^\nu V(h)}{dh^\nu} \right) . \tag{2.3}$$

Taking (1.3) into account, this expression gives

$$W_\nu(h) = \sum_{i=0}^{n-\nu} \begin{pmatrix} n-\nu \\ i \end{pmatrix} h^i W_{\nu+i} \tag{2.4}$$

This relation, which for $\nu = 0$ coincides with (1.3), can also be used to calculate the $W_\nu(h)$ of $C(h)$ in terms of the $W_{\nu+i}$ of C.

3 Parallel interior convex bodies of another given body

Let C be the convex body whose supporting function is $H = H(u)$. We consider the infinite number of hyperplanes whose distance at the origin O is, for each direction u,

$$H^*(u) = H(u) - h, \tag{3.1}$$

where h is a positive constant.

The envelope of all these hyperplanes will be a hypersurface $C(-h)$ that has at each point the same normal as the hypersurface of C at the corresponding point. Therefore, to the lines of curvature of C will correspond the lines of curvature of $C(-h)$, and the principal radii of curvature R_i^* of $C(-h)$ will be constrained by those of C by the relation

$$R_i^* = R_i - h \quad (i = 1, 2, \ldots, n-1). \tag{3.2}$$

Thus, the inverse of the total Gauss curvature of $C(-h)$ will be

$$K(-h) = (R_1 - h)(R_2 - h) \ldots (R_{n-1} - h). \tag{3.3}$$

In order for this curvature to have a constant sign, and therefore for $C(-h)$ to be the hypersurface of a convex body, it suffices that

$$h \leq \min R_i \quad \text{or} \quad h \geq \max R_i \tag{3.4}$$

If h fulfils one of these conditions, the hypersurface $C(-h)$ will also be the hypersurface of a convex body, which we also denote by $C(-h)$. This $C(-h)$ will be called the *parallel interior* convex body of C at distance h. .

If $C(-h)$ is also a convex body, then the previous relations (1.3) and (2.4) hold, simply by replacing h by $-h$. In fact, it is sufficient to replace H by $H - h$ in (1.4), as well as the R_i by $R_i - h$, and then develop according to the powers of h.

4 Relations between the invariants W_ν of convex bodies with constant width

Let $H = H(u)$ be the supporting function of a convex body C. If $-u$ indicates the direction opposite to u, the supporting hyperplanes normal to this direction will be parallel, and the distance between them will be

$$a(u) = H(u) + H(-u). \tag{4.1}$$

$a(u)$ is known as the *width* of C according to the direction u.

When $a(u)$ =constant, it is said that C is a *convex body of constant width*. These bodies, which for $n = 2$ (the case of the plane) are known as *orbiform*, and which for $n = 3$ (the case of the space) are known as *spheriform*, possess notable properties (Bonnesen-Fenchel, §15).

From the definition one may deduce that every convex body of constant width a is the parallel interior of itself at distance a; that is, $C(-a) = C(a)$. Consequently, the invariants $W_\nu(-a)$ corresponding to $C(-a)$ must be equal to the W_ν corresponding to C; only the sign can have changed. In order to see this difference in signs we observe that, assuming H and the R_i are positive, then according to (1.4) the W_ν will also be positive. However, the $W_\nu(-a)$ will have the sign that results from substituting $H - a$ and $R_i - a$ for H and R_i in (1.4). It is known (Bonnesen-Fenchel, p. 128) that for a convex body of constant width, the principal radii of curvature corresponding to the homologous lines of curvature at two opposite points have a sum equal to the width a; hence $a \geq R_i$ $(i = 1, 2, \ldots, n-1)$, and since $a \geq H$ also holds, it follows from (1.4) that the sign of $W_\nu(-a)$ is that of $(-1)^{n-\nu}$. Thus

$$W_\nu(-a) = (-1)^{n-\nu} W_\nu \tag{4.2}$$

On applying this equality to (2.4) for $h = -a$, it follows that *among the invariants W_ν of the convex bodies of constant width a of E_n there exist the relations*

$$W_\nu = \sum_{i=0}^{n-\nu} (-1)^{n-\nu+i} \binom{n-\nu}{i} a^i W_{\nu+i} \tag{4.3}$$

for $\nu = 0, 1, 2, \ldots, n$.

These $n+1$ relations are not independent. In order to check this, we write (4.3) in detail for $\nu = n, n-1, n-2, \ldots, n-\nu$ where ν is even, which gives the following relations:

$$\begin{aligned}
&0) \quad W_n = W_n \\
&1) \quad W_{n-1} = aW_n - W_{n-1} \\
&2) \quad W_{n-2} = a^2 W_n - \binom{2}{1} aW_{n-1} + W_{n-2} \\
&3) \quad W_{n-3} = a^3 W_n - \binom{3}{1} a^2 W_{n-1} + \binom{3}{2} aW_{n-2} - W_{n-3} \\
&\quad \ldots \\
&\nu) \quad W_{n-\nu} = a^\nu W_n - \binom{\nu}{1} a^{\nu-1} W_{n-1} + \binom{\nu}{2} a^{\nu-2} W_{n-2} - \cdots + W_{n-\nu}.
\end{aligned} \tag{4.4}$$

We now prove that, under the above-mentioned hypothesis that ν is an even number, the last equation written is a consequence of the preceding equations. Indeed, let us multiply equation 0) by a^ν, equation 1) by $-\binom{\nu}{1} a^{\nu-1}$, and, in general, equation i) by $(-1)^i \binom{\nu}{i} a^{\nu-i}$. Then by adding left-hand sides and by adding right-hand sides, we arrive at

$$\sum_{i=0}^{\nu} (-1)^i \binom{\nu}{i} a^{\nu-i} W_{n-i} = \sum_{j=0}^{\nu} \left(\sum_{i=0}^{i=\nu} (-1)^i \binom{\nu}{i} \binom{i}{j} \right) (-1)^j a^{\nu-j} W_{n-j}. \tag{4.5}$$

However, according to a known formula (see [1], p. 255, formula 43), this is

$$\sum_{i=0}^{i=\nu} (-1)^i \binom{\nu}{i} \binom{i}{j} = \begin{cases} 0 & \text{if } \nu \neq j \\ (-1)^j & \text{if } \nu = j \end{cases} \tag{4.6}$$

and since according to equation ν) in (4.4) the left-hand side of (4.5) is equal to $W_{n-\nu}$, it follows the identity $W_{n-\nu} = W_{n-\nu}$.

This tells us that of the $n+1$ relations (4.3) for $\nu = 0, 1, 2, \ldots, n$, all those corresponding to $n - \nu$ even can be removed, since they are a consequence of those that succeed them. In brief, we may state that: *Among the invariants W_ν of a convex body of constant width in n-dimensional Euclidean space, there exist the following independent relations:*

a) If $n = 2m$:

$$2W_{2\nu-1} + \sum_{i=1}^{n-2\nu+1} (-1)^i \binom{n-2\nu+1}{i} a^i W_{2\nu-1+i} = 0 \tag{4.7}$$

for $\nu = 1, 2, 3, \ldots, m$.

b) If $n = 2m+1$:

$$2W_{2\nu} + \sum_{i=1}^{n-2\nu} (-1)^i \binom{n-2\nu}{i} a^i W_{2\nu+i} = 0 \tag{4.8}$$

for $\nu = 0, 1, 2, 3, \ldots, m$.

5 Examples

1. For the plane $n = 2$, the only relation is

$$2W_1 - aW_2 = 0,$$

that is, since $2W_1 = L$, $W_2 = \pi$,

$$L - \pi a = 0,$$

which is the known relation between the length L and the width a of the plane convex figures of constant width (Bonnesen-Fenchel, p.131).

2. For the ordinary space, $n = 3$, the relations are

$$2W_0 - 3aW_1 + 3a^2W_2 - a^3W_3 = 0; \quad 2W_2 - aW_3 = 0,$$

which according to (1.5), (1.6), (1.7), and (1.8), can be written

$$2V - aF + \frac{2}{3}\pi a^3 = 0; \quad M - 2\pi a = 0.$$

3. For $n = 4$ we have

$$2W_1 - 3aW_2 + 3a^2W_3 - a^3W_4 = 0; \quad 2W_3 - aW_4 = 0,$$

that is, according to (1.5), (1.6), (1.7), and (1.8),

$$2F - 3aM + a^3\pi^2 = 0; \quad 4W_3 - a\pi^2 = 0.$$

References

[1] E. Netto, *Lehrbuch der Combinatorik*, B. G. Teubner, 1927, Leipzig and Berlin.

UNIVERSIDAD NACIONAL DEL LITORAL, ROSARIO, ARGENTINA

An affine invariant for convex bodies in n-dimensional space

L. A. Santaló

Received November 12, 1950

1 Introduction

Let K be a convex body in a Euclidean space of n dimensions E_n. Let P$(\xi_1, \xi_2, \ldots, \xi_n)$ be a point interior to K and $h = h(\omega)$ be the supporting function of K with respect to point P (ω denotes a point on the unit sphere). If $d\omega$ denotes the area element ($(n-1)$-dimensional) of the unit sphere corresponding to the point ω, then the integral

$$(1.1) \qquad I(P) = \frac{1}{n}\int_{\Omega} \frac{d\omega}{h^n}$$

is an invariant of K with respect to the unimodular centre-affinities of centre P [[9], p. 749][1]. $I(P)$ is precisely the measure of all the hyperplanes exterior to K, invariant as regards the centre-affinities of centre P. The surface of the unit sphere of E_n and its area are both denoted by Ω.

When P changes inside K, the minimum of $I(P)$ is an invariant of K with respect to all the unimodular affinities of E_n, which we denote by

$$(1.2) \qquad I_m = min_P I(P), \quad (P \subset K).$$

The aim of this paper is to prove that.

1)There exists only one relative minimum for $I(P)$, which is therefore the absolute minimum. If K has a centre of symmetry O, this minimum corresponds to the case $P = 0$.

2) The affine invariant I_m is related to the volume V and the affine area F_a of K by the inequalities

$$(1.3) \qquad I_m V \leq (\Omega/n)^2, \quad I_m F_a \leq \Omega^{2n}/n^{n-1},$$

in which the equals sign holds only for the case where K is an ellipsoid of n dimensions. As previously stated, in these inequalities we have

$$(1.4) \qquad \Omega = \frac{2\pi^{n/2}}{\Gamma(n/2)}.$$

Finally, we provide a geometric interpretation of the first inequality (1.3) for the case where K has a centre of symmetry, observing that this proves a conjecture by K. Mahler, which improves as far as possible the greatest upper bound of an inequality by this author.

[1] Note of the editors: Santaló considered this integral in dimension 2 in [10]. We also remark that the present paper was written in 1950 but published in the volume of *Portugaliae Math.* corresponding to 1949.

2 Study of the function $I(P)$ when P changes

The function $I(P)$ defined by (1.1) takes the infinite value for the points on the boundary of K, while for any point P interior to K it has a finite and positive value. We now find the points P interior to K for which $I(P)$ takes a stationary value (that is, its first partial derivatives vanish).

Let P_0 be a point interior to K which we take as the origin of coordinates, and let $h_0 = h_0(\omega)$ be the supporting function of K with respect to P_0. The supporting function $h = h(\omega)$ with respect to $P(\xi_1, \xi_2, \ldots, \xi_n)$ will be

$$h(\omega) = h_0(\omega) - \sum_1^n \alpha_i \xi_i, \tag{2.1}$$

where α_i are the director cosines of the direction ω.

Thus we have

$$I(P) = \frac{1}{n} \int_\Omega \frac{d\omega}{(h_0 - \sum \alpha_i \xi_i)^n}, \tag{2.2}$$

and consequently, in order for the point P to correspond to a stationary value of $I(P)$, we have

$$\frac{\partial I}{\partial \xi_i} = \int_\Omega \frac{\alpha_i d\omega}{h^{n+1}} = 0 \qquad (i = 1, 2, \ldots, n). \tag{2.3}$$

To check whether this stationary value is maximum or minimum, we observe that

$$\frac{\partial^2 I}{\partial \xi_i \partial \xi_j} = (n+1) \int_\Omega \frac{\alpha_i \alpha_j d\omega}{h^{n+2}}. \tag{2.4}$$

Therefore, if $Q(\xi_1 + \epsilon_1, \xi_2 + \epsilon_2, \ldots, \xi_n + \epsilon_n)$ is a point close to P, by Taylor's expansion, and taking into account (2.3), we have

$$\begin{aligned} I(Q) - I(P) &= \frac{n+1}{2} \sum \int_\Omega \frac{\alpha_i \alpha_j \epsilon_i \epsilon_j}{h^{n+2}} d\omega + \ldots \\ &= \frac{n+1}{2} \int_\Omega \frac{(\sum \alpha_i \epsilon_i)^2}{h^{n+2}} d\omega + \ldots \end{aligned}$$

Thus, since $h > 0$, for sufficiently small $|\epsilon_i|$, we always have $I(Q) > I(P)$; that is, *the stationary values of $I(P)$ are always minimum.*

It is thereby deduced that the minimum of $I(P)$ is unique. Indeed, according to a result by Morse [[7], especially p. 303], if all the extremal points of a function defined inside a convex body are relative minima, there exists only one of these extremal points. Thus, summarizing:

There exists only one point interior to K for which $I(P)$ takes a minimum value (which will be the absolute minimum), which is characterized by the fulfilment of conditions (2.3).

For example, if K has a centre of symmetry, the conditions (2.3) for it are obviously fulfilled, and thus: *for convex bodies with centre of symmetry, the minimum of $I(P)$ corresponds to the centre of symmetry.*

3 The affine invariant I_m and its relations with the volume and affine area

Let us denote the value (1.2) by I_m ; that is, the minimum of $I(P)$ with respect to all the points interior to K. This I_m is an invariant of K with respect to all the unimodular affinities. If K has a centre of symmetry, as already stated, I_m has the value (1.1) for the supporting function $h = h(\omega)$ referring to the centre of symmetry.

In order to compare I_m with the volume V and the affine area F_n of K (which are the other simpler affine invariants), it is necessary to recall some known results.

1°. Let K^0 be a convex body of E_n limited by a surface S^0 having at each point well-defined principle radii of curvature $R_1, R_2, \ldots, R_{n-1}$ and we write

$$D = R_1 R_2 R_3 \ldots R_{n-1}. \tag{3.1}$$

Since S^0 is convex, $R_i > 0$.

If $d\sigma$ is the ordinary area element of S^0, and $d\omega$ is the area element on the unit sphere corresponding to the direction of the normal to S^0 at the point where $d\sigma$ is considered, we have

$$d\sigma = D\, d\omega. \tag{3.2}$$

The *affine area* of S^0 is defined by

$$F_a^0 = \int_{S^0} D^{-1/(n+1)} d\sigma = \int_{\Omega} D^{-1/(n+1)} d\omega, \tag{3.3}$$

and between this affine area of S^0 and the volume V^0 of the body K^0 limited by this area, there exists Blaschke's [[2], p.198] inequality

$$(F_a^0)^{n+1} \leq \Omega^2 (nV^0)^{n-1}, \tag{3.4}$$

where the equality holds only for the case where K^0 is an ellipsoid of n dimensions.

Inequality (3.4) is due to Blaschke for the case $n = 3$, but the same proof holds almost without modification for the n-dimensional case[2]

[2]To generalize Blaschke's proof to n dimensions, it suffices to take into account the following inequality:

If $|a_{ik}|$, $|b_{ik}|$ are two positive determinants of order m, then for $p > 0$ the following are verified

$$(2^p |a_{ik} + b_{ik}|)^{1/(m+p)} \geq |a_{ik}|^{1/(m+p)} + |b_{ik}|^{1/(m+p)}. \tag{$*$}$$

For p = 0, this inequality coincides with one given by Minkowski [[5], p. 35] and can be proven in an analogous way. Indeed, the two determinants $|a_{ik}|$, $|b_{ik}|$ can always be multiplied by the same determinant with unit value such that the products are determined solely with the elements of the principle diagonal that are different from zero. Denoting these elements by a_i, b_i , we have

$$|a_{ik}| = \prod_1^m b_i, \quad |b_{ik}| = \prod_1^m b_i, \quad |a_{ik} + b_{ik}| = \prod_1^m (a_i + b_i).$$

2°. Let K and K^0 be two convex bodies of E_n limited respectively by the surfaces S, S^0. If h_ω is the supporting function of K, and $d\sigma$ is the area element of S^0, Minkowski's *first mixed volume* of K and K^0 is

$$V_1 = \frac{1}{n}\int_{S^0} h\, d\sigma, \tag{3.5}$$

and therefore there exists Minkowski's inequality

$$V_1^n > V(V^0)^{n-1}, \tag{3.6}$$

where V is the volume of K and V^0 that of K^0. The equality holds only when K and K^0 are homothetic. See, for example, Bonnesen-Fenchel [[3], p. 91].

Let us now address our problem. Let $h = h(\omega)$ be the supporting function of the convex body K referring to the point P for which $I(P)$ is minimum. As already seen, conditions (2.3) are thereby fulfilled. According to a classical result by Minkowski [[3], p.121], if conditions (2.3) are satisfied there exists a convex body K^0 limited by a surface S^0 whose principle radii of curvature $R_1, R_2, \ldots, R_{n-1}$ satisfy the condition

$$D(\omega) = R_1 R_2 \ldots R_{n-1} = h^{-(n+1)}. \tag{3.7}$$

The area element $d\sigma$ of K^0 is equal to $d\sigma = D d\omega = h^{-(n+1)} d\omega$, and therefore according to (3.5) the first mixed volume between K and K^0 is

$$V_1 = \frac{1}{n}\int_{S^0} h\, d\sigma = \frac{1}{n}\int_{\Omega} h^{-n}\, d\omega = I_m. \tag{3.8}$$

Minkowski's inequality (3.6) thus yields

$$I_m^n \geq V(V^0)^{n-1}, \tag{3.9}$$

where V is the volume of K and V^0 that of K^0.

Furthermore, the affine area of K^0, according to (3.8), is

$$F_a^0 = \int_{\Omega} D^{-1/(n+1)} d\omega = \int_{\Omega} h^{-n} d\omega = n\, I_m, \tag{3.10}$$

and thus Blaschke's inequality (3.4) gives

$$(n\, I_m)^{n+1} \leq \Omega^2 (nV^0)^{n-1}. \tag{3.11}$$

From (3.9) and (3.11), we have

$$I_m V \leq (\Omega/n)^2, \tag{3.12}$$

By writing 2^p in the form (1 + 1) (1 + 1) ··· (1 + 1) (p factors), the inequality (*) turns out to be a particular case of the general inequality

$$\left.\prod_1^s(\alpha_i+\beta_i)\right)^{1/s} > \left.\prod_1^s \alpha_i\right)^{1/s} + \left.\prod_1^s \beta_i\right)^{1/s}$$

which is well known [[5], p.39].

which is the inequality we wished to obtain.

If we wish to relate I_m with the affine area F_a, it suffices to take (3.4) into account, which combined with (3.12) gives

$$I_m^{n-1} F_a^{n-1} \leq \Omega^{2n}/n^{n-1}. \tag{3.13}$$

In both (3.12) and (3.13) the equals sign holds for n-dimensional ellipsoids (it suffices to check it for the sphere). Moreover, since in (3.11) the equality holds only for ellipsoids, and in (3.9) only for homothetic convex bodies, we have:

Between the affine invariant I_m, the volume V and the area F_a of a convex body K of E_n, there exist the inequalities (3.12) *and* (3.13), *in which the equals sign holds only for the case where K is an ellipsoid of n dimensions.*

From another point of view, and with a different interpretation, an inequality equivalent to (3.12) for $n = 3$ can be found in Blaschke [1].

4 Geometric interpretation

Consider the case in which K has a centre of symmetry O. Then, according to definition (1.1), I_m coincides with thc volume of the reciprocal polar body of K with respect to the unit sphere with centre O. Inequality (3.12) then coincides with one indicated as probable by K. Mahler [6], which proves only a much less precise bound. A better bound was obtained by A. Dvoretzki and C. A. Rogers [4]. Bound (3.12) is the best possible, since it is reached by the ellipsoid.

It would be interesting to find the optimal lower bound of the product $I_m V$, which according to K. Mahler should be

$$I_m V \geq 4^n/n!, \tag{4.1}$$

where the equality holds only for parallelpipeda. However, we have been unable to prove (3.14) by the previous method.

5 Note

Assume that K does not possess a centre of symmetry. If G is the centre of gravity of K supposedly covered by a homogeneous mass, $I(G)$ will also be an invariant by unimodular affinities of K. It is known that, in general, $I(G)$ does not coincide with I_m, as P. Pi Calleja [8] proved with an example, but the question of whether I_m can be replaced by $I(G)$ in inequalities (3.12) and (3.13) remains open.

References

[1] W. Blaschke, *Ueber affine Geometrie VII: Neue Extremaleigenschaften von Ellipse un Ellipsoid*, Bericht über die Verhunalungen der Sächsischen Akademie der Wissenschaften **69** (1917), 306–318, Leipzig.

[2] W. Blaschke, *Vorlesungen über Differentialgeometrie und geometrische Grundlagen von Einsteins Relativitätstheorie. II. Affine Differentialgeometrie*, Berlin: J. Springer. Die Grundlehren der Mathematischen Wissenschaften VII., 1923.

[3] T. Bonnesen and W. Fenchel, *Theorie der konvexen Körper*, Ergebnisse der Math. Springer, Berlin, 1934.

[4] A. Dvoretzky and C. A. Rogers, *Absolute and unconditional convergence in normed linear spaces*, Proc. Nat. Acad. Sciences of U.S.A. **36** (1950), 192–197.

[5] G. H. Hardy, J. E. Littlewood, and G. Polya, *Inequalities*, Cambridge University Press, 1934.

[6] K. Mahler, *Ein Uebertragungsprinzip für konvexe Körper*, Casopis Matematiky a Fysiky **68** (1939), 93–102.

[7] M. Morse, *Relations between the critical points of areal function of n independent variables*, Transactions of the AMS **27** (1925), 345–396.

[8] P. Pi Calleja, *Sobre la figura polar de una dada respecto de un circulo con centro en el baricentro*, Math. Notae **9** (1950), 88–93.

[9] L. A. Santaló, *Integral Geometry in Projective and Affine Spaces*, Annals of Mathematics **51** (1950), 739–755.

[10] L. A. Santaló, *Un invariante afin para curvas convexas del plano*, Math. Notae **8** (1948), 103–111.

On complete systems of inequalities among three elements of a plane convex figure

L. A. Santaló

Abstract

In order to get a better understanding of the inequalities between the fundamental elements of a convex body in space, Blaschke introduced a diagram which proved very useful for subsequent researches (Blaschke [1], Bonnesen-Fenchel [2] p.84, Hadwiger [4] p.72).

In this paper we apply a similar device for convex figures in the euclidean plane. Given three real numbers a_1, a_2, a_3 the question arises as to whether there exists a convex figure K which has these numbers as the values of three of its elements, such that $(L, F, D), (L, F, E), (F, R, r), \ldots$ where $L =$ length, $F =$ area, $D =$ diameter, $E =$ thickness, $R =$ radius of the circunscribed circle, $r =$ radius of the inscribed circle. To each triad a_1, a_2, a_3 we assign a point of the unit square; the points for which such a convex figure K exists will fill a connected domain $D(K)$. The problem is to determine $D(K)$ for each given triad of elements.

The domain $D(K)$ is determined by certain inequalities between a_1, a_2, a_3 if these inequalities form a complete system, $D(K)$ is completely determined. If some inequality is lacking, some part of the boundary of $D(K)$ remains unknown. The method gives a geometric interpretation of these inequalities and systematizes the known and the unknown results.

In this paper we consider only the elements F, L, D, E, R, r and give the diagram for seven cases. Most of the applied inequalities are known; they are included in Section 4. The inequalities and results of the cases (D, R, r), Section 9, and (E, R, r), Section 10 seem to be new.

It is clear that many other cases can be considered (Section 13); however, most of them require the discovery of new inequalities.

1 Introduction

Blaschke devised a diagram for representing convex bodies in space as points in the plane based on their three fundamental elements: the volume V, the area F and the total mean curvature M. This diagram can be seen, for example, in Bonnesen-Fenchel [2], p.84, and in a highly detailed way in Hadwiger [4], p.72. From this diagram it was seen, for example, that Minkowski's classical equalities

(1) $$M^2 \geq 4\pi F, \quad F^2 \geq 3VM,$$

from which as a consequence are deduced that

$$F^3 \geq 36\pi V^2, \quad M^3 \geq 48\pi^2 V$$

do not form a complete system. In other words, it is not sufficient that three real numbers V, F and M satisfy the relations (1) in order to ensure that there exists a convex body having such numbers as a measure of its volume, area and total mean curvature, respectively. It still lacks a further inequality, which remains unknown in the general case, and is known only in some particular cases thanks largely to work by Hadwiger [4].

Our aim is to study analogous diagrams for convex figures in the plane. If we confine ourselves to the area F and to the length L, the only inequality is the classical isoperimetric

$$L^2 - 4\pi F \geq 0, \tag{2}$$

and there is no other, since given any two real numbers F, L, related by (2), there always exists a convex figure (obviously not unique) having L for its length and F for its area. Indeed, it suffices to take a rectangle of sides a, $2b$, and to consider the figure formed by the rectangle plus two semi-circles of radius b added onto the sides of length $2b$ (Fig. 4). For this figure (which in abbreviated form we call a "rectangle terminating in semi-circles"), we have

$$L = 2a + 2\pi b, \quad F = 2ab + \pi b^2, \tag{3}$$

and therefore, given L, F, it suffices to take

$$a = \frac{1}{2}(L^2 - 4\pi F)^{\frac{1}{2}} \quad b = \frac{1}{2\pi}(L - (L^2 - 4\pi F)^{\frac{1}{2}})$$

in order to obtain a convex figure with the characteristics L, F.

Together with L and F, one may consider other elements of K, such as $D =$ diameter, $E =$ width, $R =$ radius of the circumscribed circle, and $r =$ radius of the inscribed circle. Among these elements, taken two by two, there exist certain inequalities described in Section 4. Furthermore, in these cases it is easy to check that the known inequalities form a complete system; that is, the only possible inequalities among the elements L, F, D, E, R and r , taken two by two, are those described in Section 4, all of them known.

The question changes if, instead of two elements, there exist three elements of a convex figure K, chosen from the same aforementioned L, F, D, E, R and r. Let them be denoted by a_1, a_2, a_3. The problem then arises of finding the sufficient and necessary conditions in order for there to exist, given these three real numbers a_1, a_2, a_3, a convex plane figure K having such numbers as its elements.

To this effect, it is necessary that a_1, a_2, a_3 fulfil certain inequalities. In order to show these inequalities and provide them with a clearer form, it is useful to represent for each case a diagram such as that employed by Blashcke for convex bodies. By writing $x = f_1(a_1, a_2, a_3)$, $y = f_2(a_1, a_2, a_3)$ so that x and y do not change when a similar figure is substituted for K, and where $0 \leq x \leq 1$, $0 \leq y \leq 1$, we obtain in each case a mapping of the convex figures on the points of the unit square. The points of this square for which there exists a convex figure with the characteristics a_1, a_2, a_3 will fill a certain domain, and it is precisely this domain which must be determined. Obviously, the correspondence between the points in this domain and the convex figures is not bijective, not even using a similarity, because one and the same point may correspond to convex figures with very different form.

We thereby have a diagram for each triad of elements. When the known inequalities among the elements a_1, a_2, a_3 enable the domain of the points for which there exists a convex figure K to be totally delimited, the problem is completely solved and the known inequalities form a complete system. However, if the known inequalities do not enable the said area to be totally delimited (see, for example, the case (E, R and r) and Fig. 12), it shows that inequalities are lacking. Thus the diagram serves to indicate lines of research.

In this work, we merely provide some example. Even if we confine ourselves to the elements F, L, D, E, R and r, there is still much to be taken into account (Section 13). If more elements are considered (Section 2), the number of cases grows rapidly.

Even though, for these cases under consideration here, most of the inequalities employed are known, we believe that with these diagrams an interesting systematization can be obtained, which puts the known results into order and indicates the avenues along which the study may profitably be pursued. The inequalities for the cases (D, R, r) and (E, R, r) appear to be new.

2 Main elements of a convex figure

Given a plane convex figure K, infinite characteristics of such a figure are determined. Those we consider in this work are as follows:

F = area,

L = length or its boundary perimeter,

D = diameter = maximum distance between two points in K,

E = minimum width or thickness = minimum distance between to parallel lines containing K,

R = radius of the circumscribed circle = minimum radius of the circle containing K,

r = radius of the inscribed circle = maximum radius of the circle contained by K.

Many other characteristics could also be considered. For example:

T = area of the minimum area triangle containing K,

t = area of the maximum area triangle contained by K,

I = moment of polar inertia with regard to the centre of gravity of K,

L_a = affine length of the boundary of K.

In Bonnessen-Fenchel's book [2], or in that by Polya-Szego [7], one can find many other characteristics that may be included in this general systematization.

3 Some typical convex figures

There are certain convex figures that appear frequently in problems of maxima and minima, which it would be convenient to bring together and name in order to avoid repetition.

Apart from the obvious figures of the segment, circle and equilateral triangle, these figures are as follows:

a) The Reuleaux Triangle. This is a curvilinear triangle formed from an equilateral triangle $A\ B\ C$ when both sides are replaced by arcs of circumference whose centre is the opposite vertex (Fig. 1).

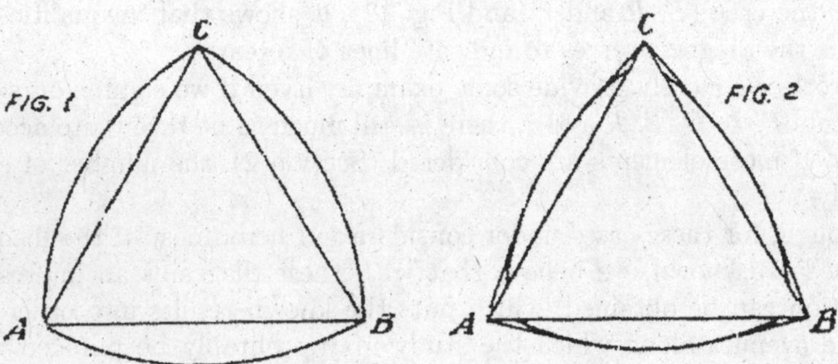

The Reuleaux triangle is a particular case of the so-called *figures of constant width*, which are those having the same width in any direction; that is, the same distance between two parallel supporting lines. For all figures of constant width we have

$$D = E, \quad L = \pi E.$$

b) Yamanouti Figures. These are obtained by taking an equilateral triangle $A\ B\ C$ and on each side placing an arc of circumference whose centre is the opposite vertex and radius less than the side of the triangle; the minimum convex envelope of the figure thus formed (obtained by tracing the tangents to the said arcs of circumference along the vertices of the triangle) is a Yamanouti figure [9] (Fig. 2).

c) A circle with two symmetric hoods. This is the figure formed by a circle plus the tangents of the circle from two exterior points A, B, belonging to a same diameter (Fig. 3). Observe that for this figure we have

$$E = r, \quad 2F = rL$$

$$L = 2(D^2 - E^2)^{\frac{1}{2}} + 2E \arcsin \frac{E}{D}.$$

d) A rectangle terminating in semi-circles. This is a rectangle with two opposite sides substituted by semi-circles (Fig. 4). For these figures we have

$$\pi r^2 - Lr + F = 0, \quad 4F = 2EL - \pi E^2.$$

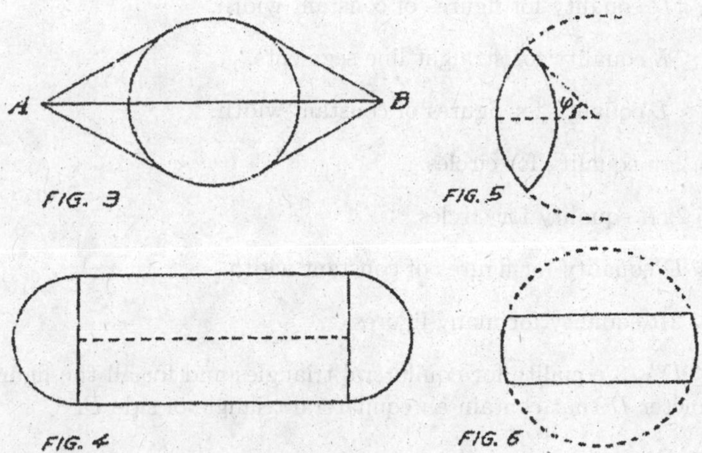

e) Symmetric lens. This is the part common to two intersecting circles of equal radius (Fig. 5). If we denote by 2φ the angle formed by the radii of one of the circles that go to the endpoints of the lens, and by a the radius of the circles, we have

$$L = 4a\varphi, \quad F = 2a^2(\varphi - \sin\varphi\cos\varphi),$$

$$D = 2R = 2a\sin\varphi, \quad E = 2r = 2a(1 - \cos\varphi).$$

f) Segment of a symmetric circle. This is the part of the circle limited by two parallel chords equidistant from the centre (Fig. 6). For these figures, the following relation holds

$$L = 2(D^2 - E^2)^{\frac{1}{2}} + 2D\arcsin(E/D),$$

$$2F = E(D^2 - E^2)^{\frac{1}{2}} + D^2\arcsin(E/D).$$

4 Known inequalities

Among the elements of a convex figure there are certain inequalities, some immediate, others more complicated, which we now describe below, indicating at the same time the figures for which the sign of equality holds. We consider only the elements L, F, D, E, R and r.

a) *Inequalities between two elements.*

(I) $4\pi F \leq L^2$ equality for circles.

(II) $4F \leq \pi D^2$ equality for circles.

(III) $E^2 \leq \sqrt{3}F$ equality for equilateral triangles.

(IV) $\pi r^2 \leq F$ equality for circles.

(V) $F \leq \pi R^2$ equality for circles.

(VI) $L \leq \pi D$ equality for figures of constant width.

(VII) $2D \leq L$ equality for straight line segments.

(VIII) $\pi E \leq L$ equality for figures of constant width.

(IX) $L \leq 2\pi r$ equality for circles.

(X) $L \leq 2\pi R$ equality for circles.

(XI) $E \leq D$ equality for figures of constant width.

(XII) $D \leq 2R$ equality for many figures.

(XIII) $R \leq D/\sqrt{3}$ equality for equilateral triangles and for all the figures of diameter D that contain an equilateral triangle of side D.

(XIV) $2r \leq D$ equality for circles.

(XV) $E \leq 2R$ equality for circles.

(XVI) $E \leq 3R$ equality for equilateral triangles.

(XVII) $2R \leq E$ equality for many figures.

(XVIII) $r \leq R$ equality for circles.

(XIX) $4R \leq L$ equality for straight line segments.

It is easy to check that these inequalities are the only ones that exist among the elements considered: L, F, D, E, R, r, taken two by two. They form a complete system. In other words, given any two elements among the six under consideration, it suffices for them to satisfy the foregoing inequalities in order for a plane convex figure having these elements to exist.

Many of the foregoing inequalities are trivial, while others are more complicated. Their proof and bibliography can be found in Bonnesen-Fenchel [2].

b) *Inequalities among more than two elements.*

(XX) $DE \leq 2F$, if $2E \leq \sqrt{3}D$ the equality holds only for triangles of base D and height E. If $2E \geq \sqrt{3}D$ then the sign < always holds.

(XXI) $2(D^2 - E^2)^{\frac{1}{2}} + 2E \arcsin(E/D) \leq L$ equality for circles with two symmetric hoods.

(XXII) $L \leq 2(D^2-E^2)^{\frac{1}{2}}+2D \arcsin(E/D)$ equality for segments of symmetric circles.

(XXIII) $2F \leq E(D^2 - E^2)^{\frac{1}{2}} + D^2$ equality for segments of symmetric circles.

(XXIV) $4F \leq 2EL - \pi E^2$ equality for rectangles terminating in semi-circles.

(XXV) $Lr \leq 2F$ equality for circles with two symmetric hoods.

(XXVI) $F \leq r(L - \pi r)$ equality for rectangles terminating in semi-circles.

(XXVII) $F \le R(L - \pi R)$ equality for circles.

(XXVIII) $8\varphi F \le L(L - 2D\cos\varphi)$, with $2\varphi D = L\sin\varphi$, equality for symmetric lenses.

All these inequalities can be found in Bonnesen-Fenchel [2] or in Sholander [8], where there is a note to the effect that the inequalities (3), (6), (5), p.81 in Bonnesen-Fenchel, must be replaced by (XXIII), (XXIV), (XXVIII) mentioned above.

On the contrary to what occurs in the case of two elements, the foregoing inequalities are far from forming a complete system.

There still remain many inequalities, and it is precisely here where the diagrams below show which these unknown inequalities are (see, for example, Sections 10 and 12). In Sections 9 and 10 we obtain new inequalities.

5 The Case of F, L, E

We begin by studying the case corresponding to the elements L, F, E; that is, the following problem: to find the conditions that three given real numbers L, F and E must satisfy in order for there to exist a convex figure in the plane that has these numbers for length, area and minimum area, respectively.
This is a complicated case that has been solved by various authors, mainly by Kubota-Hemmi [5], Sholander [8][8] and Ohmann [6]. We use the results obtained by these authors to make the corresponding diagrams. With this example, which although complicated is known, the method and its advantages are shown with the purpose of detailing the different stages that need to be overcome.

First let us write

$$x = \frac{\pi E}{L}, \quad y = \frac{E^2}{\sqrt{3}F}. \tag{1}$$

According to inequalities (VIII) and (III), each convex figure K with the elements L, F, E corresponds to a point in the unit square $0 \le x \le 1, \quad 0 \le y \le 1$. The same point corresponds to all similar figures. However, the same point may also correspond to figures that are not similar.

Consider inequality (XXIV): from this inequality and from (1) we deduce

$$y \ge \frac{4x}{\pi\sqrt{3}(2-x)}.$$

For $0 \le x \le 1$, the curve

$$\pi\sqrt{3}(2-x)y - 4x = 0$$

is an arc of hyperbole connecting the origin with the point $C(1, 4/\pi\sqrt{3})$.

The points to which a convex figure K correspond are above this curve. The points on the curve OC are those corresponding to rectangles terminating in semi-circles, changing from the segment (point O) to the circle (point C).

There still remains the upper bound. For $2\sqrt{3}E \le L$, that is, $x \le \pi/2\sqrt{3}$, Yamanouti proved that the minimum F (and therefore the maximum y) is obtained for isosceles triangles whose equal heights measure E and whose perimeter is L. For these triangles, where a is the base angle and a the base itself, we have

$$E = a \sin \alpha, \quad L = a(1 + \cos \alpha) \cos^{-1} \alpha, \quad F = (a^2/4) \tan \alpha$$

and therefore

$$x = \pi \sin \alpha \cos \alpha (1 + \cos \alpha)^{-1}, \quad y = (4/\sqrt{3}) \sin \alpha \cos \alpha.$$

These are the parametric equations (for $0 \leq \alpha\pi/3$) of the arc of curve OT, which is the upper bound of the area sought, between $0 \leq x\pi/2\sqrt{3}$. The above-mentioned isosceles triangles correspond to the points on this arc, from segment (point O) to the equilateral triangle, point $T(\pi/2\sqrt{3}, 1)$.

For $\pi/2\sqrt{3} \leq x \leq 1$, that is, $\pi E \leq L \leq 2\sqrt{3}L$, Kubota-Hemmi [5] proved that this is $2F \geq EL - \sqrt{3}E^2 \sec^2 \theta$ with θ a root of the equation $6E(\tan\theta - \theta) = L - \pi E$, such that $0 \leq \theta \leq \pi/6$. This is equivalent to the condition

$$\text{(2)} \qquad y \leq \frac{2x}{\sqrt{3}\pi - 3x \sec^2 \theta},$$

with θ a root of $6x(\tan\theta - \theta) = \pi(1 - x)$.

Taking the equality in (2), we obtain a complicated transcendent equation, but one which represents an arc such as TR in Fig. 7. For $x = 1$, we have $\theta = 0$, $y = 2/(\sqrt{3}\pi - 3)$. The points on this curve TR correspond to Yamanouti figures, which change from the equilateral triangle (point T) to the Reuleaux triangle (point R).

The points in the segment CR, for which $x = 1$, correspond to figures of constant width, from the Reuleaux triangle (point R) to the circle (point C). Recall that between the figures of fixed constant width E, the minimum of F in fact corresponds to the Reuleaux triangle, and the maximum of F corresponds to the circle (the Lebesgue-Blaschke theorem, see Bonnesen-Fenchel [2], p.132-134).

Thus we have completely delimited the area whose points correspond to convex figures in the plane with the given F, L, E related with x by (1). The problem (F, L, E) is thus completely solved. The arc TR is the most complicated.

6 The Case F, D, E

As a consequence of inequalities (XI), (II), by writing

$$x = \frac{E}{D}, \quad y = \frac{4F}{\pi D^2},$$

we have a mapping of the set of convex figures K in the unit square $0 \leq x \leq 1$, $0 \leq y \leq 1$. Inequality (XXIII) tells us that

$$x \leq \frac{2}{\pi}[x(1 - x^2)^{\frac{1}{2}} + \arcsin x]$$

is true for any $0 \leq x \leq 1$. Taking the equality in the previous inequality, we therefore have the equation of the arc OC that is the upper boundary of the domain sought of the points which are images of the convex figures (Fig.8). The points on this arc OC correspond to symmetric circle segments, which go from the segment (point O) to the circle (point C).

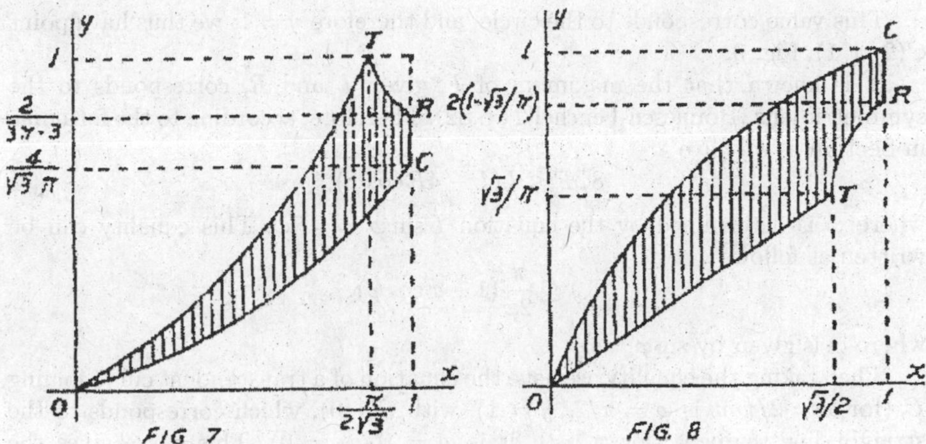

FIG 7　　FIG. 8

For the lower bound, the inequality (XX) tells us that for the interval $0 \leq x\sqrt{3}/2$ we have

$$y \geq \frac{2}{\pi}x.$$

We thus have the straight line segment OT, where T is the point $(\sqrt{3}/2, \sqrt{3}/\pi)$ corresponding to the equilateral triangle. The points in segment OT correspond to triangles with base D and height E; by fixing D and changing E, these triangles change from the segment (point O) to the equilateral triangle (point T).

Finally, for the interval $\sqrt{3}/2 \leq x \leq 1$, Kubota-Hemmi [5] prove the inequality

$$F \geq 3E[(D^2 - E^2)^{\frac{1}{2}} + E(\arcsin(E/D) - \pi/3)] - (\sqrt{3}/2)D^2,$$

that is,

$$y \geq (12/\pi)x[(1-x^2)^{\frac{1}{2}} + x(\arcsin x - \pi/3)] - 2\sqrt{3}/\pi.$$

Taking the sign = we have the equation for the arc TR. Point $R(1, 2(1-\sqrt{3}/\pi))$ corresponds to the Reuleaux triangle, while the points on the arc TR correspond to Yamanouti figures, which change from the equilateral triangle T to the Reuleaux triangle R.

The points in segment RC correspond to the figures of constant width, from the Reuleaux triangle R (minimum area) to the circle C (maximum area).

7 The Case F, L, R

As a consequence of relations (XIX) and (I), we are able to write

$$x = \frac{4R}{L}, \quad y = \frac{4\pi F}{L^2},$$

whereby $0 \leq x \leq 1, \quad 0 \leq y \leq 1$.

The obvious inequality (X) gives

$$x \geq \frac{2}{\pi}.$$

This value corresponds to the circle, and therefore $y = 1$; we thus have point $C(2/\pi, 1)$, Fig. 9.

It is known that the maximum of F, given L and R, corresponds to the symmetric lens (Bonnesen-Fenchel[2], p.82). Therefore, according to the formulae in Section 3, we have

$$8\varphi F \leq L(L - 4R\cos\varphi),$$

where φ is determined by the equation $L\sin\varphi = 4R\varphi$. This equality can be written as follows

$$y \leq \frac{\pi}{2\varphi}(1 - x\cos\varphi),$$

where φ is given by $\sin\varphi = x\varphi$.

Then taking the equality, we have the equation of a transcendent curve joining C (for $x = 2/\pi$ it is $\varphi = \pi/2$, $y = 1$), with $S(1,0)$, which corresponds to the straight line segment (for $x = 1$, it is $\varphi = 0$, $y = 0$). This curve gives the upper bound of the domain sought. All the points on this curve correspond to symmetric lenses, which change from the segment to the circle.

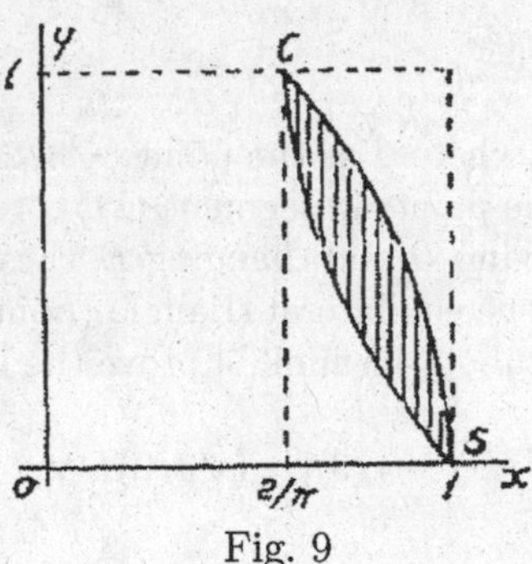

Fig. 9

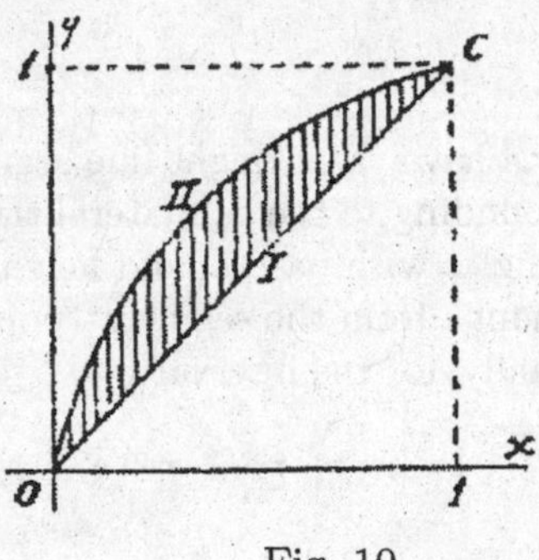

Fig. 10

The lower curve is even more complicated. According to Favard [3] [3], given L and R, the minimum of F (that is, the minimum of y and for each x) corresponds to the polygons inscribed in a circle of radius R whose sides are all equal, except for one, which is either equal in length or shorter. Although the expression of F in terms of L and R is complicated, it is well determined and enables the lower curve CS to be constructed by points.

Instead of the exact curve, in the cited reference Favard proves that a simple bound, although not very precise, is

$$F \geq R(L - 4R);$$

that is,

$$y \geq \pi x(1 - x).$$

The curve $y = \pi x(1 - x)$gives an approximation to the lower arc CS in Fig. 9, but it is not exact; for example, for $x = 2/\pi$ we have $y = 2(1 - 2/\pi) < 1$, when in fact it should be $y = 1$.

8 The Case F, L, r

This is a simple case. By writing

$$(1) \qquad x = \frac{2\pi r}{L}, \quad y = \frac{4\pi F}{L^2},$$

whereby, according to (IX) and (I), we have $0 \le x \le 1, \quad 0 \le y \le 1$.

Inequality (XXV) is written as

$$y \ge x.$$

Therefore, the diagonal OC is the lower boundary of the domain we wish to find. The points on this diagonal correspond to the circles with two symmetric hoods, which change from the segment (point O) to the circle (point C), Fig. 10.

A further important inequality is the Bonnesen (XXVI) inequality, which is written as follows

$$y \le x(2 - x).$$

Thus we have the upper arc OC that can be seen in Figure 10. The points on this arc correspond to rectangles terminating in semi-circles. The part of the plane which according to mapping (1) corresponds to convex figures is shaded part in Fig. 10.

9 The Case D, R, r

As a consequence of the evident inequalities (XXVIII) and (XII), we may write

$$x = \frac{D}{2R}, \quad y = \frac{r}{R}$$

in order always to obtain $0 \le x \le 1, \quad 0 \le y \le 1$.

Inequality (XIII) is written as follows

$$x \ge \sqrt{3}/2.$$

The points for which $x = \sqrt{3}/2$ correspond to the figures of diameter D that contain within an equilateral triangle of side D. The minimum of r for these figures corresponds to the equilateral triangle, for which $y = 1/2$. The maximum of r corresponds to the Reuleaux triangle, for which $y = \sqrt{3} - 1$ (point R). We therefore have the segment TR (Fig. 11) to which the said figures correspond, for which $\sqrt{3}R = D$ (that is, $x = \sqrt{3}/2$).

On the opposite side, $x = 1$ and may have any value. It suffices to consider the circle with two symmetric hoods, and observe that it can change from the segment S (1, 0) to the circle C (1, 1), where $D = 2\ R$ always holds.

It still remains to find the curves, RC, the upper boundary, and the lower boundary TS.

As regards the lower boundary, it is necessary to find the minimum of r given D and R. To this end, we employ the following elementary lemma:

Lemma: Between two triangles inscribed in the same circle and with a common side a, the radius of the inscribed circle is shorter for that having the shortest height with regard to side a.

In fact, if $a = BC$ and O is the centre of the inscribed circle, the angle A is fixed and so is angle $BOC = (\pi + A)/2$. When A changes, the centre O therefore describes the arc[1] of the constant angle BOC, and point A is always on the line connecting O with the middle point of the arc of circumference BC of the circle circumscribed to the triangle, which does not contain point A. Thus the distance from O to the side a, which is the radius of the inscribed circle, is shorter for the triangle in which A is at the shortest distance from a.

Once proved this lemma, let us assume a convex figure K and a circumscribed circle C_R of radius R. If $D < 2R$, there are at least three points on the boundary of C_R that belong to K and that are not on the same semi-circumference; that is, there exist three points A, B, C of K such that the triangle ABC contains the centre of C_R. Let $BC = a$ be the longest side of this triangle. The radius r of the circle inscribed in this triangle is less than or equal to the radius r of the inscribed circle of K. By fixing B, C, we can substitute A by the point A' such that $A'B = D$, $A'C \leq AC \leq D$. Then, according to the lemma, we have another triangle, of diameter D, with $r'_l \leq r_l \leq l$ and the same circumscribed circle C_R. We thus obtain a convex figure (the triangle $A'BC$) with the same D and R as K, and with a radius less than that of the inscribed circle. This process is applicable, with the sign strictly <, providing that K is not an isosceles triangle ABC with $BC = BA = D$. This figure then provides the minimum sought for r, given D and R.

For such an isosceles triangle, denoting half of the unequal angle ABC by α, we have

$$(D\cos\alpha - r)\sin\alpha = r, \quad \cos\alpha = \frac{D}{2R},$$

and therefore

$$(D^2/2R - r)(4R^2 - D^2)^{\frac{1}{2}} = 2Rr.$$

Hence, isolating r, we obtain that:

For all convex figures in the plane, among D, R and r, the following inequality is fulfilled

$$r \leq \frac{D^2(4R^2 - D^2)^{\frac{1}{2}}}{2R\,[2R + (4R^2 - D^2)^{\frac{1}{2}}]}$$

where the equality holds only for the isosceles triangle of side equal to $D \leq 2R$ inscribed within the circle of radius R.

This equality can be written thus:

$$y \leq \frac{2x^2(1 - x^2)^{\frac{1}{2}}}{1 + (1 - x^2)^{\frac{1}{2}}}.$$

Taking the equality, this is the equation for the curve TS in Fig.11, which is the lower bound of the area we are looking for.

[1] NT: Santaló uses the Spanish expression *arco capaz*, which has no short translation to English. It is the set of points from which we see a given segment with the same angle.

In order to find the upper arc RC, it is necessary to prove the following inequality:

Among D, R, r of any convex figure in the plane, the following relation holds:

$$D \geq R + r$$

where the equality holds, among others, for figures of constant width.

Indeed, according to a known theorem (Bonnesen-Fenchel [2], p.130), any convex figure K of diameter D can be completed until it forms a figure of constant width D that contains K; let this figure be K^*. Obviously, we have

$$D^* = D, \quad R^* \geq R, \quad r^* \geq r,$$

where the asterisk denotes the elements of K^*. Furthermore, for figures of constant width we have $D^* = R^* - r^*$ (Bonnesen-Fenchel [2], p.127). Hence, and from the foregoing inequalities, we obtained the statement.

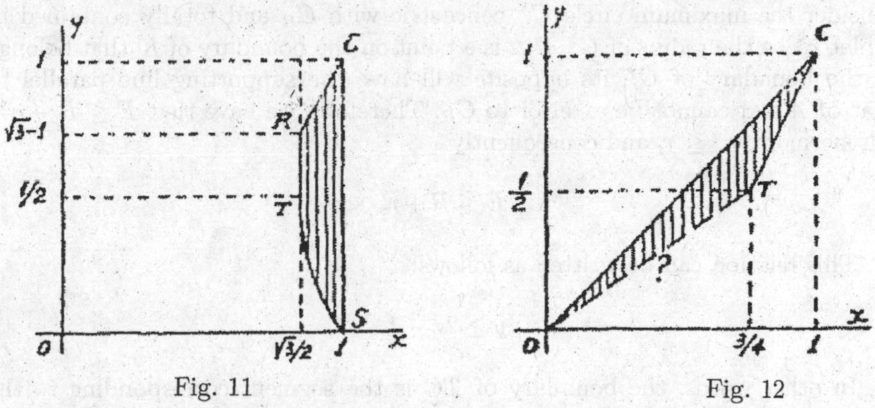

Fig. 11

Fig. 12

Examples of figures for which $D = R + r$ are obtained by taking the vertices of an equilateral triangle ABC inscribed in C_R, and a concentric interior circle C_r of radius $r > R/2$; the minimum convex envelope of C_r and the points A, B, C fulfil the condition $D = R + r$.

The relation $D \geq R + r$ can be written

$$2x \geq 1 + y;$$

that is, the upper boundary RC in Figure 11 is a segment of the straight line $y = 2x - 1$. The points on this line segment correspond to figures such as those mentioned above, which go from the Reuleaux triangle to the circumference.

10 The Case E, R, r

From the obvious relations $E \leq 2R$, $r \leq$ R, we write

$$x = \frac{E}{2R}, \quad y = \frac{r}{R},$$

whereby $0 \leq x \leq 1, \quad 0 \leq y \leq 1.$

The relation $E \geq 2r$ gives

$$x \geq y,$$

and the points on the line $y = x$ correspond to the figures $E = 2\,r$; for example, the symmetric circle segments or circles with two symmetric hoods, which change from the segment (point O) until the circle (point C) Fig.12. Thus we have the upper boundary of the area sought, which will be the diagonal OC.

The lower boundary is somewhat more complicated. Relation (XVI) gives

$$y \geq \frac{2}{3}x,$$

but here the equality holds only for equilateral triangles, for which $E = 3r$, $R = 2r$; that is, only for the point T (3/4, 1/2). For the other values of x the upper boundary is excessive, since no figures exist for which the equality holds.

Now we first find the side TC. Let C_R be the circle circumscribed to K, and consider the maximum circle C^* concentric with C_R and totally contained in K; let r^* be the radius of C^*. If A is a point on the boundary of K that belongs to the boundary of C^*, its opposite will have the supporting line parallel to that of A and cannot be exterior to C_R. Therefore, we have that $E \leq R + r^*$. Furthermore, $r^* \leq r$, and consequently

$$E \leq R + r.$$

This relation can be written as follows:

$$y \geq 2x - 1.$$

In other words, the boundary of TC is the segment corresponding to the straight line $= 2x - 1$. The points in this segment correspond to all figures for which $E = R + r$; for example, those obtained from an equilateral triangle inscribed in C_R , replacing the sides by arcs of equal circumferences whose centres are on the height corresponding to the said sides. In this way we move continuously from the equilateral triangle T to the circle C.

The problem is left unresolved for the interval $0 \leq x \leq 3/4$. A bound is that given by the line $y = (2/3)x$, but as previously stated this is not optimal, except for point T, since it is not reached by any convex figure. The optimal bound probably corresponds to the isosceles triangles that go from the segment O to the equilateral triangle T, but this remains to be proven.

11 The Case L, D, E

This is a straightforward case, solved completely by the inequalities (XXI) and (XXII) due to Kubota (see Bonnesen-Fenchel [2], p.80-81). By writing

$$x = \frac{E}{D}, \quad \frac{L}{\pi D},$$

we have $0 \leq x \leq 1, \quad 0 \leq y \leq 1$. Inequality (XXI)is written

$$y \geq \frac{2}{\pi}[(1-x^2)^{\frac{1}{2}} + \arcsin x],$$

which, by taking the equality, gives us the lower boundary of the domain sought. This curve connects the point $S(0, 2/\pi)$, which corresponds to the segments, with R (1, 1), which corresponds to all figures of constant width. The points on this curve correspond to circles with two symmetric hoods (Fig. 13).

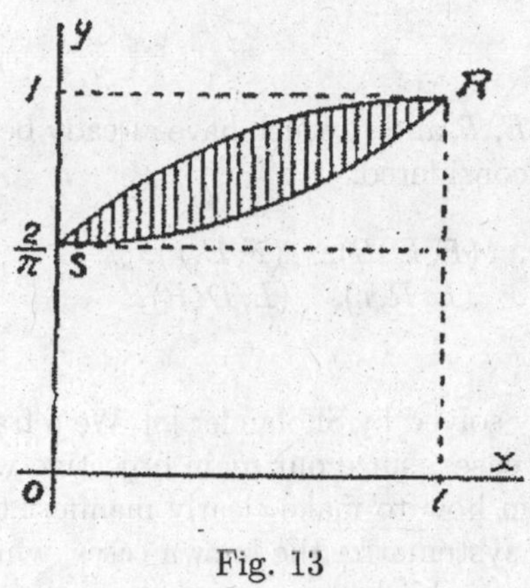

Fig. 13

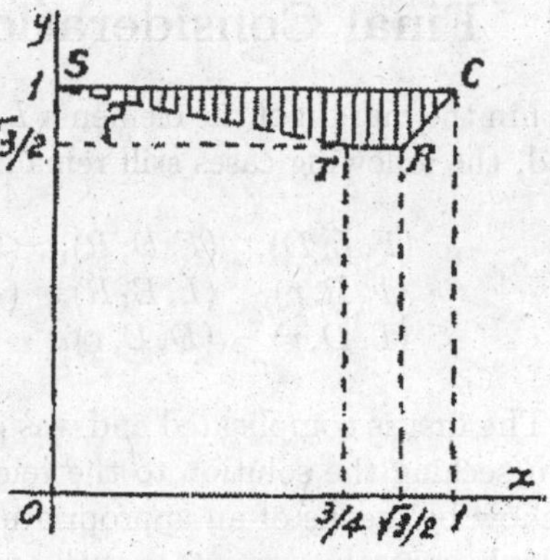

Fig. 14

Inequality (XXII) is written

$$y \leq \frac{\pi}{2}[(1-x^2)^{\frac{1}{2}} + \arcsin x],$$

and on taking the equality we obtain the upper boundary, which is another curve joining S with R (Fig.13). Its points correspond to the symmetric circle segments.

12 The Case D, E, R

Let us write

$$x = \frac{E}{2R}, \quad \frac{D}{2R}$$

whereby $0 \leq x \leq 1, \quad 0 \leq y \leq 1$. Inequality $E \leq D$ indicates that $x \leq y$. This inequality is optimal only for figures in which $E = D$; that is, for those with constant width. For these figures we have $R \leq E/\sqrt{3}$, with the equality for Reuleaux triangles, and therefore $x \geq \sqrt{3}/2$ (see Bonnesen-Fenchel [2], p.134). Consequently we have the segment joining $R(\sqrt{3}/2, \sqrt{3}/2)$, a Reuleaux triangle, with $C(1, 1)$, a circle; the points in this segment RC correspond to figures of constant width (Fig.14).

Furthermore, considering for example symmetric circle segments, $y = 1$, while x changes from 0 to 1. In other words, the segment joining $S(0, 1)$ with C is the upper boundary, reachable by the said figures and all those for which $D = 2R$ (for example, including circles with two symmetric hoods).

The equilateral triangle corresponds to point $T(3/4, \sqrt{3}/2)$. Moreover, inequality (XIII) gives $y \geq \sqrt{3}/2$; thus segment TR parallel to the axis x forms part of the boundary sought; its points correspond to figures such as Yamanouti figures, or those formed by the equilateral triangle on replacing the sides by arcs of circumference interior to the arc corresponding to the Reuleaux triangle.

The arc ST remains to be determined, and we leave it as a proposed problem. It probably corresponds to the isosceles triangles that change from the segment S to the equilateral triangle T, but still remains to be proven.

13 Final Considerations

Within the margin of the elements L, F, D, E, R and r, which have already been fixed, the following cases still remain to be considered.

$$\begin{array}{lllll} (F,L,D), & (F,D,R), & (F,D,r), & (F,E,R), & (F,E,r), \\ (F,R,r), & (L,E,R), & (L,E,r), & (L,R,r), & (L,D,R), \\ (L,D,r), & (D,E,r). & & & \end{array}$$

The first is complicated and was partially solved by Sholander [8]. We refrain from seeking the solution to the remaining cases, since our main objective was to show by means of an appropriate diagram how to make clearly manifest the difficult unresolved points as well as how to systematize the known cases, which are generally to be found scattered in various published papers.

References

[1] W. Blaschke, *Eine Frage über konvexe Körper*, Jahresberich Deutsch. Math. Ver. **25** (1916), 121–125.

[2] T. Bonnesen and W. Fenchel, *Theorie der konvexen Körper*, Ergebnisse der Math. Springer, Berlin, 1934.

[3] J. Favard, *Problemes d'extremum relatifs aux courbes covexes, I, II*, Ann. Ecole Norm. Sup. **46-47** (1929-1930), 345–369,313–324.

[4] Hugo Hadwiger, *Altes und Neues über konvexe Körper.*, (Elemente der Mathematik vom höheren Standpunkt aus. Band III.) Basel und Stuttgart: Birkhäuser Verlag, 1955.

[5] T. Kubota and D. Hemmi, *Some problems of minima concerning the oval*, J. of the Math. Soc. Japan **5** (1953), 372–389.

[6] D. Ohmann, *Extremalprobleme für konvexe Bereiche der euklidische Ebene*, Math. Zeits. **55** (1952), 347–352.

[7] G. Polya and G. Szego, *Isoperimetric inequalities in mathematical physics.*, Ann. Math. Studies, 27. Princeton: Princeton University Press. XVI, 279 p. , 1951 (English).

[8] M. Sholander, *On certain minimum problems in the theory of convex curves*, Trans. Amer. Math. Soc. **73** (1952), 139–173.

[9] M. Yamanouti, *Notes on closed convex figures*, Proc. Phys. Math. Soc. **14** (1932), 605–609, Japan.

RENDICONTI DEL CIRCOLO MATEMATICO DI PALERMO
Serie II, Tomo XXXII (1983), pp. 124-130

AN INEQUALITY BETWEEN THE PARTS INTO WHICH A CONVEX BODY IS DIVIDED BY A PLANE SECTION

L. A. SANTALÓ

A new proof is given of an inequality of J. Bokowski and E. Sperner [1] referring to the product of the volume of the two parts into which a convex body is divided by a plane. The proof, which is given for dimensions $n = 2, 3$ uses known formulas of Integral Geometry and is generalized to convex bodies of the elliptic and hyperbolic spaces.

1. Introduction.

Let K be a convex domain in the euclidean n-space E_n and let L_{n-1} be an hyperplane which divides K into two parts K_1 and K_2. Let $V(K_1)$, $V(K_2)$ denote the volumes of K_1 and K_2 respectively, D the diameter of K and σ_{n-1} the $(n-1)$-dimensional volume of the intersection $K \cap L_{n-1}$. Then, J. Bokowski and E. Sperner [1], [2] have proved the following inequality

$$V(K_1)\, V(K_2) \leq \frac{(1-2^{-n})\,(n-1)\,\omega_{n-1}}{n\,(n+1)}\, D^{n+1}\, \sigma_{n-1} \tag{1.1}$$

where ω_{n-1} denotes the volume of the $(n-1)$-dimensional unit sphere. For $n = 2, 3$ this inequality takes the form

$$F_1 F_2 \leq (D^3/4)\, \sigma_1\,, \qquad V_1 V_2 \leq (7/48)\, \pi\, D^4\, \sigma_2\,. \tag{1.2}$$

Our purpose is to give a new proof of the particular cases (1.2) and to generalize these inequalities to the elliptic and hyperbolic spaces.

2. A fundamental Lemma.

Consider the segment OA on the real line, of length a, and the segment OX of length $x \leq a$. Let $f(r)$ be an integrable non-negative function defined on the closed interval $(0, a)$, which is strictly positive ($f(r) > 0$) for $0 < r < a$. Consider the integral

$$I(x) = \int f(t_2 - t_1)\, dt_1 \wedge dt_2, \quad t_1 \in OX, \quad t_2 \in XA. \tag{2.1}$$

Then we have the following

LEMMA. *For any function $f(r)$ which satisfies the stated conditions, the integral* (2.1) *has its maximum for $x = a/2$.*

Proof. Let $F(r)$ be a primitive of $f(r)$, with $r = t_2 - t_1$, and $G(r)$ a primitive of $F(r)$. We have

$$\begin{aligned} I(x) &= \int_0^x [F(a - t_1) - F(x - t_1)]\, dt_1 = \\ &= -G(a - x) + G(0) + G(a) - G(x). \end{aligned} \tag{2.2}$$

In order that $I(x)$ have a maximum or minimum at the point x we have $I'(x) = F(a - x) - F(x) = 0$ and since $F(x)$ is an increasing function we will have $a - x = x$ and $x = a/2$. This critical value $I(a/2)$ is a maximum because $I(0) = I(a) = 0$.

3. The case $n = 2$.

We want to consider separately the cases of the euclidean, elliptic and hyperbolic planes.

a) *The euclidean plane.* Consider the line G_0 which divides K into two convex domains K_1 and K_2. Let σ_1 denote the length of the chord $G_0 \cap K$ (fig. 1). Consider the pair of points $P_1 \in K_1$, $P_2 \in K_2$ and the line G determined by

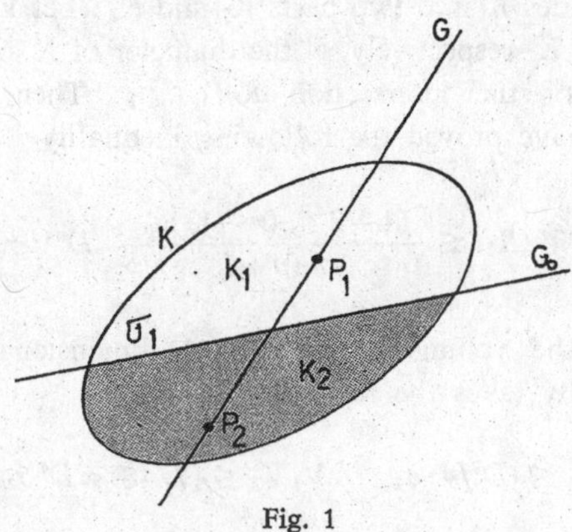

Fig. 1

them. It is well known the differential formula

$$dP_1 \wedge dP_2 = |t_2 - t_1|\, dG \wedge dt_1 \wedge dt_2 \tag{3.1}$$

where dP_1, dP_2 are the area elements of the plane at P_1, P_2, dG is the density for lines on the plane and t_1, t_2 are the abscissas of P_1, P_2 on G [3, p. 28 and 46].

Integration of both sides of (3.1) over all pairs $P_1 \in K_1$, $P_2 \in K_2$ gives: on the left side we get $F_1 F_2$ and in the right side we have the integral (2.1) for the values

$$r = t_2 - t_1, \quad f = r, \quad F = (1/2)\, r^2, \quad G = (1/6)\, r^3. \tag{3.2}$$

Therefore, denoting by a the lenght of the chord $G \cap K$, we get

$$I(x) = (1/2)\, a\, x\, (a - x), \quad I(a/2) = a^3/8. \tag{3.3}$$

Since the measure of the set of lines which cut the chord $K \cap G_0$ is equal to $2\sigma_1$ and $a \leq D$ (D = diameter of K), we have

$$F_1 F_2 = \int I(x)\, dG \leq (D^3/4)\, \sigma_1 \tag{3.4}$$

which is the first inequality (1.2)

b) *The elliptic case.* On the elliptic plane, instead of (3.1), we have [3, p. 316],

$$dP_1 \wedge dP_2 = \sin |t_2 - t_1|\, dG \wedge dt_1 \wedge dt_2. \tag{3.5}$$

We apply the fundamental lemma for the values

$$f = \sin r, \quad F = -\cos r, \quad G = -\sin r \tag{3.6}$$

and we have

$$I(x) = \sin(a - x) - \sin a + \sin x. \tag{3.7}$$

By integrating (3.5) over all pairs of points $P_1 \in K_1$, $P_2 \in K_2$ we get

$$F_1 F_2 = \int (\sin(a - x) - \sin a + \sin x)\, dG \leq \int (2 \sin(a/2) - \sin a)\, dG$$

$$= 4 \int \sin(a/2) \sin^2(a/4)\, dG \leq 8 \sin(D/2) \sin^2(D/4)\, \sigma_1 .$$

Therefore we have the following inequality

$$F_1 F_2 \leq 8 \sin (D/2) \sin^2(D/4) \sigma_1 . \tag{3.8}$$

We have applied that the measure of lines G which cut a segment of length σ_1 is equal to $2\sigma_1$, the same that in the euclidean case [3, p. 310].

c) *The hyperbolic plane.* In this case, instead of (3.1) we have [3, p. 316]

$$d P_1 \wedge d P_2 = \sinh |t_2 - t_1| \, d G \wedge d t_1 \wedge d t_2 . \tag{3.9}$$

In order to apply the lemma, we have

$$f = \sinh r, \quad F = \cosh r, \quad G = \sinh r \tag{3.10}$$

and therefore

$$\begin{aligned} I(x) &= - \sinh (a - x) + \sinh a - \sinh x, \\ I(a/2) &= 4 \sinh (a/2) \sinh^2 (a/4). \end{aligned} \tag{3.11}$$

Since the measure of lines which intersect a segment of length σ_1 is also $2\sigma_1$, [3, p. 310] we get the inequality

$$F_1 F_2 \leq 8 \sinh (D/2) \sinh^2 (D/4) \sigma_1 \tag{3.12}$$

which is the generalization to the hyperbolic plane of the first inequality of Bokowski-Sperner (1.2).

4. The case $n = 3$.

We cosider the three cases:

a) *Euclidean space.* With the customary notation we have [3, p. 237],

$$d P_1 \wedge d P_2 = (t_2 - t_1)^2 d G \wedge d t_1 \wedge d t_2 . \tag{4.1}$$

By integration over all pairs $P_1 \in K_1$, $P_2 \in K_2$, where K_1 and K_2 are now the bodies into which K is partitioned by the plane E_0, calling V_1 and V_2 the volumes of these bodies, we have

$$f = (t_2 - t_1)^2 = r^2 , \quad F = (1/3) r^3, \quad G = (1/12) r^4 \tag{4.2}$$

and

$$I(x) = (-1/12)(a-x)^4 + (1/12)a^4 - (1/12)x^4, \qquad I(a/2) = (7/96)a^4.$$

Since the measure of the set of lines which cut the plane domain $E_0 \cap K$ is $\pi\sigma_2$, where σ_2 denotes the surface area of $E_0 \cap K$ [3, p. 233], we get

$$V_1 V_2 \leq (7/96)\pi D^4 \sigma_2 \tag{4.3}$$

which is better than the second inequality of (1.2).

b) *Elliptic space.* In this case we have [3, p. 316]

$$dP_1 \wedge dP_2 = \sin^2(t_2 - t_1)\, dG \wedge dt_1 \wedge dt_2. \tag{4.4}$$

In order to apply the lemma, we have now

$$f = \sin^2 r, \quad F = (1/2)(r - \sin r \cos r), \quad G = (1/4)(r^2 - \sin^2 r) \tag{4.5}$$

and according to (2.2) we have

$$I(a/2) = (1/2)\sin^4(a/2) + (1/8)(a^2 - \sin^2 a) \tag{4.6}$$

and since the measure of the lines which cut $E_0 \cap K$ is equal to $\pi\sigma_2$ [3, p. 310], we get

$$V_1 V_2 \leq (1/8)(4\sin^4(D/2) + D^2 - \sin^2 D)\pi\sigma_2 \tag{4.7}$$

which generalizes the inequality of Bokowski-Sperner to the elliptic space.

c) *Hyperbolic space.* In this case we have [3, p. 316]

$$dP_1 \wedge dP_2 = \sinh^2(t_2 - t_1)\, dG \wedge dt_1 \wedge dt_2 \tag{4.8}$$

and therefore we have, with the notations of n. 2,

$$f = \sinh^2 r, \quad F = (1/2)(\sinh r \cosh - r), \quad G = (1/4)(\sinh^2 r - r^2) \tag{4.9}$$

and thus

$$\begin{aligned} I(x) &= (1/4)(-\sinh^2(a-x) + (a-x)^2 + \sinh^2 a - a^2 - \sinh^2 x + x^2), \\ I(a/2) &= (1/2)\sinh^4(a/2) + (1/8)(\sinh^2 a - a^2). \end{aligned} \tag{4.10}$$

Therefore, since the measure of the set of lines which intersect the set $E_0 \cap K$ is equal to $\pi \sigma_2$ [3, p. 310], we get

$$V_1 V_2 \leq (1/8)(4 \sinh^4 (D/2) + \sinh^2 D - D^2) \pi \sigma_2 \tag{4.11}$$

which is the generalization to the hyperoblic space of the second inequality (1.2).

5. A conjecture.

We have considered the case in which K is partitioned by a line (for $n = 2$) or by a plane (for $n = 3$). More general is the case of a partition of K into two sets K_1, K_2 not necessarily convex, separated by a curve (for $n = 2$) or by a surface (for $n = 3$). To apply the foregoing proof in this case we will need a lemma more general that the lemma stated in n. 2. We state it as the following conjecture:

Consider the closed interval $(0, a)$ on the real line, divided into $n + 1$ parts by the points $0 < a_1 < a_2 < \ldots < a_n < a$. Put $a_0 = 0$, $a_{n+1} = a$ and consider the sets of intervals

$$\begin{aligned} T &= \{(0, a_1), \quad (a_2, a_3), \quad (a_4, a_5), \ldots\} \\ T^* &= \{(a_1, a_2), \quad (a_3, a_4), \ldots\}. \end{aligned} \tag{5.1}$$

Consider the integral

$$I(a_1, a_2, \ldots, a_n, a) = \int_{\substack{t \in T \\ t^* \in T^*}} f(|t - t^*|) \, dt \, dt^*. \tag{5.2}$$

The conjecture is that this integral has a maximum for $n = 1$ and $a_1 = a/2$ for any integrable and non-negative function defined on the interval $(0, a)$. If it is not true, seek additional conditions for f.

In order to apply this conjecture to the generalization of the inequality (1.1) to the elliptic and the hyperbolic spaces it should be sufficient to prove it for the cases $f = \sin^n r$ and $f = \sinh^n r$.

REFERENCES

[1] Bokowski J., Sperner E., *Zerlegung konvexer Körper durch minimale Trennflächen,* J. Reine Angew. Math. 311/312, 1979, 80-100.

[2] Bokowski J., *Ungleichungen für den Inhalt von Trennflächen,* Arch. Math. (Basel) **34** 1980, 84-89.

[3] Santalò L. A., *Integral Geometry and Geometric Probability,* Encyclopedia of Math. and Applications, Addison-Wesley, Reading, 1976.

Pervenuto il 4 giugno 1982

Facultad de Ciencias
Exactas y Naturales
Universidad de Buenos Aires
(Argentina)

J.A. BARROSO editor, Aspects of Mathematics and its Applications 677

ON THE MEASURE OF LINE SEGMENTS ENTIRELY CONTAINED IN A CONVEX BODY

L.A. SANTALÓ

Academia Nacional de Ciencas Exatas, Fisicas y Naturales, Buenos Aires, Argentina

Dedicated to Leopoldo Nachbin with admiration and friendship

Let K be a convex body in the n-dimensional euclidean space $\mathbb{R}^n$. We consider the measure $M_0(l)$, in the sense of the integral geometry (i.e. invariant under the group of translations and rotations of $\mathbb{R}^n$ [6, Chap. 15]), of the set of non-oriented line segments of length l, which are entirely contained in K. This measure is related by (3.4) with the integrals I_m for the power of the chords of K. These relations allow to obtain some inequalities, like (3.6), (3.7) and (3.8) for $M_0(l)$. Next we relate $M_0(l)$ with the function $\Omega(l)$ introduced by Enns and Ehlers [3], and prove a conjecture of these authors about the maximum of the average of the random straight line path through K. Finally, for $n = 2$, $M_0(l)$ is shown to be related by (5.6) with the associated function to K introduced by W. Pohl [5]. Some representation formulas, like (3.9), (3.10) and (5.14) may be of independent interest.

1. Integrals for the Power of the Chords of a Convex Body

Let K be a convex body in the n-dimensional euclidean space $\mathbb{R}^n$. Let $\mathrm{d}G$ be the density for lines G in $\mathbb{R}^n$ in the sense of integral geometry [6, Chap. 12] and let σ denote the length of the chord $G \cap K$. The chord power integrals

$$(1.1) \qquad I_m = \int_{G\cap K \neq \emptyset} \sigma^m \, \mathrm{d}G \qquad (m \geqslant 0),$$

have been well studied [6, p. 237]. If $\mathrm{d}P_1$, $\mathrm{d}P_2$ denote the elements of volume of $\mathbb{R}^n$ at the points $P_1, P_2 \in K$ and r denotes the distance between P_1 and P_2, the integrals

$$(1.2) \qquad J_m = \int_{P_1, P_2 \in K} r^m \, \mathrm{d}P_1 \wedge \mathrm{d}P_2 \qquad (m \geqslant -(n-1)),$$

have also been considered and it is known that the relation

$$2I_m = m(m-1)J_{m-n+1} \tag{1.3}$$

holds good for $m > 1$ [6, p. 238].

For the cases $m = 0, 1$ and $m = n+1$ we have the simple formulas

$$I_0 = \tfrac{1}{2}\frac{O_{n-2}}{n-1}F, \qquad I_1 = (\tfrac{1}{2}O_{n-1})V, \qquad I_{n+1} = (\tfrac{1}{2}n(n+1))V^2, \tag{1.4}$$

where F is the surface area of K and V its volume.

We want to calculate I_m for the sphere S_r of radius r in $\mathbb{R}^n$. To this end, recalling that $\mathrm{d}G = \mathrm{d}\sigma_{n-1} \wedge \mathrm{d}O_{n-1}$ [6, (12.39)] where $\mathrm{d}\sigma_{n-1}$ is the area element of an hyperplane orthogonal to G at its intersection point with G and $\mathrm{d}O_{n-1}$ is the area element of the unit sphere at the end point of the unit vector parallel to G, we can write $\mathrm{d}G = \rho^{n-2}\,\mathrm{d}O_{n-2} \wedge \mathrm{d}\rho \wedge \mathrm{d}O_{n-1}$ and therefore we have (ρ being the distance from the center of the sphere to G)

$$\begin{aligned} I_m &= 2^{m-1}O_{n-1}O_{n-2}\int_0^r (r^2-\rho^2)^{m/2}\rho^{n-2}\,\mathrm{d}\rho \\ &= 2^{m-2}O_{n-1}O_{n-2}r^{m+n-1}\,\mathrm{B}(\tfrac{1}{2}(n-1), \tfrac{1}{2}(m+2)), \end{aligned} \tag{1.5}$$

where $\mathrm{B}(p,q) = \Gamma(p)\,\Gamma(q)/\Gamma(p+q)$ is the Beta function and O_h means the surface area of the h-dimensional unit sphere, i.e.

$$O_h = \frac{2\pi^{(h+1)/2}}{\Gamma(\frac{1}{2}(h+1))}. \tag{1.6}$$

Therefore we have

$$I_m(S_r) = \frac{2^{m-1}r^{m+n-1}\pi^{n-1/2}m\Gamma(\frac{1}{2}m)}{\Gamma(\frac{1}{2}n)\Gamma(\frac{1}{2}(m+n+1))}. \tag{1.7}$$

2. Inequalities of Hadwiger, Carleman and Blaschke for the Chord Integrals I_m

The chord integrals I_m for convex bodies in $\mathbb{R}^n$ satisfy certain inequalities. One of them is due to H. Hadwiger [4]:

(2.1)
$$2I_{n-1} \leqslant n\left(\frac{\kappa_{2n-2}}{n\kappa_{n-1}}\right)^{1/n} FV^{1-1/n} \qquad (n > 2),$$

where κ_h denotes the volume of the n-dimensional unit ball, i.e.

(2.2)
$$\kappa_h = \frac{O_{h-1}}{h} = \frac{2\pi^{h/2}}{h\Gamma(\frac{1}{2}h)},$$

and F and V denote the surface area and the volume of K respectively.

Taking into account the isoperimetric inequality

(2.3)
$$n\kappa_n^{1/n} V^{1-1/n} \leqslant F,$$

inequality (2.1) gives

(2.4)
$$4I_{n-1} \leqslant F^2 \qquad (n > 2).$$

In (2.1) and (2.4) the equality sign holds only for the sphere.

In [2] T. Carleman proved that in the plane, $n = 2$, $J_{-1} = \int r^{-1} \mathrm{d}P_1 \wedge \mathrm{d}P_2 = I_2$ has its maximum for the circle (for a given surface area) and pointed out that the same proof may be extended to showing that for convex bodies in $\mathbb{R}^n$, the integrals I_m for $m = 2, 3, \ldots, n$ have a maximum for the sphere for a given volume V. Thus, taking into account (1.7) and (2.2) we have the following set of inequalities:

(2.5)
$$I_m^n \leqslant 2^{mn-m-n+1} \pi^{(n^2/2)-mn/2} n^{m+n-1} (\Gamma(\tfrac{1}{2}n))^{m-1} \left(\frac{\Gamma(\frac{1}{2}m+1)}{\Gamma(\frac{1}{2}(m+n+1))}\right)^n V^{m+n-1},$$

for $m = 2, 3, \ldots, n$.

In [1], W. Blaschke proved that in the plane ($n = 2$) and for a given area F, the integrals I_m ($m \geqslant 4$) have its minimum for the circle. The proof is also easily extendible to $\mathbb{R}^n$, so that, taking (1.7) and (2.2) into account, we can write the new set of inequalities (for $\mathbb{R}^n$)

(2.6)
$$I_m^n \geqslant 2^{mn-m-n+1} \pi^{(n^2/2)-mn/2} n^{m+n-1} (\Gamma(\tfrac{1}{2}n))^{m-1} \left(\frac{\Gamma(\frac{1}{2}m+1)}{\Gamma(\frac{1}{2}(m+n+1))}\right)^n V^{m+n-1},$$

for $m \geqslant n+2$. In (2.5) and (2.6) the equality sign holds only for the sphere.

3. The Measure $M_0(l)$ of the Line Segments of Length l Entirely Contained in a Convex Body K in $\mathbb{R}^n$

A line segment S of given length l in $\mathbb{R}^n$ can be determined either by the line G which contains the segment and the abscissa t of the origin P of S on G, or by P and the point on the unit sphere O_{n-1} given by the direction of S. The kinematic density for sets of line segments of length l (invariant under motions in $\mathbb{R}^n$) is [6, p. 338]:

$$\mathrm{d}S = \mathrm{d}G \wedge \mathrm{d}t = \mathrm{d}P \wedge \mathrm{d}O_{n-1}. \tag{3.1}$$

Using $\mathrm{d}S = \mathrm{d}G \wedge \mathrm{d}t$ we get that the measure of the set of line segments S entirely contained in K is

$$M_0(l) = \int_{\sigma \geq l} (\sigma - l)\, \mathrm{d}G. \tag{3.2}$$

If P_1, P_2 are two points of K at a distance l, we have $\mathrm{d}P_1 \wedge \mathrm{d}P_2 = l^{n-1}\, \mathrm{d}O_{n-1} \wedge \mathrm{d}l \wedge \mathrm{d}P_1$ (up to the sign) and therefore, since we consider the measure of non-oriented segments, we have

$$\int_{P_1, P_2 \in K} l^m\, \mathrm{d}P_1 \wedge \mathrm{d}P_2 = 2 \int_0^{\mathrm{Diam}(K)} l^{m+n-1} M_0(l)\, \mathrm{d}l. \tag{3.3}$$

As a consequence of (1.3) and (3.3) we have

$$I_m = m(m-1) J_{m-n-1} = m(m-1) \int_0^{\mathrm{Diam}(K)} l^{m-2} M_0(l)\, \mathrm{d}l, \tag{3.4}$$

which holds for $m \geq 2$. In particular, for $m = 2$ we have

$$I_2 = 2 \int_0^{\mathrm{Diam}(K)} M_0(l)\, \mathrm{d}l, \tag{3.5}$$

and the first inequality (2.5) gives

$$\int_0^{\mathrm{Diam}(K)} M_0(l)\,\mathrm{d}l \leq \frac{2^{1-1/n}\pi^{(n/2)-1}n^{1+1/n}(\Gamma(\frac{1}{2}n))^{1/n}V^{(n+1)/n}}{(n+1)\Gamma(\frac{1}{2}(n+1))}, \tag{3.6}$$

where the equality sign holds only for the sphere.

For instance, for convex sets K in the plane, $n=2$, we have

$$\int_0^{\mathrm{Diam}(K)} M_0(l)\,\mathrm{d}l \leq \frac{8}{3\sqrt{\pi}} F^{3/2}, \tag{3.7}$$

where F is the surface area of K.

Taking into account the isoperimetric inequality $4\pi F \leq L^2$ and the inequality of Bieberbach $F \leq \frac{1}{4}\pi D^2$, where $D = \mathrm{diam}(K)$, we get the following inequalities (for convex sets in the plane):

$$\int_0^D M_0(l)\,\mathrm{d}l \leq \frac{L^3}{3\pi^2}, \qquad \int_0^D M_0(l)\,\mathrm{d}l \leq \tfrac{1}{3}\pi D^3, \tag{3.8}$$

with the equality sign always only for the circle.

From (3.4) we deduce that for every polynomial function of the form $f = a_2\sigma^2 + \cdots + a_h\sigma^h$ we have

$$\int_{G\cap K\neq\emptyset} f(\sigma)\,\mathrm{d}G = \int_0^D f''(\sigma)M_0(\sigma)\,\mathrm{d}\sigma. \tag{3.9}$$

By Weierstrass approximation theorem, this equality holds for every function $f(\sigma)$ having continuous derivatives $f''(\sigma)$ with the conditions $f(0) = f'(0) = 0$.

Integrating by parts the right side of (3.9) we have the following relationship

$$\int_{G\cap K\neq\emptyset} f(\sigma)\,\mathrm{d}G = -\int_0^D f'(\sigma)M_0'(\sigma)\,\mathrm{d}\sigma, \tag{3.10}$$

for every function $f(\sigma)$ having continuous derivative $f'(\sigma)$ and satisfying the condition $f(0) = f'(0) = 0$.

4. The Invariants $\Omega(l)$ of Enns–Ehlers

Denote by $K(l, \omega)$ the translate of K by a distance l in the direction ω. Enns and Ehlers [3] define $\Omega(l)$ to be the volume of $K \cap K(l, \omega)$ uniformly averaged over all directions and normalized such that $\Omega(0) = 1$. If σ denotes the length of the chord $G \cap K$, the volume of $K \cap K(l, \omega)$ is precisely $\int_{\sigma \geq l} (\sigma - l)\, \mathrm{d}\sigma_{n-1}$, where $\mathrm{d}\sigma_{n-1}$ denotes the area element on the hyperplane orthogonal to the line G which has the direction ω. Therefore, since $\mathrm{d}G = \mathrm{d}\sigma_{n-1} \wedge \mathrm{d}O_{n-1}$, where $\mathrm{d}O_{n-1}$ denotes the area element on the unit $(n-1)$-sphere corresponding to the direction ω, we have

$$\Omega(l) = \frac{2}{O_{n-1} V} \int_{\sigma \geq l} (\sigma - l)\, \mathrm{d}\sigma_{n-1} \wedge \mathrm{d}O_{n-1} = \frac{2}{O_{n-1} V} \int_{\sigma \geq l} (\sigma - l)\, \mathrm{d}G \tag{4.1}$$

and thus, according to (3.2),

$$\Omega(l) = \frac{2}{O_{n-1} V} M_0(l)\,. \tag{4.2}$$

Therefore, (3.4) gives

$$I_m = \tfrac{1}{2} m(m-1) O_{n-1} V \int_0^D l^{m-2} \Omega(l)\, \mathrm{d}l\,. \tag{4.3}$$

For instance, if $m = n + 1$, taking (1.4) into account, we have

$$\int_0^D l^{n-1} \Omega(l)\, \mathrm{d}l = \frac{V}{O_{n-1}}, \tag{4.4}$$

according to a result of Enns and Ehlers [3, (8)].

If a 'random secant' is defined by a point in the interior of K and by a direction (the point and direction have independent uniform distribution), the k-th moment of a random secant is (using (3.1))

$$E(\sigma^k) = \frac{2}{O_{n-1}V}\int \sigma^k \, dP \wedge dO_{n-1} = \frac{2}{O_{n-1}V}\int \sigma^{k+1}\, dG \tag{4.5}$$

$$= \frac{2k(k+1)}{O_{n-1}V}\int_0^D \sigma^{k-1} M_0(\sigma)\, d\sigma .$$

Thus, according to (3.4) we have

$$E(\sigma^k) = \frac{2}{O_{n-1}V} I_{k+1} ,$$

and the inequalities (2.5) give

$$E(\sigma^k) \leqslant \frac{2^{k-k/n} n^{(k/n)+1} (\Gamma(\frac{1}{2}n))^{(k+n)/n} \Gamma(\frac{1}{2}(k+3))}{\pi^{(k+1)/2} \Gamma(\frac{1}{2}(k+n+2))} V^{k/n} , \tag{4.6}$$

which holds for $k = 1, 2, \ldots, n-1$ and the equality sign holds only for the sphere. For the sphere of radius r we have $V = (2\pi^{n/2}/n\Gamma(\frac{1}{2}n))r^n$ and therefore

$$E(\sigma^k) = \frac{2^k n\Gamma(\frac{1}{2}n)\Gamma(\frac{1}{2}(k+3))}{\pi^{1/2}\Gamma(\frac{1}{2}(k+n+2))} r^k , \tag{4.7}$$

as is well-known (Enns–Ehlers [3]).

In particular (4.6) implies that of all n-dimensional convex bodies K of volume V, $E(\sigma)$ is maximized for the n-sphere. This proves a conjecture of Enns and Ehlers [3].

The inequalitiies (2.6) can be written

$$E(\sigma^k) \geqslant \frac{2^{k-k/n} n^{1+k/n} (\Gamma(\frac{1}{2}n))^{1+k/n} \Gamma(\frac{1}{2}(k+3))}{\pi^{(k+1)/2} \Gamma(\frac{1}{2}(k+n+2))} V^{k/n} , \tag{4.8}$$

valid for $k = n+1, n+2, \ldots$. The equality sign holds only for the sphere.

For the plane, $n = 2$, if F denotes the area of K, we have

$$E(\sigma) \leqslant \frac{8F^{1/2}}{3\pi^{3/2}} , \qquad E(\sigma^2) = \frac{3}{\pi} F , \tag{4.9}$$

and therefore, of all the convex sets of area F, the variance

$$E(\sigma^2)-(E(\sigma))^2 \geqslant \frac{27\pi^2-64}{9\pi^3} F$$

is minimized for the circle (as conjectured by Enns–Ehlers [3]). The conjecture that the variance is also minimized for the sphere if $n>2$ remains open.

5. The Associated Functions $A(\sigma)$ of W. Pohl

In this section we consider only the case of the plane, $n=2$. In each line G we choose a point $X(x,y)$ and the unit vector $e(\cos\theta, \sin\theta)$ corresponding to its direction. Consider the differential form $\omega = \langle \mathrm{d}X, e\rangle = \cos\theta \cdot \mathrm{d}x + \sin\theta \cdot \mathrm{d}y$. Then we have $\mathrm{d}\omega = -\sin\theta \cdot \mathrm{d}\theta \wedge \mathrm{d}x + \cos\theta \cdot \mathrm{d}\theta \wedge \mathrm{d}y = \mathrm{d}G$ (according to [6, (3.11)]). W. Pohl [5] defines the associated function $A(\sigma)$ to the convex curve ∂K by

$$A(\sigma) = \int_{\partial K} \omega = \int_{\partial K} \cos\alpha \, \mathrm{d}s\,, \tag{5.1}$$

where the integral of ω extends to the non-oriented lines (X, e), $X \in \partial K$, that determine on the convex set K a chord of length σ and in the last integral α denotes the angle between the tangent to ∂K and G at the point X corresponding to the element of the arc $\mathrm{d}s$.

A simple geometric description of $A(\sigma)$, at least for small values of σ, is the following [5]: Let ∂K_0 be the curve envelope of the chords of K of length σ. Then $A(\sigma)$ is length of ∂K_0. For instance, for a circle of diameter D we have

$$A(\sigma) = \pi(D^2-\sigma^2)^{1/2}\,. \tag{5.2}$$

Notice that our $A(\sigma)$ is one half of that of Pohl, which considers oriented lines.

Let $M_1(l)$ be the measure of the set of non-oriented line segments of length l such that one end point is inside K and one outside K. Then we have [5]

(5.3) $$M_1(l) = 2 \int_0^l A(\sigma)\, d\sigma .$$

On the other side, using the kinematic density $dS = dG \wedge dt$, we have

(5.4) $$M_1(l) = 2 \int_{\sigma \geq l} l \, dG + 2 \int_{\sigma \leq l} \sigma \, dG$$

and by virtue of (3.2) we get

(5.5) $$M_0 + \tfrac{1}{2} M_1 = \pi F ,$$

where F is the surface area of K. From (5.3) and (5.5) we have

(5.6) $$M_0 = \pi F - \int_0^l A(\sigma)\, d\sigma ,$$

and

(5.7) $$A(\sigma) = -M_0'(\sigma) .$$

The relation (5.6) can be applied to compute the measure $M_0(l)$ of non-oriented line segments of length $l \leq D$ entirely contained in a circle of diameter D. Namely, from (5.2) we have

(5.8) $$M_0(l) = \pi F - \pi \int_0^l (D^2 - \sigma^2)^{1/2}\, d\sigma$$

$$= \tfrac{1}{4}\pi \left(\pi D^2 - 2l(D^2 - l^2)^{1/2} - 2D^2 \arcsin\left(\frac{l}{D}\right)\right) ,$$

as is well known [6, p. 90].

Integrating by parts in (3.4) and taking into account (5.7), we get (for convex sets in the plane and $m \geq 1$)

(5.9) $$I_m = m \int_0^D \sigma^{m-1} A(\sigma)\, \mathrm{d}\sigma ,$$

where D is the diameter of K. This expression for the chord integrals I_m (for convex sets on the plane) is due to Pohl [5]. For $m = 1$ we have

(5.10) $$\int_0^D A(\sigma)\, \mathrm{d}\sigma = \pi F .$$

For $m = 2$, according to (2.5) we get the inequality

(5.11) $$\int_0^D \sigma A(\sigma)\, \mathrm{d}\sigma \leqslant \frac{8}{3\sqrt{\pi}} F^{3/2} .$$

For $m = 3$ we have

(5.12) $$\int_0^D \sigma^2 A(\sigma)\, \mathrm{d}\sigma = F^2 ,$$

and for $m > 3$,

(5.13) $$\int_0^D \sigma^{m-1} A(\sigma)\, \mathrm{d}\sigma \geqslant \frac{2^{m-1} \pi^{1-m/2} \Gamma(\frac{1}{2} m)}{\Gamma(\frac{1}{2}(m+3))} F^{(m+1)/2} .$$

In (5.11) and (5.13) the equality sign holds only for the circle.

From (3.10) and (5.7) we have

(5.14) $$\int_{G \cap K \neq \emptyset} f(\sigma)\, \mathrm{d}G = \int_0^D f'(\sigma) A(\sigma)\, \mathrm{d}\sigma ,$$

which holds for every function $f(\sigma)$ having a continuous derivative $f'(\sigma)$ and satisfying the conditions $f(0) = f'(0) = 0$.

The relation between the invariant $\Omega(\sigma)$ of Enns–Ehlers and the associated function $A(\sigma)$ of Pohl, according to (4.2) and (5.6) is

$$A(\sigma) = -\pi F \Omega'(\sigma)\,. \tag{5.15}$$

References

[1] W. Blaschke, Eine isoperimetrische Eigenschaft des Kreises, Math. Z. 1 (1918) 52–57.

[2] T. Carleman, Ueber eine isoperimetrische Aufgabe und ihre physikalischen Anwendungen, Math. Z. 3 (1919) 1–7.

[3] E.G. Enns and P.F. Ehlers, Random paths through a convex region, J. Appl. Probab. 15 (1978) 144–152.

[4] H. Hadwiger, Ueber zwei quadratische Distanzintegrale für Eikörper, Arch. Math. (Basel) 3 (1952) 142–144.

[5] W. Pohl, The probability of linking of random closed curves, In: Geometry Symp. Utrecht 1980. Lecture Notes in Math. 894 (Springer, Berlin, 1981) 113–126.

[6] L.A. Santaló, Integral Geometry and Geometric Probability, Encyclopedia Math. Appl. 1 (Addison–Wesley, Reading, Mass., 1976).

Journal of Microscopy, Vol. 151, *Pt* 3, *September* 1988, *pp*. 229–233.
Received 15 *September* 1987; *revised* 5 *April* 1988; *accepted* 8 *April* 1988

Affine integral geometry and convex bodies

by L. A. SANTALÓ, *Departamento de Matemáticas, Facultad de Ciencias Exactas y Naturales, Ciudad Universitaria (Nuñez),* 1428 *Buenos Aires, Argentina*

KEY WORDS. Affine transformations, r-flats, invariant densities, parallel hyperplanes, convex bodies, breadth, n-ellipsoids, n-parallelepipeds, n-simplices.

SUMMARY

Given a convex body Q in Euclidean n-dimensional space, the affine invariant measure of the set of pairs of parallel hyperplanes containing Q is an affine invariant $\mathcal{J}(Q)$ of Q, which, for ellipsoids, parallelepipeds and possibly for simplices of any dimensions, is proportional to V^{-1}, where V represents the volume of Q. Consequently, for the kind of bodies mentioned, it is possible to estimate V^{-1} from its breadth measured in uniform random directions. If the boundary of Q is of class C^2, we obtain a set of affine invariants $\mathcal{J}_h(Q)$ (h = any integer) which is easily calculable for ellipsoids. In particular, $\mathcal{J}_{-n}(Q)$ coincides with $\mathcal{J}(Q)$ and $\mathcal{J}_{-(n+1)}(Q)$ is the affine invariant measure of all pairs of parallel hyperplanes that 'support' Q as described by Firey (1972, 1985), Schneider (1978, 1979) and Weil (1979, 1981). For a general convex body Q the values of $\mathcal{J}_h(Q)$ cannot be expressed in terms of simple metric invariants (such as volume, surface area, breadth, width) and this justifies the study in the last section of certain inequalities between them.

1. INTRODUCTION

Given a convex body Q in Euclidean n-dimensional space E_n, whose boundary ∂Q is of class C^2, we obtain the set (4.2) of affine invariants $\mathcal{J}_h(Q)$, where h is any integer (positive or negative). The most important case corresponds to $h=-n$, since then the invariant $\mathcal{J}_{-n}(Q)=\mathcal{J}(Q)$ (2.4) is well defined for all convex bodies Q, without any smoothness condition for ∂Q, and coincides with the affine invariant measure of the set of pairs of parallel hyperplanes containing Q (up to a constant factor). This invariant $\mathcal{J}(Q)$ may be easily computed for n-ellipsoids and n-parallelepipeds and also for triangles in the plane and tetrahedrons in E_3. The values of $\mathcal{J}$ for n-simplices seem to be rather complicated. Formulas (2.5), (2.6) and (2.7) can be used to estimate the volume V of Q in terms of the breadth of Q in random directions (for n-ellipsoids, n-parallelepipeds, triangles in E_2 and tetrahedrons in E_3). The invariant $\mathcal{J}_{-(n+1)}(Q)$ coincides, up to a constant factor, with the affine invariant measure of the set of pairs of parallel hyperplanes that 'support' Q, in the sense of the measure of support flats introduced by Firey (1972, 1985), Schneider (1978, 1979) and Weil (1979, 1981). At the end of the paper we include some inequalities for the invariants $\mathcal{J}_h$, but they are far from being a complete system. The group of affine transformations considered is that of unimodular affinities, i.e. the affinities that are volume-preserving.

2. THE AFFINE INVARIANT $\mathcal{J}$ FOR CONVEX BODIES

It is well known that the sets of linear spaces L_r (r-flats), $r=0, 1, 2, \ldots, n-1$, in Euclidean space E_n, have an invariant density under the group of rigid motions in E_n. With these invariant densities we can measure the sets of L_r which intersect a fixed convex body Q and the results are precisely, up to a constant factor, the so-called 'quermassintegrale' $W_r(Q)$ introduced by Minkowski (see, for instance, Santaló, 1976, Chap. 14).

Instead of the group of rigid motions it may be of interest to consider any other continuous group G operating in E_n such that it transitively transforms the r-flats L_r. In this paper we will consider the group of unimodular (volume preserving) affinities, i.e. the group of transformations of E_n into itself represented by a matrix equation of the form

$$x'=Ax+B \tag{2.1}$$

where A is an $n\times n$ matrix with determinant 1 and B is an $n\times 1$ matrix.

It is well known that, except for points ($r=0$), the r-flats L_r have no invariant density under the affine group (2.1). However, the pairs of parallel hyperplanes $\mathscr{P}$ have an invariant density given by (Santaló, 1976, Chap. 11):

$$\mathrm{d}\mathscr{P}=\frac{\mathrm{d}O_{n-1}\,\mathrm{d}p_1\,\mathrm{d}p_2}{|p_1-p_2|^{n+2}} \tag{2.2}$$

where $\mathrm{d}O_{n-1}$ denotes the area element on the $(n-1)$-dimensional unit sphere corresponding to the direction perpendicular to the parallel hyperplanes and p_1 and p_2 are the distances from the origin to the hyperplanes. By integration of (2.2) we get the measure of all pairs of parallel hyperplanes completely enclosing a given convex body Q. The result is

$$m(\mathscr{P};\,Q\subset\mathscr{P})=\frac{1}{n(n+1)}\int_{O_{n-1}}\frac{\mathrm{d}O_{n-1}}{\Delta^n} \tag{2.3}$$

where the integral is extended over the $(n-1)$-dimensional unit sphere and Δ denotes the breadth of Q corresponding to the direction determined by $\mathrm{d}O_{n-1}$, i.e. the distance between the two parallel hyperplanes that support Q and are perpendicular to the direction defined by the area element $\mathrm{d}O_{n-1}$.

The measure (2.3) gives rise to the following affine invariant of Q:

$$\mathcal{J}(Q)=\int_{O_{n-1}}\frac{\mathrm{d}O_{n-1}}{\Delta^n} \tag{2.4}$$

Because of its affine invariance, $\mathcal{J}$ has the same value for the n-ball as for any n-ellipsoid of the same volume. Therefore, we readily obtain the following value for any n-ellipsoid:

$$\mathcal{J}(n\text{-ellipsoid})=\frac{O_{n-1}^2}{2^n nV} \tag{2.5}$$

where $O_{n-1}=2\pi^{n/2}/\Gamma(n/2)$ is the surface area of the $(n-1)$- dimensional unit sphere (boundary of the n-ball) and V is the volume of the n-ellipsoid.

Similarly, since all n-parallelepipeds of the same volume are affine-equivalent, a direct computation for the n-cube gives the following general value:

$$\mathcal{J}(n\text{-parallelepiped})=\frac{2^n}{(n-1)!\,V} \tag{2.6}$$

For the plane ($n=2$) and the space ($n=3$) by direct computation we have:

$$\mathcal{J}(\text{triangle})=\frac{3}{F},\quad \mathcal{J}(\text{tetrahedron})=\frac{10}{9V} \tag{2.7}$$

where F denotes the area of the triangle. The value of $\mathcal{J}$ for n-simplices remains to be derived.

From (2.5) it follows that the volume V of an ellipsoid in E_n can be estimated by means of the breadth Δ measured in uniform random directions, i.e. as an unbiased estimator of V^{-1} (for n-ellipsoids) we can take

$$\text{est}(V^{-1}) = \frac{2^n n}{O_{n-1} m} \sum_{1}^{m} \Delta_i^{-n}$$

where $\Delta_i (i = 1, 2, \ldots, m)$ are the breadths of the n-ellipsoid in random directions.

From (2.6) and (2.7) similar estimators can be deduced for the volume of parallelepipedal particles in E_n and triangular particles in E_2 or tetrahedral particles in E_3.

3. PARALLEL HYPERPLANES THAT 'SUPPORT' A CONVEX BODY

Following the ideas of Firey (1972, 1985), Schneider (1978, 1979) and Weil (1979, 1981) we consider the affine invariant measure of the set of parallel hyperplanes that 'support' a convex body Q, i.e. such that both parallel hyperplanes are supporting hyperplanes of Q.

Recall that in the affine differential geometry of convex bodies, the affine breadth Δ_a of a convex body Q at the point $x \in \partial Q$ is defined as being equal to the affine distance from x to the contact point with ∂Q of the supporting hyperplane of Q that is parallel to the supporting hyperplane at x (Blaschke, 1923, p. 110; Süss, 1927). The affine breadth Δ_a and the metric breadth Δ at a point x are related by the equality

$$\Delta_a = K^{-1/(n+1)} \Delta \tag{3.1}$$

where K is the metric Gaussian curvature of ∂Q at the point $x \in \partial Q$ (product of the $n-1$ principal curvatures). Here we assume that ∂Q is of class C^2.

For each point $x \in \partial Q$ let us consider the affine breadth Δ_a increased to $\Delta_a + \varepsilon$ by a constant ε. According to (3.1) the metric breadth Δ transforms into $\Delta + \varepsilon K^{1/(n+1)}$ and therefore the affine invariant $\mathcal{J}$ (2.4) becomes

$$\int_{O_{n-1}} [\Delta + \varepsilon K^{1/(n+1)}]^{-n} \, dO_{n-1} \tag{3.2}$$

The limit of the difference

$$\frac{1}{\varepsilon} \int_{O_{n-1}} \{\Delta^{-n} - [\Delta + \varepsilon K^{1/(n+1)}]^{-n}\} \, dO_{n-1}$$

as $\varepsilon \to 0$ may be considered the affine invariant measure of all pairs of parallel hyperplanes that 'support' Q. The result, up to a constant factor, is

$$\mathcal{J}^{\star}(Q) = \int_{O_{n-1}} \Delta^{-(n+1)} K^{1/(n+1)} \, dO_{n-1} \tag{3.3}$$

which is another affine invariant of Q.

4. OTHER AFFINE INVARIANTS AND SOME INEQUALITIES

Assume that ∂Q is of class C^2. The element of affine surface area $d\sigma_a$ of ∂Q is related to the metric curvature K and the element of metric surface area $d\sigma$ by (see, e.g., Petty, 1985, p. 113)

$$d\sigma_a = K^{1/(n+1)} \, d\sigma = K^{-n/(n+1)} \, dO_{n-1} \tag{4.1}$$

where we have used that $d\sigma = K^{-1} \, dO_{n-1}$. Taking (3.1) and (4.1) into account, we have, for any convex body Q in E_n, the set of affine invariants

$$\mathcal{J}_h(Q) = \int_{\partial Q} \Delta_a^h \, d\sigma_a = \int_{\partial Q} \Delta^h K^{(1-h)/(n+1)} \, d\sigma = \int_{O_{n-1}} \Delta^h K^{-(h+n)/(n+1)} \, dO_{n-1} \tag{4.2}$$

which are defined for any integer h (positive or negative).

The most interesting case corresponds to $h=-n$, since then $\mathcal{J}_{-n}$ does not depend on K and coincides with $\mathcal{J}$ as defined in (2.4). The affine invariant $\mathcal{J}^\star$ (see (3.3)) corresponds to $h=-(n+1)$.

The invariants $\mathcal{J}_h$ can be easily computed for the n-ball of volume V and the result, being equal for any n-parallelepiped of volume V, gives

$$\mathcal{J}_h(n\text{-ellipsoid}) = 2^{(hn-h+2)/(n+1)} n^{(n+2h-1)/(n+1)} \pi^{n(1-h)(n+1)} (\Gamma(n/2))^{2(h-1)/(n+1)} V^{(2h+n-1)/(n+1)}$$

The values of $\mathcal{J}_h$ for a general convex body Q cannot be expressed in terms of simple metric invariants of Q (such as volume, surface area, breadth, width) however, we can give some inequalities which bound its possible values.

(a) Let $p(\sigma)$ denote the supporting function of Q with respect to an interior point of Q taken as origin and let $p^\star(\sigma)$ denote that of its oppposite point, σ indicating a point of ∂Q. Then, from (4.2) we have

$$\mathcal{J}_1 = \int_{\partial Q} [p(\sigma)+p^\star(\sigma)]\, d\sigma = nV + nV^\star \tag{4.3}$$

where $V^\star$ is a mixed volume of Q and its symmetrical with respect to the origin (precisely the mixed volume $V_{(1)}(Q, -Q)$ in the notation of Bonnesen & Fenchel (1934, p. 40)). The classical inequalities of Minkowski (Bonnesen & Fenchel, 1934, pp. 91, 105) give $nV \geqslant V^\star \geqslant V$ and therefore from (4.3) we deduce

$$2nV \leqslant \mathcal{J}_1 \leqslant n(n+1)V \tag{4.4}$$

where the first equality holds only if Q is centrally symmetric. It is not known for which convex bodies the equality on the right is attained, probably for n-simplices only.

(b) The invariant $\mathcal{J} = \mathcal{J}_{-n}$ satisfies the following known inequalities (Santaló, 1976, p. 189)

$$\frac{2^n}{n!(n-1)!V} < \mathcal{J} \leqslant \frac{O_{n-1}^2}{2^n nV} \tag{4.5}$$

where the upper bound is attained if and only if Q is an ellipsoid. The lower bound is not the best possible and its best value is not known for $n>2$. For $n=2$ we have $\mathcal{J} \geqslant 3/V$ and the equality characterizes triangles (in this case V means the surface area of the triangle).

(c) According to (4.2) $\mathcal{J}_0$ is the affine area of Q and the known affine isoperimetric inequality is expressed (Petty, 1985)

$$\mathcal{J}_0^{n+1} \leqslant n^{n-1} O_{n-1}^2 V^{n-1} \tag{4.6}$$

with equality if and only if Q is an n-ellipsoid.

(d) Other trivial inequalities, a consequence of the classical Schwarz inequalities, are

$$\mathcal{J}_{p+q}^2 \leqslant \mathcal{J}_{2p} \mathcal{J}_{2q} \tag{4.7}$$

for any integers p, q.

The preceding inequalities between the invariants $\mathcal{J}_h$ are far from forming a complete system and it should be interesting to find other inequalities of the same type for the affine invariants $\mathcal{J}_h$.

ACKNOWLEDGMENT

The author is very grateful to a referee for some pertinent and valuable comments.

REFERENCES

Blaschke, W. (1923) *Vorlesungen über Differentialgeometrie II*, Springer, Berlin.
Bonnesen, T. & Fenchel, W. (1934) *Theorie der konvexen Körper*. Springer, Berlin.

Firey, W. (1972) An integral-geometric meaning of lower order area functions of convex bodies. *Mathematika*, **19**, 205–212.
Firey, W. (1985) The integral-geometric densities of tangent flats to convex bodies. Discrete geometry and convexity. *Ann. N.Y. Acad. Sci.* **440**, 92–96.
Petty, C.M. (1985) Affine isoperimetric problems. Discrete geometry and convexity, *Ann. N.Y. Acad. Sci.* **440**, 113–127.
Santaló, L.A. (1976) *Integral Geometry and Geometric Probability*. Addison-Wesley, Reading, Mass.
Schneider, R. (1978) Curvature measures of convex bodies. *Ann. Mat. Pura Appl.* **116**, 101–134.
Schneider, R. (1979) *Integralgeometrie*. Vorlesungen, Freiburg.
Süss, W. (1927) Eifläche konstanter Affinbreite. *Math. Ann.* **96**, 251–260.
Weil, W. (1979) Kinematic integral formulas for convex bodies. In: *Contributions to Geometry* (ed. by J. Tölke & J. M. Wills), pp. 60–76. Birkhäuser, Basel.
Weil, W. (1981) Berührung konvexer Körper durch q-dimensionale Ebenen. *Resultate Math.* **4**, 84–101.

Part IV. Affine Geometry, with comments by K. Leichtweiss

Santaló was an especially versatile geometer. Therefore it is not surprising that he was also occupied with problems on affine geometry. Hereby, in the papers under review, we have to distinguish between

a) *Affine invariant theory*

and

b) *Affine differential geometry.*

a) Here [50.3] seems to be one of the most typical articles in Santaló's scientific work. It was written at the time when Santaló stayed at the Institute of Advanced Studies in Princeton and deals with the foundation of Integral Geometry in projective and affine spaces.

In this discipline we mainly have to do with Geometrical Probability measures, formed by integration of certain densities which are assumed to be invariant against the action of the projective group P_n resp. the centro-equiaffine group A_n in n-space. In his paper, following ideas of S. S. Chern, Santaló successfully uses E. Cartan's calculus of exterior differential forms because our densities simply may be defined as normed n-fold exterior products of Pfaffian forms. These densities refer to special geometrical objects in the n-space like single linear subspaces or finite sets of such subspaces, fulfilling suitable conditions for their dimensions, or the space of the whole group.

The results are sometimes surprising: Unlike the Euclidean case, the space of single linear subspaces of the n-space does not possess an invariant density in the projective as well as in the affine case! Elsewhere one may easily compute the corresponding densities and transform them into a convenient form.

Finally, we should not forget to mention a very interesting application of Santaló's theory to a simplification of a proof of the following theorem of Minkowski-Hlawka, belonging to the theory of numbers: *If Q is an n-dimensional star-shaped domain of volume $< \zeta(n)$ then there exists a lattice of determinant* 1 *such that Q does not contain any lattice point* $\neq 0$.

The paper [62.2] is a continuation of [50.3] for finite sets of parallel affine subspaces of the equiaffine n-space. Here the invariant densities, if existing, may be also written in a simple metrical form and lead to certain affine invariants of a convex body

Totally different is the determination and geometric interpretation of affine invariants in [47,1] for 1. two curves with common tangent line and 2. two curves intersecting each other with different tangent lines in the plane and analogously 3. two surfaces with common tangential plane and 4. two surfaces with common straight tangent line where two different tangent planes are intersecting in the 3-space. The different results follow by lengthy purely analytic computations. They are compared with known results for projective invariants of C. C. Hsiung and others.

b) The classical equiaffine differential geometry for a smooth curve C in 3-space in the sense of W. BLASCHKE depends on its structure equation $x'''' + kx'' + tx' = 0$ where the primes denote derivation with respect to the affine arc length, given by $(x', x'', x''') = 1$. Hereby occur the tangent vector x', the affine principal vector x'', the affine binormal vector of A. WINTERNITZ $x''' + \frac{k}{4}x'$, the affine curvature k and the affine torsion t. These entities are all joint with C in an equivariant manner.

Santaló has completed in [46.3] their geometric interpretation with the help of suitable osculating figures. For instance, the straight line which connects the point x_0 of C with the point at infinity of the osculating cubical parabola of C at x_0 has the direction of the affine binormal of WINTERNITZ of C at x_0. Nowadays these techniques are no longer so interesting because of the concept of a curve as an affine immersion of K. NOMIZU and T. SASAKI.

Finally, in [60.1], Santaló was involved with affine surface theory. Already W. BLASCHKE had discovered some formal analogies between the Euclidean and the equiaffine differential geometry. In this context one has to understand the analogy between J. STEINER's Euclidean parallel surfaces and the affine parallel surfaces $S^{(\lambda)}$ with the representation $x^{(\lambda)} := x + \lambda y$ ($\lambda = \text{const} > 0$, y affine normal vector) which was investigated by Santaló. Hereby the main result says that the volume of a parallel affine surface $S^{(\lambda)}$ takes the same form as polynomial in λ as in the Euclidean case with (now) the affine area, the integrated affine mean curvature and the integrated affine total curvature as coefficients. They are up to certain integers the mixed volumes of the body bounded by $S^{(0)}$ and the body bounded by the curvature image $C^{(0)}$ of $S^{(0)}$. All these things represent a nice application of E. MUELLER's so-called *relative geometry* to the pair $\{S^{(0)}, C^{(0)}\}$.

KURT LEICHTWEISS (Stuttgart)

A GEOMETRICAL CHARACTERIZATION FOR THE AFFINE DIFFERENTIAL INVARIANTS OF A SPACE CURVE

L. A. SANTALÓ

1. **Introduction.** Let $x=x(s)$ be the vector equation of a space curve C with the affine arc length s as parameter. It is known that $x(s)$ satisfies a differential equation of the following form [**1**, p. 73; **3**, p. 76][1]

$$x'''' + kx'' + tx' = 0, \tag{1.1}$$

where the primes represent derivatives with respect to s. The vector x' is the *tangent vector* and the vectors x'' and x''' are called the *affine principal normal* and the *affine binormal*, respectively, of the curve C at the point considered. The vectors x', x'', x''' with the initial point at the point x of the curve C constitute the *affine fundamental trihedral* at x and they satisfy the following relation [**1**, p. 72; **3**, p. 78]

$$(x', x'', x''') = 1. \tag{1.2}$$

The plane determined by the point x and the edges x', x'' of the affine fundamental trihedral is the *osculating plane* at x; the plane determined by x and the edges x'', x''' is the *affine normal plane* and the plane determined by x and the edges x', x''' is the *affine rectifying plane* of the curve C at the point x.

Sometimes it is convenient to use the vector $kx'/4+x'''$ which is called the *binormal of Winternitz* [**1**, p. 76]. The invariants k and t (functions of the affine arc length s) are called the *affine curvature* and the *affine torsion* respectively.

For the affine fundamental trihedral and for k and t some geometrical characterizations have been given by Blaschke [**1**, chap. 3], Salkowski [**3**, p. 76] and Haack [**2**]. The purpose of the present paper is to give a new geometrical construction for the mentioned elements, which we believe is simpler than those previously obtained.

2. **Geometrical elements associated to an ordinary point of a space curve.** Let us consider the space curve C represented by the vector equation $x=x(s)$ (s=affine arc length) in the neighborhood of the point $x_0=x(0)$. If we denote by $x_0^{(i)}$ the derivatives $d^{(i)}x/ds^i$ at the point $s=0$, since x_0', x_0'', x_0''' are not coplanar (by (1.2)), any point y of the space can be expressed in the form

Received by the editors January 17, 1946.

[1] Numbers in brackets refer to the references cited at the end of the paper.

$$(2.1) \qquad y = x_0 + \xi_1 x_0' + \xi_2 x_0'' + \xi_3 x_0'''$$

where ξ_1, ξ_2, ξ_3 may be considered as the coordinates of the point y with respect to the affine fundamental trihedral.

For the points $x(s)$ of the curve C we have

$$x(s) = x_0 + s x_0' + \frac{1}{2} s^2 x_0'' + \frac{1}{3!} s^3 x_0'' + \frac{1}{4!} s^4 x_0''' + \cdots$$

and taking into account the relation (1.1) we obtain

$$(2.2) \qquad \begin{aligned} \xi_1 &= s - \frac{1}{4!} t s^4 - \frac{1}{5!} t' s^5 + \cdots, \\ \xi_2 &= \frac{1}{2} s^2 - \frac{1}{4!} k s^4 - \frac{1}{5!} (k' + t) s^5 + \cdots, \\ \xi_3 &= \frac{1}{3!} s^3 - \frac{1}{5!} k s^5 + \cdots, \end{aligned}$$

where $k' = dk/ds$, $t' = dt/ds$.

Let us now consider the following geometrical elements associated to the curve C at the point x_0:

(a) *The quadric cone K*. By K we denote the quadric cone determined by the tangent of C at the point x_0 and the parallel lines through x_0 to the tangents of C at four neighboring points as each of these points independently approaches x_0 along C. This quadric cone K has been considered by Haack [2] and its equation in terms of the coordinates ξ_i is [2, p. 159]

$$(2.3) \qquad \xi_2^2 + k\xi_3^2 - 2\xi_1\xi_3 = 0.$$

(b) *The osculating cubical parabola Q*. A twisted cubic which has the plane at infinity as osculating plane is called a cubical parabola. The parametric equations of a cubical parabola have the following form

$$(2.4) \qquad \xi_i = a_i + b_i u + c_i u^2 + d_i u^3 \qquad (i = 1, 2, 3).$$

In order to find the cubical parabola which has contact of the highest order with the curve C at the point x_0, let us put

$$(2.5) \qquad s = u + m_1 u^2 + m_2 u^3 + m_3 u^4 + \cdots$$

in the equations (2.2). We obtain

$$\xi_1 = u + m_1 u^2 + m_2 u^3 + \left(m_3 - \frac{1}{4!} t\right) u^4 + \cdots,$$

$$\xi_2 = \frac{1}{2} u^2 + m_1 u^3 + \left(m_2 + \frac{1}{2} m_1^2 - \frac{1}{4!} k \right) u^4 + \cdots ,$$

$$\xi_3 = \frac{1}{3!} u^3 + \frac{1}{2} m_1 u^4 + \cdots .$$

If we take the values

$$m_1 = 0, \quad m_2 = \frac{1}{4!} k, \quad m_3 = \frac{1}{4!} t$$

we see that the cubical parabola

$$(2.6) \qquad \xi_1 = u + \frac{1}{4!} k u^3, \qquad \xi_2 = \frac{1}{2} u^2, \qquad \xi_3 = \frac{1}{3!} u^3$$

is the only cubical parabola which has a fourth order contact with the curve C at the point x_0. Consequently the equations (2.6) are the parametric equations of the osculating cubical parabola Q of C at x_0.

The quadric cone K and the twisted cubic Q are affinely connected to the curve C at the point x_0. We shall use also the following cone K^* which is projectively connected with C.

(c) *The osculating quadric cone* K^*. With K^* we denote the unique quadric cone with its vertex at the point x_0 which has seven-point contact with C at the point x_0. In order to find the equation of K^* let us substitute the expansions (2.2) in the general quadratic equation

$$(2.7) \qquad A\xi_1^2 + B\xi_2^2 + C\xi_3^2 + D\xi_1\xi_2 + E\xi_1\xi_3 + F\xi_2\xi_3 = 0$$

and then, demanding that this equation be satisfied identically in s as far as the terms in s^6, we find

$$A = D = F = 0, \qquad 3B + 2E = 0, \qquad 10C - 15kB - 3kE = 0.$$

Hence, the osculating quadric cone K^* is given by the equation

$$(2.8) \qquad 20\xi_2^2 + 21k\xi_3^2 - 30\xi_1\xi_3 = 0.$$

3. **Geometrical characterization of the affine fundamental trihedral of a space curve at an ordinary point.** From (2.6) we deduce that the straight line which connects the point x_0 of the curve C with the point at infinity of the osculating cubical parabola Q at x_0 has the direction of the vector

$$\frac{1}{4} k x_0' + x_0''',$$

that is, coincides with the binormal of Winternitz (§1).

The plane determined by this binormal and the tangent line of C at x_0 is the affine rectifying plane and the polar line of this affine rectifying plane with respect to the cone K turns out to be the affine principal normal line.

If we let $\xi_2=0$ in the equation (2.3) we find that the intersection of the affine rectifying plane with the quadric cone K is composed of the tangent line and of the straight line given by the equations $\xi_2=0$, $k\xi_3-2\xi_1=0$, that is, the line which has the direction of the vector $kx_0'/2+x_0'''$.

Since the vectors x_0', $kx_0'/4+x_0'''$ and $kx_0'/2+x_0'''$, x_0''' are harmonically separated it turns out that the affine binormal is the line harmonic conjugate to the intersection line (different from the tangent) of the cone K with the affine rectifying plane with respect to the tangent line and the binormal of Winternitz. The plane determined by the affine binormal and the affine principal normal is the affine normal plane. Thus the following theorem is established:

At an ordinary point x_0 of a space curve C let us consider the osculating cubical parabola Q and the cone K defined in §2. *The direction of the point at infinity of Q gives the binormal of Winternitz of C at x_0. This binormal determines with the tangent line the affine rectifying plane, whose polar line with respect to the cone K is the affine principal normal of C at x_0. The harmonic conjugate of the intersection line (different from the tangent) of K with the affine rectifying plane with respect to the pair formed by the tangent line and the binormal of Winternitz is the affine binormal line of C at x_0.*

Instead of the cone K and the cubical parabola Q in order to give a geometrical characterization of the affine fundamental trihedral connected with the space curve C at the point x_0, we can use only the cones K and K^* (§2). For this purpose let us seek the straight line which passes through the point x_0 and has the same polar plane with respect to both cones K and K^*. If we use the coordinates ξ_1, ξ_2, ξ_3 of (2.1) the polar plane with respect to K of the line which connects x_0 with the point ξ_1^0, ξ_2^0, ξ_3^0 is given by the equation

$$\overset{0}{\xi}_3\xi_1 - \overset{0}{\xi}_2\xi_2 - (k\overset{0}{\xi}_3 - \overset{0}{\xi}_1)\xi_3 = 0 \tag{3.1}$$

and the polar plane of the same line with respect to the cone K^* is

$$15\overset{0}{\xi}_3\xi_1 - 20\overset{0}{\xi}_2\xi_2 - (21k\overset{0}{\xi}_3 - 15\overset{0}{\xi}_1)\xi_3 = 0. \tag{3.2}$$

In order that the planes (3.1) and (3.2) coincide, we must have either $\xi_2{}^0=\xi_3{}^0=0$ or $\xi_1{}^0=\xi_3{}^0=0$. The first solution corresponds to the tangent line whose common polar plane with respect to the cones K and K^* is the osculating plane. The second solution corresponds to the affine principal normal whose polar plane with respect to the cones K and K^* is the affine rectifying plane. Then, upon excluding the tangent line, the affine principal normal is the only line which has the same polar plane with respect to the quadric cones K and K^*.

We have seen that the affine rectifying plane intersects the cone K, excluding the tangent line, in the line R of the vector $kx_0'/2+x_0'''$. Analogously we find that the affine rectifying plane intersects the cone K^* besides the tangent in the line R^* of the vector $7kx_0'+10x_0'''$. Consequently the cross ratio of the tangent T, affine binormal B and the lines R and R^* has the value

$$\lambda = (TBRR^*) = \left(\infty, 0, \frac{1}{2}k, \frac{7}{10}k\right) = \frac{7}{5}.$$

Thus we have obtained:

Let K and K^* be the quadric cones attached to the space curve C at the point x_0 defined in §2. *The affine principal normal of C at x_0 is the only line which has the same polar plane with respect to both cones K and K^*; its common polar plane is the affine rectifying plane. Let R and R^* be the intersection lines, the tangents excluded, of the rectifying plane with the cones K and K^* respectively; the affine binormal is the line B such that the cross ratio $(TBRR^*)$ has the value* 7/5.

4. Geometrical interpretation of the affine curvature. If we substitute the expressions (2.6) in (2.3) we find that the cubical parabola Q has with the cone K four coincident points in x_0 and two other intersection points corresponding to the values of u satisfying the equation

$$ku^2 = 6. \tag{4.1}$$

Let us put

$$u_1 = \left(\frac{6}{k}\right)^{1/2}, \qquad u_2 = -\left(\frac{6}{k}\right)^{1/2}. \tag{4.2}$$

The corresponding intersection points of Q with K will be the points

$$A_i\left(\xi_1 = u_i + \frac{1}{4!}ku_i^3,\ \xi_2 = \frac{1}{2}u_i^2,\ \xi_3 = \frac{1}{3!}u_i^3\right) \qquad (i = 1, 2). \tag{4.3}$$

The straight line which connects A_1 and A_2 cuts the osculating plane in the point $\xi_1=\xi_3=0$, $\xi_2=3k^{-1}$. That gives the following new characterization of the affine principal normal: the affine principal normal is the line which connects the point x_0 of the curve with the intersection point of the osculating plane with the straight line A_1A_2.

The osculating plane of Q at A_i, taking into account (4.1), is

$$6u_i^2\xi_1 - 12u_i\xi_2 + 3\xi_3 - 2u_i^3 = 0 \tag{4.4}$$

and it cuts the edges of the affine fundamental trihedral in the following points

$$y_1 = x_0 + \frac{1}{3}u_ix_0', \qquad y_2 = x_0 - \frac{1}{6}u_i^2x_0'', \qquad y_3 = x_0 + \frac{2}{3}u_i^3x_0'''.$$

The absolute value of the volume of the tetrahedron whose vertices are the points x_0, y_1, y_2, y_3, taking into account (4.1) and (1.2), is

$$V = \frac{1}{6}(y_1 - x_0,\, y_2 - x_0,\, y_3 - x_0) = \frac{4}{3}k^{-3}. \tag{4.5}$$

Then we have the theorem:

Let us consider the points A_i $(i=1, 2)$ in which the osculating cubical parabola Q intersects the cone K. The osculating plane of Q at any one of the points A_i determines with the affine fundamental trihedral a tetrahedron whose volume V is related to the affine curvature k by (4.5).

5. **Geometrical interpretation of the affine torsion.** We now give a geometrical interpretation of the affine torsion t. For this purpose let us seek the intersection line of two consecutive affine normal planes. The vector equation of the affine normal plane of C at x_0 is

$$y(\lambda, \mu) = x_0 + \lambda x_0'' + \mu x_0''' \tag{5.1}$$

and the analogous equation for the affine normal plane at the point $x(s)$, taking into account the relation (1.1), takes the form

$$\begin{aligned} \bar{y}(\bar\lambda, \bar\mu) &= x(s) + \bar\lambda x''(s) + \bar\mu x'''(s) \\ &= x_0 + \left[(1 - t\bar\mu)s - \frac{1}{2}(t\bar\lambda + t'\bar\mu)s^2 + \cdots\right]x_0' \\ &\quad + \left[\bar\lambda - k\bar\mu s + \frac{1}{2}(1 - k\bar\lambda - (k' + t)\bar\mu)s^2 + \cdots\right]x_0'' \\ &\quad + \left[\bar\mu + \bar\lambda s - \frac{1}{2}k\bar\mu s^2 + \cdots\right]x_0'''. \end{aligned} \tag{5.2}$$

In order to find the intersection line of the planes (5.1) and (5.2) we must write

$$1 - t\bar{\mu} - \frac{1}{2}(t\bar{\lambda} + t'\bar{\mu})s + \cdots = 0,$$

$$\bar{\lambda} - k\bar{\mu}s + \frac{1}{2}(1 - k\bar{\lambda} - (k' + t)\bar{\mu})s^2 + \cdots = \lambda,$$

$$\bar{\mu} + \bar{\lambda}s - \frac{1}{2}k\bar{\mu}s^2 + \cdots = \mu$$

and by elimination of $\bar{\lambda}$ and $\bar{\mu}$ we find

$$(1 - \mu t) + \frac{1}{2}(\lambda t + \mu t')s + \cdots = 0.$$

For $s=0$ we have $\mu=1/t$; hence the intersection line of two consecutive affine normal planes is the line $y(\lambda)=x_0+\lambda x_0''+t^{-1}x_0'''$. Then, the intersection point of the affine binormal at the point x_0 with the consecutive affine normal plane or, what is the same, the intersection point of the affine binormal with the developable surface enveloped by the affine normal planes, is the point $B\equiv x_0+t^{-1}x_0'''$.

Let us now trace through the point B the parallel line to the tangent of C at the point x_0; if we call E its intersection point with the quadric cone K we find that it is

$$E \equiv x_0 + \frac{1}{2}kt^{-1}x_0' + t^{-1}x_0' .$$

Taking into account (1.2) for the volume of the tetrahedron whose vertices are the points x_0, B, A_1 (or A_2) and E we have the expression

$$V^* = \frac{1}{6}\begin{vmatrix} 0 & 0 & 0 & 1 \\ 0 & 0 & t^{-1} & 1 \\ \xi_1 & \xi_2 & \xi_3 & 1 \\ \frac{1}{2}kt^{-1} & 0 & t^{-1} & 1 \end{vmatrix}$$

where the values of ξ_i $(i=1, 2, 3)$ are given by (4.3) and (4.2). Consequently we have

$$V^* = \frac{1}{4}t^{-2}. \tag{5.3}$$

Hence we have obtained the following geometrical interpretation for the affine torsion t of a space curve:

Through the intersection point B of the affine binormal at the point x_0 *of a space curve C with the developable surface enveloped by the affine normal planes, we trace the parallel line to the tangent of C at* x_0. *Let E be the point in which this parallel line intersects the quadric cone K. If* A_i *is any one of the points in which the osculating cubical parabola Q intersects the cone K, the volume of the tetrahedron whose vertices are the points* x_0, B, A_i, E *is related to the affine torsion of C at* x_0 *by the relation* (5.3).

References

1. W. Blaschke, *Vorlesungen über Differentialgeometrie*, vol. 2, Berlin, 1923.
2. W. Haack, *Eine geometrische Deutung der Affin-Invarianten einer Raumkurve*, Math. Zeit. vol. 38 (1934) pp. 155–162.
3. E. Salkowski, *Affine Differentialgeometrie*, Berlin and Leipzig, 1934.

Universidad Nacional del Litoral, Rosario

AFFINE INVARIANTS OF CERTAIN PAIRS OF CURVES AND SURFACES

By L. A. Santaló

1. **Introduction.** For two curves in a plane or two surfaces in ordinary space various projective invariants have been given by Mehmke, Bouton, Segre, Buzano, Bompiani, Hsiung and others (see Bibliography).

Obviously each projective invariant is also an affine invariant, that is, an invariant with respect to the group of affine transformations. However in certain cases there are affine invariants which are not projective invariants. The purpose of the present paper is to study these cases giving affine invariants, as well as their affine and metrical characterization, for the following cases:

(a) two curves in a plane having a common tangent at two ordinary points (§§2, 3);

(b) two curves in a plane intersecting at an ordinary point (§§4, 5);

(c) two surfaces in ordinary space having a common tangent plane at two ordinary points (§§6, 7);

(d) two surfaces in ordinary space having a common tangent line but distinct tangent plane at two ordinary points (§§8, 9).

For the cases (a), (b) of plane curves we shall consider the neighborhoods of the second and the third order of the curves at the considered points. For the cases (c), (d) of two surfaces in ordinary space we shall consider only the neighborhoods of the second order of the surfaces at the considered points.

2. **Affine invariants of two plane curves having a common tangent at two ordinary points.** Suppose that O and O_1 are two ordinary points of two plane curves C and C_1 respectively, so that OO_1 is the common tangent. Let h be the distance OO_1. If we choose a cartesian coordinate system in such a way that the point O be the origin and the line OO_1 be the x-axis, the power series expansions of the two curves in the neighborhood of the points O and O_1 may be written in the form

$$C: \quad y = ax^2 + bx^3 + \cdots, \tag{2.1}$$

$$C_1: \quad y = a_1(x-h)^2 + b_1(x-h)^3 + \cdots, \tag{2.2}$$

where we suppose $a, a_1 \neq 0$.

In order to find the affine invariants of the elements of the second and the third order of the curves C, C_1 in the neighborhood of O, O_1 we have to consider the most general affine transformation which leaves the point O and the x-axis invariant:

$$x = \alpha X + \beta Y, \qquad y = \mu Y, \tag{2.3}$$

Received March 28, 1946.

where α, β, μ are arbitrary constants. By this transformation the point $O_1(h, 0)$ is carried into the point whose coordinates are h/α, 0; consequently if we call H the distance between the transformed points from O and O_1 we have

$$\alpha = h/H. \tag{2.4}$$

Let us substitute (2.3) in (2.1) and (2.2). We find two equations of the form

$$Y = AX^2 + BX^3 + \cdots, \qquad Y = A_1(X - H)^2 + B_1(X - H)^3 + \cdots,$$

where the first coefficients A, B, A_1, B_1 are given by the following system:

$$\begin{aligned} \mu A &= \alpha^2 a, & \mu B &= 2\alpha\beta a A + \alpha^3 b, \\ \mu A_1 &= \alpha^2 a_1, & \mu B_1 &= 2\alpha\beta a_1 A_1 + \alpha^3 b_1. \end{aligned} \tag{2.5}$$

Eliminating α, β, μ from this system and (2.4) we find that *there are two independent affine invariants determined by the neighborhoods of the second and the third order of the curves C and C_1 at the points O, O_1, which are*

$$I_1 = a/a_1, \qquad I_2 = ha(b_1/a_1^2 - b/a^2). \tag{2.6}$$

By interchanging a, b and a_1, b_1 in the invariant I_2 we find the invariant $I_3 = -I_2/I_1$ which is a consequence of I_1 and I_2.

The invariant I_1 is determined by the neighborhoods of the second order of C, C_1 at the points O, O_1 and the invariant I_2 is determined by the neighborhoods of the third order.

3. Metrical and affine characterization of the invariants I_1 and I_2. If the coordinate system to which the expansions (2.1) and (2.2) of C and C_1 are referred is chosen orthogonal, it is easily verified that the radius of curvature r of C and its derivative $r' = dr/ds$ with respect to the arc length s of C at the point O are given by the formulas

$$r = \frac{1}{2a}, \qquad r' = -\frac{3b}{2a^2}. \tag{3.1}$$

Similarly the radius of curvature r_1 of C_1 and its derivative $r_1' = dr_1/ds_1$ with respect to the arc length s_1 of C_1 at the point O_1 are

$$r_1 = \frac{1}{2a_1}, \qquad r_1' = -\frac{3b_1}{2a_1^2}. \tag{3.2}$$

From (3.1), (3.2) we deduce that the invariants (2.6) are expressed metrically by the formulas

$$I_1 = \frac{r_1}{r}, \qquad I_2 = \frac{h}{3r}(r' - r_1'). \tag{3.3}$$

In order to give an affine characterization of the invariant I_1 let us cut the curves C and C_1 by the parallel line to OO_1 given by the equation $y = \epsilon$. (We suppose $a > 0$; in case $a < 0$ we take $y = -\epsilon$.) If C and C_1 in the neighborhood of O and O_1 are situated at different sides of the line OO_1 we consider the two parallel lines $y = \pm\epsilon$. The area f bounded by this parallel line and the curve C in the neighborhood of O has the value

$$f = \frac{4}{3} a^{-\frac{1}{2}} \epsilon^{3/2} + o(\epsilon^{3/2}) \tag{3.4}$$

where $o(\epsilon^{3/2})$ means a function of ϵ such that $o(\epsilon^{3/2})/\epsilon^{3/2}$ tends to zero with ϵ.

Similarly the area f_1 bounded by the same line $y = \epsilon$ (or the symmetrical one $y = -\epsilon$) and the curve C_1 in the neighborhood of O_1 has the value

$$f_1 = \frac{4}{3} a_1^{-\frac{1}{2}} \epsilon^{3/2} + o(\epsilon^{3/2}). \tag{3.5}$$

From (3.4) and (3.5) we deduce

$$I_1 = a/a_1 = \lim_{\epsilon \to 0} (f_1/f)^2. \tag{3.6}$$

Since the quotient of two areas is an invariant with respect to affine transformations, (3.6) gives an affine interpretation of the invariant I_1.

Here let us make the following remark. It is known that the ratio between two areas situated on parallel planes is an invariant with respect to affine transformations in ordinary space. Consequently, if we consider two plane curves situated on parallel planes which have parallel tangents at two ordinary points O, O_1 by considering a plane which intersects both curves at a distance ϵ from the respective tangent, from (3.4) and (3.5) we deduce that *the ratio between the radii of curvature at O and O_1 of two curves situated on parallel planes with parallel tangents at O and O_1 respectively, is invariant with respect to affine transformations of the space.* This property will be used in §7.

In order to give an affine characterization of the invariant I_2 let us consider the osculating parabola of C at the point O. The equation of this parabola is found to be

$$y = ax^2 + \frac{b}{a} xy + \frac{b^2}{4a^3} y^2. \tag{3.7}$$

The diameter of this parabola which passes through O is called the *affine normal* of C at O [13; 49] and is the line

$$y = -\frac{2a^2}{b} x. \tag{3.8}$$

Similarly the affine normal of C_1 at O_1 is the line

$$y = -\frac{2a_1^2}{b_1} (x - h). \tag{3.9}$$

The polar line of the point O_1 with respect to the osculating parabola (3.7) is

$$2a^2hx + (bh - a)y = 0. \tag{3.10}$$

The intersection point M of the affine normal (3.9) of C_1 and the affine normal (3.8) of C has the abscissa

$$\xi_0 = \frac{hb}{a^2(b/a^2 - b_1/a_1^2)}, \tag{3.11}$$

and the intersection point N of the affine normal (3.9) of C_1 and the polar line (3.10) of O_1 with respect to the osculating parabola (3.7) has the abscissa

$$\xi_1 = \frac{(bh - a)h}{ha^2(b/a^2 - b_1/a_1^2) - a}. \tag{3.12}$$

Hence, taking into account the value (2.6) of I_2 we see that the ratio $D = NO_1/NM$ has the value

$$D = \frac{h - \xi_1}{\xi_0 - \xi_1} = - I_2 .$$

This formula gives an affine characterization of the invariant I_2, that is: *the affine invariant I_2 is equal, except for sign, to the ratio of the segments that the affine normal of C at O, the polar line of O_1 with respect to the osculating parabola of C at O and the tangent OO_1 determine on the affine normal of C_1 at O_1.*

From this result we deduce that $I_2 = 0$ is the necessary and sufficient condition that the affine normals of C and C_1 at O and O_1 respectively are parallel lines.

4. Affine invariants of two curves in a plane intersecting at an ordinary point. Let us consider two plane curves C, C_1 intersecting at a point O and having distinct tangents t and t_1 at O. Let us suppose that the point O is an ordinary point on both C and C_1 and choose a cartesian coordinate system so that t is the x-axis and t_1 is the y-axis. Then the curves C and C_1 may be represented by expansions of the form

$$C: \quad y = ax^2 + bx^3 + \cdots, \tag{4.1}$$

$$C_1: \quad x = a_1y^2 + b_1y^3 + \cdots, \tag{4.2}$$

where $a, a_1 \neq 0$.

The most general affine transformation which leaves the point O and the tangents t, t_1 invariant is represented by the equations

$$x = \alpha X, \qquad y = \mu Y, \tag{4.3}$$

where α, μ are arbitrary constants.

By means of this transformation the equations (4.1) and (4.2) of the curves C and C_1 transform into two other equations of the same form, namely,

$$Y = AX^2 + BX^3 + \cdots, \qquad X = A_1Y^2 + B_1Y^3 + \cdots,$$

where the coefficients A, B, A_1, B_1 are given by the following system:

$$\mu A = \alpha^2 a, \qquad \mu B = \alpha^3 b, \qquad \alpha A_1 = \mu^2 a_1, \qquad \alpha B_1 = \mu^3 b_1. \tag{4.4}$$

Eliminating α, μ from these equations we find that *there are two independent affine invariants determined by the neighborhood of the third order of the curves C and C_1, namely,*

$$i_1 = \frac{a_1 b^2}{a^3 b_1}, \qquad i_2 = \frac{a b_1^2}{a_1^3 b}. \tag{4.5}$$

Instead of these invariants it is more convenient to introduce the following ones:

$$I_1^* = i_1^2 i_2 = \frac{b^3}{a_1 a^5}, \qquad I_2^* = i_1 i_2^2 = \frac{b_1^3}{a a_1^5}. \tag{4.6}$$

5. Metric and affine characterization of the invariants I_1^* and I_2^*. Let us call ω the angle which forms the tangents t, t_1 of the curves C, C_1 at the point O. The radius of curvature r of C and its derivative $r' = dr/ds$ with respect to the arc length s at the point O are found, on performing simple calculation, to be given by the formulas

$$r = \frac{1}{2a \sin \omega}, \qquad r' = -\frac{3(b - 2a^2 \cos \omega)}{2a^2 \sin \omega}, \tag{5.1}$$

from which we deduce

$$a = \frac{1}{2r \sin \omega}, \qquad b = \frac{3 \cos \omega - r' \sin \omega}{6r^2 \sin^2 \omega}. \tag{5.2}$$

By analogy if r_1 and r_1' are the radius of curvature and its derivative with respect to the arc length s_1 of C_1 at the point O_1 we find

$$a_1 = \frac{1}{2r_1 \sin \omega}, \qquad b_1 = \frac{3 \cos \omega - r_1' \sin \omega}{6r_1^2 \sin^2 \omega}. \tag{5.3}$$

From (5.2) and (5.3) we deduce that the affine invariants (4.6) have the following metric characterization:

$$I_1^* = \frac{8}{27} \frac{r_1}{r} (3 \cos \omega - r' \sin \omega)^3, \qquad I_2^* = \frac{8}{27} \frac{r}{r_1} (3 \cos \omega - r_1' \sin \omega)^3. \tag{5.4}$$

In order to give an affine characterization of I_1^* and I_2^* we shall consider the osculating parabola Q of C at the point O, which in the present case of oblique axis is given also by the equation (3.7) and the osculating parabola Q_1 of C_1 at the point O, given by the equation

$$x = a_1 y^2 + \frac{b_1}{a_1} xy + \frac{b_1^2}{4a_1^3} x^2. \tag{5.5}$$

On the tangent t_1 of C_1 at O (y-axis of our coordinate system) we have the point A_1 in which t_1 intersects Q and the pole A_2 of the tangent t with respect to the osculating parabola Q_1. An easy calculation proves that the distances from A_1 and A_2 to the point O are

$$OA_1 = \frac{4a^3}{b^2}, \qquad OA_2 = \frac{a_1}{b_1}, \tag{5.6}$$

respectively.

Similarly we have on the tangent t (x-axis of our coordinate system) the intersection point B_1 of t with the osculating parabola Q_1 and the pole B_2 of t_1 with respect to the osculating parabola Q, whose distances to the point O are

$$OB_1 = \frac{4a_1^3}{b_1^2}, \qquad OB_2 = \frac{a}{b}. \tag{5.7}$$

Since the ratio in which a point divides a line segment is an invariant with respect to the group of affine transformations, from (5.6) and (5.7) we deduce that the ratios

$$\rho_1 = \frac{OA_2}{OA_1} = \frac{a_1 b^2}{4a^3 b_1}, \qquad \rho_2 = \frac{OB_2}{OB_1} = \frac{ab_1^2}{4ba_1^3}. \tag{5.8}$$

are affine invariants. The invariants I_1^* and I_2^* are expressed by means of ρ_1 and ρ_2 by the formulas

$$I_1^* = 64\rho_1^2\rho_2, \qquad I_2^* = 64\rho_1\rho_2^2 \tag{5.9}$$

which give an affine characterization of I_1^* and I_2^*.

We conclude with the remark that the condition $I_1^* = 0$, which is equivalent to $b = 0$, signifies that the affine normal of C at the point O (which in the present case of oblique axis is given also by the equation (3.8)) coincides with the y-axis, that is, with the tangent t_1 of C_1 at O. Similarly, the condition $I_2^* = 0$ which is equivalent to $b_1 = 0$, signifies that the affine normal of C_1 at O coincides with the x-axis, that is, with the tangent of C at O.

6. **Affine invariants of two surfaces in ordinary space having a common tangent plane at two ordinary points.** Let S and S_1 be two surfaces in ordinary space having a common tangent plane at two ordinary points O and O_1. Let h be the distance OO_1. If we choose a cartesian coordinate system in such a way that the point O be the origin, the line OO_1 be the x-axis and the common tangent

plane to S, S_1 at O, O_1 be the plane $z = 0$, the power series expansions of the two surfaces in the neighborhood of O and O_1 may be written in the form

$$S: \quad z = ax^2 + bxy + cy^2 + \cdots, \tag{6.1}$$

$$S_1: \quad z = a_1(x - h)^2 + b_1(x - h)y + c_1y^2 + \cdots. \tag{6.2}$$

In order to find the affine invariants determined by the neighborhoods of the second order of S, S_1 at the points O, O_1 we have to consider the most general affine transformation which leaves the point O, the x-axis and the plane $z = 0$ invariant, namely,

$$\begin{aligned} x &= \alpha_1 X + \beta_1 Y + \gamma_1 Z \\ y &= \phantom{\alpha_1 X + {}} \beta_2 Y + \gamma_2 Z \\ z &= \phantom{\alpha_1 X + \beta_2 Y + {}} \gamma_3 Z \end{aligned} \tag{6.3}$$

where α_1, β_1, γ_1, β_2, γ_2, γ_3 are arbitrary constants.

By means of this transformation the equations (6.1) and (6.2) of S and S_1 transform into two other equations of the same form, namely,

$$Z = AX^2 + BXY + CY^2 + \cdots,$$

$$Z = A_1(X - H)^2 + B_1(X - H)Y + C_1Y^2 + \cdots,$$

where $H = h/\alpha_1$ is the distance between the transformed points from O and O_1 and the coefficients are given by the following system:

$$\begin{aligned} \gamma_3 A &= \alpha_1^2 a, & \gamma_3 A_1 &= \alpha_1^2 a_1, \\ \gamma_3 B &= 2\alpha_1\beta_1 a + \alpha_1\beta_2 b, & \gamma_3 B_1 &= 2\alpha_1\beta_1 a_1 + \alpha_1\beta_2 b_1, \\ \gamma_3 C &= \beta_1^2 a + \beta_1\beta_2 b + \beta_2^2 c, & \gamma_3 C_1 &= \beta_1^2 a_1 + \beta_1\beta_2 b_1 + \beta_2^2 c_1. \end{aligned} \tag{6.4}$$

According to the vanishing or non-vanishing of the two coefficients a and a_1 it is necessary to distinguish four cases:

Case I. $a \neq 0$, $a_1 \neq 0$. In this case the line OO_1 does not coincide with any one of the asymptotic tangents of the surfaces S, S_1 at O, O_1 respectively. Elimination of the coefficients α_1, β_1, γ_1, β_2, γ_2, γ_3 of the affine transformation (6.3) from equations (6.4) implies that *the affine invariants determined by the neighborhoods of the second order of the two surfaces S, S_1 at the points O, O_1 are the following*:

$$J_{11} = \frac{a_1}{a}, \qquad J_{12} = \frac{4a_1c_1 - b_1^2}{4a_1^2(b_1/a_1 - b/a)^2}, \qquad J_{13} = \frac{4ac - b^2}{4a^2(b/a - b_1/a_1)^2}. \tag{6.5}$$

Case II. $a = 0$, $a_1 \neq 0$. In this case the line OO_1 is an asymptotic tangent of the surface S at O but does not coincide with an asymptotic tangent of S_1 at O_1. Elimination of the coefficients α_1, β_1, γ_1, β_2, γ_2, γ_3 from equations

(6.4) implies in this case that *the affine invariants determined by the neighborhoods of the second order of the surfaces S, S_1 at the points O, O_1 are the following*:

$$(6.6) \qquad J_{21} = \frac{1}{b^2}(bb_1 - 2ca_1), \qquad J_{22} = \frac{a_1}{b^4}(a_1c^2 - cbb_1 + b^2c_1).$$

Instead of J_{21} and J_{22} it is more convenient to replace J_{22} by the following invariant

$$(6.7) \qquad \bar{J}_{22} = J_{21}^2 - 4J_{22} = \frac{b_1^2 - 4a_1c_1}{b^2}$$

so that as invariants for this case II we shall consider J_{21} and $\bar{J}_{22}$.

Case III. $a \neq 0$, $a_1 = 0$. In this case the line OO_1 is an asymptotic tangent of S_1 at O_1 but is not an asymptotic tangent of S at O. Similarly, as in the foregoing case, we obtain the invariants

$$(6.8) \qquad J_{31} = \frac{1}{b_1^2}(bb_1 - 2c_1a), \qquad \bar{J}_{32} = \frac{b^2 - 4ac}{b_1^2}.$$

Case IV. $a = 0$, $a_1 = 0$. In this case the line OO_1 is an asymptotic tangent of both surfaces S and S_1 at O and O_1 respectively. Elimination of the coefficients of the affine transformation (6.3) from equations (6.4) gives in this case only one affine invariant, namely,

$$(6.9) \qquad J_{41} = \frac{b}{b_1}.$$

7. Metric and affine characterization of the invariants J_{ij}. For the purpose of giving a metric characterization of the invariants J_{ij} of the foregoing number, let us suppose that the coordinate system to which the expansions (6.1) and (6.2) are referred is an orthogonal cartesian coordinate system. Let us consider separately the four cases mentioned in the foregoing number.

Case I. $a \neq 0$, $a_1 \neq 0$. Let r and r_1 be the radii of curvature at the points O and O_1 of the plane curves in which the normal plane $y = 0$ intersects with the surfaces S, S_1 respectively; and K and K_1 the total curvatures of S, S_1 at the points O, O_1. Then it can be readily demonstrated that

$$(7.1) \qquad r = \frac{1}{2a}, \qquad r_1 = \frac{1}{2a_1}, \qquad K = 4ac - b^2, \qquad K_1 = 4a_1c_1 - b_1^2.$$

Furthermore if ω is the angle which the conjugate direction at O of the tangent OO_1 of the surface S forms with the tangent OO_1 and similarly ω_1 is the angle between OO_1 and its conjugate direction at O_1 on the surface S_1, we have

$$(7.2) \qquad 2\cot\omega = -b/a, \qquad 2\cot\omega_1 = -b_1/a_1.$$

From (7.1), (7.2) and (6.5) we obtain the following metrical characterization of the affine invariants J_{11}, J_{12}, J_{13}:

$$(7.3) \qquad J_{11} = \frac{r}{r_1}, \qquad J_{12} = \frac{K_1 r_1^2}{4(\cot \omega_1 - \cot \omega)^2}, \qquad J_{13} = \frac{K r^2}{4(\cot \omega - \cot \omega_1)^2}.$$

From these formulas it follows that the ratio of the total curvatures at O and O_1,

$$K/K_1 = J_{13} J_{12}^{-1} J_{11}^{-2},$$

is also an affine invariant of the two surfaces S, S_1. It may be remarked that the affine invariant $(K/K_1)(r/r_1)^{4/3} = J_{13} J_{12}^{-1} J_{11}^{-2/3}$ is also a projective invariant obtained by Hsiung [10].

We now proceed to give an affine characterization of the invariants J_{11}, J_{12}, J_{13}. We have seen (§3) that the ratio of the radii of curvature is an affine invariant for two plane curves with a common tangent line OO_1. Let us consider the plane curves C, C_1 in which the common normal plane to the surfaces S, S_1 through OO_1 intersects the surfaces S, S_1; let r, r_1 be the radii of curvature of C, C_1 at the points O, O_1. By an affine transformation the common normal plane to S, S_1 through the line OO_1 transforms into another plane which forms a certain angle, say θ, with the common normal plane to the transformed surfaces. If r^* and r_1^* are the radii of curvature of the transformed curves from C and C_1 at the transformed points from O, O_1, by §3, we have $r/r_1 = r^*/r_1^*$. Furthermore, if R and R_1 are the radii of curvature at the transformed points from O, O_1 of the curves in which the common, normal plane to the transformed surfaces from S, S_1 intersects these transformed surfaces, by Meusnier's theorem we have $r^* = R \cos \theta$, $r_1^* = R_1 \cos \theta$. Consequently $r/r_1 = r^*/r_1^* = R/R_1$ and the affine invariance of $J_{11} = r/r_1$ is proved geometrically.

In order to find an affine characterization of the invariant J_{12}, let us consider the conjugate tangent t at O of the tangent OO_1 of the surface S. From the equation (6.1) of S we deduce that the equation of t is $y = -(2a/b)x$. On the other hand the asymptotic tangents of S_1 at O_1 are

$$(7.4) \qquad y = \frac{1}{2c_1}(-b_1 \pm (b_1^2 - 4a_1c_1)^{\frac{1}{2}})(x - h), \qquad z = 0.$$

A simple calculation shows that the tangent t intersects these asymptotic tangents in the points A_1, A_2 such that the invariant J_{12} can be expressed in terms of the ratio $\rho = OA_1/OA_2$ by the following formula

$$(7.5) \qquad J_{12} = -\frac{1}{4}\left(\frac{1-\rho}{1+\rho}\right)^2$$

which gives an affine characterization of the invariant J_{12}.

A similar expression holds for the third affine invariant J_{13} expressed in terms of the ratio of the segments that the asymptotic tangents of S at O determine on the conjugate line at O_1 of the tangent O_1O of the surface S_1.

Case II. $a = 0$, $a_1 \neq 0$. In this case, with the same notations as in the foregoing case, we have

$$(7.6) \qquad b^2 = -K, \qquad b_1 = -\frac{1}{r_1}\cot\omega_1, \qquad a_1 = \frac{1}{2r_1},$$

and if M means the mean curvature of S at O, since $a = 0$, it is known that $c = M$. Hence the invariants J_{21} and $\bar{J}_{22}$ have the following metrical significance:

$$(7.7) \qquad J_{21} = \frac{(-K)^{\frac{1}{2}} + M\tan\omega_1}{Kr_1\tan\omega_1}, \qquad \bar{J}_{22} = \frac{K_1}{K}.$$

We now proceed to give an affine characterization of these invariants. Let us consider the points P, Q in which the asymptotic tangent of S at O distinct of OO_1 (given by the equations $y = -(b/c)x$, $z = 0$) intersects the asymptotic tangents of S_1 at O_1 (given by the equations (7.4)). A ready calculation shows that the ratio $\bar{J}_{22}/J_{21}^2$ is expressed in terms of the ratio $\rho = OP/OQ$ by the formula

$$(7.8) \qquad \frac{\bar{J}_{22}}{J_{21}^2} = \left(\frac{1-\rho}{1+\rho}\right)^2.$$

On the other hand, let us consider the line t_1 (given by the equations $y = -(2a_1/b_1)(x - h)$, $z = 0$), which is the harmonic conjugate of the common tangent OO_1 with respect to the asymptotic tangents of the surface S_1 at the point O_1; and its parallel line (given by the equations $y = -(2a_1/b_1)x$, $z = 0$) through the point O. The normal planes to S, S_1 through these parallel lines intersect S and S_1 respectively in two plane curves whose radii of curvature at O, O_1 have the ratio

$$(7.9) \qquad \rho_1 = \frac{4a_1c_1 - b_1^2}{2(2a_1c - bb_1)} = \frac{\bar{J}_{22}}{2J_{21}}.$$

From (7.8) and (7.9) it follows that

$$J_{21} = 2\rho_1(1+\rho)^2(1-\rho)^{-2}, \qquad \bar{J}_{22} = 4\rho_1^2(1+\rho)^2(1-\rho)^{-2}.$$

These formulas give an affine characterization of the invariants J_{21} and $\bar{J}_{22}$.

Case III. $a \neq 0$, $a_1 = 0$. This case is completely similar to the case II by interchanging S and S_1.

Case IV. $a = 0$, $a_1 = 0$. In this case, according to (7.1) and (6.9) the only affine invariant can be written

$$(7.10) \qquad J_{41} = \left(\frac{K}{K_1}\right)^{\frac{1}{2}};$$

hence, the square of the affine invariant J_{41} is equal to the ratio of the total curvatures of the surfaces S and S_1 at O and O_1 respectively.

To give an affine characterization of J_{41} let us consider the asymptotic tangent u of S at O distinct of OO_1 (given by the equations $y = -(b/c)x$, $z = 0$) and the asymptotic tangent u_1 of S_1 at O_1 distinct of OO_1 (given by the equations $y = -(b_1/c_1)(x - h)$, $z = 0$). These asymptotic tangents intersect each other in the point A whose coordinates are

$$(7.11) \qquad x = h(b_1/c_1)(b_1/c_1 - b/c)^{-1}, \qquad y = -h(b/c)(b_1/c_1)(b_1/c_1 - b/c)^{-1}.$$

We first suppose $b/c - b_1/c_1 \neq 0$. Let us consider the point B such that $2O_1B = O_1A$. The line OB is affinely connected with the configuration of the surfaces S, S_1. The normal plane to S at O through the line OB and the normal plane to S_1 at O_1 through the parallel line to OB through O_1 intersect S and S_1 respectively in two plane curves whose radii of curvature at O and O_1 have the ratio

$$(7.12) \qquad \frac{2b_1}{b} = 2J_{41}^{-1}.$$

Suppose now that $b/c - b_1/c_1 = 0$, so that the asymptotic tangents u and u_1 are parallel. In this case we may consider any pair of parallel normal planes through O and O_1 respectively; let $y = \lambda x$ and $y = \lambda(x - h)$ be these planes. The ratio between the radii of curvature at O and O_1 of the plane curves in which these planes intersect S and S_1 respectively, has the value

$$(7.13) \qquad \frac{b_1 + \lambda c_1}{b + \lambda c} = \frac{b_1}{b} = J_{41}^{-1}.$$

We have remarked in §3 that the ratio of the radii of curvature at two points O, O_1 of two plane curves situated on parallel planes with parallel tangent lines at the points O, O_1 is an invariant with respect to affine transformations of the space. Consequently (7.12) and (7.13) give an affine characterization of the invariant J_{41}.

8. Affine invariants of two surfaces in ordinary space having a common tangent line but distinct tangent planes at two ordinary points. Let now S and S_1 be two surfaces in ordinary space having a common tangent line but distinct tangent planes at two ordinary points O, O_1. Let h be the distance OO_1.

If we choose a cartesian coordinate system in such a way that the point O be the origin, the tangent plane to S at O be the plane $z = 0$ and the tangent plane to S_1 at O_1 be the plane $y = 0$, the power series expansions of the two surfaces in the neighborhood of O and O_1 may be written in the form

$$(8.1) \qquad S: \quad z = ax^2 + bxy + cy^2 + \cdots,$$

$$(8.2) \qquad S_1: \quad y = a_1(x - h)^2 + b_1(x - h)z + c_1z^2 + \cdots.$$

In order to find the affine invariants determined by the neighborhoods of the second order of S and S_1 at the points O and O_1 we have to consider the most general affine transformation which leaves the point O and the planes $y = 0$, $z = 0$ invariants:

$$x = \alpha_1 X + \beta_1 Y + \gamma_1 Z$$

$$(8.3) \qquad y = \beta_2 Y$$

$$z = \gamma_3 Z,$$

where α_1, β_1, γ_1, β_2, γ_3 are arbitrary constants.

By means of this transformation the equations (8.1) and (8.2) transform into two other equations of the same form, namely,

$$Z = AX^2 + BXY + CY^2 + \cdots,$$

$$Y = A_1(X - H)^2 + B_1(X - H)Z + C_1Z^2 + \cdots,$$

where

$$(8.4) \qquad H = h/\alpha_1$$

is the distance between the transformed points from O and O_1, and the coefficients A, B, C, A_1, B_1, C_1 are given by the following system:

$$\gamma_3 A = \alpha_1^2 a, \qquad \beta_2 A_1 = \alpha_1^2 a_1,$$

$$(8.5) \qquad \gamma_3 B = 2\alpha_1\beta_1 a + \alpha_1\beta_2 b, \qquad \beta_2 B_1 = 2\alpha_1\gamma_1 a_1 + \alpha_1\gamma_3 b_1,$$

$$\gamma_3 C = \beta_1^2 a + \beta_1\beta_2 b + \beta_2^2 c, \qquad \beta_2 C_1 = \gamma_1^2 a_1 + \gamma_1\gamma_3 b_1 + \gamma_3^2 c_1 .$$

According to the vanishing or non-vanishing of the two coefficients a and a_1 it is necessary to distinguish four cases.

Case I. $a \neq 0$, $a_1 \neq 0$. In this case the line OO_1 does not coincide with any one of the asymptotic tangents of the surfaces S, S_1 at the points O and O_1 respectively. Eliminating α_1, β_1, γ_1, β_2, γ_3 from the system (8.5) and (8.4) we find that *there are two independent affine invariants determined by the neighborhoods of the second order of the surfaces S, S_1 at the points O, O_1, namely,*

$$(8.6) \qquad J_{11}^* = \frac{a_1^2}{a^2} h^2(4ac - b^2), \qquad J_{12}^* = \frac{a^2}{a_1^2} h^2(4a_1c_1 - b_1^2).$$

Case II. $a = 0$, $a_1 \neq 0$. In this case the line OO_1 is an asymptotic tangent of the surface S at O, but it is not an asymptotic tangent of S_1 at O_1. Eliminating α_1, β_1, γ_1, β_2, γ_3 from the system (8.5) and (8.4) we see that *in this case the only affine invariant determined by the neighborhoods of the second order of the surfaces S, S_1 at the points O, O_1 is*

$$(8.7) \qquad J_{21}^* = (b_1^2 - 4a_1c_1)b^2h^4.$$

Case III. $a \neq 0$, $a_1 = 0$. The line OO_1 is an asymptotic tangent of S_1 at O_1 but is not an asymptotic tangent of S at O. Similarly as in the foregoing case the only affine invariant is found to be

(8.8) $$J^*_{31} = (b^2 - 4ac)b_1^2h^4.$$

Case IV. $a = 0$, $a_1 = 0$. The line OO_1 is an asymptotic tangent of both surfaces S, S_1 at the points O, O_1. Eliminating α_1, β_1, γ_1, β_2, γ_3 from the equations (8.5) and (8.4) we find that *in this case there is only one affine invariant determined by the neighborhoods of the second order of the surfaces S, S_1 at O, O_1, namely,*

(8.9) $$J^*_{41} = bb_1h^2.$$

9. Metric and affine characterization of the invariants J^*_{ij}. We consider separately the four cases mentioned in the foregoing number.

Case I. Let us call ω the angle between the tangent planes to S and S_1 at the points O and O_1 respectively, that is, the angle between the planes $y = 0$, $z = 0$ of our coordinate system to which are referred the equations (8.1) and (8.2) of S and S_1 respectively.

Let r be the radius of curvature at O of the plane curve in which the tangent plane to S_1 at O_1 (that is, the plane $y = 0$) intersects the surface S, and K the total curvature of S at O. Similarly, let r_1 be the radius of curvature at O_1 of the plane curve in which the tangent plane to S at O (that is, the plane $z = 0$) intersects the surface S_1, and K_1 the total curvature of S_1 at O_1. Then, if the plane $x = 0$ is chosen to be orthogonal to the x-axis, it can be readily demonstrated that

(9.1) $$2r = 1/a, \qquad 2r_1 = 1/a_1,$$
$$K = (4ac - b^2)\sin^2\omega, \qquad K_1 = (4a_1c_1 - b_1^2)\sin^2\omega.$$

From (9.1) and (8.6) we obtain the following metrical characterization of the affine invariants J^*_{11} and J^*_{12}:

(9.2) $$J^*_{11} = h^2\frac{r^2K}{r_1^2\sin^2\omega}, \qquad J^*_{12} = h^2\frac{r_1^2K_1}{r^2\sin^2\omega}.$$

From these formulas it follows in particular that the expression

(9.3) $$J^*_{11}J^*_{12} = h^4\frac{KK_1}{\sin^4\omega}$$

is an affine invariant. Buzano has proved that this invariant (9.3) is also a projective invariant [4]. Similarly

(9.4) $$J^*_{11}/J^*_{12} = (K/K_1)(r/r_1)^4$$

is an affine invariant. For the case $h = 0$ Hsiung has proved that this invariant (9.4) is also a projective invariant [7].

We now proceed to give an affine characterization of the affine invariants J^*_{11} and J^*_{12}. For this purpose let us consider the conjugate tangent t of the tangent OO_1 on the surface S and the conjugate tangent t_1 of the tangent O_1O on the surface S_1. Since we suppose that OO_1 is not an asymptotic tangent for S or S_1, t and t_1 are straight lines distinct from OO_1 whose equations are

$$(9.5) \qquad t: \quad y = -(2a/b)x, \qquad z = 0,$$

$$(9.6) \qquad t_1: \quad y = 0, \qquad z = -(2a_1/b_1)(x - h).$$

The one-parameter family of paraboloids which contain t and t_1 and are generated by lines which cut the line OO_1 and are parallel to the plane determined by the directions of t and t_1, is given by

$$(9.7) \qquad \left(x + \frac{b}{2a} y + \frac{b_1}{2a_1} z\right)(\lambda z + y) - \lambda h z = 0$$

where λ is the variable parameter.

The tangent lines at O of the curve in which the general paraboloid (9.7) intersects the surface S are given by

$$(9.8) \qquad \left(\frac{b}{2a} - \lambda hc\right)y^2 + (1 - \lambda hb)xy - \lambda hax^2 = 0, \qquad z = 0.$$

The necessary and sufficient condition that these two tangent lines coincide, that is to say, that the intersection curve of the paraboloid (9.7) and the surface S has a cusp at O, is

$$(9.9) \qquad 1 - \lambda^2 h^2 (4ac - b^2) = 0.$$

This equation gives two values of λ, say λ_1 and λ_2, each of which gives a paraboloid with the property that its intersection curve with the surface S has a cusp at O. Let us consider one of these paraboloids, for instance the paraboloid which corresponds to $\lambda = \lambda_1$. The tangent plane to this paraboloid at the point of infinity of the line OO_1 is the plane $y + \lambda_1 z = 0$ which intersects the surface S in the plane curve

$$(9.10) \qquad y + \lambda_1 z = 0, \qquad z = ax^2 - \lambda_1 bxz + \lambda_1^2 cz^2 + \cdots,$$

and the surface S_1 in the plane curve

$$(9.11) \qquad y + \lambda_1 z = 0, \qquad -\lambda z = a_1(x - h)^2 + b_1(x - h)z + c_1 z^2 + \cdots.$$

If r and r_1 are the radii of curvature of the plane curves (9.10) and (9.11) at the points O and O_1 respectively it is easily verified that

$$(9.12) \qquad \frac{r}{r_1} = -\frac{a_1}{a\lambda_1}.$$

Consequently, according to (9.9) and (8.6) we have

$$(9.13) \qquad \left(\frac{r}{r_1}\right)^2 = J_{11}^*.$$

Since the ratio r/r_1 according to §3 is an invariant with respect to affine transformations, the formula (9.13) gives an affine characterization of the invariant J_{11}^*. Evidently, by symmetry, the affine invariant J_{12}^* has an affine characterization entirely similar to the preceding one.

Cases II, III, IV. In these cases, according to (9.1) where we put $a = 0$, $a_1 \neq 0$; $a \neq 0$, $a_1 = 0$ or $a = a_1 = 0$, the invariants J_{21}^*, J_{31}^* and J_{41}^* have the following metrical characterization:

$$J_{21}^* = J_{31}^* = J_{41}^{*2} = h^4 \frac{KK_1}{\sin^4 \omega},$$

that is, they coincide with the projective invariant given by Buzano [4]. Several projective characterizations of this invariant have been given by Bompiani [2].

10. **Summary.** In this paper we compute the simplest affine invariants of certain sets of curves and surfaces. They are:

(1) Two plane curves having a common tangent at two ordinary points. Two invariants are found, determined respectively by the neighborhoods of the second and the third order. These invariants are not projective invariants. The simplest projective invariants are determined by the neighborhoods of the fourth order (see [5]).

(2) Two intersecting plane curves at an ordinary point. Two invariants are found, determined by the neighborhoods of the third order. They are not projective. The simplest projective invariants are determined by the neighborhoods of the fourth order (see [6]).

(3) Two surfaces with common tangent plane at two ordinary points, with: (I) common tangent line in general position; three invariants are found J_{11}, J_{12}, J_{13} of which the combination $J_{13}J_{12}^{-1}J_{11}^{-2/3}$ is projective. (II, III) common tangent asymptotic on one surface; two invariants are found, not projective. (IV) common tangent asymptotic on both surfaces; one invariant is found, not projective. These invariants are determined by the neighborhoods of the second order. The mentioned projective invariant is the only one (see [8] and [12]).

(4) Two surfaces with common tangent line but distinct tangent planes at two ordinary points. Same division into cases as in (3). (I) Two invariants are found, J_{11}^*, J_{12}^* of which the product is projective; this projective invariant is the only one (see [4]). (II, III, IV) One invariant is found, which is projective. These invariants are determined by the neighborhoods of the second order. It is supposed that the distance h between the two points is $\neq 0$. For $h = 0$ there is in case (I) only one affine invariant J_{11}^*/J_{12}^*, which is also the only projective invariant (see [7]); in cases (II, III, IV) there are neither affine nor projective invariants determined by the neighborhoods of the second order.

Bibliography

1. W. Blaschke, *Vorlesungen über Differentialgeometrie* II, Berlin, 1923.
2. E. Bompiani, *Invarianti proiettivi di una particolare coppia di elementi superficiali del secondo ordine*, Bollettino della Unione Matematica Italiana, vol. 14(1935), pp. 237–243.
3. C. L. Bouton, *Some examples of differential invariants*, Bulletin of the American Mathematical Society, vol. 4(1898), pp. 313–322.
4. P. Buzano, *Invariante proiettivo di una particolare coppia di elementi di superficie*, Bollettino della Unione Matematica Italiana, vol. 14(1935), pp. 93–98.
5. C. C. Hsiung, *Projective differential geometry of a pair of plane curves*, this Journal, vol. 10(1943), pp. 539–546.
6. C. C. Hsiung, *Theory of intersection of two plane curves*, Bulletin of the American Mathematical Society, vol. 49(1943), pp. 786–792.
7. C. C. Hsiung, *An invariant of intersection of two surfaces*, Bulletin of the American Mathematical Society, vol. 49(1943), pp. 877–880.
8. C. C. Hsiung, *Projective invariants of a pair of surfaces*, this Journal, vol. 10(1943), pp. 717–720.
9. C. C. Hsiung, *Projective invariants of intersection of certain pairs of surfaces*, Bulletin of the American Mathematical Society, vol. 50(1944), pp. 437–441.
10. C. C. Hsiung, *A projective invariant of a certain pair of surfaces*, this Journal, vol, 12(1945), pp. 441–443.
11. M. Mascalchi, *Un nuovo invariante proiettivo di contatto di due superficie*, Bollettino della Unione Matematica Italiana, vol. 13(1934), pp. 45–49.
12. R. Mehmke, *Ueber zwei, die Krümmung von Curven und das Gauss' sche Krümmungsmaass von Flächen betreffende charakteristische Eigenschaften der linearen Punkt-Transformationen*, Zeitschrift für Mathematik und Physik, vol. 36(1891), pp. 206–213.
13. E. Salkowski, *Affine Differentialgeometrie*, Berlin und Leipzig, 1934.
14. C. Segre, *Su alcuni punti singolari delle curve algebrische, e sulla linea parabolica di una superficie*, Atti della Reale Accademia dei Lincei, (5), vol. 6(1897), pp. 168–175.

Mathematical Institute,
Rosario, Argentina.

ANNALS OF MATHEMATICS
Vol. 51, No. 3, May, 1950

INTEGRAL GEOMETRY IN PROJECTIVE AND AFFINE SPACES

By L. A. Santaló

(Received December 27, 1948)

Introduction

The fundamental concepts of integral geometry in a general space in which a transitive Lie group of automorphisms has been defined were given by Chern [3]. Our purpose is to apply these fundamental concepts in order to study the integral geometry in projective and affine spaces and to deduce from this study some geometrical consequences.

In §1 we give a brief summary of the ideas of Chern, who was the first to apply the Cartan's theory of Lie groups to the integral geometry; only the condition (1.11) is slightly different from that given by Chern. In §2 we study the integral geometry in projective space. In §3 we consider the unimodular center affine group and its integral geometry and in §4 we apply the obtained results in order to give an elementary proof of a theorem of Minkowski-Hlawka belonging to the geometry of numbers. Finally, in §5, we indicate the main results of integral geometry in unimodular affine space.

1. On the measure of sets of geometrical elements with respect to a given group of Lie

Let E be a space of points x in which an r-parameter Lie group G_r of automorphisms has been defined. Let $a^1, a^2, \cdots, a^2$ be the parameters of G_r and

$$\omega_1(a, da), \quad \omega_2(a, da), \cdots, \omega_r(a, da) \tag{1.1}$$

its relative components (see Cartan [1] p. 79), that is, a set of r linearly independent Pfaffian forms invariant under the first group of parameters of G_r. These relative components satisfy the equations of structure of Maurer-Cartan

$$\omega_i' = \sum_{j,k=1}^{r} c_{jk}^i[\omega_j\omega_k], \qquad i = 1, 2, \cdots, r \tag{1.2}$$

where c_{jk}^i are the constants of structure of G_r, and ω_i' denotes the exterior derivative of the form ω_i. The square brackets denote exterior multiplication.

Let H be a geometrical element of E depending upon h parameters. By a geometrical element we understand a set of points of E which may be determined by a finite number of parameters. For instance, if E is the 3-dimensional euclidean space, points, straight lines, quadrics, are geometrical elements. In general, any figure transformed of a given fixed figure F by $s \in G_r$ is a geometrical element, because it may be determined by the parameters of s.

Let us assume that the subgroup of G_r, which leaves invariant the geometrical element H, is a continuous subgroup g_{r-h} depending upon $r - h$ parameters. In the space of parameters, g_{r-h} will be represented by an $(r - h)$-dimensional variety which we shall represent by the same notation g_{r-h}. The variety g_{r-h} and its transformed sg_{r-h}, by the operations s of the group of parameters of G_r, fill up the whole space of parameters and have the property that no two of them can have common point without being identical. Thus the varieties sg_{r-h} are the integral varieties of a completely integrable Pfaffian system. Furthermore the totality of varieties sg_{r-h} is invariant with respect to the first group of parameters; consequently the left hand sides of the Pfaffian system are linear combinations with constant coefficients of the relative components $\omega_1, \cdots, \omega_r$. Since the relative components are determined up to a linear transformation with constant coefficients, we may suppose the Pfaffian system which integral varieties are sg_{r-h} to be

$$(1.3) \qquad \omega_1 = 0, \qquad \omega_2 = 0, \cdots, \omega_h = 0.$$

Let

$$(1.4) \qquad \varphi_1(a^1, \cdots, a^2) = \alpha^1, \cdots, \varphi_h(a^1, \cdots, a^r) = \alpha^h$$

be h independent first integrals of (1.3). That means that to each set of constants $\alpha^1, \cdots, \alpha^h$ corresponds an integral variety sg_{r-h}. Thus we can take $\alpha^1, \cdots, \alpha^h$ as coordinates of the variety sg_{r-h}. In the original space E, to each sg_{r-h} corresponds a geometrical element sH, transformed of H by s. Conversely, to each sH of E corresponds a variety sg_{r-h}, consequently $\alpha^1, \cdots, \alpha^h$ may also be considered as coordinates of the geometrical element sH.

By "density" of the elements sH we shall mean an exterior differential form of order h of the form

$$(1.5) \qquad dH = f(\alpha^1, \cdots, \alpha^h)\,[d\alpha^1 d\alpha^2 \cdots d\alpha^h]$$

such that its value be invariant under the group G_r, i.e., under the first group of parameters. Since G_r is transitive with respect to the elements sH, this density, if it exists, is unique up to a constant factor. The measure of a set of elements sH will then be the integral of dH extended over the set.

Being independent first integrals of (1.3), the differentials $d\alpha^i$ are linearly independent combinations of $\omega_1, \cdots, \omega_h$ and we get, by exterior multiplication, an expression of the form

$$(1.6) \qquad [d\alpha^1 \cdots d\alpha^h] = \Delta(a^1, \cdots, a^2)[\omega_1 \cdots \omega_h], \qquad \Delta \neq 0$$

or

$$(1.7) \qquad [\omega_1 \cdots \omega_h] = (1/\Delta)[d\alpha^1 \cdots d\alpha^h].$$

The left hand side is invariant under the first group of parameters; therefore, in order that (1.5) be a density, we must have, up to a constant factor, $f = 1/\Delta(a)$, that is, $\Delta(a)$ must be a function of $\alpha^1, \cdots, \alpha^h$ only (condition of Chern [3]). The density is then defined up to a constant factor by

$$(1.8) \qquad dH = [\omega_1\omega_2 \cdots \omega_h].$$

The foregoing condition of Chern is, in general, not easy to apply. In many cases it is more convenient to express it in the following equivalent form.

The exterior differential form $[\omega_1 \cdots \omega_h]$ is always invariant under the first group of parameters. However, it is not always a density because it depends on the r parameters a^i and, though each point a determines an element sH, to each sH corresponds all points a of the corresponding variety sg_{r-h}. In order that (1.8) be a density its value does not change when the points are displaced on the varieties sg_{r-h}. That is, if we consider an h-dimensional variety V_h in the space of parameters, in order that (1.8) be a density, it is necessary and sufficient that the integral

$$\int_{V_h} [\omega_1 \cdots \omega_h] \tag{1.9}$$

be invariant when each point of V_h displaces on the variety sg_{r-h} which passes through it. That is equivalent to saying that the integral (1.9) is zero when extended over any closed h-dimensional variety (observe that a general V_h intersects the varieties sg_{r-h} in points only). The generalized Stokes formula (see, for instance [2] p. 40) says that if V_{h+1} is any $(h + 1)$-dimensional domain and ∂V_{h+1} is its boundary, it is

$$\int_{\partial V_{h+1}} [\omega_1 \cdots \omega_h] = \int_{V_{h+1}} [\omega_1 \cdots \omega_h]'. \tag{1.10}$$

Since ∂V_{h+1} is closed and of dimension h, the last integral must be zero for any integration domain V_{h+1}; consequently the integrand must be zero, and we get:

A necessary and sufficient condition for (1.8) *to be a density for the elements sH is*

$$[\omega_1 \cdots \omega_h]' = 0. \tag{1.11}$$

As a first and immediate application of this result, let us consider the case in which H is a geometrical element such that the subgroup of G_r, which leaves it invariant, reduces to the identity. In this case H depends on r parameters and to each element sH corresponds an unique transformation s of the group of parameters; the varieties sg_{r-h} are the points of the space of parameters and the density (1.8) is formed by the exterior product of all relative components ω_i. Having taken into account the equations of structure (1.2) the condition (1.11) is obviously satisfied in this case. We shall write

$$dG_r = [\omega_1 \cdots \omega_r] \tag{1.12}$$

and following Poincaré and Blaschke the density dG_r will be called the "cinematic density" of the group G_r. Thus we have: *the cinematic density of a group always exists.*

Let us now consider the case in which the subgroup g_{r-h}, which leaves H invariant, is a discrete group. In this case g_{r-h} will be represented in the space of parameters by a set of $(r - h)$-dimensional varieties, without common point, and congruent with respect to g_{r-h}. Analogously as in the continuous case, each of these varieties will be represented by a Pfaffian system of the form (1.3). The density for sets of elements sH will then be $dH = [\omega_1 \cdots \omega_h]$, assuming that (1.11) is satisfied. The only thing we have into account is that in the present case to each partial variety which composes sg_{r-h} corresponds the same geometrical element sH, so that, in order to measure a set of different elements sH, the domain of integration must be considered in the space of the factor group G_r/g_{r-h}. In §4 we shall see an example in which g_{r-h} is discrete.

2. Integral Geometry in Projective Space

Let P_n be the real n-dimensional projective space. Following E. Cartan, [1] p. 75, a set of $n + 1$ real numbers $x^0, \cdots, x^n$ (not all equal to zero) will be called an "analytic point", represented by x, whose coordinates are the numbers x^i. To each analytic point x corresponds the "geometric point" whose homogenous coordinates are x^i.

The projective group G_r ($r = n(n + 2)$), which we shall denote by $\mathfrak{P}$, may be represented by

$$(x^k)' = \sum_{i=0}^{n} a_i^k x^i \qquad (k = 0, 1, \cdots, n) \tag{2.1}$$

with the condition

$$| a_i^k | = 1. \tag{2.2}$$

That means that each projective transformation is determined by $n + 1$ analytic points $a_i(a_i^0, \cdots, a_i^n)$, $i = 0, 1, \cdots, n$ which satisfy the condition (2.2). Instead of (2.2) it will be more convenient to write

$$| a_0 a_1 \cdots, a_n | = 1 \tag{2.3}$$

where the left hand side represents the determinant formed with the coordinates of the analytic points a_i.

The $r = n(n + 2)$ relative components of the projective group are defined by the equations (Cartan [1] p. 84)

$$da_i = \sum_{k=0}^{n} \omega_{ik} a_k, \qquad i = 0, 1, \cdots, n. \tag{2.4}$$

From (2.3) and (2.4) we deduce

$$\omega_{ik} = | a_0 \cdot a_1 \cdots a_{k-1} da_i a_{k+1} \cdots a_n | \tag{2.5}$$

with the condition, obtained by differentiation of (2.3),

$$\sum_{i=0}^{n} \omega_{ii} = 0. \tag{2.6}$$

The equations of structure are obtained by taking the exterior derivative of (2.4) and taking into account the relations (2.4) themselves. The result is

$$(\omega_{ij})' = \sum_{k=0}^{n} [\omega_{ik} \omega_{kj}]. \tag{2.7}$$

We have now all the necessary elements to study the integral geometry in P_n.

Let us consider the geometrical elements sH defined by a completely integrable Pfaffian system

(2.8) $$\omega_{i_1j_1} = 0, \quad \omega_{i_2j_2} = 0, \cdots, \omega_{i_hj_h} = 0.$$

This is the system (1.3) of §1, which means that the integral varieties sg_{r-h} of (2.8) correspond to the geometrical elements sH obtained applying a projective transformation to one, say H, of them.

A necessary and sufficient condition in order that the elements sH possess an invariant density is given by (1.11); that is

(2.9) $$[\omega_{i_1j_1} \cdots \omega_{i_hj_h}]' = 0.$$

We want to see the form which takes this condition in the particular case of the projective group. We have, according to (2.7),

$$[\omega_{i_1j_1} \cdots \omega_{i_hj_h}]' = \sum_{m=1}^{h} (-1)^{m-1}[\omega_{i_1j_1} \cdots \omega_{i_{m-1}j_{m-1}} \sum_{l=0}^{n} [\omega_{i_ml}\omega_{lj_m}]\omega_{i_{m+1}j_{m+1}} \cdots \omega_{i_hj_h}].$$

Since the system (2.8) is completely integrable, the theorem of Frobenius, [1] p. 193, says that at least one of the forms ω_{i_ml}, ω_{lj_m} for any m and l belongs to the set (2.8). Therefore it is

(2.10) $$[\omega_{i_1j_1} \cdots \omega_{i_hj_h}]' = (-1)^h [\omega_{i_1j_1} \cdots \omega_{i_hj_h} \sum_{m=1}^{h} (\omega_{j_mj_m} - \omega_{i_mi_m})].$$

Thus, if we set $dH = [\omega_{i_1j_1} \cdots \omega_{i_hj_h}]$, we have the following

LEMMA 2.1. *In order that the geometrical elements defined by the system* (2.8) *possess an invariant density with respect to the projective group, it is necessary and sufficient that*

(2.11) $$\left[dH \cdot \sum_{m=1}^{h} (\omega_{j_mj_m} - \omega_{i_mi_m}) \right] = 0.$$

Let us apply this lemma in order to see if the linear subspaces S_h of dimension h $(h < n)$ possess a density. The geometrical element H is now a particular S_h and the set of geometrical elements sH is the set of all h-dimensional subspaces in P_n, since each of them may be obtained from S_h by a suitable projectivity. Consider the S_h defined by the analytic points $a_0, \cdots, a_h$, that is, the linear subspace defined by the parametric equations

(2.12) $$x^i = \sum_{k=0}^{h} \lambda^k a_k^i, \qquad i = 0, 1, \cdots, n.$$

The subgroup g of projectivities which leave this S_h invariant is characterized by the condition that the points $a_0, \cdots, a_h$ remain in S_h. Consequently the differentials da_i $(0 \leqq i \leqq h)$ must be linear combinations of $a_0, \cdots, a_h$ and (2.4) gives

(2.13) $$\omega_{ik} = 0 \quad \text{for} \quad \begin{cases} i = 0, 1, \cdots, h \\ k = h+1, \cdots, n. \end{cases}$$

This is the Pfaffian system which corresponds to the linear subspaces S_h; it is composed of $(h+1)(n-h)$ equations. According to the lemma, a density for S_h will exist if and only if

$$\left[\prod_{\substack{i=0,\cdots,h\\k=h+1,\cdots,n}} \omega_{ik}\left(\sum_{l=h+1}^{n}\omega_{ll}-\sum_{l=0}^{h}\omega_{ll}\right)\right]=0. \tag{2.14}$$

From (2.6) we deduce

$$\sum_{l=h+1}^{n}\omega_{ll}=-\sum_{l=0}^{h}\omega_{ll}.$$

Consequently the left hand side of (2.14) takes the form

$$-2\left[\prod_{\substack{i=0,\cdots,h\\k=h+1,\cdots,n}} \omega_{ik}\cdot\sum_{l=0}^{h}\omega_{ll}\right]. \tag{2.15}$$

Since between the relative components ω_{ik} there exists only the relation (2.14), the exterior product (2.15) cannot be zero. Consequently we have:

THEOREM 2.1. *The linear subspaces have no invariant density with respect to the projective group.*

In other words: it is not possible to define a measure, given by an integral of a form like (1.5), for sets of linear subspaces which will be invariant under the projective group.

Let us now consider as geometrical elements sets of linear subspaces $S_{h_1}+S_{h_2}+\cdots+S_{h_m}$, without common point and satisfying the condition

$$h_1+h_2+\cdots+h_m+m\leqq n+1. \tag{2.16}$$

We may take

$$\begin{array}{llll} & S_{h_1} & \text{defined by the analytic points} & a_0, a_1, \cdots, a_{h_1} \\ & S_{h_2} & \text{"} \quad\quad \text{"} & a_{h_1+1}, \cdots, a_{h_1+h_2+1} \\ (2.17) & S_{h_3} & \text{"} \quad\quad \text{"} & a_{h_1+h_2+2}, \cdots, a_{h_1+h_2+h_3+2} \\ & \cdots\cdots & & \\ & S_{h_m} & \text{"} \quad\quad \text{"} & a_{h_1+\cdots+h_{m-1}+m-1}, \cdots, a_{h_1+\cdots+h_m+m-1}. \end{array}$$

The subgroup of projectivities which leave $S_{h_1}+S_{h_2}+\cdots+S_{h_m}$ invariant is characterized by the condition that the differentials da_{i_s} are linear combinations of a_{i_s} for $h_1+\cdots+h_s+s\leqq i_s\leqq h_1+\cdots+h_{s+1}+s$ and $0\leqq s\leqq m-1$. Consequently, according to (2.4) we must have $\omega_{ij}=0$ for all pairs i, j between the limits

$$\begin{array}{ll} & 0\leqq i\leqq h_1, \quad h_1+1\leqq j\leqq n; \\ & h_1+1\leqq i\leqq h_1+h_2+1, \quad 0\leqq j\leqq h_1, \quad h_1+h_2+2\leqq j\leqq n; \\ (2.18) & \cdots\cdots\cdots \\ & h_1+\cdots+h_{m-1}+m-1\leqq i\leqq h_1+\cdots+h_m+m-1, \\ & 0\leqq j\leqq h_1+\cdots+h_{m-1}+m-2, \quad h_1+\cdots+h_m+m\leqq j\leqq n \end{array}$$

In order to apply the lemma, we observe that, having taken into account (2.5), for each row of (2.17) the sum of ω_{jj} is equal, up to the sign, to the corresponding sum of ω_{ii}. Consequently condition (2.11) is written

$$\left[\prod \omega_{ij} \cdot \sum_{l=0}^{h_1+\cdots+h_m+m-1} \omega_{ll}\right] = 0 \tag{2.19}$$

where Π is extended over all ω_{ij} with the conditions (2.18). Since $i \neq j$ in this product, we find that (2.19) holds only if

$$\sum_{l=0}^{h_1+\cdots+h_m+m-1} \omega_{ll} = 0. \tag{2.20}$$

Since (2.6) is the only relation between the relative components ω_{ij}, (2.20) holds only if

$$h_1 + h_2 + \cdots + h_m + m = n + 1. \tag{2.21}$$

Then we have:

THEOREM 2.2. *In order that the elements* $(S_{h_1} + S_{h_2} + \cdots + S_{h_m})$ *composed by m linear subspaces of dimensions* h_i, *without common point and satisfying the relation* (2.16), *have an invariant density with respect to the projective group, it is necessary and sufficient that the relation* (2.21) *holds. In this case, assumed each* S_{h_i} *defined by the points indicated in* (2.17), *the density is given by the exterior product of the forms* ω_{ij} *(given by* (2.5)) *where i, j are submitted to the conditions* (2.18).

For instance, on the straight line, $n = 1$, the pairs of points have a density with respect to the projective group, which is given by $(\xi - \eta)^{-2}[d\xi d\eta]$, where ξ, η are the non-homogenous coordinates of the points.

For the case $m = 2$, the foregoing result, following a completely different way, was obtained by Varga [7].

Finally let us observe that the cinematic density for the projective group $\mathfrak{P}$ is expressed by the exterior product of all independent ω_{ij}, given by (2.5), that is

$$d\mathfrak{P} = \left[\prod_{i,j=0}^{n}{}' \omega_{ij}\right] \tag{2.22}$$

where the accent denotes that the factor ω_{nn} is excluded, because according to (2.6) it is not independent from the others.

Taking into account (2.5) and setting

$$dA_i = [da_i^0 da_i^1 \cdots da_i^n] \qquad i = 0, 1, \cdots, n \tag{2.23}$$

we may also write, after applying a known property about adjoint determinants, [5] p. 78.

$$d\mathfrak{P} = \left[\prod_{i=0}^{n-1} dA_i \cdot \sum_{j=0}^{n} (-1)^j a_n^j \, da_n^0 \cdots da_n^{j-1} \, da_n^{j+1} \cdots da_n^n\right]. \tag{2.24}$$

That is: the measure of a set of projectivities defined by the analytic points a_i ($i = 0, 1, \cdots, n$) with the condition (2.3), is given, up to a constant factor, by the integral of the exterior differential form (2.24) extended over the set.

3. The Group of Unimodular Center-Affine Transformations

Let us now consider the n-dimensional affine space and in it the group $\mathfrak{A}$ of affine transformations of modulo 1, which leaves invariant the origin $O(0, 0, \cdots, 0)$ (unimodular center-affine group). If $x^1, x^2, \cdots, x^n$ are the non-homogeneous coordinates of the point x and $a_i^1, a_i^2, \cdots, a_i^n$ those of the points a_i ($i = 1, 2, \cdots, n$)which determine the center-affine transformation, the equations of the group may be written

$$(x^i)' = \sum_{k=1}^{n} a_k^i x^k, \qquad i = 1, 2, \cdots, n \tag{3.1}$$

with the condition

$$| a_1 a_2 \cdots a_n | = 1. \tag{3.2}$$

In the present section there is no more distinction between analytic and geometric points; since the coordinates are non-homogeneous, all points are geometric ones.

Analogously to the case of the projective group, the relative components are now defined by the equations

$$da_i = \sum_{j=1}^{n} \omega_{ij} a_j, \qquad i = 1, 2, \cdots, n \tag{3.3}$$

which, having into account (3.2) give

$$\omega_{ij} = | a_1 a_2 \cdots a_{j-1} da_i da_{j+1} \cdots a_n | \tag{3.4}$$

with the relation

$$\sum_{i=1}^{n} \omega_{ii} = 0. \tag{3.5}$$

The equations of structure are

$$(\omega_{ij})' = \sum_{k=1}^{n} [\omega_{ik} \omega_{kj}]. \tag{3.6}$$

For the linear subspaces which pass through the origin o, the group $\mathfrak{A}$ in the n-dimensional space, coincides with the projective group in the $(n - 1)$-dimensional space, assuming each S_h in the affine space as equivalent to a S_{h-1} in the projective space. Consequently, Theorems 2.1 and 2.2 may now be announced:

THEOREM 3.1. *The linear subspaces which pass through the origin have no invariant density (or measure) with respect to the unimodular center-affine group A. The elements ($S_{h_1} + S_{h_2} + \cdots + S_{h_m}$) composed of m linear subspaces of dimension*

h_i, passing through the origin and having no other common point, have an invariant density with respect to $\mathfrak{A}$ if and only if the condition

$$h_1 + h_2 + \cdots + h_m = n \tag{3.7}$$

holds.

In the last case, if each S_{h_i} is determined by the points $a_{h_1+\cdots+h_{i-1}+1}$, $a_{h_1+\cdots+h_{i-1}+2}, \cdots, \cdots, a_{h_1+h_2+\cdots+h_i}$, the density for sets of elements $(S_{h_1} + S_{h_2} + \cdots + S_{h_m})$ is given by the exterior product of the differential forms ω_{ij}, given by (3.4), corresponding to the values i, j between the limits

$$\begin{aligned} &h_1 + \cdots + h_{i-1} + 1 \leqq i \leqq h_1 + \cdots + h_i, \\ &1 \leqq j \leqq h_1 + \cdots + h_{i-1}, \qquad h_1 + \cdots + h_i + 1 \leqq j \leqq n \end{aligned} \tag{3.8}$$

for $i = 1, 2, \cdots, m$.

EXAMPLE. On the plane, $n = 2$, it is not possible to define an invariant measure with respect to $\mathfrak{A}$ for sets of straight lines through the origin. However this measure exists for sets of pairs of straight lines, because (3.7) holds for $h_1 = h_2 = 1$. If the pair of straight lines is determined by the angles φ_1, φ_2 which they form with the x-axis, it is easily found that the density takes the value $d(S_1 + S_2) = \sin^{-2}(\varphi_2 - \varphi_1)[d\varphi_1 d\varphi_2]$.

Let us now see if density exists for linear subspaces S_h which do not pass through the origin. We may consider as fixed subspace S_h, that which contains the point a_1 and is parallel to oa_i ($i = 2, 3, \cdots, h + 1$). If this S_h is assumed fixed, the differentials da_i ($i = 1, 2, \cdots, h + 1$) must be linear combinations of $a_2, \cdots a_{h+1}$. Consequently, (3.3) gives

$$\begin{aligned} &\omega_{i1} = 0 \qquad \text{for} \qquad i = 1, 2, \cdots, h + 1 \\ &\omega_{ij} = 0 \qquad \text{for} \qquad i = 1, \cdots, h + 1, j = h + 2, \cdots, n. \end{aligned} \tag{3.9}$$

According to §1, in order that the sets of S_h have a density we must have

$$\left[\prod_{i=1,\cdots,h+1} \omega_{i1} \cdot \prod_{\substack{i=1,\cdots,h+1 \\ j=h+2,\cdots,n}} \omega_{ij} \right]' = 0, \tag{3.10}$$

or, according to the Lemma 2.1, which is also applicable in this case because the equations of structure (2.7) and (3.6) have the same form in both cases, we have

$$\left[\prod_{i=1,\cdots,h+1} \omega_{i1} \cdot \prod_{\substack{i=1,\cdots,h+1 \\ j=h+2,\cdots,n}} \omega_{ij} \cdot \sum_{l=1}^{h+1} \omega_{ll} \right] = 0. \tag{3.11}$$

This condition is only satisfied if

$$\omega_{22} + \omega_{33} + \cdots + \omega_{h+1,h+1} \equiv 0 \pmod{\omega_{11}} \tag{3.12}$$

and since (3.5) is the only one relation between the relative components, (3.12) holds only if $h = 0$ or $h + 1 = n$.

For $h = 0$ the density has the value $dS_0 = [\omega_{11} \cdots \omega_{1n}]$. If the points are assumed to be determined by their coordinates $a_1^1, a_1^2, \cdots, a_1^n$ taking into account (3.4) and (3.2) we get $dS_0 = [da_1^1 da_1^2 \cdots da_1^n]$ that is, the density for points equals the element of volume, which is an obvious result, since the center-affine transformations are assumed volume preserving.

For $h = n - 1$, if each hyperplane is determined by one, say a_1, of its points and the directions oa_i $(i = 2, \cdots, n)$ the density has the value

$$dS_{n-1} = [\omega_{11}\omega_{21} \cdots \omega_{n1}]. \tag{3.13}$$

In order to give a geometrical interpretation of this density we proceed as follows.

Let b be the point on the unit hypersphere of center o such that the radius ob is perpendicular to the hyperplane determined by the points o, a_i $(i = 2, 3, \cdots, n)$; that is, the point b determines the direction normal to S_{n-1}. If $b^1, b^2, \cdots, b^n$ are the coordinates of b, the element of area on the unit hypersphere corresponding to the point b is expressed by

$$dB = \frac{[db^1 \cdots db^{i-1}\, db^{i+1} \cdots db^n]}{(-1)^i b^i} \tag{3.14}$$

where the right side is independent of i. Let $b_2, b_3, \cdots, b_n$ be n-unit orthogonal vectors with the origin o on the hyperplane determined by $o, a_2, \cdots, a_n$. If we set

$$(b_i\, db) = -(b\, db_i) = \sum_{k=1}^{n} b_i^k\, db^k, \qquad i = 2, \cdots, n \tag{3.15}$$

will be

$$\left[\prod_{i=2}^{n} (b\, db_i)\right] = \sum_{i=1}^{n} \beta_i [db^1 \cdots db^{i-1}\, db^{i+1} \cdots db^n] \tag{3.16}$$

where β_i is the complementary minor of b^i in the determinant $|\, bb_2 \cdots b_n \,| = 1$. Therefore, taking into account (3.14) and (3.15) we get (in absolute value)

$$dB = \left[\prod_{i=2}^{n} (b\, db_i)\right]. \tag{3.17}$$

On the other hand, since the vectors b_i and oa_i are on the same hyperplane we may write

$$a_i = \sum_{k=2}^{n} \lambda_i^k b_k, \qquad i = 2, 3, \cdots, n \tag{3.18}$$

and consequently

$$\left[\prod_{i=2}^{n} (b\, da_i)\right] = |\, \lambda_i^k \,| \left[\prod_{k=2}^{n} (b\, db_k)\right] = |\, \lambda_i^k \,|\, dB. \tag{3.19}$$

Since $(ba_i) = 0$ for $i = 2, 3, \cdots, n$, from (3.3) we deduce $(bda_i) = \omega_{i1}(ba_1) = \omega_{i1}\rho_1 \cos\theta$, where ρ_1 is the length oa_1 and θ the angle which forms oa_1 with ob;

that is, if p means the distance from o to the hyperplane S_{n-1}, we have $p = \rho_1 \cos \theta$ and consequently from (3.19) we deduce

$$\left[\prod_{i=2}^{n} \omega_{i1}\right] p^{n-1} = |\lambda_i^k| \, dB. \tag{3.20}$$

Moreover, from (3.3) we deduce $(bda_1) = \omega_{11}(a_1b) = \omega_{11}p$ and since $(a_1b) = p$, we have $(bda_1) = dp - (a_1db)$; if, furthermore, we take into account $[(a_1db)dB] = 0$, we get

$$[\omega_{11}dB]\, p = [dpdB]. \tag{3.21}$$

On the other hand, since the parallelepiped spanned by $oa_i (i = 1, 2, \cdots n)$ has volume 1, it is $|\lambda_i^k|\, p = 1$, and therefore (3.13), (3.20) and (3.21) give

$$dS_{n-1} = p^{-(n+1)}[dpdB]. \tag{3.22}$$

This is the wanted geometrical interpretation for dS_{n-1}. We may summarize the foregoing results in the following

THEOREM 3.2. *The points and the hyperplanes are the only linear subspaces which have an invariant density with respect to the unimodular center-affine group* $\mathfrak{A}$. *The density for points equals the element of volume. The density for hyperplanes is given by* (3.22) *where* p *denotes the distance from the origin to the hyperplane and* dB *denotes the element of area on the unit* n*-dimensional sphere corresponding to the point which gives the direction normal to the hyperplane.*

For instance, if $p = p(B)$ is the support function with respect to the origin o, of a convex body in the n-dimensional space, which contains o, the measure of all hyperplanes exterior to the convex body, invariant with respect to $\mathfrak{A}$, is given by

$$M(S_{n-1}) = (1/n) \int p^{-n}\, dB \tag{3.23}$$

the integral extended over the whole surface of the n-dimensional sphere.

Finally, we want to give a geometrical interpretation for the cinematic density of $\mathfrak{A}$.

According to §1, the cinematic density of $\mathfrak{A}$ is given by

$$d\mathfrak{A} = \left[\prod_{i,k=1}^{n}{}' \omega_{ik}\right] \tag{3.24}$$

where the accent denotes that ω_{nn} is excluded.

From (3.4) and (3.2) we deduce

$$\left[\prod_{k=1}^{n} \omega_{ik}\right] = [da_i^1\, da_i^2 \cdots da_i^n] = dA_i \tag{3.25}$$

where dA_i denotes the element of volume corresponding to the point a_i. According to (3.4) we have

$$\omega_{nk} = \alpha_k^1 da_n^1 + \alpha_k^2 da_n^2 + \cdots + \alpha_k^n da_n^n \tag{3.26}$$

where α_k^i denotes the algebraic complement of a_k^i in the determinant $| a_1 a_2 \cdots a_n | = 1$. Consequently

$$\left[\prod_{k=1}^{n-1} \omega_{nk}\right] = \sum_{j=1}^{n} \bar{\alpha}_n^j [da_n^1 \cdots da_n^{j-1} \, da_n^{j+1} \cdots da_n^n] \tag{3.27}$$

where $\bar{\alpha}_n^j$ denotes the algebraic complement of α_n^j in the adjoint determinant $| \alpha_i^j |$. By a known theorem (see, for instance, Kowalewski [5] p. 80) it is $\bar{\alpha}_n^j = (-1)^{n+j} a_n^j$. Consequently we have

$$\left[\prod_{k=1}^{n-1} \omega_{nk}\right] = \sum_{j=1}^{n} (-1)^{n+j} \, a_n^j [da_n^1 \cdots da_n^{j-1} \, da_n^{j+1} \cdots da_n^n]. \tag{3.28}$$

The right hand side of this expression is equal to n times the volume of the elementary cone, which projects from o the element of volume corresponding to the point a_n. If we represent it by dV_{a_n} from (3.24), (3.25) and (3.28) we get

$$d\mathfrak{A} = n[dA_1 dA_2 \cdots dA_{n-1} dV_{a_n}] \tag{3.29}$$

which is the geometrical interpretation for $d\mathfrak{A}$ we want to obtain.

This cinematic density can also be expressed in another form, which will be useful in the next section. It is based in the following

LEMMA 3.1. *Let $a_1, a_2, \cdots a_{n-1}$ be $n - 1$ points in the n-dimensional space; let dA_i be the element of volume corresponding to a_i and $d\bar{A}_i$ the element of $(n - 1)$-dimensional volume corresponding to a_i in the hyperplane S_{n-1} determined by the points $o, a_1, \cdots, a_{n-1}$. If dB denotes the element of area of the n-dimensional unit sphere corresponding to the point b such that the radius ob is normal to S_{n-1}, and $V(a_1 \cdots a_{n-1})$ represents the volume of the $(n - 1)$-dimensional parallelepiped spanned by the vectors oa_i $(i = 1, 2, \cdots, n - 1)$, then (in absolute value)*

$$[dA_1 dA_2 \cdots dA_{n-1}] = V(a_1 \cdots a_{n-1})[d\bar{A}_1 d\bar{A}_2 \cdots d\bar{A}_{n-1} dB]. \tag{3.30}$$

PROOF. Let ob_i $(i = 1, 2, \cdots, n - 1)$ be $n - 1$ orthogonal unit vectors contained in S_{n-1}, and w_{ik} $(k = 1, 2, \cdots, n - 1)$ be $n - 1$ orthogonal unit vectors orthogonal to a_i $(i = 1, 2, \cdots, n - 1)$ with $ow_{i,n-1} = ob$. If ρ_i is the length of oa_i, formula (3.17) applied to oa_i gives

$$dA_i = \left[\prod_{l=1}^{n-1} (w_{il} \, da_i) \, d\rho_i\right] \rho_i^{n-1} \tag{3.31}$$

and, by the same formula,

$$d\bar{A}_i = \left[\prod_{l=1}^{n-2} (w_{il} \, da_i) \, d\rho_i\right] \rho_i^{n-2}. \tag{3.32}$$

If it is,

$$a_i = \sum_{k=1}^{n-1} \lambda_i^k b_k, \qquad i = 1, 2, \cdots, n - 1 \tag{3.33}$$

will be

$$(3.34) \qquad da_i = \sum_{k=1}^{n-1} d\lambda_i^k b_k + \sum_{k=1}^{n-1} \lambda_i^k \, db_k$$

and

$$(3.35) \qquad (b \, da_i) = (w_{i,n-1} \, da_i) = \sum_{k=1}^{n-1} \lambda_i^k (b \, db_k).$$

Consequently, by exterior multiplication we get

$$(3.36) \qquad \left[\prod_{i=1}^{n-1} (w_{i,n-1} \, da_i)\right] = |\lambda_i^k| \left[\prod_{k=1}^{n-1} (b \, db_k)\right] = |\lambda_i^k| \, dB.$$

From (3.31), (3.32) and (3.36) we deduce

$$(3.37) \qquad [dA_1 \cdots dA_{n-1}] = |\lambda_i^k| \rho_1 \cdots \rho_{n-1} [d\bar{A}_1 \cdots d\bar{A}_{n-1} dB]$$

and since $|\lambda_i^k| \rho_1 \cdots \rho_{n-1} = V(a_1 \cdots a_{n-1})$, formula (3.30) is proved.

Let us now observe that if $dV_{a_{n-1}}$, analogously as in (3.29), means the volume of the elementary cone which projects from o the element of volume dA_{n-1} and $d\bar{V}_{a_{n-1}}$ has the analogue meaning in the subspace S_{n-1}, under the assumption $n > 2$, it is

$$dA_{n-1} = (n/\rho_{n-1})[dV_{a_{n-1}} d\rho_{n-1}], \qquad d\bar{A}_{n-1} = (n - 1/\rho_{n-1})[d\bar{V}_{a_{n-1}} d\rho_{n-1}]$$

and therefore, (3.30) may be written

$$(3.38) \; n[dA_1 \cdots dA_{n-2} dV_{a_{n-1}}] = (n-1) V(a_1 \cdots a_{n-1}) [d\bar{A}_1 \cdots d\bar{A}_{n-2} d\bar{V}_{a_{n-1}} dB].$$

By symmetry, the cinematic density (3.29) may also be written $d\mathfrak{A} = n \, [dA_1 \cdots dA_{n-2} dV_{a_{n-1}} dA_n]$ and therefore, from (3.38) we deduce

$$(3.39) \qquad d\mathfrak{A} = (n-1) V(a_1 \cdots a_{n-1}) [d\bar{A}_1 \cdots d\bar{A}_{n-2} d\bar{V}_{a_{n-1}} dB dA_n].$$

In order to introduce the cinematic density $d\mathfrak{A}_{n-1}$ of the unimodular center-affine group in the subspace S_{n-1}, it is enough to observe that by a change of variables $a_i^k = \rho a_i^{k*}$, $\rho = (V(a_1 \cdots a_{n-1}))^{1/n-1}$, it is

$$(3.40) \qquad \begin{aligned} &(n-1)[d\bar{A}_1 \cdots d\bar{A}_{n-2} d\bar{V}_{a_{n-1}}] \\ &\qquad = (V(a_1 \cdots a_{n-1}))^{n-1} (n-1) [d\bar{A}_1^* \cdots d\bar{A}_{n-2}^* d\bar{V}_{a_{n-1}}^*]. \end{aligned}$$

In order to set in evidence the dimension of the space set now $d\mathfrak{A}_n$ instead of $d\mathfrak{A}$ and from (3.39) and (3.40) follows

$$(3.41) \qquad d\mathfrak{A}_n = (V(a_1 \cdots a_{n-1}))^n \, [dA_n dB d\mathfrak{A}_{n-1}].$$

If h is the distance from a_n to S_{n-1} it is $V(a_1 \cdots a_{n-1}) h = 1$, and $d\mathfrak{A}_n$ takes the form

$$(3.42) \qquad d\mathfrak{A}_n = h^{-n} [dA_n dB \, d\mathfrak{A}_{n-1}].$$

(3.41) and (3.42) are two recurrent geometrical interpretations of $d\mathfrak{A}_n$ (for $n > 2$) which will be useful in the next section.

For the excluded case $n = 2$, (3.29) becomes in our actual notation

$$d\mathfrak{A}_2 = 2\rho_1^2[dA_2 dB]. \tag{3.43}$$

4. A Theorem of Minkowski-Hlawka

In order to make an application of the results of the foregoing section we want to give a proof of the following theorem ammounced by Minkowski and first proved by Hlawka [4]:

If Q is an n-dimensional star domain of volume $< \zeta(n)$, then there exists a lattice of determinant 1 *such that Q does not contain any lattice point* $\neq 0$.

This theorem was also proved by C. L. Siegel [6] and A. Weil [8]. The following proof is based on the same idea as that of these authors. It is, however, more simple and exposed from a more elementary point of view.

Let us consider in the n-dimensional space S_n the lattice L_0 of points of entire coordinates, and the set of all lattices L transformed from L_0 by the group $\mathfrak{A}_n$, that is, the set of all lattices of modulo 1. In this case the subgroup Γ of $\mathfrak{A}_n$, which leaves invariant L_0, is discrete. In the space of parameters, Γ will be represented by a set of infinite isolated points. Consequently the invariant density for sets of lattices L will be the same (3.29) or (3.41) which now we will write

$$dL = (V(a_1 \cdots a_{n-1}))^n[dA_n dB d\mathfrak{A}_{n-1}] \tag{4.1}$$

under the assumption that L is determined by o and the points $a_1, \cdots a_n$.

In order to have a one-to-one correspondence between lattices L and points of $\mathfrak{A}_n$, we must consider not only the whole space $\mathfrak{A}_n$ but the space $\mathfrak{A}_n/\Gamma$. Consequently, in what follows, whenever dL appears under the integral sign, it must be understood that the integral is taken over $\mathfrak{A}_n/\Gamma$.

Let D be a given fixed domain of volume v in S_n and consider the integral

$$I = \int_{a_n \epsilon D} dL. \tag{4.2}$$

To evaluate this integral we first keep a_n fixed. For each given $\mathfrak{A}_{n-1}$, in order to obtain different lattices L, the points a_i $(i = 1, 2, \cdots, n-1)$ can only vary in the intervals $a_i + \lambda a_n$ $(0 \leqq \lambda \leqq 1)$. Setting shortly $V(a_1 \cdots a_{n-1}) = V$ and considering the points Vb, we observe that they are contained in the hyperplane inverse of the hypersphere of diameter oa_n in the inversion of center o and power of inversion 1, because if h is the distance from a_n to the hyperplane determined by o and $a_1, a_2, \cdots, a_{n-1}$ it is always $Vh = 1$. Here, b has the same meaning as in the foregoing section.

Furthermore, when the points a_i describe the intervals $a_i + \lambda a_n (0 \leqq \lambda \leqq 1)$, independently of each other the point Vb describes the $(n-1)$-dimensional parallelepiped spanned by the vectors $\alpha_i - \alpha_0$ $(i = 1, 2, \cdots, n-1)$ where

$$\alpha_0 = \{a_1 \cdots a_{n-1}\} \text{ and}$$
$$\alpha_i = \{a_1 \cdots a_{i-1}a_i + a_n a_{i+1} \cdots a_{n-1}\}$$
$$= \{a_1 \cdots a_{n-1}\} + (-1)^{n-i}\{a_1 \cdots a_{i-1}a_{i+1} \cdots a_n\}.$$

The notation $\{a_1 a_2 \cdots a_{n-1}\}$ means the vector whose components are the determinants of order $n - 1$ in the rectangular matrix formed by the coordinates of $a_1, a_2, \cdots, a_{n-1}$.

The volume of the pyramid of vertex o and basis this parallelepiped has the value $(1/n) \mid \alpha_0 \cdots \alpha_{n-1} \mid = 1/n$, according to a well known property about adjoint determinants ([5] p. 80).

On the other hand, the volume of the last mentioned pyramid is given by $(1/n) \int V^n dB$. Consequently we get that the integral of $V^n dB$ in (4.2) has value 1.

If, only for a moment, we assume that the total measure of the unimodular center-affine transformations of the space S_{n-1} has a finite value V_{n-1}, from (4.1) we get $I = V_{n-1} \int_{a_n \in D} dA_n$. If now a_n describes D, we get the value $V_{n-1}v$. In this way each lattice L has been counted as many times as lattice points of coordinates primes among themselves (we shall say primitive lattice points) are contained in D. In fact, when a_n coincides with any one of these points it originates the same lattice. Consequently if we represent by N the number of primitive lattice points of L contained in D, we have the integral formula

$$\int N \, dL = v V_{n-1} \tag{4.3}$$

where the integration is extended over the whole space $\mathfrak{A}_n/\Gamma$.

In order to introduce in (4.3) instead of N, the total number $\bar{N}$ of lattice points contained in D for each lattice L, we follow a very useful device due to Siegel [6]. Let us consider the domain $i^{-1}D$ (of volume $i^{-n}v$), homothetic of D with respect to o and ratio i^{-1} (i integer). To every lattice point contained in D, whose coordinates have the greatest common diviser i (the number of which will be represented by N_i), corresponds a primitive lattice point in $i^{-1}D$. Therefore the same formula (4.3) applied to $i^{-1}D$, gives

$$\int N_i \, dL = v V_{n-1} i^{-n}, \qquad i = 2, 3, \cdots \tag{4.4}$$

Adding (4.3) and (4.4) for $i = 2, 3, \cdots$ we get

$$\int \bar{N} \, dL = v V_{n-1} \zeta(n). \tag{4.5}$$

It remains to evaluate V_{n-1}. For this purpose we may again follow the method of Siegel [6]. Let us consider the lattice of points of coordinates multiple of

m^{-1} (m integer); if $\bar{N}_m$ is the number of lattice points of this new lattice contained in D, (4.5) gives

$$\int (\bar{N}_m/m^n)\, dL = vV_{n-1}\zeta(n). \tag{4.6}$$

When $m \to \infty$, $\bar{N}_m/m^n$ tends to the volume v of D and therefore the total measure V_n of the lattices L will be

$$V_n = V_{n-1}\zeta(n). \tag{4.7}$$

Since formula (3.41) holds only for $n > 2$, (4.7) holds for $n = 3, 4, \cdots$. For the case $n = 2$, starting from (3.43) the same foregoing calculation gives $\int NdL = v$ instead of (4.3), $\int \bar{N}dL = v\zeta(2)$ instead of (4.5) and $V_2 = \zeta(2)$ instead of (4.7). Consequently (4.7) makes sure that every V_n is finite and gives, moreover, the known result $V_n = \zeta(2)\zeta(3) \cdots \zeta(n)$.

From (4.3), (4.5), and (4.7) we obtain the following mean values. If we consider all lattices L of modulo 1:

a) *The mean value of the number of primitive lattice points contained in a given domain D of volume v is $v/\zeta(n)$.*

b) *The mean value of the number of lattice points contained in D is equal to v.*

The announced theorem of Minkowski-Hlawka is an immediate consequence of a). In fact, if $v < \zeta(n)$, the mean value of primitive lattice points is < 1 and therefore there exist lattices without primitive lattice points in D. If D is a star domain, i.e., a point set which is measurable in the Jordan sense and which contains with any point x the whole segment λx, $0 \leqq \lambda \leqq 1$, and does not contain any primitive lattice point, it does not contain any lattice point and the theorem is proved.

5. The Unimodular Affine Group

The group $\mathfrak{K}$ of unimodular affine transformations, given by

$$(x^i)' = \sum_{j=1}^{n} a_j^i x^j + a_0^i, \qquad i = 1, 2, \cdots, n \tag{5.1}$$

with

$$| a_1 a_2 \cdots a_n | = 1 \tag{5.2}$$

may be studied in exactly the same way as in §2 and §3. Setting

$$da_i = \sum_{j=1}^{n} \omega_{ij} a_j, \qquad i = 0, 1, \cdots, n \tag{5.3}$$

it is found

$$\omega_{ij} = | a_1 \cdots a_{j-1}\, da_i a_{j+1} \cdots a_n |, \qquad \sum_{i=1}^{n} \omega_{ii} = 0 \tag{5.4}$$

with the equations of structure

$$\omega'_{ij} = \sum_{l=1}^{n} [\omega_{il}\omega_{lj}]. \tag{5.5}$$

Similarly, as in §2 and §3, we get in this case the following

THEOREM 5.1. *Except the points ($h = 0$), the linear subspaces have no invariant density with respect to the unimodular affine group. The sets ($S_{h_1} + S_{h_2} + \cdots + S_{h_m}$), with $h_1 + h_2 + \cdots + h_m + m \leqq n + 1$, of linear subspaces, admits a density only when either all h_i are equal to zero, or the condition*

$$h_1 + h_2 + \cdots + h_m + m - 1 = n$$

holds.

For instance, the sets of (hyperplane + point) admits a density, which is easily found to be equal to $p^{-(n+1)} dA_0\, dBdp$, where dA_0 is the element of volume corresponding to the point, dB is the element of area on the $(n - 1)$-dimensional unit sphere corresponding to the direction normal to the hyperplane, and p is the distance from the point to the hyperplane.

Finally, if the unimodular affine transformations are assumed determined by the point a_0 and n vectors $a_i - a_0$ ($i = 1, 2, \cdots, n$), the cinematic density is given by

$$d\mathfrak{K} = [dA_0 d\mathfrak{A}] \tag{5.6}$$

where dA_0 is the element of volume corresponding to a_0 and $d\mathfrak{A}$ is the cinematic density (3.29), (3.41) for the unimodular center-affine group with a_0 as fixed point.

THE INSTITUTE FOR ADVANCED STUDY.

BIBLIOGRAPHY

[1] E. CARTAN, La théorie des groupes finis et continus et la géométrie différentielle, traitées par la méthode du repère mobile. Paris, Gauthier-Villars, 1937.

[2] E. CARTAN, Les systèmes différentiels extérieurs et leurs applications géométriques. Actualités scientifiques et industrielles n° 994, Hermann, Paris, 1945.

[3] S. CHERN, *On integral geometry in Klein spaces*. Ann. of Math. 43, 178–189 (1942).

[4] E. HLAWKA, *Zur Geometrie der Zahlen*, Math. Zeit. 49, 285–312 (1944).

[5] G. KOWALEWSKI, Einführung in die Determinantentheorie, Leipzig, Veit, 1909.

[6] C. I. SIEGEL, *A mean value theorem in geometry of numbers*, Ann. of Math. 46, 340–347 (1945).

[7] O. VARGA, *Ueber Masse von Paaren linearer Mannigfaltigkeiten im Projektiven Raum P_n*, Revista Mathemática Hispano-Americana, 10, 241–264 (1935).

[8] A. WEIL, *Sur quelques résultats de Siegel*, Summa Brasiliensis Mathematicae, 1, 21–39 (1946).

[illegible] squares.

[illegible]

With the equation [illegible]

[illegible]

[illegible] as follows:

[illegible] the condition

[illegible]

[illegible] the [illegible] of area of the [illegible] and [illegible]

[illegible]

[illegible]

References

[1] [illegible] Paris [illegible]

[2] [illegible]

[3] [illegible]

[4] [illegible]

[5] [illegible]

[6] [illegible]

[7] [illegible]

[8] [illegible]

Steiner formula for parallel surfaces in affine geometry

L. A. Santaló

Abstract

The main purpose of this note is to prove that with a convenient definition of parallel surfaces, the Steiner formula giving the volume bounded by a closed surface S^* parallel to a given orientable and closed surface S may be generalized to the affine unimodular geometry, and takes the same form (4.5) as in the metrical case (Ω, M, C being now the affine area, integrated affine mean curvature and integrated affine total curvature, respectively). From this result certain inequalities due to Blaschke, (Section 5), follow by direct application of the Brunn-Minkowski theorem on mixed volumes.

In Section 6 we consider the family of affine invariants J_n (6.4) and certain inequalities (6.8), (6.9), (6.11) between the affine area, volume and maximal affine width of a convex body.

In Sections 7 and 8 analogous questions for the case of the plane are considered.

Introduction

The main aim of this paper is to generalize the Steiner formula to affine unimodular geometry. In the metric case, this formula gives the volume bounded by a surface parallel to another closed surface S in terms of the volume V bounded by S, the area F, the total mean curvature M and the total curvature C of this latter surface; see, for example Hadwiger [[10], p.31], Blaschke [[1], p.107], and Bonnesn-Fenchel [[6], p.49].

With a convincing definition of parallel surfaces (Section 4) and employing integral formulae (Section 3), due in part to Grotemeyer [9], the result takes the same form as in the metric case [formula (4.5)]. The interest of this result resides in the fact that for the case of convex bodies it is subsequently sufficient to apply the Brunn-Minkowski general theorem on pencils of convex bodies in order to arrive at the inequalities of Section 5 among the affine invariants of a convex body; these inequalities are due to Blaschke [2], although here they appear naturally.

We likewise make use of the formulas from Section 3, which indeed are interesting in themselves, in order to obtain some affine invariants of convex bodies (Section 6) and certain of their bounds, which lead to inequalities (6.8), (6.10), (6.11) in which the "maximum affine width" of the convex body appears. These inequalities generalize to the affine case other known inequalities belonging to the metric case. Several problems remain to be solved in this regard, as we point out at the conclusion to Section 6.

In Sections 7 and 8 we solve the same problem for the case of plane curves.

Throughout this paper we refer to affine unimodular geometry; that is, geometry corresponding to the group of unimodular affinities. The works by Blaschke [3], Salkowsky [11] and Favard [8] constitute the main references.

1 Fundamental formulae of the affine unimodular geometry of surfaces

We use E. Cartan's mobile trihedral method in a way analogous to that employed by Favard [[8], Section 11, Chapter 1].

However, for our purpose it is preferable to employ a fundamental trihedron different to that used by Favard. In fact, we employ the trihedron formed by the normal affine vector $\mathbf{I_3}$ and two vectors $\mathbf{I_1}$, $\mathbf{I_2}$ tangent to the lines of affine curvature corresponding to each point $\mathbf{X}$ of the surface The three fundamental vectors are constrained by the relation

$$(\mathbf{I_1}, \mathbf{I_2}, \mathbf{I_3}) = 1, \tag{1.1}$$

where the brackets indicate the triple vector product.

In this case, by changing the fundamental trihedron in Favard's formulae [[8], pages 398 and 399], or directly as we have done elsewhere [12], for a surface S of vector equation $\mathbf{X} = \mathbf{X}(u, v)$, of class $\mathcal{C}^3$, we obtain the following Frenet formulae:

$$\begin{aligned} d\mathbf{X} &= \omega^1\mathbf{I}_1 + \omega^2\mathbf{I}_2 \\ d\mathbf{I}_1 &= \omega_1^1\mathbf{I}_1 + \omega_1^2\mathbf{I}_2 + \omega^1\mathbf{I}_3 \\ d\mathbf{I}_2 &= \omega_2^1\mathbf{I}_1 + \omega_2^2\mathbf{I}_2 + \omega^2\mathbf{I}_3 \\ d\mathbf{I}_3 &= -k_1\omega^1\mathbf{I}_1 - k_2\omega^2\mathbf{I}_2, \end{aligned} \tag{1.2}$$

where k_1 and k_2 are the principle affine curvatures and ω^i and ω_1^h are invariant Pfaff forms[1]. Since ω^1 and ω^2 are independent, the remaining may be written in the form

$$\begin{gathered} \omega_1^2 = \alpha\omega^1 + \beta\omega^2, \qquad \omega_2^1 = \gamma\omega^1 + \delta\omega^2 \\ \omega_1^1 = -\omega_2^2 = p\omega^1 + q\omega^2, \end{gathered} \tag{1.3}$$

where

$$p = -\frac{1}{2}(\beta + \delta), \qquad q = \frac{1}{2}(\alpha + \gamma). \tag{1.4}$$

The invariants p and q are related to the Pick J invariant (Blaschke [[3], p.123]) and to the invariant k, which Favard calls the "affine curvature" ([8, p.397]), by

$$p^2 + q^2 = \frac{1}{2}J = \frac{1}{4}k^2. \tag{1.5}$$

[1] As is usual, here we take k_1, k_2 with a sign different from that used in the previous work [12].

By writing that the exterior differentials of the first members of (1.2) vanish, and applying the rules of exterior differentiation, we obtain the *integrability conditions*

$$d\omega^i = \omega^k \wedge \omega_h^i, \quad d\omega_i^h = \omega_i^m \wedge \omega_m^h, \tag{1.6}$$

where the sum over crossed repeated indices, from 1 to 3, is understood, and according to (1.2) is $\omega^3 = 0, \omega_1^3 = \omega^1, \omega_2^3 = \omega^2, \omega_3^1 = -k_1\omega^1, \omega_3^2 = -k_2\omega^2, \omega_3^3 = 0$.

If for any function f, the covariant partial derivatives f_1 and f_2 are defined by the relation

$$df = f_1\omega^1 + f_2\omega^2, \tag{1.7}$$

and taking into account (1.3) and (1.4), the conditions (1.6) can be written as follows

$$\begin{aligned} 2(q_1 - p_2) &= \alpha_1 + \gamma_1 + \beta_2 + \gamma_2 = 3(\alpha\delta - \beta\gamma) \\ 2(\beta_1 - \alpha_2) &= -3(\alpha^2 + \beta^2) - (\alpha\gamma + \beta\delta) - 2k_2 \\ 2(\delta_1 - \gamma_2) &= -3(\delta^2 + \gamma^2) - (\beta\delta + \alpha\gamma) + 2k_1 \\ k_{12} &= -\gamma(k_1 - k_2) \\ k_{21} &= \beta(k_1 - k_2). \end{aligned} \tag{1.8}$$

The covariant derivatives being defined by (1.7), the first Beltrami differential parameter (as regards the metric $ds^2 = (\omega^1)^2 + (\omega^2)^2$), is written thus

$$\nabla f = f_1^2 + f_2^2, \tag{1.9}$$

and the second Beltrami differential parameter or Laplacian of f (always with respect to the metric $ds^2 = (\omega^1)^2 + (\omega^2)^2$) can be written (see Blaschke [[4], p.70)]

$$\nabla f = \frac{d(f_1\omega^2 - f_2\omega^1)}{\omega^1 \wedge \omega^2}, \tag{1.10}$$

where the symbol in the numerator denotes the exterior differentiation.

2 The fundamental invariants

As usual, we write

$$\begin{aligned} d\Omega &= \omega^1 \wedge \omega^2 = \text{ the affine area element} \\ H &= \frac{1}{2}(k_1 + k_2) = \text{ the affine mean curvature} \\ K &= k_1 k_2 = \text{ the total affine curvature.} \end{aligned} \tag{2.1}$$

It is also necessary to define the "affine distance" of a point $\mathbf{Z}$ to the point $\mathbf{X}$ of the surface S, which according to Blaschke [[3], p.110] is defined by the mixed product as

$$\omega = (\mathbf{Z} - \mathbf{X}, \mathbf{I}_1, \mathbf{I}_2), \tag{2.2}$$

or if $\mathbf{Z}$ is the origin of the coordinates

$$\omega = (-\mathbf{X}, \mathbf{I}_1, \mathbf{I}_2). \tag{2.3}$$

The volume element, which, since we are dealing with unimodular affinities is also affine invariant, consists of the elementary pyramid having its vertex at the origin and whose base is the parallelogram determined by the vectors $\omega^1\mathbf{I}_1, \omega^2\mathbf{I}_2$, and is

$$dV = -\frac{1}{3}(\mathbf{X}, \mathbf{I}_1, \mathbf{I}_2)\,\omega^1 \wedge \omega^2 = \frac{1}{3}\,\omega d\,\Omega, \tag{2.4}$$

and therefore, if S is a closed and orientable surface, the volume of the body bounded by it is

$$V = \frac{1}{3}\int_S \omega d\,\Omega. \tag{2.5}$$

We may observe that, according to definition (1.7), the covariant partial derivatives of ω are

$$\omega_1 = -(\mathbf{X}, \mathbf{I}_3, \mathbf{I}_2); \quad \omega_2 = -(\mathbf{X}, \mathbf{I}_1, \mathbf{I}_3), \tag{2.6}$$

and therefore, according to (1.10) and taking into account (1.3) and (1.4), the Laplacian is

$$\Delta\omega = 2(1 - H\omega). \tag{2.7}$$

3 Integral Formulae

Let C be a closed curve on the surface S limiting a simply connected domain D. For any Pfaff formula θ, the following Stokes formula holds

$$\int_C \theta = \int_D d\theta. \tag{3.1}$$

Let us now apply this formula to two particular cases:

a) $\theta = \omega^n(\mathbf{X}, \mathbf{I}_3, d\mathbf{X})$, where n is any integer.

Then, taking into account (2.6) and the notation (1.9), by an easy calculation we arrive at

$$\begin{aligned}\int_C \omega^n(\mathbf{X}, \mathbf{I}_3, d\mathbf{X}) \;&=\; 2\int_D \omega^n(-1 + H\omega)\,d\Omega - \\ &\int_D n\omega^{n-1}\nabla\omega d\,\Omega.\end{aligned} \tag{3.2}$$

If S is a closed and orientable surface, then by splitting it into cells and applying to each cell the previous formula and adding, we arrive at

$$2\int_S \omega^n(-1 + H\omega)\,d\Omega - \int_S n\omega^{n-1}\nabla\omega d\,\Omega = 0, \tag{3.3}$$

or, by taking into account (2.7)

$$\int_S (\omega\Delta\omega + n\nabla\omega)\omega^{n-1}d\Omega = 0, \tag{3.4}$$

For $n = 0$, identity (3.3) yields

$$\Omega = \int_S H\omega d\Omega, \tag{3.5}$$

a formula we owe to Grotemeyer [9] , which generalizes to the affine case the analogous formula of the metric case, in which ω is the supporting function.

For $n = 1$, and taking into account (2.5) and (3.4), we have

$$\begin{aligned} V &= \frac{1}{3}\int_S (H\omega^2 - \frac{1}{2}\nabla\omega)d\,\Omega \\ &= \frac{1}{3}\int_S (H\omega + \frac{1}{2}\Delta\omega)\,\omega\,d\,\Omega. \end{aligned} \tag{3.6}$$

b) $\theta = \omega^n(\mathbf{X}, \mathbf{I}_3, d\mathbf{I}_3)$ In this case, after an easy calculation, the Stokes formula gives

$$\int_C \omega^n(\mathbf{X}, \mathbf{I}_3, d\mathbf{I}_3) = \tag{3.7}$$

$$-2\int_D \omega^n(H - K\omega)d\Omega + \int_D n\omega^{n-1}(k_2\omega_1^2 + k_1\omega_2^2)d\Omega,$$

or if S is a closed and orientable surface,

$$-2\int_S \omega^n(H - K\omega)d\Omega + \int_S n\omega^{n-1}(k_2\omega_1^2 + k_1\omega_2^2)d\Omega = 0. \tag{3.8}$$

For $n = 0$, we have

$$M = \int_S Hd\Omega = \int_S K\omega d\Omega, \tag{3.9}$$

a formula also due to Grotemeyer [9], which generalizes to the affine case a classical result by Minkowski for the metric geometry of convex bodies. M is the so-called total affine mean curvature.

4 Parallel Surfaces. The Steiner Formula

Definition: Given the surface $S : \mathbf{X} = \mathbf{X}(u, v)$, we call any surface S^* of the form

$$\mathbf{X}^* = \mathbf{X} - \lambda\mathbf{I}_3 \tag{4.1}$$

an affine surface parallel to S, with the condition that the tangent planes at the corresponding points are parallel.

It is necessary to point out that this definition is not reciprocal; that is, if S^* is parallel to S, S is not necessarily parallel to S^* [2]. From (4.1) we deduce that

$$\begin{aligned} d\mathbf{X}^* &= d\mathbf{X} - d\lambda\mathbf{I}_3 - \lambda d\mathbf{I}_3 \\ &= (1+\lambda k_1)\omega^1\mathbf{I}_1 + (1+\lambda k_2)\omega^2\mathbf{I}_2 - d\lambda\mathbf{I}_3. \end{aligned}$$

In order for the tangent planes to be parallel it must be $d\lambda = 0$; that is, $\lambda =$ constant. In this case we have

$$d\mathbf{X}^* = (\omega^1)^*\mathbf{I}_1 + (\omega^2)^*\mathbf{I}_2, \tag{4.2}$$

with

$$(\omega^1)^* = (1+\lambda k_1)\omega^1; \quad (\omega^2)^* = (1+\lambda k_2)\omega^2. \tag{4.3}$$

If S is a closed and orientable surface, the volume bounded by S^*, taking the origin within S, is according to (2.4)

$$\begin{aligned} V^* &= \frac{1}{3}\int_S (\mathbf{X}^*, d_1\mathbf{X}^*, d_2\mathbf{X}^*) \\ &= \frac{1}{3}\int_S (\mathbf{X}^*, (1+\lambda k_1)\omega^1\mathbf{I}_1, (1+\lambda k_2)\omega^2\mathbf{I}_2) \\ &= \frac{1}{3}\int_S (\mathbf{X} - \lambda\mathbf{I}_3, \mathbf{I}_1, \mathbf{I}_2)(1+\lambda k_1)(1+\lambda k_2)d\Omega \\ &= \frac{1}{3}\int_S (\omega+\lambda)(1+2\lambda H + \lambda^2 K)d\Omega \\ &= \frac{1}{3}\int_S \omega d\Omega + \frac{\lambda}{3}\int_S (1+2H\omega)d\Omega \\ &+ \frac{\lambda^2}{3}\int_S (K\omega + 2H)d\Omega + \frac{\lambda^3}{3}\int_S K d\Omega. \end{aligned}$$

Taking into account (2.4), (3.5) and (3.9), and writing

$$C = \int_S K d\Omega, \tag{4.4}$$

we arrive at

$$V^* = V + \lambda\Omega + \lambda^2 M + \frac{\lambda^3}{3}C, \tag{4.5}$$

which is the generalized Steiner formula we wished to obtain.

5 Inequalities

Let us assume that S is a closed and convex surface (always of class $\mathcal{C}^3$). In this case, it is proved (Blaschke [[3], p.161]) that by taking the vectors $\mathbf{I}_3$ from

[2] It is not difficult to see that reciprocity occurs only in the case in which S is an affine sphere.

a point of origin, the endpoints of these vectors also form a convex surface ("Krümmingsbild" – curvature indicatrix). Therefore, if S^* represents the surface $\mathbf{X}-\mathbf{I}_3$ parallel to S corresponding to the value $\lambda = 1$, the surfaces $(1-\lambda)S+\lambda S^*$ are the boundaries of a linear pencil of convex bodies. Thus, according to the Brunn-Minkowski theorem, the cubic root of the volume $V(\lambda)$ given by (4.5) is a concave function of λ, which is linear if and only if S and S^* are homothetic (see, for example, Blaschke [1], Busemann [[7], p.48] and Hadwiger [10]).

Therefore, analogously to the metric case, the following cubic inequalities are obtained:

$$\Omega^3 \geq 9V^2C, \quad M^3 \geq 3VC^2, \tag{5.1}$$

and the quadratic inequalities

$$\Omega^2 \geq 3VM, \quad M^2 \geq \Omega C, \tag{5.2}$$

which are not independent, since (5.1) are deduced from (5.2). In (5.1) the signs of equality only hold if S and its directrix of curvature are homothetic surfaces. In this case, by selecting the same origin for both surfaces, $\mathbf{X} = \alpha\mathbf{I}_3$ with $\alpha =$ constant. Therefore, by differentiating and taking into account that $\mathbf{I}_1, \mathbf{I}_2$ are independent, we have $1+\alpha k_1 = 0$, $1+\alpha k_2 = 0$; that is, k_1 and k_2 are constants, and consequently, if S is convex, it must be an ellipsoid.

For the quadratic inequalities (5.2), one may check easily and directly that the equality holds for the ellipsoid. Nevertheless, one cannot deduce from the Brunn-Minkowski theorem that this surface is the only one for which the signs of equality in (5.2) hold. By a slightly different approach, Blaschke proved that this is indeed so [4], and consequently there appears a difference as regards the metric case, in which the equality in the first bound of (5.2) also holds for those bodies referred to by Minkwoski as "Kappenkörper" of the sphere.

6 Affine width of a convex body. Other affine invariants

Let $\mathbf{X}$ be a point on the convex body bounded by a convex surface S, and let $\mathbf{Y}$ be the point on S at which the tangent plane is parallel to the tangent plane in $\mathbf{X}$. According to Süss [14], the *affine distance* from $\mathbf{X}$ to $\mathbf{Y}$ is known as the *affine width* of S corresponding to the point $\mathbf{X}$; that is

$$\begin{aligned}\Delta_a(\mathbf{X}) &= (\mathbf{Y}-\mathbf{X}, \mathbf{I}_1, \mathbf{I}_2)\\ \frac{(\mathbf{Y}-\mathbf{X}, d_1\mathbf{X}, d_2\mathbf{X})}{\omega^1 \wedge \omega^2} &= \frac{(\mathbf{Y}-\mathbf{X}).\mathbf{N}}{d\Omega}d\sigma,\end{aligned} \tag{6.1}$$

where $\mathbf{N}$ is the metric normal versor of S in $\mathbf{X}$ and $d\sigma$ is the metric area element at the same point. Since $(\mathbf{Y}-\mathbf{X}).\mathbf{N}$ is precisely the metric width Δ_m of S corresponding to the normal direction $\mathbf{N}$, we have

$$\Delta_a d\Omega = \Delta_m d\sigma, \tag{6.2}$$

or, taking into account that $d\Omega = K_m^{1/4} d\sigma$, where K_m is the metric Gauss curvature of S in $\mathbf{X}$, we obtain

$$\Delta_a K_m^{1/4} = \Delta_m. \tag{6.3}$$

Relation (6.2) enables us to provide a metric interpretation of the affine invariants

$$J_n = \int_S \Delta_a^n d\Omega; \tag{6.4}$$

that is,

$$J_n = \int_S \Delta_m^n K_m^{\frac{1-n}{4}} d\sigma = \int_S \Delta_m^n K_m^{-\frac{n+3}{4}} d\, o, \tag{6.4 *}$$

where $d\, o$ is the area element on the unit sphere E.

The most simple of these invariants are

$$J_1 = \int_S \Delta_a d\Omega = \int_S \Delta_m d\sigma \tag{6.5}$$

$$J_{-3} = \int_S \frac{d\Omega}{\Delta_a^3} = \int_S \frac{d\, o}{\Delta_m^3}, \tag{6.6}$$

in whose interpretation the curvature K_m does not appear, and thus they are defined for any convex body, regardless of the regularity conditions of their surface.

If $p(\sigma)$ indicates the supporting function corresponding to the element σ and $p^*(\sigma)$ that of its opposite point, we have

$$J_1 = \int_S |p(\sigma) + p^*(\sigma)| d\sigma = 3V + 3V^*,$$

where V^* is the mixed volume of the convex body and of its symmetric body with respect to a point. Taking into account Minkowski's inequality $V^* \geq V$, we obtain

$$J_1 \geq 6V, \tag{6.7}$$

where the equality holds only for bodies with centre of symmetry.

Therefore, if $\overline{\Delta_a}$ represents the maximum affine width

$$\overline{\Delta_a}\Omega \geq 6V, \tag{6.8}$$

where the equality holds only if $\Delta_a =$ cte; that is, according to Süss [14], for the ellipsoid.

As for (6.6), we have proved elsewhere, [13], that

$$\frac{2}{3} < J_{-3}V \leq \frac{2}{3}\pi^2, \tag{6.9}$$

from where

$$2\pi^2\overline{\triangle_a^3} \geq 3V\Omega, \tag{6.10}$$

or, combining with (6.8),

$$\pi^2\overline{\triangle_a^4} \geq 9V^2\Omega, \tag{6.11}$$

where as before the equality holds only for the ellipsoid.

These inequalities (6.9), (6.10) and (6.11) generalize to the affine case the known results of the metric geometry of convex bodies, in whose case the maximum width coincides with the diameter of the body. Thus in the metric case, (6.11) corresponds to Bieberbach's inequality, which relates the diameter with the volume (Bonnesen-Fenchel [6], p.76).

Unsolved problems. It would be interesting to find in (6.7) an upper boundary for J_1 and in (6.9) a lower boundary, better than that indicated here, for the product $J_{-3}V$. In this event one might obtain inequalities between Ω, V, and the "minimum affine width". It also remains to perform the generalization to the affine case of the maximums and minimums that appear at the same time in the various affine invariants of a convex body, in a way similar to that referred to in Bonnesen-Fenchel ([6], n° 45, p.80) for the metric case.

Finally, we point out that it would also be interesting to determine if there exist better inequalities than those in (5.1) and (5.2), similar to those found by Bonnesen for the metric case [5].

7 The case of the plane

In the case of plane curves, Frenet's formulae in unimodular affine geometry for a curve $\Gamma : \mathbf{X} = \mathbf{X}(u)$ of class $\mathcal{C}^2$ are

$$d\mathbf{X} = \omega\mathbf{I}_1, \quad d\mathbf{I}_1 = \omega\mathbf{I}_2, \quad d\mathbf{I}_2 = -k\omega\mathbf{I}_1, \tag{7.1}$$

where $\mathbf{I}_1$ is the tangent vector, $\mathbf{I}_2$ the affine normal vector, ω the affine arc element and k the affine curvature (Favard [8], p.382). Furthermore, the following relation is also fulfilled

$$\mathbf{I}_1 \wedge \mathbf{I}_2 = 1. \tag{7.2}$$

By definition, an affine parallel curve Γ^* will take the form

$$\mathbf{X}^* = \mathbf{X} - \lambda\mathbf{I}_2, \tag{7.3}$$

providing that the tangents at homologous points are parallel. We have

$$d\mathbf{X}^* = (1 + \lambda k)\omega\mathbf{I}_1 - d\lambda\mathbf{I}_2,$$

and therefore, in order for the tangents to be parallel, it is necessary that $\lambda =$ constant. In this case we obtain

$$d\mathbf{X}^* = (1 + \lambda k)d\mathbf{X}.$$

If Γ is a closed curve, the area constrained by Γ^* is expressed by

$$(7.4)\qquad \begin{aligned} F^* &= \frac{1}{2}\int_{\Gamma^*} \mathbf{X}^* \wedge d\mathbf{X}^* = \frac{1}{2}\int_{\Gamma} (\mathbf{X} - \lambda \mathbf{I}_2) \wedge (1 + \lambda k) d\mathbf{X} \\ &= \frac{1}{2}\int_{S} \mathbf{X} \wedge d\mathbf{X} + \frac{1}{2}\lambda \int_{\Gamma} (k(\mathbf{X} \wedge d\mathbf{X}) - \mathbf{I}_2 \wedge d\mathbf{X}) - \frac{1}{2}\lambda^2 \int_{\Gamma} k\mathbf{I}_2 \wedge d\mathbf{X}. \end{aligned}$$

We may observe that by applying (7.1) we arrive at

$$(7.5)\qquad d(\mathbf{I}_2 \wedge \mathbf{X}) = d\mathbf{I}_2 \wedge \mathbf{X} + \mathbf{I}_2 \wedge d\mathbf{X} = -kd\mathbf{X} \wedge \mathbf{X} + \mathbf{I}_2 \wedge d\mathbf{X},$$

and since we are dealing with an exact differential, the integral of this expression along a closed curve vanishes, thus

$$(7.6)\qquad \int_{\Gamma} k\mathbf{X} \wedge d\mathbf{X} = -\int_{\Gamma} \mathbf{I}_3 \wedge d\mathbf{X} = \int_{\Gamma} \omega = L,$$

where L is the affine length of Γ. Therefore, (7.4) can be expressed as

$$(7.7)\qquad F^* = F + \lambda L + \frac{1}{2}\lambda^2 C,$$

having written

$$(7.8)\qquad C = \int_{\Gamma} k\,\omega.$$

Formula (7.7) is the generalization to the affine case of the Steiner formula giving the area limited by one curve parallel to another.

If Γ is a convex curve, so is that described by the endpoint of the vector $\mathbf{I}_3$, assuming it to be taken from a fixed origin and therefore the pencil of curves $(1-\lambda)\mathbf{X} + \lambda \mathbf{I}_3$ is a linear pencil $(0 \leq \lambda \leq 1)$, and the Brunn-Minkwoski theorem gives the inequality

$$(7.9)\qquad L^2 \geq 2CF$$

already obtained by Blaschke [2]. The sign of equality only holds if $\mathbf{X} = \alpha \mathbf{I}_2$ (α = constant); that is, differentiating, $\omega \mathbf{I}_1 = -\alpha\, k\, \omega\, \mathbf{I}_1$, which implies that k = constant, and thus that Γ is an ellipse.

8 Affine width and other affine invariants of a plane convex curve

The affine distance from the origin of co-ordinates to the point $\mathbf{X}$ of the curve Γ is defined by

$$(8.1)\qquad \varpi = \mathbf{X} \wedge \mathbf{I}_1 = \frac{\mathbf{X} \wedge d\mathbf{X}}{\omega}.$$

Therefore, the area F can be expressed by the formula

$$(8.2)\qquad F = \frac{1}{2}\int_{\Gamma} \varpi \omega,$$

and according to (7.6) the affine length is

$$L = \int_\Gamma k\omega\omega. \tag{8.3}$$

Now let Γ be a convex curve. The following expression is called the *affine width* of Γ corresponding to the point $\mathbf{X}$

$$\Delta_a = (\mathbf{X} - \mathbf{Y}) \wedge \mathbf{I}_1 = \frac{(\mathbf{X} - \mathbf{Y}) \wedge d\mathbf{X}}{\omega}, \tag{8.4}$$

where $\mathbf{Y}$ is the point of Γ at which the tangent is parallel to the tangent in $\mathbf{X}$.

Denoting by Δ_m the metric width of Γ according to the direction of the tangent in $\mathbf{X}$, and by ds the metric area element, we have $(\mathbf{X}-\mathbf{Y})\wedge d\mathbf{X} = \Delta_m ds$, and therefore the following relation holds

$$\Delta_a \omega = \Delta_m ds. \tag{8.5}$$

Moreover, since it is known that

$$\omega = \kappa^{1/3} ds, \tag{8.6}$$

where κ is the metric curvature of Γ in $\mathbf{X}$, it turns out that the affine invariants

$$J_n = \int \Delta_a^n \omega \tag{8.7}$$

are equivalent in metric terms to

$$J_n = \int_\Gamma \Delta_m^n \kappa^{\frac{1-n}{3}} ds = \int_0^{2\pi} \Delta_m^n \kappa^{-\frac{n+2}{3}} d\varphi. \tag{8.8}$$

Among this family of affine invariants, the most important are the following, since they do not depend on the curvature of Γ and are thus defined by any convex figures, regardless of the regularity conditions of their boundaries:

$$J_1 = \int_\Gamma \Delta_a \omega = \int_\Gamma \Delta_m ds \tag{8.9}$$

$$J_{-2} = \int_\Gamma \frac{\omega}{\Delta_a^2} = \int_0^{2\pi} \frac{d\varphi}{\Delta_m^2}. \tag{8.10}$$

Since $J_1 = 2F + 2F^*$, where F^* is the Minkowski mixed area of the set limited by Γ and its symmetric with respect to a point, we have

$$J_1 \geq 4F, \tag{8.11}$$

where the equality holds only for the figures with centre of symmetry.

For J_{-2} in [13], we have proved that

$$1 \leq J_{-2}F \leq \frac{\pi^2}{2}, \tag{8.12}$$

where in the second boundary the equality holds only for the ellipses. If $\overline{\triangle_a}$ is the maximum affine width of Γ, from (8.11) we may deduce that

$$\overline{\triangle_a} \geq \frac{4F}{L}, \tag{8.13}$$

and from (8.12) that

$$\pi^2 \overline{\triangle_a^2} \geq 2FL,$$

which, together with (8.13), gives

$$\pi^2 \overline{\triangle_a^3} \geq 8F^2,$$

where the sign of equality only holds for the ellipses.

The unsolved problems referred to in Section 6 are posed analogously in the plane.

9 Values of the main affine invariants for the ellipse and the ellipsoid

In many questions it is convenient to have to hand the values of the main affine invariants for the ellipse and the ellipsoid. They may be summarized as follows:

Ellipse. Starting from the area F, we have

$$L = 2\pi^{2/3}F^{1/3}, \quad C = 2\pi^{4/3}F^{-1/3}, \quad k = \pi^{2/3}F^{-2/3}, \quad \triangle_a = 2\pi^{-2/3}F^{2/3}.$$

In addition to the inequalities proved in this paper, there are also the following (with equality solely for the ellipse):

$$L^3 \leq 8\pi^2 F, \quad C^3 F \leq 8\pi^4, \quad CL \leq 4\pi^2$$

(see Blaschke [3], [2]).

Ellipsoid. Starting from the volume V, we have

$$\Omega^2 = 12\pi V, \quad M = 4\pi, \quad C^2 = \frac{64\pi^3}{3V}$$

$$K^3 = \frac{4\pi}{3V}, \quad \triangle_a^4 = \frac{9V^2}{\pi^2}.$$

Further inequalities for convex bodies, in addition to those in Section 5, are

$$M \leq 4\pi; \quad \Omega^2 \leq 12\pi V,$$

with equality only for the ellipsoid (Blaschke [3], [2]).

References

[1] W. Blaschke, *Kreis und Kugel*, Veit, Leipzig, 1916.

[2] ———, *Ueber affine Geometrie VII: Neue Extremaleigenschaften von Ellipse un Ellipsoid*, Bericht über die Verhunalungen der Sächsischen Akademie der Wissenschaften **69** (1917), 306–318, Leipzig.

[3] W. Blaschke, *Vorlesungen über Differentialgeometrie und geometrische Grundlagen von Einsteins Relativitätstheorie. II. Affine Differentialgeometrie*, Berlin: J. Springer. Die Grundlehren der Mathematischen Wissenschaften VII., 1923.

[4] W. Blaschke, *Einführung in die Differentialgeometrie. 2. Auflage.*, Die Grundlehren der Mathematischen Wissenschaften. Vol 58. Berlin-Göttingen-Heidelberg: Springer-Verlag. VII, 1960.

[5] T. Bonnesen, *Les problèmes des isopérimètres et des isépiphanes*, Collection de monographies sur la théorie des fonctions, Gauthier-Villars, Paris, 1929.

[6] T. Bonnesen and W. Fenchel, *Theorie der konvexen Körper*, Ergebnisse der Math. Springer, Berlin, 1934.

[7] Herbert Busemann, *Convex surfaces*, Interscience Tracts in Pure and Applied Mathematics, no. 6, Interscience Publishers, Inc., New York, 1958.

[8] J. Favard, *Cours de géométrie différentielle locale.*, Cahiers Scientifiques. Fasc. 24. Paris: Gauthier-Villars VIII, 553 p. , 1957.

[9] K. P. Grotemeyer, *Die Integralsätze der affinen Flüchentheorie*, Archiv der Mathematik **3** (1952), 38–43.

[10] Hugo Hadwiger, *Altes und Neues über konvexe Körper.*, (Elemente der Mathematik vom höheren Standpunkt aus. Band III.) Basel und Stuttgart: Birkhäuser Verlag, 1955.

[11] Erich Salkowski, *Affine Differentialgeometrie.*, Berlin, Leipzig: Walter de Gruyter & Co., 1934.

[12] L. A. Santaló, *Affine differential geometriy and convex bodies (Spanish)*, Math. Notae **XVI** (1957), 20–42.

[13] ———, *Two applications of the integral geometry in affin and projective spaces*, Publ. Math. Debrecen **7** (1960), 226–237, Volume in tribute to O. Varga.

[14] W. Süss, *Eifläche konstanter Affinbreite*, Mathematische Annalen **96** (1927), 251–260.

ON THE MEASURE OF SETS OF PARALLEL LINEAR SUBSPACES IN AFFINE SPACE

L. A. SANTALÓ

1. Introduction. Let E_n be the n-dimensional euclidean real space and $\mathfrak{A}$ the group of unimodular affine transformations which operates on it. It is known that the sets of linear h-spaces L_h $(0 < h < n)$ have no invariant measure with respect to $\mathfrak{A}$ **(5)**. We wish now to consider sets of elements

$$H(L_{h_1}, L_{h_2}, \ldots, L_{h_q}) \tag{1.1}$$

composed by q parallel subspaces of dimensions $h_1, h_2, \ldots, h_q$ which transform transitively by $\mathfrak{A}$. We prove the following:

THEOREM 1. *In order that sets of elements H composed by q parallel linear subspaces of dimensions $h_1, h_2, \ldots, h_q$, which transform transitively by the unimodular affine group $\mathfrak{A}$ have an invariant measure with respect to $\mathfrak{A}$, it is necessary and sufficient that the dimensions h_i be all equal,*

$$h_1 = h_2 = h_3 = \ldots = h_q = h \tag{1.2}$$

and that

$$q = n + 1 - h. \tag{1.3}$$

In § 4 we find the explicit form of this measure together with its metrical significance and in § 5 we indicate some applications to the theory of convex bodies.

2. The Unimodular affine group. (See **2**). Each unimodular affine transformation in E_n can be defined by the position of an n-frame composed of an origin P and n independent vectors $\mathbf{I}_i$ which satisfy the condition

$$|\mathbf{I}_1, \mathbf{I}_2, \ldots, \mathbf{I}_n| = 1 \tag{2.1}$$

where the left-hand side represents the determinant formed by the components of the vectors $\mathbf{I}_i$ with respect to an orthogonal frame of reference.

The relative components of the unimodular affine group $\mathfrak{A}$ are the pfaffian forms ω_{ij} defined by the relations

$$dP = \omega_{0i}\mathbf{I}_i, \qquad d\mathbf{I}_k = \omega_{ki}\mathbf{I}_i \tag{2.2}$$

where the summation convention is used, as will be done throughout.

From (2.2) and (2.1) we deduce

$$\omega_{0i} = |\mathbf{I}_1\mathbf{I}_2 \ldots \mathbf{I}_{i-1}dP\mathbf{I}_{i+1} \ldots \mathbf{I}_n|, \quad \omega_{ki} = |\mathbf{I}_1 \ldots \mathbf{I}_{i-1}d\mathbf{I}_k\mathbf{I}_{i+1} \ldots \mathbf{I}_n|. \tag{2.3}$$

Received January 4, 1960.

By exterior differentiation of (2.2) we obtain the equations of structure

$$d\omega_{0i} = \omega_{0m} \wedge \omega_{mi} \qquad d\omega_{ki} = \omega_{km} \wedge \omega_{mi} \tag{2.4}$$

and by exterior differentiation of (2.1), having into account (2.3), we get

$$\omega_{11} + \omega_{22} + \ldots + \omega_{nn} = 0. \tag{2.5}$$

3. Measure of sets of parallel linear subspaces. Let H denote a set of q independent parallel linear subspaces

$$L_{h_1}, L_{h_2}, \ldots, L_{h_q}$$

of dimensions $h_1, h_2, \ldots, h_q$ respectively. We assume that each pair of elements H transform transitively by $\mathfrak{A}$ and that the dimensions h_i are ordered in the following way

$$n > h_1 \geqslant h_2 \geqslant \ldots \geqslant h_q \geqslant 1. \tag{3.1}$$

To each H we may associate an n-frame $(P; \mathbf{I}_1, \mathbf{I}_2, \ldots, \mathbf{I}_n)$ such that the following relations hold:

$$\begin{aligned} L_{h_1} &= \text{subspace spanned by } \mathbf{I}_1, \mathbf{I}_2, \ldots, \mathbf{I}_{h_1}; \\ L_{h_2} &= \text{subspace which passes through the endpoint of } \mathbf{I}_{h_1+1} \text{ and is} \\ &\quad \text{parallel to } \mathbf{I}_1, \mathbf{I}_2, \ldots, \mathbf{I}_{h_2}; \\ L_{h_3} &= \text{subspace which passes through the endpoint of } \mathbf{I}_{h_1+2} \text{ and is} \\ &\quad \text{parallel to } \mathbf{I}_1, \mathbf{I}_2, \ldots, \mathbf{I}_{h_3}; \\ &\ldots \\ L_{h_q} &= \text{subspace which passes through the endpoint of } \mathbf{I}_{h_1+q-1} \text{ and is} \\ &\quad \text{parallel to } \mathbf{I}_1, \mathbf{I}_2, \ldots, \mathbf{I}_{h_q}. \end{aligned} \tag{3.2}$$

The assumed transitivity for the elements H with respect to $\mathfrak{A}$ gives the condition

$$h_1 + q - 1 \leqslant n. \tag{3.3}$$

In order to see if sets of elements H have an invariant measure with respect to $\mathfrak{A}$ we follow the general method **(3; 5)**. According to (3.2) and (2.2) the completely integrable system whose integral varieties correspond to the elements H is the following

$$\begin{aligned} &\omega_{0s_1} = 0 && (s_1 = h_1 + 1, \ldots, n) \\ &\omega_{i_1m_1} = 0 && (i_1 = 1, \ldots, h;\ m_1 = h_1 + 1, \ldots, n) \\ &\omega_{0s_2} + \omega_{h_1+1,s_2} = 0 && (s_2 = h_2 + 1, \ldots, h_1) \\ &\omega_{h_1+1,s'_2} = 0 && (s'_2 = h_1 + 1, \ldots, n) \\ &\omega_{i_2m_2} = 0 && (i_2 = 1, \ldots, h_2;\ m_2 = h_2 + 1, \ldots, h_1) \\ &\omega_{0s_3} + \omega_{h_1+2,s_3} = 0 && (s_3 = h_3 + 1, \ldots, h_1) \\ &\omega_{h_1+2,s'_3} = 0 && (s'_3 = h_1 + 1, \ldots, n) \\ &\omega_{i_3m_3} = 0 && (i_3 = 1, \ldots, h_3;\ m_3 = h_3 + 1, \ldots, h_2) \\ &\ldots \\ &\omega_{0s_q} + \omega_{h_1+q-1,s_q} = 0 && (s_q = h_q + 1, \ldots, h_1) \\ &\omega_{h_1+q-1,s'_q} = 0 && (s'_q = h_1 + 1, \ldots, n) \\ &\omega_{i_qm_q} = 0 && (i_q = 1, \ldots, h_q;\ m_q = h_q + 1, \ldots, h_{q-1}). \end{aligned} \tag{3.4}$$

Note that the number of equations is

$$N = n(h_1 + q) - \sum_{i=1}^{q} h_i(h_i + 1) + \sum_{i=1}^{q-1} h_i h_{i+1} \tag{3.5}$$

and coincides with the number of parameters on which H depends, as it should.

The exterior product Π of all the relative components (3.4) is an exterior differential form of order N. The integral of Π will be a measure for sets of elements H, invariant with respect to $\mathfrak{A}$, if and only if the exterior differential $d\Pi$ vanishes when the equations (2.4) and (2.5) are taken into account (see **3**; **5**). Since the system (3.4) is completely integrable, the theorem of Frobenius (**2**, p. 193) says that in the structure equations (2.4) applied to the forms (3.4), at least one of the differential forms of each term of the sum of the right belongs to (3.4). Thus, up to the sign, which is immaterial for us since we will always take the measures in absolute value, we have

$$d\Pi = \Pi \wedge \Phi \tag{3.6}$$

where

$$\Phi = \sum_{i=1}^{h_1} \omega_{ii} + \sum_{i=1}^{h_2} \omega_{ii} + \ldots + \sum_{i=1}^{h_q} \omega_{ii} - \sum_{i=h_1+1}^{n} \omega_{ii} - \sum_{i=h_2+1}^{n} \omega_{ii} - \ldots - 2\sum_{i=h_q+1}^{n} \omega_{ii}. \tag{3.7}$$

The relative components of the set (3.4) which have equal indices are

$$\omega_{h_1+1,h_1+1}, \omega_{h_1+2,h_1+2}, \ldots, \omega_{h_1+q-1,h_1+q-1}. \tag{3.8}$$

Since the relative components are only related by the equation (2.5), the condition $d\Pi = 0$ can hold only if (3.7) is equivalent to the left side of (2.5), up to a linear combination of the forms (3.8). This is possible if and only if $h_q = h_1$ and $h_1 + q - 1 = n$. Taking into account (3.1) these relations prove the stated theorem 1.

4. Metrical interpretation of the measure. If the equations (1.2) and (1.3) are satisfied, the measure for sets of elements H composed by $q = n+1-h$ parallel linear h-spaces is given, up to a constant factor, by the integral of the form Π obtained by exterior multiplication of all the pfaffian forms (3.4). The form Π is called the density, invariant with respect to $\mathfrak{A}$, for sets of elements H.

We wish now to give a metrical interpretation of Π.

Let us put

$$\Pi = \Pi_0 \wedge \Pi_1 \wedge \ldots \wedge \Pi_n \tag{4.1}$$

where

$$\Pi_i = \omega_{i,h+1} \wedge \omega_{i,h+2} \wedge \ldots \wedge \omega_{i,n} \qquad (i = 0, 1, \ldots, n). \tag{4.2}$$

Let $(P_0 \cdot \mathbf{e}_i)$ be an orthogonal frame composed of n perpendicular unit vectors $\mathbf{e}_i$ of origin P_0, such that $\mathbf{e}_1, \mathbf{e}_2, \ldots, \mathbf{e}_h$ lie on the subspace L_h determined by $P \equiv P_0, \mathbf{I}_1, \mathbf{I}_2, \ldots, \mathbf{I}_h$. We may write

$$\mathbf{I}_\alpha = \lambda_{\alpha\beta}\mathbf{e}_\beta, \qquad \mathbf{I}_\xi = \lambda_{\xi i}\mathbf{e}_i \tag{4.3}$$

where we agree to use the following ranges of indices

$$\alpha, \beta, \gamma, \ldots = 1, 2, \ldots, h; \qquad \xi, \eta, \zeta, \ldots = h+1, h+2, \ldots, n; \tag{4.4}$$
$$i, j, k, \ldots = 1, 2, \ldots, n.$$

If we put

$$d\mathbf{e}_i = \theta_{ij}\mathbf{e}_j \tag{4.5}$$

we will have

$$d\mathbf{I}_\alpha = d\lambda_{\alpha\beta}\mathbf{e}_\beta + \lambda_{\alpha\beta}d\mathbf{e}_\beta = \phi_{\alpha\beta}\mathbf{e}_\beta + \psi_{\alpha\xi}\mathbf{e}_\xi \tag{4.6}$$

$$d\mathbf{I}_\xi = d\lambda_{\xi i}\mathbf{e}_i + \lambda_{\xi i}d\mathbf{e}_i = \sigma_{\xi i}\mathbf{e}_i \tag{4.7}$$

where

$$\phi_{\alpha\beta} = d\lambda_{\alpha\beta} + \lambda_{\alpha\gamma}\theta_{\gamma\beta}, \qquad \sigma_{\xi i} = d\lambda_{\xi i} + \lambda_{\xi j}\theta_{ji}, \qquad \psi_{\alpha\xi} = \lambda_{\alpha\beta}\theta_{\beta\xi}. \tag{4.8}$$

Let us note that, according to (4.7), the volume element at the endpoint of $\mathbf{I}_\eta$ is $\sigma_{\eta 1} \wedge \sigma_{\eta 2} \wedge \ldots \wedge \sigma_{\eta n}$ and the element of $(n\text{-}h)$-dimensional volume in the $(n\text{-}h)$-space spanned by $\mathbf{e}_{h+1}, \mathbf{e}_{h+2}, \ldots, \mathbf{e}_n$ at the orthogonal projection on it of the endpoint of $\mathbf{I}_\eta$ is

$$dP_\eta = \sigma_{\eta,h+1} \wedge \sigma_{\eta,h+2} \wedge \ldots \wedge \sigma_{\eta,n}. \tag{4.9}$$

The first relation (2.2) may be written

$$dP = \omega_{0i}\mathbf{I}_i = dx_i\mathbf{e}_i \tag{4.10}$$

and from (2.3) and (4.4) we deduce

$$\omega_{0\xi} = |\mathbf{I}_1\mathbf{I}_2 \ldots \mathbf{I}_{\xi-1}(dx_i\mathbf{e}_i)\mathbf{I}_{\xi+1} \ldots \mathbf{I}_n| = \Lambda_{\xi\eta}dx_\eta \tag{4.11}$$

where $\Lambda_{\xi\eta}$ means the algebraic complement of $\lambda_{\xi\eta}$ in the determinant

$$|\mathbf{I}_1\mathbf{I}_2 \ldots \mathbf{I}_n| = \begin{vmatrix} \lambda_{11}\lambda_{12} \ldots \lambda_{1h} & 0 \ldots 0 \\ \lambda_{21}\lambda_{22} \ldots \lambda_{2h} & 0 \ldots 0 \\ \ldots & \\ \lambda_{h1}\lambda_{h2} \ldots \lambda_{hh} & 0 \ldots 0 \\ \lambda_{h+1,1}\lambda_{h+1,2} \ldots \lambda_{h+1,h} & \lambda_{h+1,h+1} \ldots \lambda_{h+1,n} \\ \ldots & \\ \lambda_{n1}\lambda_{n2} \ldots \lambda_{nh} & \lambda_{n,h+1} \ldots \quad \lambda_{nn} \end{vmatrix} = 1. \tag{4.12}$$

Therefore, by exterior multiplication of the forms (4.11), taking into account a well-known property on adjoint determinants (**4**, p. 73) we obtain

$$\Pi_0 = |\Lambda_{\xi\eta}|dx_{h+1} \wedge dx_{h+2} \wedge \ldots \wedge dx_n = D\, dP_0 \tag{4.13}$$

where we have put

(4.14) $$D = \begin{vmatrix} \lambda_{11}\lambda_{12} \dots \lambda_{1h} \\ \lambda_{21}\lambda_{22} \dots \lambda_{2h} \\ \dots \\ \lambda_{h1}\lambda_{h2} \dots \lambda_{hh} \end{vmatrix}$$

and $dP_0 = dx_{h+1} \wedge \dots \wedge dx_n$ = element of (n-h)-dimensional volume on the (n-h)-space spanned by $\mathbf{e}_{h+1}, \mathbf{e}_{h+2}, \dots, \mathbf{e}_n$ at the point P_0.

From (2.3) and (4.7) we have

(4.15) $$\omega_{\eta\xi} = |\mathbf{I}_1\mathbf{I}_2 \dots \mathbf{I}_{\xi-1}d\mathbf{I}_\eta\mathbf{I}_{\xi+1} \dots \mathbf{I}_n|$$

and therefore

(4.16) $$\Pi_\eta = |\Lambda_\xi|\sigma_{\eta,h+1} \wedge \dots \wedge \sigma_{\eta n} = D\, dP_\eta.$$

Finally, we have,

(4.17) $$\omega_{\alpha\eta} = |\mathbf{I}_1\mathbf{I}_2 \dots \mathbf{I}_{\eta-1}d\mathbf{I}_\alpha\mathbf{I}_{\eta+1} \dots \mathbf{I}_n|$$

and therefore

(4.18) $$\Pi_\alpha = |\Lambda_{\eta\xi}|\psi_{\alpha,h+1} \wedge \dots \wedge \psi_{\alpha n} = D\psi_{\alpha,h+1} \wedge \dots \wedge \psi_{\alpha n}.$$

If we introduce the density dL_{n-h} invariant with respect to rotations about P_0 (metrical density, see **(6)**), for the linear (n-h)-spaces through P_0 spanned by $\mathbf{e}_{h+1}, \dots, \mathbf{e}_n$, that is,

(4.19) $$dL_{n-h} = (\theta_{1,h+1} \wedge \theta_{2,h+1} \wedge \dots \wedge \theta_{h,h+1}) \wedge (\theta_{1,h+2} \wedge \dots \wedge \theta_{h,h+2}) \wedge \dots \wedge (\theta_{1,n} \wedge \theta_{2,n} \dots \wedge \theta_{h,n}),$$

we have, from (4.18) and (4.8)

(4.20) $$\Pi_1 \wedge \Pi_2 \wedge \dots \wedge \Pi_h = D^n dL_{n-h}.$$

Therefore, we have,

(4.21) $$\Pi = D^{2n-h+1}\, dP_0 \wedge dP_{h+1} \wedge dP_{h+2} \wedge \dots \wedge dP_n \wedge dL_{n-h}.$$

Let us now observe that the volume S of the (n-h)-dimensional simplex of vertices $P_0, P_{h+1}, P_{h+2}, \dots, P_n$, taking into account (4.3), is given by

(4.22) $$S = \frac{1}{(n-h)!} \begin{vmatrix} \lambda_{h+1,h+1} \dots \lambda_{h+1,n} \\ \lambda_{h+2,h+1} \dots \lambda_{h+2,n} \\ \dots \\ \lambda_{n,h+1} \quad \dots \lambda_{n,n} \end{vmatrix}.$$

From (4.12) and (4.22) we get

$$D = \frac{1}{(n-h)!S}$$

and (4.21) may be written in the definitive form

(4.24) $$\Pi = \frac{dP_0 \wedge dP_{h+1} \wedge \dots \wedge dP_n \wedge dL_{n-h}}{[(n-h)!S]^{2n-h+1}}.$$

Let us summarize the meaning of the terms in (4.24). Given $n - h + 1$ parallel h-spaces, we cut them by an orthogonal (n-h)-space L_{n-h} through a fixed origin; let $P_0, P_{h+1}, \ldots, P_n$ be the intersection points. Then, S is the volume of the simplex of vertices $P_0, P_{h+1}, \ldots, P_n$; each dP_i ($i = 0, h + 1, \ldots, n$) is the element of volume at P_i of L_{n-h}, and dL_{n-h} represents the metrical density for sets of (n-h)-spaces through the origin **(6)**.

5. Application to convex bodies. Let K be a convex body in E_n. It is well known that the measure of sets of linear h-spaces, invariant with respect to the group of motions, which intersect K, gives rise up to a constant factor to the metrical invariants W_h^n (= hth mixed volume of K with the unit sphere; $h = 1, 2, \ldots, n - 1$; see **(6)**).

This result is not straightforwardly generalizable to the affine geometry, because the linear subspaces of dimension $h > 0$ have no invariant measure with respect to the unimodular affine group **(5)** .However, if we consider sets of parallel h-spaces in the sense of § 3, we find that the measure of sets of elements H composed of $n - h + 1$ parallel linear h-spaces whose convex cover $C(H)$ contains K in its interior, will give an affine invariant for K. It has the form

$$(5.1)\quad M_h^n(K) = \int \Pi = [(n-h)!]^{h-2n-1} \int_{K \subset C(H)} \frac{dP_0 \wedge dP_1 \wedge \ldots \wedge dP_{n-h} \wedge dL_{n-h}}{S^{2n-h+1}}$$

where the integral is extended over all L_{n-h} orthogonal to the parallel h-spaces which constitute H, such that $K \subset C(H)$ and dP_i ($i = 0, 1, 2, \ldots, n - h$) are the volume elements in L_{n-h} at the intersection points of L_{n-h} with H.

For $h = 1, 2, \ldots, n - 1$ we get a set of $n - 1$ affine invariants which may be considered as the affine generalization of the W_h^n of the metrical case. It seems to be an interesting open question to investigate if the affine invariants M_h^n are related by inequalities of the type of those of Minkowski for the metrical invariants W_h^n. For $h = n - 1$, see **(7)**.

Let us consider the cases $n = 2$, $n = 3$.

1. *Case of the plane* ($n = 2$). According to (1.3) we have the possibility $h = 1$, $q = 2$, that is, the elements H are composed of two parallel lines. Let θ denote the angle of the direction normal to these lines and let p_0, p_1 be their distances to a fixed origin 0. The measure of the set of parallel lines which contain K in its interior gives the following affine invariant for K:

$$(5.2)\qquad M_1^2(K) = \int \frac{dp_0 \wedge dp_1 \wedge d\theta}{|p_1 - p_0|^4} = \frac{1}{6} \int_0^\pi \frac{d\theta}{\Delta^2},$$

where $\Delta = \Delta(\theta)$ denotes the width of K in the direction θ.

2. *Case of the space* ($n = 3$). According to (1.3) we have two possibilities:

(a) $h = 2$, $q = 2$; (b) $h = 1$, $q = 3$.

For the case (a) the elements H are pairs of parallel planes. If $d\Omega$ denotes the element of area on the unit sphere corresponding to the direction normal to the planes H and p_0, p_1 are their distances to a fixed origin 0, the measure of the set of pairs of parallel planes which contain a given convex body K gives the following affine invariant for K

$$M_2^3(K) = \int \frac{dp_0 \wedge dp_1 \wedge d\Omega}{|p_0 - p_1|^5} = \frac{1}{12} \int \frac{d\Omega}{\Delta^3}, \tag{5.3}$$

where $\Delta = \Delta(\Omega)$ denotes the width of K in the direction Ω.

For the case (b) the elements H are constituted by three parallel lines. If $d\Omega$ denotes the area element on the unit sphere at the point defined by the direction of these lines and dP_0, dP_1, dP_2 are the elements of area of a plane normal to the parallel lines at the corresponding intersection points, the measure of the set of three parallel lines whose convex cover contains K, gives the following affine invariant for K:

$$M_1^3(K) = \frac{1}{64} \int d\Omega \int \frac{dP_0 \wedge dP_1 \wedge dP_2}{S^6} \tag{5.4}$$

where S denotes the area of the triangle $P_0P_1P_2$. The first integration is extended over all triangles $P_0P_1P_2$ which contain the projection K_Ω of K on the plane normal to the direction Ω. The second integration is extended over half of the unit sphere.

A direct way of obtaining the invariants (5.2) and (5.3) together with certain inequalities between them and the area (volume) of K has been given in **(7)**.

References

1. T. Bonnesen and W. Fenchel, *Theorie der konvexen Körper*, Ergebnisse der Mathematik (Berlin, 1934).
2. E. Cartan, *La Théorie des groupes finis et continus et la géométrie diférentielle traités par la méthode du repère mobile* (Paris; Gauthier-Villars, 1937).
3. S. S. Chern, *On integral geometry in Klein spaces*, Ann. Math., *43* (1942), 178–189.
4. G. Kowalewski, *Einführung in die Determinantentheorie* (New York: Chelsea Publ., 1948).
5. L. A. Santalo, *Integral geometry in projective and affine spaces*, Ann. Math., *51* (1950), 739–755.
6. ——— *Sur la mesure des espaces linéaires qui coupent un corps convexe et problèmes qui s'y rattachent*, Colloque sur les questions de réalité en Géométrie (Liege, 1955).
7. ——— *Two applications of the integral geometry in affine and projective spaces*, Publications Mathematicae Debrecen, *7* (1960), 226–237.

University of Buenos Aires

Part V. Statistics and Stereology, with comments by L. M. Cruz-Orive

Stereology: a brief overview in the light of L. A. Santaló's legacy

The term *Stereology* (from the Greek $\sigma\tau\epsilon\rho\epsilon\acute{o}\varsigma$ = solid) was coined at the foundation of the International Society for Stereology in 1963 to represent *the science of three-dimensional interpretation of bidimensional images*. Over the years Stereology has evolved into a science which defines proper sampling rules, based on sections, or projections, of an object, to estimate geometric properties of the object – in short, Stereology is essentially geometric sampling.

A glimpse into the art and spirit of Stereology can be gained with an elementary example. Consider a compact set $Y \subset \mathbb{R}^3$ of unknown volume $Q > 0$. The problem is to devise a stereological design to estimate Q. Fix a convenient axis Ox and let the interval $[a, b]$ represent the orthogonal linear projection of the set Y onto this axis. Consider a plane $L_2(x)$ normal to Ox at a point of abscissa x, and set *area* $(Y \cap L_2(x)) := f(x)$. The function $f : \mathbb{R} \to (0, \infty)$ is assumed to be observable for all x with $f(x) = 0,\ x \notin [a, b]$. By elementary calculus,

$$Q = \int_a^b f(x)\,\mathrm{d}x. \tag{1}$$

The preceding integral is computed with respect to the density $\mathrm{d}x$ of points on $\mathbb{R}$ which is invariant with respect to translations; moreover the context is geometric, hence Eq. (1) may be regarded as an elementary identity of Integral Geometry. Bearing in mind that $b - a > 0$, we can write

$$Q = (b-a) \int_a^b f(x)\, \frac{\mathrm{d}x}{b-a}, \tag{2}$$

and we realise that

$$\mathbb{P}(\mathrm{d}x) = 1_{[a,b]}(x) \cdot \frac{\mathrm{d}x}{b-a} \tag{3}$$

is the element of probability of a uniform random variable X in $[a, b]$. We have simply normalized an invariant density stemming from Integral Geometry into an element of probability measure. Now we can write,

$$Q = (b-a) \int_a^b f(x)\, \mathbb{P}(\mathrm{d}x) = (b-a)\, \mathbb{E} f(X), \tag{4}$$

where $\mathbb{E}$ is the expectation or mean value operator with respect to $\mathbb{P}(\mathrm{d}x)$. Finally, let us generate a random observation X from a uniform random variable in $[a, b]$ and set

$$\widehat{Q}(X) = (b-a) \cdot f(X). \tag{5}$$

Clearly $\mathbb{E}\widehat{Q}(X) = Q$, whereby we say that $\widehat{Q}(X)$ is an unbiased estimator of Q. Before passing it to other scientists, however, the stereologist must ensure that an estimator is not only unbiased but reasonably precise. To decrease the estimator variance we may increase the number of observations, namely the sample size. In the present case we may fix a length $T \in (0, b-a)$ and consider the improved estimator

$$\widehat{Q}(U) = T \cdot \sum_{k \in \mathbb{Z}} f(UT + kT), \tag{6}$$

where U is a uniform random variable in the interval $[0, 1]$. It is easy to show that $\widehat{Q}(U)$ is also unbiased. Now the mean sample size is $(b-a)/T$, and $\operatorname{Var} \widehat{Q}(U)$ will rapidly decrease as the sample size is increased.

The foregoing example enables us to illustrate the main features of Stereology as follows.

1. A stereological design typically uses a path involving the following disciplines: Integral Geometry → Probability → Statistics → Sampling and Inference.
2. We have made no strong assumptions about the shape of the set Y. Stereology deals with geometric sets of arbitrary shape; it is assumed that bounded sets have a finite Hausdorff measure.
3. The estimator $\widehat{Q}(U)$ typically uses a few sampling points only. Stereology is therefore very different from computed tomography, in that it is concerned with estimations and not with reconstructions.
4. Our example belongs to design based Stereology: the set Y under study is fixed and we generate random observations to estimate the desired quantity Q. On the contrary, model based Stereology is concerned with stationary random sets and it is therefore more related to Stochastic Geometry. Here sampling rules are less strict, but the underlying assumptions are stronger than in the design based case. Design based Stereology is useful in biosciences when the object under study is bounded, whereas model based Stereology is often useful in materials science, geology, geography, etc. The term *model based* does however not correspond to assuming that the sets involved have a simple geometric shape (spheres, ellipsoids, cylinders, regular polyhedra, or convex bodies in general) – this refers more to *assumption based* Stereology, which has practically vanished as the stereological sampling tools became increasingly sophisticated and assumption free. This evolution into real applicability has logically motivated an increasing degree of popularity of Stereology among scientists.

Next we concentrate briefly on the history and evolution of Stereology: this will enable us to better frame the role of Luis A. Santaló in its development. There is a certain consensus in that Stereology is first encapsulated by the celebrated needle problem described and solved by George Louis Leclerc, Comte de Buffon, in 1777. On the other hand the estimator $\widehat{Q}(U)$ described above is named the *Cavalieri estimator* after Bonaventura Cavalieri, a disciple of Galileo who lived more than a century before Buffon. Stereologists have reckoned Cavalieri as the first mathematician who established a theorem for objects of an arbitrary shape – (moreover the estimator uses parallel *Cavalieri sections*). In fact, Eq. (1) was essentially known to Archimedes (and even to Chinese mathematicians) but, until Cavalieri, Greek and Middle Age mathematicians worked instead with regular geometric models almost exclusively. Notwithstanding all this, Buffon's needle problem motivated the search for a motion invariant measure of bounded figures in the plane – and hence the birth of Geometrical Probability and Integral Geometry – more effectively than Cavalieri's work. It should be noted, however, that the incorporation of sampling and statistics belongs to the 20th century.

Integral Geometry received a fundamental push in the 1930's from the Hamburg school of William J. E. Blaschke, whose disciples exerted a great influence. A distinguished member of that school was Luis A. Santaló, whose career extended for over sixty years. His 1976 book [76.1] is the undisputed reference on Integral Geometry and Geometric Probability.

Unlike physics, whose tools emerged in close contact with higher mathematics, Stereology began to develop almost independently of Integral Geometry. Progress was therefore slow until the 1970's, when Roger E. Miles unveiled the strength of combining Integral Geometry, Probability and Statistics. From then on, Santaló was fascinated by Stereology because all of a sudden he visualized his earlier mathematical work in a different, highly motivating scenario.

It is useful to distinguish between global and particle Stereology. Global Stereology is concerned with the estimation of Hausdorff measures of whole objects or of their union, for instance the total surface area of the lung alveoli, or the pore volume fraction of an aluminium alloy. A particle is a connected set separated from other particles, (e.g. a cell, a mineral grain), and particle Stereology is concerned with the estimation of the mean individual particle measure (e.g. mean cell or grain volume, either in number, or size weighted).

Early efforts in global Stereology led to the so-called *fundamental equations of Stereology*, which were ad-hoc identities connecting ratios of Hausdorff measures defined on a population of objects with the corresponding ratios observed on line or plane sections, or on projections, of the population of objects. Details can be obtained in the books of Howard & Reed (1998, 2005) and Baddeley & Jensen (2005). As pointed out by Miles in 1972, all such identities are particular cases of a general one valid for manifolds in $\mathbb{R}^n$ hit by r-plane probes, plus another one valid for bounded r-probes. In turn, the mentioned identities are direct consequences of the corresponding Crofton formulae for r-planes and for bounded r-probes equipped with the kinematic density. Such formulae were known for a long time, and they are collected in [76.1]. Global Stereology is concerned with the implementation of such fundamental equations by means of systematic sampling in one or several stages. For instance $\widehat{Q}\,(U)$ is a one-stage estimator based on systematic plane sections taken at the points $\{(U+T)\cdot k,\ k\in\mathbb{Z}\}$, which constitute a so-called *test system* in $\mathbb{R}$. In turn, for each k the area $f(UT+kT)$ could be estimated by superimposing a test system of points in $\mathbb{R}^2$, for instance a square lattice grid, uniformly at random on the corresponding section: this would lead to a two-stage estimator of Q, and so on. Early in his career Santaló was already interested in the properties of test systems, see [39.1], [40.3]. By 1956 he had polished the relevant results almost completely, see [56.4], and his account in Chapter 8 of [76.1] is a definitive masterpiece. A distinct advantage of test systems is that the probability element that solves the estimation problem depends only on the fundamental tile of the test system and not on the object under study. For instance, if we sample a set Y with a single bounded probe equipped with the kinematic density, then the normalizing constant used to construct the probability measure for the probe involves the mean Minkowski addition of Y and the symmetric reflection of the probe with respect to the origin – an undesirable fact.

Particle Stereology is at least as old as global Stereology, and even more popular, to the point that Stereology is still occasionally associated with the *tomato salad* or the *Swiss cheese* problems, which deal with the identification of particle size distributions from their planar transects. The old and abundant literature on the subject can be checked in Baddeley & Jensen's book. Naturally Santaló was also motivated by the popularity of the problem, see [43.2], [55.1], and Chapter 16 from his book [76.1]. The underlying problem boils down to estimating particle number in three dimensions, which is a 0-dimensional property. It soon appears that any attempts at estimating particle number from two dimensional sections in a general case is futile: the dimension of the probe must be at least three. This is also the case for the connectivity number. To circumvent the problem many mathematicians adopted a simple geometrical model for the particles, notably the spheres model. The problem then reduces to inverting an Abel integral equation which unfortunately turns out to be numerically ill-conditioned. This frustrating state of affairs prevailed until the mid 1980's when the disector method was published, which marked the beginning of the gradual extinction of the classical unfolding particle size distributions problems in practical Stereology.

More recently new particle sampling and measuring tools have been developed which are based in linear subspaces containing a fixed one, for instance line probes through a fixed point, planes containing a fixed axis, etc. The corresponding results concern the so called local Stereology, which proves strong for estimating not only the first order measures discussed so far, but also second order measures aimed at characterizing, at least partially, the spatial pattern of particles and other geometrical structures in space. This has proven useful in neuroscience, and in pathology for diagnasing malignant tumours. 1980's with the publication of the disector method, which marked

Santaló also explored aspects of Stochastic Geometry, even though its foundations were laid down relatively late in his career (namely in the 1970's), see [78.2], [78.3], [79.1], [80.3] and [84.1].

Disciples and personal friends of Professor Luis A. Santaló will miss him forever as a mathematician, as one of the last universalists, as a man of his time, and as a gentleman of the highest standards. The seed of his work is deeply rooted in first quality science, however, and therefore his legacy will live in the future.

L. M. Cruz-Orive, (University of Cantabria, Santander).

Sur quelques problèmes de probabilités géométriques

L. A. Santaló à Rosario, Rép. Argentina

Si on jette au hasar sur le plan où il y a un réseau uniforme (§ 2, par exemple figs. 6, 7, 8, 9) une ligne de longeur et forme quelconque on peut obtenir dans tous les cas la *valeur moyenne* du nombre de points d'intersection avec le réseau [1]. Mais dans le cas où le ligne est simplement un segment de droite pas trop long peut obtenir, dans beaucoup de cas, non seulement la valeur moyenne si non la *probabilité* d'avoir un certain nombre de points d'intersection. L'object de ce travail est de déterminer cette probabilité dans quelques cas particuliers, c'est à dire, de déterminer la probabilité de ce qu'un segment jeté au hasard sur un réseau de la forme des figures 6, 7, 8, 9 et 11 ait un nombre déterminé de points d'intersection.

1 Mesure de quelques ensembles de segments

1.1 Densité cinématique

La position d'un segment de droite orienté sur le plan est déterminée par les deux coordonnées x, y de l'une de ses extrémités et l'angle φ qu'il forme avec une direction fixe.

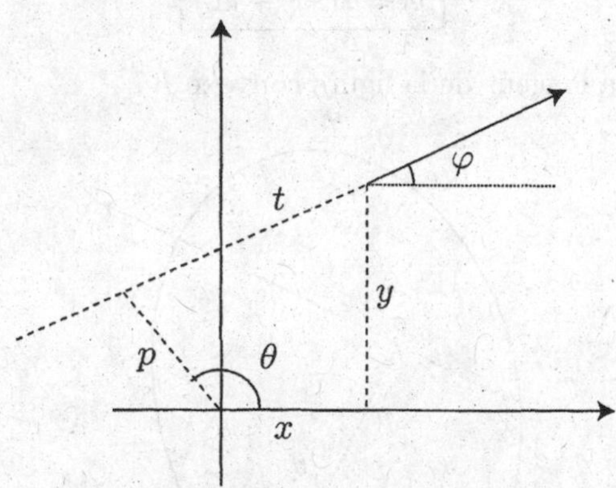

Figure 1:

Pour mesurer des ensembles des segments de la même longeur on prend

$$\mathfrak{M} = \int dx\, dy\, d\varphi, \tag{1}$$

c'ést à dire, l'intègrale sur l'ensemble consideré, de l'expression differentielle $dK = dx\,dy\,d\varphi$ qu'on appelle *densité cinématique* [2]. Par un changement de variables on voit que cette densité cinématique peut être exprimée aussi par

$$dK = dp\,d\theta\,dt, \tag{2}$$

où (fig. 1.1) p et θ sont les coordonnées normales de la droite qui contient le segment et t est la coordonnée qui fixe l'origine du segment sur cette droite.

1.2 Mesure des segments qui coupent une figure convexe

En prenant l'expression (2) de la densité cinématique, pour mesure de l'ensemble des segments de longeur l qui coupent la figure convexe K on trouve

$$\mathfrak{M} = \int dp\;d\theta\,dt = \int (\sigma + l)dp\;d\theta\,, \tag{3}$$

où σ représente la longeur de la corde que la droite du segment détérmine sur K (fig. 1.2). En appliquant deux formules connues [2]:

$$\int \sigma\,dp\,d\theta = 2\pi F, \tag{4}$$

$$\int dp\,d\theta = 2L,$$

on obtient la mesure cherchée

$$\boxed{\mathfrak{M} = 2(\pi F + lL)} \tag{5}$$

où F est l'aire, L la longeur de la figure convexe K.

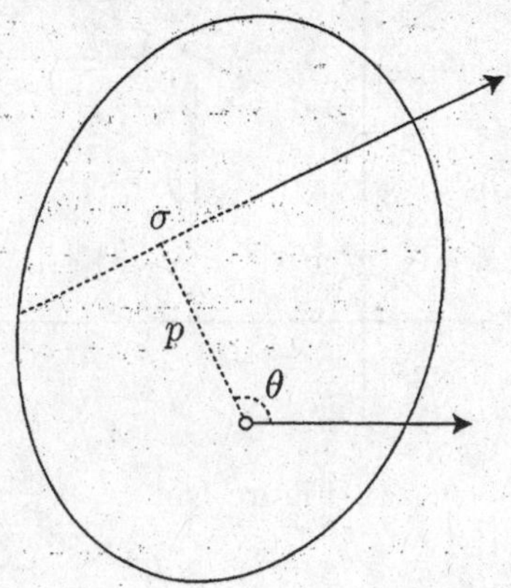

Figure 2:

1.3 Mesure des segments orientés totalement contenus dans une figure convexe

Soit una figure convexe K et un segment orienté $\vec{S}$ de longeur l. Nous voulons calculer l'integral (1) sur tout l'ensemble des segments de longeur l qui sont contenus dans K.

Prenons la forme (2) de la densité cinématique. En appellant σ la longeur de la corde que la droite contenant le segment $\vec{S}$ détermine sur K, on peut intégrer dt et on obtient

$$\mathfrak{M}_l = \int dp\, d\theta\, dt = \int_{\sigma \geq l} (\sigma - l) dp\, d\theta\,. \tag{6}$$

Pour chaque valeur de θ, si AB et CD (fig. 3) sont les cordes de longeur l correspondantes à cette direction, on aura

$$\int_{\sigma \geq l} \sigma dp\, d\theta = \int_0^{2\pi} (\text{aire } ABEDCF) d\theta\,, \tag{7}$$

$$\int_{\sigma \geq l} l dp\, d\theta = \int_0^{2\pi} (\text{aire } ABCD) d\theta\,.$$

Si on représente par Φ l'aire de la part couverte de hachures dans la figure, c'est à dire, l'aire comprise entre la figure K et le parallélogramme $ABCD$, on a donc

$$\boxed{\mathfrak{M}_l = \int_0^{2\pi} \Phi d\theta} \tag{8}$$

L'intégration du deuxième membre de cette égalité n'est pas en général facile.

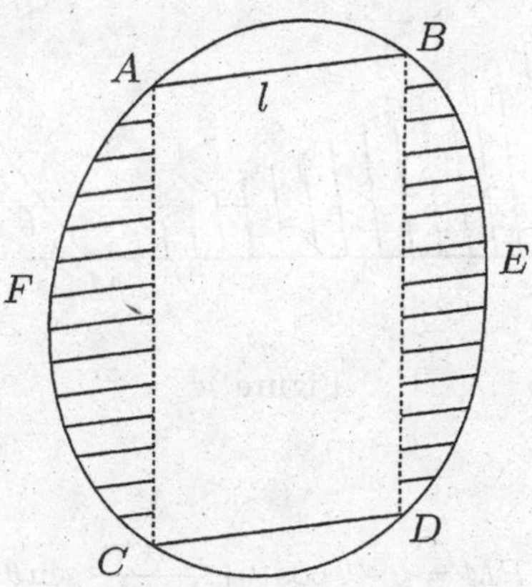

Figure 3:

Nous allons la calculer dans quelques cas particuliers.

1.4 Cas du cercle

Nous voulons calculer l'intégrale (8) dans le cas où la figure K est un cercle de rayon r. Si la longeur du segment est $l \geq 2r$, la mesure $\mathcal{M}_l$ de l'ensemble des segments qui sont à l'intérieur du cercle est évidemment nulle. Si $l \leq 2r$, on voit que l'aire Φ de (8) est constante et égale à

$$\Phi = \pi r^2 - 2r^2 \text{arcsen}\, \frac{l}{2r} - l\sqrt{r^2 - \frac{l^2}{4}}. \tag{9}$$

On a donc, pour mesure des segments de longeur $l(l \leq 2r)$ qui sont à l'intérieur du cercle de rayon r

$$\boxed{\mathfrak{M}_l = 2\pi\left(\pi r^2 - 2r^2 \text{arcsen}\, \frac{l}{2r} - l\sqrt{r^2 - \frac{l^2}{4}}\right)} \tag{10}$$

1.5 Cas du parallélogramme

Pour pouvoir calculer facilement l'intégrale (8) dans le cas ou la figure K est un parallélogramme, il faut supposer que la longeur l du segment orienté $\vec{S}$ est égale ou inférieure à la plus petite hauteur du parallélogramme. Dans ces conditions il y aura dans chaque direction segments $\vec{S}$ contenus dans K. En posant $PS = a$, $PQ = b$, pour chaque valeur de $\theta \leq \widehat{P}$ on peut calculer (fig. 4)

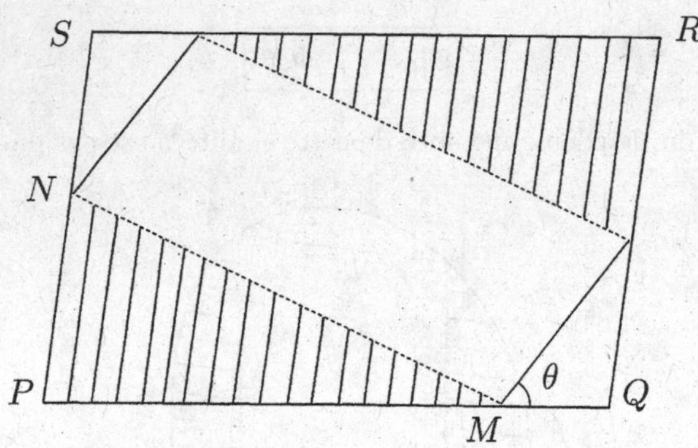

Figure 4:

$$\begin{aligned} PM &= b - l\cos\theta + \frac{l}{\text{tang}\, P}\,\text{sen}\,\theta, \\ PN &= a - \frac{l}{\text{sen}\, P}\,\text{sen}\,\theta. \end{aligned} \tag{11}$$

L'aire Φ de la formule (8) est égale à deux fois l'aire du triangle NPM, c'est à dire,

$$\Phi = \overline{PM} \cdot \overline{PN} \operatorname{sen} \widehat{P} = F - lb \operatorname{sen} \theta - la \operatorname{sen}(\widehat{P} - \theta) + l^2 \operatorname{sen} \theta \cos \theta - \frac{l^2}{\operatorname{tang} P} \operatorname{sen}^2 \theta \tag{12}$$

d'où

$$\int_0^P \Phi d\theta = \widehat{P} \cdot F + (a+b) l \cos \widehat{P} - (a+b) l + \frac{l^2}{2} \left(1 - \widehat{P} \cot \widehat{P}\right), \tag{13}$$

F représente l'aire du parallélogramme.

En faisant de même pour θ variant de $\widehat{P}$ à π on trouve

$$\int_P^\pi \Phi d\theta = (\pi - \widehat{P}) F - (a+b) l \cos \widehat{P} - (a+b) l + \frac{l^2}{2} \left[1 + (\pi - P) \cot \widehat{P}\right]. \tag{14}$$

En faisant la somme de (13) et (14) et en multipliant par deux pour obtenir l'intégration de θ à 2π, on a

$$\boxed{\mathfrak{M}_l = 2 \left[\pi F - 2(a+b) l + \frac{l^2}{2} \left(2 + (\pi - 2\widehat{P}) \cot \widehat{P}\right)\right]} \tag{15}$$

Dans le cas où le parallélogramme est un rectangle de côtés a et b et toujours dans l'hypothèse de $l \leq a$, $l \leq b$ on a

$$\boxed{\mathfrak{M}_l = 2 \left[\pi ab - 2(a+b) l + l^2\right]} \tag{16}$$

1.6 Cas du triangle

Nous supposons que l est inférieur ou égal à la plus petite des hauteurs du triangle. Pour chaque θ la valeur de l'aire Φ de(8) est l'aire du triangle AMN de la figure 5. On aura

$$\Phi = \frac{1}{2} \overline{AM} \cdot \overline{AM} \operatorname{sen} A = \frac{1}{2} \left[1 - \frac{l \operatorname{sen}(B + \theta)}{c \operatorname{sen} B}\right] bc \operatorname{sen} A. \tag{17}$$

L'intégration de 0 à π, on peut la supposer divisée en trois parties de 0 à A, de 0 à B et de 0 à C, correspondantes aux trois angles du triangle. On aura

$$\int_0^A \Phi d\theta = \frac{1}{2} bc \operatorname{sen} A \left[A + \frac{l^2}{2c^2 \operatorname{sen}^2 B} \times (A + \operatorname{sen} C \cos C + \operatorname{sen} B \cos B) - \frac{2l}{c \operatorname{sen} B} (\cos B + \cos C)\right] \tag{18}$$

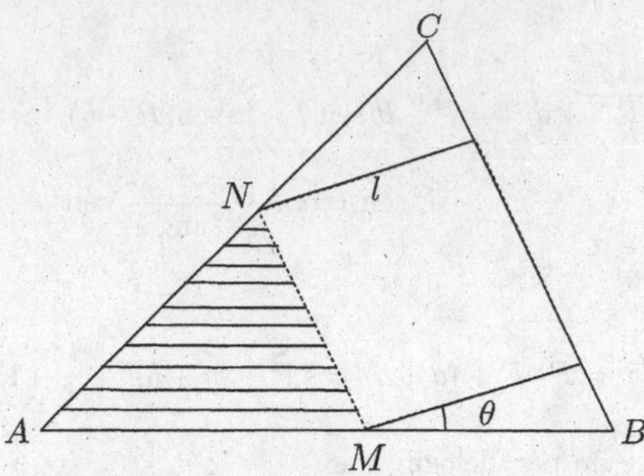

Figure 5:

Mais, dans le triangle ABC on a

$$\text{(19)}\qquad \begin{cases} 2T = bc \operatorname{sen} A = ac \operatorname{sen} B = ab \operatorname{sen} C, \\ a = b\cos C + c\cos B, \\ b = c\cos A + a\cos C, \\ c = a\cos B + b\cos A, \end{cases}$$

où T représente l'aire du triangle. Avec ces égalités, (18) s'écrit:

$$\text{(20)}\qquad \int_0^A \Phi d\theta = T\widehat{A} + \frac{1}{4}l^2\left[\frac{aA}{c\operatorname{sen} B} + \frac{a\cos C}{b} + \frac{a\cos B}{c}\right] - al(\cos B + \cos C).$$

Les relations (19) donnent en outre

$$\text{(21)}\qquad \frac{aA}{c\operatorname{sen} B} = A\cot C + A\cot B,$$

donc, finalement

$$\text{(22)}\qquad \int_0^A \Phi d\theta = T\widehat{A} + \frac{l^2}{4}\left[A\cot B + A\cot C + \frac{a\cos B}{c} + \frac{a\cos C}{b}\right] - al(\cos B + \cos C).$$

De même

$$\text{(23)}\qquad \int_0^B \Phi d\theta = T\widehat{B} + \frac{l^2}{4}\left[B\cot C + B\cot A + \frac{b\cos C}{a} + \frac{b\cos A}{c}\right] - bl(\cos C + \cos A).$$

$$(24)\qquad \int_0^C \Phi d\theta = T\widehat{C} + \frac{l^2}{4}\left[C\cot A + C\cot B + \frac{c\cos A}{b} + \frac{c\cos B}{a}\right] - cl(\cos A + \cos B).$$

En faisant la somme de (22), (23) et (24) et en multipliant par deux puisque l'intégration (8) doit s'étendre de 0 à 2π, on trouve:

$$(25)\qquad \boxed{\mathfrak{M}_l = 2\pi T - 2lL + \frac{l^2}{2}\sum_{A,B,C}((\pi - A)\cot A + 1)}$$

où $L = a + b + c$.

Comme chaque segment non orienté donne lieu à deux segments orientés, si on veut la mesure des segments *non orientés* contenus dans une figure convexe, on doit diviser par 2 les expressions antérieurs (10), (15) et (25).

2 Réseaux uniformes

Nous dirons que le plan est couvert par un *réseau uniforme* de lignes, dans lc cas où il est subdivisé par figures finites conguentes qui remplissent tout le plan. L'aire de cette figure fondamentale nous la représentons par c et la longeur de la partie de ligne qui lui appartient par u. Les figures 6, 7, 8, 9, 11 sont des exemples de réseaux uniformes. Deux positions d'une même figure sur le plan d'un réseau sont dites *différentes* dans le cas où elles ne peuvent être ammenées l'une sur l'autre par une translation du plan qui laisse invariable le réseau. Le nombre de positions différentes que peut prendre un segment de longeur l placé sur le plan d'un réseau uniforme, a une mesure finite. Pour l'obtenir il suffit de considérer l'expression (1) de la mesure et laisser occuper à l'extremité x, y tous les points de l'aire fondamentale c et dans chaque position faire varier φ de 0 à 2π. On aura donc pour *mesure totale* des positions d'un segment orienté sur le plan d'un réseau uniforme

$$(26)\qquad \boxed{\mathfrak{M}_t = \int_c dx\,dy\,d\varphi = 2\pi c}$$

On démontre aussi [1] comme généralisation dans le cas des réseaux uniformes d'une formule de Poincaré de la Géométrie Intégrale que, si n est le nombre de points d'intersection du segment $\vec{S}$ avec les ligne du réseau on a

$$(27)\qquad \boxed{\int_c n dK = 4ul}$$

où, comme toujours, l est la longeur du segment et u la longeur de la part du réseau intérieur à chaque aire fondamentale c. L'intégration de (27) est étendue sur toutes les positions différentes de $\vec{S}$.

3 Probabilités géométriques

3.1 Réseau de cercles

Soit le réseau de la figure 6. Les cercles ont le rayon r et leurs centres son les placés sur les sommets des parallélogrammes de côtés a et b. Pour ce réseau est

$$c = ab \operatorname{sen} \alpha, \qquad u = 2\pi r. \tag{28}$$

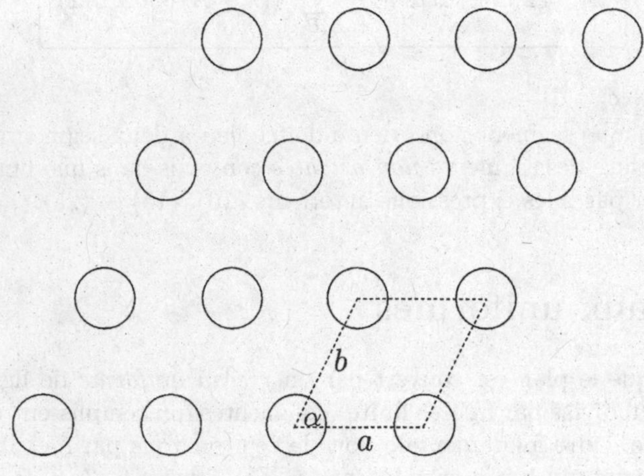

Figure 6:

Supposons un segment de longueur l tel *qu'il ne puisse couper qu'un seul des cercles du réseau.* Répresentons par $\mathfrak{M}_1$ la mesure des positions du segment dans lesquelles il a un seul point d'intersection avec les circonférences du réseau. Par $\mathfrak{M}_2$ la mesure des positions dans lesqueles il a 2 points d'intersection. Par $\mathfrak{M}_i$ la mesure des positions dans lesquelles le segment reste complètement intérieur à une circonférence et par $\mathfrak{M}_e$ la mesure des positions où le segment est extérieur à tous les cercles.

Nou allons calculer ces mesures.

La formule (26) nous donne la mesure totale des positions différentes, c'est à dire, selon (28):

$$\mathfrak{M}_t = \mathfrak{M}_1 + \mathfrak{M}_2 + \mathfrak{M}_i + \mathfrak{M}_e = 2\pi ab \operatorname{sen} \alpha. \tag{29}$$

La formule (27) donne

$$\mathfrak{M}_1 + 2\mathfrak{M}_2 = 4l \cdot 2\pi r. \tag{30}$$

La formule (5) s'écrit

$$\mathfrak{M}_1 + \mathfrak{M}_2 + \mathfrak{M}_i = 2(\pi^2 r^2 + 2\pi i l). \tag{31}$$

Finalement (10) nous donne

$$\mathfrak{M}_t = 2\pi \left[\pi r^2 - 2r^2 \arcsin \frac{l}{2r} - l\sqrt{r^2 - \frac{l^2}{4}} \right]. \tag{32}$$

Moyennant ces 4 équations on peut déterminer les 4 mesures cherchées. On trouve

$$\mathfrak{M}_1 = 4\pi^2 r^2 - 2\mathfrak{M}_i, \tag{33}$$

$$\mathfrak{M}_2 = 4\pi r l - 2\pi^2 r^2 + \mathfrak{M}_i, \tag{34}$$

$$M_e = 2\pi ab \operatorname{sen} \alpha - 2\pi^2 r^2 - 4\pi r l. \tag{35}$$

Ces mesures nous donnent les solutions de probabilités suivantes:

Un segment de longueur l est placé au hasard sur le plan du réseau des cercles de la figure 6. En supposant que ce segment ne puisse couper qu'un seul des cercles du réseau, les probabilités respectives de ce qu'il ait un point d'intersection, ou deux, qu'il soit totalement intérieur à un des cercles, ou bien soit extérieur, sont

$$\begin{cases} p_1 = \dfrac{\mathfrak{M}_1}{\mathfrak{M}_t} = \dfrac{2\pi^2 r^2 - \mathfrak{M}_i}{\pi ab \operatorname{sen} \alpha}, \\[2ex] p_2 = \dfrac{\mathfrak{M}_2}{\mathfrak{M}_t} = \dfrac{4\pi^2 r l - 2\pi^2 r^2 + \mathfrak{M}_i}{2\pi ab \operatorname{sen} \alpha}, \\[2ex] p_i = \dfrac{\mathfrak{M}_i}{\mathfrak{M}_t} = \dfrac{\mathfrak{M}_i}{2\pi ab \operatorname{sen} \alpha}, \\[2ex] p_e = \dfrac{\mathfrak{M}_e}{\mathfrak{M}_t} = \dfrac{2\pi ab \operatorname{sen} \alpha - 2\pi^2 r^2 - 4\pi r l}{2\pi ab \operatorname{sen} \alpha}. \end{cases} \tag{36}$$

Dans ces formules $\mathfrak{M}_i$ a la valeur exprimée par (32) et si $l \geq 2r$ est $\mathfrak{M}_i = 0$.

Si l'on veut seulement avoir la probabilité de ce que le segment coupe quelqu'un des cercles du plan, il sera

$$p = \frac{\mathfrak{M}_1 + \mathfrak{M}_2}{\mathfrak{M}_t} = \frac{2\pi^2 r^2 + 4\pi r l - \mathfrak{M}_i}{2\pi ab \operatorname{sen} \alpha}. \tag{37}$$

3.2 Réseau de parallélogramme

Soit le réseau de la figure 7 ou celui de la 8. Pour ces réseaux la figure fondamentale est un parallélogramme et on a

$$c = ab \operatorname{sen} \alpha, \qquad u = a + b. \tag{38}$$

Supposon qu'on jete au hasard sur le plan un segment de longueur l. *Faisons l'hypothèse que l est suffisamment petit pour ne pouvoir posséder avec le réseau que 2 points d'intersection au maximum.*

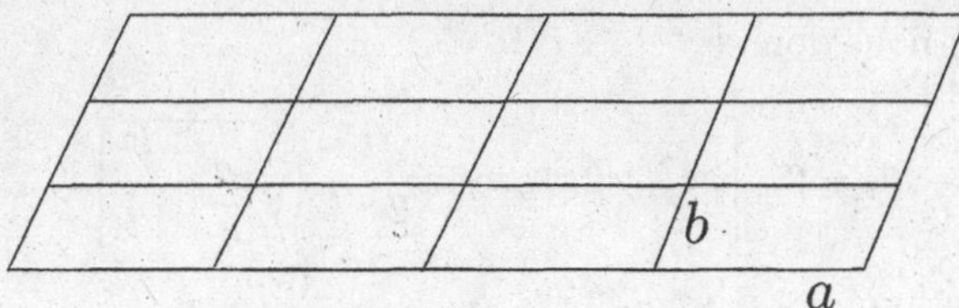

Figure 7:

En appelant $\mathfrak{M}_i$ $(i = 0, 1, 2)$ la mesure de l'ensemble de positions du segment dans lesquelles il a i points d'intersection avec le réseau, la formule (26) donne

$$\mathfrak{M}_t = \mathfrak{M}_0 + \mathfrak{M}_1 + \mathfrak{M}_2 = 2\pi ab \operatorname{sen} \alpha. \tag{39}$$

D'autre part (27) s'écrit

$$\mathfrak{M}_1 + 2\mathfrak{M}_2 = 4l(a + b). \tag{40}$$

Finalement (15) est

$$\mathfrak{M}_0 = 2 \left[\pi ab \operatorname{sen} \alpha - 2(a+b)l + \frac{l^2}{2}(2 + (\pi - 2\alpha) \cot \alpha) \right]. \tag{41}$$

Avec ces 3 équations on trouve

$$\mathfrak{M}_1 = 4(a+b)l - 2[2 + (\pi - 2\alpha) \cot \alpha] l^2, \tag{42}$$

$$\mathfrak{M}_2 = [2 + (\pi - 2\alpha) \cot \alpha] l^2. \tag{43}$$

Ainsi, *les probabilités respectives de ce que le segment ait aucun, un ou deux points d'intersection sont*

$$\begin{cases} p_0 = \dfrac{\mathfrak{M}_0}{\mathfrak{M}_t} = 1 - \dfrac{2(a+b)l}{\pi ab \operatorname{sen} \alpha} + \dfrac{l^2}{2\pi ab \operatorname{sen} \alpha}[2 + (\pi - 2\alpha) \cot \alpha] \\[2ex] p_1 = \dfrac{\mathfrak{M}_1}{\mathfrak{M}_t} = \dfrac{2(a+b)l}{\pi ab \operatorname{sen} \alpha} - \dfrac{[2 + (\pi - 2\alpha) \cot \alpha] l^2}{\pi ab \operatorname{sen} \alpha}, \\[2ex] p_2 = \dfrac{\mathfrak{M}_2}{\mathfrak{M}_t} = \dfrac{[2 + (\pi - 2\alpha) \cot \alpha] l^2}{2\pi ab \operatorname{sen} \alpha}. \end{cases} \tag{44}$$

3.3 Réseau de triangles équilatéraux

Considérons le réseau de triangles de la figure 9. La figure fondamentale de ce réseau est le losange de la figure 10 et on a

$$c = \frac{\sqrt{3}}{2} a^2, \qquad u = 3a. \tag{45}$$

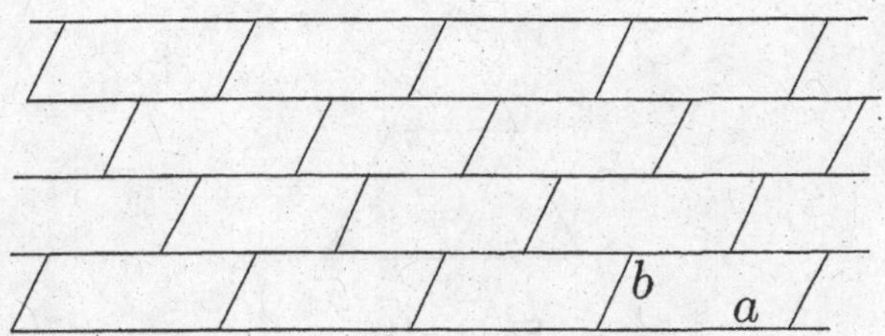

Figure 8:

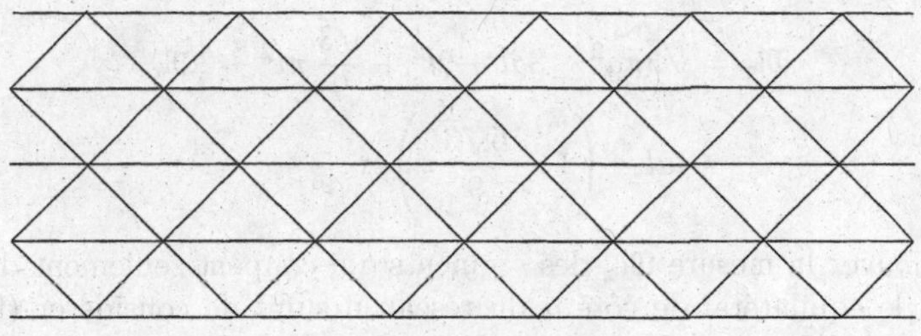

Figure 9:

Un segment de longueur l plus petite que la hauteur de ces triangles ne peut avoir avec le réseau que 0, 1, 2 ou 3 points d'intersection. Nous voulons chercher la probabilité dans chaque cas.

Représentons par $\mathfrak{M}_i$ $(i = 0, 1, 2)$ la mesure de l'ensemble de positions dans lesquelles le segment a i points d'intersection avec les côtés d'un triangle particulier du réseau.

La formule (25) nous donne:

$$\mathfrak{M}_o = \frac{\sqrt{3}}{2}\,\pi a^2 - 6al + \left(\frac{2\sqrt{3}}{3}\pi + 3\right)\frac{l^2}{2}, \tag{46}$$

où a est le côté des triangles du réseau. La mesure $\mathfrak{M}_1$ est égal à la mesure (15)

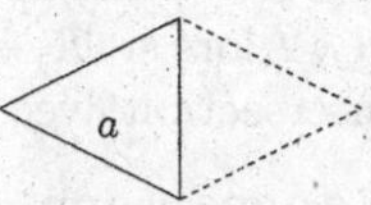

Figure 10:

des segments intérieurs au losange de côté a moins deux fois la mesure de ceux

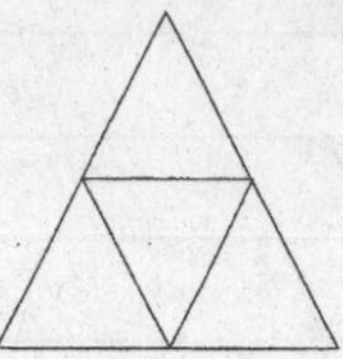

Figure 11:

intérieurs au triangle, c'est à dire,

$$(47)\qquad \mathfrak{M}_1 = \sqrt{3}\pi a^2 - 8al + 2l^2 + \frac{\sqrt{3}}{9}\pi l^2 - 2\mathfrak{M}_0$$
$$= 4al - \left(1 + \frac{5\sqrt{3}}{9}\pi\right) l^2 ,$$

Pour trouver la mesure $\mathfrak{M}_2$ des segments qui coupent seulement deux côtés d'un triangle équilatéral de côté a du réseau il suffit de considérer (fig. 11) le triangle de côté 2α dont la mesure des segments de longueur l que lui sont intérieurs, par (25), est:

$$2\sqrt{3}\pi a^2 - 12al + \left(\frac{2\sqrt{3}}{3}\pi + 3\right) \frac{l^2}{2},$$

c'est à dire,

$$(48)\qquad 4\mathfrak{M}_0 + 3\mathfrak{M}_1 + 3\mathfrak{M}_2 = 2\sqrt{3}\pi a^2 - 12al + \left(\frac{2\sqrt{3}}{3}\pi + 3\right) \frac{l^2}{2},$$

et en tenant compte de (46) et (47)

$$(49)\qquad \mathfrak{M}_2 = \left(\frac{2\sqrt{3}}{9}\pi - \frac{1}{2}\right) l^2 .$$

Dans (46) $\mathfrak{M}_0$ est la mesure des segments intérieurs à un triangle de réseau. Puisque chaque aire fondamentale se compose de deux triangles, on doit prendre $2\,\mathfrak{M}_0$. D'une façon analogue $\mathfrak{M}_1$ est la mesure des segments qui coupent un côté du triangle; il faudra prendre $3\,\mathfrak{M}_1$ pour avoir la mesure complète de ceux qui ont un point d'intersection avec le réseau. Puisque chaque figure fondamentale a 6 angles on doit prendre aussi $6\,\mathfrak{M}_2$. Alors si $\mathfrak{M}_3$ est la *mesure totale* des positions où le segment a trois points d'intersection avec le réseau, la formule (27) donne

$$(50)\qquad 3\,\mathfrak{M}_1 + 12\,\mathfrak{M}_2 + 3\,\mathfrak{M}_3 = 12al,$$

d'où

$$(51)\qquad \mathfrak{M}_3 = \left(3 - \frac{\sqrt{3}\pi}{3}\right) l^2.$$

En divisant ces mesures partials pour la mesure totale (26)

$$\mathfrak{M}_t = 2\pi \cdot \frac{1}{2}\sqrt{3}a^2 = \sqrt{3}\pi a^2, \tag{52}$$

on aura:

Les probabilités de ce qu'une tige de longueur l, jetée au hasard sur un réseau de triangles équilateraux de côté a $\left(l \leq \frac{\sqrt{3}}{2}a\right)$ ait 0, 1, 2 ou 3 points d'intersection sont, respectivament:

$$\begin{cases} p_0 = \dfrac{2\,\mathfrak{M}_o}{\mathfrak{M}_t} = 1 - \dfrac{4\sqrt{3}}{\pi}\dfrac{l}{a} + 2\left(\dfrac{\sqrt{3}}{2\pi} + \dfrac{1}{3}\right)\dfrac{l^2}{a^2}, \\ p_1 = \dfrac{3\,\mathfrak{M}_1}{\mathfrak{M}_t} = \dfrac{4\sqrt{3}}{\pi}\dfrac{l}{a} - \left(\dfrac{\sqrt{3}}{\pi} + \dfrac{5}{3}\right)\dfrac{l^2}{a^2}, \\ p_2 = \dfrac{6\,\mathfrak{M}_2}{\mathfrak{M}_t} = \left(\dfrac{4}{3} - \dfrac{\sqrt{3}}{\pi}\right)\dfrac{l^2}{a^2}, \\ p_3 = \dfrac{\mathfrak{M}_3}{\mathfrak{M}_t} = \left(\dfrac{\sqrt{3}}{\pi} - \dfrac{1}{3}\right)\dfrac{l^2}{a^2}. \end{cases} \tag{53}$$

On peut constater que

$$p_0 + p_1 + p_2 + p_3 = 1. \tag{54}$$

Si l'on veut seulement la probabilité de couper le réseau, on aura:

$$p = 1 - p_0 = \frac{4\sqrt{3}}{\pi}\frac{l}{a} - 2\left(\frac{\sqrt{3}}{2\pi} + \frac{1}{3}\right)\frac{l^2}{a}. \tag{55}$$

3.4 Réseau d'hexagones réguliers

Considérons le réseau de la figure 12 et un segment de longueur l que peut seulement couper le réseau dans 2 points au maximum, c'est à dire, l est plus petite que le côté a des hexagones.

Dans ce cas la figure fondamentale sera l'hexagone de la figure 13 avec

$$c = \frac{3\sqrt{3}}{2}a^2, \qquad u = 3a. \tag{56}$$

Pour trouver la mesure $\mathfrak{M}_o$ des segments qui sont à l'intérieur d'un hexagone du réseau il suffit de prolonguer deux paires de côtés opposés jusqu'à former un losange de côtés $2a$. Alors $\mathfrak{M}_o$ sera égal à la mesure des segments intèrieurs au losange (15), moins deux fois les segments intérieurs à un triangle équilatéral de côté a (46) moins deux fois la mesure des segments qui coupent un seul segment du réseau de triangles équilatéraux (47), moins quatre fois la mesure (49) des segments qui coupent deux côtés du même réseau. Ainsi on trouve

$$\mathfrak{M}_o = 3\sqrt{3}\pi a^2 - 12al + \left(3 - \frac{\sqrt{3}}{3}\pi\right)l^2. \tag{57}$$

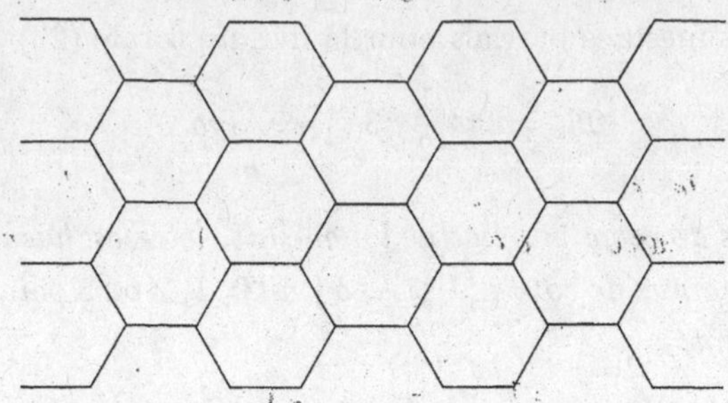

Figure 12:

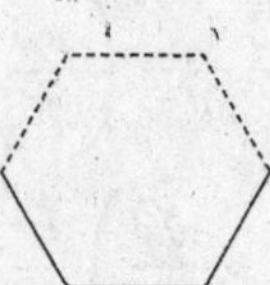

Figure 13:

La formule (27) donne

$$\mathfrak{M}_1 + 2\,\mathfrak{M}_2 = 12al, \tag{58}$$

et selon (26) et (56) on a

$$\mathfrak{M}_t = \mathfrak{M}_0 + \mathfrak{M}_1 + \mathfrak{M}_2 = 3\sqrt{3}\pi a^2. \tag{59}$$

De (57), (58) et (59) on déduit

$$\mathfrak{M}_1 = 12al - \left(6 - \frac{2\sqrt{3}}{3}\pi\right) l^2, \tag{60}$$

$$\mathfrak{M}_2 = \left(3 - \frac{\sqrt{3}}{3}\pi\right) l^2. \tag{61}$$

Donc: *Les probabilités de ce qu'une tige de longueur l jetée au hasard sur le plan du réseau hexagonal de la fig. 12 ait* 0, 1 *ou* 2 *points d'intersection avec le réseau sont*

$$\begin{cases} p_0 = \dfrac{\mathfrak{M}_0}{\mathfrak{M}_t} = 1 - \dfrac{4\sqrt{3}}{3\pi}\,\dfrac{l}{a} + \left(\dfrac{\sqrt{3}}{3\pi} - \dfrac{1}{9}\right)\dfrac{l^2}{a^2} \\ p_1 = \dfrac{\mathfrak{M}_1}{\mathfrak{M}_t} = 1 - \dfrac{4\sqrt{3}}{3\pi}\,\dfrac{l}{a} - \left(\dfrac{2\sqrt{3}}{3\pi} - \dfrac{2}{9}\right)\dfrac{l^2}{a^2} \\ p_2 = \dfrac{\mathfrak{M}_2}{\mathfrak{M}_t} = \left(\dfrac{\sqrt{3}}{3\pi} - \dfrac{1}{9}\right)\dfrac{l^2}{a^2}\,. \end{cases} \tag{62}$$

On suppose l plus petite que le côté a des hexagones.

Si l'on veut seulement avoir la probabilité de ce que le segment coupe les lignes du réseau, il sera

$$p = l - p_0 = \frac{4\sqrt{3}}{3\pi}\frac{l}{a} - \left(\frac{\sqrt{3}}{3\pi} - \frac{1}{9}\right)\frac{l^2}{a^2} \tag{63}$$

Rosario (Rép. Argentina),
Instituto de Matemàtics, Enero 1940.
(Reçu le 5, Mrs 1940)

References

[1] L. A. Santaló: Geometria Integral 31. Sobre valores medios y probabilidades geométricas, Hamburg Abhandlungen 1939.

[2] W. Blaschke, Vorlesungen über Integral geometrie I, 1936, pag. 2.

SOBRE LA DISTRIBUCION PROBABLE DE CORPUSCULOS EN UN CUERPO, DEDUCIDA DE LA DISTRIBUCION EN SUS SECCIONES Y PROBLEMAS ANALOGOS

por L. A. Santaló

Supongamos N corpúsculos de forma convexa distribuídos de manera arbitraria en el interior de un cuerpo también convexo K. Si V es el volumen de este cuerpo llamaremos *densidad media* de corpúsculos en K al cociente

$$D = \frac{N}{V} \tag{1}$$

o sea al número de ellos por unidad de volumen si la repartición fuese uniforme.

Si una recta arbitraria que atraviesa el cuerpo corta a n de estos corpúsculos y la longitud de la cuerda que determina en K es s, el cociente $\frac{n}{s}$ será la densidad media de los corpúsculos sobre la recta. Haciendo esta operación con varias rectas elegidas al azar, el objeto de esta nota consiste en ver cómo de estas distribuciones sobre las rectas se puede deducir la densidad media (1) del cuerpo entero.

En lugar de cortar por una recta se puede también cortar por varios planos y deducir de la densidad de corpúsculos en las secciones la densidad de ellos en todo el cuerpo. También consideramos los casos (§ 2, nos. 7, 9) de cortar por franjas limitadas por dos planos paralelos o bien por cilindros de sección convexa.

Como aplicación se obtiene en § 2, nº. 3 un resultado clásico de la teoría cinética de los gases.

El método de demostración consiste en utilizar algunos resultados de Probabilidades Geométricas o de Geometría Integral para los cuales remitiremos al lector principalmente a los libros de Blaschke «*Vorlesungen über Integralgeometrie*, I y II» [1] o al de Deltheil «*Probabilités géometriques*» [2] (1).

(1) Los paréntesis cuadrados se refieren a la Bibliografía citada al final.

§ 1. Caso del plano

1. *Secciones por una recta.* Supongamos primeramente una figura convexa K que contenga en su interior N corpúsculos C congruentes entre sí pero de forma cualquiera (con la sola restricción de ser también *convexos*).

Una recta cualquiera G del plano vendrá determinada por sus coordenadas polares p, ϑ o sea su distancia a un punto fijo y el ángulo de su normal con una dirección también fija. Dar una recta arbitraria en el plano significa lo mismo que dar al azar un par de números p y ϑ (p cualquier número positivo y ϑ comprendido entre 0 y 2π) y para medir un conjunto de rectas se toma la integral extendida a este conjunto de la forma diferencial [1, p. 7], [2, p. 59],

$$d\,G = d\,p \,.\, d\vartheta. \tag{2}$$

Con esta expresión si se considera una línea cualquiera de longitud l y llamamos n_1 al número de puntos en que es cortada en cada posición de la recta G se verifica [1, p. 11], [2, p. 60].

$$\int n_1\, d\,G = 2\,l \tag{3}$$

extendida la integración a todas las rectas que cortan a la línea considerada, únicas para las que es $n_1 \neq 0$.

Vamos a aplicar esta fórmula (3) a la línea formada por todos los contornos de los corpúsculos contenidos en K (fig. 1).

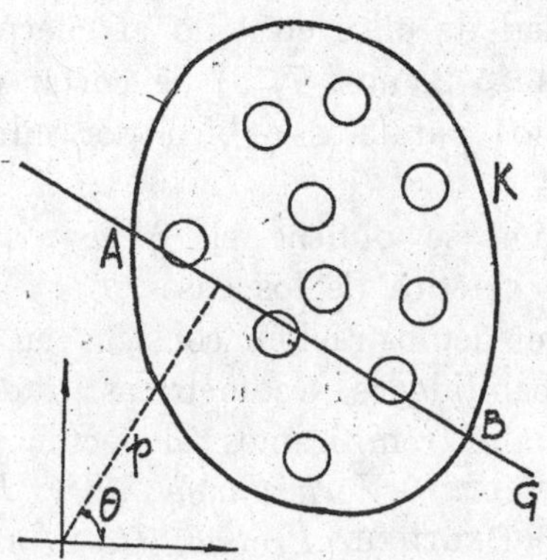

Fig. 1.

Si u es el perímetro de cada uno de ellos, la longitud total será $N u$ y si se representa por n el número de corpusculos que son cortados por la recta G, como cada uno tiene dos puntos de intersección, será

$$\int\limits_{G.K \neq 0} n\, dG = N u \tag{4}$$

donde con el símbolo $G.K \neq 0$ entendemos que la integración debe extenderse a todas las rectas que cortan a K.

Por otra parte, si se representa por U el perímetro de la figura K, por ser convexa y por tanto n_1 constante e igual a dos, la fórmula (3) nos da como medida de todas las rectas que cortan a K:

$$\int\limits_{G.K \neq 0} dG = U. \tag{5}$$

Dividiendo (4) por (5) se deduce, como *valor medio del número de corpúsculos que son cortados por una recta arbitraria G que atraviesa la figura K*:

$$\bar{n} = N \frac{u}{U}. \tag{6}$$

2. Llamando s a la longitud de la cuerda que la recta G determina en K y F al área de esta figura, es fácil ver que [1, p. 19], [2, p. 74],

$$\int\limits_{G.K \neq 0} s\, dG = \int s\, dp\, d\vartheta = \int_0^\pi F\, d\vartheta = \pi F. \tag{7}$$

Por tanto la *longitud media* de las cuerdas de K vale, dividiendo (7) por (5),

$$\bar{s} = \pi \frac{F}{U}. \tag{8}$$

3. Si se quiere ahora ver la densidad de corpúsculos sobre las rectas que atraviesan a K, bastará dividir el número medio (6) de ellos por la cuerda media (8), obteniéndose

$$\delta_G = \frac{\bar{n}}{s} = \frac{u}{\pi} \frac{N}{F}, \tag{9}$$

pero $\frac{N}{F}$ es la densidad media D de corpúsculos en la figura total K o sea el número de ellos por unidad de área si la repartición fuera uniforme, luego (9) puede escribirse

$$\delta_G = \frac{u}{\pi} D, \tag{10}$$

que es la relación que liga *la densidad media de corpúsculos sobre las cuerdas de las rectas que atraviesan a K con la densidad total de ellos.*

Por ejemplo, si los corpúsculos son circulares de radio r es

$$\delta_G = 2rD. \tag{11}$$

Esta expresión representa también el número de corpúsculos que encuentra la recta G por unidad de longitud y por tanto la *distancia media* entre ellos será la inversa, o sea

$$d_G = \frac{1}{2rD} \tag{12}$$

que nos da la *distancia media entre dos corpúsculos vecinos de los N que están distribuídos al azar en el área F.*

3. PÓLYA en un artículo muy sugestivo [5], en el que llega al mismo resultado, aunque por camino completamente distinto, da a esta cuestión la siguiente interpretación gráfica: Si desde un punto arbitrario de un bosque rodeado de árboles con sección circular de radio r distribuídos con una densidad media D, se mira en todas direcciones hasta donde alcanza la vista (fig. 2) la *distancia visible media* está dada por (12).

Tomando la inversa de la fórmula (10) tendríamos una expresión más general de esta misma distancia media para el caso de no ser las secciones circulares. (Ver el Apéndice, § 3).

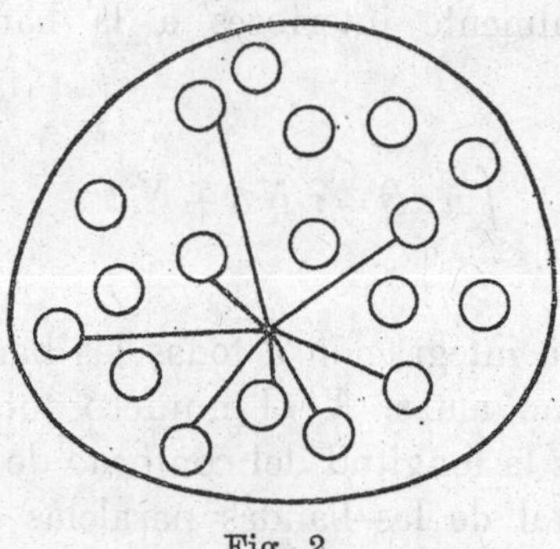

Fig. 2.

4. *Secciones por una banda paralela.* Sea el mismo problema anterior, pero ahora supongamos que se corta la figura K por una banda limitada por dos rectas paralelas a distancia Δ y de la densidad media de corpúsculos en la sección queremos deducir la densidad en la figura total (fig. 3).

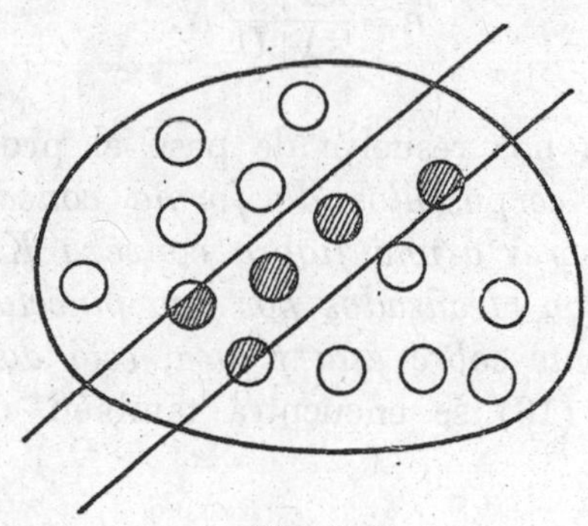

Fig. 3.

La posición de una de estas bandas paralelas queda determinada, en el plano, por las mismas coordenadas p, ϑ de una recta que puede ser, por ejemplo, su paralela media. Representando ahora por dB (densidad para conjuntos de bandas paralelas) la misma forma diferencial $dp\, d\vartheta$ de antes, o sea

$$dB = dp\, d\vartheta \qquad (13)$$

y siendo n el número de corpúsculos que en cada posición que-dan total o parcialmente interiores a la banda B, es sabido que [2]

$$\int_{B.K \neq 0} n\, dB = \pi N \Delta + N u \tag{14}$$

estando extendida la integración a todas las bandas B que cortan a K y siendo, como antes, N el número total de corpúsculos contenidos en K y u la longitud del contorno de cada uno de ellos.

La medida total de las bandas paralelas de anchura Δ que cortan a K es [7, p. 18]

$$\int_{B.K \neq 0} dB = U + \pi \Delta \tag{15}$$

luego dividiendo (14) por (15) se deduce como *valor medio del número de corpúsculos encontrados por una banda arbitraria*

$$\bar{n} = \frac{\pi\Delta + u}{\pi\Delta + U} N. \tag{16}$$

Esta fórmula nos resuelve de paso el problema siguiente: *Supuestos n corpúsculos de forma convexa cualquiera y perímetro u interiores a una figura convexa K, el número medio de ellos que son alcanzados por una pincelada de anchura* Δ *dada arbitrariamente sobre esta figura, está dado por* (16).

La fórmula (16) se encuentra también en E. GASPAR [3, p. 136].

---

(2) Sobre bandas paralelas móviles en el plano y para las fórmulas (15) y (17) se puede ver nuestra memoria [7].

La fórmula (14) se puede deducir de la de BLASCHKE [1, p. 37]

$$\int c_{01}\, dK = 2\pi\, (C_0 F_1 + C_1 F_0 + U_0 U_1) \quad (*)$$

con solo suponer que la figura K que aquí interviene, de área F_0, longitud U_0 y curvatura total C_0, se reduce a una banda de plano limitada por rectas paralelas a distancia Δ. Habrá que sustituir dK por dB, F_0 por $\frac{1}{2}\Delta$, U_0 por 1 y C_0 se anula. Como figura fija se suponen aquí los N corpúsculos, los cuales, siendo convexos, dan $C_{01} = 2\pi n$, $C_1 = 2\pi N$ y $U_1 = Nu$. Con estas sustituciones la fórmula (*) pasa a la (14) que utilizamos en el texto.

Una demostración directa de la fórmula (14) ha sido dada por E. GASPAR [3, p. 135].

5. Llamando f al área de la parte de K que queda interior a la banda B, es sabido también que [7, p. 19], [1, p. 51],

$$\int_{B.K \neq 0} f\, dB = \pi \Delta F \tag{17}$$

y por tanto el valor medio de f vale

$$\bar{f} = \frac{\pi \Delta F}{U + \pi \Delta}. \tag{18}$$

Dividiendo (16) por (18) se tendrá el *valor medio de la densidad de corpúsculos en las secciones obtenidas cortando K por B*, que valdrá

$$\delta_B = \left(1 + \frac{u}{\pi \Delta}\right) D. \tag{19}$$

Esta es la relación buscada que nos liga *la densidad total D con la densidad media de las secciones.*

§ 2. Caso del espacio

1. *Secciones por una recta.* Sea en el espacio un cuerpo convexo K conteniendo en su interior N corpúsculos iguales también convexos pero de forma cualquiera y distribuídos de una manera arbitraria. Queremos obtener la relación existente entre el número de estos corpúsculos que corta una recta al azar que atraviesa el cuerpo con el número total de ellos.

Una recta en el espacio está determinada por cuatro parámetros que pueden ser: las coordenadas x, y de su punto de intersección con un plano normal y la longitud y latitud φ y ϑ que fijan su dirección. Representando por $d\Omega$ el elemento de área sobre la esfera unidad correspondiente a la dirección de la recta, para medir conjuntos de rectas se toma [1, p. 66], [2, p. 90] la integral extendida al conjunto que sea, de la forma diferencial

$$dG = dx\, dy\, d\Omega. \tag{20}$$

Por ejemplo, llamando n_1 al número de puntos de intersección de la recta G con una superficie de área F, este número n_1 dependerá de la posición de la recta, y extendiendo la integración a todas las posiciones de la misma se verifica [1, p. 69], [2, p. 89],

$$\int n_1 \, dG = \pi F. \tag{21}$$

Consideremos la superficie formada por la suma de las superficies de los N corpúsculos, cuya área valdrá Nf, si f es el área de cada uno de ellos. Como G sólo puede tener con cada corpúsculo dos puntos de intersección, llamando n al número de corpúsculos que la recta G encuentra al atravesar K, la fórmula (21) da

$$\int_{G.K \neq 0} n \, dG = \frac{\pi}{2} N f. \tag{22}$$

Por otra parte, la medida del conjunto total de las rectas que cortan a K es (también según (21) y siendo ahora F el el área de K)

$$\int_{G.K \neq 0} dG = \frac{\pi}{2} F, \tag{23}$$

luego, dividiendo (22) por (23):

$$\bar{n} = N \frac{f}{F} \tag{24}$$

que nos da el *valor medio del número de corpúsculos que encontrará una recta arbitraria que atraviesa a K.*

2. Llamando s a la longitud de la cuerda que la recta G determina en K, es fácil ver, teniendo en cuenta (20) que [1, p. 77]

$$\int_{G.K \neq 0} s \, dG = 2\pi V \tag{25}$$

siendo V el volumen de K. De (25) y (23), por división, obtenemos como valor medio de la longitud de la cuerda s:

$$\bar{s} = \frac{4V}{F}. \tag{26}$$

Dividiendo el número medio de corpúsculos que encuentra G por la longitud media (26) de la cuerda, se tendrá la densidad media δ_G de corpúsculos sobre la recta G que será

$$\delta_G = \frac{f}{4} D, \tag{27}$$

recordando que D es la densidad $\frac{N}{V}$ de los corpúsculos en K.

Esta es, por consiguiente, *la relación existente entre la densidad de corpúsculos sobre una recta que atraviesa el cuerpo y la densidad total en el mismo.*

Dividiendo, inversamente, (26) por (24), se obtendrá la *distancia media* entre los *corpúsculos vecinos,* que valdrá

$$\bar{d} = \frac{4}{fD}. \tag{28}$$

Al decir distancia media entre corpúsculos «vecinos» entendemos, como se deduce de la manera como obtenemos la fórmula (28), que se consideran únicamente las distancias de cada corpúsculo a aquellos que son visibles desde el mismo, es decir. que se pueden unir por un segmento que no encuentra a ningún otro corpúsculo intermedio.

Si los corpúsculos son esféricos de radio r, será

$$\bar{d} = \frac{1}{\pi r^2 D}. \tag{29}$$

Pólya, en el artículo ya citado [5], obtiene también esta fórmula que interpreta de la manera siguiente: supuesta una caída de nieve formada por copos esféricos de radio r con una densidad D, la distancia visible media según todas las direcciones desde un punto envuelto por la nevada, está dada por (29). (Ver el Apéndice § 3).

3. *Aplicación a la teoría cinética de los gases.* El mismo método anterior sirve para resolver el problema siguiente: Sea un cuerpo K de área F y volumen V que contiene en su interior N corpúsculos convexos e iguales entre sí. ¿Cuál será el recorrido libre medio de un punto sin dimensiones que se mueve dentro de K en dirección arbitraria (todas igualmente probables)?

Sea G la recta sobre la cual en un momento considerado se mueve el punto (fig. 1, aunque ahora estamos en el espacio, la figura 1 sirve lo mismo). Los recorridos libres posibles serán los segmentos de esta recta limitados por dos corpúsculos sucesivos o por un corpúsculo y la pared de K. Llamando como antes s a la cuerda total que G determina en K o sea $s = AB$ y s_i a las cuerdas parciales interiores a los corpúsculos, la suma de recorridos libres posibles sobre G es

$$s - \Sigma s_i,$$

extendida la sumación al número de corpúsculos que son cortados por la recta G. En cuanto al número de recorridos libres o sea de segmentos en que la cuerda s de G queda dividida por los corpúsculos y las paredes del recipiente, es $n+1$ siendo n, como antes, el número de corpúsculos cortados por G. Extendiendo la integración a todas las rectas que cortan a K, la fórmula (25) da:

$$\int_{G.K=/=0} (s - \Sigma s_1)\, dG = 2\pi\,(V - Nv), \tag{30}$$

siendo v el volumen de cada corpúsculo. Además por la fórmula (21):

$$\int_{G.K=/=0} (n+1)\, dG = \frac{\pi}{2}\,(F + Nf) \tag{31}$$

siendo f el área de cada corpúsculo y F la del cuerpo K.

El *valor medio de los recorridos libres* se obtendrá dividiendo su suma total (30) por el número de ellos (31) o sea

$$\bar{l} = \frac{4\,(V - Nv)}{F + Nf}. \tag{32}$$

En el caso de ser los corpúsculos esféricos de radio r esta fórmula queda

$$\bar{l} = \frac{4\,(V - \frac{4}{3}N\pi r^3)}{F + 4\pi r^2 N}. \tag{33}$$

Esta fórmula se puede aplicar al caso del recorrido medio de las moléculas de un gas. Para ello basta observar que hasta ahora hemos considerado el recorrido medio de un punto sin dimensiones, si se quiere que sea el de una molécula o corpúsculo de radio r bastará suponer que ella se reduce a un punto y todas las demás duplican el radio puesto que entonces cuando el punto toque a una de estas moléculas de radio doble, es efectivamente el caso en que las dos moléculas se encontrarían. Sustituyendo en la fórmula anterior r por $2r$ se tiene pues como recorrido libre medio de las moléculas de un gas la fórmula conocida (ver por ejemplo [4, p. 34]),

$$\overline{l} = \frac{4\,(V-b)}{F+16\pi r^2 N} \tag{34}$$

indicando por b el volumen total de estas moléculas supuestas de radio $2r$.

En general para un gas a no mucha presión b y F son despreciables al lado de los demás términos de esta expresión y como valor aproximado se toma

$$\overline{l} = \frac{V}{4\pi r^2 N} = \frac{1}{4\pi r^2 D} \tag{35}$$

siendo D el número medio de moléculas por unidad de volumen.

4. *Caso de trayectorias curvas.* Recordando algunas fórmulas más de Geometría integral es fácil demostrar que el recorrido libre medio de un punto en el interior del cuerpo convexo K que contiene N corpúsculos distribuídos arbitrariamente es independiente de la forma de la trayectoria. Es decir, tiene el mismo valor dado por (32) lo mismo si la trayectoria es rectilínea, como allí se supuso, que si es circular o elíptica o cualquiera.

En efecto, supongamos que el punto debe moverse describiendo una trayectoria de forma arbitraria (pero siempre la misma) por ejemplo la representada por una curva cerrada C de longitud L (fig. 4).

Esta curva C (plana o alabeada) viene fijada en el espacio por seis coordenadas que pueden ser las tres x, y, z de uno de sus puntos, más las ϑ y φ de una dirección por este punto, más la τ de una rotación alrededor de esta dirección. Entonces es

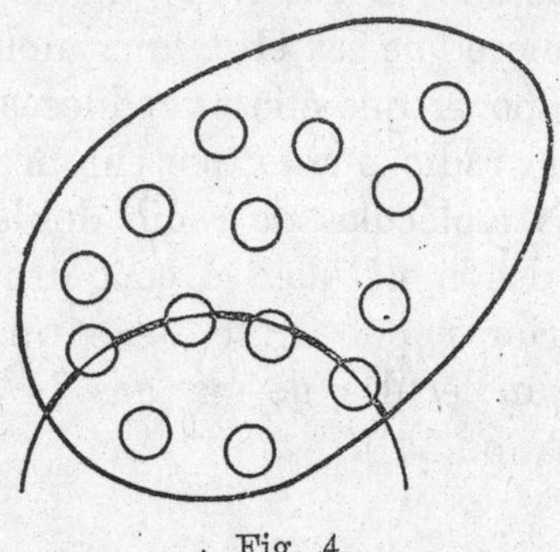

Fig. 4.

sabido que para medir un conjunto de posiciones de una tal línea se toma la integral de la llamada *densidad cinemática* que vale [1, p. 64], [6, p. 13],

$$dC = \cos\vartheta\, dx\, dy\, dz\, d\vartheta\, d\varphi\, d\tau. \qquad (36)$$

Esta medida tiene la propiedad que la caracteriza de ser independiente de la posición de los ejes coordenados de referencia, o, en otras palabras, es invariante por movimientos.

Llamando n_1 al número de puntos de intersección de esta línea con las superficies de los N corpúsculos más la del cuerpo K que los contiene, vale [6, p. 39],

$$\int n_1\, dC = 4\pi^2 (F + N f) L \qquad (37)$$

siendo como antes f el área de los corpúsculos y F la de K.

Luego si n es el número de partes, interiores a K y limitadas por dos corpúsculos o las paredes, en que c queda dividida (fig. 4), como es $n_1 = 2n$, será

$$\int n\, dC = 2\,\pi^2 (F + N f) L. \qquad (38)$$

Por otra parte, si s es la parte de C que en cada posición queda interior a K y Σs_i la suma de las partes interiores a los corpúsculos, se verifica [6, p. 42]

$$\int (s - \Sigma s_i)\, dC = 8\pi^2 (V - N v) L. \qquad (39)$$

Dividiendo la longitud total (39) de las partes en que C queda dividida por los corpúsculos y paredes de K, por la suma del número de partes (38), se tendrá la *longitud media* de ellas o sea *el recorrido libre medio para un punto cuya trayectoria tuviera la forma de la línea* C, que será

$$\bar{l} = \frac{4(V - Nv)}{F + Nf}$$

que es el mismo valor (32) obtenido para el caso de las trayectorias rectilíneas.

5. *Secciones por un plano.* Veamos ahora la relación existente entre la densidad de corpúsculos en una sección plana arbitraria del cuerpo y la densidad total. Un plano viene determinado por tres parámetros que pueden ser: su distancia p a un punto fijo y la latitud y longitud ϑ y φ de la dirección de su normal. Para medir conjuntos de planos se toma la integral, extendida al conjunto de que se trate, de la forma diferencial [1, p. 66], [2, p. 92]

$$dE = dp\, d\Omega \qquad (40)$$

siendo, como antes, $d\Omega$ el elemento de área sobre la esfera unidad correspondiente a la dirección normal al plano.

Supuestos como siempre N corpúsculos congruentes de área f, volumen v e integral de curvatura media m (3) repartidos arbitrariamente en el interior de K, si se representa por n el número de ellos que son cortados por un plano E que corta a K, es sabido [1, p. 102] que se verifica

(3) Recuérdese que se llama *integral de la curvatura media* a la integral $\frac{1}{2}\int \left(\frac{1}{r_1} + \frac{1}{r_2}\right) do$ extendida a toda la superficie del cuerpo siendo r_1 y r_2 los radios principales de curvatura correspondientes al elemento superficial do. Por ejemplo para un corpúsculo esférico de radio r sería $m = 4\pi r$. Si la superficie convexa tiene líneas de discontinuidad de la curvatura media, el valor que se debe atribuir a m es el límite de la curvatura media del cuerpo paralelo exterior a distancia ε cuando ε tiende a cero.

$$\int n\, dE = N m. \tag{41}$$

Por otra parte, la medida de todos los planos que cortan a K es [1, p. 72], [2, p. 95],

$$\int_{E.K \neq 0} dE = M \tag{42}$$

siendo M la curvatura media total de K. De (41) y (42) por división se deduce *el valor medio del número de corpúsculos que serán cortados por un plano dado al azar,* que será

$$\bar{n} = N \frac{m}{M}. \tag{43}$$

6. Llamando σ al área de la sección que el plano E determina en K es fácil deducir de la definición (40) que es [1, p. 74]

$$\int \sigma\, dE = 2\pi V \tag{44}$$

y dividiendo por (42), el *valor medio* de σ será:

$$\bar{\sigma} = 2\pi \frac{V}{M}. \tag{45}$$

Dividiendo el número medio (43) de corpúsculos que son cortados por E, por el valor medio (45) del área de la sección, se tendrá la *densidad media* de corpúsculos en las secciones de K por un plano arbitrario, que valdrá

$$\delta_E = \frac{mN}{2\pi V} = \frac{m}{2\pi} D \tag{46}$$

que relaciona *la densidad media de la sección con la densidad del conjunto.*

Si los corpúsculos son esféricos de radio r, es $m = 4\pi r$ y esta relación se reduce a

$$\delta_E = 2rD.$$

7. *Secciones por franjas paralelas.* Lo mismo dicho en el nº. 4 de § 1 para el caso del plano, se puede repetir para el espacio, suponiendo que se corta el cuerpo convexo K por una franja de espacio limitada por dos planos paralelos a distancia Δ. La posición de una de estas franjas queda determinada por la de un plano invariablemente unida a ella, de manera que como medida de un conjunto de franjas se toma la integral de la misma expresión diferencial (40) que ahora representaremos por

$$dB = dp\, d\Omega. \tag{47}$$

Llamando n al número de corpúsculos que en cada posición son alcanzados por la franja B, tiene lugar la fórmula [1, p. 101]. [8],

$$\int n\, dB = (2\pi\Delta + m)\, N \tag{48}$$

siendo, como antes, m la curvatura media total de cada corpúsculo. Como además la medida total de las franjas que cortan a K es [7, p. 34],

$$\int_{B.K \neq 0} dB = M + 2\pi\Delta, \tag{49}$$

el *valor medio del número de corpúsculos que son alcanzados por un corte de anchura Δ dado al azar en el cuerpo K* será el cociente de (48) y (49) o sea

$$\bar{n} = \frac{2\pi\Delta + m}{2\pi\Delta + M}\, N. \tag{50}$$

8. Llamando v al volumen de la parte de K que queda interior a la franja B es [7, p. 35],

$$\int v\, dB = 2\pi\Delta V, \tag{51}$$

luego, según (49), el valor medio de este volumen valdrá

$$\bar{v} = \frac{2\pi\Delta V}{M + 2\pi\Delta}. \tag{52}$$

Dividiendo (50) por (52) se obtendrá el *valor medio de la densidad de corpúsculos en las secciones de K por B,*

$$\delta_B = \left(1 + \frac{m}{2\pi\Delta}\right) D, \tag{53}$$

poniendo, como siempre, $D = \frac{N}{V}$. *Esta es la relación que liga la densidad total D con la densidad media de las secciones.* Si los corpúsculos son esféricos de radio r es $m = 4\pi r$.

9. *Secciones por cilindros convexos.* Un cilindro convexo queda determinado en el espacio por la posición de una recta paralela a sus generatrices, más un giro alrededor de esta recta. Representando por G esta recta y por τ el giro, como medida de un conjunto de cilindros se toma la integral de la expresión [6, p. 45]

$$dZ = dG\, d\tau \tag{54}$$

siendo dG la expresión dada en (20).

Con esta definición se obtiene (4) que la medida de todos los cilindros congruentes que cortan a un cuerpo convexo K es [6, p. 46]

$$\int_{Z.K \neq 0} dZ = 2\pi\,(4\pi\sigma + \pi F + \lambda M) \tag{55}$$

siendo σ el área y λ el perímetro de la sección recta del cilindro Z y F y M el área y curvatura media de K respectivamente.

Además llamando n al número de corpúsculos encontrados por este cilindro al atravesar K, tiene lugar [1, p. 101], [8],

$$\int n\, dZ = 2\pi\,(\pi f + \lambda m + 4\pi\sigma)\, N \tag{56}$$

luego, *el valor medio del número de corpúsculos encontrados por un cilindro que atraviesa al azar el cuerpo K es*

(4) Aprovecho la oportunidad para corregir que en [6] pág. 46 y 52 en esta fórmula (55) y en la (58) falta en el segundo miembro el factor 2, como es fácil darse cuenta siguiendo el razonamiento allí empleado.

$$\bar{n} = \frac{(\pi f + \lambda m + 4\pi\sigma)}{\pi F + \lambda M + 4\pi\sigma} N. \tag{57}$$

10. Llamando v al volumen de la parte de K que queda interior a Z en cada posición es [6, p. 52]

$$\int v\, dZ = 8\pi^2 \sigma V \tag{58}$$

y teniendo en cuenta (55) el valor medio de este volumen v será

$$\bar{v} = \frac{4\pi\sigma V}{4\pi\sigma + \pi F + \lambda M}. \tag{59}$$

Dividiendo por este volumen medio el número medio (57) de corpúsculos, tendremos la densidad media de ellos

$$\delta_Z = \left(1 + \frac{\pi f + \lambda m}{4\pi\sigma}\right) D$$

que es *la relación que liga la densidad media de corpúsculos en las secciones de K por cilindros congruentes arbitrarios, con la densidad total en el interior de K.*

Si los corpúsculos son esféricos de radio r habrá que sustituir $f = 4\pi r^2$, $m = 4\pi r$.

§ 3. Apéndice

En § 1, nº. 3 y § 2, nº. 2 hemos identificado la *distancia media* entre dos corpúsculos vecinos con la *distancia visible media,* considerada por Pólya, desde un punto situado en el interior del espacio que contiene a los corpúsculos. Esta identificación es lícita considerando la manera como deben entenderse dichos valores medios y que se deduce de cómo han sido obtenidos en cada caso. En el camino seguido por Pólya, se empieza por considerar un segmento de longitud x que parte de un punto fijo y se calcula luego su longitud media suponiendo que todos los corpúsculos se colocan al azar. Ello equivale a buscar la longitud media de los segmentos que se pueden colocar en la región de los corpúsculos sin que corten a ninguno de ellos. En esta forma es claro que la longitud media obtenida debe resultar igual a la distancia media obtenida por nosotros.

El camino seguido por nosotros se generaliza sin dificultad a n dimensiones. Consideremos, en efecto, como densidad para medir conjuntos de rectas la forma diferencial $dG = dP_{n-1}\, d\omega_n$, siendo dP_{n-1} el elemento de volumen de un hiperplano normal a la recta y $d\omega_n$ el elemento de área sobre la esfera unidad del espacio de n dimensiones correspondiente a la dirección de la recta. Esta expresión de dG es la generalización inmediata de (2) y (20). La medida de las rectas que cortan a un cuerpo convexo K vale

$$\int_{G.K \neq 0} dG = \frac{1}{2} \int \sigma\, d\omega_n = \frac{1}{2} \varkappa_{n-1} F, \tag{60}$$

representando por σ el área de la proyección del cuerpo K en la dirección $d\omega_n$ y poniendo 1/2 por tratarse de rectas no orientadas; la segunda integración está extendida a toda la esfera unidad n-dimensional; F es el área de K y $\varkappa_{n-1}$ el volumen de la esfera unidad del espacio de $n-1$ dimensiones([5]).

Por tanto, si f es el área de cada corpúsculo y N el número de ellos interiores al cuerpo convexo K, llamando n al número de corpúsculos cortados por la recta G en cada posición, será

$$\int n\, dG = \frac{1}{2} \varkappa_{n-1} N f. \tag{61}$$

El número de corpúsculos encontrados por una recta arbitraria será, pues, como siempre

$$\bar{n} = \frac{f}{F} N. \tag{62}$$

Por otra parte, llamando s a la longitud de la cuerda que G determina en K es

([5]) Para esta fórmula (60) recordar la llamada fórmula de CAUCHY, por ejemplo en BONNESEN-FENCHEL, *Theorie der konvexen Körper*, Berlín, 1934, pág. 48.

$$\int_{G.K=/=0} s\,dG = \frac{1}{2}\,\omega_n V \tag{63}$$

siendo ω_n el área de la esfera n-dimensional y V el volumen de K. De (63) y (60):

$$\bar{s} = \frac{\omega_n}{\varkappa_{n-1}}\,\frac{V}{F}. \tag{64}$$

Si esta longitud media la dividimos por el número medio de corpúsculos encontrados por la recta arbitraria, tendremos la *distancia media* buscada, que será

$$\bar{d} = \frac{\omega_n}{\varkappa_{n-1}}\,\frac{V}{fN}. \tag{65}$$

Si los corpúsculos son esféricos de radio r es $f = \omega_n r^{n-1}$ y poniendo como siempre $\frac{V}{N} = D$ resulta

$$\bar{d} = \frac{1}{\varkappa_{n-1}\,r^{n-1}\,D}. \tag{66}$$

El producto $\varkappa_{n-1}\,r^{n-1}$ es el volumen de la esfera de $n-1$ dimensiones de radio r. Esta fórmula (66) nos da pues *la distancia visible media desde un punto del espacio de n dimensiones supuesto rodeado por corpúsculos esféricos de radio r situados al azar*. Es fácil comprobar que (66) comprende como casos particulares a (12) y (29).

BIBLIOGRAFIA

[1] W. Blaschke, *"Vorlesungen über Integralgeometrie"* Hamburger Mathematische Einzelschriften. Nº 20-22. Leipzig und Berlin, 1936-37.

[2] R. Deltheil, *"Probabilités géométriques"* París 1926.

[3] E. Gaspar, *"Fórmulas integrales referentes a intersección de una figura plana con bandas variables"*. Publicaciones del Instituto de Matemáticas de la Facultad de Ciencias Matemáticas etc. de la Universidad N. del Litoral. Vol. II, nº 6, Rosario, 1940.

[4] L. B. Loeb, *"Kinetic Theory of Gases"* New York, 1927.

[5] G. Pólya, *"Zahlentheoretisches und Wahrscheinlichkeitstheoretisches über die Sichtweite im Walde"* Arch. der Math. und Phys. 27, p. 135-142 (1918).

[6] L. A. Santaló, *"Integralgeometrie 5. Ueber das kinematische Mass im Raum"* Actualités scientifiques et industrielles nº 357. Hermann. París, 1936.

[7] » *"Geometria integral 7. Nuevas aplicaciones del concepto de medida cinemática en el plano y en el espacio"*. Revista de la Academia de Ciencias, Madrid, 1936.

[8] » *"Geometría integral 15. Fórmula fundamental de la medida cinemática para cilindros y planos paralelos móviles"*. Abhandlungen aus dem Mathematischen Seminar der Hansischen Universität, 12, 1937.

Rosario, Julio 1943.

Reprinted from The Annals of Mathematical Statistics
Vol. XVIII, No. 1, March, 1947

ON THE FIRST TWO MOMENTS OF THE MEASURE OF A RANDOM SET

By L. A. Santaló

Universidad Nacional del Litoral, Argentina

1. Introduction. In a recent paper [3] H. E. Robbins derived general formulas for the moments of the measure of any random set X, and applied the formulas to find the mean and the variance of a random sum of intervals on a line. In subsequent papers, J. Bronowski and J. Neyman [1], using other methods, found the variance when X is a random sum of rectangles in the plane, and H. E. Robbins [4] found the variance when X is a random sum of n-dimensional intervals in n-dimensional euclidean space. In the latter paper Robbins solved also the corresponding problem for circles on the plane.

Using the methods of Robbins, our purpose in the present paper is to solve the following similar problems:

(i) Let R denote the rectangle consisting of all points (x,y) such that $0 \leq x \leq A_1$, $0 \leq y \leq A_2$, and let R' denote the larger rectangle for which $-\delta \leq x \leq A_1 + \delta$, $-\delta \leq y \leq A_2 + \delta$. Let ρ denote a rectangle of fixed dimensions, $a \times b$, but variable position in the plane. The position of ρ will be determined by the coordinates x, y of its center P and the angle φ between the side of length a and the x-axis. We suppose $(a^2 + b^2)^{\frac{1}{2}} \leq \min (A_1, A_2, \delta)$. Let a fixed number N of rectangles ρ be chosen independently with the probability density function for the coordinates (x, y, φ) of each rectangle constant and equal to $\frac{1}{2}\,\pi R'$ in the three-dimensional interval with base R' and height π and zero outside this interval. In section 3 we evaluate the first two moments of the measure of X, where X denotes the intersection of the set-theoretical sum of the N rectangles ρ with R.

(ii) Let R denote the n-dimensional interval consisting of all points $(x_1, x_2, x_3, \cdots, x_n)$ such that $0 \leq x_i \leq A_i$, $(i = 1, 2, \cdots, n)$, and let R' denote the larger interval for which $-\delta \leq x_i \leq A_i + \delta$. Let a fixed number N of n-dimensional spheres with radii r (such that $2r \leq \min (A_i, 2\delta)$) be chosen independently, with the probability density function for the centre of each n-sphere constant and equal to $1/R'$ in R' and zero outside this interval. Denoting by X the intersection of the set theoretical sum of the N n-spheres with R, we evaluate in section 4 the first two moments of the measure of X. This problem is a generalization to n-dimensional space of the case considered by Robbins for the plane ($n = 2$) in [4].

2. Preliminary formulas. Let K be an indeformable plane convex figure of variable position in the plane. The position of K may be determined by the coordinates (x, y) of a point P fixed within K and the angle φ which measures the rotation of K about P. We shall call x, y, φ the coordinates of K. The measure of a set of figures congruent with K is defined as being the integral of the differential form

$$dK = dxdyd\varphi. \tag{2.1}$$

It is readily shown that this measure does not depend on the particular point P chosen to determined the position of K[5]. For instance, the measure of the set of figures K, each of which contains in its interior a fixed point Q, has the value $2\pi F$, where F denotes the area of K; that is,

$$\int_{Q\epsilon K} dK = 2\pi F. \tag{2.2}$$

Let P_1 and P_2 be two fixed points and let l be the distance P_1P_2. The measure of the set of figures congruent with K, each of which contains both points P_1 and P_2 in its interior, will be a function of K and l, say $\mu(K, l)$. If d is the diameter of K, that is, the maximal distance between two points of K, we have $\mu(K, l) = 0$ for $l \geq d$.

Examples. Let K be a rectangle ρ of fixed dimensions $a \times b$, and let us suppose $a \leq b$. The diameter d of ρ is $d = (a^2 + b^2)^{\frac{1}{2}}$. Let $P(x, y)$ be the centre of ρ and φ the angle which forms the side of length b with the segment line P_1P_2 of length l. If we keep first φ constant, then in order that there exist positions of ρ in which it contains the segment line P_1P_2 in its interior it is necessary that

$$a - l \sin \varphi \geq 0, \qquad b - l \cos \varphi \geq 0$$

and in this case the area covered by the centres P in all these positions has the value

$$(a - l \sin \varphi)\,(b - l \cos \varphi).$$

Integrating over all permissible values of φ, we obtain

$$\mu(\rho, l) = 4 \int_{\operatorname{arc\,cos}[b/l]_1}^{\operatorname{arc\,sin}[a/l]_1} (a - l \sin \varphi)(b - l \cos \varphi)\, d\varphi \tag{2.3}$$

where we define

$$[x]_1 = \begin{cases} x \text{ if } x \leq 1 \\ 1 \text{ if } x \geq 1. \end{cases}$$

Carrying out the obvious integration in (2.3) we have

$$\mu(\rho, l) = \begin{cases} 2\pi ab - 4\, l(a + b) + 2\, l^2 & \text{for } l \leq a \leq b \\ 4(ab \operatorname{arc\,sin}\,(a/l) - \frac{1}{2}\, a^2 - bl + b(l^2 - a^2)^{\frac{1}{2}}) & \\ & \text{for } a \leq l \leq b \\ 4(ab \operatorname{arc\,sin}\,(a/l) - \operatorname{arc\,cos}\,(b/l) + b(l^2 - a^2)^{\frac{1}{2}} & \\ + a(l^2 - b^2)^{\frac{1}{2}} - \frac{1}{2}(a^2 + b^2) - \frac{1}{2}\, l^2) & \text{for } a \leq b \leq l. \end{cases} \tag{2.4}$$

As another example, let R be the rectangle consisting of all points (x, y) such that $0 \leq x \leq A_1$, $0 \leq y \leq A_2$ and let R' be the rectangle consisting of all points (x, y) such that

$$-\delta \leq x \leq A_1 + \delta, \quad -\delta \leq y \leq A_2 + \delta, \quad (a^2 + b^2)^{\frac{1}{2}} \leq \min(A_1, A_2, \delta).$$

Let us consider the set of rectangles ρ whose centers belong to R' and do not contain either P_1 or P_2, P_1 and P_2 being two fixed points which belong to R. Let l be the distance P_1P_2. According to (2.2) and the definition of $\mu(\rho, l)$ the measure of the set of rectangles ρ under consideration is

$$2\,\pi R' - 2.2\,\pi\rho + \mu(\rho, l), \tag{2.5}$$

where $R' = (A_1 + 2\,\delta)\,(A_2 + 2\,\delta)$ and $\rho = ab$.

Let K be a plane convex figure of fixed position in its plane. Let us suppose K to be translated a distance l in the direction θ, and let $F(Km, l, \theta)$ be the area of the intersection of K with the translated figure. Obviously if d is the diameter of K, $F(K, l, \theta) = 0$ for $l \geq d$. In what follows we shall consider the function

$$\Phi(K, l) = \int_0^{2\pi} F(K, l, \theta)\, d\theta. \tag{2.6}$$

Example. Let K be a rectangle R of sides A_1, A_2. Let the symbol $[x]$, as in [1], be defined by

$$[x] = \begin{matrix} x \text{ if } x \geq 0 \\ 0 \text{ if } x \leq 0. \end{matrix}$$

It is then readily seen that

$$F(R, l, \theta) = [A_1 - l \sin\theta]\,[A_2 - l\cos\theta]. \tag{2.7}$$

For our purpose the case in which $l \leq \min(A_1, A_2)$ is of interest. In this case, carrying out the immediate integrations, we obtain

$$\Phi(R, l) = 2\,\pi\, A_1A_2 - 4\,l(A_1 + A_2) + 2\,l^2. \tag{2.8}$$

Let $S_{n,r}$ be an n-dimensional sphere of radius r. $S_{n,r}$ will denote also the volume of this sphere, that is, as is known, (see [2, p. 109]),

$$S_{n,r} = \frac{(\pi r^2)^{n/2}}{\Gamma\left(\dfrac{n}{2} + 1\right)}. \tag{2.9}$$

Let us call the measure of a set of spheres $S_{n,r}$ the measure of the set of their centers. That is, if the point $P(x_1, x_2, \cdots, x_n)$ is the center of $S_{n,r}$ the measure of a set of spheres $S_{n,r}$ equals the integral extended over the set, of the differential form

$$dP = dx_1dx_2 \cdots dx_n. \tag{2.10}$$

For instance, the measure of the set of spheres $S_{n,r}$, each of which contains a fixed point Q in its interior, has the value

$$\int_{Q \in S_{n,r}} dP = S_{n,r} \tag{2.11}$$

where $S_{n,r}$ is given by (2.9).

The measure $\mu(S_{n,r}, l)$ of the set of spheres $S_{n,r}$, each of which contains totally in its interior a segment of length $l(l \leq 2r)$, equals the volume of the intersection of two-spheres $S_{n,r}$ whose centers are placed at the end points of the given segment. That is, $\mu(S_{n,r}, l)$ equals twice the volume of the spherical segment of an n-sphere of radius r and semiangle $\alpha = \text{arc cos } (l/2r)$. We will represent the volume of this spherical segment by $S_{n,r}(\alpha)$ and it may be calculated in the following way: The intersection of the n-sphere with a hyperplane at a distance x from the center is an $(n-1)$-dimensional sphere of radius $r' = (r^2 - x^2)^{\frac{1}{2}}$. Let $S_{n-1,r'}$ denote the volume of this $(n-1)$-dimensional sphere (given by the general formula (2.9)). The volume of the spherical segment, whose base has the radius $h = r \cos \alpha$, will be

$$S_{n,r}(\alpha) = \int_h^r S_{n-1,r'} \, dx.$$

Putting $x = r \cos \theta$ and substituting for $S_{n-1,r'}$ the expression given in (2.9), we obtain

$$S_{n,r}(\alpha) = \frac{\pi^{(n-1)/2} r^n}{\Gamma\left(\frac{n+1}{2}\right)} \int_0^\alpha \sin^n \theta \, d\theta = r S_{n-1,r} \int_0^\alpha \sin^n \theta \, d\theta.$$

Consequently we can write

$$\mu(S_{n,r}, l) = 2S_{n,r}(\alpha) = 2rS_{n-1,r} \int_0^\alpha \sin^n \theta \, d\theta, \tag{2.12}$$

where $S_{n-1,r}$ is the volume of the $(n-1)$-dimensional sphere of radius r and $\alpha = \text{arc cos } (l/2r)$.

In (2.12) we may substitute

$$\int_0^\alpha \sin^n \theta \, d\theta = \frac{(n-1)(n-3)\cdots 3.1}{n(n-2)\cdots 4.2} \text{arc cos } (l/2r)$$

$$- \frac{l}{2r}\left\{\frac{1}{n}\left(1 - \frac{l^2}{4r^2}\right)^{(n-1)/2} + \frac{(n-1)}{n(n-2)}\left(1 - \frac{l^2}{4r^2}\right)^{(n-3)/2} + \cdots + \frac{(n-1)(n-3)\cdots 3.1}{n(n-2)\cdots 4.2}\left(1 - \frac{l^2}{4r^2}\right)^{\frac{1}{2}}\right\} \tag{2.13}$$

for n even, and

$$\int_0^\alpha \sin^n \theta \, d\theta \, \frac{(n-1)(n-3)\cdots 4.2}{n(n-2)\cdots 3} - \frac{l}{2r}\left\{\frac{1}{n}\left(1-\frac{l^2}{4r^2}\right)^{(n-1)/2} + \frac{n-1}{n(n-2)}\left(1-\frac{l^2}{4r^2}\right)^{(n-3)/2} + \cdots + \frac{(n-1)(n-3)\cdots 4.2}{n(n-2)\cdots 5.3}\right\} \tag{2.14}$$

for n odd.

In particular, for $n = 2, 3$ we have

$$\mu(S_{2,r}, l) = 4r^2 \int_0^\alpha \sin^2 \theta \, d\theta = 2r^2 \arccos(l/2r) - \frac{1}{2} l(4r^2 - l^2)^{\frac{1}{2}} \tag{2.15}$$

$$\mu(S_{3,r}, l) = 2\pi r^3 \int_0^\alpha \sin \theta \, d\theta = \frac{4}{3}\pi r^3 - \pi r^2 l + \frac{1}{12}\pi l^3. \tag{2.16}$$

We shall now generalize the formula (2.8) to n-space.

A direction in n-space may be given by the corresponding point on the surface of the n-dimensional sphere of unit radius, that is, by the end point of the radius which is parallel to the given direction. The parametric equations of the n-sphere $\sum_1^n \xi_i^2 = 1$ are

$$\begin{aligned} \xi_1 &= \cos \varphi_1 \\ \xi_2 &= \sin \varphi_1 \cos \varphi_2 \\ \xi_3 &= \sin \varphi_1 \sin \varphi_2 \cos \varphi_3 \\ &\cdots\cdots\cdots\cdots\cdots \\ \xi_{n-1} &= \sin \varphi_1 \sin \varphi_2 \cdots \sin \varphi_{n-2} \cos \varphi_{n-1} \\ \xi_n &= \sin \varphi_1 \sin \varphi_2 \cdots \sin \varphi_{n-2} \sin \varphi_{n-1}, \end{aligned} \tag{2.17}$$

where $0 \leq \varphi_i \leq \pi$ for $i < n - 1$ and $0 \leq \varphi_{n-1} \leq 2\pi$. The element of area of this n-sphere has the value (see, [2, p. 109])

$$d\sigma = \sin^{n-2} \varphi_1 \sin^{n-3} \varphi_2 \cdots \sin \varphi_{n-2} \, d\varphi_1 d\varphi_2 \cdots d\varphi_{n-1}. \tag{2.18}$$

A direction in n-dimensional space may then be given by the $n - 1$ parameters $\varphi_1, \varphi_2, \cdots, \varphi_{n-1}$.

Given the n-dimensional interval R consisting of all points $(x_1, x_2, x_3, \cdots, x_n)$ such that $0 \leq x_i \leq A_i$ $(i = 1, 2, 3, \cdots n)$, and suppose that R is translated a distance $l(l \leq \min(A_1, A_2, A_3, \cdots, A_n))$ in the direction $(\varphi_1, \varphi_2, \cdots, \varphi_{n-1})$, the intersection of the translated interval with R is a new interval whose volume has the value $\prod_1^n (A_i - x_i)$, where $x_i = l\xi_i$ (ξ_i given by (2.17)).

Our purpose is to evaluate the integral

$$\Phi(R, l) = \int_{E_n} \prod_{1}^{n} (A_i - x_i)\, d\sigma \tag{2.19}$$

extended over the surface E_n of the n-dimensional sphere of radius unity. We shall denote by E_m either the surface of the m-dimensional sphere of radius unity or its area, given, as is known [2, p. 110] by

$$E_m = \frac{2\pi^{m/2}}{\Gamma\left(\dfrac{m}{2}\right)}. \tag{2.20}$$

Because of the symmetry, the coefficients of all the products $A_{i_1} A_{i_2} A_{i_3} \cdots A_{i_{n-k}}$ have the same value

$$\alpha_k = (-1)^k \int_{E_n} x_1 x_2 \cdots x_k \, d\sigma.$$

The integral extended over the whole surface E_n equals 2^n times the integral extended over the portion for which $\xi_i \geq 0$. Hence, taking into account (2.17) and (2.18) we get

$$\begin{aligned} \alpha_k &= (-1)^k 2^k l^k E_{n-k} \int_0^{\pi/2} \cdots \int_0^{\pi/2} \sin^{n+k-3}\varphi_1 \cos\varphi_1 \sin^{n+k-5}\varphi_2 \cos\varphi_2 \\ &\qquad \cdots \sin^{n-k-1}\varphi_k \cos\varphi_k \, d\varphi_1\, d\varphi_2 \cdots d\varphi_k \\ &= (-1)^k \frac{2^k l^k E_{n-k}}{(n+k-2)(n+k-4)\cdots(n+k-2k)} \end{aligned} \tag{2.21}$$

for $k = 1, 2, \cdots, n-1$. For $k = n$ we find that

$$\begin{aligned} \alpha_n &= (-1)^n 2^n l^n \int_0^{\pi/2} \cdots \int_0^{\pi/2} \sin^{2n-3}\varphi_1 \cos\varphi_1 \\ &\qquad \cdots \sin\varphi_{n-1} \cos\varphi_{n-1}\, d\varphi_1\, d\varphi_2 \cdots d\varphi_{n-1} \\ &= (-1)^n \frac{2^n l^n}{(2n-2)(2n-4)\cdots 4.2}. \end{aligned} \tag{2.22}$$

Hence, we have the following general formula

$$\begin{aligned} \Phi(R, l) &= A_1 A_2 \cdots A_n E_n + (-1)^n \frac{2^n l^n}{(2n-2)(2n-4)\cdots 4.2} \\ &\quad + \sum_{k=1}^{n-1} (-1)^k \Big(\sum_{i_1, i_2, \cdots, i_{n-k}} A_{i_1} A_{i_2} \cdots A_{i_{n-k}} \Big) \\ &\qquad \frac{2^k l^k E_{n-k}}{(n+k-2)(n+k-4)\cdots(n+k-2k)}. \end{aligned} \tag{2.23}$$

In particular, for $n = 2$ this result coincides with (2.8). For $n = 3$ we have

$$\Phi(R, l) = 4\pi A_1A_2A_3 - l^3 - 2\pi l(A_1A_2 + A_1A_3 + A_2A_3) + \tfrac{8}{3}l^2(A_1 + A_2 + A_3). \tag{2.24}$$

3. First problem. We can now solve the first problem (i) stated in the introduction. Denoting by the same letters either sets or their measures, we consider, as in [1] and [4], the set Y of points of R that do not belong to X. We have identically:

$$X + Y = R. \tag{3.1}$$

The general method of Robbins [3] taking into account (2.2), gives immediately the first moments

$$E(Y) = R\left(1 - \frac{\rho}{R'}\right)^N, \qquad E(X) = R\left\{1 - \left(1 - \frac{\rho}{R'}\right)^N\right\}, \tag{3.2}$$

where $R = A_1A_2$, $R' = (A_1 + 2\delta)(A_2 + 2\delta)$, $\rho = ab$.

Our remaining problem is that of evaluating the second moment of X. Let x_i, y_i, φ_i $(i = 1, 2, 3, \cdots, N)$ be the coordinates of the N rectangles ρ (section 2) and let us put, as in (2.1), $d\rho_i = dx_idy_id\varphi_i$. Let $P(x, y)$ and $P_0(x_0, y_0)$ be two points which belong to R and let us put $dP = dx\,dy$, $dP_0 = dx_0dy_0$. Let us consider the following multiple integral

$$J = \int \frac{dP\,dP_0\,d\rho_1\,d\rho_2 \cdots d\rho_N}{(2\pi R')^N} \tag{3.3}$$

extended over the sets of rectangles ρ_i (congruent with ρ) such that x_i, y_i belongs to R', $0 \leq \varphi_i \leq 2\pi$, and do not contain either P or P_0. That is, the domain of integration of J is defined by

$$\begin{gathered} -\delta \leq x_i \leq A_1 + \delta, \qquad -\delta \leq y_i \leq A_2 + \delta, \qquad 0 \leq \varphi_i \geq 2\pi, \\ P \in R, \quad P_0 \in R, \quad P \notin \rho_i, \quad P_0 \notin \rho_i, \qquad (i = 1, 2, \cdots, N). \end{gathered} \tag{3.4}$$

In order to calculate J, we can first keep the rectangles ρ_i fixed; the points P and P_0 can then vary independently over the set of points Y. That gives

$$J = \int_{(x_i,y_i)\in R'} \frac{Y^2\,d\rho_1\,d\rho_2 \cdots d\rho_N}{(2\pi R')^N} = E(Y^2). \tag{3.5}$$

We can now reverse the order of integration, an operation which is obviously justified in this case. Keeping P and P_0 fixed, we can vary each rectangle ρ_i over the set of positions in which it does not contain either P or P_0; letting l denote the distance PP_0, we have, according to (2.5),

$$J = \int_{P\in R, P_0\in R} \left(1 - \frac{4\pi\rho - \mu(\rho,l)}{2\pi R'}\right)^N dPdP_0. \tag{3.6}$$

In order to evaluate this integral we divide it into two parts $J = J_1 + J_2$, according as $0 \leq l \leq d$ or $d \leq l \leq D$, where $d = (a^2 + b^2)^{\frac{1}{2}}$ and $D = (A_1^2 + A_2^2)^{\frac{1}{2}}$. In the interval $0 \leq l \leq d$ we introduce the new variables of integration $l. \theta$ related to x, y, x_0, y_0 by

$$x_0 = x + l \cos \theta, \qquad y_0 = y + l \sin \theta \tag{3.7}$$

whence

$$\frac{\partial(x, y, x_0, y_0)}{\partial(x, y, l, \theta)} = l.$$

In terms of the new variables we have

$$J_1 = \int \left(1 - \frac{4\pi p - \mu(p, l)}{(2\pi R')}\right)^N l\, dl\, dP\, d\theta.$$

In this integral the point P can vary over the intersection of R with the figure obtained by translating R a distance l in the direction θ; that is, the integration of dP gives the function $F(R, l, \theta)$ defined in section 2. According to (2.6) we therefore have

$$J_1 = \int_0^d \left(1 - \frac{4\pi\rho - \mu(\rho, l)}{2\pi R'}\right)^N \Phi(R, l) l\, dl, \tag{3.8}$$

where $\mu(\rho, l)$ is given by (2.4) and $\Phi(R, l)$ by (2.8).

In order to evaluate J_2 we observe that in the interval $d \leq l \leq D$ $\mu(\rho, l) = 0$ and we have

$$J_2 = \left(1 - \frac{2\rho}{R'}\right)^N \int_{d \leq l \leq D} dP\, dP_0 = \left(1 - \frac{2\rho}{R'}\right)^N \left\{ \int^{0 \leq l \leq d} dP\, dP_0 - \int_{0 \leq l \leq d} dP\, dP_0 \right\}.$$

Further we have

$$\int_{0 \leq l \leq D} dP\, dP_0 = R^2 \tag{3.9}$$

and with the change of variables (3.7) and the formula (2.8) we find that

$$\int_{0 \leq l \leq d} dP\, dP_0 = \int_0^d \Phi(R, l) l\, dl = \pi A_1 A_1\, d^2 - \frac{4}{3} (A_1 + A_2)\, d^3 + \frac{1}{2} d^4. \tag{3.10}$$

Collecting (3.8), (3.9), (3.10) and taking into account (3.5) we have

$$\begin{aligned} E(Y^2) = {} & \int_0^d \left(1 - \frac{4\pi\rho - \mu(\rho, l)}{2\pi R'}\right)^N \Phi(R, l) l\, dl \\ & + \left(1 - \frac{2\rho}{R'}\right)^N \{R^2 - \pi A_1 A_2\, d^2 + \tfrac{4}{3}(A_1 + A_2)\, d^3 - \tfrac{1}{2} d_4\}, \end{aligned} \tag{3.11}$$

where $\rho = ab$, $R = A_1A_2$, $R' = (A_1 + 2\delta)(A_2 + 2\delta)$, $\mu(\rho, l)$ is given by (2.4) and $\Phi(R, l)$ by (2.8).

For the variance of X and of Y, we have by (3.1) and (3.2)

$$\sigma_X^2 = E(X^2) - E^2(X) = E(Y^2) - E^2(Y)$$

$$= \int_0^d \left(1 - \frac{4\pi\rho - \mu(\rho, l)}{2\pi R'}\right)^N \Phi(R, l) l\,dl + \left(1 - \frac{2\rho}{R'}\right)^N$$

$$\cdot \{R^2 - \pi A_1 A_2 d^2 + \tfrac{4}{3}(A_1 A_2) d^3 - \tfrac{1}{2} d^4\} - R^2\left(1 - \frac{\rho}{R'}\right)^{2N},$$

which completes the solution of our first problem stated in the introduction.

4. Second problem. In order to solve the second problem (ii) stated in the introduction we will follow the same method of the preceding section.

Let X be the intersection of the set theoretical sum of the N n-dimensional spheres $S_{n,r}$ of radius r with the n-interval R. Let us call Y the set of those points of R that do not belong to X, that is,

$$X + Y = R. \tag{4.1}$$

The general method of Robbins gives immediately

$$E(Y) = R\left(1 - \frac{S_{n,r}}{R'}\right)^N, \quad E(X) = R\left\{1 = \left(1 - \frac{S_{n,r}}{R'}\right)^N\right\} \tag{4.2}$$

where $R = \prod_1^n A_i$, $R' = \prod_1^n (A_i + 2\delta)$, and $S_{n,r}$ is given by (2.9).

We now proceed to calculate $E(Y^2)$. For this purpose let $Q_1(y_1^1, y_2^1, \cdots, y_n^1)$ and $Q_2(y_1^2, y_2^2, \cdots, y_n^2)$ be two points which belong to R and $P_i(x_1^i, x_2^i, \cdots, x_n^i)$ be the centers of the N spheres $S_{n,r}$. Let us put

$$dQ_i = dy_1^i dy_2^i \cdots dy_n^i, (i = 1, 2), \quad dP_i = dx_1^i dx_2^i \cdots dx_n^i, (i = 1, 2, \cdots, N). \tag{4.3}$$

Consider the integral

$$J = \int \frac{dQ_1 dQ_2 dP_1 dP_2 \cdots dP_N}{R'^N} \tag{4.4}$$

extended over the domain defined by

$$Q_1 \in R, \quad Q_2 \in R, \quad P_i \in R', \quad \overline{Q_1P_i} > r, \quad \overline{Q_2P_i} > r, \qquad (i = 1, 2, \cdots, N).$$

If we keep $P_1, P_2, P_3, \cdots, P_N$ fixed, each point Q_1, Q_2 can vary independently over the set Y; consequently we have

$$J = \int_{P_i \in R'} \frac{Y^2 dP_1 dP_2 \cdots dP_N}{R'^N} = E(Y^2). \tag{4.5}$$

On the other hand, if we keep Q_1 and Q_2 fixed, the integral of each dP_i gives

$R' - 2\,S_{n,r} + \mu(S_{n,r}\,,\,l)$ where $\mu(S_{n,r}\,,\,l)$ is given by (2.12) and $l = \overline{Q_1Q_2}$. Hence we have

$$J = \int_{Q_1\epsilon R, Q_2\epsilon R} \left(1 - \frac{2S_{n,r} - \mu(S_{n,r}\,,\,l)}{R'}\right)^N dQ_1\, dQ_2\,. \tag{4.6}$$

In order to calculate this integral we split it into two parts $J = J_1 + J_2$, according as $0 \leq l \leq 2r$ or $2r \leq l \leq D$, where $D = (\sum_1^n A_i^2)^{\frac{1}{2}}$. In the interval $0 \leq l \leq 2r$ we introduce the new variables of integration $l, \varphi_1\,, \varphi_2\,, \cdots, \varphi_{n-1}$ related to $y_1^1\,, y_2^1\,, \cdots y_n^1, y_1^2\,, y_2^2\,, \cdots, y_n^2$ by

$$y_i^2 = y_i^1 + l\xi_i\,, \qquad (i = 1, 2, \cdots, n), \tag{4.7}$$

where ξ_i is given in (2.17). It is found that

$$\frac{\partial(y_1^1\,, y_2^1\,, \cdots, y_n^1\,, y_1^2\,, y_2^2\,, \cdots, y_n^2)}{\partial(y_1^1\,, y_2^1\,, \cdots, y_n^1\,, l, \varphi_1\,, \cdots \varphi_{n-1})} = l^{n-1} \sin^{n-2} \varphi_1 \sin^{n-3} \varphi_2 \cdots \sin \varphi_{n-2}\,.$$

Hence we have,

$$dQ_1 dQ_2 = l^{n-1}\, dQ_1 d\sigma dl, \tag{4.8}$$

where $d\sigma$ denotes the element of area of the n-dimensional sphere of unit radius, given by (2.18). The same method used in section 3 gives

$$J_1 = \int_0^{2r} \left(1 - \frac{2S_{n,r} - \mu(S_{n,r}\,,\,l)}{R'}\right)^N \Phi(R, l) l^{n-1}\, dl, \tag{4.9}$$

where $\Phi(R, l)$ is given by (2.23).

In the interval $2r \leq l \leq D$ $\mu(S_{n,r}\,,\,l) = 0$ and we have

$$\begin{aligned} J_2 &= \left(1 = \frac{2S_{n,r}}{R'}\right)^N \int_{2r \leq l \leq D} dQ_1 dQ_2 = \left(1 = \frac{2S_{n,r}}{R'}\right)^N \\ &\quad \cdot \left\{\int_{0 \leq l \leq D} dQ_1 dQ_2 - \int_{0 \leq l \leq 2r} dQ_1 dQ_2\right\}. \end{aligned} \tag{4.10}$$

Now we have

$$\int_{0 \leq l \leq D} dQ_1\, dQ_2 = R^2 \tag{4.11}$$

and with the change of variables (4.7) we readily find that

$$\int_{0 \leq l \leq 2r} dQ_1\, dQ_2 = \int_0^{2r} \Phi(R, l) l^{n-1}\, dl. \tag{4.12}$$

Collecting (4.9), (4.10), (4.11), (4.12) and taking into account (4.5) and (2.23) we have

$$E(Y^2) = \int_0^{2r}\left(1 - \frac{2S_{n,r} - \mu(S_{n,r}, l)}{R'}\right)^N \Phi(R, l)l^{n-1}\, dl$$

$$(4.13)\quad + \left(1 - \frac{2S_{n,r}}{R'}\right)^N \left\{R^2 - \frac{2^n r^n}{n} RE_n - (-1)^n \frac{2^{3n} r^{2n}}{2n(2n-2)\cdots 4.2}\right.$$

$$- \sum_{k=1}^{n-1} (-1)^k \Big(\sum_{i_1, i_2, \cdots, i_{n-k}} A_{i_1} A_{i_2} \cdots A_{i_{n-k}}\Big)$$

$$\left.\cdot \frac{2^{n+2k} E_{n-k} r^{n+k}}{(n+k)(n+k-2)\cdots(n+k-2k)}\right\}$$

where $R = \prod_1^n A_i$, $R' = \prod_1^n (A_i + 2\delta)$; $S_{n,r}$ is given by (2.9), E_m by (2.20), $\mu(S_{n,r}, l)$ by (2.12) and $\Phi(R, l)$ by (2.23). In particular, for $n = 2$, we obtain the value given by Robbins [3, (30)], by use of (2.8), (2.15) and the equations $S_{2,n} = \pi r^2$, $E_i = 2$. For $n = 3$, the case of ordinary space it follows from (2.16), (2.24) and the equations $S_{3,r} = \frac{4}{3}\pi r^3$, $E_3 = 4\pi$, $E_2 = 2\pi$, that

$$E(Y^2) = \int_0^{2r}\left(1 - \frac{16\pi r^3 + 12\pi r^2 l - \pi l^3}{12R'}\right)^N \Big(4\pi R - l^3 - 2\pi(A_1A_2 + A_1A_3$$

$$(4.14)\quad + A_2A_3)l + \frac{8}{3}(A_1 + A_2 + A_3)l^2\Big) l^2\, dl + \left(1 - \frac{8\pi r^3}{3R'}\right)^N \Big\{R^2$$

$$- \frac{32}{3}\pi R r^3 + 8\pi(A_1A_3 + A_2A_3 + A_2A_3)r^4$$

$$- \frac{256}{15}(A_1 + A_2 + A_3)r^5 + \frac{32}{3}r^6\Big\}.$$

In this case the exact evaluation is easy if one expands the binomial under the sign of the integral and integrates term by term.

From (4.1) we see that $\sigma_X^2 = E(X^2) - E^2(X) = E(Y^2) - E^2(Y)$. Thus, from (4.2) and (4.13) we obtain immediately the second moment $E(X^2)$ and the variance σ_X^2 of X.

5. Remark. In the second problem we can substitute the n-intervals R and R' by concentric n-dimensional spheres. The problem may then be stated as follows:

Let $S_{n,a}$ denote a fixed n-dimensional sphere of radius a and $S_{n,a+\delta}$ the concentric n-dimensional sphere of radius $a + \delta$. $S_{n,a}$ and $S_{n,a+\delta}$ shall also denote the corresponding volumes. Let a fixed number N of n-dimensional spheres with radii r ($r \leq \min(a, \delta)$) be chosen independently with the probability density function for the center of each $S_{n,r}$ constant and equal to $1/S_{n,a+\delta}$ in $S_{n,a+\delta}$ and zero outside this n-sphere. Let X denote the intersection of the set-theoretical sum of the N n-spheres with $S_{n,a}$; we wish to evaluate the first two moments of the measure of X.

It suffices to observe that in this case we have

$$\Phi(S_{n,a}, l) = \mu(S_{n,a}, l)E_n = 2a\, S_{n-1,a}E_n \int_0^\alpha \sin^n\theta\, d\theta \tag{5.1}$$

where $S_{n-1,a}$ is the volume of the $(n-1)$-dimensional sphere of radius a and $\alpha = \text{arc cos } (l/2a)$.

The same method used in section 4 gives

$$E(Y) = S_{n,a}\left(1 - \frac{S_{n,r}}{S_{n,a+\delta}}\right)^N, \qquad E(X) = S_{n,a}\left\{1 - \left(1 - \frac{S_{n,r}}{S_{n,a+\delta}}\right)^N\right\}, \tag{5.2}$$

$$\begin{aligned} E(Y^2) = &\int_0^{2r}\left(1 - \frac{2S_{n,r} - \mu(S_{n,r}, l)}{S_{n,a+\delta}}\right)^N \Phi(S_{n,a}, l)l^{n-1}\, dl \\ &+ \left(1 - \frac{2S_{n,r}}{S_{n,a+\delta}}\right)^N \left\{S_{n,a}^2 - \int_0^{2r} \Phi(S_{n,a}, l)l^{n-1}\, dl\right\}, \end{aligned} \tag{5.3}$$

where $\Phi(S_{n,a}, l)$ is given by (5.1).

In particular, for $n = 2$, by use of (5.1), (2.15) and the indefinite integrals

$$\int \text{arc cos } (l/2a)l\, dl = (\tfrac{1}{2}l^2 - a^2) \text{ arc cos } (l/2a) - \tfrac{1}{4}\, l(4a^2 - l^2)^{\frac{1}{2}} + \text{constant},$$

$$\begin{aligned} \int l^2(4a^2 - l^2)^{\frac{1}{2}}\, dl = -\tfrac{1}{4}l(4a^2 - l^2)^{\frac{3}{2}} + \tfrac{1}{2}a^2l(4a^2 - l^2)^{\frac{1}{2}} \\ + 2a^4 \text{ arc sin } (l/2a) + \text{constant} \end{aligned}$$

we find that

$$\begin{aligned} E(Y^2) = 2\pi \int_0^{2r} &\left(1 - \frac{2\pi r^2 - 2r^2 \text{ arc cos } (l/2r) + \frac{1}{2}l(4r^2 - l^2)^{\frac{1}{2}}}{\pi(a+\delta)^2}\right. (2a^2 \text{ arc cos}(l/2a) \\ &\left. - \tfrac{1}{2}l(4a^2 - l^2)^{\frac{1}{2}}\right) l\, dl + \left(1 - \frac{2r^2}{(a+\delta)^2}\right)^N \left\{\pi^2 a^4 - 2\pi\left(2a^2(2r^2 - a^2) \text{ arc cos}\left(\frac{r}{a}\right)\right.\right. \\ &\left.\left. - 3a^2\, r(a^2 - r^2)^{\frac{1}{2}} + \pi a^4 + 2r(a^2 - r^2)^{\frac{3}{2}} - a^4 \text{ arc sin}(r/a)\right)\right\}. \end{aligned}$$

For $n = 3$, we have by (5.1) and 2.16)

$$\begin{aligned} E(Y^2) = 4\pi \int_0^{2r} &\left(1 - \frac{16r^3 + 12r^2 l - l^3}{16(a+\delta)^3}\right)^N (\tfrac{4}{3}\pi a^3 - \pi a^2 l + \tfrac{1}{12}\pi l^3)l^2\, dl \\ &+ 4\pi\left(1 - \frac{2r^3}{(a+\delta)^3}\right)^N \left\{\tfrac{4}{9}\pi a^6 - \tfrac{32}{9}\pi a^3 r^3 + 4\pi a^2 r^4 - \tfrac{8}{9}\pi r^6\right\}. \end{aligned}$$

From (5.2) and (5.3) with the use of the relation $\sigma_X^2 = E(X^2) - E^2(X) = E(Y^2) - E^2(Y)$ we obtain immediately the second moment $E(X^2)$ and the variance σ_X^2 of X.

REFERENCES

[1] J. Bronowski and J. Neyman, "The variance of the measure of a two-dimensional random set," *Annals of Math. Stat.*, Vol. 16 (1945), pp. 330–341.

[2] R. Deltheil, *Probabilitiés géométriques*, Gauthier-Villars, París, 1926.

[3] H. E. Robbins, "On the measure of a random set," *Annals of Math. Stat.*, Vol. 15 (1944), pp. 70–74.

[4] H. E. Robbins, "On the measure of a random set, II." *Annals of Math. Stat.*, Vol. 16 (1945), pp. 342–347.

[5] L. A. Santaló, "Sobre la medida cinemática en el plano", *Abhandlungen aus dem Mathematisches Seminar der Hamburgische Universität*, Vol. 11 (1936), pp. 222–236.

On the distribution of sizes of corpuscles contained in a body from the distribution of their sections or projections

L. A. Santaló

1 Introduction

The following problem has been studied on several occasions:

Consider an opaque convex body containing in its interior a large number of randomly distributed spheres whose radii obey a certain law of distribution. If this body is intersected by a plane, circles of different sizes are obtained as sections of spheres, the radii of which will also obey a certain law of distribution. The problem resides in how to express the distribution of the radii of the spheres by means of the radii of the circles in the plane section.

If, instead of intersecting by a plane, we consider a straight line passing through the body, chords on that line are obtained which are determined by the line in the spheres, and the analogous problem consists in deducing the distribution of the radii of the spheres based on the distribution of these chords.

This problem has recently been studied by W. P. Reid [6] who cites previous works by E. Scheil [7] and R. L. Fullman [5]. One may also refer to a prior work by S. D. Wicksell [8], who studied and solved the same problem.

The following problem is of an analogous type:

Consider a transparent convex body containing in its interior randomly distributed opaque plates with variable shape and size whose areas (irrespectively of their shape) obey a certain law of distribution. If the entire body is projected over a plane in a random direction, convex figures of variable area are obtained as a projection of the plates, and the area of these figures will also obey another certain law of distribution. The problem consists in relating these two area distributions.

Instead of plane plates, the problem may also be studied for the case of small rods or segments of variable length with a certain law of distribution.

In this paper, we consider the first problem referred to above in a more general way, assuming that instead of spheres the body contains convex corpuscles of arbitrary but similar shape, such that their size depends on a single parameter λ (similarity ratio) whose law of distribution is required in terms of the law of distribution of either the areas of the sections formed by an arbitrary plane, or of the lengths of the chords determined by an arbitrary line.

We check to see if for corpuscles having a shape that differs little from that of the sphere it is possible to obtain an integral equation of the Abel type, which is therefore solvable.

We also study the second problem regarding the projections, which leads analogously to easily solvable integral equations.

I. The problem of sections

2 Known formulae

Let us recall some integral formulae regarding convex bodies, which are required for what follows. When in these formulae we refer to randomly distributed lines or planes, our context is the theory of geometric probabilities (Deltheil [4]) or that of integral geometry (Blaschke [1]).

a) The measure of the lines G in the plane that intersect a convex figure in the plane is equal to the length L of the boundary.

b) The mean value of the lengths α of the chords that a line G in the plane determines in a convex figure in the plane is equal to

$$\bar{\alpha} = \pi S/L, \tag{2.1}$$

where S s the area and L the length of the convex figure.

c) The measure of the lines G intersecting a convex body in space is equal to $(\pi/2)S$, where S is the area of the convex body.

d) The measure of the planes intersecting a convex body in space is equal to the integral of the mean curvature M of the convex body; that is,

$$M = \frac{1}{2}\int_S \left(\frac{1}{R_1} + \frac{1}{R_2}\right) dS, \tag{2.2}$$

where R_1, R_2 are the principle radii of curvature, dS is the area element at the corresponding point, and the integration is extended to the whole surface S of the body. For instance, for a sphere of radius R this is

$$M(\text{sphere}) = 4\pi R. \tag{2.3}$$

For a plane convex body (that is, a disk or convex plane plate considered in the space), if the length of its boundary is L, the measure is

$$M = \frac{\pi}{2}L. \tag{2.4}$$

e) The mean value of the lengths α of the chords that a line determines in a convex body in space is

$$\bar{\alpha} = \frac{4V}{S}, \tag{2.5}$$

where V is the volume and S the surface of the body.

f) The mean value of the areas α of the plane sections of a convex body is

$$\bar{\alpha} = \frac{2\pi V}{M}. \tag{2.6}$$

g) If a convex body K with an integral of mean curvature M_K and surface S_K is contained within another convex body Q with characteristics analogous to M_Q and S_Q, respectively, the probability of a plane intersecting Q also intersecting K is $p_1 = M_K/M_Q$, and the probability of a line intersecting Q also intersecting K is $p_2 = S_K/S_Q$.

h) The mean value of the area of orthogonal projections of a convex body taking arbitrary planes is $\bar{S} = S/4$, where as usual S is the area of the body [2, p. 67].

3 Distribution of the areas of the plane sections of a convex body

Let K be a convex body in space and σ the area of a plane section. Given an arbitrary plane intersecting K, we denote by $\varphi(\sigma)d\sigma$ the probability that the area of the section is between σ and $\sigma + d\sigma$. This function $\varphi(\sigma)$ is, in general, difficult to calculate[1]. From it one knows that if σ_m is the maximum area of all the plane sections, then the following relations must be fulfilled:

$$\int_0^{\sigma_m} \varphi(\sigma)d\sigma = 1, \quad \int_0^{\sigma_m} \sigma\varphi(\sigma)d\sigma = \frac{2\pi V}{M}, \tag{3.1}$$

the latter being equivalent to 2, f).

Let us consider the homothetic convex body K_λ of K with homothety ratio λ. Denoting by $\varphi(\sigma, \lambda)$ the probability function of the areas σ for K_λ (such that $\varphi(\sigma) = \varphi(\sigma, 1)$), and taking into account that to each plane section with area σ in K there corresponds another plane section with area $\lambda^2\sigma$ in $K_\lambda$we obtain

$$\varphi(\sigma, \lambda)d(\lambda^2\sigma) = \varphi\left(\frac{\sigma}{\lambda^2}\right) d\sigma,$$

that is,

$$\varphi(\sigma, \lambda) = \frac{1}{\lambda^2}\varphi\left(\frac{\sigma}{\lambda^2}\right). \tag{3.2}$$

The case of the sphere: Let us consider a unit sphere. The maximum area of the plane sections is $\sigma_m = \pi$. The probability of a plane section having its radius between r and $r + dr$ is equal to $|dx|$, where $x^2 = 1 - r^2$; that is, $r(1-r^2)^{-\frac{1}{2}}\, dr$. Since $\sigma = \pi r^2$, $d\sigma = 2\pi r\, dr$, the probability of having its area between σ and $\sigma + d\sigma$ is

$$2[\pi(\pi - \sigma)]^{-\frac{1}{2}} d\sigma,$$

thus,

$$\varphi(\sigma) = \frac{1}{2\sqrt{\pi}\sqrt{\pi - \sigma}}. \tag{3.3}$$

[1] An interesting problem would be to calculate it for simple bodies: cube, tetrahedron, cylinder.

According to (3.2) for a sphere of radius λ, the probability of a plane section of the sphere being between σ and $\sigma + d\sigma$ is

$$\varphi(\sigma, \lambda) = \frac{1}{2\lambda\sqrt{\pi}\sqrt{\pi\lambda^2 - \sigma}},$$

and if $\sigma_m = \pi\lambda^2$, this can also be written as follows:

$$\varphi(\sigma, \lambda) = \frac{1}{2\lambda\sqrt{\pi}\sqrt{\sigma_m - \sigma}}. \tag{3.4}$$

4 The problem of sections: sections formed by a plane

Let us now consider the first problem referred to in the introduction: However, instead of spheres let us assume convex corpuscles K of arbitrary but similar shape. Their size is therefore determined by a single parameter λ (similarity ratio).

Let Q be a large convex body in whose interior the corpuscles K are distributed randomly. Denoting by M_λ the integral of mean curvature of the surface of corpuscles of similarity ratio λ, in accordance with 2, g), the probability of a plane that intersects Q also intersecting a particular corpuscle is M_λ/M_Q. Let us denote by $F(\lambda)d\lambda$ the number of corpuscles by a unit of volume of Q, whose λ is between λ and $\lambda + d\lambda$. The total number of corpuscles of this type will be $VF(\lambda)d\lambda$ (where V is the volume of Q), and the mean value of the number of corpuscles intersected by an arbitrary plane E whose λ is between $\lambda + d\lambda$ is

$$(M_\lambda/M_Q)VF(\lambda)d\lambda. \tag{4.1}$$

This mean value, multiplied by the probability $\varphi(\sigma, \lambda)d\lambda$ of 3, will yield the mean value of the number of plane sections of the corpuscles whose λ is between λ and $\lambda + d\lambda$, and whose area is between σ and $\sigma + d\sigma$. By integrating this expression to all the values of λ; that is, between$\lambda = \sqrt{\sigma/\sigma_m}$ (since in order for there to be a section of area σ, $\sigma \leq \sigma_m\lambda^2$) and $\lambda = \infty$, we obtain the mean value of the number of sections of the corpuscles formed by an arbitrary plane E whose area is between σ and $\sigma + d\sigma$, which is

$$d\sigma \int_{\sqrt{\sigma/\sigma_m}}^{\infty} (M_\lambda/M_Q)VF(\lambda)\varphi(\sigma, \lambda)\, d\lambda. \tag{4.2}$$

We denote by $f(\sigma)d\sigma$ the number of sections by unit of area in the intersecting plane E, and whose area is between σ and $\sigma + d\sigma$. According to (2.6), the total number of these sections in the whole area of section Q with plane E is $(2\pi V/M_Q)f(\sigma)d\sigma$, and therefore, equalling with (4.2), taking into account (3.2) and simplifying, we obtain

$$\int_{\sqrt{\sigma/\sigma_m}}^{\infty} \frac{1}{\lambda}\varphi\left(\frac{\sigma}{\lambda^2}\right) F(\lambda)d\lambda = \frac{2\pi}{M}f(\sigma), \tag{4.3}$$

where $M = M_1$ and bearing in mind that $M_\lambda = \lambda M_1$.

This equation (4.3) *is that relating the distribution* $f(\sigma)$ *of the areas of the sections of the corpuscles in plane* E *with the distribution* $F(\lambda)$ *of the sizes of the corpuscles in the body* Q.

The practical problem consists in general in determining $F(\lambda)$ if $f(\sigma)$ is known; (4.3) is then an integral equation. In order to solve it, the function $\varphi(\sigma)$ is required, which must be calculated in each case according to the shape of the corpuscles.

The case of spherical corpuscles: If the corpuscles are spheres of variable radius, then taking λ as equal to the radius, and according to (3.3), we have

$$\varphi\left(\frac{\sigma}{\lambda^2}\right) = \frac{\lambda}{2\sqrt{\pi}\sqrt{\pi\lambda^2 - \sigma}}, \tag{4.4}$$

and since for the unit sphere $M = 4\pi$ and $\sigma_m = \pi$, we arrive at the integral equation

$$\int_{\sqrt{\sigma/\pi}}^{\infty} \frac{F(\lambda)d\lambda}{\sqrt{\pi\lambda^2 - \sigma}} = \sqrt{\pi} f(\sigma). \tag{4.5}$$

In order to reduce this equation to the typical Abel form, it suffices to perform the change of variables $\pi\lambda^2 = s$, thereby obtaining

$$\int_{\sigma}^{\infty} \frac{F_1(s)ds}{\sqrt{s - \sigma}} = f_1(\sigma), \tag{4.6}$$

where

$$F_1(s) = \frac{F(\sqrt{s/\pi})}{\sqrt{s}}, \qquad f_1(\sigma) = 2\pi f(\sigma). \tag{4.7}$$

Since $f_1(\infty) = 0$, the solution to (4.6) is (see, for example, Hilbert-Courant [3, p. 158],

$$F_1(s) = -\frac{1}{\pi}\int_{s}^{\infty} \frac{f_1'(\sigma)}{\sqrt{\sigma - s}}\, d\sigma, \tag{4.8}$$

where the prime denotes the derivative with respect to σ. Returning now to the original $F(\lambda)$, $f(\sigma)$, is written as follows:

$$F(\lambda) = -2\sqrt{\pi}\lambda \int_{\pi\lambda^2}^{\infty} \frac{f'(\sigma)}{\sqrt{\sigma - \pi\lambda^2}}\, d\sigma. \tag{4.9}$$

In this case of spherical corpuscles in which the plane sections are circles, it is convenient to use the radii of these sections instead of the areas, replacing $f(\sigma)$ by the distribution function $f(r)$ of the radii. We thus obtain the relations

$$\sigma = \pi r^2, \qquad f(\sigma)d\sigma = g(r)dr,$$

whereby

$$f(\sigma) = \frac{g(r)}{2\pi r},$$

and therefore

$$f'(\sigma) = \frac{1}{4\pi^2 r}\left(\frac{g(r)}{r}\right)',$$

where the last prime denotes the derivative with respect to r, and thereby (4.9) becomes

$$(4.10) \qquad F(\lambda) = -\frac{\lambda}{\pi}\int_\lambda^\infty \left(\frac{g(r)}{r}\right)\frac{dr}{\sqrt{r^2-\lambda^2}},$$

which is the result obtained by Wicksell [8].

The case of approximately spherical corpuscles: In order to write equation (4.3), it is first necessary to calculate $\varphi(\sigma)$, which can easily be done, as seen above for the sphere. However, this calculation is more complicated where other shapes of corpuscles are concerned. Notwithstanding, for convex corpuscles with a shape that differs little from that of the sphere, by analogy with (3.4) an expression with the following form can be taken:

$$(4.11) \qquad \varphi(\sigma) = \frac{a}{(\sigma_m - \sigma)^\mu},$$

where σ_m is the maximum area of the plane sections of the corpuscle, and the parameters a, μ are available in order to obtain the best possible approximation. For $\sigma \geq \sigma_m$, we may write $\varphi(\sigma) = 0$.

In order to calculate a, μ, we have equations (3.1), which give

$$\frac{a}{1-\mu}\sigma_m^{1-\mu} = 1, \qquad \frac{a}{(1-\mu)(2-\mu)}\sigma_m^{2-\mu} = \frac{2\pi V}{M},$$

whereby

$$(4.12) \qquad a = \frac{p}{\sigma_m^p}, \qquad \mu = 1 - p,$$

in which for the sake of brevity

$$(4.13) \qquad p = \frac{M\sigma_m}{2\pi V} - 1.$$

Observe that since in any convex body, $\sigma_m \geq F/4$ and $MF \geq 12\pi V$, we have

$$p \geq \frac{MF}{8\pi V} - 1 \geq \frac{1}{2},$$

and therefore

$$\mu = \frac{1}{2}.$$

For corpuscles in which a law of the type (4.11) is applicable with the values of a, μ given by (4.12), (4.13), the integral equation (4.3) is also reducible to the Abel (generalized) type, and thus is easy to solve.

Indeed, by substituting (4.11) in (4.3), we obtain

$$aM\int_{\sqrt{\sigma/\sigma_m}}^\infty \frac{\lambda^{2\mu-1}F(\lambda)}{(\sigma_m\lambda^2-\sigma)^\mu}\,d\lambda = 2\pi f(\sigma).$$

By writing

$$\sigma_m \lambda^2 = s, \quad F_1(s) = (s/\sigma_m)^{\mu-1} F(\sqrt{s/\sigma_m}), \quad f_1(\sigma) = \frac{4\pi\sigma_m}{aM} f(\sigma)$$

we further obtain

$$\int_\sigma^\infty \frac{F_1(s)}{(s-\sigma)^\mu}\, ds = f_1(\sigma),$$

which is an equation of the Abel generalized type whose solution, if $f_1(\infty) = 0$, is [3, p. 159],

$$F_1(s) = -\frac{\sin \mu\pi}{\pi} \int_s^\infty \frac{f_1'(\sigma)}{(\sigma - s)^{1-\mu}}\, d\sigma,$$

or by returning to the original $F(\lambda)$ and $f(\sigma)$

$$F(\lambda) = -\frac{4\sigma_m \lambda^{2(1-\mu)} \sin \mu\pi}{aM} \int_{\sigma_m\lambda^2}^\infty \frac{f'(\sigma)}{(\sigma - \sigma_m\lambda^2)^{1-\mu}}\, d\sigma, \tag{4.14}$$

which for the case of the sphere $\left(\mu = \frac{1}{2},\ a = \frac{1}{2\sqrt{\pi}}\right)$ can be reduced to (4.9).

The case of equal corpuscles: If all the corpuscles are equal and there are on average N such corpuscles by a unit of volume of Q, then by considering $F(\lambda)$ as N times a Dirac delta function, the same equation (4.3) yields

$$f(\sigma) = \frac{MN}{2\pi\lambda_0}\, \varphi\left(\frac{\sigma}{(\lambda_0^2}\right),$$

where λ_0 is the value corresponding to the single size of the corpuscles.

5 Sections formed by a straight line

As in Section 4, let us consider a convex body Q containing in its interior randomly distributed corpuscles of variable size, but which are similar, such that their size is determined by the similarity ratio.

As before, the function that determines the sizes of the corpuscles is denoted by $F(\lambda)$. If a line G intersects the body Q, different chords intersecting with the corpuscles are obtained on the line. Now let σ be the variable length of these chords, and denote analogously to the previous number by $f(\sigma)$ the function such that $f(\sigma)d\sigma$ is the number of chords by a unit of length of the intersection of G with Q, whose length is between σ and $\sigma + d\sigma$. The problem consists in relating $F(\lambda)$ with $f(\sigma)$.

First observe that if, instead of by unit length, we wish to find the number (mean value) of chords with a length of between σ and $\sigma + d\sigma$ contained within all the intersection of G with Q, then according to (2.5) we have

$$(4V/S_Q) f(\sigma) d\sigma. \tag{5.1}$$

Consider a corpuscle K_1 corresponding to $\lambda = 1$. Let $\varphi(\sigma)d\sigma$ be the probability that a line G intersecting K_1 will determine a chord of length between σ

and $\sigma + d\sigma$ in the corpuscle. For a corpuscle corresponding to the general value λ, the analogous probability (by the same reasoning that led to (3.2)) will be

$$\varphi(\sigma, \lambda) = \frac{1}{\lambda}\varphi\left(\frac{\sigma}{\lambda}\right). \tag{5.2}$$

Let G be an arbitrary line intersecting Q. The probability of this line also intersecting a particular corpuscle K_λ, according to 2 g) is S_λ/S_Q , and therefore the mean value of the number of corpuscles intersected by G whose λ is between λ and $\lambda + d\lambda$ is (analogously to (4.1)) $(S_\lambda/S_Q)VF(\lambda)d\lambda$.

If σ_m is the maximum chord that a line can determine in a corpuscle K_1 of $\lambda = 1$, analogously to (4.2) we now have that the mean value of the number of chords determined by the corpuscles on line G and whose length is between σ and $\sigma + d\sigma$ is as follows:

$$d\sigma \int_{\sigma/\sigma_m}^{\infty} \frac{S_\lambda}{S_Q} VF(\lambda)\varphi(\sigma, \lambda)\, d\lambda, \tag{5.3}$$

a value which, as already seen above, is also given by (5.1). Therefore, equalling and taking into account that $S_\lambda = \lambda^2 S$ ($S =$ the area of the corpuscle corresponding to $\lambda = 1$) and (5.2), we have that the equation relating $F(\lambda)$ with $f(\sigma)$ is as follows:

$$\int_{\sigma/\sigma_m}^{\infty} \lambda\varphi\left(\frac{\sigma}{\lambda}\right) F(\lambda)\, d\lambda = \frac{4}{S} f(\sigma). \tag{5.4}$$

The case of spherical corpuscles: For a unit sphere, the probability of an arbitrarily drawn straight line determining in that sphere a chord with a length between σ and $\sigma + d\sigma$ is easy to calculate; to wit,

$$\varphi(\sigma)d\sigma = \frac{\sigma}{2} d\sigma. \tag{5.5}$$

Hence, equation (5.4) is written (where $\sigma_m = 2$, $S = 4\pi$)

$$\int_{\sigma/2}^{\infty} F(\lambda)d\lambda = \frac{2}{\pi\sigma} f(\sigma),$$

and whose solution is immediate:

$$F(\lambda) = -\frac{1}{\pi}\left(\frac{f(2\lambda)}{\lambda}\right)'. \tag{5.6}$$

An analogously simple result is obtained if convex chords with a shape not unlike the sphere are considered, for which one may take for $\varphi(\sigma)$ an expression of the more general form

$$\varphi(\sigma) = (a\sigma)^\mu \tag{5.7}$$

but of the same type as (5.5).

The parameters a and μ are determined by conditions analogous to (3.1), which in this case, according to (2.5), are

$$\int_0^{\sigma_m} \varphi(\sigma)d\sigma = 1, \quad \int_0^{\sigma_m} \sigma\varphi(\sigma)d\sigma = \frac{4V}{S}, \tag{5.8}$$

where V is the volume and S the area of a corpuscle corresponding to $\lambda = 1$. Thus,

$$\mu = \frac{\sigma_m S - 8V}{4V - \sigma_m S}, \qquad a = [(\mu+1)\sigma_m^{-(\mu+1)}]^{\frac{1}{\mu}} \tag{5.9}$$

and equation (5.4) takes the form

$$\int_{\sigma/\sigma_m}^{\infty} \lambda^{1-\mu} F(\lambda) d\lambda = \frac{4}{(a\sigma)^{\mu} S}, f(\sigma),$$

and hence,

$$F(\lambda) = \frac{4\lambda^{\mu-1} S}{(a\sigma_m)^{\mu} S} \left(\frac{f(\sigma_m \lambda)}{\lambda^{\mu}} \right), \tag{5.10}$$

which generalizes result (5.6).

The case of equal corpuscles: If all the corpuscles are equal and correspond to the value λ_0 of λ, the general equation (5.4), considering $F(\lambda)$ as N times a Dirac delta function, gives

$$f(\sigma) = \frac{\lambda_0 S N}{4} \varphi\left(\frac{\sigma}{\lambda_0}\right),$$

where N is the number of corpuscles by a unit volume of Q .

6 Plane corpuscles intersected by a plane

Now let us assume that the body Q contains plane convex corpuscles; that is, plane plates or disks with variable size, but as usual similar, such that their size is fixed by the similarity ratio λ.

With the same meaning as before for $F(\lambda)$, intersecting now by a plane E we obtain segments of variable length σ as sections of the disks, where as usual $f(\sigma)$ is the probability function of the lengths of these segments (by a unit of area); we wish to find the expression enabling $F(\lambda)$ to be calculated from $f(\sigma)$.

Consider a disk K_1 corresponding to $\lambda = 1$, and denote by $\varphi(\sigma)d\sigma$ the probability of an arbitrary plane E that intersects K_1 determining a chord with a length between σ and $\sigma + \sigma$. For a disk of similarity ratio λ, this probability is equal to

$$\varphi(\sigma, \lambda) = \frac{1}{\lambda}\varphi\left(\frac{\sigma}{\lambda}\right). \tag{6.1}$$

Hereafter we proceed exactly as in 4 to arrive at equation (4.2), with the sole difference that we now have (6.1) instead of (3.2), and we write $M = (\pi/2)L$, where L is the length of the boundary of the disks corresponding to $\lambda = 1$, whereby instead of (4.3) we now have

$$\int_{\sigma/\sigma_m}^{\infty} \varphi\left(\frac{\sigma}{\lambda}\right) F(\lambda)\, d\lambda = \frac{4}{L} f(\sigma). \tag{6.2}$$

If $\varphi(\sigma)$ is known, this integral equation enables $F(\lambda)$ to be calculated from $f(\sigma)$.

The case of circular disks. If the disks making up the plane corpuscles contained in Q are circular, function $\varphi(\sigma)$ is easy to calculate. Let us take λ as the variable radius of these disks. For $\lambda = 1$, the problem consists in finding the probability that a plane intersecting a unit disk in the space does so according to a chord of length between λ and $\lambda + d\lambda$. By the theory of geometric probabilities, it is immediately deduced that this probability is the same as that of a straight line in the plane determining in a unit disk a chord in the same conditions, and is therefore equal to

$$\varphi(\sigma) d\sigma = \frac{\sigma d\sigma}{2\sqrt{4-\sigma^2}} \tag{6.3}$$

for $\sigma < 2$, and $\varphi(\sigma) = 0$ for $\sigma \geq 2$.

By substituting this expression in (6.2), and taking into account that $L = 2\pi$, $\sigma_m = 2$, we have

$$\int_{\sigma/2}^{\infty} \frac{F(\lambda)}{\sqrt{4\lambda^2 - \sigma^2}}\, d\lambda = \frac{4}{\pi\sigma} f(\sigma), \tag{6.4}$$

which can be expressed in the Abel form

$$\int_{\infty}^{\infty} \frac{F_1(s)}{\sqrt{s-x}}\, ds = f_1(x) \tag{6.5}$$

by writing

$$s = 4\lambda^2, \qquad x = \sigma^2$$

$$F_1(s) = \frac{F\left(\frac{1}{2}\sqrt{s}\right)}{4\sqrt{s}}, \quad f_1(x) = \frac{4}{\pi\sqrt{x}} f(\sqrt{x}).$$

Equation (6.5) is the same as (4.6); thus its solution is (4.8), which on again introducing the functions $F(\lambda)$ yields

$$F(\lambda) = \frac{32\lambda}{\pi^2} \int_{2\lambda}^{\infty} \left(\frac{f(\sigma)}{\sigma}\right)' \frac{ds}{\sqrt{\sigma^2 - 4\lambda^2}}. \tag{6.6}$$

The case of equal corpuscles: As in the foregoing cases, if the plane corpuscles are all equal and correspond to value λ_0 of the parameter, equation (6.2) gives

$$f(\sigma) = \frac{LN}{4} \varphi\left(\frac{\sigma}{\lambda_0}\right),$$

where N is the number of corpuscles by a unit of volume of Q.

II. The problem of projections

7 Projections of segments

Let Q be a convex body in space of area S and volume V. Assume that its interior contains a high number of randomly distributed line segments with variable length λ. Let $F(\lambda)d\lambda$ be the average number of segments by a unit of volume whose length is between λ and $\lambda + d\lambda$.

Let us assume that Q is projected onto a plane in an arbitrary direction. The segments will be projected according to segments of variable length a contained in the projection area of Q. Let $f(a)da$ be the number of segments projected by a unit of area, whose length is between a and $a + da$.

The problem consists in finding the equation relating $F(\lambda)$ with $f(a)$, in order to be able to determine one of these functions if the other is known.

First consider the following elementary problem:

A segment in space with length λ is orthogonally projected onto a plane chosen at random. Find the probability that the length of the projected segment will be between a and $a + da$.

Denoting by 0 the angle between the segment and the normal to the plane on which it is projected, $a = \lambda \sin 0$. The probability of this length of projection being between a and $a + da$, taking into account: a) that the favourable cases are those in which the segment in the space forms an angle between 0 and $\theta d\theta$ ($d\theta$ being deduced from $da = \lambda \cos \theta d\theta$) with the normal to the plane, and its measure therefore being $2\pi \sin \theta d\theta$, and b) the possible cases, the set of all the directions from a point in the space being 2π, then this probability is $\sin \theta d\theta$; that is, equal to

$$\varphi(a)da = \frac{ada}{\lambda\sqrt{\lambda^2 - a^2}} \tag{7.1}$$

for $a < \lambda$, and $\varphi(a = 0)$ for $a \geq \lambda$.

According to the definitions of $F(\lambda)$ and $f(a)$, and taking into account the mean value in 2 h), we thereby arrive at the relation sought:

$$\int_a^\infty \frac{F(\lambda)d\lambda}{\lambda\sqrt{\lambda^2 - a^2}} = \frac{S_Q}{4aV_Q} f(a), \tag{7.2}$$

where V_Q is the volume of Q and S_Q is the area of its surface.

This equation can easily be reduced to one of the Abel type, and is therefore easy to solve. The solution is as follows:

$$F(\lambda) = -\frac{S_Q\lambda^2}{2\pi V_Q} \int_\lambda^\infty \left(\frac{f(a)}{a}\right)' \frac{da}{\sqrt{a^2 - \lambda^2}}. \tag{7.3}$$

If all the segments have the same length and there are on average N by a unit of volume, then considering $F(\lambda)$ as N times the Dirac delta function, (7.2) gives

(7.4) $$f(a) = \frac{4aV_Q N}{\lambda S_Q \sqrt{\lambda^2 - a^2}}$$

for $a < \lambda$, and $f(a) = 0$ for $a > \lambda$.

8 Projected plane corpuscles

Assume that a convex body Q contains randomly distributed plane corpuscles of variable area; let $F(\sigma)d\sigma$ be the number of these corpuscles by a unit of volume, whose area is between σ and $\sigma + d\sigma$. By projecting orthogonally onto a plane, we have convex figures with variable areas a ; let $f(a)da$ be the number of such figures by a unit of projected area, whose area is between a and $a + da$.

If θ is the angle between the normal to the plane of projection and the normal to a plane corpuscle of area σ, then $a = \sigma \cos\theta$, and therefore by the same reasoning as for the previous number the probability that the projected area will be between a and $a + da$ is

(8.1) $$\varphi(a)da = \frac{da}{\sigma}$$

for $a < \sigma$, and $\varphi(a) = 0$ for $a \geq \sigma$; that is, φ is constant and equal to $1/\sigma$ for $a < \sigma$.

The relation between $F(\sigma$ and $f(a)$ is therefore

(8.2) $$\int_a^\infty \frac{F(\sigma)}{\sigma}\, d\sigma = \frac{S_Q}{4V_Q} f(a)$$

where, as before, S_Q is the area and V_Q the volume of Q.

The calculation of $F(\sigma)$ from $f(a)$ is thus performed by the formula

(8.3) $$F(\sigma) = -\frac{\sigma S_Q}{4V_Q} f'(\sigma).$$

If the sets all have the same area σ , and there are N of such sets by a unit of volume of Q , equation (8.2) can be reduced to

$$f(a) = \frac{4aV_Q N}{\sigma S_Q}$$

for $a < \sigma$, and $f(a) = 0$ for $a \geq \sigma$. Since the right-hand side of the equation does not depend on a, it means that $f(a)$ is constant for $a < \sigma$ and is thus equal to zero for $a \geq \sigma$.

References

[1] W. Blaschke, *Vorlesungen über Integralgeometrie*, Leipzig und Berlin, Vol. I and II, 1936–1937, 3te Aufl.

[2] T. Bonnesen and W. Fenchel, *Theorie der konvexen Körper*, Ergebnisse der Math. Springer, Berlin, 1934.

[3] R. Courant and D. Hilbert, *Methods of mathematical physics. Vol. I*, Interscience Publishers, Inc., New York, N.Y., 1953.

[4] R. Deltheil, *Sur la théorie des probabilités géométriques*, Ann. Fac. Sci. Toulouse Sci. Math. Sci. Phys. (3) **11** (1919), 1–65.

[5] R. L. Fullman, *Measurement of particles sizes in opaque bodies*, J. Metals, **5** (1953), 447–452.

[6] W. P. Reid, *Distribution of sizes of spheres in a solid from a study of slices of the solid*, J. of Math. and Phys. **34** (1955), 95–102.

[7] E. Scheil, *Die berechnung der anzahl und gröszenverteilung kugelförmiger körpen mit hilfe der durch einen ebenen schnitt erhaltenen schnittktreise,*, Ztsch. anorg. allgem. Chem. **201** (1931), 259–264.

[8] S. D. Wicksell, *The corpuscle problem. a mathematical, study of a biometric problem*, Biometrica **17** (1925), 84–99.

Universidad Nacional de la Plata and Comisión Nacional de la Energía Atómica. Buenos Aires (Argentina)

J. Appl. Prob. **9**, 140–157 (1972)
Printed in Israel

AVERAGES FOR POLYGONS FORMED BY RANDOM LINES IN EUCLIDEAN AND HYPERBOLIC PLANES

L. A. SANTALÓ, *University of Buenos Aires*
I. YAÑEZ, *University of Madrid*

Abstract

We consider a countable number of independent random uniform lines in the hyperbolic plane (in the sense of the theory of geometrical probability) which divide the plane into an infinite number of convex polygonal regions. The main purpose of the paper is to compute the mean number of sides, the mean perimeter, the mean area and the second order moments of these quantities of such polygonal regions. For the Euclidean plane the problem has been considered by several authors, mainly Miles [4]–[9] who has taken it as the starting point of a series of papers which are the basis of the so-called stochastic geometry.

GEOMETRICAL PROBABILITY; RANDOM LINES; RANDOM POLYGONS; PLANE OF CONSTANT CURVATURE; HYPERBOLIC PLANE; CONVEX DOMAINS; POISSON LINE PROCESS

1. Introduction

Consider the Euclidean plane uniformly covered by random lines which will divide the plane into an infinite number of convex polygonal regions. This set of random polygonal regions was first studied by Goudsmit [2] who obtained the mean number of sides, the mean perimeter, the mean area and the mean area-squared of the polygons. More general results were obtained later by Miles ([4], [5]) and Richards [10]. Interesting generalizations to Euclidean n-dimensional space have been established by Miles ([6], [7], [8] and [9]).

In [12] one of the present authors studied the same problem for the hyperbolic plane. He considers first the regions into which a fixed circle of radius r is divided by n random lines and then takes the limit of the expected values corresponding to these regions as n and r tend to infinity in such a way that n/r tends to a finite constant.

This procedure is satisfactory for the Euclidean plane. However, for the hyperbolic plane a more detailed study is necessary. Consider, for instance, the plane divided into an infinite set of convex polygons by a countable number of independent random uniform lines and a circle of radius r placed on it (Figure 1). We may consider the mean area $E_r^*(A)$ of the *regions* into which the circle is

Received in revised form 19 March 1971.

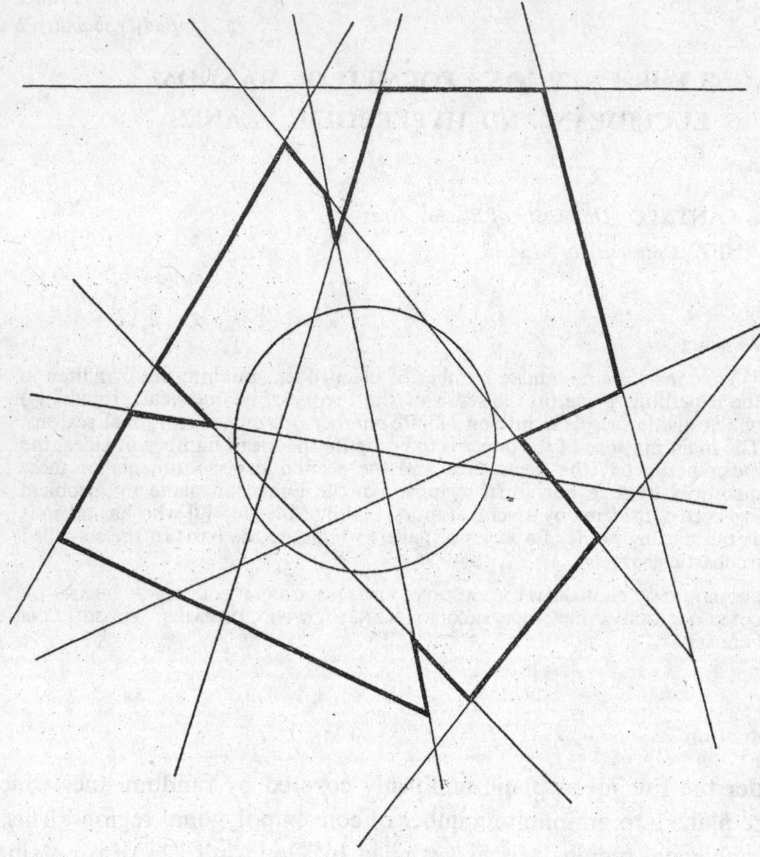

Figure 1

partitioned and the mean area $E_r(A)$ of the *polygons* having at least one point in common with the circle. In the case of Figure 1 we have the "empiric averages" $E_r^*(A) = \pi r^2/13$ and $E_r(A) = F/13$, where F is the total area of the polygons having at least one point in common with the circle. In the Euclidean plane $E_r^*(A)$ and $E_r(A)$ tend to the same limit as $r \to \infty$, while in the hyperbolic plane both limits have different values (which we will denote by $E^*(A)$ and $E(A)$ respectively). This distinction was missing in [12], where only the mean values E^* were considered. The difference arises from the fact that in the Euclidean plane the edge effects on the boundary of the circle may be disregarded and in the hyperbolic plane they may not.

In this paper we take the problem from the beginning and consider, from a general point of view, the plane of constant curvature $k \leqq 0$, which will be denoted by $H(k)$. For $k = 0$ we have the Euclidean plane and for $k = -1$ we have the hyperbolic plane.

The lemmas of Section 3 show that, instead of the circle of the example above, we may take any convex domain which expands to the whole plane; the limits of the expected values do not depend on the shape of these convex domains.

2. Random lines in the Euclidean and hyperbolic plane: compilation of known formulas

The formulas and results of this section can be seen in [11].

Consider the plane of constant curvature $k \leqq 0$. Let p, ϕ be the polar coordinates (or "geodesic" polar coordinates) of the foot of the perpendicular from the origin to the line G (Figure 2). Then, the density for the lines is

$$dG = \cos k^{\frac{1}{2}} p \, dp \wedge d\phi. \tag{2.1}$$

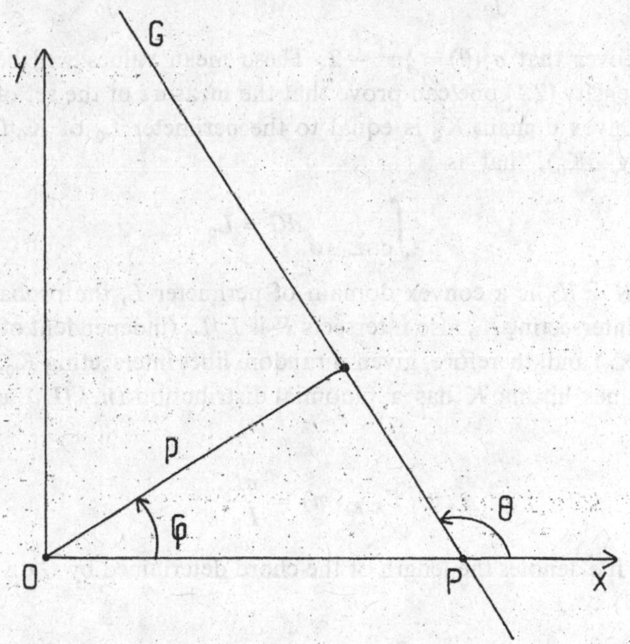

Figure 2

The measure of a set of lines is defined as the integral of dG over the set and it is the unique, up to a constant factor, which is invariant under motions in $H(k)$.

For the Euclidean plane, $k = 0$, we have

$$dG = dp \wedge d\phi \tag{2.2}$$

and for the hyperbolic plane, $k = -1$, since $\cos ix = \cosh x$, we have

$$dG = \cosh p \, dp \wedge d\phi \qquad (p \geqq 0, \quad 0 \leqq \phi \leqq 2\pi). \tag{2.3}$$

If the line G is determined by the abscissa x of the intersection point P of G with a fixed line Ox through the origin O (Figure 2) and the angle θ between G and Ox, an easy change of coordinates yields

$$dG = \sin\theta \, dx \wedge d\theta \qquad (0 \leqq \theta \leqq \pi). \tag{2.4}$$

This expression for the density of lines holds for any k. However, for $k = -1$, there are lines G (forming a set of measure ∞) which do not intersect Ox, so that the coordinate system x, θ is not admissible for all the lines of the hyperbolic plane.

From (2.4) we deduce the following mean values referring to the angle θ between a fixed line and a random line G which intersects the line:

$$E(\theta) = \tfrac{1}{2} \int_0^\pi \theta \sin\theta \, d\theta = \tfrac{1}{2}\pi, \quad E(1/\sin\theta) = \tfrac{1}{2} \int_0^\pi d\theta = \tfrac{1}{2}\pi. \tag{2.5}$$

Notice, moreover that $\sigma^2(\theta) = \frac{1}{4}\pi^2 - 2$. These mean values will be used later.

With the density (2.1) one can prove that the measure of the set of lines which intersect a convex domain K_0 is equal to the perimeter L_0 of K_0 (= length of the boundary ∂K_0), that is

$$\int_{G \cap K_0 \neq \varnothing} dG = L_0 . \tag{2.6}$$

Hence, if $K \subset K_0$ is a convex domain of perimeter L, the probability that a random line intersecting K_0 also intersects K is L/L_0 (independent of the position of K within K_0) and, therefore, given n random lines intersecting K_0, the number m of these lines hitting K has a binomial distribution $(n, L/L_0)$ and its mean value is

$$E_{K,K_0}(m) = \frac{nL}{L_0}. \tag{2.7}$$

Moreover, if s denotes the length of the chord determined by G on K_0, we have ([7], page 25)

$$\int_{G \cap K_0 \neq \varnothing} s \, dG = \pi F_0, \tag{2.8}$$

where F_0 is the area of K_0. Hence, the mean value of s is

$$E_{K_0}(s) = \pi F_0 / L_0. \tag{2.9}$$

Assuming that K_0 expands to the whole plane $H(k)$, in a sense that will be made precise in the next section, and that $n \to \infty$ in such a way that

$$\frac{n}{L_0} \to \tfrac{1}{2}\lambda, \quad \lambda = \text{constant} \tag{2.10}$$

the number of lines intersecting K is Poisson distributed with parameter $\frac{1}{2}\lambda L$, i.e., the probability that K is intersected by exactly m lines is (independently of the position of K in the plane)

$$P_m = (m!)^{-1}(\tfrac{1}{2}\lambda L)^m e^{-\lambda L/2}, \tag{2.11}$$

and the mean value of m is

$$E(m) = \tfrac{1}{2}\lambda L. \tag{2.12}$$

One says that a Poisson line system is generated in $H(k)$ or, following Miles [6], and taking (2.3) into account, that we have in $H(k)$ an isotropic homogeneous Poisson line process corresponding to a Poisson point process of intensity $\frac{1}{2}\lambda \cosh p$ in the (p, ϕ)-strip $(0 \leqq p \leqq \infty,\ 0 \leqq \phi \leqq 2\pi)$.

If K reduces to a line segment of unit length we have $L = 2$; hence, λ is equal to the mean number of lines which are intersected by an arbitrary segment of unit length. As a consequence we have that the points of intersection of an arbitrary line with the lines of the system of random lines constitute a Poisson process of parameter λ.

Other formulas that we shall use referring to planes of constant curvature are the following. The arc element ds on $H(k)$ has the form

$$ds^2 = d\rho^2 + k^{-1}\sin^2 k^{\frac{1}{2}}\rho\, d\phi^2, \tag{2.13}$$

where ρ, ϕ are geodesic polar coordinates. The element of area at the point $P(\rho, \phi)$ writes

$$dP = \frac{\sin k^{\frac{1}{2}}\rho}{k^{\frac{1}{2}}}\, d\rho \wedge d\phi. \tag{2.14}$$

From (2.13) and (2.14) it follows that the perimeter and the area of a circle of radius ρ are, respectively

$$L_C = \frac{2\pi}{k^{\frac{1}{2}}}\sin k^{\frac{1}{2}}\rho, \quad F_C = \frac{2\pi}{k}(1 - \cos k^{\frac{1}{2}}\rho). \tag{2.15}$$

Between the perimeter L and the area F of a convex domain K we have the isoperimetric inequality

$$L^2 - 4\pi F + kF^2 \geqq 0, \tag{2.16}$$

where the equality sign holds if and only if K is a circle.

Finally let us recall the following formula of Gauss-Bonnet for planes of constant curvature. Let K be a domain of $H(k)$ bounded by a single curve ∂K. If ∂K is smooth and κ_g denotes its geodesic curvature, the classical Gauss-Bonnet formula gives

(2.17) $$\int_{\partial K} \kappa_g ds = 2\pi - kF.$$

If K is a convex polygon with N vertices and θ_h $(h = 1, 2, \cdots, N)$ are its interior angles, then the formula of Gauss-Bonnet takes the form

(2.18) $$-kF = (N-2)\pi - \sum_{h=1}^{N} \theta_h$$

For (2.17) and (2.18) see, for instance, Guggenheimer ([3], page 283).

3. Two lemmas

Consider first the Euclidean plane $k = 0$. Let $K_0(t)$ be a family of convex domains depending upon the parameter t. Let $F_0(t)$ be the area and $L_0(t)$ the perimeter of $K_0(t)$. Assume that for any point P of the plane, there is a value t_P of t such that, for all $t > t_P$, we have $P \in K_0(t)$. That means that $K_0(t)$ expands over the whole plane $H(k)$ as $t \to \infty$.

Lemma 1. In the Euclidean plane we have

(3.1) $$\lim_{t\to\infty} \frac{F_0(t)}{L_0(t)} = \infty$$

independently of the shape of $K_0(t)$.

Proof. Let $C(t)$ be the greatest circle contained in $K_0(t)$ and let $R(t)$ be its radius. Let O be the center of $C(t)$ and let $h_t = h_t(\phi)$ be the support function of $K_0(t)$ with respect to the origin O. We have

(3.2) $$F_0(t) = \tfrac{1}{2} \int_{\partial K_0} h_t(\phi)\, ds_t,$$

where ds_t is the arc element of ∂K_0 at the contact point of the support line perpendicular to the direction ϕ. Since $h_t(\phi) \geqq R(t)$ we get $F_0(t) \geqq \frac{1}{2} R(t) L_0(t)$ and because $R(t) \to \infty$ as $t \to \infty$ we get (3.1). Notice that the limit of the ratio F_0/L_0^2 may have any value $\leqq 1/4\pi$ depending on the shape of $K_0(t)$. Notice, also, that (3.1) is not necessarily true for non-convex domains.

Let us now consider the hyperbolic plane $H(-1)$. With the same conditions as above, the isoperimetric inequality (2.16) gives

(3.3) $$\lim_{t\to\infty} \frac{F_0(t)}{L_0(t)} \leqq 1.$$

For simplicity, in the case of the hyperbolic plane instead of convex domains, we shall always restrict ourselves to the so-called h-convex domains, or domains which are convex with respect to horocycles (i.e., such that for each pair of points A, B belonging to the domain, the entire segments of the two horocycles AB also belong to the domain) (see [13]). Any h-convex domain is convex, but the converse is not true. If the boundary ∂K_0 is smooth, the necessary and sufficient condition for h-convexity is that the curvature of ∂K_0 (geodesic curvature) satisfies the condition $\kappa_g \geqq 1$. For instance, the circles are all h-convex. The Gauss-Bonnet formula (2.17) then gives $\lim_{t\to\infty} (F_0/L_0) \geqq 1$ and hence, taking (3.3) into account we have that for all h-convex domains which expand to the whole hyperbolic plane,

$$\lim_{t\to\infty} \frac{F_0(t)}{L_0(t)} = 1. \tag{3.4}$$

Further, for h-convex domains the diameter D_0 and the perimeter L_0 satisfy the inequality $L_0 \geqq 4 \sinh\frac{1}{2}D_0$ (see [14]) and thus

$$\lim_{t\to\infty} \frac{D_0(t)}{L_0(t)} = 0. \tag{3.5}$$

Though we have proved (3.4) and (3.5) for h-convex domains with a smooth boundary, since any convex domain may be approximated by convex domains with smooth boundaries and F_0, L_0, D_0 are continuous functionals, it follows that (3.4) and (3.5) hold for any h-convex domain. We conjecture that (3.4) and (3.5) hold for any family of convex domains (not necessarily h-convex) which expand to the whole hyperbolic plane. However the proof seems to be rather involved, so we shall restrict attention to h-convex domains. As a matter of fact it would be sufficient to consider the family of circles of radius t, but we think that it is worthwhile to point out the independence of the shape of $K_0(t)$ for the limits of Sections 4 and 5.

We can state the following result.

Lemma 2. In the hyperbolic plane, given a family of h-convex domains $K_0(t)$ such that $K_0(t)$ expands to the whole plane as $t \to \infty$, then the Relations (3.4) and (3.5) hold.

Notice that (3.1) and (3.4) can be condensed into

$$\lim_{t\to\infty} \frac{F_0(t)}{L_0(t)} = |k|^{-\frac{1}{2}}. \tag{3.6}$$

Using (2.15), one can easily verify (3.6) when $K_0(t)$ is a circle of radius t.

4. First mean values

Let K_0 be a convex domain of $H(k)$. Let F_0 be the area and L_0 the perimeter of K_0. Let $G_i (i = 1, 2, \cdots, n)$ be n lines which intersect K_0 and let V be the number of intersection points $G_i \cap G_j$ which are inside K_0. We want to compute the integral

$$I = \int_{G_h \cap K_0 \neq \varnothing} V \, dG_1 \wedge dG_2 \wedge \cdots \wedge dG_n \tag{4.1}$$

extended over all the lines G_h which intersect K_0 $(h = 1, 2, \cdots, n)$. Let V_{ij} be the function of G_i, G_j which is equal to 1 if $G_i \cap G_j \in K_0$ and is equal to 0 otherwise (set $V_{ii} = 0$ for completeness). We have $V = \sum_{i<j} V_{ij}$ and

$$I = \tfrac{1}{2}n(n-1) \int V_{ij}\, dG_1 \wedge dG_2 \wedge \cdots \wedge dG_n$$

$$(4.2) \qquad = \tfrac{1}{2}n(n-1)L_0^{n-2} \int V_{ij}\, dG_i \wedge dG_j = \tfrac{1}{2}n(n-1)\, L_0^{n-2}\, 2 \int s_i dG_i$$

$$= n(n-1)\pi F_0 L_0^{n-2},$$

where s_i denotes the length of the chord $G_i \cap K_0$ and the integrals are extended over all the lines which intersect K_0.

Since, by (2.6)

$$(4.3) \qquad \int_{G_h \cap K \neq \varnothing} dG_1 \wedge dG_2 \wedge \cdots \wedge dG_n = L_0^n,$$

we obtain the following result:

Given independently and at random n lines which intersect a convex domain K_0, the expected number of intersection points of these lines which are interior to K_0, is

$$(4.4) \qquad E_{K_0,n}(V) = \pi n(n-1)\, \frac{F_0}{L_0^2}.$$

Note. Let $K_0(t)$ be a family of convex (or h-convex) domains which expand to the whole $H(k)$ as $t \to \infty$. For the Euclidean plane the limit of $E_{K_0(t),n}(V)$ as $t \to \infty$ depends on the shape of $K_0(t)$ and we can only affirm that it is at most $\frac{1}{4}n(n-1)$. On the other hand, for the hyperbolic plane, according to (3.4), the limit is 0, independently of the shape of $K_0(t)$ (assumed h-convex).

The same method leads to the evaluation of

$$(4.5) \qquad I_2 = \int_{G_h \cap K_0 \neq \varnothing} V^2 dG_1 \wedge dG_2 \wedge \cdots \wedge dG_n.$$

In fact we have $V^2 = (\sum_{i<j} V_{ij})^2 = \sum_{i<j} V_{ij}^2 + 2\sum V_{ij} V_{kl}$, where in the second sum the range of indices assumes $i < j$, $k < l$ and the cases $i = k, j = l$ are excluded. Thus

$$I_2 = \int V\, dG_1 \wedge \cdots \wedge dG_n + 2 \int \sum V_{ij} V_{kl}\, dG_1 \wedge \cdots \wedge dG_n$$

$$(4.6) \qquad = n(n-1)\pi F_0 L_0^{n-2} + 2L_0^{n-4} \binom{n}{4} 3 \int V_{ij} V_{kl}\, dG_i \wedge dG_j \wedge dG_k \wedge dG_l$$

$$+ 2L_0^{n-3} \binom{n}{3} 3 \int V_{ij} V_{il}\, dG_i \wedge dG_j \wedge dG_l,$$

where in the first integral of the last term the factor 3 arises from the possibilities $V_{ij}V_{kl}$, $V_{ik}V_{jl}$, $V_{il}V_{jk}$ (assuming $i<j<k<l$) and in the second integral the factor 3 arises from the possibilities $V_{ij}V_{il}$, $V_{ij}V_{jl}$, $V_{il}V_{jl}$. Performing the integration we get

$$(4.7)\quad I_2 = n(n-1)\pi F_0 L_0^{n-2} + 24\binom{n}{4}\pi^2 F_0^2 L_0^{n-4} + 24\binom{n}{3} L_0^{n-3} \int_{G\cap K_0 \neq \varnothing} s^2 dG,$$

where s is the length of the chord $G \cap K_0$. Division by (4.3) yields

$$(4.8)\quad E_{K_0.n}(V^2) = \pi n(n-1)\frac{F_0}{L_0^2} + 24\pi^2\binom{n}{4}\frac{F_0^2}{L_0^4} + 24\binom{n}{3} L_0^{-3} \int_{G\cap K_0\neq\varnothing} s^2 dG.$$

If D_0 is the diameter of K_0 we have $\int s^2 dG \leqq D_0 \int s\, dG = \pi F_0 D_0$ and therefore

$$(4.9)\quad E_{K_0.n}(V^2) \leqq \pi n(n-1)\frac{F_0}{L_0^2} + 24\pi^2\binom{n}{4}\frac{F_0^2}{L_0^4} + 24\binom{n}{3}\pi\frac{F_0 D_0}{L_0^3}.$$

These formulas are valid for any convex domain of $H(k)$. Assume now, that in the case $k<0$, K_0 is h-convex. Furthermore suppose that K_0 is dependent on a parameter t and that $K_0(t)$ expands to the whole plane as $t\to\infty$. Assume further that n increases with t in such a way that

$$(4.10)\quad \lim_{t\to\infty} \frac{n(t)}{L_0(t)} = \tfrac{1}{2}\lambda,$$

where λ is a constant. According to Section 2 we get an isotropic homogeneous Poisson line process of parameter λ, i.e., such that λ is equal to the mean number of lines intersecting an arbitrary segment of unit length.

For the random variable V/F_0, depending on t (and therefore on n, by (4.10)) we have, using (4.9) and (3.5),

$$(4.11)\quad \lim_{t\to\infty} E_{K_0.n}\left(\frac{V}{F_0}\right) = \lim_{t\to\infty} \frac{\pi n(n-1)}{L_0^2} = \tfrac{1}{4}\pi\lambda^2$$

and

$$(4.12)\quad \lim_{t\to\infty} \sigma^2_{K_0,n}\left(\frac{V}{F_0}\right) = \lim_{t\to\infty}\left[E_{K_0,n}\left(\frac{V^2}{F_0^2}\right) - E^2_{K_0,n}\left(\frac{V}{F_0}\right)\right] = 0.$$

It follows that, for any convex (or h-convex) domain $K_0(t)$ which expands to the whole plane we have, with probability one,

$$(4.13)\quad \lim_{t\to\infty} \frac{V}{F_0} = \tfrac{1}{4}\pi\lambda^2.$$

From (4.13), (4.10) and (3.6) we get, with probability one,

$$(4.14)\quad \lim_{t\to\infty} \frac{n}{V} = \lim_{t\to\infty} \frac{n}{L_0}\frac{L_0}{F_0}\frac{F_0}{V} = \frac{2|k|^{\frac{1}{2}}}{\pi\lambda}.$$

Note, for later use, that (4.13) gives the mean number of intersection points of two lines per unit area in $H(k)$ and therefore the probability that a random element of area $d\sigma$ in the plane contains an intersection point of two lines, is

$$\frac{1}{4}\pi\lambda^2 d\sigma. \tag{4.15}$$

5. Mean values for the regions into which a convex (or h-convex) domain is divided by random lines

Consider the plane of constant curvature k and a convex (or h-convex) domain $K_0 = K_0(t)$ in it. We desire to study some mean values concerning the regions into which K_0 is divided by n random lines G_i which intersect K_0. Assume that there are not three or more lines intersecting in a common point of K_0 (according to the measure defined in Section 2, these lines form a set of measure zero). The chords $G_i \cap K_0$ and the boundary ∂K_0 form a plane graph which has $V + 2n$ vertices (V vertices which are interior to K_0 and $2n$ vertices on ∂K_0). The number of sides of the graph is clearly $\frac{1}{2}(4V + 6n) = 2V + 3n$. Therefore, calling P the number of regions into which K_0 is partitioned by the random lines, the classical Euler's formula (regions − sides + vertices = 1) gives $P - 2V - 3n + V + 2n = 1$ and we get

$$P = V + n + 1. \tag{5.1}$$

For instance, in the case of Figure 1 we have $n = 6$, $V = 6$, $P = 13$, number of sides = 30.

The equality (5.1) and (4.4) give

$$E_{K_0\,n}(P) = \pi n(n-1)\frac{F_0}{L_0^2} + n + 1, \tag{5.2}$$

which is *the mean value of the number of regions into which a convex domain K_0 is partitioned by n random lines which cross it.*

Assuming that $K_0(t)$ expands to the whole plane $H(k)$ as $t \to \infty$ with the Condition (4.10), the mean number of regions per unit area will be (using (4.13) and (3.6))

$$\lim_{t\to\infty} \frac{P}{F_0} = \lim_{t\to\infty}\left(\frac{V}{F_0} + \frac{n}{L_0}\frac{L_0}{F_0} + \frac{1}{F_0}\right) = \frac{1}{4}\pi\lambda^2 + \frac{1}{2}\lambda|k|^{\frac{1}{2}} \tag{5.3}$$

and hence the limit of the mean area A of the regions into which K_0 is partitioned is

$$E^*(A) = \frac{4}{\pi\lambda^2 + 2\lambda|k|^{\frac{1}{2}}}. \tag{5.4}$$

As in (4.11) and (4.14) the limit (5.3) (and hence (5.4)), like the following limits (5.6) and (5.7) are limits "in probability", i.e., the equality occurs with probability one.

Let now N_i be the number of vertices of the ith region ($i = 1, 2, \cdots, P$). We have

$$\sum_{i=1}^{P} N_i = 4V + 4n \tag{5.5}$$

and therefore the limit of the mean number of vertices of a region with the boundary regions included, is

$$E^*(N) = \lim_{t\to\infty} \frac{4V + 4n}{P} = \lim_{t\to\infty} \frac{4V + 4n}{V + n + 1} = 4. \tag{5.6}$$

Finally, if s_i is the length of the chord $G_i \cap K_0$, the sum of the perimeters of the regions, for a given set of lines $G_1, G_2, \cdots, G_n$ which cut K_0 is $2\sum s_i + L_0$ ($i = 1, 2, \cdots, n$) and thus

$$E^*(L) = \lim_{t\to\infty} \frac{2\sum s_i + L_0}{P} = \lim_{t\to\infty} \frac{2\sum s_i/n + L_0/n}{V/n + 1 + 1/n} \tag{5.7}$$

and, since $\lim_{t\to\infty}(L_0/n) = 2/\lambda$, $\lim_{t\to\infty}(V/n) = \pi\lambda/2|k|^{\frac{1}{2}}$, $\lim_{t\to\infty}[\sum s_i/n] =$ limit of the mean length of the chords $G_i \cap K_0 = \lim_{t\to\infty}(\pi F_0/L_0) = \pi/|k|^{\frac{1}{2}}$, we get

$$E^*(L) = \frac{4\pi\lambda + 4|k|^{\frac{1}{2}}}{\pi\lambda^2 + 2\lambda|k|^{\frac{1}{2}}}. \tag{5.8}$$

The limit second order moments $E^*(A^2)$, $E^*(AL)$, $E^*(L^2), \cdots$ for regions, are known for the Euclidean plane, because they coincide with the second order moments for polygons and they have been given by Miles ([4], [5]). For the hyperbolic plane we only know $E^*(A^2)$ and we have $E^*(A^2) < \infty$ if and only if $\lambda > \frac{1}{2}$(cf. [12]). Moreover, for the hyperbolic plane we do not know if these limit second order moments are dependent or not on the shape of the expanding domain $K_0(t)$.

6. Mean values for polygons determined by random lines

Consider the Poisson line system described in Sections 2 and 4, which partitions $H(k)$ into an aggregate T of random convex polygons. Our object is now to investigate some mean values of certain quantities Z attached to each polygon, the basic ones being the area A, the perimeter L and the number of sides (or vertices) N.

To this end, the natural way should be to consider first the subaggregate of polygons T_t having at least one point in common with $K_0(t)$ and then make $t \to \infty$. However in this case it seems to be difficult to compute directly the empiric averages $\sum_t Z/\sum_t 1$ ($\sum_t Z =$ sum of Z-values for the polygons of T_t) and we must follow an indirect method which makes necessary the introduction of certain assumptions.

Let $F_{Z,t}(z)$ (z is a particular value of Z) be the empiric distribution function of Z for the finite aggregate T_t ($F_{Z,t}(z)$ depends on the way of selecting the random polygons). Then we make the assumption that for each Z there exists a distribution function $F_Z(z)$ such that $F_{Z,t}(z)$ tends almost surely to $F_Z(z)$ as $t \to \infty$ and, moreover, the empiric averages $\Sigma_t Z / \Sigma_t 1$ converge almost surely to the ergodic mean $\int Z F_Z(dz)$.

For details about these assumptions and their proof for the Euclidean plane, together with some deep reasons for their plausibility in general, see Miles ([4], [5], [6] and [8]).

We will first consider the following mean values, depending of the way of selecting the random polygons.

(a) Select at random a point on the plane and consider the value of Z of the polygon which contains the point. The corresponding mean value will be denoted by $E_A(Z)$, $Z = A, L, N$.

(b) Select a vertex at random (i.e., an intersection point of two lines) and, with probability $\frac{1}{4}$, select one of the four polygons having this vertex. This procedure gives rise to the mean values which we will denote by $E_N(Z)$.

(c) Select at random a point on one of the lines and then, with probability $\frac{1}{2}$, select one of the two polygons which contain the point in its boundary. We call $E_L(Z)$ the corresponding mean value.

Of course these mean values presuppose that the stated selections are meaningful, in particular, the assumption that it is meaningful to select a random vertex of T.

We proceed to compute these mean values.

Method (a). Given a random line G, the intersection points of G with the lines $G_1, G_2, \cdots$ of the Poisson line system which determines T are distributed according to a Poisson distribution of parameter λ. Therefore, given a random point P on the plane and a random line G through P, the probability that the distance from P to the first intersection point of G with T lies in the interval ρ $\rho + d\rho$ ($\rho \geqq 0$), is

$$\lambda e^{-\lambda\rho} d\rho. \tag{6.1}$$

According to (2.14) the area of the polygon which contains P will be

$$A = \int_0^{2\pi} \int_0^{\rho} \frac{\sin k^{\frac{1}{2}}\rho}{k^{\frac{1}{2}}} d\rho \wedge d\phi = k^{-1} \int_0^{2\pi} (1 - \cos k^{\frac{1}{2}}\rho) d\phi, \tag{6.2}$$

and therefore, by (6.1),

$$E_A(A) = k^{-1} \int_0^{2\pi} \int_0^{\infty} \lambda e^{-\lambda\rho} (1 - \cos k^{\frac{1}{2}}\rho) d\rho \wedge d\phi. \tag{6.3}$$

Performing the integration we get

(6.4) $$E_A(A) = \infty \text{ for } \lambda \leqq |k|^{\frac{1}{2}},\ E_A(A) = \frac{2\pi}{\lambda^2 + k} \quad \text{for } \lambda > |k|^{\frac{1}{2}}.$$

In order to compute $E_A(L)$ we apply (2.13) and calling α the angle of G with the side of the polygon at the intersection point (Figure 3), we have

(6.5) $$L = \int_0^{2\pi} \frac{\sin k^{\frac{1}{2}}\rho}{k^{\frac{1}{2}}} \frac{1}{\sin \alpha} d\phi.$$

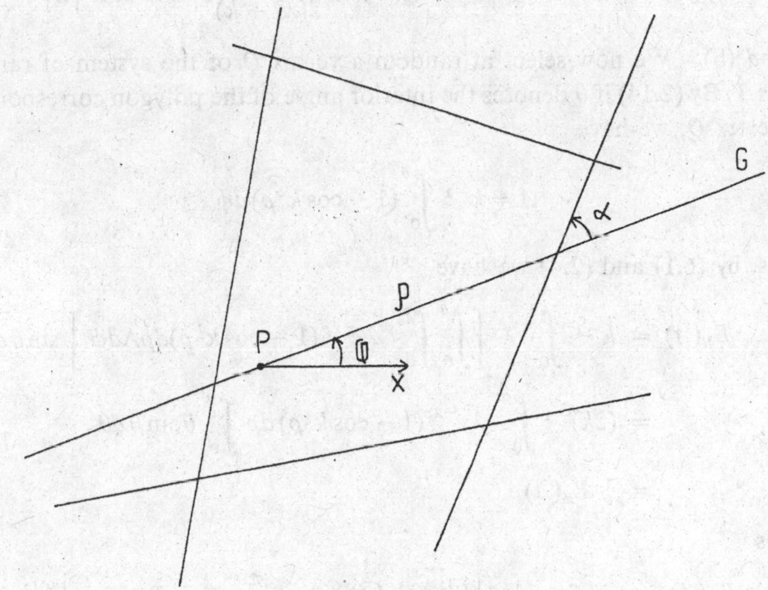

Figure 3

Since ρ and α are independent, taking (2.5) into account, we have

(6.6) $$\begin{aligned} E_A(L) &= \int_0^{2\pi} E\left(\frac{\sin k^{\frac{1}{2}}\rho}{k^{\frac{1}{2}}}\right) E\left(\frac{1}{\sin \alpha}\right) d\phi \\ &= (\pi\lambda/2k^{\frac{1}{2}}) \int_0^{2\pi}\int_0^{\infty} e^{-\lambda\rho} \sin k^{\frac{1}{2}}\rho \, d\rho \wedge d\phi, \end{aligned}$$

and therefore

(6.7) $$E_A(L) = \infty \text{ for } \lambda \leqq |k|^{\frac{1}{2}},\ E_A(L) = \frac{\pi^2\lambda}{\lambda^2 + k} \quad \text{for } \lambda > |k|^{\frac{1}{2}}.$$

We now proceed to compute $E_A(N)$. Consider the random variable $u(d\sigma(Q))$ associated to each area element $d\sigma(Q)$ of the plane, defined as 1 if the area element $d\sigma(Q)$ contains a vertex of T and the segment PQ does not intersect any line of T, and 0 otherwise. According to (4.15) and since the probability that a segment

PQ of length ρ does not contain any point in a Poisson process with parameter λ is equal to $e^{-\lambda\rho}$, we have

$$E_A(N) = \tfrac{1}{4} \int_0^{2\pi} \int_0^{\infty} \pi\lambda^2 e^{-\lambda\rho} d\sigma = \tfrac{1}{2}k^{-\frac{1}{2}}\pi^2\lambda^2 \int_0^{\infty} e^{-\lambda\rho} \sin k^{\frac{1}{2}}\rho d\rho \tag{6.8}$$

or

$$E_A(N) = \infty \text{ for } \lambda \leq |k|^{\frac{1}{2}},\ E_A(N) = \frac{\pi^2\lambda^2}{2(\lambda^2+k)} \quad \text{for } \lambda > |k|^{\frac{1}{2}}. \tag{6.9}$$

Method (b). We now select at random a vertex Q of the system of random polygons T. By (2.14) if θ denotes the interior angle of the polygon corresponding to the vertex Q, we have

$$A = k^{-1} \int_0^{\theta} (1 - \cos k^{\frac{1}{2}}\rho)\, d\phi \tag{6.10}$$

and thus, by (6.1) and (2.4) we have

$$\begin{aligned} E_N(A) &= k^{-1} \int_0^{\theta} \tfrac{1}{2} \left[\int_0^{\theta} \int_0^{\infty} \lambda e^{-\lambda\rho}(1-\cos k^{\frac{1}{2}}\rho) d\rho \Lambda d\phi \right] \sin\theta\, d\theta \\ &= (2k)^{-1} \int_0^{\infty} \lambda e^{-\lambda\rho}(1-\cos k^{\frac{1}{2}}\rho)\, d\rho \int_0^{\pi} \theta \sin\theta\, d\theta \\ &= \tfrac{1}{4}\, E_A(A) \end{aligned} \tag{6.11}$$

and thus

$$E_N(A) = \infty \text{ for } \lambda \leq |k|^{\frac{1}{2}},\ E_N(A) = \frac{\pi}{2(\lambda^2+k)} \quad \text{for } \lambda > |k|^{\frac{1}{2}}. \tag{6.12}$$

To compute $E_N(L)$ we observe that the sides opposite to the chosen vertex Q, by similar considerations as above, give the term $\frac{1}{4}E_A(L)$ and the sides which are adjacent to Q give the mean length $1/\lambda$. Thus we have, $E_N(L) = \infty$ for $\lambda \leq |k|^{\frac{1}{2}}$ and

$$E_N(L) = \tfrac{1}{4}E_A(L) + 2/\lambda = \frac{(\pi^2+8)\lambda^2 + 8k}{4\lambda(\lambda^2+k)} \quad \text{for } \lambda > |k|^{\frac{1}{2}}. \tag{6.13}$$

Similarly, we get $E_N(N) = \infty$ for $\lambda \leq |k|^{\frac{1}{2}}$ and

$$E_N(N) = \tfrac{1}{4}E_A(N) + 3 = \frac{(\pi^2+24)\lambda^2 + 24k}{8(\lambda^2+k)} \quad \text{for } \lambda > |k|^{\frac{1}{2}}, \tag{6.14}$$

where the term $E_A(N)$ stands for the vertices which are opposite to Q and the term 3 stands for the two vertices adjacent to Q and for Q itself.

Method (c). By similar considerations as above we get the remaining mean values:

(6.15) $$E_L(A) = \tfrac{1}{2}E_A(A) = \frac{\pi}{\lambda^2 + k}, \quad \lambda > |k|^{\frac{1}{2}}$$

(6.16) $$E_L(L) = \tfrac{1}{2}E_A(L) + 2/\lambda = \frac{(\pi^2 + 4)\lambda^2 + 4k}{2\lambda(\lambda^2 + k)}, \quad \lambda > |k|^{\frac{1}{2}}$$

(6.17) $$E_L(N) = \tfrac{1}{2}E_A(N) + 2 = \frac{(\pi^2 + 8)\lambda^2 + 8k}{4(\lambda^2 + k)}, \quad \lambda > |k|^{\frac{1}{2}}.$$

For $\lambda \leqq |k|^{\frac{1}{2}}$ these mean values are all ∞.

Notice that in the case of the hyperbolic plane, $k = -1$, in order to obtain a finite value of the means E_A, E_N, E_L we must have $\lambda > 1$.

Alternative proof. We are indebted to the referee for the remark that all the preceding mean values can be deduced from $E_A(A)$ using a very ingenious device due to Miles [5]. Indeed, select a random vertex of T, say Q, and denote by $P_1, \cdots, P_4$, in order about Q, the four polygons having Q as common vertex. Then $P_{1234} = P_1 \cup P_2 \cup P_3 \cup P_4$ is a random polygon selected by method (a); $P_{12} = P_1 \cup P_2$, $P_{34} = P_3 \cup P_4$ are random polygons selected by method (c) and P_1, P_2, P_3, P_4 are random polygons selected by method (b). Using these remarks one can deduce some relations between the means E_A, E_N, E_L which determine all of them from $E_A(A)$. Though this method is more elegant and shorter, the proofs given above are perhaps more natural and straightforward.

7. The second moments

Let $dF_A(A,N,L)$ denote the probability that the polygon chosen according to method (a) has area between A and $A + dA$, perimeter between L and $L + dL$ and has N vertices. Similarly, let $dF_N(A,N,L)$ and $dF_L(A,N,L)$ denote the analogous probabilities for the polygons chosen with methods (b) and (c). We have made the assumption that all such probabilities exist.

Let $dF(A,N,L)$ denote the probability that a random polygon of T has area between A and $A + dA$, perimeter between L and $L + dL$ and has N vertices. Here the method of sampling a random polygon has the following sense: we consider first the finite number of polygons which intersect a convex (or h-convex) domain $K_0(t)$, all equally likely; for this finite set the probability $dF_t(A,N,L)$ has a well-defined meaning and then $dF(A,N,L)$ is the limit value of this probability as $t \to \infty$, so that $K_0(t)$ expands to the whole plane $H(k)$. Our assumption is that this limit exists almost surely.

Since the probability that a randomly chosen point on the plane belongs to a polygon of area A is proportional to A (i.e., the probability of choosing a polygon of area A by method (a) is proportional to A), we have the equation

(7.1) $$dF_A(A,N,L) = \frac{A\,dF(A,N,L)}{E(A)},$$

and similarly

$$(7.2)\qquad dF_N(A,N,L)=\frac{N\,dF(A,N,L)}{E(N)},\quad dF_L(A,N,L)=\frac{L\,dF(A,N,L)}{E(L)}.$$

Multiplying (7.1) and (7.2) by A, L, N and integrating over all values of A, N, L we get

$$(7.3)\qquad E(A^2)=E(A)\,E_A(A),\quad E(LA)=E(A)\,E_A(L),\ E(NA)=E(A)\,E_A(N),$$

$$(7.4)\qquad E(AN)=E(N)\,E_N(A),\ E(LN)=E(N)E_N(L),\ E(N^2)=E(N)\,E_N(N),$$

$$(7.5)\qquad E(AL)=E(L)\,E_L(A),\ E(L^2)=E(L)\,E_L(L), E(NL)=E(L)\,E_L(N).$$

From these relations we deduce

$$(7.6)\qquad \begin{aligned} E(A)\,E_A(L) &= E(L)\,E_L(A),\\ E(N)\,E_N(A) &= E(A)\,E_A(N),\\ E(L)\,E_L(N) &= E(N)\,E_N(L), \end{aligned}$$

which yield the identity

$$(7.7)\qquad E_A(L)\,E_N(A)\,E_L(N)=E_L(A)\,E_A(N)\,E_N(L)$$

which can be directly verified from the results of Section 6.

The identity (7.7) shows that the Equations (7.6) are not independent and thus they are not sufficient for computing the mean values $E(A)$, $E(N)$, $E(L)$. It seems very plausible that a new relation must be given by the Gauss-Bonnet formula (2.18). Indeed calling A_i, N_i the area and the number of vertices of the polygons having at least one point in common with $K_0(t)$ $(i=1,2,\cdots,P(t))$, using (2.18) we have

$$(7.8)\qquad -kA_i=(N_i-2)\pi-\sum_{h=1}^{N_i}\theta_{ih},\quad (i=1,2,\cdots,P(t))$$

where θ_{ih} $(h=1,2,\cdots,N_i)$ are the interior angles of the ith polygon. Since the total angle at a point is 2π we may accept that, almost surely,

$$(7.9)\qquad \sum_{i,h}\theta_{ih}\ \text{is equivalent to}\ \tfrac{1}{2}\pi\sum_i N_i\ \text{as}\ t\to\infty.$$

This assumption supposes that "edge effects" are negligible, a fact that is surely true for the Euclidean plane, but which should be interesting to prove for the hyperbolic plane. Accepting (7.9), then (7.8) gives

$$(7.10)\qquad \pi E(N)+2kE(A)=4\pi.$$

This equation, together with the system (7.6) and the values of Section 6 give the result

$$(7.11)\quad E(A)=\frac{4\pi}{\pi^2\lambda^2-2|k|},\ E(L)=\frac{4\pi^2\lambda}{\pi^2\lambda^2-2|k|},\ E(N)=4+\frac{8|k|}{\pi^2\lambda^2-2|k|}$$

for $\lambda>(2|k|)^{\frac{1}{2}}/\pi$, and $E(A)$, $E(N)$, $E(L)=\infty$ for $\lambda\leqq(2|k|)^{\frac{1}{2}}/\pi$. Since $k\leqq 0$ in the expressions above we can put $|k|=-k$.

Taking (7.11) into account, the Equations (7.3), (7.4) and (7.5) give the following second order moments. For $\lambda>1$,

$$(7.12)\quad E(A^2)=\frac{8\pi^2}{(\pi^2\lambda^2-2|k|)(\lambda^2+k)},$$

$$(7.13)\quad E(AL)=\frac{4\pi^3\lambda}{(\pi^2\lambda^2-2|k|)(\lambda^2+k)},$$

$$(7.14)\quad E(AN)=\frac{2\pi^3\lambda^2}{(\pi^2\lambda^2-2|k|)(\lambda^2+k)},$$

$$(7.15)\quad E(L^2)=\frac{2\pi^2[(\pi^2+4)\lambda^2+4k]}{(\pi^2\lambda^2-2|k|)(\lambda^2+k)},$$

$$(7.16)\quad E(NL)=\frac{\pi^2\lambda[(\pi^2+8)\lambda^2+8k]}{(\pi^2\lambda^2-2|k|)(\lambda^2+k)},$$

$$(7.17)\quad E(N^2)=\frac{\pi^2\lambda^2[(\pi^2+24)\lambda^2+24k]}{2(\pi^2\lambda^2-2|k|)(\lambda^2+k)},$$

and for $\lambda\leqq 1$ the six second order moments become ∞.

Notes. 1. For the Euclidean plane $k=0$, the values (7.11) coincide with the values (5.4), (5.6) and (5.8), that is, we have $E^*=E$. These values agree with those given by Miles [4]. Also the second order moments (7.12),$\cdots$,(7.17), for $k=0$ agree with those given by Miles [4].

2. For the Euclidean plane the first and second order moments of A, N, L are finite for any $\lambda>0$. For the hyperbolic plane the first order moments are finite if and only if $\lambda>2^{\frac{1}{2}}/\pi=0.450$, and the second order moments are finite if and only if $\lambda>1$. We do not know if these critical values of λ increase with higher order moments.

The authors are indebted to the referee for many valuable suggestions and improvements.

References

[1] FELLER, W. (1966) *An Introduction to Probability Theory and its Applications.* Vol. II. J. Wiley, New York.

[2] GOUDSMIT, S. A. (1945) Random distributions of lines in a plane. *Rev. Modern Phys.* **17**, 321–322.

[3] GUGGENHEIMER, H. (1963) *Differential Geometry.* McGraw Hill, New York.

[4] MILES, R. E. (1964) Random polygons determined by random lines in a plane. *Proc. Nat. Acad. Sci.* **52**, 901–907.

[5] MILES, R. E. (1964) Random polygons determined by random lines in a plane (II). *Proc. Nat. Acad. Sci.* **52**, 1157–1160.

[6] MILES, R. E. (1970) A synopsis of Poisson flats in Euclidean spaces. *Izv. Akad. Nauk Armjan. SSR Ser. Mat.* V, **3**, 263–285.

[7] MILES, R. E. (1969) Poisson flats in Euclidean spaces, Part I: A finite number of random uniform flats. *Adv. Appl. Prob.* **1**, 211–237.

[8] MILES, R. E. (1971) Poisson flats in Euclidean spaces, Part II: Homogeneous Poisson flats and the complementary theorem. *Adv. Appl. Prob.* **3**, 1–43.

[9] MILES, R. E. (1970) On the homogeneous planar Poisson point process. *Math. Biosciences* **6**, 85–127.

[10] RICHARDS, P. I. (1964) Averages for polygons formed by random lines. *Proc. Nat. Acad. Sci.* **52**, 1160–1164.

[11] SANTALÓ, L. A. (1953) *Introduction to Integral Geometry*. Hermann, Paris.

[12] SANTALÓ, L. A. (1966) Valores medios para polígonos formados por rectas al azar en el plano hiperbólico. *Rev. Mat. y Fis. Teor. Univ. Tucumán* **16**, 29–43.

[13] SANTALÓ, L. A. (1967) Horocycles and convex sets in hyperbolic plane. *Arch. Math. (Basel)* **18**, 529–533.

[14] SANTALÓ, L. A. (1969) Convexidad en el plano hiperbólico. *Rev. Mat. y Fis. Teor. Univ. Tucuman* **19**, 173–183.

J. Appl. Prob. **15**, 494–501 (1978)
Printed in Israel

PLATE AND LINE SEGMENT PROCESSES

N. A. FAVA AND
L. A. SANTALÓ, *University of Buenos Aires*

Abstract

Random processes of convex plates and line segments imbedded in R^3 are considered in this paper, and the expected values of certain random variables associated with such processes are computed under a mean stationarity assumption, by resorting to some general formulas of integral geometry.

GEOMETRICAL PROBABILITY; INTEGRAL GEOMETRY; RANDOM SETS; CONVEX PLATES; LINE SEGMENTS; POISSON DISTRIBUTION

1. Introduction

Random processes of diverse geometric figures have been studied in several papers in view of their applications. Coleman (1974), Parker and Cowan (1976) and more recently Berman (1977) are some of the authors who have solved several problems regarding this kind of process.

In the first part of this paper we consider a random process of convex plates in R^3, computing the expected values of certain random variables associated with such a process under a fairly general assumption of stationarity which is obviously satisfied in the case of the strongly stationary Poisson process. The following section is devoted to line segment processes of the type considered by Parker and Cowan (1976). In this connection, we show how the formulas they state for the three-dimensional case, as well as those they derive for the plane, can be obtained from the now classical results of integral geometry as exposed, for example, in the work of Santaló (1936), (1976).

In the last section, we consider a mixed process of plates and of line segments. By 'plates' we shall always understand convex plates.

2. Plate processes

Let us consider in R^3 a random process of oriented plates depending on a non-negative random parameter ρ, in such a way that any two plates which correspond to the same value of ρ are congruent.

Suppose also that each plate of the family contains a 'distinguished' point H, such that any two congruent plates, if carried to coincide, the distinguished points coincide as well.

Received 21 June 1977; revision received 26 October 1977.

We shall assume that our plate process can be decomposed into four mutually independent processes, namely:

(i) The distinguished points H, which form a point process with the property that the expected value of the random variable $H(A)$ = the number of points H within the Borel set $A \subset R^3$ is invariant under translations (in keeping with usage, we call this property 'mean stationarity').

(ii) The unit normal vectors u to each plate, which we shall assume to be mutually independent and uniformly distributed on the unit spherical surface S^2, with density $du/4\pi$, where du is the area element of S^2.

(iii) The angles φ formed by an oriented segment in each plate with an oriented segment in the plane containing the plate. We assume that these angles are mutually independent and uniformly distributed on the interval $[0, 2\pi)$.

(iv) The values of ρ for each plate of the process, which we shall assume to be mutually independent with a common distribution function $F(\rho)$ concentrated in $[0, \infty)$ and such that if $f = f(\rho)$ is the area and $s = s(\rho)$ the perimeter of a plate with parameter ρ, then

$$E(f) = \int_0^\infty f(\rho)dF(\rho) \quad \text{and} \quad E(s) = \int_0^\infty s(\rho)dF(\rho) \tag{1}$$

are both finite.

Three further assumptions are:

(v) the points H can be labelled according to their distance to the origin — for two or more points at an equal distance, a systematic method for continuing the labelling is chosen;

(vi) the process of points H is almost surely orderly, that is, $\Pr\{H(\{x\}) = 0 \text{ or } 1 \text{ for every } x \in R^3\} = 1$;

(vii) the expected value of $H(A)$ is finite if A is bounded.

We remark that this model formulation is completely analogous to the one given by Parker and Cowan (1976).

Since each plate $\mathcal{P}$ of the family can be identified with the ordered set (H, u, φ, ρ), our plate process is equivalent to a point process in the product space

$$Z = R^3 \times S^2 \times [0, 2\pi) \times [0, \infty) \tag{2}$$

which we call the associated point process (APP).

Since $EH(\cdot)$ is a measure on the Borel subsets of R^3, our first assumption entails the equation

$$EH(A) = \lambda m(A), \tag{3}$$

where m stands for Lebesgue measure and λ is a constant called 'intensity' of the point process (i).

From now on, we shall denote by A an arbitrary convex set in R^3.

In order to compute the expected value of the random variable $N(A)$ = the number of plates of the process which intersect with A, we consider on the σ-field of Borel subsets of Z the product measure

$$d\mathcal{P} = \left(\frac{\lambda}{8\pi^2}\right) dH du d\varphi dF(\rho), \tag{4}$$

whose integral over any Borel set $U \subset Z$ represents the average number of points of the APP within U. By $dH = dxdydz$, we denote the volume element of R^3.

The measure (4) can be written in the form

$$d\mathcal{P} = \left(\frac{\lambda}{8\pi^2}\right) dK dF(\rho), \tag{5}$$

where $dK = dHdud\varphi$ is the kinematic density of integral geometry (Santaló (1976), Chapter 15). Hence, we may compute the expected value of $N(A)$ as follows:

$$EN(A) = \int_{\mathcal{P}\cap A\neq\emptyset} d\mathcal{P} = \frac{\lambda}{8\pi^2}\int_0^\infty dF(\rho)\int_{\mathcal{P}\cap A\neq\emptyset} dK,$$

where the last integral extends over the region of all plates $\mathcal{P}$ with parameter ρ that intersect with A.

It is known (Santaló (1976), Formulas (13.66) and (15.63) for $n = 3$) that the value of the last integral is $8\pi^2 V_A + \pi^2 F_A s + 4\pi M_A f$, where V_A, F_A and M_A represent the volume, the surface area and the mean curvature of A. Therefore

$$\begin{aligned} EN(A) &= \frac{\lambda}{8\pi^2}\int_0^\infty \{8\pi^2 V_A + \pi^2 F_A s(\rho) + 4\pi M_A f(\rho)\} dF(\rho) \\ &= \lambda V_A + \frac{1}{8}\lambda F_A E(s) + (\lambda M_A/2\pi)E(f). \end{aligned} \tag{6}$$

We shall denote by $\alpha = \alpha(H, u, \varphi, \rho)$ the area of the intersection of the plate $\mathcal{P}$ with A.

If we wish to compute the expected value of the random variable $S(A)$ = total area within A of all plates of the process which intersect with A, we have

$$ES(A) = \int \alpha d\mathcal{P} = \frac{\lambda}{8\pi^2}\int_0^\infty dF(\rho)\int \alpha dK.$$

On the other hand, the formula (15.20) of Santaló (1976) in the particular case under consideration, gives us the value

$$\int \alpha dK = 8\pi^2 V_A f(\rho),$$

where $V_A = m(A)$ is, as before, the volume of A. Hence,

$$ES(A) = \lambda V_A E(f). \tag{7}$$

For our next computation we shall make use of the following hypothesis:

(viii) if A and B are disjoint Borel sets in R^3, $E[H(A)H(B)] = EH(A) \cdot EH(B)$. The need for this assumption is carefully discussed in Parker and Cowan (1976).

For each pair $\mathcal{P}_1$ and $\mathcal{P}_2$ of different plates, we shall denote by $l(A \cap \mathcal{P}_1 \cap \mathcal{P}_2)$ the length of the intersection within parentheses.

In order to calculate the expected value of the random variable $L(A)$ = total length within A of the intersection of each pair of different plates of the process, we consider on the cartesian product $Z^2 = Z \times Z$ the product measure $d\mathcal{P}_1 d\mathcal{P}_2$, where $d\mathcal{P}_1$ and $d\mathcal{P}_2$ denote the measure (5).

The diagonal of Z^2 being a set of measure zero, we have

$$EL(A) = \frac{1}{2} \int l(A \cap \mathcal{P}_1 \cap \mathcal{P}_2) d\mathcal{P}_1 d\mathcal{P}_2, \tag{8}$$

where $\mathcal{P}_i$ is the plate corresponding to the point $(H_i, u_i, \varphi_i, \rho_i)$, $i = 1, 2$.

Using the factorization

$$d\mathcal{P}_i = \frac{\lambda}{8\pi^2} dK_i dF(\rho_i),$$

where $dK_i = dH_i du_i d\varphi_i$ is the kinematic density for $\mathcal{P}_i$, we can represent the right-hand member of (8) in the form

$$\frac{\lambda^2}{128\pi^4} \int_0^\infty dF(\rho_2) \int_0^\infty dF(\rho_1) \int l(A \cap \mathcal{P}_1 \cap \mathcal{P}_2) dK_1 dK_2.$$

On the other hand, the formula (15.22) of Santaló (1976) gives for the last integral the value $16\pi^5 V_A f(\rho_1) f(\rho_2)$. Hence, it follows that

$$EL(A) = \frac{1}{8} \pi \lambda^2 V_A E(f)^2. \tag{9}$$

Next, we consider the random variable $T(A)$ = the number of intersections within A of every three different plates of the process.

To this end, we introduce on the cartesian product $Z^3 = Z \times Z \times Z$ the product measure $d\mathcal{P}_1 d\mathcal{P}_2 d\mathcal{P}_3$, where each $d\mathcal{P}_i$ denotes the measure (5) and we assume:

(ix) if A, B and C are disjoint Borel sets in R^3, then $E[H(A)H(B)H(C)] = EH(A)EH(B)EH(C)$. Under this assumption, the expected value of $T(A)$ is given by

$$ET(A) = \frac{1}{6} \int_{A \cap \mathcal{P}_1 \cap \mathcal{P}_2 \cap \mathcal{P}_3 \neq \emptyset} d\mathcal{P}_1 d\mathcal{P}_2 d\mathcal{P}_3, \tag{10}$$

where $\mathcal{P}_i$ corresponds to the point $(H_i, u_i, \varphi_i, \rho_i)$, $i = 1, 2, 3$.

By factorizing each measure $d\mathcal{P}_i$ through the corresponding kinematic density dK_i, we get for the right-hand member of (10) the value

$$\frac{\lambda^3}{3072\pi^6}\int_0^\infty\int_0^\infty\int_0^\infty dF(\rho_1)dF(\rho_2)dF(\rho_3)\int_{A\cap\mathcal{P}_1\cap\mathcal{P}_2\cap\mathcal{P}_3\neq\emptyset} dK_1dK_2dK_3.$$

The value of the last integral can be obtained from the formula (15.22) of Santaló (1976). The result is $64\pi^7 V_A f(\rho_1)f(\rho_2)f(\rho_3)$. Hence

$$ET(A)=\frac{\pi}{48}\lambda^3 V_A E(f)^3. \tag{11}$$

In passing, we remark that since each member of Equations (7) and (11) defines a measure on the Borel sets $A\subset R^3$, it follows that they hold true for any Borel set A.

It may also be useful to remark that our assumptions (viii) and (ix) are both satisfied in the case of the Poisson process, whose definition is given below.

The Poisson process. If we suppose, in addition to (i) through (vii), that the point process (i) is a homogeneous Poisson process of intensity λ, then the random variable $N(A)$ has a Poisson distribution with the expectation given by (6).

Let us call D the distance from the origin to the nearest plate of the process. If we take for A the ball of radius r centered at the origin, then $P\{D>r\}=P\{N(A)=0\}$. But for this particular A we have

$$V_A=\frac{4\pi r^3}{3},\quad F_A=4\pi r^2,\quad M_A=4\pi r,$$

so that the expected value of $N(A)$ is the number

$$\mu=\frac{4\pi\lambda r^3}{3}+\frac{\pi\lambda r^2}{2}E(s)+2\lambda rE(f). \tag{12}$$

Hence, under the Poisson assumption,

$$P\{D>r\}=e^{-\mu}$$

with μ given by (12). From this, we readily obtain for the probability density function of D the expression

$$\lambda e^{-\mu}\{4\pi r^2+\pi rE(s)+2E(f)\}.$$

Examples. (a) If all plates of the process are similar, we may distinguish in each one of them the center of homothety H and we can take for ρ the ratio of similarity of each plate with respect to a fixed plate of the family. With this choice of the random parameter, we have $f(\rho)=f_1\rho^2$ and $s(\rho)=s_1\rho$, where $f_1=f(1)$ is the area and $s_1=s(1)$ the perimeter of the fixed plate. In this example we have $E(f)=f_1E(\rho^2)$ and $E(s)=s_1E(\rho)$.

(b) If the distribution function of the random parameter ρ is concentrated on a single point, then the plate process becomes a process of congruent plates.

3. Line segment processes

Let us consider in R^3 a random process of oriented line segments with variable length. We shall assume that this process can be decomposed into three mutually independent processes, namely:

(i) The mid points M of each segment, which form a mean stationary process of intensity σ, that is, if $M(A)$ denotes the number of mid points within the Borel set A, then $EM(A) = \sigma m(A)$.

(ii) The unit vectors u in the direction of each segment, which we assume to be mutually independent and uniformly distributed over the unit spherical surface S^2, with density $du/4\pi$, where du is the area element of S^2.

(iii) The lengths l of each segment, which we suppose mutually independent with a common distribution $G(l)$ concentrated in $[0, \infty)$.

Except for the minor hypothesis of orientation, these are the basic assumptions of the model formulated by Parker and Cowan (1976).

Since each segment $\mathcal{S}$ can be identified with the ordered set (M, u, l), the segment process is equivalent to a point process in the product space $W = R^3 \times S^2 \times [0, \infty)$, which we call, as before, the associated point process (APP).

If we wish to compute the expected value of the random variable $N(A)$ = the number of segments of the process which intersect with the convex set A, we are led to consider on the σ-field of Borel subsets of W the product measure

$$d\mathcal{S} = \frac{\sigma}{4\pi}\, dMdudG(l), \tag{13}$$

where $dM = dxdydz$ represents the Lebesgue measure in R^3.

To conform to the kinematic fundamental formula of integral geometry, we introduce on the Borel subsets of the product space $W_1 = W \times [0, 2\pi)$ the product measure

$$d\mu = d\mathcal{S}d\varphi, \tag{14}$$

where $d\varphi$ is the Lebesgue measure on $[0, 2\pi)$.

The measure (14) can be written in the form

$$d\mu = \frac{\sigma}{4\pi}\, dKdG(l), \tag{15}$$

where dK is the kinematic density of Santaló (1976), Chapter 15 for $n = 3$.

Hence, we have

$$2\pi EN(A) = \int_{A\cap\mathcal{S}\neq\emptyset} d\mu = \frac{\sigma}{4\pi}\int_0^\infty dG(l)\int_{A\cap\mathcal{S}\neq\emptyset} dK.$$

It is known (Santaló (1976), Formulas (13.65) and (15.63) for $n=3$) that the value of the last integral is $8\pi^2 V_A + 2\pi^2 F_A l$. Therefore

$$EN(A) = \sigma V_A + \frac{\sigma}{4} F_A E(l). \tag{16}$$

As to the random variable $L(A)$ = total length within A of all segments of the process which intersect with A, let us denote by $t(A \cap \mathcal{S})$ the length of the intersection within parentheses. Then

$$\int_{W_1} t(A \cap \mathcal{S}) d\varphi d\mathcal{S} = \frac{\sigma}{4\pi} \int_{W_1} t(A \cap \mathcal{S}) dK dG(l)$$

and by applying Fubini's theorem in the last relation, we get

$$2\pi EL(A) = \frac{\sigma}{4\pi} \int_0^\infty dG(l) \int t(A \cap \mathcal{S}) dK.$$

According to Formula (15.20) of Santaló (1976), the value of the last integral is $8\pi^2 V_A l$. Hence

$$EL(A) = \sigma V_A E(l). \tag{17}$$

For the same reasons as before, the last formula holds for any Borel set A.

4. Mixed process

Let us consider in R^3 a mixed process consisting of two mutually independent processes of plates and of line segments of the types defined in the preceding sections.

Keeping the previous notations, we consider the product space $W_1 \times Z$ endowed with the product measure

$$d\mu d\mathcal{P} = d\varphi d\mathcal{S} d\mathcal{P} = \frac{\lambda\sigma}{32\pi^3} dG(l) dF(\rho) dK_s dK_p,$$

where dK_s and dK_p stand respectively for the kinematic density of $\mathcal{S}$ and of $\mathcal{P}$.

To compute the expected value of the random variable $I(A)$ = the number of segment-plate intersections within the convex set A, we have the formula

$$EI(A) = \int_{A \cap \mathcal{S} \cap \mathcal{P} \neq \emptyset} d\mathcal{S} d\mathcal{P}.$$

Therefore, by integrating $d\mu d\mathcal{P}$ over the set $A \cap \mathcal{S} \cap \mathcal{P} \neq \emptyset$, we get

$$2\pi EI(A) = \frac{\lambda\sigma}{32\pi^3} \int_0^\infty \int_0^\infty dG(l) dF(\rho) \int_{A \cap \mathcal{S} \cap \mathcal{P} \neq \emptyset} dK_s dK_p.$$

On the other hand, by virtue of Formula (15.22) of Santaló (1976), the value of the last integral is $32\pi^4 V_A lf$. Hence

$$(18) \qquad EI(A) = \tfrac{1}{2}\lambda\sigma V_A E(l)E(f).$$

Final remarks

(1) Note that for Equations (16), (17) and (18) the hypothesis of orientation that we attached to the segment process is irrelevant, and that it was adopted for ease of reference only.

(2) The same general formulas of integral geometry can be used to study line segment processes in the plane. The treatment there is slightly simpler, for one gets directly the kinematic density in R^2.

Acknowledgement

The authors are indebted to the referee for several remarks which have contributed to improve the presentation.

References

[1] Berman, M. (1977) Distance distributions associated with Poisson processes of geometric figures. *J. Appl. Prob.* **14**, 195–199.

[2] Coleman, R. (1974) The distance from a given point to the nearest end of one member of a random process of linear segments. In *Stochastic Geometry*, ed. E. F. Harding and D. G. Kendall, Wiley, New York, 192–201.

[3] Parker, P. and Cowan, R. (1976) Some properties of line segment processes. *J. Appl. Prob.* **13**, 96–107.

[4] Santaló, L. A. (1936) Integralgeometrie 5. *Exposés de Géométrie*, Hermann et Cie, Paris.

[5] Santaló, L. A. (1976) Integral geometry and geometric probability. In *Encyclopedia of Mathematics and its Applications*. Addison Wesley, London.

Z. Wahrscheinlichkeitstheorie verw. Gebiete
50, 85–96 (1979)

Zeitschrift für
Wahrscheinlichkeitstheorie
und verwandte Gebiete

Random Processes of Manifolds in R^n

N.A. Fava and L.A. Santaló

Department of Mathematics, University of Buenos Aires, Ciudad Universitaria,
Nuñez, Buenos Aires (Argentina)

Dedicated to Professor Leopold Schmetterer
on the occasion of his 60th Birthday

Summary. We compute the expected values of certain random variables associated with a random process of manifolds in R^n by inserting certain general formulas of integral geometry into the definition of the moment measures of a point process.

1. Introduction

Increasing interest in stochastic processes of geometric figures has arisen recently in view of their applications and their connections with geometry. Berman [1], Coleman [2], Fava and Santaló [3], and Parker and Cowan [8] are some of the authors who have dealt with problems closely connected with the subject of this paper.

In this article we show how some previous results concerning processes of geometric figures in the plane and in three dimensional space can be studied in a unified manner in the more general context of R^n, by taking into account certain invariants linked to each manifold, namely, the integrals of mean curvature and the Euler-Poincaré characteristic. This is done by inserting some general formulas of integral geometry into the definition of the moment measures of a point process.

Acknowledgement. We are deeply indebted to Klaus Krickeberg for helping us to overcome several flaws in our original manuscript. In particular, we are indebted to him for the general outline included in Sect. 4 and for drawing our attention to his paper [6].

2. Geometric Preliminaries

All manifolds appearing in the sequel are assumed to have a finite simplicial decomposition and the symbol $\chi(\Sigma)$ is used to denote the Euler-Poincaré characteristic of Σ. Recall that the Euler-Poincaré characteristic is a topological invariant which has the value zero for the empty set, and for a compact manifold Σ^n of dimension n which has a finite simplicial decomposition with α_i simplexes of dimension i, equals

$$\chi(\Sigma^n)=\alpha_0-\alpha_1+\ldots+(-1)^n\alpha_n.$$

That is, $\chi(\Sigma^n)$ can be computed simply by counting the simplexes in any simplicial decomposition of Σ^n. For instance, for a simplex of dimension n or equivalently, for a topological ball D^n in R^n, we have $\chi(D^n)=1$ and for a topological sphere $S^{n-1}=\partial D^n$ in R^n we have $\chi(S^{n-1})=1-(-1)^n$. For a 2-dimensional torus T in R^3 we have $\chi(T)=0$. In general, for a closed surface Σ_g^2 of genus g (i.e. a compact manifold of dimension 2 homeomorphic to a 2-dimensional sphere with g handles) we have $\chi(\Sigma_g^2)=2(1-g)$. If Σ^n is composed of m disjoint compact manifolds Σ^{ni} of dimension n, then $\chi(\Sigma^n)=\chi(\Sigma^{n1})+\chi(\Sigma^{n2})+\ldots+\chi(\Sigma^{nm})$.

Concerning the Euler-Poincaré characteristic from the point of view which is of interest in Integral geometry, see Hadwiger [5], Lefschetz [7] and Groemer [4].

With reference to the integrals of mean curvature $M_i(\Sigma)$ $(i=0,1,\ldots,n-1)$ they are well defined for hypersurfaces of class C^2 in R^n in terms of their principal curvatures, see [10] or [5]. For compact manifolds of dimension n in R^n (which will be called "bodies") the integrals of mean curvature refer to the boundary $\partial\Sigma^n$. For manifolds of dimension less than $n-1$ and for non smooth manifolds, the integrals of mean curvature have to be computed by the familiar device of considering the integrals for the parallel set to a distance ε and letting ε tend to zero [10].

For the ordinary space, $n=3$, the following cases are of interest: (a) Σ^3 = convex polyhedron; then $M_0(\Sigma^3)=M_0(\partial\Sigma^3)$ = surface area of Σ^3, $M_1(\Sigma^3)=M_1(\partial\Sigma^3)=(1/2)\,\Sigma(\pi-\alpha_i)\,a_i$, where a_i are the lengths of the edges and α_i the corresponding dihedral angles, the sum being extended over all edges of Σ^3. (b) Σ^2 = convex plate of area f and perimeter u; then $M_0=2f$, $M_1=(\pi/2)\,u$, $M_2=4\pi$. (c) Σ^1 = linear segment of length s; then $M_0=0$, $M_1=\pi s$, $M_2=4\pi$.

Throughout the paper, we shall denote by

$$O_k=\frac{2\pi^{(k+1)/2}}{\Gamma((k+1)/2)}$$

the surface area of the k-dimensional unit sphere S^k; thus $O_0=2$, $O_1=2\pi$, $O_2=4\pi$, $O_3=2\pi^2$, etc.

By R^n we mean the n-dimensional euclidean space.

3. Processes of Manifolds in R^n. Processes of Bodies

Let us consider in R^n a random process of compact manifolds, the location of each manifold being given by a point H lying in it and an orthonormal n-frame composed of the origin H and a set of orientation vectors $u_{n-1},\ldots,u_1$, where u_k

is a point of the k-dimensional unit sphere S^k, while the shape of the manifold is given by an element ρ of a certain probability space $(\mathfrak{S}, Q)$, where $\mathfrak{S}$ is a locally compact Hausdorff space with a countable base which we call the shape space, in such a way that two manifolds of the family which have assigned the same value of ρ are congruent.

We assume that the manifolds of a given realization can be enumerated according to the distance of the corresponding point H to the origin and that this enumeration is measurable (for two or more points at an equal distance a systematic method for continuing the enumeration is chosen).

We shall also assume that (for an arbitrary realization) our process can be decomposed into $n+1$ mutually independent processes, namely,

(i) the points H corresponding to the different manifolds of the realization, which form a point process in R^n with the property that the expected value of the random variable $N(A)$=number of points H within the Borel set $A \subset R^n$ is invariant under translations in R^n and finite if A is bounded.

(ii) for each fixed k, the unit vectors u_k corresponding to the different elements of the realization, which we assume to be mutually independent and uniformly distributed on the unit sphere S^k, with density du_k/O_k, where du_k stands for the area element of $S^k (k=1, 2, \ldots, n-1)$.

(iii) the sequence of values of the "shape parameter" ρ corresponding to the different elements of the realization, which we assume to be mutually independent and distributed in $\mathfrak{S}$ according to the same probability law Q.

Finally, we assume that if $V(\rho)$ is the volume, $M_i(\rho)$ the i-th integral of mean curvature and $\chi(\rho)$ the Euler-Poincaré characteristic of a manifold which has assigned the value ρ of the shape parameter, then the mean values

$$E(V) = \int_{\mathfrak{S}} V(\rho)\, Q(d\rho), \quad E(M_i) = \int_{\mathfrak{S}} M_i(\rho)\, Q(d\rho) \qquad (i=1, 2, \ldots, n-1),$$

$$E(\chi) = \int_{\mathfrak{S}} \chi(\rho)\, Q(d\rho)$$

are all finite.

4. The Mathematical Model

We recall that a point process (more precisely its law P) in a locally compact second countable Hausdorff space Z is a probability measure in the space $\mathscr{M}$ of all point measures in Z, i.e., locally finite sums of δ-measures. Denoting by $\mu^{\otimes h} = \mu \otimes \ldots \otimes \mu$ the h-fold product of $\mu \in \mathscr{M}$, the h-th moment measure of P is the mixture

$$\nu^{(h)} = \int_{\mathscr{M}} \mu^{\otimes h} P(d\mu)$$

provided that this exists as a locally finite measure in Z^h, in which case P is called an h-th order process. Explicitly, this means that the functional defined on $\mathscr{M}$ by

$$\zeta_f^{(h)}(\mu) = \mu^{\otimes h}(f) \qquad (\mu \in \mathscr{M}), \tag{1}$$

where f is a Borel function on Z^h such that (1) exists for every $\mu \in \mathcal{M}$ (for example, f bounded with compact support or $f \geqq 0$), is a random variable on the probability space $(\mathcal{M}, P)$ and

$$\nu^{(h)}(f) = \int_{\mathcal{M}} \mu^{\otimes h}(f)\, P(d\mu) = E(\zeta_f^{(h)}). \tag{2}$$

A point $z \in Z$ such that $\mu(\{z\}) \geqq 2$ is called a multiple point of the realization μ. In this paper we shall be dealing exclusively with simple processes, that is, processes having almost surely no multiple points. Accordingly, every realization μ, except for a set of P-measure zero, can be identified with its support (a countable set without accumulation points), each point of which carries a mass equal to one.

If f is a function on Z^h of the tipe described above, then $\zeta_f^{(h)}(\mu) = \mu^{\otimes h}(f)$ is the sum of the values $f(z_1, z_2, \ldots, z_h)$ over the set of all h-tuples $(z_1, z_2, \ldots, z_h)$ formed from the "points" of the realization μ. We consider two particular cases:

(a) $h = 1$; then $\zeta_f(\mu) = \zeta_f^{(1)}(\mu)$ is the sum of the values $f(z)$ over the set of all points z of the realization μ;

(b) $h = 1$ and $f = 1_A$ = the indicator function of a bounded Borel set $A \subset Z$; then $\zeta_A(\mu) = \zeta_{1_A}(\mu) = \mu(A)$ is the number of points of μ falling into A, and

$$\nu^{(1)}(A) = E(\zeta_A) = \int_{\mathcal{M}} \mu(A)\, P(d\mu)$$

is the average number of such points, i.e., $\nu^{(1)}$ is the intensity measure of P.

Returning to the random process described in the preceding section, since each manifold of the family under consideration can be identified through a one to one correspondence with the ordered set $z = (H, u_{n-1}, \ldots, u_1, \rho)$, mathematically we are dealing with a point process in the locally compact space $Z = R^n \times S^{n-1} \times \ldots \times S^1 \times \mathfrak{S}$ and the assumptions we made in that section imply that the first moment measure $\nu^{(1)}$ of this point process is given by

$$\nu^{(1)}(dz) = \frac{\lambda}{O_{n-1} \cdots O_1} dH\, du_{n-1} \ldots du_1\, Q(d\rho) \tag{3}$$

where λ is a positive constant and $dH = dx_1 \ldots dx_n$ is the volume element of R^n. Let us prove this fact.

Recalling that $N(A)$ denotes the number of points H within the Borel set $A \subset R^n$ and taking into account that $EN(.)$ is a measure in R^n, our hypothesis (i) of Sect. 3 entails the equation

$$EN(A) = \lambda V(A),$$

where λ is a positive constant and $V(A)$ the volume of A.

Let us consider a rectangular set $A \times B \subset Z$, where $A \subset R^n$ is bounded, $B = B_{n-1} \times \ldots \times B_1 \times B_0$ with $B_k \subset S^k (k = 1, 2, \ldots, n-1)$, $B_0 \subset \mathfrak{S}$ and all sets involved are Borel sets. To compute the conditional expectation of the random variable $\zeta_{A \times B}$ (see notation above) given that $N(A) = m$, let us write $z_1, \ldots, z_m$ to denote those points of the realization whose projections on R^n fall into A. Setting z_j

$=(H_j,\omega_j)$, where $H_j\in A$ and $\omega_j\in S^{n-1}\times\ldots\times S^1\times\mathfrak{S}$, we have $\zeta_{A\times B}=1_B(\omega_1)+\ldots+1_B(\omega_m)$.

Hence, taking expected values

$$E(\zeta_{A\times B}|N(A)=m)=m\Pr\{\omega_j\in B\},$$

that is

$$E(\zeta_{A\times B}|N(A))=N(A)\int_B\frac{du_{n-1}}{O_{n-1}}\ldots\frac{du_1}{O_1}Q(d\rho)$$

and taking expectations again in the last equation, we get

$$\begin{aligned}E\zeta_{A\times B}=\nu^{(1)}(A\times B)&=\lambda V(A)\int_B\frac{du_{n-1}\ldots du_1\,Q(d\rho)}{O_{n-1}\ldots O_1}\\&=\frac{\lambda}{O_{n-1}\ldots O_1}\int_{A\times B}dH\,du_{n-1}\ldots du_1\,Q(d\rho)\end{aligned}$$

which proves our assertion.

Remark. For brevity of notation, we shall occassionally use the symbol dz instead of the strictly correct ones $\nu^{(1)}(dz)$ or $d\nu^{(1)}(z)$.

For our next computations it is of the utmost importance to observe that the measure (3) can be written in the form

$$d\nu^{(1)}=\frac{\lambda}{O_{n-1}\ldots O_1}dKQ(d\rho) \tag{4}$$

where $dK=dH\,du_{n-1}\ldots du_1$ is the so called kinematic density of integral geometry [10, Chap. 15] which, as is well known, is the unique density, up to a constant factor, which is invariant with respect to the euclidean motions of R^n.

5. Mean Values of Intersections

Assuming first that all the manifolds z of our family as well as the set $A\subset R^n$ are bodies, we regard the function f defined on Z by the formula $f(z)=\chi(A\cap z)$. Then, the random variable ζ_f which we call $X(A)$ is the sum

$$X(A)=\sum\chi(A\cap z) \tag{5}$$

where summation extends to all bodies z of the realization.

To compute the expected value of (5) we have the formula

$$\begin{aligned}EX(A)=\nu^{(1)}(f)&=\int f\,d\nu^{(1)}=\int\chi(A\cap z)\,dz\\&=\frac{\lambda}{O_{n-1}\ldots O_1}\int_{\mathfrak{S}}Q(d\rho)\int\chi(A\cap z)\,dK.\end{aligned}$$

On the other hand, according to the kinematic fundamental formula of integral geometry [10, formula (15.36)], the value of the last integral on the right hand is

$$O_1 \ldots O_{n-2}\Big[O_{n-1}\,\chi(A)\,V(\rho)+O_{n-1}\,\chi(\rho)\,V(A) + (1/n)\sum_{i=0}^{n-2}\binom{n}{i+1} M_i(A)\,M_{n-i-2}(\rho)\Big].$$

Hence

$$EX(A)=\lambda\Big[\chi(A)\,E(V)+E(\chi)\,V(A) + (1/nO_{n-1})\sum_{i=0}^{n-2}\binom{n}{i+1} M_i(A)\,E(M_{n-i-2})\Big] \tag{6}$$

Since the Euler-Poincaré characteristic of a convex body equals one, if we assume that all the bodies involved in the present context are convex, then $X(A)$ represents the number of bodies of the realization which intersect with A and formula (6) becomes

$$EX(A)=\lambda\Big[V(A)+E(V)+(1/nO_{n-1})\sum_{i=0}^{n-2}\binom{n}{i+1} M_i(A)E(M_{n-i-2})\Big] \tag{7}$$

Let us consider some particular cases of (7):

(a) For $n=3$, the integrals of mean curvature of an arbitrary convex body K are $M_0(K)=F(K)=$surface area of K and $M_1(K)=M(K)=$integral of mean curvature of the boundary of K, so that in this case we have

$$EX(A)=\lambda[V(A)+E(V)+(1/4\pi)\,F(A)\,E(M)+(1/4\pi)\,M(A)\,E(F)] \tag{8}$$

where $F=F(\rho)$ is the surface area of any body of the family which corresponds to the value ρ of the shape parameter and $E(F)=\int F(\rho)\,Q(d\rho)$.

(b) For $n=3$, if each manifold z is a convex plate we may consider it as a flattened convex body, and we only have to insert the values $F=2f=$twice the are of the plate and $M=(\pi/2)\,u$, where $u=u(\rho)$ is the perimeter of z, to get the formula

$$EX(A)=\lambda[V(A)+(1/8)\,F(A)\,E(u)+(1/2\pi)\,M(A)\,E(f)]. \tag{9}$$

(c) For $n=3$, suppose that each manifold z is a linear segment of length $s=s(\rho)$. If we think of each segment as the limit of a narrowing convex plate, we may insert the values $u=2s$ and $f=0$ in the above formula and we obtain

$$EX(A)=\lambda\,V(A)+(\lambda/4)\,F(A)\,E(s). \tag{10}$$

(d) For $n=2$, the integrals of mean curvature of a plane convex set K are $M_0(K)=u(K)=$the perimeter of K and $M_1(K)=2\pi$, thus

$$EX(A)=\lambda[f(A)+E(f)+(1/2\pi)\,u(A)\,E(u)] \tag{11}$$

where $f=f(\rho)$ is the area and $u=u(\rho)$ the perimeter of z, while $f(A)$ denotes the area of A.

(e) For $n=2$ and each z a linear segment of length $s=s(\rho)$, by inserting in (11) the values $f=0$ and $u=2s$, we obtain

$$EX(A)=\lambda f(A)+(\lambda/\pi)\,u(A)\,E(s) \tag{12}$$

in agreement with [8].

Once more we emphasize that for the validity of the last formulas (7), ..., (12) all the manifolds involved must be convex.

Turning back our attention to a process of (not necessarily convex) bodies in R^n, we fix a number q in the set $\{1,2,\dots,n-1\}$ and consider the function f defined on Z by $f(z)=M_{q-1}(A\cap z)$, where A is some fixed body. Then ζ_f is the random variable $Y_{q-1}(A)$ defined by

$$Y_{q-1}(A)=\sum M_{q-1}(A\cap z) \tag{13}$$

with summation extended over all bodies z of the realization (we complete the definition by writing $M_{q-1}(\emptyset)=0$).

To compute the expected value of (13), we have

$$EY_{q-1}(A)=\int f\,dv^{(1)}=\int M_{q-1}(A\cap z)\,dz$$
$$=\frac{\lambda}{O_{n-1}\dots O_1}\int Q(d\rho)\int M_{q-1}(A\cap z)\,dK.$$

On the other hand, the formula (15.72) of [10] gives for the last integral the value

$$O_{n-2}\dots O_1\Bigg[O_{n-1}M_{q-1}(A)\,V(\rho)+O_{n-1}V(A)\,M_{q-1}(\rho)$$
$$+\frac{(n-q)\,O_{q-1}}{O_{n-q-1}}\sum_{h=n-q}^{n-2}\frac{\binom{q-1}{q+h-n}O_{2n-h-q}O_h}{(h+1)\,O_{n-h}O_{h+q-n}}M_{h+q-n}(A)\,M_{n-2-h}(\rho)\Bigg],$$

where, for $q=1$, the last sum must be deleted. It follows from here that

$$EY_{q-1}(A)=\lambda\Bigg[M_{q-1}(A)\,E(V)+V(A)\,E(M_{q-1})$$
$$+\frac{(n-q)O_{q-1}}{O_{n-1}O_{n-q-1}}\sum_{h=n-q}^{n-2}\frac{\binom{q-1}{q+h-n}O_{2n-h-q}O_h}{(h+1)\,O_{n-h}O_{h+q-n}}M_{h+q-n}(A)\,E(M_{n-2-h})\Bigg]. \tag{14}$$

We consider two particular cases of the preceding relation for $n=3$:

1. $q=1$. Then $Y_0=\sum M_0(A\cap z)=$ total surface area of all sets $A\cap z$; $M_0(A)=F(A)=$ surface area of A; $M_0(\rho)=F(\rho)=F=$ surface area of z, and formula (14) yields

$$EY_0(A)=\lambda[F(A)\,E(V)+V(A)\,E(F)].$$

In particular, if the manifolds are plates with area $f=f(\rho)$, then $Y_0(A) = 2V_2(A)$ = twice the total area within A of all plates z which intersect with A, and we only have to write $V=0$, $F=2f$ in the last relation to get $EV_2(A) = \lambda V(A) E(f)$.

2. $q=2$. Then $Y_1(A) = \sum M_1(A \cap z)$ and we obtain

$$EY_1(A) = \lambda[M_1(A) E(V) + V(A) E(M_1) + (\pi^2/16) F(A) E(F)].$$

6. Manifolds of Dimension Less Than *n*

Let $A = A^q$ be a q-dimensional compact manifold in R^n and let us suppose that all manifolds z are compact manifolds of dimension r, where $r \leqq n$ and $r+q-n \geqq 0$. We shall denote by $\sigma_r(M^r)$ the volume of the r-dimensional manifold M^r. For a manifold M of dimension zero, $\sigma_0(M)$ denotes the number of points of M.

If f is the function defined on Z by the formula $f(z) = \sigma_{r+q-n}(A \cap z)$, then ζ_f (see the definition in section 3) is the random variable

$$V_{r+q-n}(A) = \sum \sigma_{r+q-n}(A \cap z)$$

where the summation extends over all manifolds z of the realization. Its expected value is

$$EV_{r+q-n}(A) = E(\zeta_f) = v^{(1)}(f) = \int \sigma_{r+q-n}(A \cap z)\, dz$$
$$= \frac{\lambda}{O_{n-1} \dots O_1} \int_{\mathfrak{S}} Q(d\rho) \int \sigma_{r+q-n}(A \cap z)\, dK.$$

Now, the formula (15.20) of [10] gives for the last integral on the right hand the value $(O_n \dots O_1) O_{r+q-n} (O_q O_r)^{-1} \sigma_q(A) \sigma_r(\rho)$, where $\sigma_r(\rho) = \sigma_r(z)$. Hence

$$EV_{r+q-n}(A) = \lambda \frac{O_n O_{r+q-n}}{O_q O_r} \sigma_q(A) E(\sigma_r). \tag{15}$$

If $r+q-n=0$, then V_0 denotes the number of intersection points of A with the manifolds of the realization.

Formula (15) contains many particular cases which may be useful in practical situations of the type encountered in stereology. Let us consider some examples.

(a) If $q=n$, we get $EV_r(A) = \lambda V(A) E(\sigma_r)$, where $V_r(A)$ denotes the total r-dimensional volume within A of all manifolds z which intersect with A. Recall that the manifolds z have dimension r.

(b) If $n=3$, $q=2$, $r=1$ (A is a surface of area $F(A)$ and z are curves of length $L(\rho)$), then $V_0(A)$ denotes the number of intersection points of A with the curves of the realization, and (15) gives $EV_0(A) = (\lambda/2) F(A) E(L)$.

(c) If $n=3$, $q=2$, $r=2$, (15) gives $EV_1(A) = (\lambda/4) \pi F(A) E(F)$, where $V_1(A)$ is the total length of the intersections of the surface A with the surfaces of the realization (of area $F = F(\rho)$).

7. Mean Values of Multiple Intersections

Suppose that all manifolds z of our process have the same dimension r. We shall make an additional assumption concerning the structure of the h-th moment measure of P, namely, we assume that if $A_1\ A_2, \ldots, A_h$ are disjoint Borel sets in R^n, then we have

$$E[N(A_1)\ldots N(A_h)]=EN(A_1)\ldots EN(A_h). \tag{16}$$

For example, (16) is satisfied in the case of the Poisson process as defined in the next section.

Recalling that $\sigma_i(M^i)$ denotes the volume of the i-th dimensional manifold M^i and assuming that $q+rh-nh\geqq 0$, we wish to compute the expected value of the random variable

$$W(A)=\sum \sigma_{q+rh-nh}(A\cap z_1\cap\ldots\cap z_h)$$

where A is a compact q-dimensional manifold and summation extends over all sets $\{z_1,\ldots,z_h\}$ formed by h different elements of the realization.

To this end, we start by considering the mappings π_1 and π_2 defined on Z^h by $(H,u_{n-1},\ldots,u_1,\rho)\to H$ and $(H,u_{n-1},\ldots,u_1,\rho)\to(u_{n-1},\ldots,u_1)$ respectively.

Let D_1 be the closed subset of Z^h formed by all points $(z_1,z_2,\ldots,z_h)$ such that $\pi_1(z_i)=\pi_1(z_j)$ for some pair of indexes i and j such that $1\leqq i<j\leqq h$ and similarly, let D_2 be the closed subset of Z^h formed by all points $(z_1,\ldots,z_h)$ such that $\pi_2(z_i)=\pi_2(z_j)$ for some pair of different indexes i and j. We write $D=D_1\cup D_2$ and define the function f on Z^h by

$$f(z_1,\ldots,z_h)=\begin{cases}\sigma_{q+rh-nh}(A\cap z_1\cap\ldots\cap z_h) & \text{outside } D\\ 0 \quad \text{on } D.\end{cases}$$

It is clear that almost surely

$$W(A)=\frac{1}{h!}\zeta_f^{(h)},$$

the factorial in the denominator accounting for all permutations.

We postpone the proof of the following facts:

(a) The restriction of $\nu^{(h)}$ to the open set Z^h-D equals the restriction of the h-fold product $\nu^{(1)}\otimes\ldots\otimes\nu^{(1)}$ to the same set;

(b) $\nu^{(1)}\otimes\ldots\otimes\nu^{(1)}(D)=0$.

Taking them for granted, the computation runs as follows:

$$\begin{aligned}EW(A)&=\frac{1}{h!}E(\zeta_f^{(h)})=\frac{1}{h!}\int f\,d\nu^{(h)}=\frac{1}{h!}\int_{Z^h-D} f\,d\nu^{(h)}\\ &=\frac{1}{h!}\int_{Z^h-D} f\,d(\nu^{(1)}\otimes\ldots\otimes\nu^{(1)})=\frac{1}{h!}\int_{Z^h} f(z_1,\ldots,z_h)\,dz_1\ldots dz_h\end{aligned}$$

where $dz_i = \nu^{(1)}(dz_i) = \lambda(O_{n-1} \dots O_1)^{-1} dK_i Q(d\rho_i)$, $i = 1, 2, \dots, h$. Hence

$$EW(A) = \frac{\lambda^h}{h!(O_1 \dots O_{n-1})^h} \int_{\mathfrak{S}} Q(d\rho_1) \dots \int_{\mathfrak{S}} Q(d\rho_h) \times \int \sigma_{q+rh-nh}(A \cap z_1 \cap \dots \cap z_h) \, dK_1 \dots dK_h.$$

Now, the formula (15.22) of [10] gives for the last integral the value $(O_1 \dots O_n)^h O_{q+rh-nh}(O_q O_r^h)^{-1} \sigma_q(A) \sigma_r(\rho_1) \dots \sigma_r(\rho_h)$, where $\sigma_r(\rho_i) = \sigma_r(z_i)$. Therefore

$$EW(A) = \frac{\lambda^h}{h!} \frac{O_n^h O_{q+rh-nh}}{O_q O_r^h} \sigma_q(A) [E(\sigma_r)]^h. \tag{17}$$

Next, we give the proof of (a) and (b) in the case $h = 2$.

Let $U_1 = A_1 \times B_1$ and $U_2 = A_2 \times B_2$, where A_1 and A_2 are disjoint bounded open sets in R^n, while B_1 and B_2 are rectangular subsets of $S^{n-1} \times \dots \times S^1 \times \mathfrak{S}$. To compute the conditional expectation of $\zeta^{(2)}_{U_1 \times U_2}$ given that $N(A_1) = m$ and $N(A_2) = p$, let us write $z_j = (H_j, \omega_j)$ and $z'_k = (H'_k, \omega'_k)$ to denote the points of the realization μ such that $H_j \in A_1$ $(1 \leqq j \leqq m)$ and $H'_k \in A_2 (1 \leqq k \leqq p)$. Then

$$\zeta^{(2)}_{U_1 \times U_2}(\mu) = \zeta_{U_1}(\mu) \zeta_{U_2}(\mu) = \{1_{B_1}(\omega_1) + \dots + 1_{B_1}(\omega_m)\} \{1_{B_2}(\omega'_1) + \dots + 1_{B_2}(\omega'_p)\} = \sum_{j=1}^{m} \sum_{k=1}^{p} 1_{B_1}(\omega_j) 1_{B_2}(\omega'_k).$$

Hence

$$E(\zeta^{(2)}_{U_1 \times U_2} | N(A_1) = m, N(A_2) = p) = m p \Pr(B_1) \Pr(B_2)$$

where $\Pr(B) = \int_B (O_{n-1} \dots O_1)^{-1} du_{n-1} \dots du_1 Q(d\rho)$. That is

$$E(\zeta^{(2)}_{U_1 \times U_2} | N(A_1), N(A_2)) = N(A_1) N(A_2) \Pr(B_1) \Pr(B_2).$$

Taking expected values, we get

$$\begin{aligned} \nu^{(2)}(U_1 \times U_2) &= E(\zeta^{(2)}_{U_1 \times U_2}) = E(N(A_1) N(A_2)) \Pr(B_1) \Pr(B_2) \\ &= \{EN(A_1) \Pr(B_1)\} \{EN(A_2) \Pr(B_2)\} \\ &= \nu^{(1)}(U_1) \nu^{(1)}(U_2) = \nu^{(1)} \otimes \nu^{(1)}(U_1 \times U_2) \end{aligned}$$

which proves (a).

As for (b), we may assume without loss of generality that each manifold z is a point in R^n, so that $Z = R^n$ and D_1 becomes the diagonal of Z^2.

If E is a bounded subset of D_1, let $E' = \{z : (z, z) \in E\}$. Then, from the fact that $\nu^{(1)}$ assigns the value zero to each set consisting of a single point,

$$\nu^{(1)} \otimes \nu^{(1)}(E) = \int \nu^{(1)}(E_z) \nu^{(1)}(dz) = \int_{E'} \nu^{(1)}(\{z\}) \nu^{(1)}(dz) = 0.$$

Hence $\nu^{(1)} \otimes \nu^{(1)}(\mathrm{D}_1) = 0$ and similarly for D_2. Thus $\nu^{(1)} \otimes \nu^{(1)}(D) = 0$ and the proof of (17) is complete.

Examples. 1. Assuming that $n=3$ and that each z is a surface, we may consider two particular cases:

(a) $q=3$, $h=2$. Then $W(A)$ represents the total length within A of all intersections of two different surfaces of the realization and its expected value is

$$EW(A)=(\pi\lambda^2/8)\,V(A)\,[E(f)]^2$$

where $f=f(\rho)$ denotes the surface area of z.

(b) $q=3$, $h=3$. Then $W(A)$ becomes the number of intersection points within A of every three different surfaces of the realization and its expected value is

$$EW(A)=(\pi\lambda^3/48)\,V(A)\,[E(f)]^3.$$

2. Assuming that $n=2$ and that each z is a linear segment of length $s(\rho)$, we consider the case $q=2$, $h=2$; then $W(A)$ is the number of segment-segment crossings within A, and we get the result of Parker and Cowan [8]:

$$EW(A)=(\lambda^2/\pi)\,F(A)\,[E(s)]^2,$$

where $F(A)$ denotes the area of A.

8. Processes of Poisson of Convex Manifolds

If in addition to the previously stated hypothesis, the point process (i) of Sect. 3 satisfies the following two conditions:

(a) For every finite set of disjoint Borel subsets of R^n, say $\{A_1, A_2, \ldots, A_k\}$, the random variables $N(A_1), \ldots, N(A_k)$ are mutually independent,

(b) For every bounded Borel set $A\subset R^n$, $P\{N(A)=m\}=[\lambda V(A)]^m(m!)^{-1}\exp(-\lambda V(A))$ $(m=0,1,2,\ldots)$ then we say that our process of manifolds P is a Poisson process of intensity λ.

In this section we assume that both (a) and (b) hold and that the arbitrary set A as well as the manifolds z are convex.

Under these assumptions, it is easy to see that the random variable $X(A)$ = the number of manifolds which intersect with A has a Poisson distribution with the expectation given by (7). Calling D the distance from the origin to the nearest manifold of the process, if we take for A the ball of radius r centered at the origin, then $P\{D>r\}=P\{X(A)=0\}$. But for this particular A, we have

$$V(A)=(O_{n-1}/n)\,r^n, \quad M_i(A)=O_{n-1}\,r^{n-i-1}\,(i=0,1,\ldots,n-1)$$

so that the expected value of $X(A)$ is the number

$$\psi=\lambda\left[\frac{O_{n-1}}{n}r^n+E(V)+\frac{1}{n}\sum_{i=0}^{n-2}\binom{n}{i+1}r^{n-i-1}\,E(M_{n-i-2})\right]. \tag{18}$$

Hence, under the Poisson assumption,

$$P\{D>r\}=\exp(-\psi)$$

with ψ given by (18). It follows from here that the probability density function of the random variable D is

$$\lambda \exp(-\psi)\left[O_{n-1} r^{n-1} + \sum_{i=0}^{n-2} \binom{n-1}{i+1} E(M_{n-i-2}) r^{n-i-2}\right], \quad r>0.$$

For the ordinary space ($n=3$) we have three possibilities:

(a) Each manifold z is a convex body. Then $M_0(z)=F(\rho)=$surface area of z, $M_1(z)=M_1(\rho)=$first integral of mean curvature and $\psi=\lambda[4\pi r^3/3+E(V)+E(M_1)r^2+E(F)r]$.

(b) Each manifold z is a convex plate. Then $M_0(z)=2f$, where $f=f(\rho)=$area of z; $M_1(z)=(\pi/2)u$, where $u=u(\rho)=$perimeter of z and $\psi=\lambda[4\pi r^3/3+(\pi r^2/2)E(u)+2rE(f)]$.

(c) Each set z is a linear segment of length $s=s(\rho)$. Then $M_0(z)=M_0(\rho)=0$, $M_1(z)=M_1(\rho)=\pi s$ and $\psi=\lambda[4\pi r^3/3+\pi r^2 E(s)]$.

References

1. Berman, M.: Distance distributions associated with Poisson processes of geometric figures. J. Appl Probability **14**, 195–199 (1977)
2. Coleman, R.: The distance from a given point to the nearest end of one member of a random process of linear segments. In Stochastic Geometry. Ed. Harding and Kendall pp. 192–201. New York: Wiley 1974
3. Fava, N., Santaló, L.A.: Plate and line segment processes, J. Appl. Probability **15**, 494–501 (1978)
4. Groemer, H.: Eulersche Characteristic, Projectionen und Quermassintegrale. Math. Ann. **198**, 23–56 (1972)
5. Hadwiger, H.: Vorlesungen über Inhalt, Oberfläche und Isoperimetrie. Berlin-Heidelberg-New York: Springer 1957
6. Krickeberg, K.: Moments of Point-processes. Stochastic Geometry. Ed. Harding and Kendall, pp. 89–113. New York: Wiley 1974
7. Lefschetz, S.: Introduction to Topology. Princeton University Press 1949
8. Parker, P., Cowan, R.: Some properties of line segment processes. J. Appl. Probability **13**, 96–107 (1976)
9. Santaló, L.A.: Integralgeometrie 5, Exposés de Géométrie. Paris: Hermann 1936
10. Santaló, L.A.: Integral Geometry and Geometric Probability, Encyclopedia of Mathematics and its Applications. Reading, Mass.: Addison-Wesley 1976

Received February 24, 1978; in revised form March 17, 1979

STOCHASTICA, Vol. IV, nº 1 (1980)

RANDOM LINES AND TESSELLATIONS IN A PLANE

L. A. Santaló

ABSTRACT

Our purpose is the study of the so called "mixed random mosaics", formed by superposition of a given tessellation, not random, of congruent convex polygons and a homogeneous Poisson line process. We give the mean area, the mean perimeter and the mean number of sides of the polygons into which such mosaics divide the plane.

1. Introduction

Lines in the euclidean plane E_2 are parametrized by (p,θ), the polar coordinates of the foot of the perpendicular from the origin to the line. The density element for lines, invariant under euclidean motions, is $dp\wedge d\theta$. The measure of the set of lines intersecting a convex set K is equal to the length L of the boundary ∂K. For a line segment of length b this measure is equal to 2b, since the segment must be considered as a flattened convex set.

The standard homogeneous Poisson line process of intensity λ is that line process corresponding to a homogeneous Poisson point process of constant intensity λ in the strip $\{(p,\theta); 0\leqslant p<\infty, 0\leqslant\theta<2\pi\}$. The fundamental property of this line process is that the number m of lines of the process hitting a convex set K has a Poisson distribution of intensity λL. Moreover, each line of the plane intersects the lines of the process in a linear homogeneous point process of Poisson of intensity 2λ. This property also holds for the intersection of a line of the process with the other lines of the process. For all these questions see, for instance Kendall-Moran [9], Solomon [18], or [17].

A Poisson system of random lines of intensity λ partitiones the plane into a random tessellation (or mosaic) of simple convex polygons with almost surely each vertex being a vertex of four polygons of the tessellation. These random polygons were first studied by Goudsmidt [6] who obtained the mean number of sides, the mean perimeter, the mean area and the mean area squared of the polygons. More general results were obtained later by Miles [10], [12] and Richards [15]. Interesting and fundamental questions referring to ergodicity and edge effects have been carefully treated by Cowan [3], [4] and Ambartzumian [1], [2] (see also Miles [11], [13]).

Our purpose is the study of mixed random tesselations of the plane, originated when a Poisson system of random lines is superposed on a preexistent tessellation, not random, formed by compact, convex, congruent polygons which cover the whole plane without overlapping. For instance, figs.1, 2 represent the case of tessellations of congruent pentagons crossed by a random Poisson system of lines.

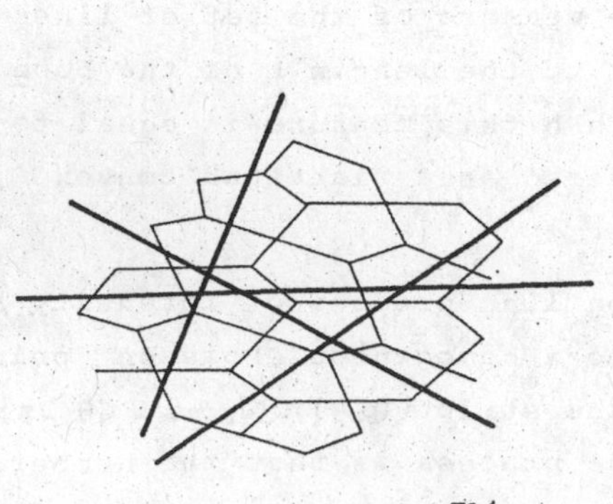

FIG.1

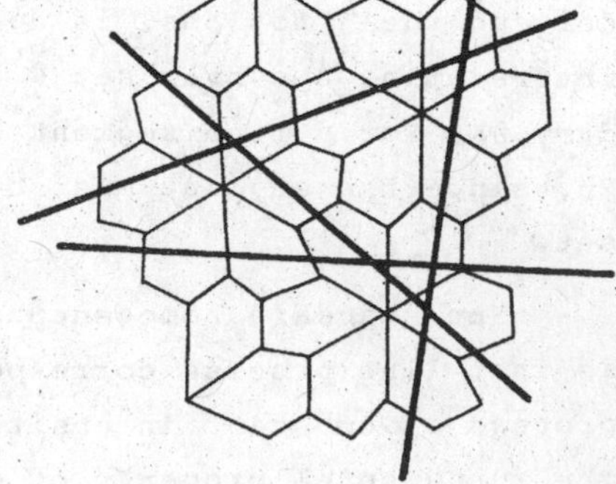

FIG.2

We obtain the mean are (5.3), the mean number of sides (5.4) and the mean perimeter (5.5) of the resulting polygons. It seems to be an interesting open question to find the second order moments of these characteristics for that kind of mixed mosaics.

2. Tessellations of compact, congruent, convex polygons.

We use the term tessellation for any arrangement of bounded, convex, congruent polygons (called the fundamental polygons or cells of the tessellation) fitting together so as to cover the whole plane without overlapping. Examples of tessellations and their relations to group theory can be seen in the books of Coxeter [5], Guggenheimer [8] or in Grünbaum-Shephard [7].

Assume a given tessellation T whose fundamental polygons have area f, perimeter u and number of sides n (equal to the number of vertices). It is known that the only possible values of n are 3,4,5,6. Consider a circle Q(R) of radius R (which we will assume sufficiently large) and let $\nu(\partial Q)$ be the number of fundamental polygons which are intersected by the boundary ∂Q and $\nu(Q)$ the number of fundamental polygons within Q. Then we have

$$(2.1) \qquad \lim_{R\to\infty} \frac{\nu(\partial Q)}{\nu(Q)} = 0.$$

For a proof, notice that if D denotes the diameter of a fundamental polygon, we have (for large R),

$$\nu(\partial Q) \leqslant \frac{\pi(R+D)^2 - \pi(R-D)^2}{f} = \frac{4\pi RD}{f}, \quad \nu(Q)\ f \geqslant \pi(R-D)^2$$

and (2.1) follows.

This relation (2.1) or the weaker one $\nu(\partial Q)/\pi R^2 \to 0$ (as $R\to\infty$) makes possible to eliminate the "edge effects" in some passage to the limit which we shall perform later. It is worthy to note that in the hyperbolic plane these "edge effects" are not negligible, so that the passage from a finite region to the whole plane in some tessellation problems, must be treated with care (see [16]).

Let now n_k denote the number of vertices of each fundamental polygon which are surrounded by k faces of the tessellation ($k \geqslant 3$) We shall need the following identities

$$\sum n_k = n \, , \qquad \sum n_k \left(\frac{1}{2} - \frac{1}{k}\right) = 1 \tag{2.2}$$

where the sums are extended over all values of k. The first equality is nothing else than the definition of n_k and the second is an easy consequence of the Euler relation vertices-sides+faces=1, applied to the bounded planar graph formed by the edges of the tessellation within Q for $R \to \infty$.

3. Some results of stochastic geometry

Consider a circle $Q=Q(R)$ of radius R. The random variable $m=m(Q)=$ number of lines of a given homogeneous Poisson line process of intensity λ hitting Q has a Poisson $2\pi R\lambda$-distribution (see Miles [10] or Santaló [17]; notice that in [17] we use $\lambda/2$ instead of the present λ).

Therefore we have the following moments

$$\begin{aligned} &E(m) = 2\pi R\lambda \, , \quad E(m^2) = 2\pi R\lambda + (2\pi R\lambda)^2 \\ &E(m^3) = 2\pi R\lambda + 3(2\pi R\lambda)^2 + (2\pi R\lambda)^3 \\ &E(m^4) = 2\pi R\lambda + 7(2\pi R\lambda)^2 + 6(2\pi R\lambda)^3 + (2\pi R\lambda)^4 \end{aligned} \tag{3.1}$$

Moreover, we know that for m random lines (in the sense of geometrical probability) which meet a circle Q, the mean number of intersection points n_P which are inside Q is [17, p. 53]

$$E(n_P | m) = m(m-1)/4 \tag{3.2}$$

and thus, applying (3.1)

(3.3) $$E(n_P) = EE(n_P | m) = \pi^2 R^2 \lambda^2 .$$

We also know that [17, p. 54]

(3.4) $$E(n_P^2 | m) = \frac{1}{2}\binom{m}{2} + \frac{3}{2}\binom{m}{4} + \frac{16}{\pi 2}\binom{m}{3}$$

and thus

(3.5) $$E(n_P^2) = \pi^2 R^2 \lambda^2 + \pi^4 R^4 \lambda^4 + \frac{64}{3} \pi R^3 \lambda^3 .$$

From (3.3) and (3.5) we have, as $R \to \infty$,

(3.6) $$\lim E\left(\frac{n_P}{\pi R^2}\right) = \pi\lambda^2 \quad , \quad \lim E\left(\frac{n_P}{\pi R^2}\right)^2 = \pi^2\lambda^4$$

so that, for any realization of the process, we have, almost sure

(3.7) $$\lim_{R\to\infty} \frac{n_P}{\pi R^2} = \pi\lambda^2 .$$

Where n_P is the number of intersections of the lines of the Poisson process within Q.

4. Tessellations and Random Lines.

Consider a fixed tessellation T with the characteristics specified in n.2, and a Poisson line process P of intensity λ superposed to it. The tessellation T and the line process P define on the plane a mixed random mosaic M. Consider a circle Q(R) of large radius R. Let n_P denote the number of intersection points of the lines of the process within Q, n_T the number of vertices of the tessellation within Q and n_{PT} the number of intersections of lines of the process with sides of the tessellation within Q. From a classical Crofton's formula of integral geometry ([17], p. 31) we know that $E(n_{PT}|m) = mL/\pi R$, where L is the total length of the sides of the tessellation within Q and m denotes the number of lines of the process intersecting Q. Neglecting edge effects, for R sufficiently large, we can take $\pi R^2/f$ as the number of fundamental polygons intersected by Q and thus we have

$$(4.1)\qquad E(n_{PT}|m) = \frac{uRm}{2f} \quad , \quad E(n_{PT}) = \frac{\pi R^2 u\lambda}{f} \quad , \quad E\left(\frac{n_{PT}}{\pi R^2}\right) = \frac{u\lambda}{f} \; .$$

For n_T and R large, with the notations of n. 2, we have

$$(4.2)\qquad \frac{n_T}{\pi R^2} = \frac{1}{f} \sum_k \frac{n_k}{k} \lambda^2 .$$

Thus, using (3.3), we have

$$(4.3)\qquad E\left(\frac{n_P n_T}{(\pi R^2)^2}\right) = \frac{\pi}{f} \sum_k \frac{n_k}{k} \lambda^2 \; .$$

The second moments involving n_P and n_{PT}, other than $E(n_P^2)$, are not easy to calculate exactly. However, we shall give some upper bounds.

Noting that n_{PT}, for each line of the process P and large R is less than the number of fundamental poligons within a rectangle of sides 2R and 2D (where D means the diameter of the fundamental polygons), we have $n_{PT} < (4RD/f)m$ and thus, using (3.1)

$$(4.4)\qquad E\left(\frac{n_{PT}}{\pi R^2}\right)^2 < \frac{32D^2}{\pi f^2} \left(\frac{\lambda}{R} + 2\pi\lambda^2\right)$$

On the other hand we have $n_P \leqslant m(m-1)/2$ so that $E(n_{PT}n_P/(\pi R^2)^2) < 2Dm^2(m-1)/\pi^2 R^3 f$ and using (3.1)

$$(4.5)\qquad E\left(\frac{n_{PT}n_P}{(\pi R^2)^2}\right) < \frac{16D}{f} \left(\frac{\lambda^2}{R} + \pi\lambda^3\right) \; .$$

The total number of vertices v of the mixed random mosaic M generated by the union of T and P which are inside Q, is $v = n_P + n_{PT} + n_T$. Therefore, using (3.3), (3.5), (4.1), (4.2), (4.3), (4.4), and

(4.5) we have

(4.6) $$E\left(\frac{v}{\pi R^2}\right) = \pi\lambda^2 + \frac{u\lambda}{f} + \frac{1}{f}\sum_k \frac{n_k}{k}$$

(4.7) $$E\left(\frac{v^2}{(\pi R^2)^2}\right) < \frac{\lambda^2}{R^2} + \pi^2\lambda^4 + \frac{64\lambda^3}{3\pi R} + \frac{32D^2}{\pi f^2}\left(\frac{\lambda}{R} + 2\pi\lambda^2\right) + \frac{1}{f^2}\sum_k\left(\frac{n_k}{k}\right)^2 + \frac{32D}{f}\left(\frac{\lambda^2}{R} + \pi\lambda^3\right) + \frac{2\pi\lambda^2}{f}\sum_k \frac{n_k}{k} + \frac{2u\lambda}{f^2}\sum_k \frac{n_k}{k}$$

and, as $R\to\infty$, we have

$$\lim_{R\to\infty} \operatorname{var}\,(v^2/(\pi R^2)^2) < \left(\frac{32\pi D}{f} - \frac{2\pi u}{f}\right)\lambda^3 + \left(\frac{64D^2}{f^2} - \frac{u^2}{f^2}\right)\lambda^2$$

5. Mean values of the area, perimeter and number of sides of the cells of a mixed random mosaic.

Consider the mixed random mosaic M of n.4. composed of the tessellation T of congruent polygons and a homogeneous Poisson line process P of intensity λ. With the notations of the foregoing paragraph we have that tha number of edges e of M within the circle Q of radius R is

(5.1) $$e = 2n_P + 2n_{PT} + \frac{1}{2}\,\frac{\pi R^2 n}{f}$$

where $n = \sum n_k$ is the number of sides of the fundamental polygons of the tessellation T. This equality disregards some "edge effects" on the boundary of Q which may be neglected for R sufficiently large. Therefore, as $R\to\infty$, we have

(5.2) $$\lim_{R\to\infty} E(e/\pi R^2) = 2\pi\lambda^2 + 2u\lambda/f + n/2f.$$

By Euler's relation, we have

$$n_P + n_{PT} + (\pi R^2/f) \sum_k (n_k/k) - e + c = 1$$

where c denotes the number of cells of M within Q. Taking expectation and using (2.2) we deduce

$$\lim_{R\to\infty} E(c/\pi R^2) = \pi\lambda^2 + (u/f)\lambda + 1/f.$$

Thus, as $R\to\infty$, we have almost surely $\lim(\pi R^2/c) = (\pi\lambda^2+u\lambda/f + 1/f)^{-1}$ and the mean area of the cells of the mixed mosaic results

(5.3) $$E(A) = \frac{f}{\pi f\lambda^2 + u\lambda + 1} .$$

For the mean number of sides of each region, we have

(5.4) $$E(N) = \lim_{R\to\infty} (2e/c) = \frac{4\pi f\lambda^2 + 4u\lambda + n}{\pi f\lambda^2 + u\lambda + 1} .$$

Note that for $n=4$ (tessellation of quadrilaterals) is $E(N)=4$, independently of λ.

In order to find the mean perimeter S we note that the total length of all sides of T within Q is $(\pi R^2/2f)u$ and the mean value of the total length of the chords that the lines of the process P intercept in Q is $\pi^2 R^2\lambda$ ([17], p. 30). Therefore the mean value of the total length of the sides of the mosaic within Q is $\pi R^2(u/2f+\pi\lambda)$ and the mean perimeter (as $R\to\infty$) is given by

(5.5) $$E(S) = \frac{2\pi f\lambda + u}{\pi f\lambda^2 + u\lambda + 1} .$$

Note that for $\lambda=0$ the mixed mosaic reduces to the tessellation

T (not random) and for $f \to \infty$ it reduces to the well known random division of the plane by a Poisson line process [10].

6. Example

Considere the tessellation of parallelograms of the fig.3.

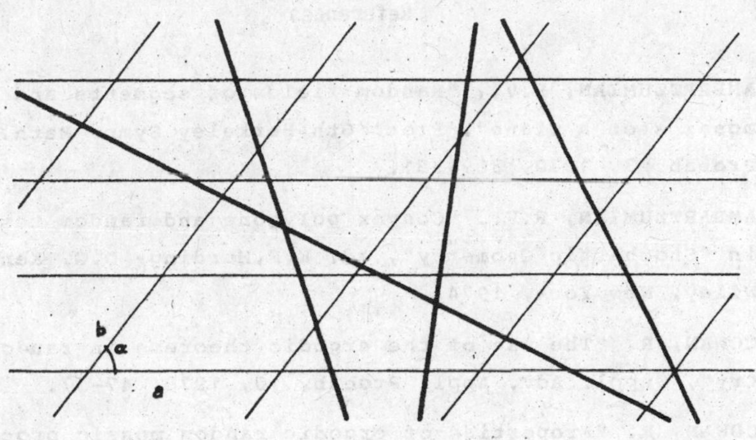

FIGURA 3

We have

$$f = a\, b \sin\alpha \;,\; u = 2(a+b)\,,\; n = n_4 = 4$$

and therefore

$$E(A) = \frac{ab \sin\alpha}{\pi ab \sin\alpha\lambda^2 + 2(a+b)\lambda + 1}\;,\quad E(N) = 4$$

$$E(S) = \frac{2\pi ab \sin\alpha\,\lambda + 2(a+b)}{\pi ab \sin\alpha\lambda^2 + 2(a+b)\lambda + 1}.$$

If $a \to \infty$ we have the plane divided by parallel lines at distance $\Delta = b \sin\alpha$. Thus, a homogeneous Poisson line process of in tensity λ determines on the plane on which are ruled parallel li nes at a distance Δ apart, a mixed random mosaic of characteristics

$$E(A) = \frac{\Delta}{\pi\Delta\lambda^2+2\lambda}, \quad E(N)= 4 , \quad E(S) = \frac{2\pi\Delta\lambda + 2}{\pi\Delta\lambda^2 + 2\lambda}.$$

References

[1] AMBARTZUMIAN, R.V., "Random fields of segments and random mosaics on a plane". Proc. 6th Berkeley Symp. Math. Statist. Probab. 3, 1970, 369-381.

[2] AMBARTZUMIAN, R.V., "Convex polygons and random tessellations", in "Stochastic Geometry", ed. E.F.Harding- D.G. Kendall, J. Wiley, New York, 1974.

[3] COWAN, R. "The use of the ergodic theorems in random geometry", Suppl. Adv. Appl. Probab. 10, 1978, 47-57.

[4] COWAN, R. "Properties of ergodic random mosaic processes", to be published in Mathematische Nachrichten, 1980.

[5] COXETER, H.S.M. "Introduction to Geometry", J. Wiley, New York, 1961.

[6] GOUDSMIDT, S. "Random distribution of lines in a plane", Rev. Modern Physics, 17, 1945, 321-322.

[7] GRÜNBAUM, B. - SHEPHARD, G.C., "Isohedral tiling of a plane by polygons", Comm. Math. Helvetici, 53, 1978, 522-571.

[8] GUGGENHEIMER, H.W., "Plane Geometry and its Groups",Holden Day, San Francisco, 1967.

[9] KENDALL, M.G. - MORAN, P.A.P. "Geometrical probability", Griffin, London, 1963.

[10] MILES, R.E., "Random polygons determined by random lines in a plane",I, Proc. Nat. Acad. Sc. USA, 52, 1964, 901-907; II, 52, 1964, 1157-1160.

[11] MILES, R.E., "Poisson flats in Euclidean spaces, II: Homogeneous Poisson flats and the complementary theorem", Adv. Appl. Prob. 3, 1971, 1 - 43.

[12] MILES, R.E., "The various aggregates of random polygons determined by random lines in a plane", Adv. in Math. 10, 1973 256-290.

[13] MILES, R.E., "The random division of space", Suppl. Adv. Prob. 1972, 243-266.

[14] MILES, R.E., "The elimination of edge effects in planar sampling". In "Stochastic Geometry", ed. E.F. Harding-D.G. Kendall, J. Wiley, New York, 1974.

[15] RICHARDS, P.I., "Averages for polygons formed by random lines", Proc. Nat. Acad. Sc. USA, 52, 1964, 1160-1164.

[16] SANTALÓ, L.A. - YAÑEZ, I., "Averages of polygons formed by random lines in euclidean and hyperbolic planes", J. Appl. Prob. 9, 1972, 140-157.

[17] SANTALÓ, L.A., "Integral Geometry and Geometric Probability" Addison Wesley Publ. Reading, 1976.

[18] SOLOMON, H., "Geometric probability", Soc. for Industrial and Appl. Math. Philadelphia, Penn., 1978.

Departamento de Matemáticas.
Universidad de Buenos Aires.
Ciudad Universitaria (Nuñez).
BUENOS AIRES, Argentina.

Math. Nachr. **117** (1984) 129–133

Mixed Random Mosaics

By L. A. Santaló of Buenos Aires

(Received September 15, 1982)

Abstract. Cowan [2] has defined random mosaics processes RMP in R^2 and has given some basic properties of them. In particular Cowan introduces the fundamental parameters α, β_k, γ_k of the process and, in terms of them, he computes the mean values of the area a, perimeter h, number of arcs w and number of vertices v of a typical polygon of the RMP. Our purpose is to consider the RMP obtained by superposition of two independent random mosaics. Then, the characteristics a_{12}, h_{12}, w_{12}, v_{12} of the resulting process are computed in terms of the characteristics a_i, h_i, w_i, v_i of each process. The case of non random tessellations mixed with random mosaics is also considered.

1. Random Mosaics

We consider homogeneous, ergodic, random mosaics processes in R^2 in the sense of Cowan [2], [1]. Let $\mathfrak{M}(\omega)$ denote a realization of a random mosaic process RMP in R^2. Consider a domain D in R^2 chosen independently of the RMP and let $M(D, \omega)$ denote the number of arcs of the process intersecting with D. We will assume that for bounded D we have $EM<\infty$.

Following Cowan [2] we call $C_k(D, \omega)$ the number of nodes within D which have k emanating arcs, $J_k(D, \omega)$ the number of junctions of order k within D and $S(D, \omega)$ the total length of arcs of $\mathfrak{M}(\omega)$ contained in D. Recall that a node is called a junction if one of the angles formed at the node is π.

Then, Cowan [2] establishes the relations

$$(1.1)\qquad \begin{aligned} ES(D)&=\alpha m(D)\\ EC_k(D)&=\beta_k m(D)\\ EJ_k(D)&=\gamma_k m(D) \end{aligned}$$

which define the fundamental parameters α, β_k, γ_k of the RMP. $m(D)$ denotes the Lebesgue measure of D.

Under very general hypothesis on the RMP, Cowan shows that the mean area a, the mean perimeter h, the mean number of arcs w and the mean number of vertices v of a "typical polygon" (suitable defined), are given by

$$(1.2)\qquad a=\frac{2}{\Sigma\,(k-2)\,\beta_k},\quad h=\frac{4\alpha}{\Sigma\,(k-2)\,\beta_k}$$
$$w=\frac{2\Sigma k\beta_k}{\Sigma\,(k-2)\,\beta_k},\quad v=\frac{2\Sigma\,(k\beta_k-\gamma_k)}{\Sigma\,(k-2)\,\beta_k}.$$

We assume that the RMP considered satisfies the necessary conditions in order that (1.2) be true.

2. Intersection of a Random Mosaic with a Rectifiable Curve

Let K be a rectifiable curve of length L, placed at random on the plane of the random mosaic $\mathfrak{M}$. By "placed at random" we understand that in order to measure a set of positions of K we take the integral of the differential form $dPd\varphi$ (kinematic density of the Integral Geometry [7, Chap. 6]), where P is a fixed point with respect to K and φ denotes a rotation about P. We want to calculate the average of the number $N\,(K\cap\mathfrak{M}(\omega))$ of the intersection points of K with the arcs of $\mathfrak{M}(\omega)$.

Let d be the diameter of K and take a disk B_r of radius $r\gg d$. We represent by B_r the disk and the set of arcs of $\mathfrak{M}(\omega)$ contained in it. Let $S(B_r,\omega)$ be the total length of the arcs of $\mathfrak{M}(\omega)$ within B_r. The so called Poincares formula of the integral geometry [7, p. 111] gives

$$(2.1)\qquad \int N\,(K\cap B_r(\omega))\,dPd\varphi=4LS(B_r,\omega)$$

where the integral on the left is extended to all $P\in R^2$, $0\leqq\varphi\leqq 2\pi$.

From this equality we obviously deduce

$$\frac{2\pi\cdot\pi\,(r-d)^2}{2\pi\cdot\pi r^2}\int\limits_{P\in B_{r-d}}\frac{N\,(K\cap B_r)}{2\pi\cdot\pi\,(r-d)^2}\,dPd\varphi\leqq 4L\,\frac{S(B_r,\omega)}{2\pi\cdot\pi r^2}$$
$$\leqq\frac{2\pi\cdot\pi\,(r+d)^2}{2\pi\cdot\pi r^2}\int\limits_{P\in B_{r+d}}\frac{N\,(K\cap B_r)}{2\pi\cdot\pi\,(r+d)^2}\,dPd\varphi\,.$$

By $r\to\infty$ the left and right sides of these inequalities converge with probability one to the same quantity $EN\,(K\cap\mathfrak{M})$ and $S(B_r,\omega)/\pi r^2\to\alpha$ almost sure (see Cowan [2]) so that we have

$$(2.2)\qquad EN\,(K\cap\mathfrak{M})=\frac{2\alpha L}{\pi}\,.$$

Notice that, by (1.1), α is the mean length of the arcs of $\mathfrak{M}$ by unit area. Therefore, if $\mathfrak{M}$ reduces to a non random tessellation of congruent fundamental polygons of area a and perimeter L_0, we have $\alpha=L_0/2a$ and $EN\,(K\cap\mathfrak{M})=LL_0/\pi a$, as is well known [7, p. 134)]

3. Intersections of Two Random Mosaics

Let $\mathfrak{M}_1(\omega_1)$, $\mathfrak{M}_2(\omega_2)$ be two realizations of two RMP. Let $B_r(\omega_2)$ denote the part of $\mathfrak{M}_2(\omega_2)$ which is contained in the disk B_r of radius r. According to (2.2) we have

$$EN\,(B_r(\omega_2)\cap\mathfrak{M}_1)=\frac{2\alpha_1 S(B_r,\,\omega_2)}{\pi} \tag{3.1}$$

where α_1 is the parameter α corresponding to $\mathfrak{M}_1$. Dividing by πr^2 and taking the limit by $r\to\infty$, we get, almost surely (Cowan [2, (3.3)])

$$EN\,((\mathfrak{M}_2\cap\mathfrak{M}_1)/\text{unit area})=\frac{2\alpha_1\alpha_2}{\pi}\,. \tag{3.2}$$

If D is any domain of Lebesgue measure $m(D)$ and N_{12} denote the number of points of the intersection $\mathfrak{M}_1\cap\mathfrak{M}_2$ within D, (3.2) can be written

$$EN_{12}(D)=\frac{2\alpha_1\alpha_2}{\pi}\,m(D)\,. \tag{3.3}$$

4. Mixed Random Mosaics

Equation (3.3) together with (1.1) makes possible the computation of the first moments (1.2) for the typical polygons of the mixed random mosaic $\mathfrak{M}_1\cup\mathfrak{M}_2$. result of superposing $\mathfrak{M}_1$ and $\mathfrak{M}_2$.

Suppose that we superpose two random mosaics realizations $\mathfrak{M}_1(\omega_1)$ and $\mathfrak{M}_2(\omega_2)$. Let α_i, β_k^i, γ_k^i be the fundamental parameters of $\mathfrak{M}_i$ $(i=1,\,2)$ defined by (1.1). The result is a new random mosaic whose fundamental parameters α, β_k, γ_k are related to those of $\mathfrak{M}_1$, $\mathfrak{M}_2$ (almost sure) by the equations

$$\begin{aligned}\alpha_1&=\alpha_1+\alpha_2\\ \beta_k&=\beta_k^1+\beta_k^2,\quad k\neq 4\\ \beta_4&=\beta_4^1+\beta_4^2+\nu_{12}\\ \gamma_k&=\gamma_k^1+\gamma_k^2\,.\end{aligned} \tag{4.1}$$

These relations follow immediately from the definitions. The new parameter ν_{12} is defined by the relation $EN_{12}(D)=\nu_{12}m(D)$ and therefore, according to (3.3) we have

$$\nu_{12}=\frac{2\alpha_1\alpha_2}{\pi}\,. \tag{4.2}$$

From (4.1), (4.2) and (1.2) we can compute the first moments a_{12}, h_{12}, w_{12}, v_{12} of the characteristics of a typical polygone of the mixed mosaic $\mathfrak{M}_1\cup\mathfrak{M}_2$. The results are the following:

$$a_{12}=\frac{2\pi a_1 a_2}{2\pi\,(a_1+a_2)+h_1 h_2} \tag{4.3}$$

9*

$$h_{12}=\frac{2\pi\,(a_1h_2+a_2h_1)}{2\pi\,(a_1+a_2)+h_1h_2}$$

$$w_{12}=\frac{2\pi\,(w_1a_2+w_2a_1)+4h_1h_2}{2\pi\,(a_1+a_2)+h_1h_2}$$

$$v_{12}=\frac{2\pi\,(a_2v_1+a_1v_2)+4h_1h_2}{2\pi\,(a_1+a_2)+h_1h_2}.$$

5. Examples

1. If $\mathfrak{M}_1$ and $\mathfrak{M}_2$ are homogeneous POISSON line processes (POISSON random mosaics) of parameters λ_1, λ_2 we know that (see [7, p. 57] and [4], [5])

$$(5.1)\qquad a_i=\frac{4}{\pi\lambda_i^2}\,,\quad h_i=\frac{4}{\lambda_i}\,,\quad w_i=v_i=4\,,$$

and the first moments of the characteristics of the random mosaic $\mathfrak{M}_1\cup\mathfrak{M}_2$ results to be equal to the first moments of the characteristics of the homogeneous POISSON random mosaic of lines of parameter $\lambda_1+\lambda_2$, as it should be a priori.

2. For a random mosaic of VORONOI type [7, p. 20], [1], [5] we have

$$(5.2)\qquad a_1=1/\lambda_1,\quad h_1=4/\lambda_1^{1/2},\quad w_1=v_1=6$$

and for a random mosaic of DELAUNAY ([6, chap. 8], [5]) we have

$$(5.3)\qquad a_1=1/2\lambda_1,\quad h_1=32/9\pi\sqrt{\lambda_1},\quad w_1=v_1=3\,.$$

By (4.3) we can compute the first moments of the characteristics of the random mosaics obtained by superposing two mosaics of any pair of mosaics of VORONOI, POISSON or DELAUNAY. Some relations between these characteristics can be easily obtained. For instance, if $\mathfrak{M}_1$ and $\mathfrak{M}_2$ are VORONOI mosaics we allways have

$$(5.4)\qquad 6>w_{12}>4+2\pi/(\pi+4)=4{,}87\,\ldots$$

3. Assume that $\mathfrak{M}_1$ is a non random tessellation (i.e. an arrangement of congruent polygons fitting together so as to cover the whole plane without overlapping (see COXETER [3, chap. 4]). Then, a_1, h_1, $w_1=v_1$ are the area, perimeter and number of sides of a polygon of the tessellation (fundamental domain) and (4.3) applies for the mixed mosaics generated by a non random tessellation and a random mosaic superposed to it [8].

4. Let $\mathfrak{M}_1$, $\mathfrak{M}_2$ be both non random tessellations. Assume that $\mathfrak{M}_1$ is fixed and $\mathfrak{M}_2$ moving in the plane without deformation, with the kinematic density of Integral Geometry [7]. Then $\mathfrak{M}_1\cup\mathfrak{M}_2$ is a random mosaic and its characteristics are also given by (4.3)

5. The extension of (4.3) to the case of more than two RMP is straightforward.

References

[1] R. Cowan, The use of the ergodic theorems in random geometry, Suppl. Adv. Appl. Prob. **10** (1978) 47 – 57
[2] –, Properties of ergodic random mosaic processes, Math. Nachr. **97** (1980) 89–102
[3] H. S. M. Coxeter, Introduction to Geometry, John Wiley, New York 1961
[4] R. E. Miles, Random polygons determined by random lines in a plane, I and II, Proc. Nat. Acad. Sc. U.S.A. **52** (1964) 901–907; 1157 – 1160
[5] –, On the homogeneous planar Poisson process, Math. Biosc. **6** (1970) 85 – 127
[6] C. A. Rogers, Packing and covering, The University Press, Cambridge 1964
[7] L. A. Santaló, Integral Geometry and Geometric Probability, Encyclop. Math. and Appl. Addison-Wesley, Reading 1976
[8] –, Random lines and tessellations in a plane, Stochastica, IV (1980) 3 – 13

Ciudad Universitaria (Nuñez)
Departamento de Matematica
Buenos Aires
Argentina

7 Comments on Some of Santaló's Papers

To get an idea of the current influence of Santaló's work, we include below, in chronological order, comments on some of his articles, made by specialists in their respective fields.

Commentary on article

[36.2] Some problems referring to geometrical probabilities

Neither 1936, nor the following years, were productive ones for science and culture in Spain. That's why this small pearl which Professor Santaló offers us is a surprising and welcome gift from that period. In it, making use of kinematics density as a basic tool, in simple language and without superfluous literature, he deals with the problems of random coverage of a convex figure on the plane and the random volume occupied in a convex body in space. The average value for the uncovered surface and the unoccupied volume is obtained.

The passage of time has given us interesting applications for the random coverage problem. To mention just one pair of examples, the coverage of the circumference through random arcs or the surface of a sphere through random circles is applied in Biology to the study of the inactivation of a virus through the action of antibodies; the coverage of mobile telephone networks is studied and simulated through an application of the problem of surface coverage. These and other current applications of the problem highlight the peculiarity of the application that Professor Santaló mentions in point 2 of §1: the calculation of the surface destroyed in hitting the bull's eye in target practice, on the assumption that all the bullets hit it.

F. Montes (University of València, Spain).

Commentary on article

[43.2] On the probable distribution of corpuscles in a body, deduced from the distribution of its sections, and analogous problems

In this paper Santaló considers a convex body K with volume V and N convex figures distributed at random inside it. Let $D = N/V$. The aim is to determine D using measurements of sections of K with random figures of a known shape (lines, bands, cylinders). Firstly, he considers the case of a planar K. It is shown that if

K is crossed by a random line, the mean density of the corpuscles on the chord determined on it is uD/π, where u is the perimeter of the figures. He extends this result to bands of width δ. Analogous results are obtained in the case of the space, and other quantities closely related to this density are deduced; in particular, a generalization of some results given by Pólya (*Arch. Math. Phys.* 1918). Particularly interesting is an application to the Kinetic Theory of Gases. He gives the value of the mean free path of the molecules of a gas, generalizing a known result of Loeb (Kinetic Theory of Gases, New York, 1927). These results were generalized by Dumas-Dumas-Golse (*J. Statist. Phys.* 87 (1997), 943–950) and used by Golse in Lorentz gases (*Proc. ICM. Madrid*, III (2006), 183–203). Also [43.2] was used by F. Boca and A. Zaharescu in [*Comm. Math. Phys.* **269** (2007), 425–471] for the study of the distribution of the free path lengths in the periodic 2-dimensional Lorentz gas in the small-scatterer limit. This work by Santaló becomes fundamental, since it is considered to be the basis of Stereology, a field where Santaló had made several significant contributions. Stereology is one of the fields where the theoretic results on Integral Geometry and Stochastic Geometry are applied most efficiently. It is an important and efficient tool in many applications, such as petrography, material science and bioscience, including histology, bone anatomy and neuroanatomy.

A. Simó (University of Castellón, Spain).

Commentary on article

[46.1] Convex Regions on the n-dimensional spherical surface

The question which motivates the main result included in this article, (an extension to the sphere of Jung's theorem), lies in the problem of covering a finite set of points on the plane with the smallest possible circle, and was posed by J. J. Sylvester in 1857, (see [*Philosophical Magazine*, **20**, (1860), 203–222]). The mathematical interest in the n- dimensional problem is centered around Jung's inequality, which relates the diameter of a given (finite or infinite) bounded set K, (defined as the least upper bound for the finite distances between its points), to the radius of the circumscribed sphere to K. A dual question consists in relating the diameter of this bounded set K with the radius of the inscribed sphere in K.

The statement and proof of this result in the case of plane and finite sets was presented by H. W. E. Jung, in [*Journal für de Reine Angew. Math.* **123**, (1901), 241–257], and [*Journal für de Reine Angew. Math.*, **137**, (1909), 310–313]. L. M. Blumenthal and G. E. Wahlin proved in [*Bull. A.M.S.*, **47**, (1941), 771–777] the corresponding generalization to infinite sets in E^n. In this paper, they also mentioned a relation between the breadth of a convex set, (defined as the minimum distance of two parallel supporting lines of it), and its inscribed radius, obtained by W. Blaschke in 1914, (see [*Jahresbericht der Deutschen Math. Ver.*, **23**, (1914), 369–374]). The generalization of this last result to convex sets in E^n was published in 1922 by P. Steinhagen, in [*Abh. Math. Sem. Univ. Hamburg* **1** (1922), 15–26].

What Santaló did to extend Jung's Theorem to the sphere was to bound from below the diameter D of any set $K \subseteq S_1^n$ lying on an hemisphere by the corresponding diameter of an equilateral spherical simplex having the same circumradius R as K, (Theorem 1). To do that, Santaló computed the diameter of this simplex, (Lemma 1), showing that is the smallest among all the n-dimensional spherical simplexes on S_1^n whose circumscribed radius is R and which contain the center of its circumscribed sphere, (Lemma 2). The second main theorem (Theorem 2) constitutes the extension to the n-sphere of Steinhagen-Blaschke's result and it is obtained by duality from Theorem 1. We can find also a brief but precise description of all the results of this paper in Schenider's report about Santaló's contributions to Convex Geometry.

There are articles dealing with extensions of Jung's theorem to wider contexts. See, for example, the generalization of Jung's theorem to bounded regions in finite dimensional Banach spaces, published by F. Bohnenblust in [*Ann. of Math.*, **39**, no. 2, (1938), 301–308]. Recently, we have a version of Jung's theorem for the spheres and hyperbolic spaces due to B. V. Dekster, in [*Acta Math. Sci. Hungar.*, **67**, no. 4, (1995), 315–331]. In 1997, two versions of Jung's theorem were proved for bounded sets in a metric space with curvature bounded above in the sense of Alexandrov, (see Dekster's work [*Proc. Amer. Math. Soc.*, **125**, no. 8, (1997), 2425–2433], and the work due to U. Lang and V. Schroeder [*Ann. Glob. Anal. and Geom.*, **15**, 1997, 263–275]).

Finally, and as far as applications are concerned, the use of simplexes in Santaló's proof of Jung's theorem in the sphere becomes a useful tool in the fields of optimization theory and operational research. As an example of this, in Betke's paper [Discrete and Computational Geom., **32**, 2004, 317–338] a polynomial procedure to solve the feasibility problem is described, which is directly inspired by the results of Santaló's paper reviewed here.

V. Palmer (University of Castellon, Spain).

Commentary on article

[49.1] Integral Geometry in three-dimensional spaces of constant curvature

In [49.1] Santaló found a formula relating the volume of a hyperbolic tetrahedra with the measure of the set of hyperplanes intersecting it, using methods of Integral Geometry. See also [76.1], p. 311. E. Suárez-Peiró in [*Pacific J. Math.* **194** (2000), 229–255] generalizes Santaló's formula to hyperbolic simplices of higher dimension.

Given a n-simplex in hyperbolic n space $\Delta \subset \mathbf{H}^n$ and given a fixed vertex $v_0 \in \Delta$, E. Suárez-Peiró constructs a subset Δ^P of the de Sitter sphere, that can be viewed as the set of hyperplanes that intersect Δ, up to a subset of measure zero. This construction is made in the de Sitter sphere, where points correspond naturally to oriented hyperplanes, so that a distinguished vertex is required for not counting twice a hyperplane with opposite orientations. More precisely Δ^P is the set of points in the de Sitter sphere which, viewed as oriented hyperplanes, their intersect Δ and that the halfspace corresponding to orientation contains the fixed vertex v_0. The volume $vol_n(\Delta^P)$ is called the dual volume of Δ.

Santaló's equality in the three dimensional case says that, for any compact 3-simplex Δ in hyperbolic space,

$$Vol_3(\Delta) + Vol_3(\Delta^P) = \frac{1}{2}\sum_F(\pi - \alpha_F)Vol_1(F),$$

where the sum runs over all edges F of Δ, $Vol_1(F)$ denotes its length and α_F its dihedral angle.

Suárez-Peiró generalizes this formula to higher dimensions. Here, given a face F of a simplex $\Delta \subset \mathbf{H}^n$, θ_F denotes the volume of its polar face F, and in codimension two it is precisely the exterior dihedral angle.

Theorem (Suárez-Peiró) *Let Δ be a hyperbolic n-simplex. Then its dual volume is*

$$Vol_n(\Delta^P) = \sum_{k=0}^{[\frac{n-1}{2}]} ((-1)^k \cdot c_{2k+1} \cdot \sum V_{2k+1}(F) \cdot \theta_F$$

where the sum runs over all odd dimensional faces of Δ, $c_n = 1$ and

$$c_i = \frac{Vol(\mathbf{S}^n)}{Vol(\mathbf{S}^i) \cdot Vol(\mathbf{S}^{n-i-1})}, \quad \text{for } 0 \leq i \leq n-1.$$

When the dimension n is odd, this formula relates the volume of the simplex with the volume of its dual:

$$\begin{aligned} &Vol_n(\Delta^P) + (-1)^{\frac{n+1}{2}} Vol_n(\Delta) \\ &= \sum_{k=0}^{\frac{n-3}{2}} (-1)^k \cdot c_{2k+1} \cdot \sum_{dim(F)=2k+1} Vol_{2k+1}(F) \cdot \theta_F. \end{aligned}$$

When the dimension n is even, this formula relates the dual volume of the simplex with the volume of its boundary $\partial\Delta$, i.e. its faces of codimension one:

$$\begin{aligned} &Vol_n(\Delta^P) + (-1)^{\frac{n}{2}} \cdot c_{n-1} \cdot Vol_{n-1}(\partial\Delta) \\ &= \sum_{k=0}^{\frac{n-4}{2}} (-1)^k \cdot c_{2k+1} \cdot \sum_{dim(F)=2k+1} Vol_{2k+1}(F) \cdot \theta_F. \end{aligned}$$

This formula is proved by means of a new Schläfli formula for simplices in semi-riemannian spaces of constant curvature. In the 1850's L. Schläfli found a differential formula that computes the variation of volume of a spherical simplex when its faces are perturbed, in terms of the variation of dihedral angles and the length of codimension two faces [J. Milnor Collected Papers, **1**, (1994)]. Schläfli proved his formula in any dimension, and in 1936 H. Kneser [*Deutsche Math.* **1** (1936), 337–340] gave a new proof that easily generalizes to the hyperbolic case. Here the proof of Kneser is adapted to semi-riemannian spaces.

This work was realized when E. Suárez-Peiró was Ph. D. student under the supervision of J. M. Montesinos. He had obtained a differential equation for tetrahedron in de Sitter space by differentiating the formula of Santaló and then applying the usual Schläfli formula for hyperbolic tetrahedra [Unpublished manuscript contained in a letter from J. M. Montesinos to L. A. Santaló, 18–6–1993]. With this unpublished result in mind, Montesinos suggested the strategy of this paper, that became a chapter of Suárez-Peiró's thesis.

J. Porti (Autonomus University of Barcelona, Spain)

Commentary on articles

[52.1] Measure of sets of geodesics in a Riemannian space and applications to integral formulas in Elliptic and Hyperbolic spaces

[76.1] Integral Geometry and Geometric Probability

Santaló's formula (see [52.1] and [76.1] chap.19) describes the Liouville measure on the unit tangent bundle of a Riemannian manifold in terms of the geodesic flow and the measure of a codimension one submanifold.

Let M be an n-dimensional Riemannian manifold, $\pi : UM \to M$ be the unit tangent bundle of M, du the Liuoville measure on UM, and $g_t : UM \to UM$ the geodesic flow. One way to define du is to start with the standard contact form α and define $du = \alpha \wedge d\alpha^{n-1}$. Liuoville's theorem says that du is invariant under the geodesic flow g_t (since α is). Locally du is just the product measure $dm \times dv$ where dm is the Riemannian volume form and dv is the standard volume form on the unit $n-1$ sphere.

For any (locally defined) codimension 1 submanifold $N \subset M$ let dx be the Riemannian volume element of the submanifold. We let $SN = \pi^{-1}(N) \subset UM$, and for each $x \in N$ we let N_x be a unit normal to N at x. Then we have a smooth map $G : SN \times R \to UM$ given by $G(v, t) = g_t(v)$. Santaló's formula says:

$$G^*(du) = | < v, N_x > | \mathrm{d}t\, \mathrm{d}v\, \mathrm{d}x$$

The formula is used to convert integrals over subsets $Q \subset UM$ of the unit tangent bundle to iterated integrals first over a fixed unit speed geodesic (say parameterized on $I(\gamma) \subset R$), and then over the space Γ of geodesics which are parameterized by their intersections with a fixed codimension one submanifold and endowed with the measure

$$d\gamma = | < v, N_x > | \mathrm{d}v\, \mathrm{d}x.$$

i.e.

$$\int_Q f(u)du = \int_{\gamma\in\Gamma} \int_{I(\gamma)} f(\gamma'(t))\mathrm{d}t\, \mathrm{d}\gamma$$

One of the most important applications is to the study of Riemannian manifolds with smooth boundary. In this case we let $N = \partial M$, let N_x be the inwardly pointing unit normal vector, and $U^+\partial M = \{v \in SN| < v, N_x >> 0\}$. For any $u \in UM$ we let

$l(u) = sup\{t \geq 0 | g_t(u) \text{ is defined}\}$. Note that $l(u) = \infty$ means that $g_t(u)$ is defined for all $t > 0$. We let $\overline{UM} = \{u \in UM | l(-u) < \infty\} \cup U^+\partial M$, i.e. $u \in \overline{UM}$ means $u = g_t(v)$ for some some $v \in U^+\partial M$ and some $t \geq 0$. In this setting Santaló's formula takes the form:

$$\int_{\overline{UM}} f(u)du = \int_{U^+\partial M} \int_0^{l(v)} f(g_t(v))dt < v, N_x > \mathrm{d}v\,\mathrm{d}x.$$

One immediate application, by simply putting $f(u) = 1$, is:

$$Volume(\overline{UM}) = C_1(n) \int_{U^+\partial M} l(v) < v, N_x > \mathrm{d}v\,\mathrm{d}x.$$

Since the Liuoville measure is locally a product measure, in the special case $\overline{UM} = UM$ this says $Volume(M) = C_2(n) \int_{U^+\partial M} l(v) < v, N_x > \mathrm{d}v\,\mathrm{d}x$.

The formula has been used often to prove isoperimetric and rigidity results. A sample of such applications can be found in C. Croke, ([*Comment. Math. Helv.* **59** (1984), 187–192], [*Ann. Sci. École Norm. Sup.* **13** (1980), 419–435]), and in C. Croke, N. Dairbekov and V. Sharafutdinov, *Trans. Amer. Math. Soc.*, to appear.

The description of Santaló's formula for the timelike geodesic flow for Lorentzian surfaces is given in L. Andersson, M. Dahl, and R. Howard, *Trans. Amer. Math. Soc.* **348** (1996), 2307–2329.

C. Croke (University of Pennsylvania, USA).

Commentary on article

[59.1] On complete systems of inequalities between elements of a plane convex figure

In 1916 Blaschke raised the problem of finding a complete system of inequalities for the volume, the integral mean curvature and the surface area of a convex body in the 3-dimensional Euclidean space; he realized that the well-known inequalities of Minkowski and the isoperimetric inequality were not enough to determine a complete system. Many mathematicians have contributed to this problem (Hadwiger, Bieri, Groemer, Sangwine-Yager, ...), but it is still open.

This paper opened up a new line of research. In fact, Santaló extended Blaschke's problem and the tools developed by Blaschke to inequalities relating any finite collection of geometric magnitudes. In particular, he considered the area, the perimeter, the diameter, the minimal width, the circumradius and the inradius of a planar convex set and considered inequalities relating any three of these magnitudes. Of the 20 possible cases he solved 6. In the last ten years, due mainly to the work of Hernández Cifre, Salinas and Borozcky Jr. ([*Amer. Math. Monthly* **107** (2000), 893–900], [*Discrete Comput. Geom.* **23** (2000), 381–388], [*Monatsch Math.* **138** (2003), 95–110]) 7 other cases were solved. And there are still 7 open cases.

The more recent important contributions are due to Brandenberg ([Ph. D. Dissertation, Technische Universität Munchen 2002]) and to Ting and Keller

([*Discrete Comput. Geom.* **33** (2005), 369–393]). They attacked the challenging problem of finding complete systems of inequalities for inequalities relating collections of four geometric magnitudes. Many new geometric inequalities have been found as a consequence of this work.

S. SEGURA-GÓMIS (University of Alicante, Spain).

Commentary on article

[63.1] A relation between the mean curvatures of parallel convex bodies in spaces of constant curvature

In this paper Santaló studies the mean curvature integrals of parallel convex sets in constant curvature spaces. More precisely, he finds polynomial expressions of these integrals that take the same value on all parallel sets of a convex body. For instance, if $L(\lambda), M_1(\lambda)$ are respectively the length and the total geodesic curvature of the parallel set $Q(\lambda)$ at distance $\lambda \geq 0$ from a convex set Q in the plane of constant curvature K, then $L^2(\lambda) + K^{-1}M_1^2(\lambda)$ is shown to be independent of λ. This is used to give a proof of the isoperimetric inequality in constant curvature planes.

For a convex body Q in the 3-dimensional space of constant curvature K, the area $F(\lambda)$ and the total mean curvature $M_1(\lambda)$ of the parallel body $Q(\lambda)$ make the expression $M_1^2 + KF(\lambda)^2 - 4\pi F(\lambda)$ constant on λ. As an application, Santaló shows the following isoperimetric inequality in case $K > 0$

$$M_1^2 + KF^2 - 4\pi F > 0. \tag{1}$$

The last paragraph of the paper raises the question to generalize the previous inequality to the space of negative curvature. Next we show that (1) cannot hold in hyperbolic space. Indeed, let Q_r be a 2-dimensional disc of radius r contained in a geodesic plane of hyperbolic 3-space (of curvature K). In this case $F(Q_r) = 4\pi(-K)^{-1}(\cosh(\sqrt{-K}r) - 1)$, and $M_1(Q_r) = \pi^2(-K)^{-1/2}\sinh(\sqrt{-K}r)$. Now, substitution in (1) gives a contradiction for r big enough.

It would be interesting to find polynomial relations between the mean curvature integrals of parallel convex bodies in constant curvature spaces of higher dimensions. This has been done for Euclidean spaces in R. Masó and A. M. Naveira, *Bull. Soc. Math. Roumanie* **43(91)** (2000), 299–311.

A. M. NAVEIRA (University of Valencia, Spain)
G. SOLANES (Autonomous University of Barcelona, Spain)

Commentary on articles

[67.2] Horocycles and convex sets in the hyperbolic plane

[68.1] Horospheres and convex bodies in hyperbolic space

[69.1] Convexity in the hyperbolic plane

[72.1] Averages for polygons formed by random lines in Euclidean and hyperbolic planes

[80.1] Notes on the Integral Geometry in the hyperbolic plane

The study of h-convex sets started by L. A. Santaló in [67.2, 68.1], and in particular the estimates on the quotient area/length for bounded h-convex sets in the hyperbolic plane [72.1], has had a big influence on recent research. Answering a question of Santaló in [72.1], E. Gallego and A. Reventós ([*J. Differential Geom.* **21** (1985), 63–72]) constructed a family of sequences of h-convex sets expanding over the whole space with limit of the quotient area/length of any value bigger than 1. A. M. Naveira and A. Tarrío ([*Arch. Math.* (Basel) **68** (1997), 514–519]) extended the estimates of Santaló–Yáñez to higher odd dimensions with some restrictions on the expansion.

In a series of papers, A. Borisenko, E. Gallego, V. Miquel, A. Reventós, G. Solanes and V. Vlasenko ([*Illinois J. Math.* **43** (1999), 61–78], [*Mat. Fiz. Anal. Geom.* **6** (1999), 223–233], [*Geom. Dedicata* **76** (1999), 275–289], [*Dopov. Nats. Akad. Nauk Ukr. Mat. Prirodozn. Tekh. Nauki* 2000, 10–14], [*J. London Math. Soc.* (2) **64** (2001), 161–178], [*Differential Geom. Appl.* **14** (2001), 267–280], [*Ann. Global Anal. Geom.* **21** (2002), 191–202]) extended the estimates in full generality to any dimension, and also gave estimates for λ-convex sets in Hadamard manifolds of this quotient and also of the quotient (total k-th mean curvature)/area.

In the course of the proof of the above estimates, other interesting estimates on the maximal distance from any interior point of the h-convex set to its boundary and for the angle between the normal vector to the boundary and the radial vector were obtained. In turn, all these estimates have been useful for applications in other problems, for example the study of the volume-preserving mean curvature flow of bounded h-convex hypersurfaces of the hyperbolic space by E. Cabezas-Rivas and V. Miquel ([*Indiana Univ. Math J.* **56** (2007) –2086]).

On the other hand, the idea of doing Integral Geometry in the hyperbolic space using horospheres was also raised in [67.2, 68.1] and [80.1] for dimensions 2 and 3, and had its continuation in a systematic study of this geometry by A. M. Naveira and A. Tarrío ([Arch. Math. (Bassel) **68** (1997), 514-519]), and E. Gallego, A. M. Naveira and G. Solanes ([*Geom. Dedicata* **103** (2004), 103–114], [*Differential Geom. Appl.* **22** (2005), no. 3, 315–325], [*Israel J. Math.* 145 (2005), 271–284], [*Trans. Amer. Math. Soc.* **358** (2006), no. 3, 1105–1115]). In these studies new formulas for the measure of horospheres and λ-convex sets intersecting convex bodies were found.

E. Gallego (Autonomous University of Barcelona, Spain).
V. Miquel (University of València, Spain).

Commentary on article

[69.2] On some geometric inequalities in the style of Fáry

The total absolute curvature $K(\Sigma)$ of a compact hypersurface Σ_{n-1} in $\mathbb{R}^n$, introduced by Chern and Lashof in [*Amer. J. Math.*, **79** (1957), 306–318], can be defined as

$$K(\Sigma) = \int_{\Sigma_{n-1}} |G(x)| \mathrm{d}x,$$

where $G(x)$ denotes the Gauss curvature of Σ at x. The well known Chern-Lashof inequality (previously proven by Milnor in [Junior thesis, Princeton University, 1950. Reprinted in *Collected papers*, Vol. I, Publish or Perish, (1994)]) states that $K(\Sigma)$ is bounded below by the topology of Σ, namely by the sum of its Betti numbers.

In the paper [69.2] Santaló looks for *geometric* bounds of the total absolute curvature. Precisely, the pursued inequality is

$$r^{n-1} K(\Sigma) \geq F(\Sigma), \tag{1}$$

where Σ has $(n-1)$-dimensional volume F, and it is contained in a ball of radius r. Unfortunately, as remarked later by Chakerian (MR.39#3395) and by Santaló himself (Zbl.0176.19606), the proof given in [69.2] is not valid. The argument is based on comparing the sets of critical values of the projections onto planes of different dimensions. The overlooked point is the fact that these sets present singularities. The same problem already occurred in Fáry's paper [*Acta Sci. Math. Szeged*, **12** (1950), 117–124], where he considered similar constructions in the space of dimension three.

This way, the inequality (1) remained as an interesting conjecture. In fact, up to the present only the cases $n = 2, 3$ have been solved. The plane case was previously solved in the above mentioned paper by Fáry. For surfaces in $\mathbb{R}^3$ the inequality (1) was first established by Aminov [*Ukrain. Geometr. Sb. Vyp.* **18** (1975), 3–15]. A similar proof can be found in Burago and Zalgaller's book [Grundlehren der Mathematischen Wissenschaften, 285. Springer-Verlag, Berlin, 1988]. Combined with the Minkowski type inequalities of Shahin [*Proc. Amer. Math. Soc.* **19** (1968), 609–613], the same method can be used in higher dimensions to get

$$r^2 K(\Sigma) \geq |M_{n-3}(\Sigma)|. \tag{2}$$

where $M_i(\Sigma)$ denotes the integral of the i-th order mean curvature. More easily, the classical Minkowski formula gives $rK \geq |M_{n-2}|$. A stronger type of inequalities, which also appeared in [69.2] and still remain open, are the following

$$r^{n-m-1} K(\Sigma) \geq M_m^+(\Sigma) \qquad m = 2, \ldots, n-1, \tag{3}$$

where M_m^+ is the average of the $(n-m-1)$-dimensional volumes of the critical sets of projections onto all the planes of dimension $n-m$. It is worth noting that $M_m^+ = M_m$ if Σ is convex, but in the non-convex case M_m^+ is bigger than the integral of the absolute value of the m-th mean curvature. For instance, M_1^+ is the integral of the absolute values of the normal curvatures (see Langevin and Shifrin [*Amer. J. Math.* **104** (1982), 553–605]).

Quite recently, Ekholm and Kutschebauch [*Math. Scand.* **96** (2005), 224–242.] obtained an inequality for singular plane curves, which fills the gap in Santaló's proof of (1) in the case of surfaces. This improves the result of Aminov, since their method applies to surfaces with (stable) singularities. Moreover (altough this is not noted in Ekholm and Kutschebauch's paper) it can be used to prove

$$r^2 K(\Sigma) \geq M^+_{n-3}(\Sigma) \tag{4}$$

for a hypersurface $\Sigma_{n-1} \subset \mathbb{R}^n$. Note that this implies (2) and is a particular case of (3).

Finally, in the paper [69.2] Santaló conjectures

$$r^m M^+_m(\Sigma) \geq F(\Sigma). \tag{5}$$

This is an interesting problem, open in the case $m > 1$ (for $m = 1$ this is a consequence of the Minkowski formula).

G. Solanes (Autonomous University of Barcelona, Spain).

Commentary on article

[72.1] Averages for polygons formed by random lines in Euclidean and hyperbolic planes

A stationary isotropic Poisson line process tessellates the plane into convex polygons. A typical polygon of this tessellation is known as Poisson polygon and depends only on the line process intensity $\lambda > 0$. Its (random) characteristics as the number of vertices, number of edges or area content have been studied intensively. The mean values and variances were obtained by Miles in the early 70s, as well as higher-dimensional cases, and the anisotropic case has been considered by Matheron and later by Mecke, Schneider, Weil and others. We remark that to our knowledge the complete distribution of the Poisson polygon is not yet known.

Santaló and Yañez considered a Poisson polygon as well, although not in the Euclidean plane, but in a plane of constant negative curvature (mainly the hyperbolic plane with curvature -1). Though the problems look very similar, the situation is completely different and its solution demands a good knowledge of Hyperbolic Geometry, which can hardly be supported by visual intuition as in the Euclidean case. The basic technique needed is the Integral Geometry in the Hyperbolic plane developed by Santaló in *Introduction to Integral Geometry*, Hermann, Paris 1953. For example, while the invariant measure over lines in the Euclidean plane is $dp\, d\phi$, (p, ϕ) being the polar coordinates of the foot of the perpendicular from the origin to the line, in the hyperbolic plane it is $\cosh p\, dp\, d\phi$. Another important difference is that, having a family $K(t)$ of convex domains expanding over the whole plain as $t \to \infty$, we do not have $\lim_{t\to\infty} \frac{\text{area}(K(t))}{\text{perimeter}(K(t))} = \infty$ as in the Euclidean space. The limit is less or equal to 1 in the hyperbolic plane, and only after restriction to special convex domains called h-convex (convex with respect to horocycles) the limit can be shown to be equal to 1.

The authors derive the mean and variance for the number of intersection points of the line process with a convex domain, and also the mean number of vertices, mean edge length and mean area of the Poisson polygon in the Hyperbolic Geometry, together with some second moments. For example, the mean area of a Poisson polygon in the plane of constant curvature k is $\mathbb{E}A = \frac{4\pi}{\pi^2\lambda^2 - 2|k|}$ under the assumption $\lambda > \sqrt{2|k|}\pi$, and infinite otherwise.

J. RATAJ (Mathematical Institute of Charles University, Czech Republic).

Commentary on articles

[78.2] Plate and line segment processes

[79.1] Random processes on manifolds in $\mathbb{R}^n$

In concord with the main object of Stereology, namely, the knowledge of the composition and spatial or volumetrical structure of an heterogeneous material by studying its plane sections, in [78.2] we study an stochastical process of figures in the plane, according to a model of geometrical stochastic process introduced by P. Parker and R. Cowan (*J. Appl. Probability* **13** (1976), 96–107) for the case of a process of segments. The most important technical resources for that purpose are the formulas related with the kinematic density of Integral Geometry. The classical examples of these aleatory processes appear in petrography, in metallurgy, in the microscopic observation of histologic cutting and in the study of porous materials. [79.1] represents a generalization of the preceding one to Euclidean spaces of arbitrary dimension.

N. FAVA (University of Buenos Aires, Argentina).

Commentary on article

[80.3] Random lines and tessellations in the plane

Santaló determines mean geometric characteristics of planar "mixed random mosaics", which are obtained here as superpositions of a deterministic and a random tessellation. The deterministic tessellation consists of a partition of the plane into compact, convex and congruent n-sided polygons having area f and perimeter length u. The random tessellation is generated by a stationary (homogeneous) and isotropic Poisson line process with intensity λ (λ is here the mean number of lines of the process that hit a disk with unit perimeter).

Combining geometric identities and relations for Poisson line processes, he first shows that the number of cells c in a circular window of radius R satisfies

$$\lim_{R\to\infty} E\frac{c}{\pi R^2} = \pi\lambda^2 + \frac{u\lambda}{f} + \frac{1}{f} =: \overline{c}. \tag{1}$$

This has a clear geometric interpretation; if we, for instance, count the cells by enumerating their upper right vertices. The three terms in (1) correspond to the number of cells whose upper right vertices either lies in the interior, or on an edge, or coincides with a vertex of a cell of the deterministic tiling, respectively. Santaló then shows that the mean area $E(A)$ of the typical cell is just $1/\overline{c}$, and its mean number of edges $E(N)$ and mean perimeter $E(S)$ are given by

$$E(N) = (4\pi\lambda^2 + 4u\lambda/f + n/f)/\overline{c}, \qquad E(S) = (2\pi\lambda + u/f)/\overline{c},$$

all of which allows for clear geometrical interpretations.

The idea of superimposing tessellations was later extended in [L. A. Santaló, *Math. Nachr.* **117**, 1984, 129–133] and [J. Mecke, *Math. Operationsforsch. Statist., Ser. Statist.* **15**, 1984, 437–442] to the case where both tessellations are random.

M. KIDERLEN (University of Karlsruhe, Germany)

8 Book Reviews

8.1 Some Books Reviewed by L. A. Santaló

Santaló possessed a global vision of the many special branches of geometry. The many book reviews he wrote are a testimony to this, and some of these appear below.

Theory and applications of distance geometry
Blumenthal, Leonard M.
Oxford, at the Clarendon Press, 1953. xi+347 pp.
MR0054981

Since Menger's work [*Math. Ann.* **100** (1928), 75–163], the field known as metric topology or distance geometry (i.e., geometry of the subgroup of homeomorphisms for which the distance of two points is an invariant) has been greatly extended in content and application. A first survey was done by the author in his excellent volume on distance geometries [*Univ. of Missouri Studies* **13**, (1938)]. The present book furnishes a detailed, self-contained and in many aspects a very complete treatment of the theory.

Part I deals with semimetric and metric spaces, starting from the preliminary notions and examples. The concepts of betweenness and convexity are introduced, as well as the definition and characteristic properties of metric segments. A chapter is devoted to curve theory in metric spaces; the concepts of arc length, geodesic arcs and different definitions of curvature and torsion are given and discussed. There is no reference to the work of Busemann on metric methods in Finsler spaces.

Part II deals with the distance geometry in Euclidean (E_n) and Hilbert ($\mathcal{H}$) spaces. A fundamental problem of distance geometry is the so-called "space problem"; it seeks necessary and sufficient metric conditions that an arbitrary distance space of a specified class $\{S\}$ must satisfy in order that it may be congruent with a member of a given subclass of $\{S\}$. This problem is solved for the class of semimetric spaces and the subclass of the single space E_n and for the class of separable semimetric spaces and the subclass of the single $\mathcal{H}$. The conditions are given in terms of the so-called Cayley-Menger determinants. Other noteworthy metric characterizations of E_n and $\mathcal{H}$ are given; for instance, a finitely compact, convex, externally convex semimetric space with the weak euclidean four-point property is congruent with a Euclidean space of finite dimension. Part II contains also the study of the congruence indices of some Euclidean subsets (a regular polygon, a conic, a circle) with respect to the class of semimetric spaces and with respect to the class of subsets of E_n. A space R is said to have congruence indices (n, k) with respect to a class $\{S\}$ of spaces provided any space S of $\{S\}$ containing more than $n + k$ pairwise distinct points, is congruently imbeddable in R whenever each n of its points, not necessarily pairwise distinct, has that property. When the congruence indices are $(n, 0)$, n is said to be the "congruence order" of R with respect to $\{S\}$. A typical theorem proved in this direction is the following: the circular disc of radius r has congruence order three with respect to subsets of E_2.

Part III contains the distance geometry of the non-Euclidean spaces (spherical $S_{n,r}$ of dimension n and radius r, elliptic $\mathcal{E}_{n,r}$ and hyperbolic $\mathcal{H}_{n,r}$). Some determinants, generalization of those of Cayley-Menger for Euclidean spaces, play here a fundamental role in order to solve the space problem and some characterizations for such spaces. Some needed properties of these spaces, not readily found in the literature, are developed in the book. It is found that these properties do not characterize the non-Euclidean spaces; they give rise to the definition of certain semimetric spaces with identical distance geometry to the non-Euclidean spaces. For instance, the hyperbolic cosine function that is characteristic of hyperbolic space may be replaced by any one of a large class of functions without essential change in the distance geometry of the space. Some spherical theorems of Helly type are considered in detail, as well as the congruence indices of hemispheres and small caps. The metric geometry of the elliptic space $\mathcal{E}_{n,r}$ is particularly attractive and interesting. The necessary distinction between congruence and superposability gives rise to noteworthy questions referring to the existence of equilateral subsets in $\mathcal{E}_{n,r}$, solved only for $n \leq 3$. A chapter is devoted to the following problems: a) to find necessary and sufficient conditions for superposability of two congruent subsets of $\mathcal{E}_{n,r}$; b) to determine when a given congruence between two subsets of $\mathcal{E}_{n,r}$ may be extended to the whole space. The case of the elliptic plane is considered in detail; certain freely movable configurations are introduced which are useful in order to obtain the congruence order of $\mathcal{E}_{2,r}$ with respect to the class of all semimetric spaces. The analogous congruence order of $\mathcal{E}_{n,r}$ for $n > 2$ is not known.

Part IV is devoted to the applications of distance geometry. Three illustrations are given: a) Applications to determinant theory (some properties of determinants of the Cayley-Menger type, difficult to prove in a purely algebraic way, are neatly proved when translated to its metrical meaning); b) Applications to sets of linear inequalities (provided by the results referring to congruence indices of hemispheres and small caps and by the intersection theorems for convex subsets of $S_{n,r}$); c) Applications to lattice theory (a normed lattice L may be made into a metric space $D(L)$ by attaching to each pair of elements x, y of L the "distance" $(x, y) = |x + y| - |xy|$; then, metric concepts and relations in $D(L)$ such as betweenness, congruent imbedding,... lead to concepts and relations in L). Applications to calculus of variations are not given.

The book is very well written; the exposition is lucid and the proofs are carefully arranged. It contains mainly the personal work of the author on the subject. Every chapter is well provided with exercises and concludes with references to the literature dealing with the material discussed. A course in general topology and a course in abstract algebra is all the preparation recommendable, though not indispensable, for a thorough understanding of the author's exposition.

L. A. Santaló

Non-Euclidean Geometry (3rd ed.)
Coxeter, H. S. M.
Mathematical Expositions, no. 2. University of Toronto Press,
Toronto, Ont., 1957. xv+309 pp.
MR0087965

The 3rd edition of this excellent book differs from the first [published in 1942; MR0006835 (4,50a)] only by the following additions, put together as a new chapter. a) A description of the two families of mid-lines or Hjelmslev lines of two given lines, coplanar or skew; b) an elementary derivation of the basic formulae of spherical trigonometry (Napier chain) and hyperbolic trigonometry (Engel chain); c) the computation of the Gaussian curvature of the elliptic and hyperbolic planes when they are referred to isometric or geodesic coordinates; d) a theorem on quadratic forms related with the formulae for distance and angle in non-Euclidean Geometry (Chap. XII); e) a proof of Schläfli's formula for the differential of the volume of a tetrahedron; f) a brief historical survey of construction problems in non-Euclidean Geometry by means of ruler, compasses, hypercompass and horocompass, including recent bibliography.

L. A. Santaló

Integral Geometry (Romanian)
Stoka, Marius I.
Editura Academiei Republicii Socialiste România, Bucharest 1967 239 pp.
MR0217747

The classical theory of geometrical probabilities, originated in the work of Buffon (1777), Barbier (1860), Crofton (1868), Czuber (1884) and nicely exposed in the book of R. Deltheil [Probabilités géométriques, Gauthier-Villars, Paris, 1926], seems to be having a revival in recent years, in its physical and statistical applications [M. G. Kendall and P. A. P. Moran, Geometrical probability, Hafner, New York, 1963; MR0174068 (30 #4275)] as well as in its geometrical consequences (e.g., the book under review). When this last aspect predominates, the theory takes the name of Integral Geometry (following Blaschke) and its tools and methods have proved to be very useful, as much a source of examples in group theory and measure theory as in the geometry of convex bodies. The present book is a clear exposition and compilation of many results in this field since the initial work of Blaschke.

Let X_n be the real number space of coordinates $x_1, x_2, \cdots, x_n$ and let $\mathcal{F}_q$ be a q-dimensional family of p-dimensional varieties V_p in X_n; let $\alpha_1, \alpha_2, \cdots, \alpha_q$ be the parameters which determine the elements of $\mathcal{F}_q$. A Lie group G of automorphisms of X_n is called a group of invariance of $\mathcal{F}_q$ if for any $g \in G$ and $V_p \in \mathcal{F}_q$, $g V_p \in \mathcal{F}_q$. Then the mapping $V_p \mapsto g V_p$ defines a group of automorphisms G^* in the parameter space $\alpha_1, \cdots, \alpha_q$, and any form $\omega = \varphi(\alpha_1, \cdots, \alpha_q)\, d\alpha_1 \wedge \cdots \wedge d\alpha_q$ invariant with respect to G^* is called a "density" in $\mathcal{F}_q$ (the integral of ω defines a measure in $\mathcal{F}_q$). In Chapters I and II the author gives different conditions for the existence and uniqueness of such an invariant density when $\mathcal{F}_q$ and G are given. Chapter III deals with sets of N points P_i in Euclidean n-space. The group G is that of Euclidean motions and the invariant density becomes $\omega = dP_1 \wedge dP_2 \wedge \cdots \wedge dP_N$, where dP_i denotes the volume element at P_i. To calculate the integral of ω over particular sets, the author gives a useful formula due to Crofton and applies it to calculate the mean value of the distance between two points in the sphere. Chapter IV deals with the Integral Geometry of the Euclidean plane, treating the standard subjects: density for lines, Crofton's formulas (including some formulas of Masotti Biggiogero and the reviewer), kinematic density, the principal formula of Blaschke and applications to isoperimetric problems and Hadwiger's theorems. Chapter V is devoted to the Integral Geometry of Euclidean 3-space (densities for lines and planes, Herglotz's formula, kinematic density, random rotations). Several applications are given, for instance, to the problem of determining the distribution of the sizes of particles embedded in a body from the measurement of the figures formed by their intersections with random planes. Chapter VI presents a detailed study of Integral Geometry in the projective and centro-affine unimodular planes, the latter with applications to convex figures. Analogous topics for 3-dimensional projective and centro-affine unimodular spaces are treated in Chapter VII. The final Chapter VIII deals with Integral Geometry on surfaces, mainly confined to surfaces of constant curvature; it contains several of the author's results in this subject, and applications to integral formulas and inequalities pertaining to convex figures.

The book contains no exercises, but it reads well and provides a grounding in those essentials that it covers which will prepare the interested reader for further study in the original papers listed in the almost exhaustive bibliography given at the end.

L. A. Santaló

Integral Geometry (French)
Stoka, Marius I.
Mémorial des Sciences Mathématiques, Fasc. 165
Gauthier-Villars Éditeur, Paris (1968) 65 pp.
MR0231336

This book is not a translation of the previous book by the author published with the same title in Romanian [Editura Acad. R.S.R., Bucharest, 1967; MR0217747 (36 #836)]. Although most of its chapters are contained in the Romanian book, the present monograph is mainly devoted to the foundations and general principles of Integral Geometry according to the point of view of the author. Chapters I and II give conditions in order that an r-parameter Lie group G_r which acts as a transformation group of an n-dimensional space X_n have an integral invariant (in this case G_r is said to be measurable). If G_r leaves invariant a q-dimensional family F_q of p-dimensional varieties V_p, and $\phi(\alpha^i)$ is a function integral invariant by G_r, then the differential form $\phi(\alpha^i)d\alpha^1 \wedge \cdots \wedge d\alpha^q$ is said to be a density for the family of varieties. If this density is unique, up to a constant factor, then F_q is said to be measurable. Some general theorems on measurability of families of varieties under a given transitive group are given. Chapters III and IV apply the general theory to the cases of the Euclidean plane and Euclidean 3-dimensional space. For instance: (a) If F_2 is the family of lines of the plane, then F_2 is not measurable for the projective group G_8 and it is measurable for the group of motions G_3; (b) if F_3 is the family of circles of the plane, then F_3 is measurable for the affine group G_6. Chapter V deals with Integral Geometry on a 2-dimensional Riemannian space of constant curvature. The author gives the 2 and 3 parameter families of curves which are measurable under the group of motions of the space. In particular, the family of geodesic curves is measurable; some applications of this fact are given. The book is nicely written, and though it is essentially a collection of previous papers of the author, they are here presented in a uniform and clear style.

L. A. Santaló

The Differential Geometry of congruences in elliptic space
(Romanian)
Roşca, Radu
Editura Academiei Republicii Socialiste România, Bucharest 1969 208 pp.
MR0259777

The book contains a systematic exposition of the Differential Geometry of line congruences in three-dimensional elliptic space $P_e{}^3$, using throughout the methods of É. Cartan and his exterior differential calculus. Though the subject belongs to classical Differential Geometry, its study is worthwhile and instructive mainly because of the richness in several aspects of the geometry of $P_e{}^3$ and its applications to different topics (groups, kinematics). Chapter 1 contains definitions and first theorems on differential forms and the method of moving frames in $P_e{}^3$. Chapter 2 introduces the fundamental formulae for congruences in $P_e{}^3$ and defines the normal and diagonal congruences associated to a given congruence. Chapter 3 studies properties of the derived congruences by right and left parallelism. Chapter 4 deals with pseudospherical and pseudospherical normal congruences. Chapters 5, 6 and 7 are devoted to some special congruences (normal, isotropic, Ribaucour, Vincensini, Guichard) and associated surfaces (Bianchi, Clifford). Chapters 8, 9 and 10 treat pairs of congruences (conformal pairs of Tihockiĭ and Finikov, pairs of Thybaut, pairs J), including the original work of the author on these subjects. Chapters 11 and 12 are devoted to a type of double cyclical congruences, called K_2 congruences (normal congruences to a minimal surface), and to certain recent results on configurations, nets and sequences of conjugate nets in $P_e{}^3$. Chapter 13 is a survey of the theory of congruences in elliptic n-dimensional space. The book is clearly written and the topics are well selected, so that it will surely be useful as a source of reference as well as complement to standard courses on Differential Geometry.

L. A. Santaló

Introduction to global Differential Geometry[4] (Portuguese)

do Carmo, Manfredo Perdigao
Instituto de Matemática Pura e Aplicada,
Conselho Nacional de Pesquisas, Rio de Janeiro 1970. v+159 pp.
MR0275325

This is a small nice book on selected topics of the Differential Geometry in the large of surfaces embedded in E_3. The book is carefully written, with well selected examples to clarify the theory and to show why certain hypotheses in the theorems are really necessary. The contents are the following. (1) Rigidity of the sphere (following S. S. Chern [*Duke Math. J.* **12** (1945), 270–290; MR0012492 (7,29f)]). (2) Completeness of surfaces: theorem of Hopf and Rinow. (3) First and second variation of the arc length: theorem of Bonnet (no previous knowledge of the calculus of variations is necessary for the understanding of this and the next chapter). (4) Jacobi fields and conjugate points (applications to surfaces with $K \leq 0$). (5) Covering surfaces: theorems of Hadamard: (i) a complete, simply connected surface with $K \leq 0$ has a complete covering which is diffeomorphic with the Euclidean plane; (ii) a compact surface with $K > 0$ is diffeomorphic to a sphere. (6) Surfaces with $K = 0$ (a complete surface in E_3 with $K = 0$ is a cylinder or a plane (the proof is analogous to that of W. S. Massey [*Tôhoku Math. J.* **14** (1962), 73–79; MR0139088 (25 #2527)])). (7) Jacobi's conditions for shortest arcs between two points of a surface. (8) Hilbert's theorem about the impossibility of embedding isometrically a surface with $K = -1$ in E_3 without a singularity (the proof uses the same basic tool that was used by Hilbert). The book is very suitable as a complement of a first course of Differential Geometry.

L. A. SANTALÓ

Geometric Probability

Solomon, Herbert
Regional Conference Series in Applied Mathematics, **28**.
Society for Industrial and Applied Mathematics, Philadelphia, Pa., 1978.
MR0488215

During the 1960's, after the publication of the lucid monograph of M. G. Kendall and P. A. P. Moran [*Geometrical Probability*, Hafner, New York, 1963; MR0174068 (30 #4275)] and the fundamental papers of R. E. Miles [*Proc. Nat. Acad. Sci.* U.S.A. **52** (1964), 901–907; MR0168000 (29 #5265); ibid. **52** (1964), 1157–1160; MR0169258 (29 #6510); *Advances in Appl. Probability* **1** (1969), 211–237; MR0259977 (41 #4606); ibid. **3** (1971), 1–43; MR0287619 (44 #4822)], the theory of Geometric Probability enjoyed a resurgence of interest. Several problems arose and many interesting

[4] Almost all topics analysed in this book were included and extended by the author in [*Differerential Geometry of curves and surfaces*, Prentice-Hall, Inc. 1976. viii+503 pp.] and [*Riemannian Geometry*, Birkhäuser, Boston, MA, 1992. xiv+300 pp.].

results were obtained that have deep connections with other branches of pure and applied mathematics. Moreover, methods and tools of Geometric Probability became essential for the so-called stereology [see *Stereology* (Proc. Second Internat. Congr., Chicago, 1967); Springer, New York, 1967]. The present monograph, which contains a thorough and clear presentation of selected topics, will surely stimulate interest in this field – a field in which elementary concepts of geometry, probability and groups mix together, yielding harmonious results, interesting in themselves and in various fields of applications.

Following a discussion of the historical development, Chapter 1 deals with Buffon's needle problem and its extensions, with interesting comments on experimental designs and statistical estimation procedures for π. Chapter 2 takes up density and measure for random geometric elements (points, lines, pairs of points in the plane). Chapter 3 examines in detail the number of intersections of random lines in the plane and their angles of intersection, treating questions of variance, higher moments and distribution of the number of intersections. Several problems arise pertaining to the distribution of the area A, the perimeter L and the number of sides N of the polygons formed by random lines in the plane. Finding the exact distribution of A, L, N is still an open question, but the author gives simple derivations of the second order moments and some third order moments, recapturing classical results of Miles and pointing out some empirical distribution functions obtained by Monte Carlo methods. The case of anisotropic lines (sets of lines with density $dp \wedge dF(\theta)$, where $F(\theta)$ is the cumulative distribution of θ) is also considered, and its relation to highway traffic flow models is pointed out. Chapter 4 deals with problems of covering the circumference of a circle by random arcs (moments, distribution and asymptotic distribution coverages) and with the still unsolved problem of covering the surface of a sphere with randomly placed circular caps. Chapter 5 gives the classical Crofton theorems on fixed points and mean values and applies them to a detailed treatment of Sylvester's four-points problem and its extension to three-space. The last chapter discusses several results on random chords in the circle and the sphere, some of them motivated by biological problems (position of chromosomes on a karyograph). Six different randomization models by which random chords in a circle can be formed are considered, and for each of them the mean and variance of the total number of intersections within the circle is derived. The last problem, also thoroughly discussed, is that of finding the distribution of the length of a random chord of a three-dimensional sphere under eight different randomization models.

The book is well organized and carefully written. Some recent results are included and unsolved questions are pointed out to encourage further research. The exposition throughout is clear and concise so that the whole monograph is very enjoyable reading.

L. A. Santaló

Convexity (Spanish)
Toranzos, Fausto A.; Nanclares, Jorge H.
Cursos, Seminarios y Tesis del PEAM, No. 4. PEAM,
Universidad del Zulia, Maracaibo, 1978. iv+155 pp.
MR0513876

This is a nice expository brochure devoted to the foundations of convexity and to certain applications of this theory to different branches of pure and applied mathematics. Chapter 1 contains the basic concepts and definitions (convex hull and some metric and topological properties of convex sets). Chapter 2 deals with separation theorems, complementary sets, hyperplanes of support, extreme points, the Krein-Milman theorem and polarity. Chapter 3 is devoted to the theorem of Helly and its generalizations, with applications to some combinatorial properties of convex sets (separation theorems of Kirchberger and Krasnosel'skii). Chapter 4 is entitled "Convexity and optimization" and contains some applications of convexity (convex polytopes) to linear programming, with general notions on convex functions (Fenchel's theorem). the last chapter, 5, contains some selected topics on convex sets, some of them classical (Blaschke convergence theorem, sets of constant breadth, Borsuk's problem) and others, less classical, which offer perspectives for research (mainly topics related to star-shaped sets). The book is carefully written in a very clear and well- motivated fashion.

L. A. Santaló

Stochastic Geometry[5] (German)
Stoyan, Dietrich; Mecke, Joseph
Scientific Paperbacks, Mathematics/Physics Series, 275.
Akademie-Verlag, Berlin, 1983. 132 pp.
MR0727292

This paper is an excellent mise au point of the foundations and present problems of Stochastic Geometry, written by two authors who know the subject very well and have made important contributions to it. The name "Stochastic Geometry" was proposed by E. F. Harding and D. G. Kendall [Stochastic geometry, edited by Harding and Kendall, Wiley, London, 1974; MR0350792 (50 #3284)] as a result of the introduction – initiated by R. E. Miles and pursued by R. Davidson and K. Krickeberg – of stochastic processes in classical geometrical probability and Integral Geometry. Section 1 provides the needed basic notions and tools on set theory, topology, convex sets, measure and integration theory. Section 2 deals with the theory of random closed sets and their statistics, mainly following G. Matheron

[5] The contents of this monography was extended by the same authors in [*Stochastic Geometry and its aplications*. With a foreword by D. G. Kendall. Wiley Series in Probability and Mathematical Statistics. John Wiley & Sons, Ltd., Chichester, 1987. 345 pp.], and a revised version was published in Russian and German [*Geometric Probabilities and Stochastic Geometry*. With R. V. Ambartzumjan. Akademie Verlag, Berlin, 1993. 400 pp.].

[Random sets and Integral Geometry, Wiley, New York, 1975; MR0385969 (52 #6828)]. Accurate definitions of the basic concepts are given (independence, stationarity, isotropy, covariance, covariance function, correlation function) and the most significant formulas and theorems are stated, mostly without proof. Reference is made to the theory of image analysis and mathematical morphology in the sense of Matheron and J. Serra . Section 3 deals with general point processes and marked point processes: definitions, moments, generating functionals, Palm distribution, operations (thinning and clustering), and statistics for stationary point processes. Section 4 is devoted to random locally finite measures in $\mathbb{R}^d$. In Section 5 the rather abstract concepts above are applied to the study of the Boole model and the book suddenly takes on an attractive geometric flavour which it preserves until the end. Several formulas and examples on the Boole model with convex grains are given. Section 6 is a fine study of some models of point processes and their statistics (Poisson, Cox, Neyman-Scott and others). Section 7 deals with line and fibre processes on the plane and their intersections with fixed or random lines and other planar line or fibre processes. Section 8 is a very nice introduction to random mosaics, with accurate definitions and the statement, without proof, of their principal mean values and characteristics (especially Poisson and Voronoï mosaics in the plane and in 3-space). The last section (Section 9) is an excellent survey on stereology (classical stereological formulas for the germ-grain model and for collections of spheres). The book, very well written, is very condensed, so that it is not easy to read, but the great deal of information it contains makes worthwhile any effort of the reader. Many of the statements are given without proof, but an accurate and well-selected bibliography will help the reader who desires more information or wishes to be acquainted with more details. For all those who are working in Stochastic Geometry and related fields – students or specialists – the book will be very useful as a guide and good reference.

L. A. Santaló

Theory of convex bodies
Bonnesen, T., Fenchel, W.
Moscow, Idaho (USA): BCS Associates. IX, 1987. 172 pp.
Zbl 0628.52001

The book is a faithful translation of the classical 'Theorie der konvexen Körper' of the authors originally published in German (1934; Zbl 0008.07708) and reprinted (revised) in (1974; Zbl 0277.52001). For more than fifty years the book has been the main reference for all research on convexity and convex bodies, a field that during this period has greatly developped in itself and has very much increased its applications to many branches of mathematics. The contents is the following: basic concepts, centroids and convex hull, boundary points and support planes, representation by convex functions, linear combination of convex bodies, approximation of convex bodies, number and figures associated with convex bodies, integral formulas for the volume and mixed volumes, symmetrization, inequalities, Brunn-Minkowski theory with applications, determination of convex bodies by curvature functions, special convex bodies. In an excellent and concise form the different topics are presented with detailed proofs and precise indication of corresponding bibliography, which is practically complete until 1934. The usefulness of the book is still in force, so that the present translation will be wellcome for all mathematicians (students and research workers) interested in convexity.

L. A. SANTALÓ

Factorization calculus and Geometric Probability

Ambartzumian, R. V.

Encyclopedia of Mathematics and its Applications, 33. Cambridge University Press, Cambridge, 1990. xii+286 pp. ISBN: 0-521-34535-9
MR1075011

The book may be considered as an interesting continuation of the reviewer's book Integral Geometry and Geometric Probability [Addison-Wesley, Reading, MA, 1976; MR0433364 (55 #6340)] and the author's book Combinatorial Integral Geometry [Wiley, New York, 1982; MR0679133 (84g:60017)] with noteworthy complements to their content and systematic derivations towards those topics in Geometric Probability which are now known as Stochastic Geometry. We quote from the preface: "Traditionally, Integral Geometry considers only finite sets of geometrical elements (lines, planes, etc.) and measures in the spaces of such sets which should be invariant with respect to an appropriate group acting in the basic space (which is $\mathbb{R}^n$ in most of the present book). If the group contains translations of $\mathbb{R}^n$, then the measures in question are necessarily totally infinite and cannot be normalized to become probability measures. Yet a step towards countably infinite sets of geometric elements changes the situation: spaces of such sets admit probability measures which are invariant and these measures are numerous. The step from finite sets to countably infinite sets directly transfers an integral geometrician into the domain of probability. The vast field of inquiry that opens up surely deserves attention by virtue of the mathematical elegance of its problems and as a potentially rich source of models for applied sciences."

Chapter 1 is devoted to the so-called Cavalieri principle which allows one to replace many Jacobian calculations by clear geometrical reasoning, and other prerequisites, such as Lebesgue and Haar factorization of measures, parametrization maps, metrics and convexity, and the classical problem of finding the invariant measure of the sets of lines separating two planar convex domains. Chapter 2 ("Measures invariant with respect to translations") analyzes the spaces of directed and nondirected lines and planes in $\mathbb{R}^2$ and $\mathbb{R}^3$, the spaces of segments and flats and product spaces with their random measures with invariant first and second moment measures (the Davidson theorem). In Chapter 3 ("Measures invariant with respect to Euclidean motions") the method of Haar factorization is applied to measures of several situations: rotations, geodesic lines on the sphere, kinematic measure, pairs of points, lines and planes in $\mathbb{R}^3$, triads of points and lines, measures in shape spaces.

Chapter 4 is devoted to Haar measures in groups of affine transformations on $\mathbb{R}^2$ and $\mathbb{R}^3$, triads and quadruples of points in $\mathbb{R}^2$ and quintuples in $\mathbb{R}^3$, together with the invariant measure in the space of tetrahedral shapes and shapes of quintuples in $\mathbb{R}^3$. Derivation of Haar measures on groups based on knowledge of Haar measures on their subgroups simplifies the treatment. Chapter 5 treats the combinatorics of lines in $\mathbb{R}^2$ and planes in $\mathbb{R}^3$, in particular the so-called flag representation of bounded centrally symmetric convex bodies in $\mathbb{R}^3$ (a complete presentation can be found in the author's text cited above). Chapter 6 ("Basic integrals") presents several integrations useful for the study of geometrical processes related with the Euclidean motion invariant Integral Geometry (Pleijel identities, isoperimetric inequalities, analysis of sets of segments by means of moving test segments, random chord length distribution for convex polygons). Particularly noteworthy are the given measures on the family of sets $A(x) = \{$triangular shapes for which the minimal interior angle is greater than $x\}$.

The second half of the book is devoted to random point processes and related questions of Stochastic Geometry, always from the original point of view (factorization calculus) successfully followed throughout by the author. Chapter 7 gives the basic notions on random point processes, treating in detail the case of point processes on the linc (k-subsets of a linear interval, finite sets on $[a, b)$, consistent families, point processes of Poisson and Shepp) and introducing the notion of marked point processes and moment measures with interesting examples carefully selected (segment processes, random mosaics). Averaging in the space of realizations is also applied to examples with useful results concerning the nodes in segment processes and random mosaics in $\mathbb{R}^2$. Chapter 8 goes deep into the author's idea of connecting Palm distributions with Integral Geometry, with a base on Haar factorization. Various definitions of Palm distributions are given with applications and examples referring to point, cluster and renewal processes. The main goal of Chapter 9 ("Poisson-generated geometrical processes") is to derive size and shape distributions of typical configurations generated by Poisson processes, in particular random mosaics generated by Poisson line processes (distribution of the polygon randomly chosen from the Poisson line mosaic if the weight given to each polygon is the numer of vertices, the perimeter length or the area of the polygon). Several mean values of random variables depending on random polygons are treated in detail and similar ideas are applied to the study of Voronoï mosaics. In Chapter 10, the last chapter, the author considers geometrical processes of a general nature (not necessarily Poisson), namely line processes, random mosaics and Boolean models for disc processes with the analysis of the induced one-dimensional marked point or segment processes with angular marks, giving some basic relations which have stereological significance. Several averages of combinatorial decompositions of general line processes remain valid for a broad class of random mosaics, not necessarily generated by line processes which apart from the motion invariance are determined by further explicitly given conditions. The technique is also applied to domain processes and new possible expansions are indicated (for instance, to line and plane processes in $\mathbb{R}^3$) with examples showing that the motion invariance often imposes rather heavy restrictions on the distributions of intersection processes.

The book is written in the characteristic style of the author, full of new and interesting ideas and, although in appearance the prerequisites are only standard courses on analysis and probability, it is not easy to read. It must be carefully thought out in many of its details, but the effort is well compensated by the great deal of information and new ways of thinking it supplies.

L. A. SANTALÓ

Geometry of surfaces

Stillwell, John

Universitext. New York: Springer-Verlag. xi, 1992. 216 pp.

Zbl 0752.53002

The book is essentially devoted to the geometry of surfaces of constant curvature, exploring the interplay between geometry and topology and introducing in the basic notions of curvature, group actions and covering spaces. The contents is the following: 1. The Euclidean plane (isometries and its classification). 2. Euclidean surfaces (cylinder, torus, Klein bottle, quotient surfaces, the covering isometry group, Killing-Hopf theorem). 3. The sphere (isometries, stereographic projection, groups of isometries, the elliptic plane). 4. The hyperbolic plane (the pseudosphere, the half plane and the disc models, classification of isometries). 5. Hyperbolic surfaces (construction of hyperbolic surfaces from polygons, geometric realization of compact surfaces). 6. Paths and geodesics (normal forms of compact orientable surfaces, Euler characteristic, paths and homotopy, the fundamental group, genus, closed geodesic paths). 7. Planar and spherical tesselations (fundamental regions, triangle tesselations). 8. Tessellations of compact surfaces (orbifolds and desingularization).

The prerequisites are a little linear algebra, calculus as far as hyperbolic functions, basic group theory and basic topology. Surely the book will stimulate the interest in areas, which arise in a wide variety of situations ranging from geometry and analytic functions to physics. The book is carefully written and contains valuable exercises and informal discussions with historical remarks at the end of each chapter.

L. A. SANTALÓ

Integral Geometry in Statistical Physics. Percolation, complex fluids and the structure of the universe (German)

Mecke, Klaus

Reihe Physik. 25. Frankfurt/Main: Harri Deutsch.
München: Univ. iv, 1993. 213 pp.

Zbl 0863.52005

The purpose of this book is to show many and important applications of Integral Geometry to statistical physics particularly to percolation, complex fluids and structure of the universe. The first half (Chapters 1,2) is devoted to integral and

Stochastic Geometry and the rest (Chapters 3-6) to statistical physics, mainly to the distribution of grains or holes in porous media.

Chapter 1 summarizes the main concepts of Integral Geometry. The quermass-integral or Minkowski functionals $W_\nu(A)$, $\nu = 1, 2, \ldots, d$ of any set A of dimension d are defined and the following Hadwiger's theorem is stated: Every nonnegative, isometry-invariant, monotone and additive functional $F(A)$ can be written in the form $F(A) = \sum_0^d c_\nu W_\nu(A)$, where the c_ν are nonnegative constants depending on F; this theorem is basis for applications to the physics which the author has in mind. Then, the fundamental kinematic and the generalized Crofton's formulas are given, besides the Integral Geometry in lattices and some applications to stereology.

Chapter 2 expounds the basic concepts of Stochastic Geometry: the Boolean or germ-grain model, Voronoï tessellations, mean values of Minkowski functionals, oriented cylinders, correlation functions and the Poisson-Gauss process.

Chapter 3 is a short introduction to the Euler characteristic and its relation with percolation (when a structure changes from a collection of many disconnected parts into basically one big comglomarate, one says that percolation occurs).

Chapter 4 applies the preceding concepts to describe the structure of the universe, that is, a Boolean model and Minkowski functionals are used for finding the distribution of the galaxics and clusters.

Chapter 5: Given a configuration A of bodies, because of the energy $E(A)$ is a functional which is motion-invariant, continuous and additive; by the theorem of Hadwiger, $E(A)$ is a linear combination of the Minkowski functionals of A. This is the bridge between Stochastic Geometry and statistical physics, which the author applies to discuss some properties of microemulsions, for instance a water-amphiphil or water-oil-amphiphil system, as well a microemulsion as a system of boundaries of surfaces. Also the following items are considered: morphology of liquid models, structure functions, lattice models and perspectives on other methods and applications.

In Chapter 6, two renormalization groups are presented, that of Migdal-Kadanoff and that of approximation by finite cells. Moreover, the duration of the structure functions in the microemulsions and the Monte Carlo simulation are discussed. The book ends with an extensive bibliography. In summary, this text will be useful not only to specialists in statistical physics but also to those in neighboring fields.

L. A. Santaló

The kinematic formula in Riemannian homogeneous spaces
Howard, Ralph
Mem. Amer. Math. Soc. 106, 1993. no. 509, vi+69 pp.
MR1169230

Let G be a Lie group and K a closed subgroup of G. If M, N are compact submanifolds of G/K, then one of the main purposes of Integral Geometry is to compute integrals of the type

$$\int_G I(M \cap gN)\Omega_G(g), \quad g \in G,$$

called "kinematic formulas", where I is an integral invariant and Ω_G is the invariant measure on G. Section 1 is a clear summary of the results obtained, which are a set of interesting and very general kinematic formulas, including as particular cases most of the known formulas of this kind seen from new angles and perspectives. In Section 2 the so-called "basic integral formula" is proven, which relates the integral of any Borel measurable function of h on $M \times N$ with the integral over G of the integral of h on $M \cap gN$. In Section 3 of the paper the basic integral formula is applied to the case $I(M \cap gN) = \mathrm{vol}(M \cap gN)$, and examples are given of how the "transfer principle" (Section 7) can be used to compute the constants occurring in the kinematic formulas. Noteworthy are the examples of homogeneous spaces (in particular CP^n) where the measures are different from the Riemannian volume of the submanifolds and where these measures are not unique, so that the choice of the measure to be used is determined by the type of the integral geometric formula to be proven. Of particular interest is the appendix to this section, entitled "Cauchy-Crofton formulas and invariant volumes", which clarifies the different integral formulas according to the chosen integral volumes for submanifolds of CP^{2n}. In Section 4 the author gives a very general definition of integral invariants for p-dimensional submanifolds M of R^n, which are the integrals I^{Q_α} of some homogeneous invariant polynomials Q_α, $p+1 \leq \alpha \leq n$, defined on the second fundamental form of the submanifold M. Then the kinematic formula is written as

$$\int_G I^P(M \cap gN)\Omega_G(g) = \sum_\alpha I^{Q_\alpha}(M) I^{R_\alpha}(N),$$

where R_α is the analogous invariant polynomial defined for the q- dimensional submanifold N. These invariant polynomials and the corresponding integral invariants will be important in other branches of geometry and topology. The form of kinematic formulas in G/K does not depend on the full group G but only on the invariant theory of the isotropy subgroup K, a property which leads to the so-called "transfer principle" which allows the author to translate kinematic formulas proven for G/K to any other homogeneous space with an isotropy group equivalent to K.

Sections 5, 6 and 7 contain the lemmas needed to prove the stated results of Section 4. Many of them are interesting in themselves and lead to a restatement of the kinematic formula in terms of certain algebraic definitions. Section 8 applies the results obtained to spaces of constant sectional curvature. The last two sections are devoted to elegant proofs of the Chern-Federer kinematic formula and the formula of Weyl on tubes, giving an algebraic characterization of the polynomials appearing as integrands of both formulas.

The paper is very important and will be very useful as a reference for future work on kinematic formulas. Moreover, mainly for the geometry of $C\mathbf{P}^n$, it contains new ideas and problems which will surely be the source of new and interesting developments.

L. A. SANTALÓ

Stochastic Geometry[6]
Weil, Wolfgang; Wieacker, John A.
Handbook of Convex Geometry, Vol. A, B, 1391–1438,
North-Holland, Amsterdam, 1993.
MR1243013

The paper is an excellent survey on numerous aspects and results in Geometric Probability and Stochastic Geometry, with historical remarks and an extensive biography. The paper surveys the most recent developments in this area and is structured as follows. (1) Random points in a convex body. Let $K \subset \mathbf{R}^d$ be a convex body with interior points and let $X_1, \cdots, X_n$ be n independently and uniformly distributed random points in K. Then the convex hull $Q_n = \text{conv}\{X_1, \cdots, X_n\}$ is a random polytope P^d and for any measurable nonnegative function $P^d \to [0, \infty]$, $g \circ Q_n$ is a random variable and we write $E_n(g)$ for the expected value of $g \circ Q_n$. The explicit computation of $E_n(g)$ for some functionals g is complicated and the moments have been considered by several authors (Kingman, Buchta, Müller, Affentrangen), obtaining, for instance, that the rth normalised moment $M_r(n, K)$ is minimal if and only if K is an ellipsoid. The conjecture that $M_1(n, K)$ is maximal if and only if K is a simplex is still open. General results have been obtained on the asymptotic behaviour of $E_n(g)$ as $n \to \infty$ (Rényi and Sulanke, Efron, Wieacker, Bárány, Schneider). (2) Random flats intersecting a convex body. Instead of considering the convex hull of random points, one may consider the intersection of random closed half- spaces generated by random hyperplanes (Rényi and Sulanke, Ziezold, Schneider) and obtain interesting results. (3) Random convex bodies. Some results of the following type are stated. Let $X_1, \cdots, X_n$ be a sequence of independent, identically distributed random convex bodies (such that $Ed(X_1, \{0\}) < \infty$); then almost surely $n^{-1}(X_1 + \cdots + X_n) \to E(X_1)$ as $n \to \infty$ (Artstein and Vitale, Araujo and Giné). logarithm, ergodic theorems) are mentioned. (4) Random sets. Some properties and theorems referring to closed sets (RACS) are stated, including various results of Choquet, Kendall, Matheron, Weil, Wieacker, Zähle. (5) Point processes. The general version of point processes following the approaches of Kendall and Matheron and Weil and Wieacker, among others, is given. Further results on Poisson processes are mentioned and a wide variety of formulas are given from the classical

[6] Although this publication is not a book we consider interesting to include this review.

and modern literature. (6) Random surfaces. A random closed set X is called a random m-surface if X is a locally countable m-rectifiable closed subset of $\mathbf{R}^d$ and $E\lambda_m(X \cap K) < \infty$ for all convex sets in $\mathbf{R}^d$. The most important real parameter of X is the density $D_m(X)$ defined by $E\lambda_m(X \cap A) = D_m(X)/\lambda_d(A)$, for any Borel set A in $\mathbf{R}^d$. The work of Wieacker is treated in detail and several additional results are established. (7) Random mosaics. Classical examples obtained from a stationary Poisson point process (Voronoĭ, Delaunay) have been investigated mainly by Miles, Matheron, Stoyan, Mecke, Zähle and Cowan. The paper includes some extensions of these results and related questions. (8) Stereology. The so-called fundamental formulae of stereology are mentioned, given different possibilities for unbiased estimation of the quermass densities. However, the determination of variances is in general still open.

L. A. Santaló

8.2 Reviews of Some Books by L. A. Santaló

Santaló wrote more than 20 books. We reproduce the reviews of those more interesting and more directly related to its investigation work.

Integral Geometry in spaces of constant curvature[7] (Spanish)
Santaló, Luis A.
Repub. Argentina. Publ. Comision Nac. Energia Atomica.
Ser. Mat. 1, 1952. no. 1, 68 pp.
MR0051544

This paper deals with the Integral Geometry in a space of constant curvature of dimension n. It is divided into two parts, Part I concerned with the densities of linear subspaces and integral formulas derived therefrom and Part II with the kinematic density. These densities are determined explicitly by E. Cartan's method of moving frames. The total measures of several homogeneous spaces are computed, such as the measure of all linear subspaces of a given dimension through a fixed point. Many integral formulas are given, of which we mention the following as samples. 1) The integral over the density of r-dimensional linear subspaces of the area of intersection of such a linear subspace with a fixed regular q-dimensional submanifold C_q is equal to a constant multiple of the area of C_q. This generalizes a well-known formula of Cauchy and it is of interest to notice that it is independent of the curvature of the space. 2) Generalizations of the classical Crofton formulas. 3) Integrals of the

[7] Most of the topics analysed here were exposed in the ICM of Cambridge, 1950, (see [50.2]), and some misprints were corrected later in [62.3]. These topics were included and extended in the second part of [76.1].

elementary symmetric functions of the principal curvatures of a hypersurface and integral formulas for the mean values of such invariants when a hypersurface is cut by a linear subspace. Cases 2) and 3) are carried out only in Euclidean space.

The main result in Part II is the kinematic formula in a space of constant curvature, which contains many formulas of Integral Geometry as special or limiting cases.

S. Chern

Introduction to Integral Geometry
Santaló, L. A.
Actualités Sci. Ind., no. 1198, Publ. Inst. Math. Univ. Nancago II. Herman et Cie, Paris, 1953. 127 pp.
MR0060840

The book consists of three parts. The first deals with the Euclidean plane and treats the following topics: density and measure for sets of points (it is emphasized from the outset that the densities are to be considered as obtained by exterior, rather than ordinary, multiplication of differentials). Density and measure for sets of straight lines, sets of pairs of points, sets of pairs of straight lines, kinematic measure, sets of segments, sets of rectifiable curves, fundamental formulas of Blaschke, applications (for instance, the isoperimetric inequality), lattices of figures. The author remarks in the introduction regarding the concepts of this first part: "However elementary and simple they may appear, we think it worthwhile to keep them always in mind to achieve a better understanding of the latest generalizations." This part is obviously most useful (besides being pretty), and it is merely a sad reflection on an at present very prevalent way of writing, that the author feels obliged to justify his procedure.

The second part deals with the geometry on surfaces; its subheadings are: density for sets of geodesics, geodesics which intersect a given curve, kinematic density on surfaces, Integral Geometry on surfaces of constant curvature. The author implies that, for instance, density for geodesics is (except for surfaces with constant curvature) not as well justified as for the plane, because there are no motions. Actually, the standard procedure in Differential Geometry of considering local concepts (like angle) as given by the local Euclidean Geometry imposes this density.

The third part deals with general Integral Geometry and treats the topics: properties of Lie groups (and any others, the moving frames and relative components of E. Cartan, exterior differentiation, Pfaffian forms), density and measure in homogeneous spaces. This general theory is then applied to the following special cases: the group of central affinities in the plane, the unimodular group in the plane, the projective group, the generalized Poincaré formula in the plane, Integral Geometry in the plane of Cayley, the group of motions in n-dimensional Euclidean space.

There is a bibliography at the end of each major topic. Considering that there are only 123 pages of text, very much material is covered; nevertheless the book is unusually readable, and will serve its purpose very well.

H. BUSEMANN

Non-Euclidean geometries (Spanish)
Santaló, Luis A.
Editorial Universitaria de Buenos Aires, Buenos Aires 1961. 64 pp.
MR0177336

The author gives a very clear and easily readable representation to the development, the terms and most important sets of non-Euclidean Geometry. Higher previous knowledge is not presupposed; the most remarkable characteristics and differences in relation to Euclidean Geometry are emphasized clearly. The historical development following only the axiomatic method is lit up. The first chapter is dedicated to Euclide's *Elements* and in particular to the fifth postulate (parallel axiom) and its different forms. In the second chapter non- Euclidean Geometry of Gauss, Lobachevsky and Bolyai is axiomatically introduced. In this first part under reference notions and theorems are more outlined by references to the literature than proved in detail. Two further chapters bring an outline of Projective Geometry and prepare F. Klein entrance to non-Euclidean Geometry. In three further chapters – the main part of the monography – the descriptive characteristics, the measurement of angles and segments, the notion of area and special curves (generalization of the notion of circle) in non-Euclidean Hyperbolic Geometry are treated. Finally in a further chapter other well-known models of non-Euclidean Geometry (projective interpretation of elliptical geometry, geometry on the pseudosphere and the Poincaré model) are treated. In the last chapter relation with other geometries and the development of geometry in general are obtained. In small space the author presents plenty of results and an excellent overview of the structures and propositions of non-Euclidean Geometry. The concise and elegant proofs of projective (synthetic) geometry are used in a competent way and many interesting questions are refered to briefly.

H. R. MÜLLER

Integral Geometry and Geometric Probability
Santaló, Luis A.
(With a foreword by Mark Kac)
Encyclopedia of Mathematics and its Applications, Vol. 1.
Addison-Wesley Publishing Co., 1976. xvii+404 pp.
MR0433364

Geometric Probability and Integral Geometry began as one subject: the study of probability measures of sets of geometric objects such as points, lines, and motions.

It was epitomized by the pioneering work of Morgan Crofton in the second half of the last century. Afterwards the topics separated: writings of E. Czuber (1884), H. Poincaré (1912) and R. Deltheil (1926) emphasized probability questions; on the other hand, the seminal ideas of G. Herglotz' 1933 Göttingen lectures matured into the Integral Geometry of W. Blaschke's Hamburg School (1935–1939) of which the author is a distinguished representative. In Integral Geometry, appropriate group invariance requirements determine the most interesting measures; these notions have been applied to geometric questions such as isoperimetric problems. Since the time of Blaschke's school, Integral Geometry has grown steadily. Separately, and principally in the last 20 years, Geometric Probability and its relative, Stochastic Geometry, initiated by papers of the late R. Davidson (1968–1970), have resumed a flourishing development, often motivated by practical applications.

The author's monograph covers all these topics, in their current states, with encyclopedic fullness and pedagogic skill. By so doing, it promises to become both a standard textbook and the basic reference for years to come. Yet its greatest service may be the re-unification of Integral Geometry with modern Geometric Probability, to the benefit of both subjects.

There are four parts to the book. The first, plane Integral Geometry, introduces the notions, methods and applications of the subject in its simplest and richest setting, the Euclidean plane. This part includes a chapter on Poisson processes in the plane. General Integral Geometry forms the second part which contains: differential forms, Lie groups and invariant measures; densities and associated measures in homogeneous spaces; the specializations to affine groups and motion groups treated by É. Cartan's method of moving frames. These first two parts give an enlarged and updated version of the author's earlier monograph [*Introduction to Integral Geometry*, Actualités Sci. Indust., No. 1198, Hermann, Paris, 1953; MR0060840 (15,736d)]. A detailed survey of Integral Geometry in Euclidean n-space makes up the third part, with a chapter on statistical applications including stereology: determination of particle size distributions in space from the size distributions in random sections. The final part is called Integral Geometry in spaces of constant curvature; its first two chapters cover non-Euclidean Integral Geometry; the last chapter deals with Integral Geometry in a variety of spaces of non-constant curvature. This includes, under the section titles: Integral Geometry (in the sense) of Gelfand, a survey of Radon problems which require the reconstruction of functions from their integrals over manifolds of a given family. Each chapter consists of a development of the principal ideas, with proofs, and a final section in smaller type of notes, providing a comprehensive character to the work by outlining related, subsidiary developments. In some cases, exercises are added, at times with solutions. This enables the author to give both a survey, in the style of the Ergebnisse series, as well as a textbook for readers with suitable background. Here a caution is needed: sometimes the author omits precise assumptions for the validity of theorems and formulas (for example, the general kinematic formula of H. Federer, described on p. 276).

The bibliography contains more than 700 items. Taken with the bibliographies of references 51, 152, 274, 335, 367, 428, 429 and 647, it appears to be nearly complete through 1974, with many items from 1975–1976. There is a thorough author index and subject index. The work deserves better production quality: some linguistic and punctuation slips show poor editing and the frequency of misprints indicates inadequate proofreading.

Mark Kac describes the spirit of this work in his foreword: "Probability is measure theory with a 'soul' which in this case is provided by ... the most ancient and noble of all mathematical disciplines, namely Geometry".

W. J. Firey

The History of Aeronautics (Spanish)
Santaló, L. A.
Espasa-Calpe Argentina, 1946.

Some years ago, during an informal conversation about his life, Santaló expressed to me his interest in aviation and its history. In fact, he told me that during the Spanish Civil War he was doing his military service in Alcantarilla, Murcia, where he was working on some calculations for the Spanish Republican Air Force, and since this left him with a lot of spare time, he pursued his curiosity by learning more about the scientific basis of aviation. Prior to the appearance of his long book [46.6], he had already published two short papers on the subject [42.12 and 45.14].

In the introduction to this book, he describes how throughout history two types of invention can be distinguished. Some of them, such as the radiotelephone, were neither deliberately sought or predicted by humanity, which suddenly found itself in possession of them without having thought previously of their possibility.

Other inventions, on the other hand, were the obsession of many preceding generations which, each one with its own available the means and perspectives, attempted to develop these ideas. The most typical example of this type of invention are flying machines. The most obvious proof of man's desire to fly is that it is expressed in all ancient mythologies and religions.

It is perhaps possible to state that aviation has been the most deeply felt desire running through all successive generations.

In this book, Santaló presents us with a "*History of Aviation*" covering the entire range from the first primitive urges as expressed in mythology, through the early successive frustrated attempts of which we have knowledge, down to the period in which the book was written. In short, this history is perfectly summarized in the introduction that Santaló provides in this work:

The first part is devoted to what we might call the prehistory of aeronautics, including the largely fantastic projects, many with more literary than scientific worth, which occurred before man managed to actually leave the ground by means of devices of his own construction.

The second part concerns the study of the very first airborne vehicles; air balloons, in which man first left the ground, the first atmospheric observations were made and the first flights were undertaken.

The third part deals with the history of the first steerable airborne craft: dirigible balloons. With these dirigibles, two features were introduced that were to prove indispensable: combustion engines for the production of energy, and propellers, to convert this energy into the driving force required to move such craft through the atmospheric ocean.

With these two features, it was not long before the tenacity of those convinced of the possibility of flight with devices heavier than air was able to convert their dreams into reality. The first aircraft emerged, and it is their history that provides the content for the fourth part of the book.

The fifth part, necessarily brief due to the nature of the work, is devoted to the history of military aviation. This deserves a chapter to itself, since it deals with the military application of one of the principles of aeronautics, and as a result of the military operations carried out by aircraft during a certain period, the exact state of the development of aeronautics in itself at that time may be ascertained.

In the sixth part, flying machines other than aeroplanes (helicopters, autogyros, gliders) are studied, while in a separate chapter the history of the parachute is considered, a development which has been of great assistance in aeronautics since its invention and practical use.

As an appendix, and prompted by the inertia of the preceding historical overview, we have been unable to resist the temptation of delving into the uncertain territory of future possibilities, the chief of which is a brief exploration of the current perspectives as regards interplanetary travel.

This is an attractive and highly readable book, fully accessible to the public at large. Despite having been published many years ago, it enables the reader to understand deeply the subsequent development of aeronautics, which has become one of the modern cornerstones in the development of humanity.

A. M. NAVEIRA

Psychology of invention in the mathematical field
by *Jacques Hadamard*
Princeton University Press 1945 (Dover 1954)
Translated to Spanish by
L. A. Santaló
Ed. Espasa Calpe 1947. 234 pp.

This book was inspired by a talk given by H. Poincaré at the Psychological Society of Paris in 1908 and is based on a series of talks given by the author at the École Libre des Hautes Études in New York in 1943.

The subjects it deals with are mainly the role of the unconscious in mathematical discovery, the subsequent conscious work, the different types of mathematical intelligence, and paradoxical cases of intuition, etc.

It is accompanied by two addenda: the first consists of surveys on the working methods of mathematicians, previously published in Enseignement Mathématiques, **IV**, 1902, **VI**, 1904, by E. Claparède and T. Flournoy. Here we find curious questions such as: Do you think that certain rules of hygiene such as diet, mealtimes, rest periods, etc., are useful for mathematicians?, and Have you ever noticed periods of excitation and stimulation followed by others of depression and inability to work?

The second addendum consists of a letter by Albert Einstein written in reply to a questionnaire containing questions similar to those above and analyzing his own mechanisms of thought.

While this is not a work written by Santaló, we include it here as relevant, since he himself thought it worthwhile to translate the book in 1947.

A. REVENTÓS

9 On the Correspondence between Santaló and Vidal-Abascal

Marcelo Santaló (Luís Santaló's brother) studied Mathematical Sciences in Madrid, where one of his fellow students was Enrique Vidal-Abascal. It appears that Marcelo acted as a bridge for what was to be a long and exceptional human and scientific relationship between Vidal-Abascal and Santaló, who was three years younger than Vidal-Abascal.

Thanks to Enrique Vidal Costa, Vidal-Abascal's son, we have been granted access to several letters (most of them handwritten) that Santaló sent to his friend Vidal-Abascal. These letters, written in Spanish, cover the years 1936 to 1983, although it is clear that the correspondence in full was far more voluminous and extensive.

The first of these letters is dated Monday, January 1st, 1936, and in it Santaló provides Vidal-Abascal with details about course that were to be given in Madrid (possibly doctoral courses). Of particular interest is information concerning Santaló's recommendations about the bibliography for "Higher Geometry", in which he tells Vidal-Abascal that the book they were following was Blaschke's "Differentialgeometrie".

As already mentioned in "A Short Biography of L.A. Santaló", Santaló travelled to the Argentina Republic with the moral support of Julio Rey Pastor, who was to be instrumental in introducing Santaló to this country. This distinguished Spanish mathematician of the first half of the 20th century often spent long periods in Buenos Aires and is mentioned in this first letter.

On his arrival in Argentina, Santaló settled in Rosario (Litoral Province) where he worked at the Instituto de Matemáticas, which had just been created and was headed by B. Levi[8], and where he was to remain for several years.

From a letter dated 1947, we know that Vidal Abasacal sent Santaló a copy of an article entitled "Parallel Curves on Surfaces of Constant Curvature", which Santaló submitted for publication to the Journal of the Unión Matemática Argentina with the remark that there was a lot of material in the *Math. Notae Journal*. Santaló also said in the following paragraph that he was moving to Buenos Aires: "*At last, after a thousand and one difficulties in the search for lodgings, which seemed an almost insoluble problem, I found an apartment that though tiny is enough for myself, my wife and my daughter. The address is . . .*"

This apartment was to be his home base for the rest of his life. Santaló also wrote: "*In the meantime, I received the offprint of your work in the Revista Hispano-Americana;*

[8] In a conversation in Buenos Aires between Rey Pastor, Cortés Pla (Dean of the Faculty of Engineering at the Universidad del Litorial) and Levi-Civita, the latter recommended Beppo Levi to the other two as Director of the Institute.

I think it is very nice, especially the generalization of Holditch's theorem to the case of surfaces of constant curvature."

We have access to two letters from 1948. In the first, Santaló congratulates Vidal-Abascal for his results on parallel curves. He was most likely referring to the article that Vidal-Abascal published in the Proceedings of the American Mathematical Society, and says that he is going to send Vidal-Abascal some of his own published papers. We find the following quotation particularly interesting: "*The Bonnesen-Fenchel cannot be found here. Luckily I hung on to my copy I had before, since it's difficult to get hold of, and as well as being magnificent and vital for convex figures it includes all the bibliography up to 1934.*"

There are also several references to the personal life of the families of both mathematicians.

In 1948, Santaló moved to Princeton, where he spent a long time at that prestigious university. From a letter he wrote to Vidal-Abascal in November of the same year, Vidal-Abascal's high mathematical level and aspirations are patently clear. Indeed, from the first paragraph one deduces that Vidal-Abascal intended to go to Princeton himself (perhaps to work with Allendoerfer). Santaló writes: "*I get the impression from your letter that you think that perhaps a letter from me might help you in your plans to spend some time here. That's why I include a letter written "with another respect", should you find that it may be useful. If it does not meet your requirements you can tear it up and tell me frankly what it is I can do. I'd be pleased to do anything on your behalf. These last few months I finished a work on the same subject you propose and also entitled "Integral Geometry on Surfaces", but you mustn't trouble yourself about that because there is unlikely to be much coincidence. The paper is going to be published at Duke and I'll send you an offprint when it comes out. You'll see that, although the proof seems different to me, your result you sent me on the measure of geodesics intersecting a curve; that is, $\int n dG = 2L$ (dG = the density of geodesics, L the boundary of the curve, n = number of intersecting points) and for the "convex curves" ($n = 2$) gives $\int n dG = 2L$, already published in two previous works: W. Blaschke "Integralgeometrie 11", Zur variationsrechnung Abhandlungen Hamburg,* **11**, *1936, and M. Haimovici "Géométrie intégrale sur les surfaces curves", Annales Scientifiques de l'Université de Jassy. In the same work, Blaschke also obtains the formula $\int \sigma dG = \pi F$."*

We extract the following paragraph from a letter dated March, 1949: "*Next week I set off South again; once more to Argentina. Sorting through my papers I found your last letter; I can't remember whether or not I replied to it. I'll say sorry, just in case. You ask about a book for studying fiber bundles. There is no such book. Alexandroff's "Topology" is difficult to get hold of; it's terribly expensive and doesn't have anything on the subject you want, which is all contained in odd papers, mainly in the Annals of Mathematics. We've studied these things here at the Veblen Seminar with Allendoerfer and in recent months with Chern, who is here once again for a stay. It has been a good time and I'm sorry I have to leave. The months just fly by. I don't know when my job at Duke is going to happen. All the American journals are so swamped with articles that it takes more than a year for a paper to be published. As you thought, in order to generalize Poincaré's Formula to surfaces I take as congruent curves those having equal $\rho g(s)$. Just now I've been doing the analogy for manifolds of n-dimensions. It's not hard, but I can't find applications to the formulas obtained, and a formula without* *applications isn't very interesting. For example, by defining a density dG for geodesics in a n-dimensional manifold, given a manifold V_{n-1}, one obtains $\int N dG = cte \cdot F$, $F =$ area. Poincaré's formula can also be generalized, but I'd like to find applications. I'll carry on writing. Let me know what you get."*

We would like to draw attention to two conclusions that seem evident. Santal's unflagging interest in finding applications for his theoretical researches, and Vidal-Abascal's keenness for studying the latest theories; for example, fiber bundles in the 1940s.

In 1949, Vidal-Abascal is awarded the Alfonso X (The Wise) Prize from the Spanish Higher Council of Scientific Research. Wishing to publish a paper worthy of such a distinction, Vidal-Abascal sent Santaló a draft and asked him to write a prologue. Santaló declined the honour with the following phrase: "*I received your paper when I arrived in Princeton in May, and what with one thing and another time has passed and I still haven't replied until now. First let me thank you for sending it and congratulations on the award. Although I'd be delighted to do it, I don't think it would be appropriate for me to write the prologue.*"

We believe that Santaló declined because of the political situation in Spain at that time, since he would not have wished to have placed his friend in a compromising situation (recall that Santaló was a political refugee).

In the same letter, Santaló goes on to say: "*Many thanks for the kind words you said about me in the paper! It is very interesting and I hope you continue with your work in this field. Cartan's work is an inexhaustable vein and I believe that there is much that can be developed from the geometric point of view. On my arrival here I accepted a contract from the Universidad de la Plata, so now I have definitively left Rosario.*"

In April 1952, Santaló wrote another letter containg some interesting things. It is not easy to read, since the paper makes it almost illegible. Vidal-Abascal rewrote this letter, even down to copying Santaló's signature! In this letter, Santaló says: "*I was delighted to receive your letter and to learn that you finally managed to get to Lausanne. I hope your health improves and that you can use the time to accumulate problems that you can work on later in Spain. I envy you being able to be with De Rham: his work is absolutely up to date and promises many things.*

As for what you say about my work, that seems fine to me. If $Adx + Bdy$ takes the same value in A and B, on integrating, since the directions are opposed, they vanish. This is analogous for more dimensions, for if the volume V of the connection of the sphere (indispensable for Stokes' Theorem to hold, I believe), on intersecting by sgr $\rightarrow h$, the orientations are opposite at the points of intersection (one exit corresponds to each entry), save for excetional cases of tangency, which have no influence. What you say would be true if we took the absolute value of $[\omega_1 \ldots \omega_n]$.

Your interpretation as integral invariant is of course fine, but I'm not sure if it works. As I say, this seems okay to me, but in the same work "Integral Geometry in Projective and Affine Spaces", when speaking about the Minkowski-Hlawka Theorem, I think there's an inconsistency of mine that should be put right (a pass to the limit under the integral sign). Since I don't get the Zentralblatt here, I'd be grateful if you could let me known what it says if you have the time. There's nothing about that in the Math. Review. If you haven't got time to copy it, just tell me – if you remember when you write – if the Zentralblatt has anything to say on this.

I'm glad your still working on these things; it's very likely that your idea of duality will give good results, but as you say, when the geodesics can have several common points, it gets complicated. I've just been working on Integral Geometry in Hermitian spaces, which may have applications to Algebraic Geometry. I'm also studying Lichnerowicz's notes, which tie in with De Rham and Hodge's work. I didn't know that you also liked to paint; why don't you send me a picture to show as a model of mathematical painting?"

This letter ends with a note in the margin that says: "*I don't remember if I thanked you for the citations from me you included in your prize-winning work on "Integral Geometry in Curved Surfaces". Many thanks* !! "

From the letter dated in Buenos Aires on December 2nd, 1952, we select the following: "*I'm glad you could spend some time with Varela Gil and also with A. Weil; I owe many ideas to Weil and we became good friends in Chicago. He is very friendly, although his mordant character makes him argue with people. However, we've always got on very well together.*

As regards what you say about geodesics, of course, once obtained the invariant $\Sigma[d\rho_i du^i]$ *for* 2*-dimensional sets, any power* $(\Sigma[d\rho_i du^i])^h$, $(h \leq n-1)$ *(the power extended as exterior multiplication of h equal factors) will have the same invariance properties, and therefore can be used as the density for h-dimensional sets of geodesics; except for a constant factor, this density seems to be unique. This is precisely something that I don't think has been done: calculating measures of certain h-dimensional sets of geodesics (congruences, for* $n = 3$ *for example). In a paper of mine that appeared in the Summa Brasiliensis I do this for* $h = n - 1$*, but not for* $h < n - 1$*. For example, in the* 3*-dimensional case, the measure is a congruence of geodesics* $(h = 2)$*. Perhaps there are interesting properties (it's possible, although I don't know if it is known, that the measure of a congruence of normals could be the area of the normal surface; if it's not a congruence of normals, then I don't know). Send me everything you publish, because it interests me very much, and I'll do the same for you.*"

The paper Santaló refers to is [52.1], which has had a great impact, as witnessed by the fact that Santaló devotes an almost entire section to it in the last chapter of [76.1], and also for its broad implications in higher research into the same subject (see, for example, Croke's commentary in this volume).

En 1955 Vidal-Abascal was preparing the publication of his book "*Introduction to Differential Geometry*", Dossat, Madrid 1956. In this regard, some interesting paragraphs referring to this work appear in a letter Santaló wrote in the same year: "*I'm liking the book more and more. The last part with Cartan's method is magnificent in both clarity and content; it all goes together very well and with the added notes the amount of material is very great. I have no doubt that it will be a big success.*

As for the Addendum on Affine Geometry, after much thought I think it would be better not to do it. I've been trying to write it, believing that it would be easy, but now I realise that if it were going to be brief, as it should be, it wouldn't be understood very well and wouldn't be worthwhile; on the other hand if I tried to clarify and extend it, it would take up too many pages. That's why I think it's better to leave it for another time. If a second edition comes out later we might be able to do something between the two of us. I'm sorry, but I think the book is fine as it is, and if anything were to be added it would have to be of the same high quality, something fully worked out.

*As for the last material you sent me, I have nothing to remark, except that I see on page 247 you have cited Beltrami's famous 1868 paper in English (*An attempt to interpret the non-Euclidean Geometry*). It seems to me that if you still had time it would be better either to translate the title into Spanish or include the original in Italian, which I cannot put my hands on right now to send you but which I have somewhere scattered about. Some days ago I had them sent you Rey Pastor-Balanzat's and my "Analytic Geometry". It's just something didactic aimed at the university graduate market. I'm not very pleased with it, but it's not all bad."*

The letter dated December 8th, 1955, is less interesting from a mathematical point of view. Nevertheless, it is worth pointing out Santaló's reference to the almost simultaneous publication of two books in Spanish on Differential Geometry: the first by Vidal-Abascal mentioned above, and the second a translation of D. J. Struick's "Introduction to Differential Geometry".

In November 1967, an International Congress on Differential Geometry was held in Santiago de Compostela (Spain) organized by Vidal-Abascal. Santaló was one of the distinguished specialists invited to attend. In his talk, he gave the definition of "absolute total curvatures" of a submanifold in Euclidean space. In two letters written by Santaló from Buenos Aires in December of the same year, he congratulates his friend on the success of the Congress.

In December 1972, Vidal-Abascal paid a visit to the universities of Buenos Aires and Bahía Blanca in Argentina, as well as other cultural centres in the country. From a letter sent by Santaló dated November 7th, 1972, it is clear that he helped Vidal-Abascal in the organization of this trip

In a further letter dated October 5th, 1976, Santaló writes: "*As for this end, I'm on the verge of retirement age. I'm gradually dropping what's known as research and devoting myself more to teaching and education, from the theoretical perspective really, because actual teaching is starting to bore me. I've corrected the proofs of a book on Integral Geometry and Geometric Probabilities, to be published by Addison-Wesley. They say it will come out in December. Of course, I'll send you a copy of the first edition, or get them to send you one. I think it has turned out all right, although it's difficult to say. I'm realizing that I'm a bit antiquated, although it seems that now there's a revival of Classical Mathematics. We'll see.*"

With the thirty years of hindsight since the publication of [76.1], Santaló's modesty is once more evident when it comes to judging his outstanding scientific work stretching over the last forty years, an achievement that is also reflected in this book.

As we know, Santaló became interested in the problems surrounding the teaching of Mathematics in the 1970s. According to what Santaló's family told Naveira in November, 2007, this interest arouse out of Santaló's deep concern for the situation in which the teaching of this speciality found itself in the Argentine Republic, a state of affairs that also affected one of his own daughters!

In the same letter he relates how he attended the 3rd International Congress on the Teaching of Mathematics held in Karlsruhe (Germany) in August 1976. There he met up with Enrique Vidal Costa (Vidal-Abascal's son), who was a teacher of Mathematics at the Institute of Secondary Education in Girona (Santaló's native city). This meeting revived many memories of Santaló's youth, which he mentions in the letter. For its human value, it is worth quoting the following paragraph:

"*You might not see your son, but today I also write to him to thank him for a letter he wrote that was published in "La Vanguardia"*[9] *in Barcelona, in which he speaks of our meeting in Germany and sings my praises. This gave the family great pleasure, which was also gratifying for me. We connoisseurs all know each other very well and the occurrence is of no great importance. What a coincidence that your son happens to be in Girona! We spoke a lot about my town and his way of looking at things interested me very much; I suppose that for him it's a backwater, but he is no doubt doing and will do valuable work.*"

The last letter, a copy of which is available to us, is dated November 19th, 1983, and in it Santaló tells Vidal-Abascal that he has been made Professor Emeritus at the University of Buenos Aires and now confines himself to supervising seminars and research work.

Both in the correspondence quoted here and in that omitted, there are many other comments of a personal nature regarding the families of Santaló and Vidal-Abascal, as well as references to other mathematicians and mutual friends with whom they had maintained relationships, both at a personal and professional level.

Reading the correspondence placed at our disposal, one arrives at the conclusion that Santaló and Vidal-Abascal were in great harmony. During his stay in Hamburg, Santaló acquired a sound mathematical training in those subjects in which he was subsequently to work to develop the theory of Integral Geometry, and which were the hallmark of almost all his scientific endeavours. On the other hand, Vidal-Abascal, who possessed exceptional foresight, was obliged to train himself, far removed from the main centres of research, and whose boldness and far-sightedness already in the 1940s and 50s was prompting him to break away from the secular backwardness under which he had been brought up, continued to live for Spanish science, in particular Mathematics.

Although Santaló became an Argentinian citizen, he never forgot his Spanish origins, and every time the opportunity presented itself he always helped the development of scientific-mathematical development in Spain to the best of his ability, both in the general sense, and in particular the work that Vidal-Abascal initiated in Santiago de Compostela.

A. M. Naveira.
A. Reventós.

[9] A Spanish national daily.

10 Some Remarks on Two Classifications of Santaló's Papers

From the data obtained through MathSciRev, Zentralblatt für Mathematik and the bibliographical sources placed a our disposal by his family, Santaló's referenced production according to our classification comes to 192 publications amounting to approximately 2,000 pages.

For editorial reasons, the present work cannot exceed a certain number of pages, and therefore a judicious selection has been made of Santaló's scientific papers, which seemed of the greatest interest when seen in the light of scientific content and homogeneity. The choice made by the Editors, assisted by the different respective specialists in their fields, may therefore be seen in Chapter 5.

In March, 2008, Naveira and Reventós met Professor A. Córdoba from the Universidad Autónoma de Madrid at the CRM in Barcelona, where they spoke to him about the project for the "Selected Papers of Santaló", which they were working on since April 2006, and as a result of which the selection was made. Córdoba informed them that in the late 1990s, when he was President of the RSME Scientific Committee, he and Professor Fernández Pérez, also of the Universidad Autónoma de Madrid, had already put forward the idea of publishing a selection of Santaló's work. To that end, Córdoba and Fernández Pérez got in touch with Santaló through Professor Feferman. Santaló duly sent them a list of 151 scientific articles which he himself had rated on a scale of 2 to 10. However, for a variety of reasons, the project never got off the ground.

Among the research papers on the following list of Santaló's scientific production, those chosen for inclusion are marked with (*), and are accompanied by the rating he accorded to his own scientific production.

34.1(2), 34.2(3), 34.3(3), 35.1(10), 35.2(6), 36.1*(7), 36.2(6), 36.3(10), 36.4(10), 36.5(9), 37.1(8), 39.1(4), 40.1*(9), 40.2*(10), 40.3(8), 40.4(8), 40.5(5), 41.1*(7), 41.2*(7), 41.3*(8), 41.4*(8), 41.5(6), 41.6(5), 41.7, 41.8(6), 41.11(9), 42.1*(10), 42.2*(8), 42.3*(10), 42.4(4), 42.5(6), 42.6(4), 42.7(6), 42.8, 42.9(7), 42.10, 42.11, 43.1*(10), 43.2*(8), 43.3, 43.4(7), 43.5(8), 43.6(2), 43.7, 44.1*(8), 44.2, 44.3, 44.4, 44.5(6), 44.6(8), 44.7(8), 44.8, 45.1*(8), 45.2*(8), 45.3, 45.4, 45.5, 45.6(6), 45.7(8), 45.8(8), 45.9(7), 45.10(5), 46.1*(10), 46.2*(8), 46.3*(9), 46.4(7), 46.5(8), 46.7, 46.8(8), 46.9, 46.10(6), 47.1*(9), 47.2*(8), 47.3*(6), 47.4, 47.5(7), 47.6(8), 48.1(8), 48.2(8), 49.1*(8), 49.2*(9), 49.3*(10), 49.4(7), 49.5, 50.1*(9), 50.2*(9), 50.3*(10), 50.4(7), 50.5(9), 50.6(8), 50.7(3), 50.8, 51.2(8), 51.3(8), 51.4(7), 51.5(6), 51.6, 52.1*(8), 52.2*(10), 52.3(7), 52.4(8), 52.5(7), 53.2, 53.3, 54.1*(8), 54.2(8), 54.3(7), 54.4(8), 54.5(8), 54.6(9), 54.7, 54.8, 55.1*(9), 55.2(8), 55.5(6), 55.6(8), 56.1*(8), 56.3(8), 56.4(9), 57.1(9), 57.2(7), 57.3(9), 57.4(4), 59.1*(9), 59.2(9), 60.1*(8), 60.2*(10), 60.3(8), 62.1*(8), 62.2*(10), 62.3*(7), 62.4(9), 62.5, 63.1*(8), 65.2(8),

66.1*(9), 66.2*(6), 66.4(7), 67.1*(10), 67.2*(9), 67.3(8), 67.4(9), 67.5(6), 68.1*(9), 68.2, 68.3, 69.1*(10), 69.2*(8), 69.3(2), 70.1*(10), 70.2(7), 72.1*(8), 72.2(9), 72.3(7), 74.1*(10), 74.2(8), 75.1*(9), 75.2(8), 76.3(7), 77.1, 78.1*(6), 78.2*(8), 78.3(7), 78.4(3), 79.1*(8), 80.1*(7), 80.2*(9), 80.3*(8), 80.4(6), 82.1(8), 83.1*(9), 83.3, 84.1*(10), 84.3, 85.1(7), 85.2(8), 86.1*(9), 86.2, 87.1(8), 88.1*(8), 88.2, 88.3(9), 89.2, 91.1, 95.2.

The following remarks are included as of possible interest to the reader:

a. Of the 18 articles that Santaló awarded 10 points, 14 were chosen for our selection, with the exception of papers 35.1, 36.3, 36.4 and 49.3, which amounts to 78%. Article 36.3 was not included due to lack of space. The content of 35.1 did not appear to fit clearly into any of the Chapters of the Selection. Many of the results in 36.4 appear in subsequent papers, as well as being a little too long for inclusion.

b. Santaló provided us with 27 articles with a rating of 9 points, from which we have selected 15 (56%). Papers [36.5, 41.11, 57.1, 57.3] are all of great interest, but we were obliged to omit them for lack of space. The results in [50.5, 54.6 and 56.4] are basically included in 76.1. Papers [59.2, 67.4 and 72.2] are not included for reasons of homogeneity with the contents of the Selection. Article [88.3] consists of the content of a talk and also seemed to be too long; furthermore, paper [84.1], which is included, deals with the same subject.

c. In our opinion, all the articles appearing with a score of 8 seemed worthy of inclusion in the Selection, but for reasons of editing we are unable to include them all. In fact, we have chosen 21 from a total of 51, which amounts to 42%.

d. From the 23 papers with a score of 7 we have chosen five, which is 22%. Article [36.1] is the content of Santaló's doctoral thesis and is chosen for its undoubtable historical interest. This same criterion also determined the selection of [41.1; 41.2 and 62.3]. Many of the fundamental results in [62.3] appear explicitly in [76.1]. Paper [80.1] is of great scientific value because of its subsequent influence, see, for example, the comments by Gallego and Miquel.

e. From the articles with a rating of 6 points, we have selected [47.3, 66.2 and 78.1]; that is, 20%. [47.3] is included on the recommendation of our scientific advisors. [66.2] seemed interesting in itself, as well as for its applications to geodesic mosaics in the hyperbolic plane. Likewise, [78.1] seemed interesting since it belongs to the Integral Geometry on surfaces research line, of which very little is known if surfaces of constant curvature are excepted.

f. None of the other articles scored by Santaló have been chosen for the Selection.

A. M. Naveira.
A. Reventós.

11 Authorization

Springer-Verlag wish to thank to the original publishers of the following papers by L. A. Santaló for permission to reprint them here.

The numbers following each source correspond to the numeration of the articles.

Reprinted from Math. Notae © by Instituto de Matemática "Beppo Levi", Rosario, Argentina: [42.1, 49.1, 59.1, 62.3, 78.2].

Reprinted from Bull. Amer. Math. Soc. © by The American Mathematical Society [45.2, 44.1, 46.3].

Reprinted from Proc. Amer. Math. Soc. © by The American Mathematical Society [50.1, 68.1].

Reprinted from Rev. Un. Mat. Argentina © by The Unión Matemática Argentina [43.2, 62.1, 63.1, 45.1, 66.1].

Reprinted from Arch. Math. (Basel) © by Birkhäuser [67.2].

Reprinted from Univ. Nac. Tucumán Rev. © by The Faculty of Sciences Ex. and Tech., National University of Tucumán, [42.2, 60.1, 66.2, 69.1].

Reprinted from Amer. J. Math. © by the Johns Hopkins University Press. [41.3, 52.2, 69.2].

Reprinted from Symposia Math., Inst. Naz. di Alta Matematica © by INdAM [74.1].

Reprinted from Rev. Acad. Ci. Ex. Fis. Nat. © by The Real Academia Ciencias Exactas, F´ısicas y Naturales, Madrid, [36.1].

Reprinted from Duke Math. J. © by Duke Math. J. [42.1, 43.1, 49.2, 47.1].

Reprinted from Proc. Internat. Congress of Math. © by The American Mathematical Society, [50.2].

Reprinted from Summa Brasil. Math. © by The Sociedade Brasileira de Matemàtica, [52.1].

Reprinted from Univ. Politec. Torino. Rend. Sem. Mat. © by Sem. Mat. Univ. Pol. Torino, [54.1, 80.2].

Reprinted from Rend. Circ. Mat. Palermo © by Circolo Matematico di Palermo, [83.1].

Reprinted from Rev. Fac. Sci., Univ. Istanbul © by Istanbul Technical University [56.1].

Reprinted from Canad. J. Math. © by CMS [62.2].

Reprinted from Ann. Mat. Pura Appl. © by Annali di Matematica Pura ed Applicata, [75.1].

Reprinted from Portugaliae Math. © by The Sociedade Portuguesa de Matemàtica, [46.2, 49.3, 80.1].

Reprinted from Aspects of Math. and its Appl. © by Elsevier Science Publ., North-Holland, [86.1].

Reprinted from Tohöku Math. J. © by Tohöku Mathematical Journal, [40.1, 41.1, 41.2].

Reprinted from Ann. of Math. © by The Annals of Mathematics [46.1, 50.3].

Reprinted from J. of Microscopy © by Royal Microscopical Society [88.1].

Reprinted from Rev. Mat. Hisp. Amer. © by The Real Sociedad Matemática Española, [36.2].

Reprinted from Ann. Math. Statist. Statistics © by Institute of Mathematical Statistics, [47.2].

Reprinted from Trabajos Estadist. © by The Sociedad Española Investigación Operativa, [55.1]

Reprinted from J. Appl. Probab © by Applied Probability Trust, [72.1, 78.2].

Reprinted from Stochastica © by The Universidad Politécnica Cataluña , [80.3].

Reprinted from Math. Nachr. © by –[84.1].

We would also like to thank the Instituto de España for their permission to publish the illustrations in page 15.

12 Acknowledgements

The Editors wish to express their sincere gratitude to individuals and institutions for all their help, both in the scientific and financial sense, as well as for their moral support, which together have enabled us to compile this edition of the Selected Papers of Professor Santaló. First, the support provided by Mme. Hilda Rossi, Santaló's widow, as well as that of his daughters, Tessy, Alicia and Claudia. Their help has been of great value in the accomplishment of our work. A special thanks goes to the distinguished Professor Simon K. Donaldson (Royal Society Research Professor and 1986 Fields Medal Winner), who kindly accepted to write a preface for this publication. In the scientific field, we are also grateful to Professors L. M. Cruz-Orive, R. Langevin, K. Leichtweiss, R. Schneider and E. Teufel, who at our request provide short introductions to each chapter. We likewise wish to express our gratitude to the following institutions:

- Universidad de Valencia (UVEG)[10].
- DGI (Spain) and FEDER Project MTM 2007-65852.
- Universitat Autònoma de Barcelona (UAB)[11].
- Institut d'Estudis Catalans (IEC), Barcelona.
- Departamento de Geometría y Topología de la UVEG.
- Departament de Matemàtiques de la UAB.
- Departament de Matemàtiques de la Universitat Jaume I, (UJI), Castellón.
- Departamento de Matemáticas de la Universidad Nacional del Centro de la Provincia de Buenos Aires (UNCPBA), Tandil, Argentina.
- Departamento de Matemáticas de la Universidad de Buenos Aires (UBA).
- Centre de Recerca Matemàtica (CRM), Barcelona.
- Societat Catalana de Matemàtiques (SCM), Barcelona.
- Cátedra Lluís Santaló d'Aplicacions de la Matemàtica (CLS), Universidad de Girona (UdG).
- Familia Ares Rodríguez, Buenos Aires.

The financial support received has been vital too the preparation of these Selected Papers.

At all times we have been fortunate to count on the invaluable moral support of Real Academia de Ciencias Exactas, Físicas y Naturales (RACEFN), Madrid;

[10] Moreover, A. M. Naveira was in sabbatical year from this University during the preparation of this Selecta.

[11] Moreover, A. Reventós was in sabbatical period from this University during the preparation of this Selecta.

Real Sociedad Matemática Española (RSME), Madrid; Instituto de España (IE), Madrid.

The legal representatives of the above-mentioned institutions have also provided advice and assistence in the solution of many problems. In particular, we would like to mention, among others, to Professors Andradas, Cordoba and Gil-Medrano (RSME), Castellet and Bruna (CRM), Barceló (CLS), and Casacuberta and Perelló (SCM).

Our thanks also go to the valuable scientific contributions from the authors of the "Comments" that appear in section 7 of these Selected Papers, as likewise to Professors Abardia (UAB), Hernández Cifre (UM) and Carreras (UVEG) who provided help in other aspects of the work.

It is only fair also to mention Ms. Catriona Byrne, Editorial Director for Mathematics at Springer, who from the moment that we presented her with the idea of producing such a publication was always available to help.

The efforts made in gathering together and analyzing the scientific production and dessemination work of Professor Santaló has essentially been facilitated by the specialized bibliographic services of the Universities of Barcelona, Buenos Aires, Girona, Rosario and Valencia.

Finally, we would like to thank Mr. J. Palmer for his careful translation of papers 40.2, 41.4, 42.2, 42.3, 45.1, 46.2, 47.3, 49.1, 49.3, 55.1, 59.1, 60.1, 62.1, 62.3, 63.1, 66.1, 66.2, 67.1, and 78.1, as well as for his general revision of the whole text.

Printed in the United States
By Bookmasters

Printed in the United States
By Bookmasters